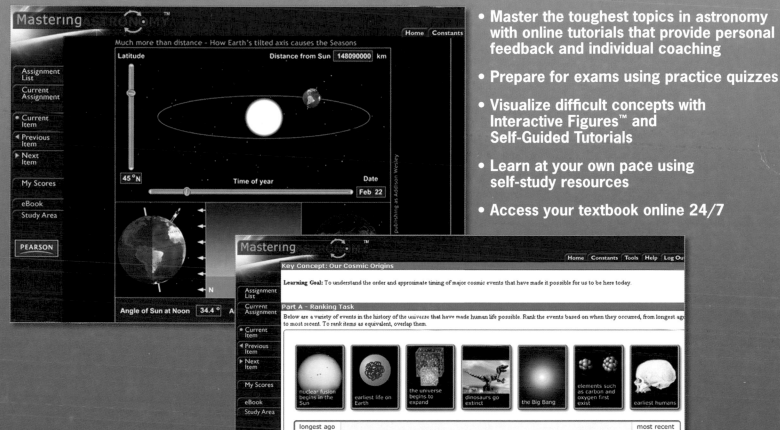

LAUNCH YOUR STUDENTS ON THE PATH
TO COSMIC DISCOVERY AND SUCCESS

Based on the most up-to-date astronomical research, ***The Essential Cosmic Perspective,*** **Fifth Edition** retains all of the features that have made this text so popular with new features to help students learn about the process of science and how to interpret visual data. The **Fifth Edition** focuses on building student appreciation of science by helping them understand how astronomers know what they know about the universe, and how information can be gleaned from looking at and interpreting images. The textbook package also includes updated supplements to support the book's pedagogy, making it the most effective text in the one-semester astronomy market.

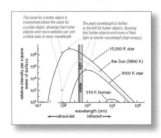

7.3   Mars: A Victim of Planetary Freeze-Drying

• What geological features tell us that water once flowed on Mars?

Dry riverbeds, eroded craters, and chemical analysis of Martian rocks all show that water once flowed on Mars, though any periods of rainfall seem to have ended at least 3 billion years ago. Mars today still has water ice

How do your students
identify what is important?
See pages W-2 and W-3 of this walkthrough.

HALLMARK OF SCIENCE   A
phenomena that rely
geometry to explain th

How do you give your students
an understanding of the process of science?
See pages W-4 and W-5 of this walkthrough.

How do your students
understand visual information?
See pages W-6 and W-7 of this walkthrough.

MA™

How do you
assess your students?
See pages W-8 and W-9 of this walkthrough.

The Essential Cosmic Perspective's clear and directed learning path enhances students' understanding of key concepts. Using proven ideas from education research, each chapter is built around explicit learning goals, which are reinforced throughout the text.

Motivational Learning Goals begin each chapter to focus students on the most important concepts ahead.

7

earth and the terrestrial worlds

learning goals

7.1 Earth as a Planet
- Why is Earth geologically active?
- What processes shape Earth's surface?
- How does Earth's atmosphere affect the planet?

7.2 The Moon and Mercury: Geologically Dead
- Was there ever geological activity on the Moon or Mercury?

7.3 Mars: A Victim of Planetary Freeze-Drying
- What geological features tell us that water once flowed on Mars?
- Why did Mars change?

7.4 Venus: A Hothouse World
- Is Venus geologically active?
- Why is Venus so hot?

7.5 Earth as a Living Planet
- What unique features of Earth are important for life?
- How is human activity changing our planet?
- What makes a planet habitable?

- Was there ever geological activity on the Moon or Mercury?

- Why is Venus so hot?

188

One section, one learning goal—each section is written to address a single learning goal, guiding students through each chapter with a focus on the key concepts.

7.2 The Moon and Mercury: Geologically Dead

In the rest of this chapter, we will investigate the histories of the terrestrial worlds, with the ultimate goal of learning how and why Earth became unique. We'll start in this section with the two worlds that have the simplest histories: the Moon and Mercury (Figure 7.16).

The simple histories of the Moon and Mercury are a direct consequence of their small sizes. Both these worlds are considerably smaller than Venus, Earth, or Mars, so they long ago lost most of their internal heat, leaving them without any energy source to power ongoing geological activity. Small size also explains their lack of significant atmospheres: Their gravity is too weak to hold gas for long periods of time, and without ongoing volcanism they lack the outgassing needed to replenish gas lost in the past.

Moon   Mercury

Figure 7.16
Similar views of the Moon and Mercury, shown to scale. The Mercury photo was obtained during the January 2008

- **Was there ever geological activity on the Moon or Mercury?**

- **Was there ever geological activity on the Moon or Mercury?**

200   Part III Learning from Other Worlds

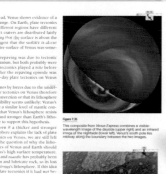

Venus lacks any similar features. Instead, Venus shows evidence of a very different type of global geological change. On Earth, plate tectonics reshapes the surface gradually, so that different regions have different ages. On Venus the relatively few impact craters are distributed fairly uniformly over the entire planet, suggesting that the surface is about the same age everywhere. Crater counts suggest that the surface is about 750 million years old. Apparently, the entire surface of Venus was somehow "repaved" at that time.

We do not know how much of the repaving was due to tectonic processes and how much was due to volcanism, but both probably were important. It is even possible that plate tectonics played a role before and during the repaving, only to stop after the repaving episode was over. Either way, the nature of present-day plate tectonics on Venus poses a major mystery.

Figure 7.35
This composite from Venus Express combines a visible-wavelength image of the dayside (upper right) and an infrared image of the nightside (lower left). Venus's south pole lies midway along the boundary between the two images.

- Why is Venus so hot?

- **Why is Venus so hot?**

Chapter 7 Earth and the Terrestrial Worlds 213

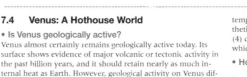

summary of key concepts

7.1   Earth as a Planet

- **Why is Earth geologically active?**

Internal heat drives geological activity, and Earth retains internal heat because of its relatively large size for a terrestrial world. This heat causes mantle **convection** and keeps Earth's lithosphere thin, ensuring active surface geology. It also keeps part of Earth's core melted, and circulation of this molten metal creates Earth's magne...

- **What processes shape Earth**
The four major geological process volcanism, **tectonics**, and **erosion**. Earth has experienced many impacts, but most craters have been erased by other processes. We owe the existence of our atmosphere and oceans to volcanic **outgassing**. A special type of tectonics—**plate tectonics**—shapes much of Earth's surface. Ice, water, and wind drive rampant erosion on our planet.

- **How does Earth's atmosphere affect the planet?**
Two crucial effects are (1) protecting the surface from dangerous solar radiation—ultraviolet is absorbed by ozone and X rays are absorbed high in the atmosphere—and (2) the **greenhouse effect**, without which the surface temperature would be below freezing.

7.2   The Moon and Mercury: Geologically Dead

- **Was there ever geological activity on the Moon or Mercury?**
Both the Moon and Mercury had some volcanism and tectonics when they were young. However, because of their small sizes, their interiors long ago cooled too much for ongoing geological activity.

7.3   Mars: A Victim of Planetary Freeze-Drying

- **Was there ever geological activity on the Moon or Mercury?**

chemical analysis of Martian rocks all show that water once flowed on Mars, though any periods of rainfall seem to have ended at least 3 billion years ago. Mars today still has water ice underground and in its polar caps and could possibly have pockets of underground liquid water.

- **Why did Mars change?**
Mars's atmosphere must once have been thicker with a stronger greenhouse effect, so change must have occurred due to loss of atmospheric gas. Much of the lost gas probably was stripped away by the solar wind, after Mars lost its magnetic field and protective magnetosphere. Mars also lost water, because solar ultraviolet light split water molecules apart and the hydrogen escaped to space.

7.4   Venus: A Hothouse World

- **Is Venus geologically active?**
Venus almost certainly remains geologically active today. Its surface shows evidence of major volcanic or tectonic activity in the past billion years, and it should retain nearly as much internal heat as Earth. However, geological activity on Venus differs from that on Earth in at least two key ways: lack of erosion and lack of plate tectonics.

- **Why is Venus so hot?**
Venus's extreme surface heat is a result of its thick, carbon dioxide atmosphere, which creates a very strong greenhouse effect. The reason Venus has such a thick atmosphere has to do with its distance from the Sun... like those on Earth, ...ide dissolved in water and became locked away in **carbonate** rock. Carbon dioxide remained in Venus's atmosphere, creating a **runaway greenhouse effect**.

- **Why is Venus so hot?**

7.5   Earth as a Living Planet

- **What unique features of Earth are important for life?**
Unique features of Earth on which we depend for survival are (1) surface liquid water, made possible by Earth's moderate

temperature; (2) atmospheric oxygen, a product of photosynthetic life; (3) plate tectonics, driven by internal heat; and (4) climate stability, a result of the **carbon dioxide cycle**, which in turn requires plate tectonics.

- **How is human activity changing our planet?**

CO₂ concentration over the past 400,000 years

The global average temperature has risen about 0.8°C over the past hundred years, accompanied by an even larger rise in the atmospheric CO₂ concentration—a result of fossil fuel burning and other human activity. The current CO₂ concentration is higher than at any time in the past million years, and climate models indicate that this higher concentration is indeed the cause of **global warming**.

- **What makes a planet habitable?**
We can trace Earth's habitability to its relatively large size and its distance from the Sun. Its size keeps the internal heat that allowed volcanic outgassing to lead to our oceans and atmosphere, and also drives the plate tectonics that helps regulate our climate through the carbon dioxide cycle. Its distance from the Sun is neither too close nor too far, thereby allowing liquid water to exist on Earth's surface.

Visual summaries at the end of each chapter consolidate understanding with answers to the learning goals presented in text and figures to reinforce learning.

How do you give your students
AN UNDERSTANDING OF THE PROCESS OF SCIENCE?

Cosmic Context two-page figures combine text and illustrations into accessible and coherent visual summaries that will help improve your students' understanding of essential topics. Five figures have been added to the **Fifth Edition,** covering the Copernican revolution, seasons, global warming, the H–R diagram, and dark matter and dark energy.

HALLMARKS OF SCIENCE

HALLMARK OF SCIENCE   **A scientific model must seek explanations for observed phenomena that rely solely on natural causes.** The ancient Greeks used geometry to explain their observations of planetary motion.

Hallmarks of Science call attention to the process of science and how it applies to the topics discussed in the figure.

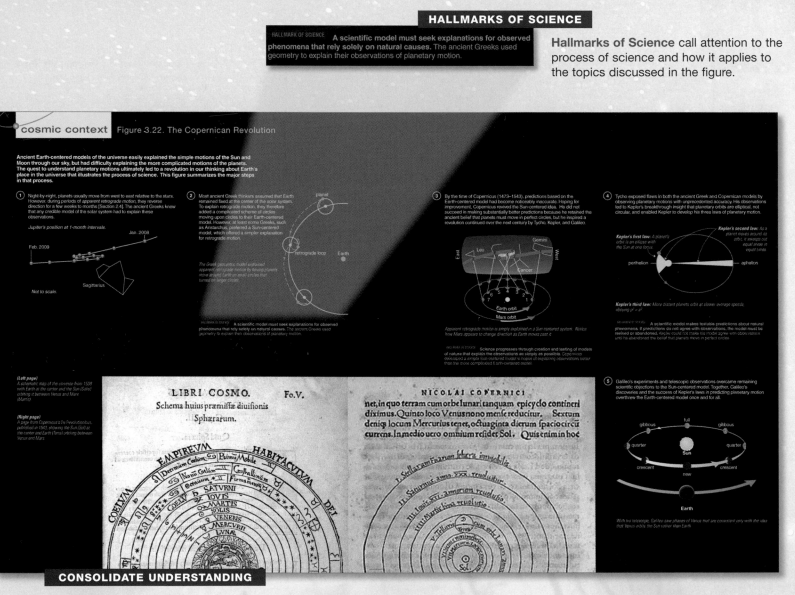

CONSOLIDATE UNDERSTANDING

Key Features of the Cosmic Context figures include:

• A brief introductory paragraph that gives a "30-second message" about the central theme of the figure

• Numbered steps that help students navigate through the information

• A central illustration that ties together and reinforces key concepts in the chapter or part

• Annotated figures that work in conjunction with the central illustration to give students a comprehensive visual summary of what they have learned

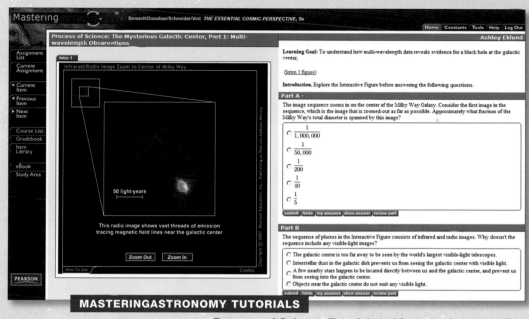

MA™ **N**ew Process of Science tutorials ask students to solve problems that make them think deeply about how astronomers use observational data and models to learn about the universe. These tutorials reinforce the book's emphasis on the process of science for a seamless integration of the text and media.

MASTERINGASTRONOMY TUTORIALS

Process of Science Tutorials in MasteringAstronomy™ challenge students to think like a scientist in answering interactive online questions.

---

larger
35. What is the most common kind of particle in the solar wind, in which the overall charge is neutral? (a) proton (b) electron (c) helium nucleus
36. Which of these things poses the greatest hazard to communications satellites? (a) photons from the Sun (b) solar magnetic fields (c) particles from the Sun

Process of Science

Examining How Science Works

37. *Inside the Sun.* Scientists claim to know what is going on inside the Sun, even though we cannot directly observe the Sun's interior. What is the basis for these claims, and how do they relate to the hallmarks of science outlined in Section 3.4?
38. *The Solar Neutrino Problem.* Early solar neutrino experiments detected only about a third of the number of neutrinos predicted by the theory of fusion in the Sun. Why didn't scientists simply abandon their models at this point? What features of the Sun did the model get right? What alternatives were there for explaining the mismatch between the predictions and the observations?

Investigate Further

In-Depth Questions to Increase Your Understanding

Short-Answer/Essay Questions

39. *The End of Fusion I.* Describe what would happen in the Sun if fusion reactions abruptly ceased.
40. *The End of Fusion II.* If fusion reactions in the Sun were to suddenly cease, would we be able to tell? If so, how?
41. *A Really Strong Force.* How would the interior temperature of the Sun be different if the strong force that binds nuclei together were 10 times stronger?
42. *Measuring the Sun's Rotation.* Suppose you observe a sunspot from Earth, taking a picture of the Sun each day. Using these photos, you measure the time it takes for the sunspot to return to the position it had in the first picture. Now suppose a friend of yours standing on Pluto does the same thing. Would you and your friend measure the same time for the sunspot to return to where it started? Explain your answer.

Quantitative Problems

Be sure to show all calculations clearly and state your final answers in complete sentences.

47. *The Color of the Sun.* The Sun's average surface temperature is about 5800 K. Use Wien's law (see Cosmic Calculations 5.1) to calculate the wavelength of peak thermal emission from the Sun. What color does this wavelength correspond to in the visible-light spectrum? Why do you think the Sun appears white or yellow to our eyes?
48. *The Color of a Sunspot.* The typical temperature of a sunspot is about 4000 K. Use Wien's law (see Cosmic Calculations 5.1) to calculate the wavelength of peak thermal emission from a sunspot. What color does this wavelength correspond to in the visible-light spectrum? How does this color compare with that of the Sun?
49. *Solar Mass Loss.* Estimate how much mass the Sun will lose through fusion reactions during its 10-billion-year life. You can simplify the problem by assuming the Sun's energy output remains constant. Compare the amount of mass lost with Earth's mass.
50. *Pressure of the Photosphere.* The gas pressure of the photosphere changes substantially from its upper levels to its lower levels. Near the top of the photosphere the temperature is about 4500 K and there are about 1.6×10^{16} gas particles per cubic centimeter. In the middle the temperature is about 5800 K and there are about 1.0×10^{17} gas particles per cubic centimeter. At the bottom of the photosphere the temperature is about 7000 K and there are about 1.5×10^{17} gas particles per cubic centimeter. Compare the pressures of each of these layers and explain the reason for the trend in pressure that you find. How do these gas pressures compare with Earth's atmospheric pressure at sea level? (*Hint:* See Cosmic Calculations 10.1.)
51. *The Lifetime of the Sun.* The total mass of the Sun is about 2×10^{30} kg, of which about 75% was hydrogen when the Sun formed. However, only about 13% of this hydrogen ever becomes available for fusion in the core. The rest remains in layers of the Sun where the temperature is too low for fusion.
    a. Based on the given information, calculate the total mass of hydrogen available for fusion over the lifetime of the Sun.

END-OF-CHAPTER QUESTIONS

The Process of Science is also reinforced in targeted end-of-chapter questions.

How do your students
UNDERSTAND VISUAL INFORMATION?

Enhanced visual pedagogy makes this already student-friendly text even more accessible through the use of Visual Skills Check activities, Cosmic Context process and summary figures, carefully crafted annotations, stepped zoom-in figures, and multi-wavelength images.

VISUAL SKILLS CHECK

Visual Skills Check questions help students check their understanding of the many different types of visual information used in astronomy.

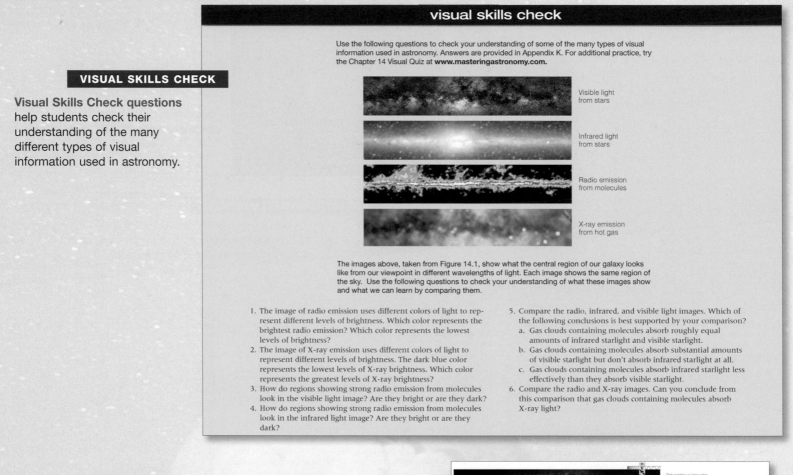

visual skills check

Use the following questions to check your understanding of some of the many types of visual information used in astronomy. Answers are provided in Appendix K. For additional practice, try the Chapter 14 Visual Quiz at **www.masteringastronomy.com**.

Visible light from stars

Infrared light from stars

Radio emission from molecules

X-ray emission from hot gas

The images above, taken from Figure 14.1, show what the central region of our galaxy looks like from our viewpoint in different wavelengths of light. Each image shows the same region of the sky. Use the following questions to check your understanding of what these images show and what we can learn by comparing them.

1. The image of radio emission uses different colors of light to represent different levels of brightness. Which color represents the brightest radio emission? Which color represents the lowest levels of brightness?
2. The image of X-ray emission uses different colors of light to represent different levels of brightness. The dark blue color represents the lowest levels of X-ray brightness. Which color represents the greatest levels of X-ray brightness?
3. How do regions showing strong radio emission from molecules look in the visible light image? Are they bright or are they dark?
4. How do regions showing strong radio emission from molecules look in the infrared light image? Are they bright or are they dark?

5. Compare the radio, infrared, and visible light images. Which of the following conclusions is best supported by your comparison?
   a. Gas clouds containing molecules absorb roughly equal amounts of infrared starlight and visible starlight.
   b. Gas clouds containing molecules absorb substantial amounts of visible starlight but don't absorb infrared starlight at all.
   c. Gas clouds containing molecules absorb infrared starlight less effectively than they absorb visible starlight.
6. Compare the radio and X-ray images. Can you conclude from this comparison that gas clouds containing molecules absorb X-ray light?

Figure 14.1 interactive photo

Spiral arms in the Galaxy M51. This photo from the Hubble Space Telescope shows M51's two magnificent spiral arms along with a smaller galaxy that is currently interacting with one of those arms. Notice that the spiral arms are much bluer in color than the central bulge. Because massive blue stars live only for a few million years, the relative blueness of the spiral arms tells us that stars must be forming more actively within them than elsewhere in the galaxy. (The large image shows a region roughly 90,000 light-years across.)

Dark patches on inner edge of spiral arm show where gas clouds are packing together ... and compression of these clouds triggers star formation in the arm.

Blue specks are young stars that formed in the spiral arm.

Flow of gas and stars through spiral arm

Red patches are ionization nebulae around the hottest, youngest stars.

600 km

1 km

5 cm

Figure 8.21 interactive figure

This sequence zooms in on the *Huygens* landing site on Titan. Left: a global view taken by the orbiting *Cassini* spacecraft. Center: an aerial view from the descending probe. Right: a surface view taken by the probe after landing; the "rocks," which are 10–20 centimeters across, are presumably made of ice.

ZOOM-IN FIGURES

Zoom-in figures help students understand the scale of astronomical objects shown in the book.

Interactive Figure™ and Interactive Photo™ icons throughout the text direct students to dynamic versions of key book figures on MasteringAstronomy.™ Using these online applets, students can manipulate factors such as time, wavelength, scale, and perspective to increase their understanding.

INTERPRETING FIGURES FROM BOOK

Annotations serve as the voice of the instructor to explain key aspects of complex figures and photos.

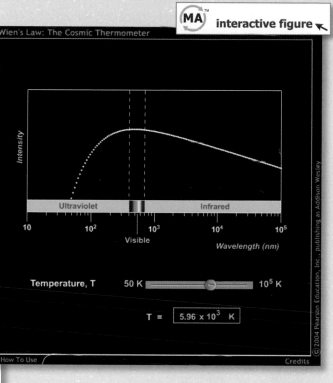

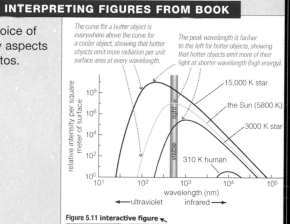

Figure 5.11 interactive figure

This graph of idealized thermal radiation spectra demonstrates the two laws of thermal radiation: (1) Each square meter of a hotter object's surface emits more light at all wavelengths; (2) hotter objects emit photons with a higher average energy. Notice that the graph uses power-of-10 scales on both axes, so that we can see all the curves even though the differences between them are quite large.

INTERPRETING PHOTOS FROM BOOK

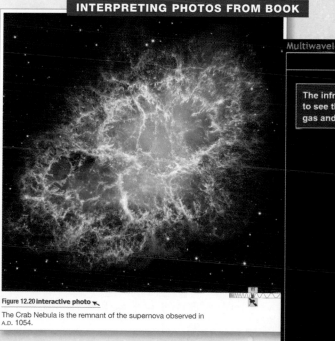

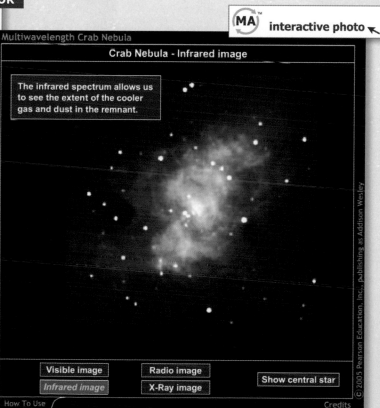

Figure 12.20 interactive photo

The Crab Nebula is the remnant of the supernova observed in A.D. 1054.

Wavelength icons help students interpret what is being shown in each photo and what information can be gleaned from it.

How do you
ASSESS YOUR STUDENTS?

MasteringAstronomy™ is the most widely used and advanced astronomy tutorial and assessment system available. It has established an unparalleled database of students' most common wrong answers nationwide to provide each student with feedback specific to his or her misconceptions and hints when they get stuck.

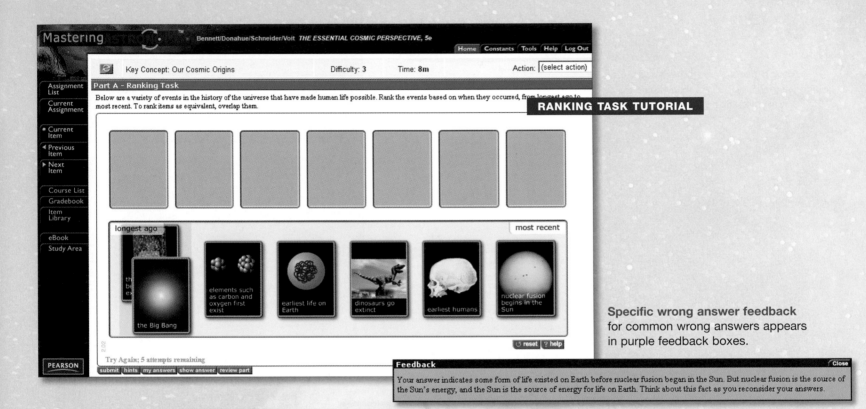

Specific wrong answer feedback for common wrong answers appears in purple feedback boxes.

UNIQUELY POWERFUL GRADEBOOK AND DIAGNOSTICS

The MasteringAstronomy Gradebook allows you to see at-a-glance where your students are having difficulties. Shades of pink instantly highlight struggling students and challenging assignments.

Mastering
ASTRONOMY™

MasteringAstronomy provides students with access to a wealth of book-specific self-study resources in the **Study Area**, which they can use whether or not you assign homework.

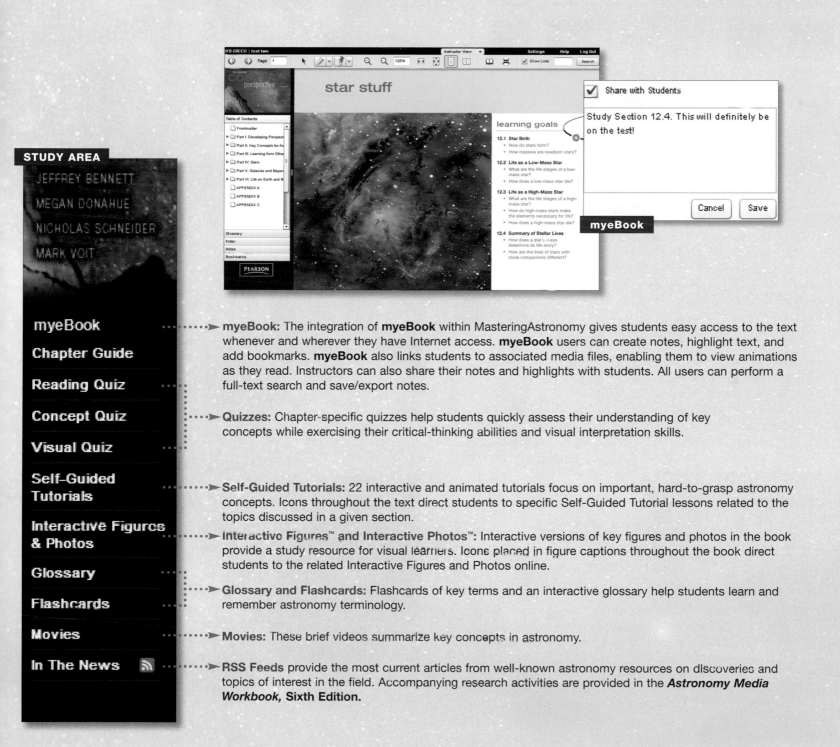

STUDY AREA

JEFFREY BENNETT

MEGAN DONAHUE

NICHOLAS SCHNEIDER

MARK VOIT

myeBook

Chapter Guide

Reading Quiz

Concept Quiz

Visual Quiz

Self-Guided Tutorials

Interactive Figures & Photos

Glossary

Flashcards

Movies

In The News

myeBook: The integration of **myeBook** within MasteringAstronomy gives students easy access to the text whenever and wherever they have Internet access. **myeBook** users can create notes, highlight text, and add bookmarks. **myeBook** also links students to associated media files, enabling them to view animations as they read. Instructors can also share their notes and highlights with students. All users can perform a full-text search and save/export notes.

Quizzes: Chapter-specific quizzes help students quickly assess their understanding of key concepts while exercising their critical-thinking abilities and visual interpretation skills.

Self-Guided Tutorials: 22 interactive and animated tutorials focus on important, hard-to-grasp astronomy concepts. Icons throughout the text direct students to specific Self-Guided Tutorial lessons related to the topics discussed in a given section.

Interactive Figures™ and Interactive Photos™: Interactive versions of key figures and photos in the book provide a study resource for visual learners. Icons placed in figure captions throughout the book direct students to the related Interactive Figures and Photos online.

Glossary and Flashcards: Flashcards of key terms and an interactive glossary help students learn and remember astronomy terminology.

Movies: These brief videos summarize key concepts in astronomy.

RSS Feeds provide the most current articles from well-known astronomy resources on discoveries and topics of interest in the field. Accompanying research activities are provided in the *Astronomy Media Workbook,* Sixth Edition.

Instructor and Student Supplements

FOR INSTRUCTORS

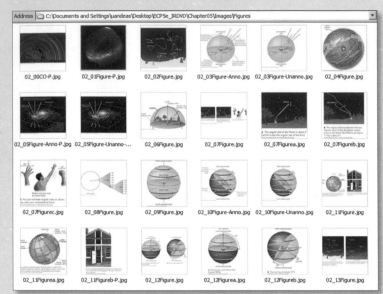

Address | C:\Documents and Settings\uandeas\Desktop\ECP5e_IRDVD\Chapter05\Images\Figures

02_00CO-P.jpg   02_01Figure-P.jpg   02_02Figure.jpg   02_03Figure-Anno.jpg   02_03Figure-Unanno.jpg   02_04Figure.jpg

02_05Figure-Anno-P.jpg   02_05Figure-Unanno-...   02_06Figure.jpg   02_07Figure.jpg   02_07Figurea.jpg   02_07Figureb.jpg

02_07Figurec.jpg   02_08Figure.jpg   02_09Figure.jpg   02_10Figure-Anno.jpg   02_10Figure-Unanno.jpg   02_11Figure.jpg

02_11Figurea.jpg   02_11Figureb-P.jpg   02_12Figure.jpg   02_12Figurea.jpg   02_12Figureb.jpg   02_13Figure.jpg

Instructor Resource DVD ▲
978-0-321-58161-7 | 0-321-58161-X
This DVD includes high resolution JPEGs of all images from the
book for improved in-class projection, Interactive Figures and
Photos™ based on figures from the book, informative applets and
animations, pre-built PowerPoint® Lecture Outlines, and Clicker
Quizzes based on the book and book-specific media resources.

Test Bank (IRC Download)
978-0-321-50612-2 | 0-321-50612-X
The Test Bank contains Multiple Choice, True/False, and Short
Answer questions, plus a new set of Process of Science questions
for each chapter. TestGen® and Microsoft® Word files are available
for download from the Instructor Resource Center (IRC).

Instructor Guide (IRC Download)
978-0-321-58160-0 | 0-321-58160-1
The Instructor Guide contains teaching tips, sample syllabi, and answers
to all of the end-of-chapter questions. Microsoft Word files are
available for download from the Instructor Resource Center (IRC).

Cosmos DVD Boxed Set
© 2000 • 978-0-8053-8572-4 • 0-8053-8572-X
Arguably the most popular science television
program ever produced, Carl Sagan's Cosmos series
has captivated more than 600 million reviewers.
The entire updated series is available on
DVD for all adopting professors. The
Instructor Guide includes a lesson plan
showing how to integrate specific video
segments into your classes.

FOR STUDENTS

Starry Night Pro™ 6 Student DVD
978-0-321-53973-1 | 0-321-53973-7
This best-selling planetarium software lets you escape the Milky
Way and travel within 700 million light years of space. View more
than 16 million stars in stunningly realistic star fields. Zoom in on
thousands of galaxies, nebulae, and star clusters. Move through
200,000 years of time to see key celestial events in our dynamic
and ever-changing universe. Blast off from Earth and see the
motions of the planets from a new perspective. Hailed for its
breath taking realism, powerful features, and intuitive interface,
Starry Night Pro lives up to its reputation as astronomy software's
brightest…night after night.

Voyager: SkyGazer, College Edition v4.0
Built specifically for students studying introductory astronomy,
the updated Voyager: SkyGazer, College Edition v4.0, combines

exceptional planetarium software
with informative pre-packaged
tutorials. Based on the popular
Voyager software, this dual-platform
CD-ROM is included at no additional
charge with all new copies of the
textbook. You will find recommended
SkyGazer activities at the end of each
chapter in the text and in the
Astronomy Media Workbook.

Lecture Tutorials for Introductory Astronomy, Second Edition
978-0-132-39226-6 | 0-132-39226-7
Funded by the National Science Foundation, Lecture Tutorials for
Introductory Astronomy is designed to help make large lecture-

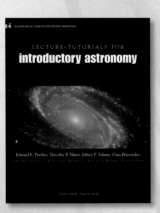

format courses more interactive with
easy-to implement student activities
that can be integrated into existing
course structures. The Second Edition
contains nine new activities that focus
on planetary science, system related
topics, and the interactions of light and
matter. Each of the 38 Lecture
Tutorials challenges students with a
series of carefully designed questions
that spark classroom discussion,
engage students in critical reasoning,
and require no equipment.

For a complete list of supplements, visit the web catalog at **www.pearsonhighered.com**

the essential

cosmic

perspective

FIFTH EDITION

JEFFREY BENNETT
University of Colorado at Boulder

MEGAN DONAHUE
Michigan State University

NICHOLAS SCHNEIDER
University of Colorado at Boulder

MARK VOIT
Michigan State University

Addison-Wesley

San Francisco   Boston   New York   Cape Town
Hong Kong   London   Madrid   Mexico City   Montreal
Munich   Paris   Singapore   Sydney   Tokyo   Toronto

Publisher: James Smith
Executive Editor: Nancy Whilton
Development Manager: Michael Gillespie
Associate Development Editor: Ashley Eklund
Senior Media Producer: Deb Greco
Media Producer: David Huth
Senior Marketing Manager: Scott Dustan
Associate Director of Production: Erin Gregg
Managing Editor: Corinne Benson
Production Supervisor: Mary O'Connell
Production Service: Lifland et al., Bookmakers
Composition: Progressive Information Technologies
Design: Sandra N. Salamony
Cover Design: Sandra N. Salamony
Manufacturing Buyer: Jeffrey Sargent
Director, Image Resource Center: Melinda Patelli
Manager, Rights and Permissions: Zina Arabia
Manager, Cover Visual Research & Permissions: Karen Sanatar
Image Permission Coordinator: Elaine Soares
Printer and Binder: Quebecor Versailles
Cover Printer: Phoenix Color
Cover Photograph (Milky Way over Utah): Wally Pacholka

Library of Congress Cataloging-in-Publication Data

The essential cosmic perspective / Jeffrey Bennett ... [et al.]. -- 5th ed.
      p. cm.
   ISBN 978-0-321-58088-7
   1. Astronomy--Textbook. I. Bennett, Jeffrey O.

   QB43.3.E87 2009
   520--dc22

                              2008044433

ISBN 10-digit 0-321-58088-5; 13-digit 978-0-321-58088-7 (Student edition)
ISBN 10-digit 0-321-58089-3; 13-digit 978-0-321-58089-4 (Professional copy)

Addison-Wesley
is an imprint of

PEARSON

www.pearsonhighered.com   1 2 3 4 5 6 7 8 9 10— QWV—12 11 10 09 08

dedication

TO ALL WHO HAVE EVER WONDERED about the mysteries of the universe. We hope this book will answer some of your questions—and that it will also raise new questions in your mind that will keep you curious and interested in the ongoing human adventure of astronomy.

And, especially, to Michaela, Emily, Sebastian, Grant, Nathan, Brooke, and Angela. The study of the universe begins at birth, and we hope that you will grow up in a world with far less poverty, hatred, and war so that all people will have the opportunity to contemplate the mysteries of the universe into which they are born.

brief contents

detailed contents

preface

WE HUMANS HAVE GAZED into the sky for countless generations. We have wondered how our lives are connected to the Sun, Moon, planets, and stars that adorn the heavens. Today, through the science of astronomy, we know that these connections go far deeper than our ancestors ever imagined. This book tells the story of modern astronomy and the new perspective, *The Essential Cosmic Perspective,* that astronomy gives us on ourselves and our planet.

This book grew out of our experience teaching astronomy to both college students and the general public over the past 30 years. During this time, a flood of new discoveries fueled a revolution in our understanding of the cosmos but had little impact on the basic organization and approach of most astronomy textbooks. We felt the time had come to rethink how to organize and teach the major concepts in astronomy to reflect this revolution in scientific understanding. This book is the result.

Who Is This Book for?

The Essential Cosmic Perspective is designed as a textbook for college courses in introductory astronomy, but is suitable for anyone who is curious about the universe. We assume no prior knowledge of astronomy or physics, and the book is especially written for students who do not intend to major in mathematics or science.

We have tailored *The Essential Cosmic Perspective* to one-semester survey courses in astronomy by carefully selecting the most important topics and presenting them with only as much depth as can be realistically learned in one semester. This book may also be used for two-semester astronomy sequences, though instructors of such courses may wish to consider our more comprehensive book, *The Cosmic Perspective.*

About the Fifth Edition

The underlying philosophy, goals, and structure of *The Essential Cosmic Perspective* remain the same as in past editions, but we have thoroughly updated the text and made a number of other improvements. Here, briefly, is a list of the significant changes you'll find in this fifth edition:

- *Fully Updated Science:* Astronomy is a fast-moving field, and numerous new developments have occurred since the prior edition was published. The topics updated in this edition include

  - New developments in the study of extrasolar planets and planetary systems

  - Discussion of the IAU decision to create a new "dwarf planet" category

  - New results and images from spacecraft exploring our solar system, including *Phoenix* on Mars, *Cassini* at Saturn, *MESSENGER* at Mercury, and more

  - The latest observational evidence for dark matter and dark energy

  - Recent results from the Spitzer Space Telescope, Hubble Space Telescope, and Chandra X-ray Observatory

  - And much more

- *New Cosmic Context 2-page Visual Summaries:* We have added 5 new, 2-page *Cosmic Context* spreads, and improved the existing 12 spreads from the fourth edition. You'll find one of these at the end of each of the 6 parts of the book, and the rest appear within the main bodies of various chapters.

- *New "See It for Yourself" Feature:* Within the bodies of chapters, we have added an occasional feature called "See It for Yourself" that asks students to try a simple observation or experiment that will help them understand the chapter material.

- *New "Process of Science" Questions:* Long-time users know that our text has the process of science deeply integrated throughout the textbook. To help students focus on this important concept, in each chapter we have added two or more "process of science" questions to the end-of-chapter exercise sets.

- *New "Visual Skills Check" Questions:* Each chapter's end-of-chapter exercises concludes with a new set of questions designed to help students build their skills at interpreting the many types of visual information used in astronomy.

- *Enhanced MasteringAstronomy™ (www.masteringastronomy .com):* We have reached the point where *The Essential Cosmic Perspective* is no longer just a textbook; rather it is a "learning package" that combines a printed book with deeply integrated, interactive media that we have developed to support every chapter of our book. For students, MasteringAstronomy™ provides a wealth of tutorials and activities to build understanding, while quizzes and exercises allow them to test what they've learned. For instructors, MasteringAstronomy™ provides the unprecedented ability to quickly build, post, and automatically grade pre- and post-lecture diagnostic tests, weekly homework assignments, and exams of appropriate difficulty, duration, and content coverage. It also provides the ability to record detailed information on the step-by-step work of every student directly into a powerful and easy-to-use gradebook, and to evaluate results with a sophisticated suite of diagnostics. Among the changes you'll find to the MasteringAstronomy™ site for this edition are

- Updated quizzes in the self-study area, including a Reading Quiz, Concept Quiz, and Visual Quiz for each chapter

- New Interactive Figures and Photos in the self-study area

- Nearly 200 automatically graded tutorials, more than 50 of which are entirely new for this edition, along with most of the end-of-chapter questions available for online homework assignment and automatic grading and diagnostics

- At least one new tutorial per chapter that focuses specifically on helping students understand the process of science

- RSS feeds from a variety of notable astronomy publications

- A fully customizable interactive eBook with embedded links to multimedia and glossary terms

Themes of *The Essential Cosmic Perspective*

The Essential Cosmic Perspective offers a broad survey of our modern understanding of the cosmos and of how we have built that understanding. Such a survey can be presented in a number of different ways. We have chosen to interweave a few key themes throughout the book, each selected to help make the subject more appealing to students who may never have taken any formal science courses and who may begin the course with little understanding of how science works. Our book is built around the following five key themes:

- *Theme 1: We are a part of the universe and thus can learn about our origins by studying the universe.* This is the overarching theme of *The Essential Cosmic Perspective,* as we continually emphasize that learning about the universe helps us understand ourselves. Studying the intimate connections between human life and the cosmos gives students a reason to care about astronomy and also deepens their appreciation of the unique and fragile nature of our planet.

- *Theme 2: The universe is comprehensible through scientific principles that anyone can understand.* We can understand the universe because the same physical laws appear to be at work in every aspect, on every scale, and in every age of the universe. Moreover, while professional scientists generally have discovered the laws, anyone can understand their fundamental features. Students can learn enough in one or two terms of astronomy to comprehend the basic reasons for many phenomena that they see around them—ranging

from seasonal changes and phases of the Moon to the most esoteric astronomical images that appear in the news.

- *Theme 3: Science is not a body of facts but rather a process through which we seek to understand the world around us.* Many students assume that science is just a laundry list of facts. The long history of astronomy shows that science is a process through which we learn about our universe—a process that is not always a straight line to the truth. That is why our ideas about the cosmos sometimes change as we learn more, as they did dramatically when we first recognized that Earth is a planet going around the Sun rather than the center of the universe. In this book, we continually emphasize the nature of science so that students can understand how and why modern theories have gained acceptance and why these theories may change in the future.

- *Theme 4: A course in astronomy is the beginning of a lifelong learning experience.* Building upon the prior themes, we emphasize that what students learn in their astronomy course is not an end but a beginning. By remembering a few key physical principles and understanding the nature of science, students can follow astronomical developments for the rest of their lives. We therefore seek to motivate students to continue to participate in the ongoing human adventure of astronomical discovery.

- *Theme 5: Astronomy affects each of us personally with the new perspectives it offers.* We all conduct the daily business of our lives with reference to some "world view"—a set of personal beliefs about our place and purpose in the universe that we have developed through a combination of schooling, religious training, and personal thought. This world view shapes our beliefs and many of our actions. Although astronomy does not mandate a particular set of beliefs, it does provide perspectives on the architecture of the universe that can influence how we view ourselves and our world, which can potentially affect our behavior. In many respects, the role of astronomy in shaping world views may represent the deepest connection between the universe and the everyday lives of humans.

Pedagogical Principles of *The Essential Cosmic Perspective*

No matter how an astronomy course is taught, it is very important to present material according to a clear set of pedagogical principles. The following list briefly summarizes the major pedagogical principles that we apply throughout this book. (The *Instructor Guide* describes these principles in more detail.)

- *Stay focused on the big picture.* Astronomy is filled with interesting facts and details, but they are meaningless unless they fit into a big picture view of the universe. We therefore take care to stay focused on the big picture (essentially the themes discussed above) at all times. A major benefit of this approach is that although students may forget individual facts and details after the course is over, the big picture framework should stay with them for life.

- *Always provide context first.* We all learn new material more easily when we understand why we are learning it. We therefore begin the book (in Chapter 1) with a broad overview of modern understanding of the cosmos so that students know what they will be studying in the rest of the book. We maintain this "context first" approach throughout the book by always telling students what they will be learning, and why, before diving into the details.

- *Make the material relevant.* It's human nature to be more interested in subjects that seem relevant to our lives. Fortunately, astronomy is filled with ideas that touch each of us personally. By emphasizing our personal connections to the cosmos, we make the material more meaningful, inspiring students to put in the effort necessary to learn it.

- *Emphasize conceptual understanding over the "stamp collecting" of facts.* If we are not careful, astronomy can appear to be an overwhelming collection of facts that are easily forgotten when the course ends. We therefore emphasize a few key concepts that we use over and over again. For example, the laws of conservation of energy and conservation of angular momentum (introduced in Section 4.3) reappear throughout the book, and we find that the wide variety of features found on the terrestrial planets can be understood through just a few basic geological processes. Research shows that, long after the course is over, students are far more likely to retain such conceptual ideas than individual facts or details.

- *Proceed from the more familiar and concrete to the less familiar and abstract.* It's well known that children learn best by starting with concrete ideas and then generalizing to abstractions. The same is true for many adults. We therefore always try to "build bridges to the familiar"—that is, to begin with concrete or familiar ideas and then gradually develop more general principles from them.

- *Use plain language.* Surveys have found that the number of new terms in many introductory astronomy books is larger than the number of words taught in many first-year foreign language courses. This means that most books are teaching astronomy in what looks to students like a foreign language! It is much easier for students to understand key astronomical concepts if they are explained in plain English without resorting to unnecessary jargon. We have gone to great lengths to eliminate jargon as much as possible or, at minimum, to replace standard jargon with terms that are easier to remember in the context of the subject matter.

- *Recognize and address student misconceptions.* Students do not arrive as blank slates. Most students enter our courses not only lacking the knowledge we hope to teach but often holding misconceptions about astronomical ideas. Therefore, to teach correct ideas, we must also help students recognize the paradoxes in their prior misconceptions. We address this issue in a number of ways, most overtly with Common

Misconceptions boxes. These summarize commonly held misconceptions and explain why they cannot be correct.

The Topical (Part) Structure of *The Essential Cosmic Perspective*

The Essential Cosmic Perspective is organized into six broad topical areas (the six parts in the table of contents), each approached in a distinctive way designed to help maintain the focus on the themes discussed earlier. Here, we summarize the guiding philosophy through which we have approached each topic.

Every part concludes with a two-page Cosmic Context figure, which ties together into a coherent whole the diverse ideas covered in the individual chapters.

Part I: Developing Perspective (Chapters 1–3)

Guiding Philosophy: Introduce the big picture, the process of science, and the historical context of astronomy.

The basic goal of these chapters is to give students a big picture overview and context for the rest of the book and to help them develop an appreciation for the process of science and how science has developed through history. Chapter 1 offers an overview of our modern understanding of the cosmos, thereby giving students perspective on the entire universe. Chapter 2 provides an introduction to basic sky phenomena, including seasons and phases of the Moon, and a perspective on how phenomena we experience every day are tied to the broader cosmos. Chapter 3 discusses the nature of science, offering a historical perspective on the development of science and giving students perspective on how science works and how it differs from nonscience.

The Cosmic Context for Part I appears on pp. 82–83.

Part II: Key Concepts for Astronomy (Chapters 4–5)

Guiding Philosophy: Bridges to the familiar.

These chapters lay the groundwork for understanding astronomy through what is sometimes called the "universality of physics"—the idea that a few key principles governing matter, energy, light, and motion explain both the phenomena of our daily lives and the mysteries of the cosmos. Chapter 4 covers the laws of motion, the crucial conservation laws of angular momentum and energy, and the universal law of gravitation. Chapter 5 covers the nature of light and matter, spectra, and telescopes.

The Cosmic Context for Part II appears on pp. 140–141.

Part III: Learning from Other Worlds (Chapters 6–9)

Guiding Philosophy: Learning about Earth by learning about other planets in our solar system and beyond.

This set of chapters begins with a broad overview of the solar system and discussion of solar system formation in Chapter 6; this chapter also includes discussion of extrasolar planets. The next three chapters focus respectively on the terrestrial planets, the jovian planets, and the small bodies of the solar system. Note that Part III is essentially independent of Parts IV and V, and thus can be covered either before or after them.

The Cosmic Context for Part III appears on pp. 282–283.

Part IV: Stars (Chapters 10–13)

Guiding Philosophy: We are intimately connected to the stars.

These are our chapters on stars and stellar life cycles. Chapter 10 covers the Sun in depth, so that it can serve as a concrete model for building an understanding of other stars. Chapter 11 describes the general properties of stars, how we measure these properties, and how we classify stars using the H-R diagram. Chapter 12 covers stellar evolution, tracing the birth-to-death lives of both low- and high-mass stars. Chapter 13 covers the end points of stellar evolution: white dwarfs, neutron stars, and black holes.

The Cosmic Context for Part IV appears on pp. 384–385.

Part V: Galaxies and Beyond (Chapters 14–17)

Guiding Philosophy: Present galaxy evolution in a way that parallels the teaching of stellar evolution, integrating cosmological ideas in the places where they most naturally arise.

These chapters cover galaxies and cosmology. Chapter 14 presents the Milky Way as a paradigm for galaxies in much the same way that Chapter 10 uses the Sun as a paradigm for stars. Chapter 15 presents the variety of galaxies, how we determine key parameters such as galactic distances and age, and current understanding of galaxy evolution. Chapter 16 focuses on dark matter and dark energy and their role in the fate of the universe. Chapter 17 covers the theory of the Big Bang. Throughout these chapters, we integrate cosmological ideas as they arise. For example, we cover Hubble's law in Chapter 15 because of its importance to the cosmic distance scale and to our understanding of what we see when we look at distant galaxies.

The Cosmic Context for Part V appears on pp. 498–499.

Part VI: Life on Earth and Beyond (Chapter 18)

Guiding Philosophy: The study of life on Earth helps us understand the search for life in the universe.

This part consists of a single chapter. It may be considered optional, to be used as time allows. Those who wish to teach a more detailed course on astrobiology may consider the text *Life in the Universe*, Second Edition by Bennett and Shostak.

The Cosmic Context for Part VI appears on pp. 534–535.

Pedagogical Features of *The Essential Cosmic Perspective*

Alongside the main narrative, *The Essential Cosmic Perspective* includes a number of pedagogical devices designed to enhance student learning:

- **Learning Goals** Presented as key questions, motivational learning goals begin every chapter, and every section of every chapter is carefully written to address the specific learning goal in the title. This helps students stay focused on the big picture and stay motivated by the understanding they will gain.

- **Chapter Summary** The end-of-chapter summary offers a concise review of the learning goal questions, helping reinforce student understanding of key concepts from the chapter. Thumbnail figures are included to remind students of key illustrations and photos in the chapter.

- **Highlighted "Essential Points"** These call attention to key points and help students find the relevant discussion in the text.

- **Annotated Figures** Key figures in each chapter now include the research-proven technique of "annotation"— carefully crafted text placed on the figure (in blue) to guide students through interpreting graphs, following process figures, and translating between different representations.

- **Cosmic Context Two-Page Visual Summaries** These new two-page figures pull together related ideas in spectacular visual summaries.

- **Wavelength/Observatory Icons** For astronomical photographs (or astronomy art that may be confused with photographs), simple icons identify the wavelength band; whether the image is a photo, artist's impression, or computer simulation; and whether the image came from ground-based or space-based observations.

- **MasteringAstronomy™ Self-Guided Tutorials** Lessons from within the highly acclaimed self-guided tutorials on www.masteringastronomy.com are referenced above specific section titles to direct students to targeted, self-paced help.

- **Think About It** This feature, which appears throughout the book as short questions integrated into the narrative, gives students the opportunity to reflect on important new concepts. It also serves as an excellent starting point for classroom discussions.

- **NEW! See It for Yourself** This feature, which appears throughout the book as short questions integrated into the narrative, gives students the opportunity to conduct simple observations or experiments that will help them understand key concepts.

- **Common Misconceptions** These boxes address popularly held but incorrect ideas related to the chapter material.

- **Special Topic Boxes** These boxes contain supplementary discussion topics related to the chapter material but not prerequisite to the continuing discussion.

- **Cosmic Calculations Boxes** These boxes contain optional mathematics, set in the margin of the text.

- **The Big Picture** Every chapter narrative ends with this feature. It helps students put what they've learned in the chapter into the context of the overall goal of gaining a broader perspective on ourselves and our planet.

- **End-of-Chapter Questions** Each chapter includes an extensive set of exercises that can be used for study, discussion, or assignment. NEW in this edition: All of the end-of-chapter exercises are organized into the following subsets:

  - Review Questions: Questions that students should be able to answer from the reading alone.

  - Does It Make Sense? (or similar title): A set of short statements for each of which students are expected to determine whether the statement makes sense, and to explain why or why not. These exercises are generally easy once students understand a particular concept, but very difficult otherwise; thus, they are an excellent probe of comprehension.

  - Quick Quiz: A short multiple-choice quiz that allows students to check their progress.

  - Process of Science Questions: New to this edition, essay and discussion questions that ask students to reflect on how science is done and how astronomers have learned about the universe over time.

  - Short-Answer/Essay Questions: Questions that go beyond the Review Questions in asking for conceptual interpretation.

  - Quantitative Problems: Problems that require some mathematics, usually based on topics covered in the Cosmic Calculations boxes.

  - Discussion Questions: Open-ended questions for class discussions.

  - Web Projects: Online research projects designed for independent study.

  Nearly all end-of-chapter questions are available at www.masteringastronomy.com for online homework assignment and automatic grading and diagnostics.

- **NEW! Visual Skills Check** Each chapter ends with a set of questions designed to help students build their skills at interpreting the many types of visual information used in astronomy. Answers to these questions appear in Appendix K.

- **Glossary** A detailed glossary makes it easy for students to look up important terms.

- **Appendixes** The appendixes include a number of useful references and tables, including key constants (Appendix A), key formulas (Appendix B), key mathematical skills (Appendix C), and numerous data tables and star charts (Appendixes D–J), plus the answers to the Visual Skills Check questions (Appendix K).

MasteringAstronomy™—A New Paradigm in Astronomy Teaching

What is the single most important factor in student success in astronomy? Both research and common sense reveal the same answer: *study time*. No matter how good the teacher, or how good the textbook, students learn only when they spend adequate time studying. Unfortunately, limitations on resources for grading have prevented most instructors from assigning much homework despite its obvious benefits to student learning. And limitations on help and office hours have made it difficult for students to make sure they use self-study time effectively. That, in a nutshell, is why we have created MasteringAstronomy™. For students, it provides the first adaptive-learning, online system to coach them *individually*—responding to their errors with specific, targeted feedback, and providing hints for partial credit to help them when they get stuck. For professors, MasteringAstronomy™ provides the unprecedented ability to automatically monitor and record students' step-by-step work and evaluate the effectiveness of assignments and exams. As a result, we believe that MasteringAstronomy™ will create a paradigm shift in the way astronomy courses are taught: For the first time, it will be possible, even in large classes, to ensure that each student spends his or her study time on optimal learning activities outside of class.

MasteringAstronomy™ provides students with a wealth of self-study resources, including interactive tutorials targeting the most difficult concepts of the course, interactive versions of key figures and photos, and quizzes and other activities for self-assessment covering every chapter and every week. For professors, MasteringAstronomy™ provides the first library of tutoring activities and assessment problems pretested and informed by students nationally. You can choose from more than 4,500 activities and problems to automatically assign, grade, and track: pre- and post-lecture diagnostic quizzes, tutoring activities, end-of-chapter problems, even test bank questions. You can find a walk-through of the major features of MasteringAstronomy™ in the front of this book, though of course the best way to become familiar with it is to spend some time on the Web site. We invite you to visit www.masteringastronomy.com to see it for yourself.

Finally, in a world where every publisher walks into a professor's office and claims that their Web site is better than anyone else's, we'd like to point out four reasons why you'll discover that MasteringAstronomy™ really does stand out from the crowd:

- MasteringAstronomy™ has been built specifically to support the structure and pedagogy of *The Essential Cosmic Perspective*. You'll find the same concepts emphasized in the book and the Web site, using the same terminology and the same pedagogical approaches. This type of consistency ensures that students focus on the concepts, without the risk of becoming confused by different presentations.

- Nearly all MasteringAstronomy™ content has been developed either directly by *The Essential Cosmic Perspective* author team or in close collaboration with outstanding educators including Jim Dove, Ed Prather, Tim Slater, Daniel Lorenz, Jonathan Williams, Lauren Jones, and others. The direct involvement of book authors ensures that you can expect the same high level of quality in our Web site that you have come to expect in our textbook.

- The MasteringAstronomy™ platform uses the same unique student-driven engine as the highly successful MasteringPhysics™ product (the most widely adopted physics tutorial and assessment system), developed by a group led by MIT physicist David Pritchard. This robust platform gives instructors unprecedented power not only to tailor content to their own courses, but also to evaluate the effectiveness of assignments and exams, thereby enabling instructors to adapt their lectures, assignments, and tests to the students' needs as the course progresses.

- With *The Essential Cosmic Perspective*, Fifth Edition, we are excited to be offering two new features on Mastering-Astronomy™:

  - **RSS Feeds**, which provide the latest articles from a number of astronomy publications and organizations.

  - **Interactive myeBook**, a fully customizable eBook. Users can search for words or phrases, create notes, highlight text, bookmark sections, click on definitions to key terms, and launch Self-Guided Tutorials and Interactive Figures and Photos™ as they read. Professors also have the ability to annotate the text for their course and hide chapters not covered in their syllabi.

All new textbooks come with a free, 1-year subscription to MasteringAstronomy™. Students buying a used book can purchase access online at www.masteringastronomy.com.

Additional Supplements for *The Essential Cosmic Perspective*

The Essential Cosmic Perspective is much more than just a textbook. It is a complete package of teaching, learning, and assessment resources designed to help both teachers and students. In addition to MasteringAstronomy™, the following supplements are available with this book:

- FREE with all new books: ***Voyager: SkyGazer v4.0 College Edition™ CD-ROM*** Based on *Voyager IV*, one of the world's most popular planetarium programs, *SkyGazer*

makes it easy for students to learn constellations and explore the wonders of the sky through interactive exercises. The *SkyGazer* CD is packaged free with all new copies of this book. Accompanying activities are available in LoPresto's *Astronomy Media Workbook*, Sixth Edition.

- ***NEW! Starry Night™ Pro Software*** Now available as an additional option with *The Essential Cosmic Perspective, Starry Night*™ has been acclaimed as the world's most realistic desktop planetarium software, and is available as an additional bundle. Ask your Pearson sales representative for details.

- ***In-Class Active-Learning Tutorials (ISBN 0-8053-8296-8)*** This workbook by Marvin DeJong provides fifty 20-minute in-class tutorial activities to choose from. Designed for use in large lecture classes, these activities are also suitable for labs. Students can complete these short tutorials on their own or in peer-learning groups. Each activity targets specific learning objectives such as understanding Newton's laws, understanding Mars's retrograde motion, tracking stars on the H-R diagram, or comparing the properties of planets.

- ***NEW! Astronomy Media Workbook, Sixth Edition (ISBN 0-321-55627-5)*** Now offered on CD-ROM as an additional bundling option, the *Astronomy Media Workbook* by Michael LoPresto includes a wide selection of in-depth activities based on the Interactive Figures™ and RSS Feeds on MasteringAstronomy™, and *Voyager: SkyGazer v4.0* planetarium software. These thought-provoking projects are suitable for labs or homework assignments.

Instructor-Only Supplements

Several additional supplements are available for instructors only. Contact your local Addison-Wesley sales representative to find out more about the following supplements:

- ***Instructor Resource DVD-ROM*** This DVD-ROM provides a wealth of lecture and teaching resources, including high-resolution JPEGs of all images from the book for in-class projection, Interactive Figures and Photos™ based on figures from the book, informative applets and animations, pre-built PowerPoint® Lecture Outlines, and PRS-enabled Clicker Quizzes based on the book and book-specific interactive media.

- ***Clickers in the Astronomy Classroom (ISBN 0-8053-9616-0)*** This 100-page handbook by Douglas Duncan provides everything you need to know to successfully introduce or enhance your use of CRS (clicker) quizzing in your astronomy class—the research-proven benefits, common pitfalls to avoid, and a wealth of thought-provoking astronomy questions for every week of your course.

- ***Instructor Guide (ISBN 0-321-58160-1)*** This guide contains a detailed overview of the text, sample syllabi for courses of different emphasis and duration, suggestions for teaching strategies, answers or discussion points for all Think About It and See It for Yourself questions in the text, solutions to end-of-chapter problems, and a detailed reference guide summarizing media resources available for every chapter and section in the book. Word files can be downloaded from the Instructor Resource Center (www.pearsonhighered.com/irc).

- **Carl Sagan's *Cosmos* (DVD or Video)** The *Best of Cosmos* and the complete, revised, enhanced, and updated *Cosmos* series are available free to qualified adopters of *The Essential Cosmic Perspective*. A week-by-week guide of segments to include in your course is provided in the *Instructor Guide*.

- ***Test Bank (ISBN 0-321-58162-8)*** The *Test Bank* includes hundreds of multiple-choice, true/false, and short answer questions, plus a new set of Process of Science questions for each chapter. TestGen® and Word files can be downloaded from the Instructor Resource Center.

Acknowledgments

A textbook may carry author names, but it is the result of hard work by a long list of committed individuals. We could not possibly list everyone who has helped, but we would like to call attention to a few people who have played particularly important roles. First, we thank our editors and friends at Addison-Wesley, who have stuck with us through thick and thin, including Adam Black, Nancy Whilton, Linda Davis, Michael Gillespie, Ashley Eklund, Scott Dustan, and Mary O'Connell. Special thanks to our past and present production teams, especially Joan Marsh, Mary Douglas, Brandi Nelson, and Sally Lifland; our art and design team, Mark Ong, Judy and John Waller, and Dartmouth Publishing; our interactive media team, Jim Dove, Lauren Jones, David Huth, and Cadre Design; and our MasteringAstronomy™ team, Deb Greco, Andi Pascarella, Katie Foley, and Caroline Power.

We've also been fortunate to have an outstanding group of reviewers, whose extensive comments and suggestions helped us shape the book. We thank all those who have reviewed the book in various stages, including

Marilyn Akins, *Broome Community College*
Christopher M. Anderson, *University of Wisconsin*
John Anderson, *University of North Florida*
Peter S. Anderson, *Oakland Community College*
Keith Ashman, *University of Missouri—Kansas City*
Simon P. Balm, *Santa Monica College*
Nadine Barlow, *Northern Arizona University*
John Beaver, *University of Wisconsin at Fox Valley*
Peter A. Becker, *George Mason University*
Timothy C. Beers, *Michigan State University*
Jim Bell, *Cornell University*
Priscilla J. Benson, *Wellesley College*
Bernard W. Bopp, *University of Toledo*
Sukanta Bose, *Washington State University*

David Brain, *University of California Berkeley Space Sciences Laboratory*
David Branch, *University of Oklahoma*
John C. Brandt, *University of New Mexico*
James E. Brau, *University of Oregon*
Jean P. Brodie, *UCO/Lick Observatory, University of California, Santa Cruz*
James Brooks, *Florida State University*
Daniel Bruton, *Stephen F. Austin State University*
Amy Campbell, *Louisiana State University*
Eugene R. Capriotti, *Ohio State University*
Eric Carlson, *Wake Forest University*
David A. Cebula, *Pacific University*
Supriya Chakrabarti, *Boston University*
Kwang-Ping Cheng, *California State University Fullerton*
Dipak Chowdhury, *Indiana University—Purdue University Fort Wayne*
Chris Churchill, *New Mexico State University*
Josh Colwell, *University of Colorado*
Anita B. Corn, *Colorado School of Mines*
Philip E. Corn, *Red Rocks Community College*
Kelli Corrado, *Montgomery County Community College*
John Cowan, *University of Oklahoma*
Kevin Crosby, *Carthage College*
Christopher Crow, *Indiana University—Purdue University Fort Wayne*
Manfred Cuntz, *University of Texas at Arlington*
Christopher De Vries, *California State University Stanislaus*
John M. Dickey, *University of Minnesota*
Bryan Dunne, *University of Illinois, Urbana-Champaign*
Suzan Edwards, *Smith College*
Robert Egler, *North Carolina State University at Raleigh*
Paul Eskridge, *Minnesota State University*
David Falk, *Los Angeles Valley College*
Timothy Farris, *Vanderbilt University*
Robert A. Fesen, *Dartmouth College*
Tom Fleming, *University of Arizona*
Douglas Franklin, *Western Illinois University*
Sidney Freudenstein, *Metropolitan State College of Denver*
Martin Gaskell, *University of Nebraska*
Richard Gelderman, *Western Kentucky University*
Harold A. Geller, *George Mason University*
Donna Gifford, *Pima Community College*
Mitch Gillam, *Marion L. Steele High School*
Bernard Gilroy, *The Hun School of Princeton*
David Graff, *U.S. Merchant Marine Academy*
Richard Gray, *Appalachian State University*
Kevin Grazier, *Jet Propulsion Laboratory*
Robert Greeney, *Holyoke Community College*
Henry Greenside, *Duke University*
Alan Greer, *Gonzaga University*
John Griffith, *Lin-Benton Community College*
David Griffiths, *Oregon State University*
David Grinspoon, *University of Colorado*
John Gris, *University of Delaware*
Bruce Gronich, *University of Texas at El Paso*

Thomasana Hail, *Parkland University*
Jim Hamm, *Big Bend Community College*
Charles Hartley, *Hartwick College*
J. Hasbun, *University of West Georgia*
Joe Heafner, *Catawba Valley Community College*
Scott Hildreth, *Chabot College*
Mark Hollabaugh, *Normandale Community College*
Richard Holland, *Southern Illinois University, Carbondale*
Joseph Howard, *Salisbury University*
James Christopher Hunt, *Prince George's Community College*
Richard Ignace, *University of Wisconsin*
James Imamura, *University of Oregon*
Douglas R. Ingram, *Texas Christian University*
Assad Istephan, *Madonna University*
Bruce Jakosky, *University of Colorado*
Adam G. Jensen, *University of Colorado*
Adam Johnston, *Weber State University*
Lauren Jones, *Gettysburg College*
William Keel, *University of Alabama*
Julia Kennefick, *University of Arkansas*
Steve Kipp, *University of Minnesota, Mankato*
Kurtis Koll, *Cameron University*
Ichishiro Konno, *University of Texas at San Antonio*
John Kormendy, *University of Texas at Austin*
Eric Korpela, *University of California, Berkeley*
Kevin Krisciunas, *Texas A&M*
Ted La Rosa, *Kennesaw State University*
Kristine Larsen, *Central Connecticut State University*
Ana Marie Larson, *University of Washington*
Stephen Lattanzio, *Orange Coast College*
Larry Lebofsky, *University of Arizona*
Patrick Lestrade, *Mississippi State University*
Nancy Levenson, *University of Kentucky*
David M. Lind, *Florida State University*
Abraham Loeb, *Harvard University*
Michael LoPresto, *Henry Ford Community College*
William R. Luebke, *Modesto Junior College*
Darrell Jack MacConnell, *Community College of Baltimore City*
Marie Machacek, *Massachusetts Institute of Technology*
Loris Magnani, *University of Georgia*
Steven Majewski, *University of Virginia*
Phil Matheson, *Salt Lake Community College*
John Mattox, *Fayetteville State University*
Marles McCurdy, *Tarrant County College*
Stacy McGaugh, *University of Maryland*
Barry Metz, *Delaware County Community College*
William Millar, *Grand Rapids Community College*
Dinah Moche, *Queensborough Community College of City University, New York*
Stephen Murray, *University of California, Santa Cruz*
Zdzislaw E. Musielak, *University of Texas at Arlington*
Charles Nelson, *Drake University*
Gerald H. Newsom, *Ohio State University*
Brian Oetiker, *Sam Houston State University*
John P. Oliver, *University of Florida*
Stacy Palen, *Weber State University*

Russell L. Palma, *Sam Houston State University*
Bryan Penprase, *Pomona College*
Eric S. Perlman, *University of Maryland, Baltimore County*
Peggy Perozzo, *Mary Baldwin College*
Charles Peterson, *University of Missouri, Columbia*
Cynthia W. Peterson, *University of Connecticut*
Jorge Piekarewicz, *Florida State University*
Lawrence Pinsky, *University of Houston*
Stephanie Plante, *Grossmont College*
Jascha Polet, *California State Polytechnic University, Pomona*
Matthew Price, *Oregon State University*
Harrison B. Prosper, *Florida State University*
Monica Ramirez, *Aims College, Colorado*
Christina Reeves-Shull, *Richland College*
Todd M. Rigg, *City College of San Francisco*
Elizabeth Roettger, *DePaul University*
Roy Rubins, *University of Texas at Arlington*
Carl Rutledge, *East Central University*
Bob Sackett, *Saddleback College*
Rex Saffer, *Villanova University*
John Safko, *University of South Carolina*
James A. Scarborough, *Delta State University*
Britt Scharringhausen, *Ithaca College*
Ann Schmiedekamp, *Pennsylvania State University, Abington*
Joslyn Schoemer, *Denver Museum of Nature and Science*
James Schombert, *University of Oregon*
Gregory Seab, *University of New Orleans*
Larry Sessions, *Metropolitan State College of Denver*
Ralph Siegel, *Montgomery College, Germantown Campus*
Philip I. Siemens, *Oregon State University*
Caroline Simpson, *Florida International University*
Paul Sipiera, *William Harper Rainey College*
Earl F. Skelton, *George Washington University*
Michael Skrutskie, *University of Virginia*
Mark H. Slovak, *Louisiana State University*
Norma Small-Warren, *Howard University*
Dale Smith, *Bowling Green State University*
James R. Sowell, *Georgia Technical University*
Kelli Spangler, *Montgomery County Community College*
John Spencer, *Lowell Observatory*
Darryl Stanford, *City College of San Francisco*
George R. Stanley, *San Antonio College*
John Stolar, *West Chester University*
Jack Sulentic, *University of Alabama*
C. Sean Sutton, *Mount Holyoke College*
Beverley A. P. Taylor, *Miami University*
Brett Taylor, *Radford University*
Donald M. Terndrup, *Ohio State University*
Frank Timmes, *School of Art Institute of Chicago*
David Trott, *Metro State College*
David Vakil, *El Camino College*
Trina Van Ausdal, *Salt Lake Community College*
Licia Verde, *University of Pennsylvania*
Nicole Vogt, *New Mexico State University*
Darryl Walke, *Rariton Valley Community College*
Fred Walter, *State University of New York, Stony Brook*

James Webb, *Florida International University*
Mark Whittle, *University of Virginia*
Paul J. Wiita, *Georgia State University*
Lisa M. Will, *Mesa Community College*
Jonathan Williams, *University of Florida*
J. Wayne Wooten, *Pensacola Junior College*
Scott Yager, *Brevard College*
Andrew Young, *Casper College*
Arthur Young, *San Diego State University*
Min S. Yun, *University of Massachusetts, Amherst*
Dennis Zaritsky, *University of California, Santa Cruz*
Robert L. Zimmerman, *University of Oregon*

Historical Accuracy Reviewer—Owen Gingerich, *Harvard–Smithsonian*

In addition, we thank the following colleagues who helped us clarify technical points or checked the accuracy of technical discussions in the book:

Nahum Arav, *University of Colorado*
Phil Armitage, *University of Colorado*
Thomas Ayres, *University of Colorado*
Cecilia Barnbaum, *Valdosta State University*
Rick Binzel, *Massachusetts Institute of Technology*
Howard Bond, *Space Telescope Science Institute*
David Brain, *University of California Berkeley Space Sciences Laboratory*
Humberto Campins, *University of Florida*
Robin Canup, *Southwest Research Institute*
Clark Chapman, *Southwest Research Institute*
Kelly Cline, *Carroll College*
Josh Colwell, *University of Colorado*
Mark Dickinson, *National Optical Astronomy Observatory*
Jim Dove, *Metropolitan State College of Denver*
Harry Ferguson, *Space Telescope Science Institute*
Andrew Hamilton, *University of Colorado*

Todd Henry, *Georgia State University*
Dennis Hibbert, *Everett Community College*
Dave Jewitt, *University of Hawaii*
Hal Levison, *Southwest Research Institute*
Mario Livio, *Space Telescope Science Institute*
J. McKim Malville, *University of Colorado*
Mark Marley, *New Mexico State University*
Linda Martel, *University of Hawaii*
Kevin McLin, *University of Colorado*
Michael Mendillo, *Boston University*
Rachel Osten, *National Radio Astronomy Observatory*
Bob Pappalardo, *University of Colorado*
Bennett Seidenstein, *Arundel High School*
Michael Shara, *American Museum of Natural History*
Bob Stein, *Michigan State University*
Glen Stewart, *University of Colorado*
John Stolar, *West Chester University*
Jeff Taylor, *University of Hawaii*
Dave Tholen, *University of Hawaii*
Nick Thomas, *MPI/Lindau (Germany)*
Dimitri Veras, *University of Colorado*
John Weiss, *University of Colorado*
Francis Wilkin, *Union College*
Don Yeomans, *Jet Propulsion Laboratory*

Finally, we thank the many people who have greatly influenced our outlook on education and our perspective on the universe over the years, including Tom Ayres, Fran Bagenal, Forrest Boley, Robert A. Brown, George Dulk, Erica Ellingson, Katy Garmany, Jeff Goldstein, David Grinspoon, Robin Heyden, Don Hunten, Geoffrey Marcy, Joan Marsh, Catherine McCord, Dick McCray, Dee Mook, Cheri Morrow, Charlie Pellerin, Carl Sagan, Mike Shull, John Spencer, and John Stocke.

Jeff Bennett
Megan Donahue
Nick Schneider
Mark Voit

about the authors

Jeffrey Bennett

Megan Donahue

JEFFREY BENNETT holds a B.A. (1981) in biophysics from the University of California, San Diego, and an M.S. and Ph.D. (1987) in astrophysics from the University of Colorado, Boulder. He has taught at every level from preschool through graduate school, including more than 50 college classes in astronomy, physics, mathematics, and education. He served 2 years as a visiting senior scientist at NASA headquarters, where he created NASA's "IDEAS" program, started a program to fly teachers aboard NASA's airborne observatories (including the hopefully soon-to-be-flying SOFIA), and worked on numerous educational programs for the Hubble Space Telescope and other space science missions. He also proposed the idea for and helped develop both the Colorado Scale Model Solar System on the CU-Boulder campus and the *Voyage* Scale Model Solar System on the National Mall in Washington, D.C. (He is pictured here with the model Sun.) In addition to this astronomy textbook, he has written college-level textbooks in astrobiology, mathematics, and statistics; two books for the general public, *On the Cosmic Horizon* (Pearson Addison-Wesley, 2001) and *Beyond UFOs* (Princeton University Press, 2008); and an award-winning series of children's books that includes *Max Goes to the Moon, Max Goes to Mars, Max Goes to Jupiter,* and *Max's Ice Age Adventure.* When not working, he enjoys participating in masters swimming and in the daily adventures of life with his wife, Lisa; his children, Grant and Brooke; and his dog, Cosmo. His personal Web site is www.jeffreybennett.com.

MEGAN DONAHUE is an associate professor in the Department of Physics and Astronomy at Michigan State University. Her current research is mainly on clusters of galaxies: their contents—dark matter, hot gas, galaxies, active galactic nuclei—and what they reveal about the contents of the universe and how galaxies form and evolve. She grew up on a farm in Nebraska and received a B.A. in physics from MIT, where she began her research career as an X-ray astronomer. She has a Ph.D. in astrophysics from the University of Colorado, for a thesis on theory and optical observations of intergalactic and intracluster gas. That thesis won the 1993 Trumpler Award from the Astronomical Society for the Pacific for an outstanding astrophysics doctoral dissertation in North America. She continued postdoctoral research in optical and X-ray observations as a Carnegie Fellow at Carnegie Observatories in Pasadena, California, and later as an STScI Institute Fellow at Space Telescope. Megan was a staff astronomer at the Space Telescope Science Institute until 2003, when she joined the MSU faculty. Megan is married to Mark Voit, and they collaborate on many projects, including this textbook and the raising of their children, Michaela, Sebastian, and Angela. Between the births of Sebastian and Angela, Megan qualified for and ran the 2000 Boston Marathon. These days, Megan runs, orienteers, and plays piano and bass guitar whenever her children allow it.

Nicholas Schneider

Mark Voit

NICHOLAS SCHNEIDER is an associate professor in the Department of Astrophysical and Planetary Sciences at the University of Colorado and a researcher in the Laboratory for Atmospheric and Space Physics. He received his B.A. in physics and astronomy from Dartmouth College in 1979 and his Ph.D. in planetary science from the University of Arizona in 1988. In 1991, he received the National Science Foundation's Presidential Young Investigator Award. His research interests include planetary atmospheres and planetary astronomy, with a focus on the odd case of Jupiter's moon Io. He enjoys teaching at all levels and is active in efforts to improve undergraduate astronomy education. Off the job, he enjoys exploring the outdoors with his family and figuring out how things work.

MARK VOIT is an associate professor in the Department of Physics and Astronomy at Michigan State University. He earned his B.A. in astrophysical sciences at Princeton University and his Ph.D. in astrophysics at the University of Colorado in 1990. He continued his studies at the California Institute of Technology, where he was a research fellow in theoretical astrophysics, and then moved on to Johns Hopkins University as a Hubble Fellow. Before going to Michigan State, Mark worked in the Office of Public Outreach at the Space Telescope, where he developed museum exhibitions about the Hubble Space Telescope and was the scientist behind NASA's HubbleSite. His research interests range from interstellar processes in our own galaxy to the clustering of galaxies in the early universe. He is married to coauthor Megan Donahue, and they try to play outdoors with their three children whenever possible, enjoying hiking, camping, running, and orienteering. Mark is also author of the popular book *Hubble Space Telescope: New Views of the Universe.*

how to succeed
in your astronomy course

Using This Book

Each chapter in this book is designed to make it easy for you to study effectively and efficiently. To get the most out of each chapter, you might wish to use the following study plan:

- A textbook is not a novel, and you'll learn best by reading the elements of this text in the following order:

  1. Start by reading the Learning Goals and the introductory paragraphs at the beginning of the chapter so that you'll know what you are trying to learn.

  2. Next, get an overview of the key concepts by studying the illustrations and reading their captions. The illustrations highlight almost all of the major concepts, so this "illustrations first" strategy gives you an opportunity to survey the concepts before you read about them in depth. You will find the *Cosmic Context* figures to be especially useful. Also look for the Interactive Figure icons—when you see one, go to the MasteringAstronomy™ Web site (www.masteringastronomy.com) to try the interactive version.

  3. Read the chapter narrative, but save the boxed features (Common Misconceptions, Special Topics, Cosmic Calculations) to read later. As you read, make notes on the pages to remind yourself of ideas you'll want to review later. Avoid using a highlight pen; underlining with pen or pencil is far more effective, because it forces you to take greater care and therefore helps keep you alert as you study. Be careful to underline selectively—it won't help you later if you've underlined everything.

  4. After reading the chapter once, go back through and read the boxed material. You should read all of the Common Misconceptions and Special Topics boxes; whether you choose to read the Cosmic Calculations is up to you and your instructor. Also watch for the MasteringAstronomy™ tutorial icons throughout the chapter; if a concept is giving you trouble, go to the MasteringAstronomy™ site to try the relevant tutorial.

  5. Then turn your attention to the Chapter Summary. The best way to use the summary is to try to answer the Learning Goal questions for yourself before reading the short answers given in the summary.

- After completing the reading as described above, start testing your understanding with the end-of-chapter exercises. A good way to begin is to make sure you can answer all of the Review Questions; if you don't know an answer, look back through the chapter until you figure it out. Then test your understanding a little more deeply by trying the "Does It Make Sense?", Quick Quiz, and Visual Skill Check questions.

- You can further check your understanding and get feedback on difficulties by trying the online quizzes at www.masteringastronomy.com. Each chapter has three quizzes: a Reading Quiz, a Concept Quiz, and a Visual Quiz. Try the Reading Quiz first. Once you clear up any difficulties you have with it, try the Concept and Visual quizzes.

- If your course has a quantitative emphasis, work through all of the examples in the Cosmic Calculations before trying the quantitative problems for yourself. Remember that you should always try to answer questions qualitatively before you begin plugging numbers into a calculator. For example, make an order-of-magnitude estimate of what your answer should be so that you'll know your calculation is on the right track, and be sure that your answer makes sense and has the appropriate units.

- If you have done all the above, you will have already made use of numerous resources on the MasteringAstronomy™ Web site (www.masteringastronomy.com). Don't stop there; visit the site again and make use of other resources that will help you further build your understanding. These resources have been developed specifically to help you learn the most important ideas in your astronomy course, and they have been extensively tested to make sure they are effective. They really do work, and the only way you'll gain their benefits is by going on the Web site and using them.

If Your Course Is	Times for Reading the Assigned Text (per week)	Times for Homework Assignments (per week)	Times for Review and Test Preparation (average per week)	Total Study Time (per week)
3 credits	2 to 4 hours	2 to 3 hours	2 hours	6 to 9 hours
4 credits	3 to 5 hours	2 to 4 hours	3 hours	8 to 12 hours
5 credits	3 to 5 hours	3 to 6 hours	4 hours	10 to 15 hours

The Key to Success: Study Time

The single most important key to success in any college course is to spend enough time studying. A general rule of thumb for college classes is that you should expect to study about 2 to 3 hours per week *outside* of class for each unit of credit. For example, based on this rule of thumb, a student taking 15 credit hours should expect to spend 30 to 45 hours each week studying outside of class. Combined with time in class, this works out to a total of 45 to 60 hours spent on academic work—not much more than the time a typical job requires, and you get to choose your own hours. Of course, if you are working while you attend school, you will need to budget your time carefully.

As a rough guideline, your studying time in astronomy might be divided as shown in the table at the top of this page. If you find that you are spending fewer hours than these guidelines suggest, you can probably improve your grade by studying longer. If you are spending more hours than these guidelines suggest, you may be studying inefficiently; in that case, you should talk to your instructor about how to study more effectively.

General Strategies for Studying

- Don't miss class. Listening to lectures and participating in discussions is much more effective than reading someone else's notes. Active participation will help you retain what you are learning.

- Take advantage of the resources offered by your professor, whether it be e-mail, office hours, review sessions, online chats, or simply finding opportunities to talk to and get to know your professor. Most professors will go out of their way to help you learn in any way that they can.

- Budget your time effectively. Studying 1 or 2 hours each day is more effective, and far less painful, than studying all night before homework is due or before exams.

- If a concept gives you trouble, do additional reading or studying beyond what has been assigned. And if you still have trouble, ask for help: You surely can find friends, peers, or teachers who will be glad to help you learn.

- Working together with friends can be valuable in helping you understand difficult concepts. However, be sure that you learn *with* your friends and do not become dependent on them.

- Be sure that any work you turn in is of *collegiate quality:* neat and easy to read, well organized, and demonstrating mastery of the subject matter. Although it takes extra effort to make your work look this good, the effort will help you solidify your learning and is also good practice for the expectations that future professors and employers will have.

Preparing for Exams

- Study the Review Questions, and rework problems and other assignments; try additional questions to be sure you understand the concepts. Study your performance on assignments, quizzes, or exams from earlier in the term.

- Study the relevant online tutorials and chapter quizzes available at www.masteringastronomy.com.

- Study your notes from lectures and discussions. Pay attention to what your instructor expects you to know for an exam.

- Reread the relevant sections in the textbook, paying special attention to notes you have made on the pages.

- Study individually *before* joining a study group with friends. Study groups are effective only if every individual comes prepared to contribute.

- Don't stay up too late before an exam. Don't eat a big meal within an hour of the exam (thinking is more difficult when blood is being diverted to the digestive system).

- Try to relax before and during the exam. If you have studied effectively, you are capable of doing well. Staying relaxed will help you think clearly.

foreword: the meaning of *the cosmic perspective*

by Neil deGrasse Tyson

© Neil deGrasse Tyson

Astrophysicist Neil deGrasse Tyson is the Frederick P. Rose Director of New York City's Hayden Planetarium at the American Museum of Natural History. He has written numerous books and articles, hosts the PBS series NOVA scienceNOW, *and was named one of the "Time 100"—Time Magazine's list of the 100 most influential people in the world. He contributed this essay about the meaning of "The Cosmic Perspective," abridged from his 100th essay written for* Natural History *magazine.*

> Of all the sciences cultivated by mankind, Astronomy is acknowledged to be, and undoubtedly is, the most sublime, the most interesting, and the most useful. For, by knowledge derived from this science, not only the bulk of the Earth is discovered . . . ; but our very faculties are enlarged with the grandeur of the ideas it conveys, our minds exalted above [their] low contracted prejudices.
>
> —James Ferguson, *Astronomy Explained Upon Sir Isaac Newton's Principles, and Made Easy To Those Who Have Not Studied Mathematics* (1757)

LONG BEFORE ANYONE knew that the universe had a beginning, before we knew that the nearest large galaxy lies two and a half million light-years from Earth, before we knew how stars work or whether atoms exist, James Ferguson's enthusiastic introduction to his favorite science rang true.

But who gets to think that way? Who gets to celebrate this cosmic view of life? Not the migrant farm worker. Not the sweatshop worker. Certainly not the homeless person rummaging through the trash for food. You need the luxury of time not spent on mere survival. You need to live in a nation whose government values the search to understand humanity's place in the universe. You need a society in which intellectual pursuit can take you to the frontiers of discovery, and in which news of your discoveries can be routinely disseminated.

When I pause and reflect on our expanding universe, with its galaxies hurtling away from one another, embedded with the ever-stretching, four-dimensional fabric of space and time, sometimes I forget that uncounted people walk this Earth without food or shelter, and that children are disproportionately represented among them.

When I pore over the data that establish the mysterious presence of dark matter and dark energy throughout the universe, sometimes I forget that every day—every twenty-four-hour rotation of Earth—people are killing and being killed. In the name of someone's ideology.

When I track the orbits of asteroids, comets, and planets, each one a pirouetting dancer in a cosmic ballet choreographed by the forces of gravity, sometimes I forget that too many people act in wanton disregard for the delicate interplay of Earth's atmosphere, oceans, and land, with consequences that our children and our children's children will witness and pay for with their health and well-being.

And sometimes I forget that powerful people rarely do all they can to help those who cannot help themselves.

I occasionally forget those things because, however big the world is—in our hearts, our minds, and our outsize atlases—the universe is even bigger. A depressing thought to some, but a liberating thought to me.

Consider an adult who tends to the traumas of a child: a broken toy, a scraped knee, a schoolyard bully. Adults know that kids have no clue what constitutes a genuine problem, because inexperience greatly limits their childhood perspective.

As grown-ups, dare we admit to ourselves that we, too, have a collective immaturity of view? Dare we admit that our thoughts and behaviors spring from a belief that the world revolves around us? Part the curtains of society's racial, ethnic, religious, national, and cultural conflicts, and you find the human ego turning the knobs and pulling the levers.

Now imagine a world in which everyone, but especially people with power and influence, holds an expanded view of our place in the cosmos. With that perspective, our problems would shrink—or never arise at all—and we could celebrate our earthly differences while shunning the behavior of our predecessors who slaughtered each other because of them.

• • •

Back in February 2000, the newly rebuilt Hayden Planetarium featured a space show called "Passport to the Universe," which took visitors on a virtual zoom from New York City to the edge of the cosmos. En route the audience saw Earth, then the solar system, then the 100 billion stars of the Milky Way galaxy shrink to barely visible dots on the planetarium dome.

I soon received a letter from an Ivy League professor of psychology who wanted to administer a questionnaire to visitors, assessing the depth of their depression after viewing the show. Our show, he wrote, elicited the most dramatic feelings of smallness he had ever experienced.

How could that be? Every time I see the show, I feel alive and spirited and connected. I also feel large, knowing that the goings-on within the three-pound human brain are what enabled us to figure out our place in the universe.

Allow me to suggest that it's the professor, not I, who has misread nature. His ego was too big to begin with, inflated by delusions of significance and fed by cultural assumptions that human beings are more important than everything else in the universe.

In all fairness to the fellow, powerful forces in society leave most of us susceptible. As was I . . . until the day I learned in biology class that more bacteria live and work in one centimeter of my colon than the number of people who have ever existed in the world. That kind of information makes you think twice about who—or what—is actually in charge.

From that day on, I began to think of people not as the masters of space and time but as participants in a great cosmic chain of being, with a direct genetic link across species both living and extinct, extending back nearly 4 billion years to the earliest single-celled organisms on Earth.

• • •

Need more ego softeners? Simple comparisons of quantity, size, and scale do the job well.

Take water. It's simple, common, and vital. There are more molecules of water in an eight-ounce cup of the stuff than there are cups of water in all the world's oceans. Every cup that passes through a single person and eventually rejoins the world's water supply holds enough molecules to mix 1,500 of them into every other cup of water in the world. No way around it: some of the water you just drank passed through the kidneys of Socrates, Genghis Khan, and Joan of Arc.

How about air? Also vital. A single breathful draws in more air molecules than there are breathfuls of air in Earth's entire atmosphere. That means some of the air you just breathed passed through the lungs of Napoleon, Beethoven, Lincoln, and Billy the Kid.

Time to get cosmic. There are more stars in the universe than grains of sand on any beach, more stars than seconds have passed since Earth formed, more stars than words and sounds ever uttered by all the humans who ever lived.

Want a sweeping view of the past? Our unfolding cosmic perspective takes you there. Light takes time to reach Earth's observatories from the depths of space, and so you see objects and phenomena not as they are but as they once were. That means the universe acts like a giant time machine: the farther away you look, the further back in time you see—back almost to the beginning of time itself. Within that horizon of reckoning, cosmic evolution unfolds continuously, in full view.

Want to know what we're made of? Again, the cosmic perspective offers a bigger answer than you might expect. The chemical elements of the universe are forged in the fires of high-mass stars that end their lives in stupendous explosions, enriching their host galaxies with the chemical arsenal of life as we know it. We are not simply in the universe. The universe is in us. Yes, we are stardust.

• • •

Again and again across the centuries, cosmic discoveries have demoted our self-image. Earth was once assumed to be astronomically unique, until astronomers learned that Earth is just another planet orbiting the Sun. Then we presumed the Sun was unique, until we learned that the countless stars of the night sky are suns themselves. Then we presumed our galaxy, the Milky Way, was the entire known universe, until we established that the countless fuzzy things in the sky are other galaxies, dotting the landscape of our known universe.

The cosmic perspective flows from fundamental knowledge. But it's more than just what you know. It's also about having the wisdom and insight to apply that knowledge to assessing our place in the universe. And its attributes are clear:

- The cosmic perspective comes from the frontiers of science, yet is not solely the provenance of the scientist. It belongs to everyone.

- The cosmic perspective is humble.

- The cosmic perspective is spiritual—even redemptive—but is not religious.

- The cosmic perspective enables us to grasp, in the same thought, the large and the small.

- The cosmic perspective opens our minds to extraordinary ideas but does not leave them so open that our brains spill out, making us susceptible to believing anything we're told.

- The cosmic perspective opens our eyes to the universe, not as a benevolent cradle designed to nurture life but as a cold, lonely, hazardous place.

- The cosmic perspective shows Earth to be a mote, but a precious mote and, for the moment, the only home we have.

- The cosmic perspective finds beauty in the images of planets, moons, stars, and nebulae but also celebrates the laws of physics that shape them.

- The cosmic perspective enables us to see beyond our circumstances, allowing us to transcend the primal search for food, shelter, and sex.

- The cosmic perspective reminds us that in space, where there is no air, a flag will not wave—an indication that perhaps flag waving and space exploration do not mix.

- The cosmic perspective not only embraces our genetic kinship with all life on Earth but also values our chemical kinship with any yet-to-be discovered life in the universe, as well as our atomic kinship with the universe itself.

• • •

At least once a week, if not once a day, we might each ponder what cosmic truths lie undiscovered before us, perhaps awaiting the arrival of a clever thinker, an ingenious experiment, or an innovative space mission to reveal them. We might further ponder how those discoveries may one day transform life on Earth.

Absent such curiosity, we are no different from the provincial farmer who expresses no need to venture beyond the county line, because his forty acres meet all his needs. Yet if all our predecessors had felt that way, the farmer would instead be a cave dweller, chasing down his dinner with a stick and a rock.

During our brief stay on planet Earth, we owe ourselves and our descendants the opportunity to explore—in part because it's fun to do. But there's a far nobler reason. The day our knowledge of the cosmos ceases to expand, we risk regressing to the childish view that the universe figuratively and literally revolves around us. In that bleak world, arms-bearing, resource-hungry people and nations would be prone to act on their "low contracted prejudices." And that would be the last gasp of human enlightenment—until the rise of a visionary new culture that could once again embrace the cosmic perspective.

1

our place in the universe

learning goals

1.1 Our Modern View of the Universe

- What is our place in the universe?
- How did we come to be?
- How can we know what the universe was like in the past?
- Can we see the entire universe?

1.2 The Scale of the Universe

- How big is Earth compared to our solar system?
- How far away are the stars?
- How big is the Milky Way Galaxy?
- How big is the universe?
- How do our lifetimes compare to the age of the universe?

1.3 Spaceship Earth

- How is Earth moving in our solar system?
- How is our solar system moving in tho Milky Way Galaxy?
- How do galaxies move within the universe?
- Are we ever sitting still?

Far from city lights on a clear night, you can gaze upward at a sky filled with stars. Lie back and watch for a few hours, and you will observe the stars marching steadily across the sky. Confronted by the seemingly infinite heavens, you might wonder how Earth and the universe came to be. If you do, you will be sharing an experience common to humans around the world and in thousands of generations past.

Modern science offers answers to many of our fundamental questions about the universe and our place within it. We now know the basic content and scale of the universe. We know the age of Earth and the approximate age of the universe. And, although much remains to be discovered, we are rapidly learning how the simple ingredients of the early universe developed into the incredible diversity of life on Earth.

In this first chapter, we will survey the content and history of the universe, the scale of the universe, and the motions of Earth. We'll develop a "big picture" perspective of our place in the universe that will provide a base on which we can build a deeper understanding in the rest of the book.

1.1   Our Modern View of the Universe

If you observe the sky carefully, you can see why most of our ancestors believed that the heavens revolved about a stationary Earth. The Sun, Moon, planets, and stars appear to circle around our sky each day, and we cannot feel the constant motion of Earth as it rotates on its axis and orbits the Sun. It therefore seems quite natural to assume that we live in an Earth-centered, or *geocentric*, universe.

Nevertheless, we now know that Earth is a planet orbiting a rather average star in a vast universe. The historical path to this knowledge was long and complex. In later chapters, we'll see that many ancient beliefs made sense in their day and changed only when people were confronted by strong evidence to the contrary. We'll also see how the process of science enabled us to acquire this evidence and to learn that we are connected to the stars in ways our ancestors never imagined. First, however, it's useful to have a general picture of the universe as we know it today.

• What is our place in the universe?

Figure 1.1 illustrates our place in the universe with what we might call our "cosmic address." Earth is a planet in our **solar system**, which consists of the Sun and all the objects that orbit it: the planets and their moons, and countless smaller objects including rocky *asteroids* and icy *comets*.

Our Sun is a star, just like the stars we see in our night sky. The Sun and all the stars we can see with the naked eye make up only a small part of a huge, disk-shaped collection of stars called the **Milky Way Galaxy**. A **galaxy** is a great island of stars in space, containing from a few hundred million to a trillion or more stars. The Milky Way Galaxy is relatively

Universe

approx. size: 10^{21} km

Local Supercluster

approx. size: 3×10^{19} km

Local Group

approx. size: 10^{18} km

Milky Way Galaxy

Solar System
(*not to scale*)

approx. size: 10^{10} km

Earth

approx. size: 10^4 km

Figure 1.1
Our cosmic address.

large, containing more than 100 billion stars. Our solar system is located a little over halfway from the galactic center to the edge of the galactic disk.

Billions of other galaxies are scattered throughout space. Some galaxies are fairly isolated, but many others are found in groups. Our Milky Way, for example, is one of the two largest among about 40 galaxies in the **Local Group**. Groups of galaxies with more than a few dozen members are often called **galaxy clusters**.

We live on one planet orbiting one star among more than 100 billion stars in the Milky Way Galaxy, which in turn is one of billions of galaxies in the universe.

On a very large scale, observations show that galaxies and galaxy clusters appear to be arranged in giant chains and sheets with huge voids between them; the background of Figure 1.1 shows this large-scale structure. The regions in which galaxies and galaxy clusters are most tightly packed are called *superclusters*, which are essentially clusters of galaxy clusters. Our Local Group is located in the outskirts of the Local Supercluster.

Together, all these structures make up our **universe**. In other words, the universe is the sum total of all matter and energy, encompassing the superclusters and voids and everything within them.

think about it Some people think that our tiny physical size in the vast universe makes us insignificant. Others think that our ability to learn about the wonders of the universe gives us significance despite our small size. What do *you* think?

• How did we come to be?

According to modern science, we humans are newcomers in an old universe. We'll devote much of the rest of this textbook to studying the scientific evidence that backs up this idea. To help prepare you for this study, let's look at a quick overview of the scientific history of the universe, as summarized in Figure 1.2 (pp. 6–7).

The Big Bang and the Expanding Universe Telescopic observations of distant galaxies show that the entire universe is **expanding**, meaning that the average distances between galaxies are increasing with time. This fact implies that galaxies must have been closer together in the past, and if we go back far enough, we must reach the point at which the expansion began. We call this beginning the **Big Bang**, and from the observed rate of expansion we estimate that it occurred about 14 billion years ago. The three cubes in the upper left corner of Figure 1.2 represent the expansion of a small piece of the universe over time.

The rate at which galaxies are moving apart suggests that the universe was born about 14 billion years ago, in the event we call the Big Bang.

The universe as a whole has continued to expand ever since the Big Bang, but on smaller scales the force of gravity has drawn matter together. Structures such as galaxies and galaxy clusters occupy regions where gravity has won out against the overall expansion. That is, while the universe as a whole continues to expand, individual galaxies and their contents do *not* expand. This idea is also illustrated by the three cubes in Figure 1.2. Notice that as the region as a whole grew larger, the matter within it clumped into galaxies and galaxy clusters. Most galaxies, including our own Milky Way, probably formed within a few billion years after the Big Bang.

Stellar Lives and Galactic Recycling Within galaxies like the Milky Way, gravity drives the collapse of clouds of gas and dust to form stars and planets. Stars are not living organisms, but they nonetheless go through "life cycles." A star is born when gravity compresses the material in a cloud to the point where the center becomes dense and hot enough to generate energy by **nuclear fusion**, the process in which lightweight atomic nuclei smash together and stick (or fuse) to make heavier nuclei. The star "lives" as long as it can generate energy from fusion and "dies" when it finally exhausts its usable fuel.

Stars are born in interstellar clouds, produce energy and new elements through nuclear fusion, and release those new elements in interstellar space when they die.

In its final death throes, a star blows much of its content back out into space. In particular, massive stars die in titanic explosions called *supernovae*. The returned matter mixes with other matter floating between the stars in the galaxy, eventually becoming part of new clouds of gas and dust from which future generations of stars can be born. Galaxies therefore function as cosmic recycling plants, recycling material expelled from dying stars into new generations of stars and planets. This cycle is illustrated in the lower right of Figure 1.2. Our own solar system is a product of many generations of such recycling.

Basic Astronomical Objects, Units, and Motions

This box summarizes a few key astronomical definitions introduced in this chapter and used throughout the book.

Basic Astronomical Objects

star A large, glowing ball of gas that generates heat and light through nuclear fusion in its core. Our Sun is a star.

planet A moderately large object that orbits a star and shines primarily by reflecting light from its star. According to a definition approved in 2006, an object can be considered a planet only if it: (1) orbits a star; (2) is large enough for its own gravity to make it round; and (3) has cleared most other objects from its orbital path. An object that meets the first two criteria but has *not* cleared its orbital path, like Pluto, is designated a *dwarf planet*.

moon (or satellite) An object that orbits a planet. The term *satellite* is also used more generally to refer to any object orbiting another object.

asteroid A relatively small and rocky object that orbits a star. Asteroids are officially considered part of a category known as *small solar system bodies*.

comet A relatively small and ice-rich object that orbits a star. Like asteroids, comets are considered *small solar system bodies*.

Collections of Astronomical Objects

solar system The Sun and all the material that orbits it, including the planets, dwarf planets, and small solar system bodies. Although the term *solar system* technically refers only to our own star system (*solar* means "of the Sun"), it is often applied to other star systems as well.

star system A star (sometimes more than one star) and any planets and other materials that orbit it. Roughly half of all star systems contain two or more stars.

galaxy A great island of stars in space, containing from a few hundred million to a trillion or more stars, all held together by gravity and orbiting a common center.

cluster (or group) of galaxies A collection of galaxies bound together by gravity. Small collections (up to a few dozen galaxies) are generally called *groups*, while larger collections are called *clusters*.

supercluster A gigantic region of space where many individual galaxies and many groups and clusters of galaxies are packed more closely together than elsewhere in the universe.

universe (or cosmos) The sum total of all matter and energy—that is, all galaxies and everything between them.

observable universe The portion of the entire universe that can be seen from Earth, at least in principle. The observable universe is probably only a tiny portion of the entire universe.

Astronomical Distance Units

astronomical unit (AU) The average distance between Earth and the Sun, which is about 150 million kilometers. More technically, 1 AU is the length of the semimajor axis of Earth's orbit.

light-year The distance that light can travel in 1 year, which is about 9.46 trillion kilometers.

Terms Relating to Motion

rotation The spinning of an object around its axis. For example, Earth rotates once each day around its axis, which is an imaginary line connecting the North Pole to the South Pole.

orbit (revolution) The orbital motion of one object around another. For example, Earth orbits around the Sun once each year.

expansion (of the universe) The increase in the average distance between galaxies as time progresses. Note that while the universe as a whole is expanding, individual galaxies and their contents (as well as groups and clusters of galaxies) do *not* expand.

Throughout this book we will see that human life is intimately connected with the development of the universe as a whole. This illustration presents an overview of our cosmic origins, showing some of the crucial steps that made our existence possible.

(1) **Birth of the Universe:** The expansion of the universe began with the hot and dense Big Bang. The cubes show how one region of the universe has expanded with time. The universe continues to expand, but on smaller scales gravity has pulled matter together to make galaxies.

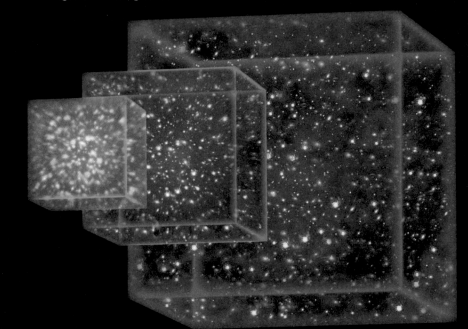

(4) **Earth and Life:** By the time our solar system was born, 4½ billion years ago, about 2% of the original hydrogen and helium had been converted into heavier elements. We are therefore "star stuff," because we and our planet are made from elements manufactured in stars that lived and died long ago.

2 **Galaxies as Cosmic Recycling Plants:** The early universe contained only two chemical elements: hydrogen and helium. All other elements were made by stars and recycled from one stellar generation to the next within galaxies like our Milky Way.

Stars are born in clouds of gas and dust; planets may form in surrounding disks.

Stars shine with energy released by nuclear fusion, which ultimately manufactures all elements heavier than hydrogen and helium.

Massive stars explode when they die, scattering the elements they've produced into space.

3 **Life Cycles of Stars:** Many generations of stars have lived and died in the Milky Way.

One light-year (ly) is defined as the distance that light can travel in 1 year. This distance is fixed because light always travels at the same speed—the *speed of light*, which is 300,000 km/s (186,000 mi/s).

We can calculate the distance represented by a light-year by recalling that

$$distance = speed \times time$$

For example, if you travel at a speed of 50 km/hr for 2 hours, you will travel 100 km. To find the distance represented by 1 light-year, we simply multiply the speed of light by 1 year:

$$1 \text{ light-year} = (\text{speed of light}) \times (1 \text{ yr})$$

Because we are given the speed of light in units of kilometers per second but the time as 1 year, we must carry out the multiplication while converting 1 year into seconds. You can find a review of unit conversions in Appendix C; here, we show the result for this particular case:

$$
\begin{aligned}
1 \text{ light-year} &= \left(300{,}000 \, \frac{km}{s}\right) \times (1 \text{ yr}) \\
&= \left(300{,}000 \, \frac{km}{s}\right) \times \left(1 \, yr \times 365 \, \frac{day}{yr}\right. \\
&\quad \left. \times \, 24 \, \frac{hr}{day} \times 60 \, \frac{min}{hr} \times 60 \, \frac{s}{min}\right) \\
&= 9{,}460{,}000{,}000{,}000 \text{ km}
\end{aligned}
$$

That is, 1 light-year is equivalent to 9.46 trillion km, which is easier to remember as almost 10 trillion km.

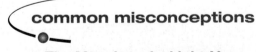

Stars Manufacture the Elements of Earth and Life The recycling of stellar material is connected to our existence in an even deeper way. By studying stars of different ages, we have learned that the early universe contained only the simplest chemical elements: hydrogen and helium (and a trace of lithium). We and Earth are made primarily of other elements, such as carbon, nitrogen, oxygen, and iron. Where did these other elements come from? Evidence shows that these elements were manufactured by stars—some through the nuclear fusion that makes stars shine, and others through nuclear reactions accompanying the explosions that end stellar lives.

> We are "star stuff"—made of material that was manufactured in stars from the simple elements born in the Big Bang.

By the time our solar system formed, about $4\frac{1}{2}$ billion years ago, earlier generations of stars had already converted about 2% of our galaxy's original hydrogen and helium into heavier elements. Therefore, the cloud that gave birth to our solar system was made of about 98% hydrogen and helium and 2% other elements. That 2% may seem a small amount, but it was more than enough to make the small rocky planets of our solar system, including Earth. On Earth, some of these elements became the raw ingredients of simple life forms, which ultimately blossomed into the great diversity of life on Earth today.

In summary, most of the material from which we and our planet are made was created inside stars that lived and died before the birth of our Sun. As astronomer Carl Sagan (1934–1996) said, we are "star stuff."

• How can we know what the universe was like in the past?

You may wonder how we can claim to know anything about what the universe was like in the distant past. It's possible because we can actually see into the past by studying light from distant stars and galaxies.

Light travels extremely fast by earthly standards. The speed of light is 300,000 kilometers per second, a speed at which it would be possible to circle Earth nearly eight times in just 1 second. Nevertheless, even light takes time to travel the vast distances in space. For example, light takes about 1 second to reach Earth from the Moon, and about 8 minutes to reach Earth from the Sun. Light from stars takes many years to reach us, so we measure distances to stars in units called **light-years**. One light-year is the distance that light can travel in 1 year—about 10 trillion kilometers, or 6 trillion miles (see Cosmic Calculations 1.1). Note that a light-year is a unit of *distance*, not time.

Because light takes time to travel through space, we are led to a remarkable fact: **The farther away we look in distance, the further back we look in time**. For example, the brightest star in the night sky, Sirius, is about 8 light-years away, which means its light takes about 8 years to reach us. When we look at Sirius, we are seeing it not as it is today but as it was about 8 years ago.

> Light takes time to travel the vast distances in space. When we look deep into space, we also look far into the past.

The effect is more dramatic at greater distances. The Andromeda Galaxy (also known as M31) lies about 2.5 million light-years from Earth. Figure 1.3 is therefore a picture of how this galaxy looked about

2.5 million years ago, when early humans were first walking on Earth. We see more distant galaxies as they were even further back into the past.

It's also amazing to realize that any "snapshot" of a distant galaxy is a picture of both space and time. For example, because the Andromeda Galaxy is about 100,000 light-years in diameter, the light we see from the far side of the galaxy must have left on its journey to us 100,000 years before the light from the near side. Figure 1.3 therefore shows different parts of the galaxy spread over a time period of 100,000 years. When we study the universe, it is impossible to separate space and time.

see it for yourself Use a star chart (see Appendix J) to find the Andromeda Galaxy in the night sky. At a dark site, you can see a glow from its central regions with your naked eye; otherwise, you can find it with binoculars. Note that you are seeing light that traveled freely through space for some 2.5 million years before reaching your eyes. If students on a planet in the Andromeda Galaxy were looking at the Milky Way now, what would they see? Could they know that we exist here on Earth? Explain.

• Can we see the entire universe?

The fact that looking deep into space means looking far back in time allows us to observe how the universe has changed through time (Figure 1.4). Remember that we think the universe is about 14 billion years old. If we look at a galaxy that is 7 billion light-years away, we see it as it looked 7 billion years ago—which means we see it as it was when the universe was half its current age. If we look at a galaxy that is 12 billion light-years away, we see it as it was 12 billion years ago, when the universe was only 2 billion years old. Thus, by looking to great distances, we can see what the universe looked like when it was younger. The remarkable photo that opens this chapter (p. 1), taken by the Hubble Space Telescope, shows some galaxies that are more than 12 billion light-years away.

Even with the most powerful telescopes imaginable, there's a limit to how far we can see into space: In a universe that is 14 billion years old, we cannot possibly see anything more than 14 billion light-years away. Beyond that—say, at a distance of 15 billion light-years—we'd be looking back to a time before the universe existed, which means there's nothing to see.

Figure 1.3

The Andromeda Galaxy (M31). When we look at this galaxy, we see light that has been traveling through space for 2.5 million years. The inset shows its location in the constellation Andromeda.

Figure 1.4

The farther away we look in space, the further back we look in time. The age of the universe therefore puts a limit on the size of the observable universe—the portion of the entire universe that we could observe in principle.

Far: We see a galaxy 7 billion light-years away as it was 7 billion years ago—when the universe was half its current age of 14 billion years.

Farther: We see a galaxy 12 billion light-years away as it was 12 billion years ago—when the universe was only about 2 billion years old.

The limit of our observable universe: Light from nearly 14 billion light-years away shows the universe as it looked shortly after the Big Bang, before galaxies existed.

Beyond the observable universe: We cannot see anything farther than 14 billion light-years away, because light has not had enough time to reach us.

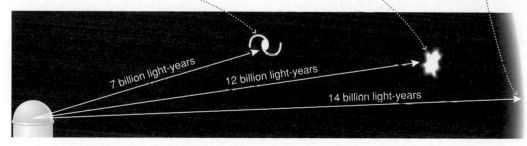

Because the universe is about 14 billion years old, we cannot observe light coming from anything more than 14 billion light-years away.

Our **observable universe**—the portion of the entire universe that we can potentially observe—therefore consists only of objects that lie within 14 billion light-years of Earth. This fact does not put any limit on the size of the *entire* universe, which may be far larger than our observable universe. We simply have no hope of seeing or studying anything beyond the bounds of our observable universe. Anything that exists beyond our observable universe is so far away that its light has not yet had time to reach us.

(MA) Scale of the Universe Tutorial, Lessons 1–3

1.2 The Scale of the Universe

The numbers in our description of the size and age of the universe probably have little meaning for you—after all, they are literally astronomical—but understanding them is crucial to the interpretation of nearly every topic in astronomy. In this section, we will try to give meaning to astronomical distances and times.

• How big is Earth compared to our solar system?

Illustrations and photo montages often make our solar system look like it is crowded with planets and moons, but the reality is far different. One of the best ways to develop perspective on cosmic sizes and distances is to imagine our solar system shrunk down to a scale that would allow you to walk through it. The Voyage scale model solar system in Washington, D.C., makes such a walk possible (Figure 1.5). The Voyage model shows the Sun and the planets, and the distances between them, at *one ten-billionth* of their actual sizes and distances.

Figure 1.6a shows the Sun and planets at their correct sizes (but not distances) on the Voyage scale: The model Sun is about the size of a large grapefruit, Jupiter is about the size of a marble, and Earth is about the size of a pinhead or the ball point in a pen. You can immediately see some key facts about our solar system. For example, the Sun is far larger than any of the planets; in mass, the Sun outweighs all the planets combined by a factor of more than 1000. The planets also vary considerably in size: The storm on Jupiter known as the Great Red Spot (visible near Jupiter's lower left in Figure 1.6a) could swallow up the entire Earth.

The scale of the solar system becomes even more remarkable when you combine the sizes shown in Figure 1.6a with the distances illustrated by the map of the Voyage model in Figure 1.6b. For example, the ball point–sized Earth is located about 15 meters (16.5 yards) from the grapefruit-sized Sun, which means you can picture Earth's orbit by imagining a ball point taking a year to make a circle of radius 15 meters around a grapefruit.

Perhaps the most striking feature of our solar system when we view it to scale is its emptiness. The Voyage model shows the planets along a straight path, so we'd need to draw each planet's orbit around the model Sun to show the full extent of our planetary system. Fitting all these orbits would require an area measuring more than a kilometer on a side—an area equivalent to more than 300 football fields arranged in a grid.

Figure 1.5

This photo shows the pedestals housing the Sun (the gold sphere on the nearest pedestal) and the inner planets in the Voyage scale model solar system (Washington, D.C.). The model planets are encased in the sidewalk-facing disks visible at about eye level on the planet pedestals. The building at the left is the National Air and Space Museum.

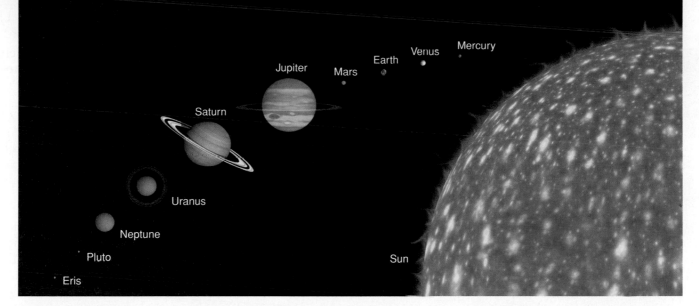

a This painting shows the scaled sizes (but not distances) of the Sun, the planets, and the two largest known dwarf planets.

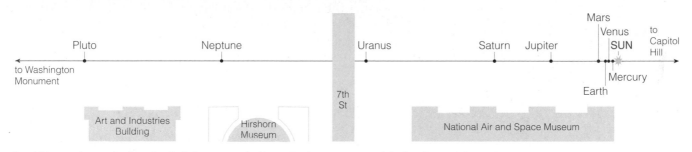

b This map shows the locations of the Sun and planets in the Voyage model; the distance from the Sun to Pluto is about 600 meters (1/3 mile). Planets are lined up in the model, but in reality each planet orbits the Sun independently and a perfect alignment never occurs.

Figure 1.6 interactive figure

The Voyage scale model represents the solar system at one *ten-billionth* of its actual size. Pluto is included in the Voyage model, which was built before the International Astronomical Union classified Pluto as a dwarf planet.

---

special topic: --

How Many Planets Are There in Our Solar System?

AS CHILDREN, WE were taught that there are nine planets in our solar system. However, as you've probably heard, in 2006 astronomers voted to demote Pluto to a dwarf *planet*, leaving our solar system with only eight official planets. Why the change, and is this really the end for Pluto as a planet?

When Pluto was discovered in 1930, it was assumed to be similar to other planets. But as we'll discuss in Chapter 9, we've since learned that Pluto is much smaller than any of the first eight planets and that it shares the outer solar system with thousands of other icy objects. Still, as long as Pluto was the largest known of these objects, most astronomers were content to leave the planetary status quo. Change was forced by the 2005 discovery of an object called Eris. Because Eris is slightly larger than Pluto, astronomers could no longer avoid the question of what objects should count as planets.

At a contentious meeting in August 2006, members of the International Astronomical Union (IAU) considered various options, ultimately voting to define a *planet* as an object that: (1) orbits a star (but is itself neither a star nor a moon); (2) is massive enough for its own gravity to give it a nearly round shape; and (3) has cleared the neighborhood around its orbit. Objects that meet the first two criteria but that have not cleared their orbital neighborhoods—including Pluto, Eris and the asteroid Ceres—are designated *dwarf planets*. The myriad objects that orbit the Sun but are too small to be round, including most asteroids and comets, make up a class called *small solar system bodies*.

It is important to remember that this new definition was established by a vote, making it politics, not science. The politics may yet change again, but the solar system will remain the same. Some people are likely to keep thinking of Pluto as a planet regardless of what professional astronomers say, much as many people still talk of Europe and Asia as separate continents even though both belong to the same land mass (Eurasia). So if you're a Pluto fan, don't despair: It's good to know the official definitions, but it's better to understand the science behind them.

Figure 1.7

The Moon is the most distant place ever visited by humans, yet on the 1-to-10-billion scale of the Voyage model it is only about 4 centimeters ($1\frac{1}{2}$ inches) from Earth. This famous photograph from the first Moon landing (*Apollo 11* in July 1969) shows astronaut Buzz Aldrin, with Neil Armstrong reflected in his visor. Armstrong was the first to step onto the Moon's surface, saying, "That's one small step for a man, one giant leap for mankind."

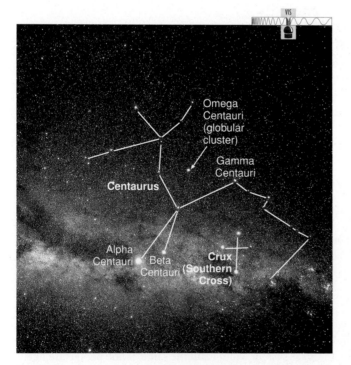

Figure 1.8

This photograph and diagram show the constellation Centaurus, visible from tropical and southern latitudes. Alpha Centauri's real distance of 4.4 light-years becomes 4400 kilometers (2700 miles) on the 1-to-10-billion Voyage scale.

Spread over this large area, only the grapefruit-size Sun, the planets, and a few moons would be big enough to notice with your eyes. The rest of it would look virtually empty (that's why we call it *space*!).

> **think about it** Earth is the only place in our solar system—and the only place we yet know of in the universe—with conditions suitable for human life. How does visualizing Earth to scale affect your perspective on human existence? How does it affect your perspective on our planet? Explain.

On a scale in which the Sun is the size of a grapefruit, Earth is the size of a ball point from a pen, orbiting the Sun at a distance of 15 meters.

Seeing our solar system to scale also helps put space exploration into perspective. The Moon, the only other world on which humans have ever stepped (Figure 1.7), lies only about 4 centimeters ($1\frac{1}{2}$ inches) from Earth in the Voyage model. On this scale, the palm of your hand can cover the entire region of the universe in which humans have so far traveled. The trip to Mars is some 200 times as far as the trip to the Moon, even when Mars is on the same side of its orbit as Earth. And while you can walk from the Sun to Pluto in just a few minutes on the Voyage scale, the *New Horizons* spacecraft that is currently making the real journey will have been in space nearly a decade by the time it finally flies past Pluto in 2015.

• How far away are the stars?

Imagine that you start at the Voyage model Sun in Washington, D.C. You walk the roughly 600-meter distance to Pluto (just over $\frac{1}{3}$ mile) and decide to keep going to find the nearest star besides the Sun. How far would you have to go?

Amazingly, you would need to walk to California. That is, on the same scale that allows you to walk from the Sun to Pluto in minutes, even the nearest stars would be more than 4000 kilometers (2500 miles) away. If this answer seems hard to believe, you can check it for yourself. A light-year is about 10 trillion kilometers, which becomes 1000 kilometers on the 1-to-10-billion scale (because 10 trillion ÷ 10 billion = 1000). The nearest star system to our own, a three-star system called Alpha Centauri (Figure 1.8), is about 4.4 light-years away. That distance becomes about 4400 kilometers (2700 miles) on the 1-to-10-billion scale, or roughly equivalent to the distance across the United States.

On the same scale on which Pluto is just a few minutes' walk from the Sun or Earth, the distance to the nearest stars is equivalent to the distance across the United States.

The tremendous distances to the stars give us some perspective on the technological challenge of astronomy. For example, because the largest star of the Alpha Centauri system is roughly the same size and brightness as our Sun, viewing it in the night sky is somewhat like being in Washington, D.C., and seeing a very bright grapefruit in San Francisco (neglecting the problems introduced by the curvature of the Earth). It may seem remarkable that we can see this star at all, but the blackness of the night sky allows the naked eye to see it as a faint dot of light. It looks much brighter through powerful telescopes, but we still cannot see any features of the star's surface.

Now, consider the difficulty of detecting *planets* orbiting nearby stars. It is equivalent to looking from Washington, D.C., and trying to find ball points or marbles orbiting grapefruits in California or beyond.

When you consider this challenge, it is remarkable to realize that we now have technology capable of finding such planets, at least in some cases [Section 6.5].

The vast distances to the stars also offer a sobering lesson about interstellar travel. Although science fiction shows and movies like *Star Trek* and *Star Wars* make such travel look easy, the reality is far different. Consider the *Voyager 2* spacecraft. Launched in 1977, *Voyager 2* flew by Jupiter in 1979, Saturn in 1981, Uranus in 1986, and Neptune in 1989. It is now bound for the stars at a speed of close to 50,000 kilometers per hour—about 100 times as fast as a speeding bullet. But even at this speed, *Voyager 2* would take about 100,000 years to reach Alpha Centauri if it were headed in that direction (which it's not). Convenient interstellar travel remains well beyond our present technology.

• How big is the Milky Way Galaxy?

The vast separation between our solar system and Alpha Centauri is typical of the separations among star systems here in the outskirts of the Milky Way Galaxy. The 1-to-10-billion scale is useless for thinking about distances beyond the nearest stars, because more distant stars would not fit on Earth on this scale. Visualizing the entire galaxy requires a new scale.

Let's further reduce our solar system scale by a factor of 1 billion (making it a scale of 1 to 10^{19}). On this new scale, each light-year becomes 1 millimeter, and the 100,000-light-year diameter of the Milky Way Galaxy becomes 100 meters, or about the length of a football field. Visualize a football field with a scale model of our galaxy centered over midfield. Our entire solar system is a microscopic dot located around the 20-yard line. The 4.4-light-year separation between our solar system and Alpha Centauri becomes just 4.4 millimeters on this scale—smaller than the width of your little finger. If you stood at the position of our solar system in this model, millions of star systems would lie within reach of your arms.

Another way to put the galaxy into perspective is to consider its number of stars—more than 100 billion. Imagine that tonight you are having difficulty falling asleep (perhaps because you are contemplating the scale of the universe). Instead of counting sheep, you decide to count stars. If you are able to count about one star each second, on average, how long would it take you to count 100 billion stars in the Milky Way? Clearly, the answer is 100 billion (10^{11}) seconds, but how long is that?

It would take thousands of years just to count out loud the number of stars in the Milky Way Galaxy.

Amazingly, 100 billion seconds turns out to be more than 3000 years. (You can confirm this by dividing 100 billion by the number of seconds in 1 year.) You would need thousands of years just to *count* the stars in the Milky Way Galaxy, and this assumes you never take a break—no sleeping, no eating, and absolutely no dying!

• How big is the universe?

As incredible as the scale of our galaxy may seem, the Milky Way is only one of roughly 100 billion galaxies in the observable universe. Just as it would take thousands of years to count the stars in the Milky Way, it would take thousands of years to count all the galaxies.

common misconceptions

Confusing Very Different Things

Most people are familiar with the terms *solar system* and *galaxy*, but people sometimes mix them up. Notice how incredibly different our solar system is from our galaxy. Our solar system is a single star system consisting of our Sun and the various objects that orbit it, including Earth and the other planets. Our galaxy is a collection of more than 100 billion star systems—so many that it would take thousands of years just to count them. Confusing the terms *solar system* and *galaxy* means making a mistake by a factor of 100 billion—a fairly big mistake!

Figure 1.9

The number of stars in the observable universe is comparable to the number of grains of dry sand on all the beaches on Earth.

Figure 1.10

The cosmic calendar compresses the 14-billion-year history of the universe into 1 year, so that each month represents a little more than 1 billion years. This cosmic calendar is adapted from a version created by Carl Sagan.

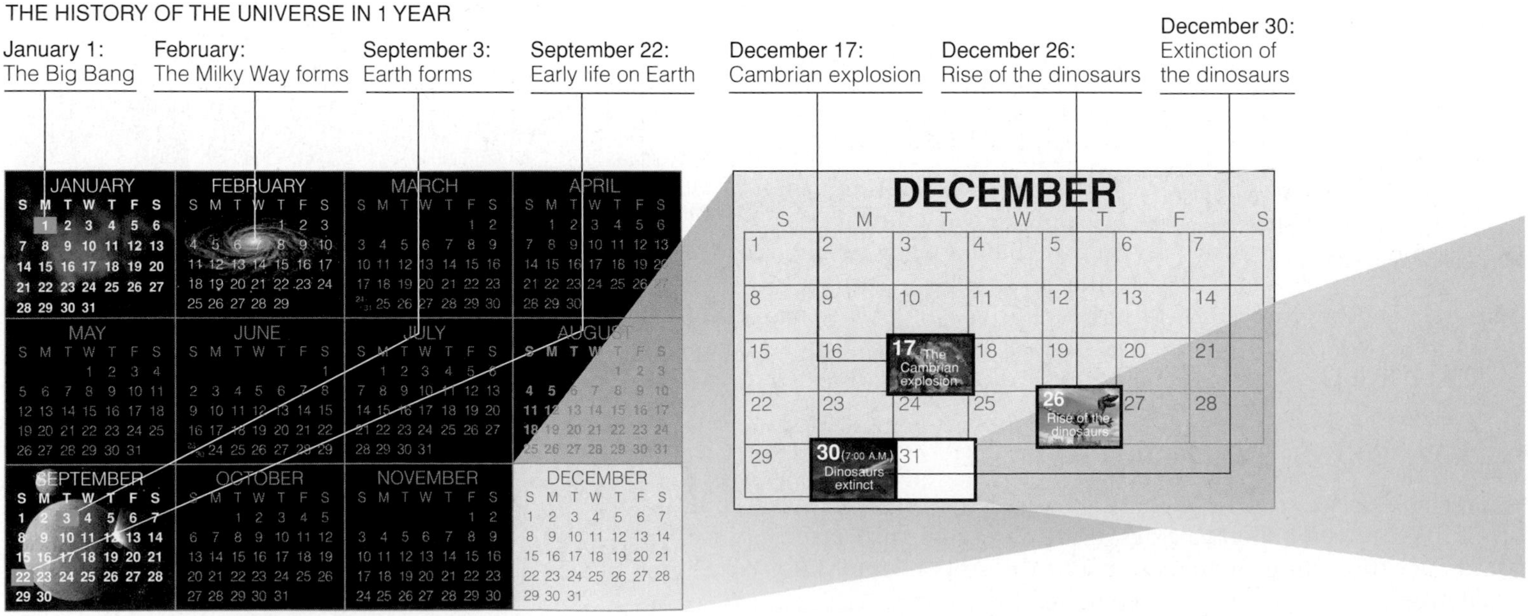

Think for a moment about the total number of stars in all these galaxies. If we assume 100 billion stars per galaxy, the total number of stars in the observable universe is roughly 100 billion × 100 billion, or 10,000,000,000,000,000,000,000 (10^{22}).

Roughly speaking, there are as many stars in the observable universe as there are grains of sand on all the beaches on Earth.

How big is this number? Visit a beach. Run your hands through the fine-grained sand. Imagine counting each tiny grain of sand as it slips through your fingers. Then imagine counting every grain of sand on the beach and continuing on to count every grain of dry sand on *every* beach on Earth. If you could actually complete this task, you would find that, roughly speaking, the number of grains of sand is comparable to the number of stars in the observable universe (Figure 1.9).

think about it Contemplate the fact that there are as many stars in the observable universe as grains of sand on all the beaches on Earth and that each star is a potential sun for a system of planets. With so many possible homes for life, do you think it is conceivable that life exists only on Earth? Why or why not?

• How do our lifetimes compare to the age of the universe?

Now that we have developed some perspective on the scale of space, we can do the same for the scale of time. Imagine the entire history of the universe, from the Big Bang to the present, compressed into a single year. We can represent this history with a *cosmic calendar*, on which the Big Bang takes place at the first instant of January 1 and the present day is just before the stroke of midnight on December 31 (Figure 1.10). For a universe that is about 14 billion years old, each month on the cosmic calendar represents a little more than 1 billion years.

On this time scale, the Milky Way Galaxy probably formed sometime in February. Many generations of stars lived and died in the subsequent cosmic months, enriching the galaxy with the "star stuff" from which we and our planet are made.

THE HISTORY OF THE UNIVERSE IN 1 YEAR

January 1:
The Big Bang

February:
The Milky Way forms

September 3:
Earth forms

September 22:
Early life on Earth

December 17:
Cambrian explosion

December 26:
Rise of the dinosaurs

December 30:
Extinction of the dinosaurs

JANUARY S M T W T F S

FEBRUARY S M T W T F S

MARCH S M T W T F S

APRIL S M T W T F S

MAY S M T W T F S

JUNE S M T W T F S

JULY S M T W T F S

AUGUST S M T W T F S

SEPTEMBER S M T W T F S

OCTOBER S M T W T F S

NOVEMBER S M T W T F S

DECEMBER S M T W T F S

DECEMBER

17 The Cambrian explosion

26 Rise of the dinosaurs

30 (7:00 A.M.) Dinosaurs extinct

Our solar system and our planet did not form until early September on this scale, or $4\frac{1}{2}$ billion years ago in real time. By late September, life on Earth was flourishing. However, for most of Earth's history, living organisms remained relatively primitive and microscopic in size. On the scale of the cosmic calendar, recognizable animals became prominent only in mid-December. Early dinosaurs appeared on the day after Christmas. Then, in a cosmic instant, the dinosaurs disappeared forever—probably due to the impact of an asteroid or a comet [Section 9.4]. In real time, the death of the dinosaurs occurred some 65 million years ago, but on the cosmic calendar it was only yesterday. With the dinosaurs gone, small furry mammals inherited Earth. Some 60 million years later, or around 9 P.M. on December 31 of the cosmic calendar, early hominids (human ancestors) began to walk upright.

If we imagine the 14-billion-year history of the universe compressed into 1 year, a human lifetime lasts only a fraction of a second.

Perhaps the most astonishing thing about the cosmic calendar is that the entire history of human civilization falls into just the last half-minute. The ancient Egyptians built the pyramids only about 11 seconds ago on this scale. About 1 second ago, Kepler and Galileo proved that Earth orbits the Sun rather than vice versa. The average college student was born about 0.05 second ago, around 11:59:59.95 P.M. on the cosmic calendar. On the scale of cosmic time, the human species is the youngest of infants, and a human lifetime is a mere blink of an eye.

1.3 Spaceship Earth

Wherever you are as you read this book, you probably have the feeling that you're "just sitting here." Nothing could be further from the truth. In fact, you are being spun in circles as Earth rotates, you are racing around the Sun in Earth's orbit, and you are careening through the cosmos in the Milky Way Galaxy. In the words of noted inventor and philosopher R. Buckminster Fuller (1895–1983), you are a traveler on *spaceship Earth*. In this section, we'll take a brief look at the motion of spaceship Earth through the universe.

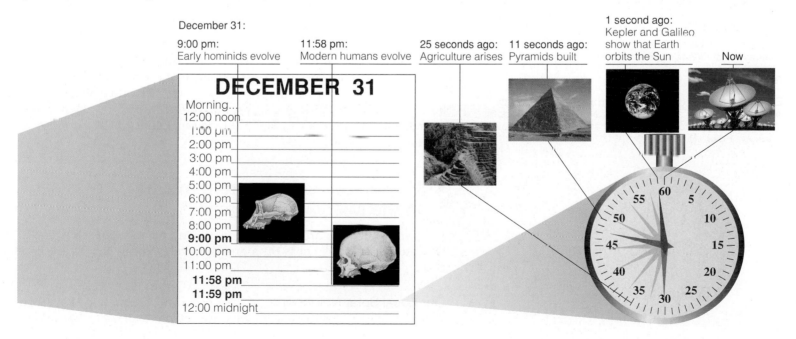

December 31:

9:00 pm:
Early hominids evolve

11:58 pm:
Modern humans evolve

25 seconds ago:
Agriculture arises

11 seconds ago:
Pyramids built

1 second ago:
Kepler and Galileo
show that Earth
orbits the Sun

Now

DECEMBER 31

Morning...
12:00 noon
1:00 pm
2:00 pm
3:00 pm
4:00 pm
5:00 pm
6:00 pm
7:00 pm
8:00 pm
9:00 pm
10:00 pm
11:00 pm
11:58 pm
11:59 pm
12:00 midnight

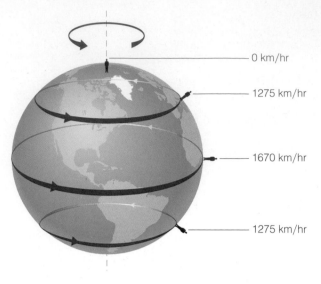

0 km/hr

1275 km/hr

1670 km/hr

1275 km/hr

Figure 1.11

As Earth rotates, your speed around Earth's axis depends on your location: The closer you are to the equator, the faster you travel with rotation. Notice that Earth rotates from west to east, which is why the Sun appears to rise in the east and set in the west.

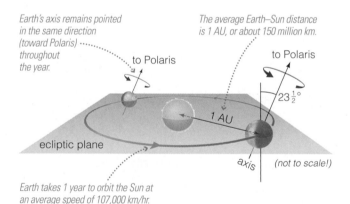

Earth's axis remains pointed in the same direction (toward Polaris) throughout the year.

to Polaris

The average Earth–Sun distance is 1 AU, or about 150 million km.

to Polaris

$23\frac{1}{2}°$

1 AU

ecliptic plane

axis   (not to scale!)

Earth takes 1 year to orbit the Sun at an average speed of 107,000 km/hr.

Figure 1.12

Earth takes a year to complete an orbit of the Sun, but its orbital speed is still surprisingly fast. Notice that Earth both rotates and orbits counterclockwise as viewed from above the North Pole.

• How is Earth moving in our solar system?

The most basic motions of Earth are its daily **rotation** (spin) and its yearly **orbit** (or *revolution*) around the Sun.

Earth rotates once each day around its axis, which is the imaginary line connecting the North Pole to the South Pole. Earth rotates from west to east—counterclockwise as viewed from above the North Pole—which is why the Sun and stars appear to rise in the east and set in the west each day. Although we do not feel any obvious effects from Earth's rotation, the speed of rotation is substantial (Figure 1.11). Unless you live very near the North or South Poles, you are whirling around Earth's axis at a speed of more than 1000 kilometers per hour (600 miles per hour)—faster than most airplanes travel.

Earth rotates once each day and orbits the Sun once each year. Its average orbital distance, called an *astronomical unit* (AU), is about 150 million kilometers.

At the same time Earth is rotating, it is also orbiting the Sun, completing one orbit each year (Figure 1.12). Earth's average orbital distance is called an **astronomical unit**, or **AU**, equivalent to about 150 million kilometers (93 million miles). Again, even though we don't feel the effects of this motion, the speed is impressive: At all times we are racing around the Sun at a speed in excess of 100,000 kilometers per hour (60,000 miles per hour), faster than any spacecraft yet launched.

As you study Figure 1.12, notice that Earth's orbital path defines a flat plane that we call the **ecliptic plane**. Earth's axis is tilted by $23\frac{1}{2}°$ from a line *perpendicular* to the ecliptic plane. This **axis tilt** happens to be oriented so that it points almost directly at a star called *Polaris*, or the *North Star*. Keep in mind that the idea of axis tilt makes sense only in relation to the ecliptic plane. That is, the idea of "tilt" by itself has no meaning in space, where there is no absolute up or down. In space, "up" and "down" mean only "away from the center of Earth (or another planet)" and "toward the center of Earth," respectively.

think about it   If there is no up or down in space, why do you think most globes have the North Pole on top? Would it be equally correct to have the South Pole on top or to turn the globe sideways? Explain.

Notice also that Earth orbits the Sun in the same direction that it rotates on its axis: counterclockwise as viewed from above the North Pole. This is not a coincidence but a consequence of the way our planet was born. As we'll discuss in Chapter 6, Earth and the other planets were born in a spinning disk of gas that surrounded our Sun when it was young, and Earth rotates and orbits in the same direction as the disk was spinning.

• How is our solar system moving in the Milky Way Galaxy?

Rotation and orbit are only part of the travels of spaceship Earth. Our entire solar system is on a great journey within the Milky Way Galaxy.

Our Local Solar Neighborhood Let's begin with the motion of our solar system relative to nearby stars in what we call our *local solar neighborhood*,

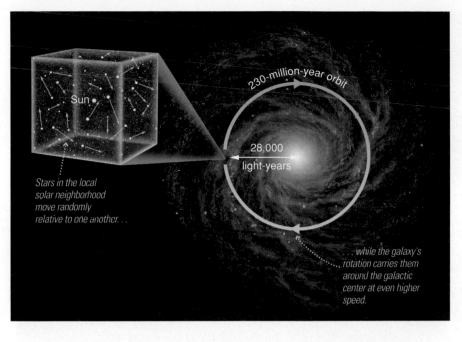

Figure 1.13

This painting illustrates the motion of our solar system within the Milky Way Galaxy. The "zoom in" box shows that stars in our local solar neighborhood move essentially at random relative to one another. At the same time, the entire galaxy is rotating, so that all these stars orbit around the center of the galaxy.

the region of the Sun and nearby stars. The small box in Figure 1.13 shows that stars within the local solar neighborhood (like the stars of any other small region of the galaxy) move essentially at random relative to one another. They also generally move quite fast. For example, we are moving relative to nearby stars at an average speed of about 70,000 kilometers per hour (40,000 miles per hour), about three times as fast as the Space Station orbits Earth.

Stars in our local solar neighborhood move in essentially random directions relative to each other.

Given these high speeds, why don't we see nearby stars racing around the sky? The answer lies in their vast distances from us. You've probably noticed that a distant airplane appears to move through the sky more slowly than one flying close overhead. Stars are so far away that even at speeds of 70,000 kilometers per hour, their motions would be noticeable to the naked eye only if we watched them for thousands of years. That is why the patterns in the constellations seem to remain fixed. Nevertheless, in 10,000 years the constellations will be noticeably different from those we see today. In 500,000 years they will be unrecognizable. If you could watch a time-lapse movie made over millions of years, you *would* see stars racing across the sky.

think about it Despite the chaos of motion in the local solar neighborhood over millions and billions of years, collisions between star systems are extremely rare. Explain why. (*Hint:* Consider the sizes of star systems, such as the solar system, relative to the distances between them.)

Galactic Rotation If you look closely at leaves floating in a stream, their motions relative to one another might appear random, just like the motions of stars in the local solar neighborhood. As you widen your view, you see that all the leaves are being carried in the same general direction by the current. In the same way, as we widen our view beyond the local solar neighborhood, the seemingly random motions of its stars give way to a simpler and even faster motion: rotation of the Milky Way Galaxy. Our solar system, located about 28,000 light-years from the

Most of the galaxy's light comes from stars and gas in the galactic disk and central bulge . . .

. . . but measurements suggest that most of the mass lies unseen in the spherical halo that surrounds the entire disk.

Figure 1.14

This painting shows an edge-on view of the Milky Way Galaxy. Study of galactic rotation shows that although most visible stars lie in the disk and central bulge, most of the mass lies in the halo that surrounds and encompasses the disk. Because this mass emits no light that we have detected, we call it *dark matter.*

galactic center, completes one orbit of the galaxy in about 230 million years. Even if you could watch from outside our galaxy, this motion would be unnoticeable to your naked eye. However, if you calculate the speed of our solar system as we orbit the center of the galaxy, you will find that it is close to 800,000 kilometers per hour (500,000 miles per hour).

The Sun and other stars in our neighborhood orbit the center of the galaxy every 230 million years, because the entire galaxy is rotating.

Careful study of the galaxy's rotation reveals one of the greatest mysteries in science. Stars at different distances from the galactic center orbit at different speeds, and we can learn how mass is distributed in the galaxy by measuring these speeds. Such studies indicate that the stars in the disk of the galaxy represent only the "tip of the iceberg" compared to the mass of the entire galaxy (Figure 1.14). Most of the mass of the galaxy seems to be located outside the visible disk, in what we call the *halo*. We don't know the nature of this mass, but we call it *dark matter* because we have not detected any light coming from it. Studies of other galaxies suggest that they also are made mostly of dark matter, which means this mysterious matter must significantly outweigh the ordinary matter that makes up planets and stars. An even more mysterious *dark energy* seems to make up much of the total energy content of the universe. We'll discuss the mysteries of dark matter and dark energy in Chapter 16.

• How do galaxies move within the universe?

The billions of galaxies in the universe also move relative to one another. Within the Local Group (see Figure 1.1), some of the galaxies move toward us, some move away from us, and at least two small galaxies (known as the Large and Small Magellanic Clouds) apparently orbit our Milky Way Galaxy. Again, the speeds are enormous by earthly standards. For example, the Milky Way is moving toward the Andromeda Galaxy at about 300,000 kilometers per hour (180,000 miles per hour). Despite this high speed, we needn't worry about a collision anytime soon. Even if the Milky Way and Andromeda Galaxies are approaching each other head-on, it will be billions of years before any collision begins.

When we look outside the Local Group, however, we find two astonishing facts recognized in the 1920s by Edwin Hubble, for whom the Hubble Space Telescope was named:

1. Virtually every galaxy outside the Local Group is moving *away* from us.
2. The more distant the galaxy, the faster it appears to be racing away.

These facts might make it sound like we suffer from a cosmic case of chicken pox, but there is a much more natural explanation: *The entire universe is expanding.* We'll save the details for later in the book (Chapter 15), but you can understand the basic idea by thinking about a raisin cake baking in an oven.

Imagine that you make a raisin cake in which the distance between adjacent raisins is 1 centimeter. You place the cake into the oven, where it expands as it bakes. After 1 hour, you remove the cake, which has expanded so that the distance between adjacent raisins has increased to 3 centimeters (Figure 1.15). The expansion of the cake seems fairly obvious. But what would you see if you lived *in* the cake, as we live in the universe?

Pick any raisin (it doesn't matter which one) and call it the Local Raisin. Figure 1.15 shows one possible choice, with three nearby raisins also labeled. The accompanying table summarizes what you would see if you lived within the Local Raisin. Notice, for example, that Raisin 1 starts out at a distance of 1 centimeter before baking and ends up at a distance of 3 centimeters after baking, which means it moves a distance of 2 centimeters away from the Local Raisin during the hour of baking. Hence, its speed as seen from the Local Raisin is 2 centimeters per hour. Raisin 2 moves from a distance of 2 centimeters before baking to a distance of 6 centimeters after baking, which means it moves a distance of 4 centimeters away from the Local Raisin during the hour. Hence, its speed is 4 centimeters per hour, or twice as fast as the speed of Raisin 1. Generalizing, the fact that the cake is expanding means that all the raisins are moving away from the Local Raisin, with more distant raisins moving away faster.

Distant galaxies are all moving away from us, with more distant ones moving faster, indicating that we live in an expanding universe.

Hubble's discovery that galaxies are moving in much the same way as the raisins in the cake, with most moving away from us and more distant ones moving away faster, implies that our universe is expanding much like the raisin cake. If you now imagine the Local Raisin as representing our Local Group of galaxies and the other raisins as representing more distant galaxies or clusters of galaxies, you have a basic picture of the expansion of the universe. Like the expanding batter between the raisins in the cake, *space* itself is growing between galaxies. More distant galaxies move away from us faster because they are carried along with this expansion like the raisins in the expanding cake. Many billions of light-years away, we see galaxies moving away from us at speeds approaching the speed of light.

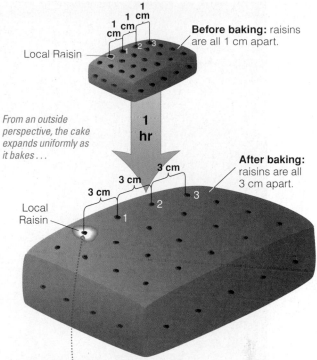

Before baking: raisins are all 1 cm apart.

Local Raisin

From an outside perspective, the cake expands uniformly as it bakes . . .

1 hr

After baking: raisins are all 3 cm apart.

Local Raisin

. . . but from the point of view of the Local Raisin, all other raisins move farther away during baking, with more distant raisins moving faster.

Distances and Speeds as Seen from the Local Raisin

Raisin Number	Distance Before Baking	Distance After Baking (1 hour later)	Speed
1	1 cm	3 cm	2 cm/hr
2	2 cm	6 cm	4 cm/hr
3	3 cm	9 cm	6 cm/hr
⋮	⋮	⋮	⋮

Figure 1.15 interactive figure ↖

An expanding raisin cake offers an analogy to the expanding universe. Someone living in one of the raisins inside the cake could figure out that the cake is expanding by noticing that all other raisins are moving away, with more distant raisins moving away faster. In the same way, we know that we live in an expanding universe because all galaxies outside our Local Group are moving away from us, with more distant ones moving faster.

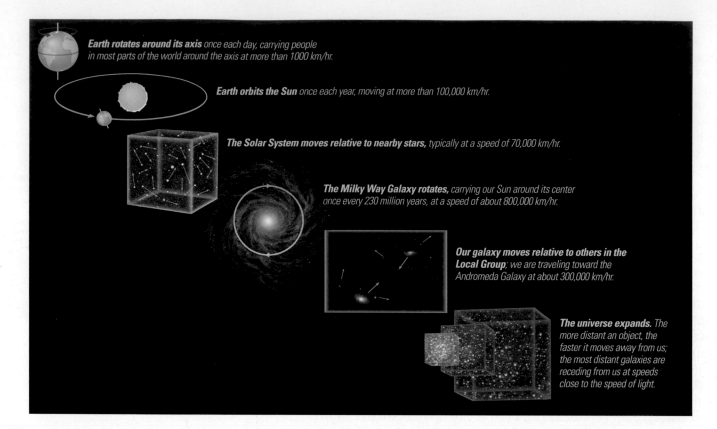

Earth rotates around its axis once each day, carrying people in most parts of the world around the axis at more than 1000 km/hr.

Earth orbits the Sun once each year, moving at more than 100,000 km/hr.

The Solar System moves relative to nearby stars, typically at a speed of 70,000 km/hr.

The Milky Way Galaxy rotates, carrying our Sun around its center once every 230 million years, at a speed of about 800,000 km/hr.

Our galaxy moves relative to others in the Local Group; we are traveling toward the Andromeda Galaxy at about 300,000 km/hr.

The universe expands. The more distant an object, the faster it moves away from us; the most distant galaxies are receding from us at speeds close to the speed of light.

Figure 1.16

This figure summarizes the basic motions of Earth in the universe, along with their associated speeds.

There's one important distinction between the raisin cake and the universe: A cake has a center and edges, but we do not think the same is true of the entire universe. Anyone living in any galaxy in an expanding universe sees just what we see—other galaxies moving away, with more distant ones moving away faster. Because the view from each point in the universe is about the same, no place can claim to be any more "central" than any other place.

It's also important to realize that, unlike the case with a raisin cake, we can't actually *see* galaxies moving apart with time—the distances are too vast for any motion to be noticeable on the time scale of a human life. Instead, we measure the speeds of galaxies by spreading their light into spectra and observing what we call *Doppler shifts* [Section 5.2]. This illustrates how modern astronomy depends both on careful observations and on using current understanding of the laws of nature to explain what we see.

• Are we ever sitting still?

We and our planet are constantly on the move through the universe, and at surprisingly high speeds.

As we have seen, we are never truly sitting still. Figure 1.16 summarizes the motions we have covered. We spin around Earth's axis at more than 1000 km/hr, while our planet orbits the Sun at more than 100,000 km/hr. Our solar system moves among the stars of the local solar neighborhood at typical speeds of 70,000 km/hr, while also orbiting the center of the Milky Way Galaxy at a speed of about 800,000 km/hr. Our galaxy moves among the other galaxies of the Local Group, while all other galaxies move away from us at speeds that increase with distance in our expanding universe. Spaceship Earth is carrying us on a remarkable journey.

---

the big picture
Putting Chapter 1 into Context

In this first chapter, we developed a broad overview of our place in the universe. As we consider the universe in more depth in the rest of the book, remember the following "big picture" ideas:

- Earth is not the center of the universe but instead is a planet orbiting a rather ordinary star in the Milky Way Galaxy. The Milky Way Galaxy, in turn, is one of billions of galaxies in our observable universe.

- We are "star stuff." The atoms from which we are made began as hydrogen and helium in the Big Bang and were later fused into heavier elements by massive stars. Stellar deaths released these atoms into space, where our galaxy recycled them into new stars and planets. Our solar system formed from such recycled matter some $4\frac{1}{2}$ billion years ago.

- Cosmic distances are literally astronomical, but we can put them in perspective with the aid of scale models and other scaling techniques. When you think about these enormous scales, don't forget that every star is a sun and every planet is a unique world.

- We are latecomers on the scale of cosmic time. The universe was already more than half its current age when our solar system formed, and it took billions of years more before humans arrived on the scene.

- All of us are being carried through the cosmos on spaceship Earth. Although we cannot feel this motion, the associated speeds are surprisingly high. Learning about the motions of spaceship Earth gives us a new perspective on the cosmos and helps us understand its nature and history.

summary of key concepts

1.1 Our Modern View of the Universe

• What is our place in the universe?

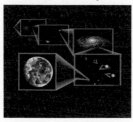
Earth is a planet orbiting the Sun. Our Sun is one of more than 100 billion stars in the **Milky Way Galaxy**. Our galaxy is one of about 40 galaxies in the **Local Group**. The Local Group is one small part of the **Local Supercluster**, which is one small part of the **universe**.

• How did we come to be?

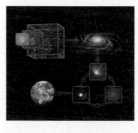
The universe began in the **Big Bang** and has been expanding ever since, except in localized regions where gravity has caused matter to collapse into galaxies and stars. The Big Bang essentially produced only two chemical elements: hydrogen and helium. The rest have been produced by stars, which is why we are "star stuff."

• How can we know what the universe was like in the past?

Light takes time to travel through space, so the farther away we look in distance, the further back we look in time. When we look billions of **light-years** away, we see pieces of the universe as they were billions of years ago.

• Can we see the entire universe?

No. The age of the universe limits the extent of our **observable universe**. Because the universe is about 14 billion years old, our observable universe extends to a distance of about 14 billion light-years. If we tried to look beyond that distance, we'd be trying to look to a time before the universe existed.

1.2 The Scale of the Universe

• How big is Earth compared to our solar system?

On a scale of 1 to 10 billion, the Sun is about the size of a grapefruit. Planets are much smaller, with Earth the size of a ball point and Jupiter the size of a marble on this scale. The distances between planets are huge compared to their sizes, with Earth orbiting 15 meters from the Sun on this scale.

• How far away are the stars?

On the 1-to-10-billion scale, it is possible to walk from the Sun to Pluto in just a few minutes. On the same scale, the nearest stars besides the Sun are thousands of kilometers away.

• How big is the Milky Way Galaxy?

Using a scale on which the Milky Way galaxy is the size of a football field, the distance to the nearest star would be only about 4 millimeters. There are so many stars in our galaxy that it would take thousands of years just to count them.

• How big is the universe?

The observable universe contains roughly 100 billion galaxies, and the total number of stars is comparable to the number of grains of dry sand on all the beaches on Earth.

• How do our lifetimes compare to the age of the universe?

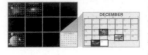

On a cosmic calendar that compresses the history of the universe into 1 year, human civilization is just a few seconds old, and a human lifetime lasts only a fraction of a second.

1.3 Spaceship Earth

• How is Earth moving in our solar system?

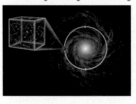

Earth **rotates** on its axis once each day and **orbits** the Sun once each year. Earth orbits at an average distance from the Sun of 1 **AU** and with an **axis tilt** of $23\frac{1}{2}°$ to a line perpendicular to the **ecliptic plane**.

• How is our solar system moving in the Milky Way Galaxy?

We move randomly relative to other stars in our local solar neighborhood. The speeds are substantial by earthly standards, but stars are so far away that their motion is undetectable to the naked eye. Our Sun and other stars in our neighborhood orbit the center of the galaxy every 230 million years, because the entire galaxy is rotating.

• How do galaxies move within the universe?

Galaxies move essentially at random within the Local Group, but all galaxies beyond the Local Group are moving away from us. More distant galaxies are moving faster, which tells us that we live in an expanding universe.

• Are we ever sitting still?

We are never truly sitting still. We spin around Earth's axis and orbit the Sun. Our solar system moves among the stars of the local solar neighborhood while orbiting the center of the Milky Way Galaxy. Our galaxy moves among the other galaxies of the Local Group, while all other galaxies move away from us in our expanding universe.

exercises and problems

For instructor-assigned homework go to **www.masteringastronomy.com**.

Mastering
ASTRONOMY™

Review Questions

Short-Answer Questions Based on the Reading

1. What do we mean by a *geocentric* universe? Contrast a geocentric view with our modern view of the universe.
2. Briefly describe the major levels of structure (such as planet, star, galaxy) in the universe.
3. What do we mean when we say that the universe is *expanding*? How does expansion lead to the idea of the *Big Bang*?
4. What did Carl Sagan mean when he said that we are "star stuff"?
5. How fast does light travel? What is a *light-year*?
6. Explain the statement *The farther away we look in distance, the further back we look in time.*

7. What do we mean by the *observable universe?* Is it the same thing as the entire universe?

8. Describe the solar system as it looks on the 1-to-10-billion scale used in the text. How far away are other stars on this same scale?

9. Describe at least one way to put the scale of the Milky Way Galaxy into perspective and at least one way to put the size of the observable universe into perspective.

10. Use the cosmic calendar to describe how the human race fits into the scale of time.

11. Define *astronomical unit, ecliptic plane*, and *axis tilt*. Explain how each is related to Earth's rotation and/or orbit.

12. What is the shape of the Milky Way Galaxy? Describe our solar system's location and motion.

13. Distinguish between our galaxy's *disk* and *halo*. Where does the mysterious *dark matter* seem to reside?

14. What key observations lead us to conclude that the universe is expanding? Use the raisin cake model to explain how these observations imply expansion.

Test Your Understanding

Does It Make Sense?

Decide whether the statement makes sense (or is clearly true) or does not make sense (or is clearly false). Explain clearly; not all of these have definitive answers, so your explanation is more important than your chosen answer.

Example: I walked east from our base camp at the North Pole.

Solution: The statement does not make sense because east has no meaning at the North Pole—all directions are south from the North Pole.

15. Our solar system is bigger than some galaxies.

16. The universe is billions of light-years in age.

17. It will take me light-years to complete this homework assignment!

18. Someday we may build spaceships capable of traveling a light-year in only a decade.

19. Astronomers discovered a moon that does not orbit a planet.

20. NASA plans soon to launch a spaceship that will photograph our Milky Way Galaxy from beyond its halo.

21. The observable universe is the same size today as it was a few billion years ago.

22. Photographs of distant galaxies show them as they were when they were much younger than they are today.

23. At a nearby park, I built a scale model of our solar system in which I used a basketball to represent Earth.

24. Because nearly all galaxies are moving away from us, we must be located at the center of the universe.

Quick Quiz

Choose the best answer to each of the following. Explain your reasoning with one or more complete sentences.

25. Which of the following correctly lists our "cosmic address" from small to large? (a) Earth, solar system, Milky Way Galaxy, Local Group, Local Supercluster, universe (b) Earth, solar system, Local Group, Local Supercluster, Milky Way Galaxy, universe (c) Earth, Milky Way Galaxy, solar system, Local Group, Local Supercluster, universe.

26. When we say the universe is *expanding*, we mean that (a) everything in the universe is growing in size. (b) the average distance between galaxies is growing with time. (c) the universe is getting older.

27. If stars existed but galaxies did not (a) we would probably exist anyway. (b) we would not exist because life on Earth depends on the light of galaxies. (c) we would not exist because we are made of material that was recycled in galaxies.

28. Could we see a galaxy that is 20 billion light-years away? (a) Yes, if we had a big enough telescope. (b) No, because it would be beyond the bounds of our observable universe. (c) No, because a galaxy could not possibly be that far away.

29. The star Betelgeuse is about 425 light-years away. If it explodes tonight (a) we'll know because it will be brighter than the full Moon in the sky. (b) we'll know because debris from the explosion will rain down on us from space. (c) we won't know about it until 425 years from now.

30. If we represent the solar system on a scale that allows us to walk from the Sun to Pluto in a few minutes, then (a) the planets would be the size of basketballs and the nearest stars would be a few miles away. (b) the planets would all be marble-size or smaller and the nearest stars would be thousands of miles away. (c) the planets would be microscopic and the stars would be light-years away.

31. The total number of stars in the observable universe is roughly equivalent to (a) the number of grains of sand on all the beaches on Earth. (b) the number of grains of sand on Miami Beach. (c) infinity.

32. The age of our solar system is about (a) one-third of the age of the universe. (b) three-fourths of the age of the universe. (c) two billion years less than the age of the universe.

33. An astronomical unit is (a) any planet's average distance from the Sun. (b) Earth's average distance from the Sun. (c) any large astronomical distance.

34. The fact that nearly all galaxies are moving away from us, with more distant ones moving faster, tells us that (a) the universe is expanding. (b) galaxies repel each other like magnets. (c) our galaxy lies near the center of the universe.

Process of Science

Examining How Science Works

35. *Earth as a Planet.* For most of human history, scholars assumed Earth was the center of the universe. Today, we know that Earth is just one planet orbiting the Sun, and the Sun is just one star in a vast universe. How did science make it possible for us to learn these facts about Earth?

36. *Thinking About Scale.* One key to success in science is finding a simple way to evaluate new ideas, and making a simple scale model is often helpful. Suppose someone tells you that the reason it is warmer during the day than at night is that the day side of Earth is closer to the Sun than the night side. Evaluate this idea by thinking about the size of Earth and its distance from the Sun in a scale model of the solar system.

37. *Looking for Evidence.* In this first chapter, we have discussed the scientific story of the universe but have not yet discussed most of the evidence that backs it up. Choose one idea presented in this chapter—such as the idea that there are billions of galaxies in the universe, or that the universe was born in the Big Bang, or that the galaxy contains more dark matter than ordinary matter—and briefly discuss the type of evidence you would want to see before accepting the idea. (*Hint*: It's okay to look ahead in the book to see the evidence presented in later chapters.)

Investigate Further

Short-Answer/Essay Questions

38. *Our Cosmic Origins.* Write one to three paragraphs summarizing why we could not be here if the universe did not contain both stars and galaxies.

39. *Alien Technology.* Some people believe that Earth is regularly visited by aliens who travel here from other star systems. For this to be true, how much more advanced than our own technology would the aliens' technology have to be? Write one to two paragraphs to give a sense of the technological difference. (*Hint:* The ideas of scale in this chapter can help you contrast the distance the aliens would have to travel with the distances we are now capable of traveling.)

40. *Stellar Collisions.* Is there any danger that another star will come crashing through our solar system in the near future? Explain.

41. *Raisin Cake Universe.* Suppose that all the raisins in a cake are 1 centimeter apart before baking and 4 centimeters apart after baking.
    a. Draw diagrams to represent the cake before and after baking.
    b. Identify one raisin as the Local Raisin on your diagrams. Construct a table showing the distances and speeds of other raisins as seen from the Local Raisin.
    c. Briefly explain how your expanding cake is similar to the expansion of the universe.

42. *The Cosmic Perspective.* Write a short essay describing how the ideas presented in this chapter affect your perspectives on your own life and on human civilization.

Quantitative Problems

Be sure to show all calculations clearly and state your final answers in complete sentences.

43. *Distances by Light.* Just as a light-year is the distance that light can travel in 1 year, we define a light-second as the distance that light can travel in 1 second, a light-minute as the distance that light can travel in 1 minute, and so on. Calculate the distance in both kilometers and miles represented by each of the following:
    a. 1 light-second
    b. 1 light-minute
    c. 1 light-hour
    d. 1 light-day

44. *Moonlight and Sunlight.* How long does it take light to travel from
    a. the Moon to Earth?
    b. the Sun to Earth?

45. *Saturn vs. the Milky Way.* Photos of Saturn and photos of galaxies can look so similar that children often think the photos show similar objects. In reality, a galaxy is far larger than any planet. About how many times larger is the diameter of the Milky Way Galaxy than the diameter of Saturn's rings? (*Data:* Saturn's rings are about 270,000 km in diameter; the Milky Way is 100,000 light-years in diameter.)

46. *Driving Trips.* Imagine that you could drive your car at a constant speed of 100 km/hr (62 mi), even across oceans and in space. How long would it take to drive
    a. around Earth's equator? (*Hint:* Use Earth's circumference of about 40,000 km.)
    b. from the Sun to Earth?
    c. from the Sun to Pluto? (*Hint:* You can find Pluto's distance in Appendix E.)
    d. to Alpha Centauri (4.4 light-years away)?

47. *Faster Trip.* Suppose you wanted to reach Alpha Centauri in 100 years.
    a. How fast would you have to go, in km/hr?
    b. How many times faster is the speed you found in (a) than the speeds of our fastest current spacecraft (around 50,000 km/hr)?

Discussion Questions

48. *Vast Orbs.* Dutch astronomer Christiaan Huygens may have been the first person to truly understand both the large sizes of other planets and the great distances to other stars. In 1690, he wrote: "How vast those Orbs must be, and how inconsiderable this Earth, the Theatre upon which all our mighty Designs, all our Navigations, and all our Wars are transacted, is when compared to them. A very fit consideration, and matter of Reflection, for those Kings and Princes who sacrifice the Lives of so many People, only to flatter their Ambition in being Masters of some pitiful corner of this small Spot." What do you think he meant? Explain.

49. *Infant Species.* In the last few tenths of a second before midnight on December 31 of the cosmic calendar, we have developed an incredible civilization and learned a great deal about the universe, but we also have developed technology through which we could destroy ourselves. The midnight bell is striking, and the choice for the future is ours. How far into the next cosmic year do you think our civilization will survive? Defend your opinion.

50. *A Human Adventure.* Astronomical discoveries clearly are important to science, but are they also important to our personal lives? Defend your opinion.

Web Projects

51. *Astronomy on the Web.* The Web contains a vast amount of astronomical information. Spend at least an hour exploring astronomy on the Web. Write two or three paragraphs summarizing what you learned from your search. What was your favorite astronomical Web site, and why?

52. *NASA Missions.* Visit the NASA Web site to learn about upcoming astronomy missions. Write a one-page summary of the mission you feel is most likely to provide new astronomical information during the time you are enrolled in this astronomy course.

53. *The Hubble Ultra Deep Field.* The photo that opens this chapter is called the Hubble Ultra Deep Field. Find the photo on the Hubble Space Telescope Web site. Learn how it was taken, what it shows, and what we've learned from it. Write a short summary of your findings.

Use the following questions to check your understanding of some of the many types of visual information used in astronomy. Answers are provided in Appendix K. For additional practice, try the Chapter 1 Visual Quiz at **www.masteringastronomy.com**.

The figure above shows the sizes of Earth and the Moon to scale; the scale used is 1 cm = 4000 km. Using what you've learned about astronomical scale in this chapter, answer the following questions. *Hint*: If you are unsure of the answers, you can calculate them using the following real values:

Diameter of Earth = 12,800 km
Earth–Moon distance = 384,000 km
Diameter of Sun = 1,400,000 km
Earth–Sun distance = 150,000,000 km

1. If you wanted to show the distance between Earth and the Moon on the same scale, about how far apart would you need to place the two photos above?
   a. 10 centimeters (about the width of your hand)
   b. 1 meter (about the length of your arm)
   c. 100 meters (about the length of a football field)
   d. 1 kilometer (a little more than a half mile)
2. Suppose you wanted to show the Sun on the same scale. About how big would it need to be?
   a. 2.5 centimeters in diameter (the size of a golf ball)
   b. 25 centimeters in diameter (the size of a basketball)
   c. 2.5 meters in diameter (about 8 feet across)
   d. 2.5 kilometers in diameter (the size of a small town)

3. About how far away from Earth would the Sun be located on this scale?
   a. 3.75 meters (about 12 feet)
   b. 37.5 meters (about the height of a 12-story building)
   c. 375 meters (about the length of four football fields)
   d. 37.5 kilometers (the size of a large city)
4. Could you use the same scale to represent the distances to nearby stars? Why or why not?

2

discovering the universe for yourself

learning goals

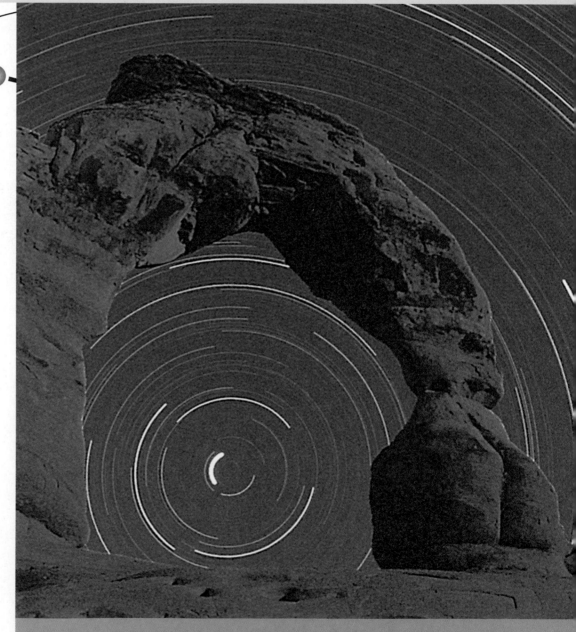

This time-exposure photograph shows star paths at Arches National Park, Utah.

This is an exciting time in the history of astronomy. A new generation of telescopes is scanning the depths of the universe. Increasingly sophisticated space probes are collecting new data about the planets and other objects in our solar system. Rapid advances in computing technology are allowing scientists to analyze the vast amount of new data and to model the processes that occur in planets, stars, galaxies, and the universe.

One goal of this book is to help *you* share in the ongoing adventure of astronomical discovery. One of the best ways to become a part of this adventure is to do what other humans have done for thousands of generations: Go outside, observe the sky around you, and contemplate the awe-inspiring universe of which you are a part. In this chapter, we'll discuss a few key ideas that will help you understand what you see in the sky.

essential preparation

1. What is our place in the universe? [Section 1.1]

2. How far away are the stars? [Section 1.2]

3. Are we ever sitting still? [Section 1.3]

2.1   Patterns in the Night Sky

Today we take for granted that we live on a small planet orbiting an ordinary star in one of many galaxies in the universe. But this fact is not obvious from a casual glance at the night sky, and we've learned about our place in the cosmos only through a long history of careful observations. In this section, we'll discuss the major features of the night sky, and how we understand them in light of our current knowledge of the universe.

• What does the universe look like from Earth?

Shortly after sunset, as daylight fades to darkness, the sky appears to fill slowly with stars. On clear, moonless nights far from city lights, more than 2000 stars may be visible to your naked eye, along with the whitish band of light that we call the *Milky Way* (Figure 2.1). As you look at the stars, your mind may group them into patterns that look like familiar shapes or objects. If you observe the sky night after night or year after year, you will recognize the same patterns of stars. These patterns have not changed noticeably in the past few thousand years.

Constellations   People of nearly every culture gave names to patterns they saw in the sky. We usually refer to such patterns as constellations, but to astronomers the term has a more precise meaning: A **constellation** is a *region* of the sky with well-defined borders; the familiar patterns of stars merely help us locate these constellations.

Bright stars help us identify constellations, which officially are *regions* of the sky.

The names and borders of the 88 official constellations [see Appendix I] were chosen in 1928 by members of the International Astronomical Union. Note that, just as every spot of land in the continental United States is part of some state, every point in the sky belongs to some constellation. For example, Figure 2.2 shows the borders of the constellation Orion and several of its neighbors.

Recognizing the patterns of just 20 to 40 constellations is enough to make the sky seem as familiar as your own neighborhood. The best way

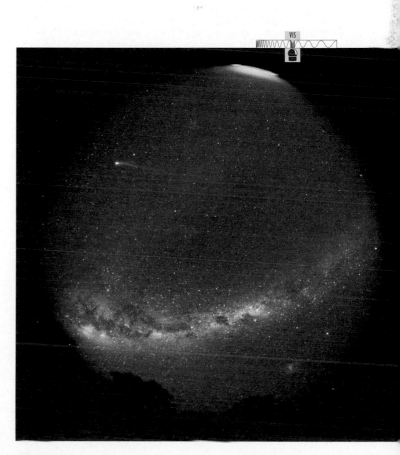

Figure 2.1

A "fish-eye" photograph of the Australian night sky. The Milky Way is the prominent whitish band with dark lanes running through it. A comet (Comet Hyakutake, in 1996) is visible near the upper left of this photo.

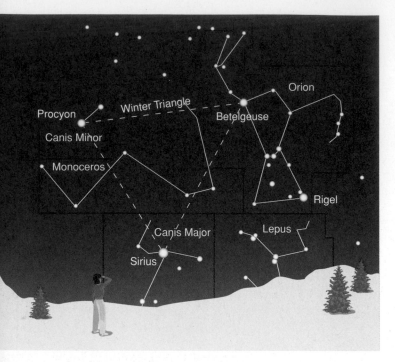

Figure 2.2

Red lines mark official borders of several constellations near Orion. Yellow lines connect recognizable patterns of stars within constellations. Sirius, Procyon, and Betelgeuse form a pattern that spans several constellations and is called the *Winter Triangle*. It is easy to see on clear winter evenings.

to learn the constellations is to go out and view them, guided by a few visits to a planetarium and star charts like the ones in the back of this book [Appendix J]. The *SkyGazer* software that comes with this book can also help you learn constellations.

The Celestial Sphere The stars in a particular constellation appear to lie close to one another but may actually be at very different distances from Earth. This illusion occurs because we lack depth perception when we look into space, a consequence of the fact that the stars are so far away [Section 1.2]. The ancient Greeks mistook this illusion for reality, imagining the stars to lie on a great **celestial sphere** that surrounds Earth (Figure 2.3a).

> All stars appear to lie on a *celestial sphere*, but in reality they lie at different distances from Earth.

We now know that Earth does not really lie in the center of a giant ball of stars, but we can still use the idea of a celestial sphere to help us understand and map the sky. For reference, we identify four special points and circles.

- The **north celestial pole** is the point directly over Earth's North Pole.

- The **south celestial pole** is the point directly over Earth's South Pole.

- The **celestial equator**, which is a projection of Earth's equator into space, makes a complete circle around the celestial sphere.

- The **ecliptic** is the path the Sun follows as it appears to circle around the celestial sphere once each year. It crosses the celestial equator at a $23\frac{1}{2}°$ angle, because that is the tilt of Earth's axis.

Just as a globe shows us the layout of Earth's surface, a model of the celestial sphere shows how the stars appear to be arranged in the sky. A typical model (Figure 2.3b) shows constellation borders, patterns of bright stars, the celestial poles, the celestial equator, and the ecliptic.

Figure 2.3

The stars appear to lie on a great celestial sphere that surrounds Earth.

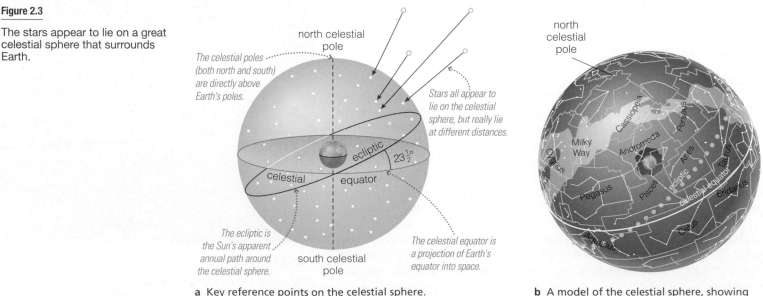

a Key reference points on the celestial sphere.

b A model of the celestial sphere, showing constellation borders and more.

The Milky Way The band of light that we call the Milky Way circles all the way around the celestial sphere, passing through more than a dozen constellations, and bears an important relationship to the Milky Way Galaxy: *It traces our galaxy's disk of stars—the galactic plane—as it appears from our location in the outskirts of the galaxy.*

The Milky Way in the night sky is our view in all directions into the disk of our galaxy.

Figure 2.4 shows this idea. The Milky Way Galaxy is shaped like a thin pancake with a bulge in the middle (see Figure 1.14). We view the universe from our location a little more than halfway out from the center of this "pancake." In any direction that we look *within* the disk of the galaxy, we see countless stars and vast clouds of interstellar gas and dust. These stars and clouds form the Milky Way in our night sky. In contrast, relatively few stars and gas clouds block our view when we look in directions pointing *away* from the galactic disk, thereby allowing us a clear view to the far reaches of the universe (as long as we use sufficiently powerful telescopes).

You'll notice that the Milky Way varies in width and has dark lanes running through it. It is widest in the direction of the constellation Sagittarius, because that is the direction in which we are looking toward the galaxy's central bulge. The dark lanes appear in regions where particularly dense interstellar clouds obscure our view of stars behind them. The Milky Way's abundant gas and dust prevent us from seeing more than a few thousand light-years into our galaxy's disk. This limitation kept much of our own galaxy hidden from view until just a few decades ago, when new technologies allowed us to peer through the clouds by recording forms of light that are invisible to our eyes (such as radio waves and X-rays [Section 5.1]).

The Local Sky The celestial sphere provides a useful way of thinking about the appearance of the universe from Earth. But it is not what we actually see when we go outside. Instead, your **local sky**—the sky as seen from wherever you happen to be standing—appears to take the shape of a hemisphere or dome. The dome shape arises from the fact that we see only half of the celestial sphere at any particular moment from any particular location, while the ground blocks the other half from view.

Figure 2.5 shows key reference features of the local sky. The boundary between Earth and sky defines the **horizon**. The point directly overhead is the **zenith**. The **meridian** is an imaginary half-circle stretching from the horizon due south, through the zenith, to the horizon due north.

We pinpoint an object in the local sky by stating its altitude above the horizon and direction along the horizon.

We can pinpoint the position of any object in the local sky by stating its **direction** along the horizon (sometimes stated as *azimuth*, which is degrees clockwise from due north) and its **altitude** above the horizon. For example, Figure 2.5 shows a person pointing to a star located in the southeast direction at an altitude of 60°. Note that the zenith has altitude 90° but no direction, because it is straight overhead.

Angular Sizes and Distances Our lack of depth perception on the celestial sphere means we have no way to judge the true sizes or separations of the objects we see in the sky. However, we can describe the *angular* sizes or separations of objects even without knowing how far away they are.

Figure 2.4

This painting shows how our galaxy's structure affects our view from Earth.

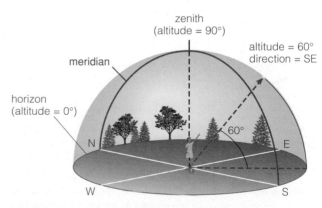

Figure 2.5

From any place on Earth, the local sky looks like a dome (hemisphere). This diagram shows key reference points in the local sky. It also shows how we can describe any position in the local sky by its altitude and direction.

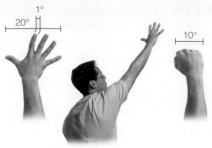

a The angular sizes of the Sun and the Moon are about 1/2°.

b The angular distance between the two "pointer stars" of the Big Dipper is about 5°.

c You can estimate angular sizes or distances with your outstretched hand.

Stretch out your arm as shown here.

Figure 2.6

We measure *angular sizes* or *angular distances*, rather than actual sizes or distances, when we look at objects in the sky.

The **angular size** of an object is the angle it appears to span in your field of view. For example, the angular sizes of the Sun and the Moon are each about $\frac{1}{2}°$ (Figure 2.6a). Note that angular size does not by itself tell us an object's true size, because angular size also depends on distance:

The farther away an object is, the smaller its angular size.

The farther away an object is, the smaller its angular size. For example, the Sun is about 400 times larger in diameter than the Moon, but it has the same angular size in our sky because it is also about 400 times farther away.

think about it Children often try to describe the sizes of objects in the sky (such as the Moon or an airplane) in inches or miles, or by holding their fingers apart and saying, "It was THIS big." Can we really describe objects in the sky in this way? Why or why not?

The **angular distance** between a pair of objects in the sky is the angle that appears to separate them. For example, the angular distance between the "pointer stars" at the end of the Big Dipper's bowl is about 5° (Figure 2.6b). You can use your outstretched hand to make rough estimates of angles in the sky (Figure 2.6c).

For more precise astronomical measurements, we subdivide each degree into 60 **arcminutes** (abbreviated ′) and subdivide each arcminute into 60 arcseconds (abbreviated ″) (Figure 2.7). For example, we read 35°27′15″ as "35 degrees, 27 arcminutes, 15 arcseconds."

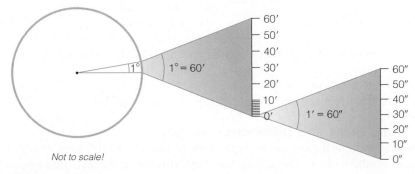

Not to scale!

Figure 2.7

We subdivide each degree into 60 arcminutes and each arcminute into 60 arcseconds.

• Why do stars rise and set?

If you spend a few hours out under a starry sky, you'll notice that the universe seems to be circling around us, with stars moving gradually across the sky from east to west. Many ancient people took this appearance at face value, concluding that we lie in the center of a universe that rotates around us each day. Today we know that the ancients had it backward: It is Earth that rotates, not the rest of the universe, and that is why the Sun, Moon, planets, and stars all move across our sky each day.

We can picture the movement of the sky by imagining the celestial sphere rotating around Earth (Figure 2.8a). From this perspective you can see how the universe seems to turn around us: Every object on the celestial sphere appears to make a simple daily circle around Earth. However, the motion can look a little more complex in the local sky, because the horizon cuts the celestial sphere in half. Figure 2.8b shows the idea for a location in the Northern Hemisphere. If you study the figure carefully, you'll notice the following key facts about the paths of various stars through the local sky:

- Stars near the north celestial pole do not rise or set; rather, they remain above the horizon and make daily counterclockwise circles around the north celestial pole. We say that such stars are **circumpolar**.

- Stars near the south celestial pole never rise above the horizon at all.

- All other stars have daily circles that are partly above the horizon and partly below it. Because Earth rotates from west to east (counterclockwise as viewed from above the North Pole), these stars appear to rise in the east and set in the west.

Earth's west-to-east rotation makes stars appear to move from east to west through the sky as they circle around the celestial poles.

The time-exposure photograph that opens this chapter (p. 26) shows a part of the daily paths of stars. Paths of circumpolar stars are visible within the arch; notice that the complete daily circles for these stars are above the horizon, although the photo shows only a portion of each circle. The north celestial pole lies at

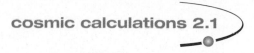
Angular Size, Physical Size, and Distance

If you hold a quarter in front of your eye, it can block your entire field of view. But as you move it farther way, it appears to get smaller and it blocks less of your view. As long as a quarter or any other object is far enough away so that its angular size is relatively small (less than a few degrees), the following formula describes the relationship between the object's angular size, physical size, and distance:

$$\frac{\text{angular size}}{360°} = \frac{\text{physical size}}{2\pi \times \text{distance}}$$

Example: The angular diameter of the Moon is about 0.5° and the Moon is about 380,000 km away. What is the Moon's physical diameter?

Solution: To solve the formula for physical size, we multiply both sides by $2\pi \times$ distance and rearrange:

$$\text{physical size} = \text{angular size} \times \frac{2\pi \times \text{distance}}{360°}$$

We now plug in the given values of the Moon's angular size and distance:

$$\text{physical size} = 0.5° \times \frac{2\pi \times 380,000 \text{ km}}{360°}$$
$$\approx 3300 \text{ km}$$

The Moon's diameter is about 3300 km. (This differs from the precise value of 3476 km because we used inexact values for the angular size and distance.)

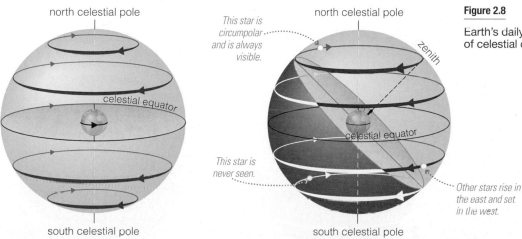

plete daily circles for these stars are above the horizon, although the photo shows only a portion of each circle. The north celestial pole lies at

a Earth rotates from west to east (black arrow), making the celestial sphere *appear* to rotate around us from east to west (red arrows).

b The local sky for a Northern Hemisphere location (40°N). The horizon slices through the celestial sphere at an angle to the equator, causing the daily circles of stars to appear tilted in the local sky. Note: It is easier to follow the star paths if you rotate the page so that the zenith points up.

Figure 2.8

Earth's daily rotation explains the apparent daily motions of celestial objects in our sky.

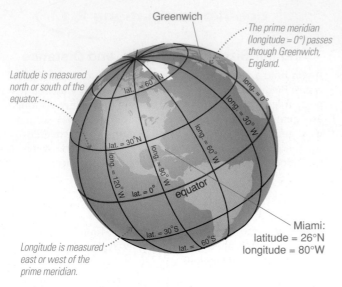

Greenwich

The prime meridian (longitude = 0°) passes through Greenwich, England.

Latitude is measured north or south of the equator.

Longitude is measured east or west of the prime meridian.

Miami:
latitude = 26°N
longitude = 80°W

Figure 2.9

We can locate any place on Earth's surface by its latitude and longitude.

the center of these circles. The circles grow larger for stars farther from the north celestial pole. If they are large enough, the circles cross the horizon, so that the stars rise in the east and set in the west. The same ideas apply in the Southern Hemisphere, except that circumpolar stars are those near the south celestial pole.

think about it Do distant galaxies also rise and set like the stars in our sky? Why or why not?

• Why do the constellations we see depend on latitude and time of year?

If you stay in one place, the basic patterns of motion in the sky will stay the same from one night to the next. However, if you travel far north or south, you'll see a different set of constellations than you see at home. And even if you stay in one place, you'll see different constellations at different times of year. Let's explore why.

Variation with Latitude **Latitude** measures north-south position on Earth, and **longitude** measures east-west position (Figure 2.9). Latitude is defined to be 0° at the equator, increasing to 90°N at the North Pole and 90°S at the South Pole. By international treaty, longitude is defined to be 0° along a line passing through Greenwich, England. Stating a latitude and a longitude pinpoints a location on Earth. For example, Miami lies at about 26°N latitude and 80°W longitude.

The constellations you see depend on your latitude, but not on your longitude.

Latitude affects the constellations we see because it affects the locations of the horizon and zenith relative to the celestial sphere. Figure 2.10 shows how this works for the latitudes of the North Pole (90°N) and Sydney, Australia (34°S). Note that although the local sky varies with latitude, it does *not* vary with longitude. For example, Charleston (South Carolina) and San Diego (California) are at about the same latitude, so people in both cities see the same set of constellations at night.

You can learn much more about how the sky varies with latitude by studying diagrams like those in Figures 2.8 and 2.10. For example, at the

Figure 2.10 interactive figure

The sky varies with latitude. Notice that the altitude of the celestial pole that is visible in your sky is always equal to your latitude.

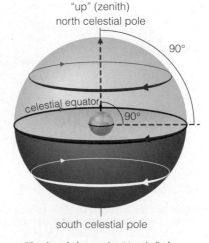

"up" (zenith)
north celestial pole

90°

celestial equator

90°

south celestial pole

a The local sky at the North Pole (latitude 90°N).

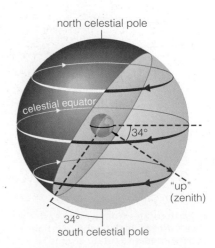

north celestial pole

celestial equator

34°

"up" (zenith)

34°

south celestial pole

b The local sky at Sydney, Australia (latitude 34°S).

North Pole, you can only see objects that lie on the northern half of the celestial sphere, and they are all circumpolar. That is why the Sun remains above the horizon for 6 months at the North Pole: The Sun lies north of the celestial equator for half of each year (see the yellow dots in Figure 2.3b), so during these 6 months, it circles the sky at the North Pole just like a circumpolar star.

The altitude of the celestial pole in your sky is equal to your latitude.

The diagrams in Figures 2.8 and 2.10 also show a fact that is very important to navigation: *The altitude of the celestial pole in your sky is equal to your latitude.* For example, if you see the north celestial pole at an altitude of 40° above your north horizon, your latitude is 40°N. Similarly, if you see the south celestial pole at an altitude of 34° above your south horizon, your latitude is 34°S. Finding the north celestial pole is fairly easy, because it lies very close to the star Polaris, also known as the North Star (Figure 2.11a). In the Southern Hemisphere, you can find the south celestial pole with the aid of the Southern Cross (Figure 2.11b).

see it for yourself What is *your* latitude? Use Figure 2.11 to find the celestial pole in your sky, and estimate its altitude with your hand as shown in Figure 2.6c. Is its altitude what you expect?

Variation with Time of Year The night sky changes throughout the year because of Earth's changing position in its orbit around the Sun. As Earth orbits, the Sun *appears* to move steadily eastward along the ecliptic, with the stars of different constellations in the background at different

Figure 2.11 interactive figure

You can determine your latitude by measuring the altitude of the celestial pole in your sky.

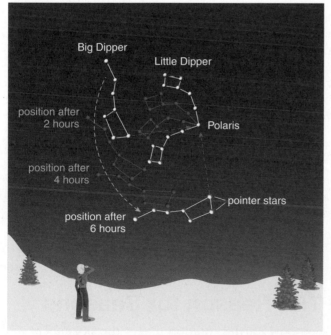

looking northward in the Northern Hemisphere

a The pointer stars of the Big Dipper point to the North Star, Polaris, which lies within 1° of the north celestial pole. The sky appears to turn *counterclockwise* around the north celestial pole.

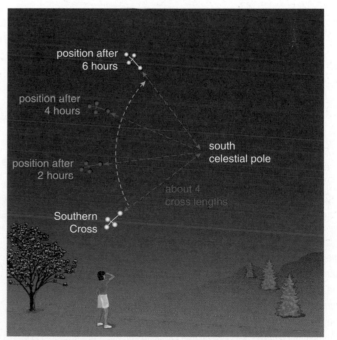

looking southward in the Southern Hemisphere

b The Southern Cross points to the south celestial pole, which is not marked by any bright star. The sky appears to turn *clockwise* around the south celestial pole.

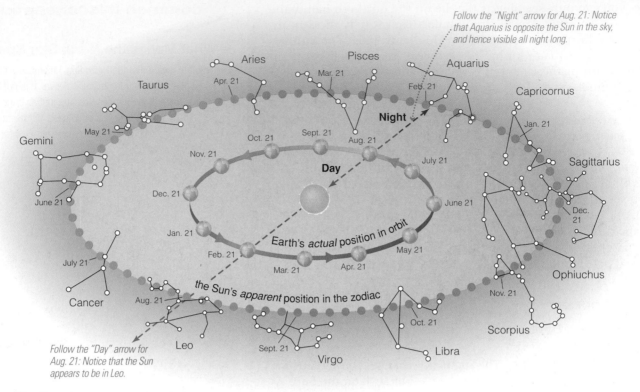

Follow the "Night" arrow for Aug. 21: Notice that Aquarius is opposite the Sun in the sky, and hence visible all night long.

Aries
Pisces
Aquarius
Taurus
Apr. 21
Mar. 21
Feb. 21
Capricornus
May 21
Night
Jan. 21
Gemini
Oct. 21
Sept. 21
Day
July 21
Sagittarius
Nov. 21
Aug. 21
Dec. 21
June 21
Jan. 21
June 21
Dec. 21
July 21
Earth's *actual* position in orbit
May 21
Feb. 21
Ophiuchus
Mar. 21
Apr. 21
Nov. 21
the Sun's *apparent* position in the zodiac
Cancer
Aug. 21
Oct. 21
Scorpius
Leo
Sept. 21
Libra
Virgo

Follow the "Day" arrow for Aug. 21: Notice that the Sun appears to be in Leo.

Figure 2.12 interactive figure

The Sun appears to move steadily eastward along the ecliptic as Earth orbits the Sun, so we see the Sun against the background of different zodiac constellations at different times of year. For example, on August 21 the Sun appears to be in Leo, because it is between us and the much more distant stars that make up Leo.

times of year. The constellations along the ecliptic make up what we call the **zodiac;** tradition places 12 constellations along the zodiac, but the official borders include a thirteenth constellation, Ophiuchus.

The constellations visible at a particular time of night change as we orbit the Sun.

Figure 2.12 shows the idea. In late August, for example, the Sun appears to be in Leo. We therefore cannot see Leo at this time (because it is in our daytime sky), but we can see Aquarius all night long because of its location opposite Leo on the celestial sphere. Six months later, in February, we see Leo at night while Aquarius is above the horizon only in the daytime.

see it for yourself Based on Figure 2.12 and today's date, in what constellation does the Sun currently appear? What constellation of the zodiac will be on your meridian at midnight? What constellation of the zodiac will you see in the west shortly after sunset? Go outside at night to confirm your answers.

common misconceptions

Stars in the Daytime

Stars may appear to vanish in the daytime and "come out" at night, but in reality the stars are always present. The reason you don't see stars in the daytime is that their dim light is overwhelmed by the bright daytime sky. You *can* see bright stars in the daytime with the aid of a telescope, and you may see stars in the daytime if you are fortunate enough to observe a total eclipse of the Sun. Astronauts can also see stars in the daytime. Above Earth's atmosphere, where there is no air to scatter sunlight, the Sun is a bright disk against a dark sky filled with stars. (However, the Sun is so bright, that astronauts must block its light if they wish to see the stars.)

MA Seasons Tutorial, Lessons 1–3

2.2 The Reason for Seasons

We have seen how Earth's rotation makes the sky appear to circle us daily and how the night sky changes as Earth orbits the Sun each year. The combination of Earth's rotation and orbit also leads to the progression of the seasons. In this section, we'll explore the reason for seasons.

• What causes the seasons?

You know that we have seasonal changes, such as longer and warmer days in summer and shorter and cooler days in winter. But why do the seasons occur? The answer is that the tilt of Earth's axis causes sunlight to fall differently on Earth at different times of year.

Figure 2.13 (pp. 36–37) shows the idea. Notice that Earth's axis remains pointed in the same direction in space (toward Polaris) throughout the year. However, as Earth orbits the Sun, the orientation of the axis *relative to the Sun* changes over the course of each orbit.

> Earth's axis points in the same direction all year round, which means its orientation *relative to the Sun* changes as Earth orbits the Sun.

For example, look at Earth on the left side of Figure 2.13. The Northern Hemisphere is tipped toward the Sun, making it summer there, while the Southern Hemisphere is tipped away from the Sun, making it winter there. Half an orbit later (when Earth is on the right side of Figure 2.13), the situation is reversed. It is winter in the Northern Hemisphere because it is tipped away from the Sun, while it is summer in the Southern Hemisphere because it is tipped toward the Sun. That is why the two hemispheres experience opposite seasons.

Figure 2.13 also shows why the days are longer and warmer in summer and shorter and cooler in winter. Sunlight strikes the summer

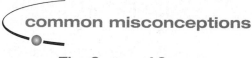

common misconceptions

The Cause of Seasons

Many people guess that seasons are caused by variations in Earth's distance from the Sun. But if this were true, the whole Earth would have to have summer or winter at the same time, and it doesn't: The seasons are opposite in the Northern and Southern Hemispheres. In fact, Earth's slightly varying orbital distance has virtually no effect on the weather. The real cause of seasons is Earth's axis tilt, which causes the two hemispheres to "take turns" being tipped toward the Sun over the course of each year.

special topic: --

How Long Is a Day?

WE USUALLY ASSOCIATE our 24-hour day with Earth's rotation, but if you measure the rotation period, you'll find that it is about 23 hours and 56 minutes (more precisely $23^h56^m4.09^s$)—or about 4 minutes short of 24 hours. What's going on?

Astronomically, we define two different types of day. Earth's 23 hour and 56 minute rotation period, which we measure by timing how long it takes any star to make one full circuit through our sky, is called a **sidereal day**; *sidereal* (pronounced *sy-dear-ee-al*) means "related to the stars." Our 24-hour day, which we call a **solar day**, is the average time it takes *the Sun* to make one circuit through the sky.

A simple demonstration shows why the solar day is about 4 minutes longer than the sidereal day. Set an object representing the Sun on a table, and stand a few steps away to represent Earth. Point at the Sun and imagine that you also happen to be pointing toward a distant star that lies in the same direction. If you rotate (counterclockwise) while standing in place, you'll again be pointing at both the Sun and the star after one full rotation. However, to show that Earth also orbits the Sun, you should take a couple of steps around the Sun (counterclockwise) as you rotate (see figure). After one full rotation, you will again be pointing in the direction of the distant star, so this rotation represents a sidereal day. But it does not represent a solar day, because you will not yet be pointing back at the Sun; you need to rotate a bit more. This "extra" bit of rotation makes a solar day longer than a sidereal day.

The only problem with this demonstration is that it exaggerates Earth's daily orbital motion. Earth takes about 365 days (1 year) to make a full 360° orbit around the Sun, which means about 1° per day. A solar day therefore represents about 361° of rotation, rather than the 360° for a sidereal day. The extra 1° rotation takes about $\frac{1}{360}$ of Earth's rotation period, which is about 4 minutes.

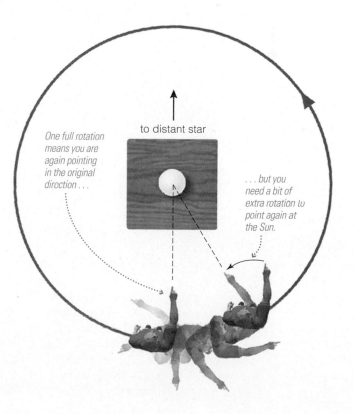

One full rotation means you are again pointing in the original direction . . .

to distant star

. . . but you need a bit of extra rotation to point again at the Sun.

Earth's seasons are caused by the tilt of its rotation axis, which is why the seasons are opposite in the two hemispheres. The seasons do *not* depend on Earth's distance from the Sun, which varies only slightly throughout the year.

① Axis Tilt: Earth's axis points in the same direction throughout the year, which causes changes in Earth's orientation *relative to the Sun.*

23½°

N

S

② Northern Summer/Southern Winter: In June, sunlight falls more directly on the Northern Hemisphere, which makes it summer there because solar energy is more concentrated and the Sun follows a longer and higher path through the sky. The Southern Hemisphere receives less direct sunlight, making it winter.

Summer (June) Solstice

The Northern Hemisphere is tipped most directly toward the Sun.

Noon rays of sunlight hit the ground at a steeper angle in the Northern Hemisphere, meaning more concentrated sunlight and shorter shadows.

Noon rays of sunlight hit the ground at a shallower angle in the Southern Hemisphere, meaning less concentrated sunlight and longer shadows.

Interpreting the Diagram

To interpret the seasons diagram properly, keep in mind:

1. Earth's size relative to its orbit would be microscopic on this scale, meaning that both hemispheres are at essentially the same distance from the Sun.

2. The diagram is a side view of Earth's orbit. A top-down view (below) shows that Earth orbits in a nearly perfect circle and comes closest to the Sun in January.

Spring Equinox

147.1 million km · *January 3*

152.1 million km

July 4

Fall Equinox

(3) Spring/Fall: Spring and fall begin when sunlight falls equally on both hemispheres, which happens twice a year: In March, when spring begins in the Northern Hemisphere and fall in the Southern Hemisphere; and in September, when fall begins in the Northern Hemisphere and spring in the Southern Hemisphere.

(4) Northern Winter/Southern Summer: In December, sunlight falls less directly on the Northern Hemisphere, which makes it winter because solar energy is less concentrated and the Sun follows a shorter and lower path through the sky. The Southern Hemisphere receives more direct sunlight, making it summer.

Spring (March) Equinox

The Sun shines equally on both hemispheres.

The variation in Earth's orientation relative to the Sun means that the seasons are linked to four special points in Earth's orbit:

Solstices *are the two points at which sunlight becomes most extreme for the two hemispheres.*

Equinoxes *are the two points at which the hemispheres are equally illuminated.*

Winter (December) Solstice

The Southern Hemisphere is tipped most directly toward the Sun.

Fall (September) Equinox

The Sun shines equally on both hemispheres.

Noon rays of sunlight hit the ground at a shallower angle in the Northern Hemisphere, meaning less concentrated sunlight and longer shadows.

Noon rays of sunlight hit the ground at a steeper angle in the Southern Hemisphere, meaning more concentrated sunlight and shorter shadows.

hemisphere at a steeper angle than it strikes the winter hemisphere. The steeper angle means sunlight is more concentrated, which is why summer tends to be warmer than winter. The steeper angle also means that the Sun follows a longer and higher path through the summer sky, which is why the days are long and midday shadows are short. You can see the annual change in how high the Sun rises in your sky by observing the Sun's position at the same time each day (Figure 2.14).

Summer occurs in your hemisphere when sunlight hits it more directly, and winter occurs when the sunlight is less direct.

Notice that the seasons on Earth are caused only by the axis tilt and *not* by any change in Earth's distance from the Sun. Although Earth's orbital distance varies over the course of each year, the variation is fairly small: Earth is only about 3% farther from the Sun at its farthest point than at its nearest. The difference in the strength of sunlight due to this small change in distance is easily overwhelmed by the effects caused by the $23\frac{1}{2}°$ axis tilt. (*Note:* Some people also misinterpret diagrams like Figure 2.13, getting the mistaken impression that orbital position can make one hemisphere closer to the Sun. The mistake happens because figures like this are not drawn to scale. If the figure were drawn to scale, Earth would be microscopic, showing that there is virtually no difference in the distances of the two hemispheres from the Sun.)

think about it Jupiter has an axis tilt of about 3°, small enough to be insignificant. Saturn has an axis tilt of about 27°, slightly greater than that of Earth. Both planets have nearly circular orbits around the Sun. Do you expect Jupiter to have seasons? Do you expect Saturn to have seasons? Explain.

• How do we mark the progression of the seasons?

Today, we use a calendar to mark the progression of the seasons. In ancient times, however, people tracked the seasons by observing the Sun's changing position in our sky. Let's explore how the Sun's changing path allows us to mark the changing of the seasons.

Solstices and Equinoxes To help us track the seasons, we define four special moments in the year, each of which corresponds to one of the four special positions in Earth's orbit shown in Figure 2.13.

- The **summer (June) solstice**, which occurs around June 21 each year, is the moment when the Northern Hemisphere receives its most direct sunlight (and the Southern Hemisphere gets its least direct sunlight).

- The **winter (December) solstice**, which occurs around December 21, is the moment when the Northern Hemisphere receives its least direct sunlight.

- The **spring (March) equinox** occurs around March 21, at the moment when the Northern Hemisphere goes from being tipped slightly away from the Sun to being tipped slightly toward the Sun.

- The **fall (September) equinox** occurs around September 22 and marks the opposite change, when the Northern Hemisphere first starts to be tipped away from the Sun.

Figure 2.14

This composite photograph shows midday images of the Sun at 7- to 11-day intervals over the course of a year, always from the same spot (the Parthenon in Athens, Greece) and at the same time of day (technically, at the same "mean solar time"). Notice the dramatic change in the Sun's midday altitude over the course of the year. The "figure 8" shape (called an *analemma*) is due to the combination of Earth's axis tilt and Earth's varying speed as it orbits the Sun.

We use the equinoxes and solstices to mark the progression of the seasons.

The exact dates and times of the solstices and equinoxes vary from year to year but stay within a couple of days of the dates given above. In fact, our modern calendar includes leap years in a pattern specifically designed to keep the solstices and equinoxes around the same dates: We generally add a day (February 29) for leap year every fourth year, but skip leap year when a century changes (for example, in the years 1700, 1800, 1900) *unless* the century year is divisible by 400 (for example, 2000). This pattern makes the average length of the calendar year match the true length of the year,* which is about 11 minutes short of $365 \frac{1}{4}$ days.

Identifying the Solstices and Equinoxes Ancient people recognized the days on which the solstices and equinoxes occur by observing the Sun in the sky. Many ancient structures were used for this purpose, including Stonehenge in England and the Sun Dagger in New Mexico [Section 3.1].

The Sun rises precisely due east and sets precisely due west *only* on the days of the spring and fall equinoxes.

The equinoxes occur on the only two days of the year on which the Sun rises precisely due east and sets precisely due west (Figure 2.15). These are also the only two days when sunlight falls equally on both hemispheres. The summer solstice occurs on the day that the Sun follows its longest and highest path through the Northern Hemisphere sky (and its shortest and lowest path through the Southern Hemisphere sky). It is therefore the day that the Sun rises and sets farther to the north than on any other day of the year, and on which the noon Sun reaches its highest point in the Northern Hemisphere sky. The opposite is true on the day of the winter solstice, when the Sun rises and sets farthest to the south and the noon Sun is lower in the Northern Hemisphere sky than on any other day of the year.

First Days of Seasons We usually say that each equinox and solstice marks the first day of a season. For example, the day of the summer solstice is usually said to be the "first day of summer." Notice, however, that the summer solstice occurs when the Northern Hemisphere has its *maximum* tilt toward the Sun. You might then wonder why we consider the summer solstice to be the beginning rather than the midpoint of summer.

In part, the choice of the summer solstice as the "first" day of summer is somewhat arbitrary. However, the choice makes sense in at least two ways. First, it was much easier for ancient people to identify the days on which the Sun reached extreme positions in the sky—such as when it reached its highest point on the summer solstice—than other days in between. Second, we usually think of the seasons in terms of weather, and the solstices and equinoxes correspond quite well with the beginnings of seasonal weather patterns. For example, although the Sun's path through the Northern Hemisphere sky is longest and highest around the time of the summer solstice, the warmest days tend to come 1 to 2 months later. To understand why, think about what happens when you heat a pot of

---

*Technically, we are referring here to the *tropical year*—the time from one spring equinox to the next. Axis precession (discussed later in this section) causes the tropical year to be slightly shorter (by about 20 minutes) than Earth's orbital period, called the *sidereal year*.

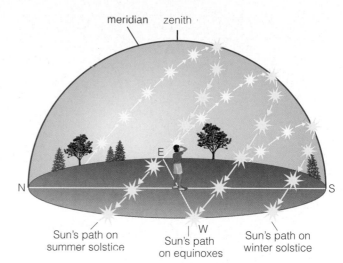

Figure 2.15

This diagram shows the Sun's path on the solstices and equinoxes for the Northern Hemisphere sky (latitude 40°N). Notice that the Sun rises exactly due east and sets exactly due west only on the equinoxes. The summer solstice occurs on the day that the Sun rises and sets farthest to the north and reaches its highest point in the sky. The winter solstice occurs on the day that the Sun rises and sets farthest to the south and traces its lowest path through the sky.

common misconceptions

High Noon

When is the Sun directly overhead in your sky? Many people answer "at noon." It's true that the Sun reaches its *highest* point each day when it crosses the meridian, giving us the term "high noon" (though the meridian crossing is rarely at precisely 12:00). However, unless you live in the Tropics (between latitudes 23.5°S and 23.5°N), the Sun is *never* directly overhead. In fact, any time you can see the Sun as you walk around, you can be sure it is *not* at your zenith. Unless you are lying down, seeing an object at the zenith requires tilting your head back into a very uncomfortable position.

cold soup. Even though you may have the stove turned on high from the start, it takes a while for the soup to warm up. In the same way, it takes some time for sunlight to heat the ground and oceans from the cold of winter to the warmth of summer. "Midsummer" in terms of weather therefore comes in late July or early August, which makes the summer solstice a pretty good choice for the "first day of summer." For similar reasons, the winter solstice is a good choice for the first day of winter, and the spring and fall equinoxes are good choices for the first days of those seasons.

Seasons Around the World The names of the solstices and equinoxes generally reflect the northern seasons, which can make things sound strange when we talk about seasons in the Southern Hemisphere. For example, when Earth is at the orbital point usually called the *summer* solstice, it is *winter* in the Southern Hemisphere. This apparent injustice to people in the Southern Hemisphere arose because the solstices and equinoxes were named by people living in the Northern Hemisphere. A similar injustice affects people living in equatorial regions. If you study Figure 2.13, you'll see that Earth's equator gets its most direct sunlight on the two equinoxes and its least direct sunlight on the solstices. People living near the equator therefore don't experience four seasons in the same way as people living at mid-latitudes. Instead, equatorial regions have rainy and dry seasons, with the rainy seasons coming when the Sun is higher in the sky.

At very high latitudes, the Sun becomes circumpolar in summer, remaining above the horizon all day long.

In addition, seasonal variations around the times of the solstices are more extreme at high latitudes. For example, Alaska has much longer summer days and winter nights than Florida. In fact, the Sun becomes circumpolar at very high latitudes (within the *Arctic* and *Antarctic Circles*) in the summer. The Sun never sets during these summer days in what we call the *land of the midnight Sun* (Figure 2.16). Of course, the name "land of noon darkness" would be more appropriate in the winter, when the Sun never rises above the horizon at these high latitudes.

Figure 2.16

This sequence of photos shows the progression of the Sun all the way around the horizon on the summer solstice at the Arctic Circle. Notice that the Sun does not set but instead skims the northern horizon at midnight. It then gradually rises higher, reaching its highest point at noon, when it appears due south.

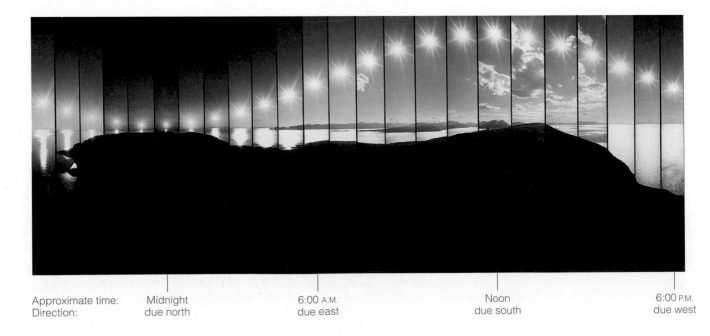

| Approximate time: | Midnight | 6:00 A.M. | Noon | 6:00 P.M. |
| Direction: | due north | due east | due south | due west |

• How does the orientation of Earth's axis change with time?

We have now discussed both daily and seasonal changes in the sky, but there are other changes that occur over longer periods of time. One of the most important of these slow changes is called **precession**, a gradual wobble that changes the orientation of Earth's axis in space.

Precession occurs with many rotating objects. You can see it easily by spinning a top (Figure 2.17a). As the top spins rapidly, you'll notice that its axis also sweeps out a circle at a slower rate. We say that the top's axis *precesses*. Earth's axis precesses in much the same way, but far more slowly (Figure 2.17b). Each cycle of Earth's precession takes about 26,000 years, gradually changing where the axis points in space. Today, the axis points toward Polaris, making it our North Star. Some 13,000 years from now, Vega will be the star closest to true north (within a few degrees). At most times, the axis does not point near any bright star.

The tilt of Earth's axis remains close to $23\frac{1}{2}°$, but the direction the axis points in space changes slowly with the 26,000-year cycle of precession.

Notice that precession does not change the *amount* of the axis tilt (which stays close to $23\frac{1}{2}°$) and therefore does not affect the pattern of the seasons. However, because the solstices and equinoxes correspond to points in Earth's orbit that depend on the direction the axis points in space, their positions in the orbit gradually shift with the cycle of precession. As a result, the constellations associated with the solstices and equinoxes change over time. For example, a couple thousand years ago the Sun appeared in the constellation Cancer on the day of the summer solstice, but it now appears in Gemini. This explains something you can see on any world map: The latitude at which the Sun is directly overhead on the summer solstice ($23\frac{1}{2}°$) is called the *Tropic of*

Figure 2.17 interactive figure

Precession affects the orientation of a spinning object's axis, but not the amount of its tilt.

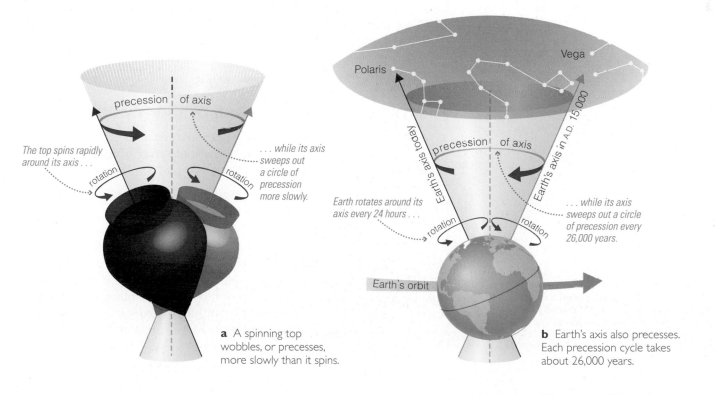

a A spinning top wobbles, or precesses, more slowly than it spins.

b Earth's axis also precesses. Each precession cycle takes about 26,000 years.

Cancer, telling us that it was named back when the Sun appeared in Cancer on the summer solstice.

Why does precession occur? It is caused by gravity's effect on a tilted, rotating object that is *not* a perfect sphere. A spinning top precesses because Earth's gravity tries to pull over its lopsided, tilted spin axis. Gravity does not succeed in pulling it over—at least until friction slows the rate of spin—but instead causes the axis to precess. The spinning Earth precesses because gravitational tugs from the Sun and Moon try to "straighten out" our planet's bulging equator, which has the same tilt as the axis. Again, gravity does not succeed in straightening out the tilt but only causes the axis to precess.

(MA) **Phases of the Moon Tutorial, Lessons 1–3**

2.3 The Moon, Our Constant Companion

Aside from the Sun, the Moon is the brightest and most noticeable object in our sky. The Moon is our constant companion in space, orbiting Earth about once every $27\frac{1}{3}$ days.

Figure 2.18 shows the Moon's orbit on the same scale we used for the model solar system in Section 1.2. Remember that on this scale, the Sun is about the size of a large grapefruit and is located about 15 meters from Earth. The entire orbit of the Moon would fit easily inside the Sun, and for practical purposes we can consider Earth and the Moon to share the same orbit around the Sun.

Like all objects in space, the Moon appears to reside on the celestial sphere. Earth's daily rotation makes the Moon appear to rise in the east and set in the west each day. In addition, because it orbits Earth, the Moon appears to move eastward from night to night through the constellations of the zodiac. Each circuit through the constellations takes the same $27\frac{1}{3}$ days that the Moon takes to orbit Earth. If you do the math, you'll see that this means the Moon moves relative to the stars by about $\frac{1}{2}°$—its own angular size—each hour. You can notice this gradual motion in just a few hours by checking the Moon's position compared to bright stars near it in the sky.

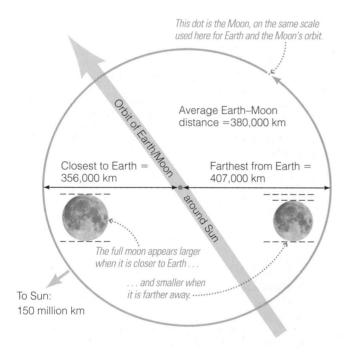

This dot is the Moon, on the same scale used here for Earth and the Moon's orbit.

Orbit of Earth/Moon around Sun

Average Earth–Moon distance =380,000 km

Closest to Earth = 356,000 km

Farthest from Earth = 407,000 km

The full moon appears larger when it is closer to Earth . . .

. . . and smaller when it is farther away.

To Sun: 150 million km

Figure 2.18

The Moon's orbit around Earth, shown on the 1-to-10-billion scale used in Section 1.2 (see Figure 1.6). The segment shown of our orbit around the Sun looks nearly straight because the distance to the Sun is so great in comparison to the size of the Moon's orbit. The inset photos contrast the relative angular size of the full moon in our sky when the Moon is at the near and far points of its orbit; of course, full moon occurs only when the Moon is opposite the Sun as seen from Earth.

• ## Why do we see phases of the Moon?

As the Moon moves through the sky, both its appearance and the time at which it rises and sets change with the cycle of **lunar phases**. The phase of the Moon on any given day depends on its position relative to the Sun as it orbits Earth.

The phase of the Moon depends on its position relative to the Sun as it orbits Earth.

The easiest way to understand the lunar phases is with the simple demonstration illustrated in Figure 2.19. Take a ball outside on a sunny day. (If it's dark or cloudy, you can use a flashlight instead of the Sun; put the flashlight on a table a few meters away and shine it toward you.) Hold the ball at arm's length to represent the Moon while your head represents Earth. Slowly spin around (counterclockwise), so that the ball goes around you just like the Moon orbits Earth. As you turn, you'll see the ball go through phases just like the Moon. If you think about what's

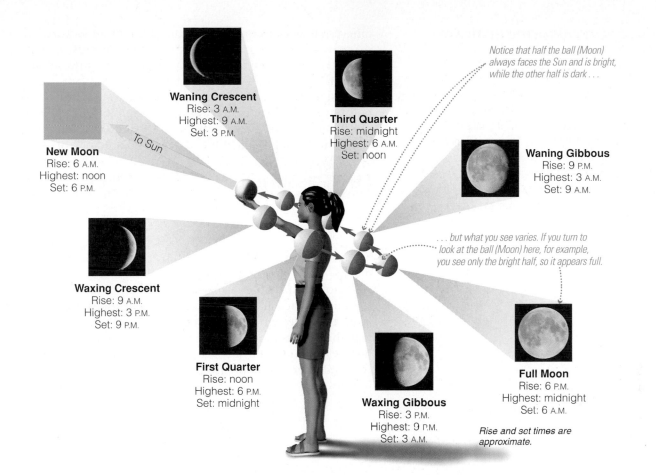

Waning Crescent
Rise: 3 A.M.
Highest: 9 A.M.
Set: 3 P.M.

Third Quarter
Rise: midnight
Highest: 6 A.M.
Set: noon

Notice that half the ball (Moon) always faces the Sun and is bright, while the other half is dark . . .

New Moon
Rise: 6 A.M.
Highest: noon
Set: 6 P.M.

To Sun

Waning Gibbous
Rise: 9 P.M.
Highest: 3 A.M.
Set: 9 A.M.

. . . but what you see varies. If you turn to look at the ball (Moon) here, for example, you see only the bright half, so it appears full.

Waxing Crescent
Rise: 9 A.M.
Highest: 3 P.M.
Set: 9 P.M.

First Quarter
Rise: noon
Highest: 6 P.M.
Set: midnight

Waxing Gibbous
Rise: 3 P.M.
Highest: 9 P.M.
Set: 3 A.M.

Full Moon
Rise: 6 P.M.
Highest: midnight
Set: 6 A.M.

Rise and set times are approximate.

Figure 2.19 interactive figure

A simple demonstration illustrates the phases of the Moon. Hold a ball at arm's length; your head represents Earth and the ball represents the Moon. As you turn, you'll see the ball go through phases just like those of the Moon. The photos show what the Moon looks like at each orbital position. (The new moon photo shows blue sky, because a new moon is always close to the Sun in the sky and hence hidden from view by the bright light of the Sun.)

happening, you'll realize that the phases of the ball result from just two basic facts:

1. Half the ball always faces the Sun (or flashlight) and therefore is bright, while the other half faces away from the Sun and therefore is dark.
2. As you look at the ball at different positions in its "orbit" around your head, you see different combinations of its bright and dark faces.

For example, when you hold the ball directly opposite the Sun, you see only the bright portion of the ball, which represents the "full" phase. When you hold the ball at its "first-quarter" position, half the face you see is dark and the other half is bright.

We see lunar phases for the same reason. Half the Moon is always illuminated by the Sun, but the amount of this illuminated half that we see from Earth depends on the Moon's position in its orbit. The photographs in Figure 2.19 show how the phases look. Each complete cycle of phases, from one new moon to the next, takes about $29\frac{1}{2}$ days—hence the origin of the word *month* (think "moonth"). This is about 2 days longer than the Moon's actual orbital period because of Earth's motion around the Sun during the time the Moon is orbiting around Earth.

The Moon's phase affects not only its appearance, but also its rise and set times.

The different phases not only look different but also rise, reach their highest points, and set at

common misconceptions

Moon in the Daytime

In traditions and stories, night is so closely associated with the Moon that many people mistakenly believe that the Moon is visible only in the nighttime sky. In fact, the Moon is above the horizon as often in the daytime as at night, though it is easily visible only when its light is not drowned out by sunlight. For example, a first-quarter moon is easy to spot in the late afternoon as it rises through the eastern sky, and a third-quarter moon is visible in the morning as it heads toward the western horizon.

common misconceptions

The "Dark Side" of the Moon

Although we see many *phases* of the Moon, we do not see many *faces*. In fact, from Earth we always see (nearly) the same face of the Moon (see Chapter 4). This leads to a common point of confusion. The term *dark side of the Moon* really should be used to mean the night side— that is, the side facing away from the Sun. Unfortunately, *dark side* traditionally meant what would better be called the *far side*—the face that never can be seen from Earth. Many people still refer to the far side as the "dark side," even though this side is not necessarily dark. For example, during new moon the far side faces the Sun and hence is completely sunlit. The only time the far side is completely dark is at full moon, when it faces away from both the Sun and Earth.

different times. For example, because full moon occurs when the Moon is opposite the Sun in the sky, the full moon must rise around sunset, reach its highest point in the sky at midnight, and set around sunrise. Similarly, because first-quarter moon occurs when the Moon is about 90° east of the Sun in our sky, it must rise around noon, reach its highest point around 6 P.M., and set around midnight. Figure 2.19 lists the approximate rise, highest point, and set times for each phase. (The exact times vary with latitude, time of year, and other factors.)

think about it Suppose you go outside in the morning and notice that the visible face of the Moon is half-light and half-dark. Is this a first-quarter or third-quarter moon? How do you know?

Notice that the phases from new to full are said to be *waxing*, which means "increasing." Phases from full to new are *waning*, or "decreasing." Also notice that no phase is called a "half moon." Instead, we see half the moon's face at first-quarter and third-quarter phases; these phases mark the times when the Moon is one-quarter or three-quarters of the way through its monthly cycle (taken to begin at new moon). The phases just before and after new moon are called *crescent*, while those just before and after full moon are called *gibbous* (pronounced with a hard *g* as in "gift").

(MA) Eclipses Tutorial, Lessons 1–3

• What causes eclipses?

The phases of the Moon are the most obvious effect of the Moon's orbit around Earth. Occasionally, however, we see something much more dramatic. The Moon and Earth cast shadows in sunlight, and these shadows can create **eclipses** when the Sun, Earth, and Moon fall into a straight line. Eclipses come in two basic types:

- A **lunar eclipse** occurs when Earth lies directly between the Sun and the Moon, so that Earth's shadow falls on the Moon.

- A **solar eclipse** occurs when the Moon lies directly between the Sun and Earth, so that the Moon's shadow falls on Earth. People living within the area covered by the Moon's shadow will see the Sun blocked or partially blocked from view.

To understand when and why eclipses occur, we must investigate the Moon's orbit of Earth a little more deeply.

Conditions for Eclipses Look again at Figure 2.19. The figure makes it look like the Sun, Earth, and Moon line up with every new and full moon. If this figure told the whole story of the Moon's orbit, we would have both a lunar and a solar eclipse every month—but we don't.

We see a lunar eclipse when Earth's shadow falls on the Moon, and a solar eclipse when the Moon blocks our view of the Sun.

The missing piece of the story in Figure 2.19 is that the Moon's orbit is slightly inclined (by about 5°) to the ecliptic plane (the plane of Earth's orbit around the Sun). To visualize

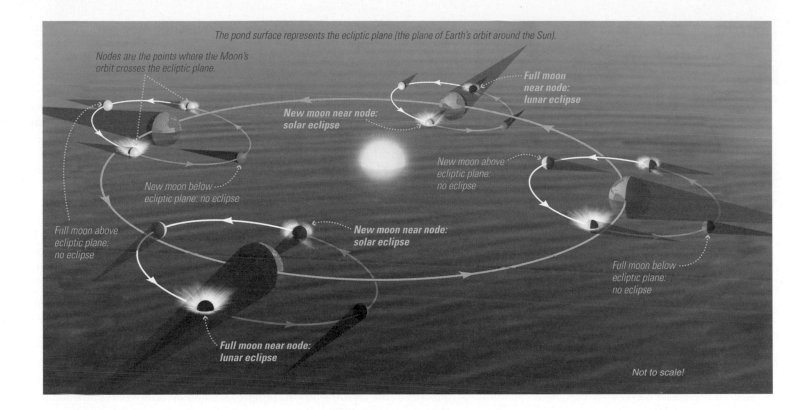

The pond surface represents the ecliptic plane (the plane of Earth's orbit around the Sun).

Nodes are the points where the Moon's orbit crosses the ecliptic plane.

Full moon near node: lunar eclipse

New moon near node: solar eclipse

New moon below ecliptic plane: no eclipse

New moon above ecliptic plane: no eclipse

Full moon above ecliptic plane: no eclipse

New moon near node: solar eclipse

Full moon below ecliptic plane: no eclipse

Full moon near node: lunar eclipse

Not to scale!

this inclination, imagine the ecliptic plane as the surface of a pond, as shown in Figure 2.20. Because of the inclination of its orbit, the Moon spends most of its time either above or below this surface. It crosses *through* this surface only twice during each orbit: once coming out and once going back in. The two points in each orbit at which the Moon crosses the surface are called the **nodes** of the Moon's orbit.

As you study Figure 2.20, notice that the nodes are aligned approximately the same way (diagonally on the page) throughout the year. As a result, the nodes lie in a straight line with the Sun and Earth only about twice each year. We therefore find the following conditions for an eclipse to occur:

1. The phase of the Moon must be full (for a lunar eclipse) or new (for a solar eclipse), since those are the only phases at which the Sun, Earth, and Moon can lie in a straight line.
2. The new or full moon must occur during one of the periods when the nodes of the Moon's orbit are aligned with the Sun and Earth, since those are the only times when the new and full moons lie in the ecliptic plane.

We see an eclipse only when a full or new moon occurs at one of the points where the Moon's orbit crosses the ecliptic plane.

Although there are two basic types of eclipse—lunar and solar—each of these types can look different depending on precisely how the shadows fall. The shadow of the Moon or Earth consists of two distinct regions: a central **umbra**, where sunlight is completely blocked, and a surrounding **penumbra**, where sunlight is only partially blocked (Figure 2.21). Therefore, an umbral shadow is totally dark, while a penumbral shadow is only slightly darker than no shadow. Let's see how this affects eclipses.

Figure 2.20 interactive figure

This illustration represents the ecliptic plane as the surface of a pond. The Moon's orbit is tilted by about 5° to the ecliptic plane, so the Moon spends half of each orbit above the plane (the pond surface) and half below it. Eclipses occur only when the Moon is at both a node (passing through the pond surface) *and* a phase of either new moon (for a solar eclipse) or full moon (for a lunar eclipse)—as is the case with the lower left and top right orbits shown.

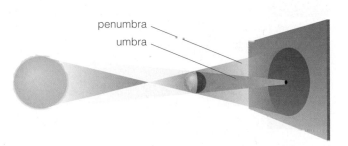

penumbra

umbra

Figure 2.21

The shadow cast by an object in sunlight. Sunlight is fully blocked in the umbra and partially blocked in the penumbra.

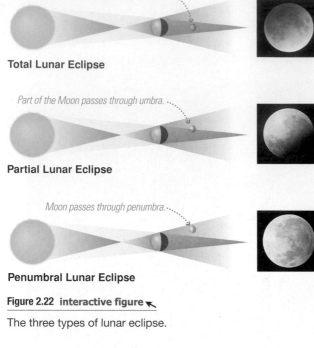

Total Lunar Eclipse

Moon passes entirely through umbra.

Part of the Moon passes through umbra.

Partial Lunar Eclipse

Moon passes through penumbra.

Penumbral Lunar Eclipse

Figure 2.22 interactive figure

The three types of lunar eclipse.

A total solar eclipse occurs in the small central region.

Moon

path of total eclipse

A partial solar eclipse occurs in the lighter area surrounding the area of totality.

Moon

path of annular eclipse

If the Moon's umbral shadow does not reach Earth, an annular eclipse occurs in the small central region.

Figure 2.23 interactive figure

The three types of solar eclipse. The diagrams show the Moon's shadow falling on Earth; note the dark central umbra surrounded by the much lighter penumbra.

Lunar Eclipses

A lunar eclipse begins at the moment when the Moon's orbit first carries it into Earth's penumbra. After that, we will see one of three types of lunar eclipse (Figure 2.22). If the Sun, Earth, and Moon are nearly perfectly aligned, the Moon will pass through Earth's umbra and we will see a **total lunar eclipse**. If the alignment is somewhat less perfect, only part of the full moon will pass through the umbra (with the rest in the penumbra) and we will see a **partial lunar eclipse**. If the Moon passes *only* through Earth's penumbra, we will see a **penumbral lunar eclipse**.

Penumbral eclipses are slightly more common that total lunar eclipses and partial lunar eclipses, but they are the least visually impressive because the full moon darkens only slightly. Earth's umbral shadow clearly darkens part of the Moon's face during a partial lunar eclipse, and the curvature of this shadow demonstrates that Earth is round. A total lunar eclipse is particularly spectacular because the Moon becomes dark and eerily red during **totality**—the time during which the Moon is entirely engulfed in the umbra. Totality typically lasts about an hour. The Moon becomes dark because it is in shadow, and red because Earth's atmosphere bends some of the red light from the Sun toward the Moon.

Solar Eclipses

We can also see three types of solar eclipse (Figure 2.23). If a solar eclipse occurs when the Moon is relatively close to Earth in its orbit, the Moon's umbra touches a small area of Earth's surface (no more than about 270 kilometers in diameter). Anyone within this area will see a **total solar eclipse**. Surrounding the region of totality is a much larger area (typically about 7000 kilometers in diameter) that falls within the Moon's penumbral shadow. Anyone within this region will see a **partial solar eclipse**, in which only part of the Sun is blocked from view. If the eclipse occurs when the Moon is relatively far from Earth, the umbra may not reach Earth's surface at all. In that case, anyone in the small region of Earth directly behind the umbra will see an **annular solar eclipse**, in which a ring of sunlight surrounds the disk of the Moon. Again, anyone in the surrounding penumbral shadow will see a partial solar eclipse.

A total solar eclipse is visible only within the narrow path that the Moon's umbral shadow makes across Earth's surface.

The combination of Earth's rotation and the orbital motion of the Moon causes the Moon's umbral and penumbral shadows to race across the face of Earth at a typical speed of about 1700 kilometers per hour. As a result, the umbral shadow traces a narrow path across Earth, and totality never lasts more than a few minutes in any particular place.

A total solar eclipse is a spectacular sight. It begins when the disk of the Moon first appears to touch the Sun. Over the next couple of hours, the Moon appears to take a larger and larger "bite" out of the Sun. As totality approaches, the sky darkens and temperatures fall. Birds head back to their nests, and crickets begin their nighttime chirping. During the few minutes of totality, the Moon completely blocks the normally visible disk of the Sun, allowing the faint *corona* to be seen (Figure 2.24). The surrounding sky takes on a twilight glow, and planets and bright stars become visible in the daytime. As totality ends, the Sun slowly emerges from behind the Moon over the next couple of hours. However, because your eyes have adapted to the darkness, totality appears to end far more abruptly than it began.

Figure 2.24

This multiple-exposure photograph shows the progression of a total solar eclipse. Totality (central image) lasts only a few minutes, during which time we can see the faint corona around the outline of the Sun. This photo was taken July 22, 1990, in La Paz, Mexico.

Predicting Eclipses   Few phenomena have so inspired and humbled humans throughout the ages as eclipses. For many cultures, eclipses were mystical events associated with fate or the gods, and countless stories and legends surround them. Much of the mystery of eclipses probably stems from the relative difficulty of predicting them.

Look again at Figure 2.20. The two periods each year when the nodes of the Moon's orbit are nearly aligned with the Sun are called **eclipse seasons**. Each eclipse season lasts a few weeks, so some type of lunar eclipse occurs during each eclipse season's full moon, and some type of solar eclipse occurs during its new moon.

If Figure 2.20 told the whole story, eclipse seasons would occur every 6 months, and predicting eclipses would be easy. For example, if eclipse seasons always occurred in January and July, eclipses would always occur on the dates of new and full moons in those months. But the figure does not show one important thing about the Moon's orbit: The nodes slowly move around the orbit. As a result, eclipse seasons occur slightly less than 6 months apart (about 173 days apart) and do not recur in the same months year after year.

The general pattern of eclipses repeats with the roughly 18-year saros cycle.

The combination of the changing dates of eclipse seasons and the $29\frac{1}{2}$-day cycle of lunar phases makes eclipses recur in a cycle of about 18 years $11\frac{1}{3}$ days. This cycle is called the **saros cycle**. Astronomers in many ancient cultures identified the saros cycle and thus could predict *when* eclipses would occur. However, the saros cycle does not account for all the complications involved in predicting eclipses. If a solar eclipse occurred today, the one that would occur 18 years $11\frac{1}{3}$ days from now would not be visible from the same places on Earth and might not be of the same type. For example, one might be total and the other only partial. No ancient culture achieved the ability to predict eclipses in every detail.

Today, we can predict eclipses because we know the precise details of the orbits of Earth and the Moon. Table 2.1 lists upcoming lunar eclipses; notice that, as we expect, eclipses generally come a little less than 6 months apart. Figure 2.25 shows paths of totality for total solar eclipses (but not for partial or annular eclipses) from 2009 to 2033, using color coding to show eclipses that repeat with the saros cycle.

TABLE 2.1   *Lunar Eclipses 2009–2012**

Date	Type	Where You Can See It
Feb. 9, 2009	penumbral	eastern Europe, Asia, Australia, Pacific, western North America
Jul. 7, 2009	penumbral	Australia, Pacific, Americas
Aug. 6, 2009	penumbral	Americas, Europe, Africa, western Asia
Dec. 31, 2009	partial	Europe, Africa, Asia, Australia
Jun. 26, 2010	partial	eastern Asia, Australia, Pacific, western Americas
Dec. 21, 2010	total	eastern Asia, Australia, Pacific, Americas, Europe
Jun. 15, 2011	total	South America, Europe, Africa, Asia, Australia
Dec. 10, 2011	total	Europe, Africa, Asia, Australia, North America
Jun. 4, 2012	partial	Asia, Australia, Americas
Nov. 28, 2012	penumbral	Europe, Africa, Asia, Australia, North America

*Dates are based on Universal Time and hence are those in Greenwich, England, at the time of the eclipse; to see an eclipse, check a news or Web source for the exact local time and date. Eclipse predictions by Fred Espenak; see NASA's Eclipse Web site.

Figure 2.25

Paths of totality for total solar eclipses, 2009 to 2033. Paths of the same color represent eclipses in successive saros cycles, separated by 18 years 11 days. For example, the 2028 eclipse occurs 18 years 11 days after the 2010 eclipse (both shown in orange). Eclipse predictions by Fred Espenak; see NASA's Eclipse Web site.

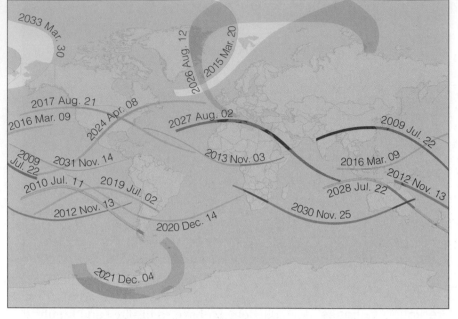

think about it Table 2.1 shows one exception to the "rule" of eclipses coming about 6 months apart: the eclipses of July 7 and August 6, 2009. How can eclipses occur a month apart like this? Should you be surprised that both these lunar eclipses are penumbral? Explain.

2.4 The Ancient Mystery of the Planets

We've now covered the appearance and motion of the stars, Sun, and Moon in the sky. That leaves us with the planets yet to discuss. As we'll soon see, planetary motion posed an ancient mystery that played a critical role in the development of modern civilization.

Five planets are easy to find with the naked eye: Mercury, Venus, Mars, Jupiter, and Saturn. Mercury is visible only infrequently, and then only just after sunset or just before sunrise because it is so close to the Sun. Venus often shines brightly in the early evening in the west or before dawn in the east. If you see a very bright "star" in the early evening or early morning, it is probably Venus. Jupiter, when it is visible at night, is the brightest object in the sky besides the Moon and Venus. Mars is often recognizable by its reddish color, though you should check a star chart to make sure you aren't looking at a bright red star. Saturn is also easy to see with the naked eye, but because many stars are just as bright as Saturn, it helps to know where to look. (It also helps to know that planets tend not to twinkle as much as stars.)

• What was once so mysterious about the movement of planets in our sky?

Ancient people carefully observed the movements of the planets among the stars, and you can observe the same motions for yourself. On any particular night, the planets move through the sky just like stars, rising in the east and setting in the west. However, if you observe planetary positions among the stars over a period of weeks or months, you'll notice that the planets wander slowly through the constellations of the zodiac. (The word *planet* comes from the Greek for "wandering star.") By itself, this wandering might not be too surprising. After all, the Sun and Moon also move among the constellations. But the planets move in what seems to be a very strange way.

Unlike the Sun and Moon, which move steadily eastward relative to the stars, the planets vary substantially in both speed and brightness as they move among the stars. Moreover, while the planets *usually* move eastward relative to the stars, they occasionally reverse course, moving westward rather than eastward through the zodiac (Figure 2.26). These periods of **apparent retrograde motion** (*retrograde* means "backward") last from a few weeks to a few months, depending on the planet.

Ancient astronomers could easily explain the daily paths of the stars through the sky by imagining that the celestial sphere was real and that it actually rotated around Earth each day. But the apparent retrograde motion of the planets posed a far greater mystery: What could cause the planets to sometimes go backward? As we'll discuss in Chapter 3, the ancient Greeks came up with some very clever ways to explain the occasional backward motion of the planets, despite being wedded to the incorrect idea of an Earth-centered universe. However, the Greek explanation was quite complex, and ultimately proven wrong.

In contrast, apparent retrograde motion has a simple explanation in a Sun-centered solar system. You can demonstrate it for yourself with the help of a friend (Figure 2.27a). Pick a spot in an open field to represent the Sun. You can represent Earth, walking counterclockwise around the Sun, while your friend represents a more distant planet (such as Mars or Jupiter) by walking counterclockwise around the Sun at a greater distance. Your friend should walk more slowly than you, because more distant planets orbit the Sun more slowly. As you walk, watch how your friend appears to move relative to buildings or trees in the distance. Although both of you always walk the same way around the Sun, your

Figure 2.26

This composite of 29 individual photos shows Mars from June through November 2003. Notice that Mars usually moves eastward (left) relative to the stars, but reverses course during its apparent retrograde motion. Note also that Mars is biggest and brightest in the middle of the retrograde loop, because that is where it is closest to Earth in its orbit. (The white dots in a line just right of center are the planet Uranus, which by coincidence was in the same part of the sky.)

Figure 2.27 interactive figure

Apparent retrograde motion—the occasional "backward" motion of the planets relative to the stars—has a simple explanation in a Sun-centered solar system.

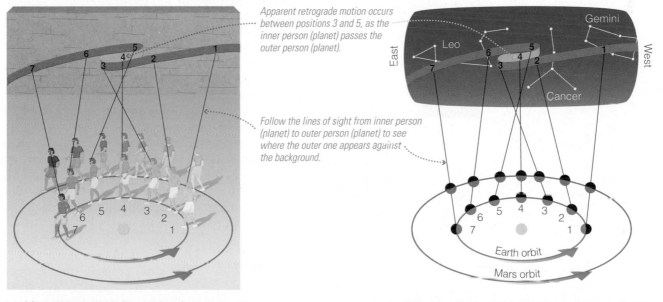

Apparent retrograde motion occurs between positions 3 and 5, as the inner person (planet) passes the outer person (planet).

Follow the lines of sight from inner person (planet) to outer person (planet) to see where the outer one appears against the background.

a This demonstration shows why planets sometimes seem to go backward relative to distant stars. Watch how your friend (in red) usually appears to move forward against the background of the building in the distance but appears to move backward as you (in blue) catch up and pass her in your "orbit."

b This diagram shows how the same idea applies to a planet. Follow the lines of sight from Earth to Mars in numerical order. Notice that Mars appears to move westward relative to the distant stars as Earth passes it in its orbit (from points 3 to 5 in the diagram).

friend will appear to move backward against the background during the part of your "orbit" at which you catch up to and pass him or her. To understand the apparent retrograde motions of Mercury and Venus, which are closer to the Sun than is Earth, simply switch places with your friend and repeat the demonstration.

A planet appears to move backward relative to the stars during the period when Earth passes it in its orbit.

This demonstration closely models actual planet motions. For example, because Mars takes about 2 years to orbit the Sun (more precisely, 1.88 years), it covers about half its orbit during the 1 year in which Earth makes a complete orbit. If you trace lines of sight from Earth to Mars from different points in their orbits, you will see that the line of sight usually moves eastward relative to the stars but moves westward during the time when Earth is passing Mars in its orbit (Figure 2.27b). Like your friend in the demonstration, Mars never actually changes direction. It only *appears* to change direction from our perspective on Earth.

• Why did the ancient Greeks reject the real explanation for planetary motion?

If the apparent retrograde motion of the planets is so readily explained by recognizing that Earth orbits the Sun, why wasn't this idea accepted in ancient times? In fact, the idea that Earth goes around the Sun was suggested as early as 260 B.C. by the Greek astronomer Aristarchus. No one knows why Aristarchus proposed a Sun-centered solar system, but the fact that it explains planetary motion so naturally probably played a role. Nevertheless, Aristarchus's contemporaries rejected his idea, and the Sun-centered solar system did not gain wide acceptance until almost 2000 years later.

Although there were many reasons why the Greeks were reluctant to abandon the idea of an Earth-centered universe, one of the most important was their inability to detect something called **stellar parallax.** Extend your arm and hold up one finger. If you keep your finger still and alternately close your left eye and right eye, your finger will appear to jump back and forth against the background. This apparent shifting, called *parallax,* occurs because your two eyes view your finger from opposite sides of your nose. If you move your finger closer to your face, the parallax increases. If you look at a distant tree or flagpole instead of your finger, you may not notice any parallax at all. This little experiment shows that parallax depends on distance, with nearer objects exhibiting greater parallax than more distant objects.

If you now imagine that your two eyes represent Earth at opposite sides of its orbit around the Sun and that your finger represents a relatively nearby star, you have the idea of stellar parallax. Because we view the stars from different places in our orbit at different times of year, nearby stars should *appear* to shift back and forth against the background of more distant stars (Figure 2.28).

The Greeks knew that stellar parallax should occur if Earth orbits the Sun, but they could not detect it.

Because the Greeks believed that all stars lie on the same celestial sphere, they expected to see stellar parallax in a slightly different way. If Earth orbited the Sun, they reasoned, at different times of year we would be closer to different parts of the celestial sphere and would notice changes in the angular separations of stars. However, no matter how hard they searched, they could

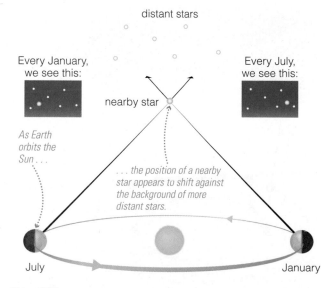

distant stars

Every January, we see this:

nearby star

Every July, we see this:

As Earth orbits the Sun . . .

. . . the position of a nearby star appears to shift against the background of more distant stars.

July

January

Figure 2.28

Stellar parallax is an apparent shift in the position of a nearby star as we look at it from different places in Earth's orbit. This figure is greatly exaggerated; in reality, the amount of shift is far too small to detect with the naked eye.

find no sign of stellar parallax. They concluded that one of the following must be true:

1. Earth orbits the Sun, but the stars are so far away that stellar parallax is not detectable to the naked eye.
2. There is no stellar parallax because Earth remains stationary at the center of the universe.

Aside from notable exceptions such as Aristarchus, the Greeks rejected the correct answer (the first one) because they could not imagine that the stars could be *that* far away. Today, we can detect stellar parallax with the aid of telescopes, providing direct proof that Earth really does orbit the Sun. Careful measurements of stellar parallax also provide the most reliable means of measuring distances to nearby stars [Section 11.1].

think about it How far apart are opposite sides of Earth's orbit? How far away are the nearest stars? Describe the challenge of detecting stellar parallax. It may help to visualize Earth's orbit and the distance to the stars on the 1-to-10-billion scale used in Chapter 1.

The ancient mystery of the planets drove much of the historical debate over Earth's place in the universe. In many ways, the modern technological society we take for granted today can be traced directly to the scientific revolution that began in the quest to explain the strange wandering of the planets among the stars in our sky. We will turn our attention to this revolution in the next chapter.

----

the big picture
Putting Chapter 2 into Context

In this chapter, we surveyed the phenomena of our sky. Keep the following "big picture" ideas in mind as you continue your study of astronomy:

- You can enhance your enjoyment of learning astronomy by observing the sky. The more you learn about the appearance and apparent motions of objects in the sky, the more you will appreciate what you can see in the universe.

- From our vantage point on Earth, it is convenient to imagine that we are at the center of a great celestial sphere—even though we really are on a planet orbiting a star in a vast universe. We can then understand what we see in the local sky by thinking about how the celestial sphere appears from our latitude.

- Most of the phenomena of the sky are relatively easy to observe and understand. The more complex phenomena—particularly eclipses and apparent retrograde motion of the planets—challenged our ancestors for thousands of years. The desire to understand these phenomena helped drive the development of science and technology.

summary of key concepts

2.1 Patterns in the Night Sky

• **What does the universe look like from Earth?**

Stars and other celestial objects appear to lie on a great **celestial sphere** surrounding Earth. We divide the celestial sphere into **constellations** with well-defined borders. From any location on Earth, we see half the celestial sphere at any given time as the dome of our **local sky**, in which the **horizon** is the boundary between Earth and sky, the **zenith** is the point directly overhead, and the **meridian** runs from due south to due north through the zenith.

• **Why do stars rise and set?**

Earth's rotation makes stars appear to circle around Earth each day. A star whose complete circle lies above our horizon is said to be **circumpolar.** Other stars have circles that cross the horizon, so they rise in the east and set in the west each day.

• **Why do the constellations we see depend on latitude and time of year?**

The visible constellations vary with time of year because our night sky lies in different directions in space as we orbit the Sun. The constellations vary with **latitude** because your latitude determines the orientation of your horizon relative to the celestial sphere. The sky does not vary with **longitude.**

2.2 The Reason for Seasons

• **What causes the seasons?**

The tilt of Earth's axis causes the seasons. The axis points in the same direction (toward Polaris) throughout the year. Thus, as Earth orbits the Sun, sunlight hits different parts of Earth more directly at different times of year.

• **How do we mark the progression of the seasons?**
The **summer** and **winter solstices** are the times during the year when the Northern Hemisphere gets its most and least direct sunlight, respectively. The **spring** and **fall equinoxes** are the two times when both hemispheres get equally direct sunlight.

• **How does the orientation of Earth's axis change with time?**

Earth's 26,000-year cycle of **precession** changes the orientation of its axis in space, although the tilt remains about $23\frac{1}{2}°$. The changing orientation of the axis does not affect the pattern of seasons, but it changes the identity of the north star and shifts the locations of the solstices and equinoxes in Earth's orbit.

2.3 The Moon, Our Constant Companion

• **Why do we see phases of the Moon?**

The **phase** of the Moon depends on its position relative to the Sun as it orbits Earth. The half of the Moon facing the Sun is always illuminated while the other half is dark, but from Earth we see varying combinations of the illuminated and dark halves.

• **What causes eclipses?**

We see a **lunar eclipse** when Earth's shadow falls on the Moon and a **solar eclipse** when the Moon blocks our view of the Sun. We do not see an eclipse at every new and full moon because the Moon's orbit is slightly inclined to the ecliptic plane. Eclipses come in different types, depending on where the dark **umbral** and lighter **penumbral** shadows fall.

2.4 The Ancient Mystery of the Planets

• **What was once so mysterious about the movement of planets in our sky?**

Planets generally appear to move eastward relative to the stars over the course of the year, but for weeks or months they reverse course in their periods of **apparent retrograde motion.** This motion occurs when Earth passes by (or is passed by) another planet in its orbit, but it posed a major mystery to ancient people who assumed Earth to be at the center of the universe.

• **Why did the ancient Greeks reject the real explanation for planetary motion?**

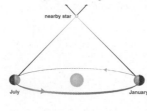

The Greeks rejected the idea that Earth goes around the Sun in part because they could not detect **stellar parallax**—slight apparent shifts in stellar positions over the course of the year. To most Greeks, it seemed unlikely that the stars could be so far away as to make parallax undetectable to the naked eye, even though that is, in fact, the case.

exercises and problems

For instructor-assigned homework go to www.masteringastronomy.com.

Review Questions

Short-Answer Questions Based on the Reading

1. What are *constellations?* How did they get their names?
2. Suppose you were making a model of the celestial sphere with a ball. Briefly describe all the things you would need to mark on your celestial sphere.
3. On a clear, dark night, the sky may appear to be "full" of stars. Does this appearance accurately reflect the way stars are distributed in space? Explain.
4. Why does the *local sky* look like a dome? Define *horizon, zenith,* and *meridian.* How do we describe the location of an object in the local sky?
5. Explain why we can measure only *angular sizes* and *angular distances* for objects in the sky. What are *arcminutes* and *arcseconds?*
6. What are *circumpolar stars?* Are more stars circumpolar at the North Pole or in the United States? Explain.
7. What are *latitude* and *longitude?* Does the local sky vary with latitude? Does it vary with longitude? Explain.
8. What is the *zodiac,* and why do we see different parts of it at different times of year?
9. Suppose Earth's axis had no tilt. Would we still have seasons? Why or why not?
10. Briefly describe what is special about the summer and winter solstices and the spring and fall equinoxes.
11. What is *precession,* and how does it affect the sky that we see from Earth?
12. Briefly describe the Moon's cycle of *phases.* Can you ever see a full moon at noon? Explain.
13. Suppose you lived on the Sun (and could ignore the heat). Would you still see the Moon go through phases as it orbits Earth? Why or why not?
14. Why don't we see an *eclipse* at every new and full moon? Describe the conditions that must be met for us to see a *solar* or *lunar eclipse.*
15. What do we mean by the *apparent retrograde motion* of the planets? Why was it difficult for ancient astronomers to explain but is easy for us to explain?
16. What is *stellar parallax?* Briefly describe the role it played in making ancient astronomers believe in an Earth-centered universe.

Test Your Understanding

Does It Make Sense?

Decide whether the statement makes sense (or is clearly true) or does not make sense (or is clearly false). Explain clearly; not all of these have definitive answers, so your explanation is more important than your chosen answer.

17. The constellation Orion didn't exist when my grandfather was a child.
18. When I looked into the dark lanes of the Milky Way with my binoculars, I saw what must have been a cluster of distant galaxies.
19. Last night the Moon was so big that it stretched for a mile across the sky.
20. I live in the United States, and during my first trip to Argentina I saw many constellations that I'd never seen before.

21. Last night I saw Jupiter right in the middle of the Big Dipper. (*Hint:* Is the Big Dipper part of the zodiac?)
22. Last night I saw Mars move westward through the sky in its apparent retrograde motion.
23. Although all the known stars appear to rise in the east and set in the west, we might someday discover a star that will appear to rise in the west and set in the east.
24. If Earth's orbit were a perfect circle, we would not have seasons.
25. Because of precession, someday it will be summer everywhere on Earth at the same time.
26. This morning I saw the full moon setting at about the same time the Sun was rising.

Quick Quiz

Choose the best answer to each of the following. Explain your reasoning with one or more complete sentences.

27. Two stars that are in the same constellation (a) must both be part of the same cluster of stars in space. (b) must both have been discovered at about the same time. (c) may actually be very far away from each other.
28. The north celestial pole is 35° above your northern horizon. This tells you that (a) you are at latitude 35°N. (b) you are at longitude 35°E. (c) you are at latitude 35°S.
29. Beijing and Philadelphia have about the same latitude but very different longitudes. Therefore, tonight's night sky in these two places (a) will look about the same. (b) will have completely different sets of constellations. (c) will have partially different sets of constellations.
30. In winter, Earth's axis points toward the star Polaris. In spring, (a) the axis also points toward Polaris. (b) the axis points toward Vega. (c) the axis points toward the Sun.
31. When it is summer in Australia, it is (a) winter in the United States. (b) summer in the United States. (c) spring in the United States.
32. If the Sun rises precisely due east, (a) you must be located at Earth's equator. (b) it must be the day of either the spring or fall equinox. (c) it must be the day of the summer solstice.
33. A week after full moon, the Moon's phase is (a) first quarter. (b) third quarter. (c) new.
34. Some type of lunar or solar eclipse (not necessarily a total eclipse) occurs (a) about once every 18 years. (b) about once a month. (c) at least four times a year.
35. If there is going to be a total lunar eclipse tonight, then you know that (a) the Moon's phase is full. (b) the Moon's phase is new. (c) the Moon is unusually close to Earth.
36. When we see Saturn going through a period of apparent retrograde motion, it means (a) Saturn is temporarily moving backward in its orbit of the Sun. (b) Earth is passing Saturn in its orbit, with both planets on the same side of the Sun. (c) Saturn and Earth must be on opposite sides of the Sun.

Process of Science

Examining How Science Works

37. *Earth-Centered or Sun-Centered?* The phenomena discussed in this chapter are all visible to the naked eye and therefore have been known throughout human history, even during the thousands of years when Earth was assumed to be at the center of the

universe. For each of the following, decide whether the phenomenon is consistent or inconsistent with a belief in an Earth-centered system. If consistent, describe how. If inconsistent, explain why, and also explain why the inconsistency did not immediately lead people to abandon the Earth-centered model.

a. The daily paths of stars through the sky
b. Seasons
c. Phases of the Moon
d. Eclipses
e. Apparent retrograde motion of the planets

38. *Shadow Phases.* Many people incorrectly guess that the phases of the Moon are caused by Earth's shadow falling on the Moon. How would you convince a friend that the phases of the Moon have nothing to do with Earth's shadow? Describe the observations you would use to show that Earth's shadow isn't the cause of phases.

Investigate Further

In-Depth Questions to Increase Your Understanding

Short-Answer/Essay Questions

39. *New Planet.* Suppose we discover a planet in another solar system that has a circular orbit and an axis tilt of 35°. Would you expect this planet to have seasons? If so, would you expect them to be more extreme than the seasons on Earth? If not, why not?

40. *Your View.*
a. Find your latitude and longitude, and state the source of your information.
b. Describe the altitude and direction in your local sky at which the north or south celestial pole appears.
c. Is Polaris a circumpolar star in your sky? Explain.

41. *View from the Moon.* Suppose you lived on the Moon, in which case you would see Earth going through phases in your sky. Assume you live near the center of the face that looks toward Earth.
a. Suppose you see a full Earth in your sky. What phase of the Moon would people on Earth see? Explain.
b. Suppose people on Earth see a full moon. What phase would you see for Earth? Explain.
c. Suppose people on Earth see a waxing gibbous moon. What phase would you see for Earth? Explain.
d. Suppose people on Earth are viewing a total lunar eclipse. What would you see from your home on the Moon? Explain.

42. *A Farther Moon.* Suppose the distance to the Moon were twice its actual value. Would it still be possible to have a total solar eclipse? Why or why not?

43. *A Smaller Earth.* Suppose Earth were smaller. Would solar eclipses be any different? If so, how? What about lunar eclipses? Explain.

44. *Observing Planetary Motion.* Find out what planets are currently visible in your evening sky. At least once a week, observe the planets and draw a diagram showing the position of each visible planet relative to stars in a zodiac constellation. From week to week, note how the planets are moving relative to the stars. Can you see any of the apparently wandering features of planetary motion? Explain.

Quantitative Problems

Be sure to show all calculations clearly and state your final answers in complete sentences.

45. *Arcminutes and Arcseconds.* There are 360° in a full circle.
a. How many arcminutes are in a full circle?
b. How many arcseconds are in a full circle?
c. The Moon's angular size is about $\frac{1}{2}$° What is this in arcminutes? In arcseconds?

46. *Find the Sun's Diameter.* The Sun has an angular diameter of about 0.5° and an average distance from Earth of about 150 million km. What is the Sun's approximate physical diameter? Compare your answer to the actual value of 1,390,000 km.

47. *Find a Star's Diameter.* The supergiant star Betelgeuse (in the constellation Orion) has a measured angular diameter of 0.044 arcsecond from Earth and a distance from Earth of 427 light-years. What is the actual diameter of Betelgeuse? Compare your answer to the size of our Sun and the Earth–Sun distance.

48. *Eclipse Conditions.* The Moon's precise equatorial diameter is 3476 km, and its orbital distance from Earth varies between 356,400 km and 406,700 km. The Sun's diameter is 1,390,000 km, and its distance from Earth ranges between 147.5 and 152.6 million km.
a. Find the Moon's angular size at its minimum and maximum distances from Earth.
b. Find the Sun's angular size at its minimum and maximum distances from Earth.
c. Based on your answers to (a) and (b), is it possible to have a total solar eclipse when the Moon and Sun are both at their maximum distances? Explain.

Discussion Questions

49. *Earth-Centered Language.* Many common phrases reflect the ancient Earth-centered view of our universe. For example, the phrase "the Sun rises each day" implies that the Sun is really moving over Earth. We know that the Sun only *appears* to rise as the rotation of Earth carries us to a place where we can see the Sun in our sky. Identify other common phrases that imply an Earth-centered viewpoint.

50. *Flat Earth Society.* Believe it or not, there is an organization called the Flat Earth Society. Its members hold that Earth is flat and that all indications to the contrary (such as pictures of Earth from space) are fabrications made as part of a conspiracy to hide the truth from the public. Discuss the evidence for a round Earth and how you can check it for yourself. In light of the evidence, is it possible that the Flat Earth Society is correct? Defend your opinion.

Web Projects

51. *Sky Information.* Search the Web for sources of daily information about sky phenomena (such as lunar phases, times of sunrise and sunset, or dates of equinoxes and solstices). Identify and briefly describe your favorite source.

52. *Constellations.* Search the Web for information about the constellations and their mythology. Write a one- to three-page report about one or more constellations.

53. *Upcoming Eclipse.* Find information about an upcoming solar or lunar eclipse. Write a one- to three-page report about how you could best observe the eclipse, including any necessary travel to a viewing site, and what you could expect to see. Bonus: Describe how you could photograph the eclipse.

Use the following questions to check your understanding of some of the many types of visual information used in astronomy. Answers are provided in Appendix K. For additional practice, try the Chapter 2 Visual Quiz at **www.masteringastronomy.com**.

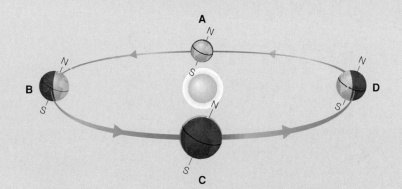

The figure above is a typical diagram used to describe Earth's seasons.

1. Which of the four labeled points (A through D) represents the beginning of summer for the Northern Hemisphere?
2. Which of the four labeled points represents the beginning of summer for the Southern Hemisphere?
3. Which of the four labeled points represents the beginning of spring for the Northern Hemisphere?
4. Which of the four labeled points represents the beginning of spring for the Southern Hemisphere?
5. Diagrams like the one shown in the figure are useful for representing seasons, but they can also be misleading because they exaggerate the sizes of Earth and the Sun relative to the orbit. If Earth were correctly scaled relative to the orbit in the figure, how big would it be?
   a. about half the size shown
   b. about 2 millimeters across
   c. about 0.1 millimeter across
   d. microscopic

The figure above (based on Figure 2.12) shows the Sun's path through the constellations of the zodiac.

6. As viewed from Earth, in which zodiac constellation does the Sun appear to be located on April 21?
   a. Leo
   b. Aquarius
   c. Libra
   d. Aries
7. If the date is April 21, what zodiac constellation will be visible on your meridian at midnight?
   a. Leo
   b. Aquarius
   c. Libra
   d. Aries
8. If the date is April 21, what zodiac constellation will you see setting in the west shortly after sunset?
   a. Scorpius
   b. Pisces
   c. Taurus
   d. Virgo

3

the science of astronomy

Today we know that Earth is a planet orbiting a rather ordinary star, in a galaxy of more than a hundred billion stars, in an incredibly vast universe. We know that Earth, along with the entire cosmos, is in constant motion. We know that, on the scale of cosmic time, human civilization has existed for only the briefest moment. How did we manage to learn these things?

It wasn't easy. Astronomy is the oldest of the sciences, with roots extending as far back as recorded history allows us to see. But the most impressive advances in knowledge have come in just the past few centuries.

In this chapter, we will trace how modern astronomy grew from its roots in ancient observations, including those of the Greeks. We'll pay special attention to the unfolding of the Copernican revolution, which overturned the ancient belief in an Earth-centered universe and laid the foundation for the rise of our technological civilization. Finally, we'll explore the nature of modern science and the distinction between science and nonscience.

essential preparation

1. What does the universe look like from Earth? [Section 2.1]

2. What was once so mysterious about planetary motion in our sky? [Section 2.4]

3. Why did the ancient Greeks reject the real explanation for planetary motion? [Section 2.4]

3.1   The Ancient Roots of Science

A common stereotype holds that scientists walk around in white lab coats and somehow think differently than other people. In reality, scientific thinking is a fundamental part of human nature. In this section, we will trace the roots of science to experiences common to nearly all people and nearly all cultures.

• In what ways do all humans use scientific thinking?

Scientific thinking comes naturally to us. By about a year of age, a baby notices that objects fall to the ground when she drops them. She lets go of a ball—it falls. She pushes a plate of food from her high chair—it falls, too. She continues to drop all kinds of objects, and they all plummet to Earth. Through her powers of observation, the baby learns about the physical world, finding that things fall when they are unsupported. Eventually, she becomes so certain of this fact that, to her parents' delight, she no longer needs to test it continually.

Scientific thinking is based on everyday observations and trial-and-error experiments.

One day somebody gives the baby a helium balloon. She releases it, and to her surprise it rises to the ceiling! Her understanding of nature must be revised. She now knows that the principle "all things fall" does not represent the whole truth, although it still serves her quite well in most situations. It will be years before she learns enough about the atmosphere, the force of gravity, and the concept of density to understand *why* the balloon rises when most other objects fall. For now, she is delighted to observe something new and unexpected.

The baby's experience with falling objects and balloons exemplifies scientific thinking. In essence, it is a way of learning about nature through

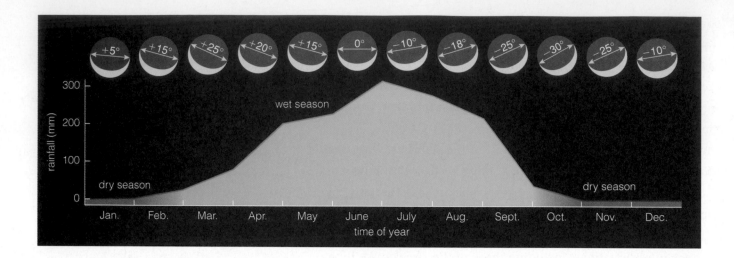

Figure 3.1

Science is rooted in careful observation of the world around us. This diagram shows how central Africans used the orientation of the waxing crescent moon to predict rainfall. The graph depicts the annual rainfall pattern in central Nigeria, and the Moon diagrams show the varying angle of the "horns" of a waxing crescent moon relative to the western horizon. (Adapted from *Ancient Astronomers* by Anthony F. Aveni.)

careful observation and trial-and-error experiments. Rather than thinking differently than other people, modern scientists are trained to organize everyday thinking in a way that makes it easier for them to share their discoveries and use their collective wisdom.

think about it Describe a few cases where you have learned by trial and error—while cooking, participating in sports, fixing something, learning on the job, or in any other situation.

Just as learning to communicate through language, art, or music is a gradual process for a child, the development of science has been a gradual process for humanity. Science in its modern form requires painstaking attention to detail, relentless testing of each piece of information to ensure its reliability, and a willingness to give up old beliefs that are not consistent with observed facts about the physical world. For professional scientists, these demands are the "hard work" part of the job. At heart, professional scientists are like the baby with the balloon, delighted by the unexpected and motivated by those rare moments when they—and all of us—learn something new about the universe.

How did astronomical observations benefit ancient societies?

We will discuss modern science shortly, but first we will explore how it arose from the observations of ancient peoples. Our exploration begins in central Africa, where people long ago learned to predict the weather with reasonable accuracy by making careful observations of the Moon. Remember that the Moon begins its monthly cycle as a crescent in the western sky just after sunset. Through long traditions of sky watching, central African societies discovered that the orientation of the crescent "horns" relative to the horizon is closely tied to local rainfall patterns (Figure 3.1).

Why did ancient people make such careful and detailed observations of the sky? In part, it was probably to satisfy their inherent curiosity. But astronomy also played a practical role for them. They used the changing positions of the Sun, Moon, and stars to keep track of the time

TABLE 3.1 *The Seven Days of the Week and the Astronomical Objects They Honor*

The correspondence between objects and days is easy to see in French and Spanish. In English, the correspondence becomes clear when we look at the names of the objects used by the Teutonic tribes who lived in the region of modern-day Germany.

Object	Teutonic Name	English	French	Spanish
Sun	Sun	Sunday	dimanche	domingo
Moon	Moon	Monday	lundi	lunes
Mars	Tiw	Tuesday	mardi	martes
Mercury	Woden	Wednesday	mercredi	miércoles
Jupiter	Thor	Thursday	jeudi	jueves
Venus	Fria	Friday	vendredi	viernes
Saturn	Saturn	Saturday	samedi	sábado

and seasons, crucial skills for people who depended on agriculture. Some cultures even learned to navigate by the Sun and stars.

Ancient people used observations of the sky to keep track of the time and seasons and as an aid in navigation.

Modern measures of time come directly from ancient observations of motion in the sky. The length of our day is the time it takes the Sun to make one full circuit of the sky. The length of a month comes from the Moon's cycle of phases [Section 2.3], and our year is based on the cycle of the seasons [Section 2.2]. The seven days of the week were named after the seven objects that could be seen with the naked eye and appeared to move among the constellations: the Sun, the Moon, and the five planets recognized in ancient times (Table 3.1).

• What did ancient civilizations achieve in astronomy?

Nearly all ancient civilizations practiced astronomy at some level. Many built remarkable structures for observing the sky. Let's explore a few of the ways that ancient societies studied the sky.

Determining the Time of Day In the daytime, ancient peoples could tell time by observing the Sun's path through the sky. Many cultures probably used the shadows cast by sticks as simple sundials. The ancient Egyptians built huge obelisks, often decorated in homage to the Sun, that probably also served as simple clocks (Figure 3.2). At night, ancient people could estimate the time from the position and phase of the Moon (see Figure 2.19) or by observing the constellations visible at a particular time of night.

We can trace the origins of our modern clock to ancient Egypt, some 4000 years ago. The Egyptians divided the daylight into 12 equal parts, and we still break the 24-hour day into 12 hours each of A.M. and P.M. The abbreviations *A.M.* and *P.M.* stand for the Latin terms *ante meridiem* and *post meridiem*, respectively, which mean "before the middle of the day" and "after the middle of the day."

Marking the Seasons Many ancient cultures built structures to help them mark the seasons. One of the oldest standing human-made structures served such a purpose: Stonehenge in southern England (Figure 3.3). Stonehenge served both as an astronomical device for keeping track of the seasons and as a social and religious gathering place.

Among the most spectacular structures used to mark the seasons was the Templo Mayor in the Aztec city of Tenochtitlán, located on the site of modern-day Mexico City (Figure 3.4). Twin temples stood on a flat-topped, 150-foot-high pyramid. From the vantage point of a royal observer watching from the opposite side of the plaza, the Sun rose directly through the notch between the temples on the equinoxes.

Many cultures aligned their buildings with the cardinal directions (north, south, east, and west), enabling them to mark the rising and setting of the Sun relative to the building orientation. Other structures were used to mark the Sun's position on special dates. For example, the ancient Anasazi people carved a 19-turn spiral—known as the *Sun Dagger*—on a vertical cliff face in Chaco Canyon, New Mexico (Figure 3.5). The Sun's rays form a dagger of sunlight that pierces the center of the carved spiral only once each year—at noon on the summer solstice.

Figure 3.2

This ancient Egyptian obelisk, which stands 83 feet tall and weighs 331 tons, resides in St. Peter's Square at the Vatican in Rome. It is one of 21 surviving obelisks from ancient Egypt, most of which are now scattered around the world. Shadows cast by the obelisks may have been used to tell time.

Figure 3.3

The remains of Stonehenge today. It was built in stages from about 2750 B.C. to about 1550 B.C.

Figure 3.4

This scale model shows the Templo Mayor and the surrounding plaza as they are thought to have looked before the Spanish conquistadores destroyed Aztec civilization. The structure was used to help mark the seasons.

Figure 3.5

The Sun Dagger. Three large slabs of rock in front of the spiral produce patterns of light and shadow that vary throughout the year and form a dagger of sunlight that pierces the center of the spiral only at noon on the summer solstice.

Lunar Calendars Some ancient civilizations paid particular attention to lunar phases and used them as the basis for calendars. Some months on a lunar calendar have 29 days and others have 30 days, so that the average matches the $29\frac{1}{2}$-day lunar cycle. A 12-month lunar calendar has only 354 or 355 days, or about 11 days fewer than a calendar based on the Sun. Such a calendar is still used in the Muslim religion. That is why the month-long fast of Ramadan (the ninth month) begins about 11 days earlier with each subsequent year.

Other lunar calendars remain roughly synchronized with solar calendars by taking advantage of an interesting coincidence: 19 years on a solar calendar is almost precisely 235 months on a lunar calendar. As a result, the lunar phases repeat on the same dates about every 19 years (a pattern known as the *Metonic cycle*). For example, there was a full moon on October 17, 2005, and there will be a full moon 19 years later, on October 17, 2024. Because an ordinary lunar calendar has only $19 \times 12 = 228$ months in a 19-year period, adding 7 extra months (to make 235) can keep the lunar calendar roughly synchronized to the seasons. The Jewish calendar does this by adding a thirteenth month in the third, sixth, eighth, eleventh, fourteenth, seventeenth, and nineteenth years of each 19-year cycle.

> Remarkable ancient achievements included accurate calendars, eclipse prediction, navigational tools, and elaborate structures for astronomical observations.

In addition to following the lunar phases, some ancient cultures discovered other lunar cycles. In the Middle East more than 2500 years ago, the ancient Babylonians achieved remarkable success in predicting eclipses, thanks to their recognition of the approximately 18-year saros cycle [Section 2.3]. The Mayans of Central America also appear to have been experts at eclipse prediction, but we know few details about their accomplishments because the Spanish conquistadores burned most Mayan writings.

Ancient Structures and Archaeoastronomy It's easy to establish the astronomical intentions of ancient cultures that left extensive written records, such as the Chinese and the Egyptians. In other cases, however, claims that ancient structures served astronomical purposes can be much more difficult to evaluate.

The study of ancient structures in search of astronomical connections is called *archaeoastronomy,* a word that combines archaeology and astronomy. Scientists engaged in archaeoastronomy usually start by evaluating an ancient structure to see whether it shows any particular astronomical alignments. For example, they may check to see whether an observer in a central location would see particular stars rise above specially marked stones, or whether sunlight enters through a window only on special days like the solstices or equinoxes. However, the mere existence of astronomical alignments is not enough to establish that a structure had an astronomical purpose; the alignments may be coincidental.

Native American Medicine Wheels—stone circles found throughout the northern plains of the United States—offer an example of the difficulty of trying to establish the intentions of ancient builders. In the 1970s, a study of the Big Horn Medicine Wheel in Wyoming (Figure 3.6) seemed to indicate that its 28 "spokes" were aligned with the rise and set of particular stars. However, later research showed that the original study had failed to take into account the motion of stars as they rise above the horizon and the way the atmosphere affects the visibility of stars at the latitude of Big Horn. In reality the spokes do *not* show any special alignments with bright stars. Moreover, if Medicine Wheels really did serve an astronomical purpose, we'd expect all of them to have been built with consistent alignments—but that is not the case.

In some cases, scientists can use other clues to establish the intentions of ancient builders. For example, lodges built by the Pawnee people in Kansas feature strategically placed holes for observing the passage of constellations that figure prominently in Pawnee folklore. The correspondence between the folklore and the structural features provides a strong case for deliberate intent rather than coincidence. Similarly, traditions of the Inca Empire of South America held that its rulers were descendents of the Sun and therefore demanded that movements of the Sun be watched closely. This fact supports the idea that astronomical alignments in Inca cities and ceremonial centers, such as the World Heritage Site of Machu Picchu (Figure 3.7), were deliberate rather than accidental.

A different type of evidence makes a convincing case for the astronomical sophistication of ancient Polynesians, who lived and traveled among the many islands of the mid- and South Pacific. Navigation was crucial to survival, because the next island in a journey usually was too distant to be seen. The most esteemed position in Polynesian culture was that of the Navigator, a person who had acquired the knowledge necessary to navigate great distances among the islands. Navigators used a combination of detailed knowledge of astronomy and equally impressive knowledge of the patterns of waves and swells around different islands (Figure 3.8).

3.2   Ancient Greek Science

Before a structure such as Stonehenge or the Templo Mayor could be built, careful observations had to be made and repeated over and over to ensure their accuracy. Careful, repeatable observations also underlie modern science. Elements of modern science were therefore present in many early human cultures. If the circumstances of history had been different, almost any culture might have been the first to develop what we consider to be modern science. In the end, however, history takes only one of countless possible paths. The path that led to modern science emerged from the ancient civilizations of the Mediterranean and the Middle East—especially from ancient Greece.

Greece gradually rose as a power in the Middle East beginning around 800 B.C. and was well-established by about 500 B.C. Its geographical location placed it at a crossroads for travelers, merchants, and armies from northern Africa, Asia, and Europe. Building on the diverse ideas brought forth by the meeting of these many cultures, ancient Greek philosophers soon began their efforts to move human understanding of nature from the mythological to the rational. Their ideas spread widely with the conquests of Alexander the Great (356–323 B.C.), who had been personally tutored by Aristotle and had a keen interest in science. Alexander founded the city of Alexandria in Egypt, and shortly after his death the city commenced work on a great research center and library. The Library of Alexandria (Figure 3.9) opened in about 300 B.C. and remained the world's preeminent center of research for some 700 years. At its peak, it may have held as many as a half million books, handwritten on papyrus scrolls. Most of these scrolls were ultimately burned, their contents lost forever.

Figure 3.6

The Big Horn Medicine Wheel in Wyoming. A study once claimed that its "spokes" have astronomically significant alignments, but later research showed the claim was in error—making this a good example of how science adapts as new data come to light.

Figure 3.7

The World Heritage Site of Machu Picchu has structures aligned with sunrise at the winter and summer solstices.

Figure 3.8

A Micronesian stick chart, an instrument used by Polynesian Navigators to represent swell patterns around islands.

a This rendering shows an artist's reconstruction of how the Great Hall may have looked in the ancient Library of Alexandria.

b A similar rendering to (a), this time showing a scroll room in the ancient library.

c The New Library of Alexandria, Egypt, which opened in 2003.

Figure 3.9

The ancient Library of Alexandria thrived for some 700 years, starting in about 300 B.C.

think about it Estimate the number of books you're likely to read in your lifetime and compare this number to the half million books once housed in the Library of Alexandria. Can you think of other ways to put into perspective the loss of ancient wisdom resulting from the destruction of the Library of Alexandria?

• Why does modern science trace its roots to the Greeks?

Greek philosophers developed at least three major innovations that helped pave the way for modern science. First, they developed a tradition of trying to understand nature without relying on supernatural explanations, and of working communally to debate and challenge each other's ideas. Second, the Greeks used mathematics to give precision to their ideas, which allowed them to explore the implications of new ideas in much greater depth than would have otherwise been possible. Third, while much of their philosophical activity consisted of subtle debates grounded only in thought and was not scientific in the modern sense, the Greeks also saw the power of reasoning from observations. They understood that an explanation could not be right if it disagreed with observed facts.

The Greeks developed models of nature that aimed to explain and predict observed phenomena.

Perhaps most important, the Greeks combined all three innovations to create **models** of nature, a practice that is still central to modern science. Scientific models differ somewhat from the models you may be familiar with in everyday life. In our daily lives, we tend to think of models as miniature physical representations, such as model cars or airplanes. In contrast, a scientific model is a conceptual representation created to explain and predict observed phenomena. For example, a model of Earth's climate uses logic and mathematics to represent what we know about how the climate works. Its purpose is to explain and predict climate changes, such as the changes that may occur with global warming. Just as a model airplane does not faithfully represent every aspect of a real airplane, a scientific model may not fully explain all our observations of nature. Nevertheless, even the failings of a scientific model can be useful, because they often point the way toward building a better model.

In astronomy, the Greeks constructed conceptual models of the universe in an attempt to explain what they observed in the sky, an effort

Columbus and a Flat Earth

A widespread myth gives credit to Columbus for learning that Earth is round, but knowledge of Earth's shape predated Columbus by nearly 2000 years. Not only were scholars of Columbus's time well aware that Earth is round, but they even knew its approximate size: Earth's circumference was first measured in about 240 B.C. by the Greek scientist Eratosthenes. In fact, a likely reason why Columbus had so much difficulty finding a sponsor for his voyages was that he tried to argue a point on which he was dead wrong: He claimed the distance by sea from western Europe to eastern Asia to be much less than the scholars knew it to be. Indeed, when he finally found a patron in Spain and left on his journey, he was so woefully underprepared that the voyage would almost certainly have ended in disaster if the Americas hadn't stood in his way.

that quickly led them past simplistic ideas of a flat Earth under a dome-shaped sky to a far more sophisticated view of the cosmos. We do not know precisely when other Greeks first began to think that Earth is round, but this idea was being taught as early as about 500 B.C. by the famous mathematician Pythagoras (c. 560–480 B.C.). He and his followers envisioned Earth as a sphere floating at the center of the celestial sphere. More than a century later, Aristotle cited observations of Earth's curved shadow on the Moon during lunar eclipses as evidence for a spherical Earth. Thus, Greek philosophers adopted a **geocentric model** of the universe (recall that *geocentric* means "Earth-centered"), with a spherical Earth at the center of a great celestial sphere.

• How did the Greeks explain planetary motion?

Greek philosophers quickly realized that there had to be more to the heavens than just a single sphere surrounding Earth. To account for the fact that the Sun and Moon each move gradually eastward through the constellations, the Greeks added separate spheres for them, with these spheres turning at different rates from the sphere of the stars. The planets also move relative to the stars, so the Greeks added additional spheres for each planet (Figure 3.10).

The difficulty with this model was that it made it hard to explain the apparent retrograde motion of the planets [Section 2.4]. You might guess that the Greeks would simply have allowed the planetary spheres to sometimes turn forward and sometimes turn backward relative to the sphere of the stars, but they did not because it would have violated their deeply held belief in "heavenly perfection." According to this idea, enunciated most clearly by Plato, heavenly objects could move only in perfect circles. But how could the planets sometimes go backward in our sky if they were moving in perfect circles?

One potential answer would have been to discard the geocentric model and replace it with a Sun-centered model, since such a model gives a simple and natural explanation for apparent retrograde motion (see Figure 2.27). While such a model was indeed proposed by Aristarchus in about 260 B.C., it never gained much support in ancient times—in part because of the lack of detectable stellar parallax [Section 2.4], but also because Aristotle and others developed (incorrect) ideas about physics that required having Earth at the center of the universe.

The Greeks came up with a number of ingenious ideas for explaining planetary motion while preserving Earth's central position and the idea that heavenly objects move in perfect circles. These ideas were refined for centuries and reached their culmination in the work of Claudius Ptolemy (c. A.D. 100–170; pronounced *tol-e-mee*). We refer to Ptolemy's model as the **Ptolemaic model** to distinguish it from earlier geocentric models.

In the Ptolemaic model, each planet moved on a small circle whose center moved around Earth on a larger circle. The essence of the Ptolemaic model was that each planet moves on a small circle whose center moves around Earth on a larger circle (Figure 3.11). (The small circle is called an *epicycle*, and the larger circle is called a *deferent*.) A planet following this circle-upon-circle motion traces a loop as seen from Earth, with the backward portion of the loop mimicking apparent retrograde motion. However, to make his model agree well with observations, Ptolemy had to include a number of other complexities, such as positioning some of the large circles slightly off-center from

Figure 3.10

This model represents the Greek idea of the heavenly spheres (c. 400 B.C.). Earth is a sphere that rests in the center. The Moon, the Sun, and the planets each have their own spheres. The outermost sphere holds the stars.

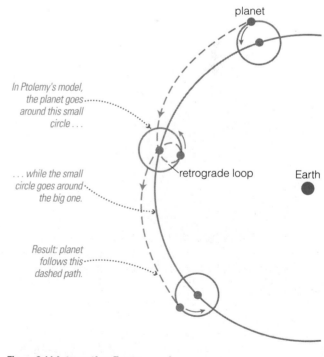

planet

In Ptolemy's model, the planet goes around this small circle . . .

. . . while the small circle goes around the big one.

retrograde loop

Earth

Result: planet follows this dashed path.

Figure 3.11 interactive figure

This diagram shows how the Ptolemaic model accounted for apparent retrograde motion. Each planet is assumed to move around a small circle that turns upon a larger circle. The resulting path (dashed) includes a loop in which the planet goes backward as seen from Earth.

Eratosthenes Measures Earth

The first accurate estimate of Earth's circumference was made by the Greek scientist Eratosthenes in about 240 B.C. Eratosthenes knew that the Sun passed directly overhead in the Egyptian city of Syene (modern-day Aswan) on the summer solstice but that on the same day the Sun came only within 7° of the zenith in the city of Alexandria. He concluded that Alexandria must be 7° of latitude north of Syene (see figure), making the north-south distance between the two cities $\frac{7}{360}$ of Earth's circumference.

Eratosthenes estimated the north-south distance between Syene and Alexandria to be 5000 stadia (the *stadium* was a Greek unit of distance), which meant

$$\frac{7}{360} \times \text{Earth's circumference} = 5000 \text{ stadia}$$

Multiplying both sides by $\frac{360}{7}$ gives us

$$\text{Earth's circumference} = \frac{360}{7} \times 5000 \text{ stadia} \approx 250,000 \text{ stadia}$$

Based on the actual sizes of Greek stadiums, we estimate that stadia must have been about $\frac{1}{6}$ km each, making Eratosthenes' estimate about $\frac{250,000}{6} = 42,000$ km—remarkably close to the actual value of just over 40,000 km.

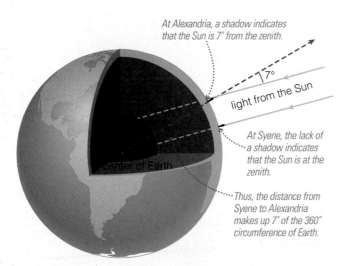

At Alexandria, a shadow indicates that the Sun is 7° from the zenith.

light from the Sun

At Syene, the lack of a shadow indicates that the Sun is at the zenith.

Thus, the distance from Syene to Alexandria makes up 7° of the 360° circumference of Earth.

This diagram shows how Eratosthenes concluded that the north-south distance from Syene to Alexandria is $\frac{7}{360}$ of Earth's circumference.

Earth. As a result, the full Ptolemaic model was mathematically quite complex, and using it to predict planetary positions required long and tedious calculations. Many centuries later, while supervising computations based on the Ptolemaic model, the Spanish monarch Alphonso X (1221–1284) is said to have complained, "If I had been present at the creation, I would have recommended a simpler design for the universe."

Despite its complexity, the Ptolemaic model proved remarkably successful: It could correctly forecast future planetary positions to within a few degrees of arc, which is about the angular extent of your hand held at arm's length against the sky. This was sufficiently accurate to keep the model in use for the next 1500 years. When Ptolemy's book describing the model was translated by Arabic scholars around A.D. 800, they gave it the title *Almagest*, derived from words meaning "the greatest compilation."

• How did Islamic scientists preserve and extend Greek science?

Much of Greek knowledge was lost with the destruction of the Library of Alexandria. That which survived was preserved primarily thanks to the rise of a new center of intellectual inquiry in Baghdad (in present-day Iraq). While European civilization fell into the period of intellectual decline known as the Dark Ages, scholars of the new religion of Islam sought knowledge of mathematics and astronomy in hopes of better understanding the wisdom of Allah. During the eighth and ninth centuries A.D., scholars working in the Muslim empire translated and thereby saved many ancient Greek works.

Around A.D. 800, the Islamic leader Al-Mamun (A.D. 786–833) established a "House of Wisdom" in Baghdad with a mission much like that of the destroyed Library of Alexandria. Using the translated Greek scientific manuscripts as building blocks, scholars in Baghdad developed the mathematics of algebra and many new instruments and techniques for astronomical observation. Most of the official names of constellations and stars come from Arabic because of the work of these scholars. If you look at a star chart, you will see that the names of many bright stars begin with *al* (e.g., Aldebaran, Algol), which simply means "the" in Arabic.

Islamic scholars preserved and extended ancient Greek scholarship, and their work helped ignite the European Renaissance.

The Islamic world of the Middle Ages was in frequent contact with Hindu scholars from India, who in turn brought knowledge of ideas and discoveries from China. Hence, the intellectual center in Baghdad achieved a synthesis of the surviving work of the ancient Greeks and that of the Indians and the Chinese. The accumulated knowledge of the Arabs spread throughout the Byzantine empire (part of the former Roman empire). When the Byzantine capital of Constantinople (modern-day Istanbul) fell to the Turks in 1453, many Eastern scholars headed west to Europe, carrying with them the knowledge that helped ignite the European Renaissance.

3.3 The Copernican Revolution

The Greeks and other ancient peoples developed many important ideas of science, but what we now think of as science arose during the European Renaissance. Within a half century after the fall of Constantinople, Polish scientist Nicholas Copernicus (1473–1543) began the work that ultimately overturned the Earth-centered Ptolemaic model.

• How did Copernicus, Tycho, and Kepler challenge the Earth-centered model?

The new ideas introduced by Copernicus fundamentally changed the way we perceive our place in the universe. The story of this dramatic change, known as the **Copernican revolution**, is in many ways the story of the origin of modern science. It is also the story of several key personalities, beginning with Copernicus himself.

Copernicus Copernicus was born in Torún, Poland, on February 19, 1473. He began studying astronomy in his late teens, and soon learned that tables of planetary motion based on the Ptolemaic model had been growing increasingly inaccurate. He therefore began a quest to find a better way to predict planetary positions.

Copernicus was aware of and adopted Aristarchus's ancient Sun-centered idea, probably because it offered such a simple explanation for the apparent retrograde motion of the planets. But he went beyond Aristarchus in working out mathematical details of the model. In the process, Copernicus discovered simple geometric relationships that allowed him to calculate each planet's orbital period around the Sun and its relative distance from the Sun in terms of Earth–Sun distance. The model's success in providing a geometric layout for the solar system further convinced him that the Sun-centered idea must be correct.

Copernicus (1473–1543)

Despite his own confidence in the model, Copernicus was hesitant to publish his work, fearing that the idea of a moving Earth would be considered absurd. However, he discussed his system with other scholars, including high-ranking officials of the Church, who urged him to publish a book. Copernicus saw the first printed copy of his book, *De Revolutionibus Orbium Caelestium* ("Concerning the Revolutions of the Heavenly Spheres"), on the day he died—May 24, 1543.

Copernicus's Sun-centered model had the right general ideas, but its predictions were not substantially better than those of Ptolemy's Earth-centered model.

Publication of the book spread the Sun-centered idea widely, and many scholars were drawn to its aesthetic advantages. Nevertheless, the Copernican model gained relatively few converts over the next 50 years, for a good reason: It didn't work all that well. The primary problem was that while Copernicus had been willing to overturn Earth's central place in the cosmos, he had held fast to the ancient belief that heavenly motion must occur in perfect circles. This incorrect assumption forced him to add numerous complexities to his system (including circles on circles much like those used by Ptolemy) to get it to make decent predictions. In the end, his complete model was no more accurate and no less complex than the Ptolemaic model, and few people were willing to throw out thousands of years of tradition for a new model that worked just as poorly as the old one.

Tycho Part of the difficulty faced by astronomers who sought to improve either the Ptolemaic model or the Copernican model was a lack of quality data. The telescope had not yet been invented, and existing naked-eye observations were not very accurate. Better data were needed, and they were provided by the Danish nobleman Tycho Brahe (1546–1601), usually known simply as Tycho (pronounced "tie-koe").

Tycho was an eccentric genius who once lost part of his nose in a sword fight with another student over who was the better mathematician. In 1563, Tycho decided to observe a widely anticipated alignment of

Tycho Brahe (1546–1601)

Figure 3.12

Tycho Brahe in his naked-eye observatory, which worked much like a giant protractor. He could sit and observe a planet through the rectangular hole in the wall as an assistant used a sliding marker to measure the angle on the protractor.

Johannes Kepler
(1571–1630)

Jupiter and Saturn. To his surprise, the alignment occurred nearly 2 days later than the date Copernicus had predicted. Resolving to improve the state of astronomical prediction, he set about compiling careful observations of stellar and planetary positions in the sky.

Tycho's fame grew after he observed what he called a *nova*, meaning "new star," in 1572 and proved that it was much farther away than the Moon. (Today, we know that Tycho saw a *supernova*—the explosion of a distant star [Section 12.3].) In 1577, Tycho observed a comet and proved that it too lay in the realm of the heavens. Others, including Aristotle, had argued that comets were phenomena of Earth's atmosphere. King Frederick II of Denmark decided to sponsor Tycho's ongoing work, providing him with money to build an unparalleled observatory for naked-eye observations (Figure 3.12). After Frederick II died in 1588, Tycho moved to Prague, where his work was supported by German emperor Rudolf II.

Over a period of three decades, Tycho and his assistants compiled naked-eye observations accurate to within less than 1 arcminute—less than the thickness of a fingernail viewed at arm's length. Despite the quality of his observations, Tycho never succeeded in coming up with a satisfying explanation for planetary motion. He was convinced that the *planets* must orbit the Sun, but his inability to detect stellar parallax [Section 2.4] led him to conclude that Earth must remain stationary. He therefore advocated a model in which the Sun orbits Earth while all other planets orbit the Sun. Few people took this model seriously.

> **Tycho's accurate naked-eye observations provided the data needed to improve the Copernican system.**

Kepler Tycho failed to explain the motions of the planets satisfactorily, but he succeeded in finding someone who could: In 1600, he hired the young German astronomer Johannes Kepler (1571–1630). Kepler and Tycho had a strained relationship, but Tycho recognized the talent of his young apprentice. In 1601, as he lay on his deathbed, Tycho begged Kepler to find a system that would make sense of his observations so "that it may not appear I have lived in vain."

Kepler was deeply religious and believed that understanding the geometry of the heavens would bring him closer to God. Like Copernicus, he believed that planetary orbits should be perfect circles, so he worked diligently to match circular motions to Tycho's data. After years of effort, he found a set of circular orbits that matched most of Tycho's observations quite well. Even in the worst cases, which were for the planet Mars, Kepler's predicted positions differed from Tycho's observations by only about 8 arcminutes.

Kepler surely was tempted to ignore these discrepancies and attribute them to errors by Tycho. After all, 8 arcminutes is barely one-fourth the angular diameter of the full moon. But Kepler trusted Tycho's careful work. The small discrepancies finally led Kepler to abandon the idea of circular orbits—and to find the correct solution to the ancient riddle of planetary motion. About this event, Kepler wrote:

> *If I had believed that we could ignore these eight minutes [of arc], I would have patched up my hypothesis accordingly. But, since it was not permissible to ignore, those eight minutes pointed the road to a complete reformation in astronomy.*

Kepler's key discovery was that planetary orbits are not circles but instead are a special type of oval called an **ellipse**. You can draw a circle by putting a pencil on the end of a string, tacking the string to a board, and

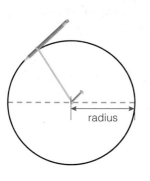

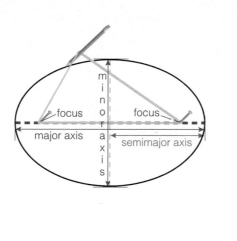

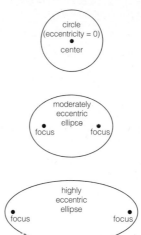

Figure 3.13 interactive figure

An ellipse is a special type of oval. These diagrams show how an ellipse differs from a circle and how different ellipses vary in their eccentricity.

a Drawing a circle with a string of fixed length.

b Drawing an ellipse with a string of fixed length.

c Eccentricity describes how much an ellipse deviates from a perfect circle.

By using elliptical orbits, Kepler created a Sun-centered model that predicted planetary positions with outstanding accuracy.

pulling the pencil around (Figure 3.13a). Drawing an ellipse is similar, except that you must stretch the string around *two* tacks (Figure 3.13b). The locations of the two tacks are called the **foci** (singular, **focus**) of the ellipse. The long axis of the ellipse is called its *major axis*, each half of which is called a **semimajor axis**; as we'll see shortly, the length of the semimajor axis is particularly important in astronomy. The short axis is called the *minor axis*. By altering the distance between the two foci while keeping the length of string the same, you can draw ellipses of varying **eccentricity**, a quantity that describes the amount by which an ellipse is stretched out compared to a perfect circle (Figure 3.13c). A circle is an ellipse with zero eccentricity, and greater eccentricity means a more elongated ellipse.

Kepler's decision to trust the data over his preconceived beliefs marked an important transition point in the history of science. Once he abandoned perfect circles in favor of ellipses, Kepler soon came up with a model that could predict planetary positions with far greater accuracy than Ptolemy's Earth-centered model. Kepler's model withstood the test of time and became accepted not only as a model of nature but also as a deep, underlying truth about planetary motion.

(MA) Orbits and Kepler's Laws Tutorial, Lessons 2–4

• What are Kepler's three laws of planetary motion?

Kepler summarized his discoveries with three simple laws that we now call **Kepler's laws of planetary motion**. He published the first two laws in 1610 and the third in 1618.

Kepler's first law: The orbit of each planet about the Sun is an ellipse with the Sun at one focus.

Kepler's first law tells us that the orbit of each planet about the Sun is an ellipse with the Sun at one focus (Figure 3.14). (There is nothing at the other focus.) In essence, this law tells us that a planet's distance from the Sun varies during its orbit. It is closest at the point called

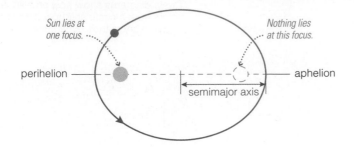

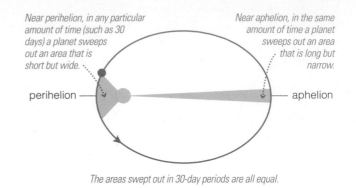

Near perihelion, in any particular amount of time (such as 30 days) a planet sweeps out an area that is short but wide.

Near aphelion, in the same amount of time a planet sweeps out an area that is long but narrow.

perihelion — aphelion

The areas swept out in 30-day periods are all equal.

Figure 3.14 interactive figure

Kepler's first law: The orbit of each planet about the Sun is an ellipse with the Sun at one focus. (The eccentricity shown here is exaggerated compared to the actual eccentricities of the planets.)

Figure 3.15 interactive figure

Kepler's second law: As a planet moves around its orbit, it sweeps out equal areas (the shaded regions) in equal times.

cosmic calculations 3.2

Kepler's Third Law

When Kepler discovered his third law ($p^2 = a^3$), he knew only that it applied to the orbits of planets about the Sun. In fact, it applies to any orbiting object as long as the following two conditions are met:

1. The object orbits the Sun *or* another star of precisely the same mass.
2. We use units of *years* for the orbital period and *AU* for the orbital distance.

(Newton extended the law to *all* orbiting objects; see Cosmic Calculations 4.1.)

Example 1: The largest asteroid, Ceres, orbits the Sun at an average distance (semimajor axis) of 2.77 AU. What is its orbital period?

Solution: Both conditions are met, so we solve Kepler's third law for the orbital period p and substitute the given orbital distance, $a = 2.77$ AU:

$$p^2 = a^3 \implies p = \sqrt{a^3} = \sqrt{2.77^3} = 4.6$$

Ceres has an orbital period of 4.6 years.

Example 2: A planet is discovered orbiting every 3 months around a star of the same mass as our Sun. What is the planet's average orbital distance?

Solution: The first condition is met, and we can satisfy the second by converting the orbital period from months to years: $p = 3$ months $= 0.25$ year. We now solve Kepler's third law for the average distance a:

$$p^2 = a^3 \implies a = \sqrt[3]{p^2} = \sqrt[3]{0.25^2} = 0.40$$

The planet orbits its star at an average distance of 0.40 AU, which is nearly the same as Mercury's average distance from the Sun.

perihelion (from the Greek for "near the Sun") and farthest at the point called **aphelion** (from the Greek for "away from the Sun"). The *average* of a planet's perihelion and aphelion distances is the length of its *semimajor axis*. We will refer to this simply as the planet's average distance from the Sun.

Kepler's second law: As a planet moves around its orbit, it sweeps out equal areas in equal times.

Kepler's second law states that as a planet moves around its orbit, it sweeps out equal areas in equal times. As shown in Figure 3.15, this means the planet moves a greater distance when it is near perihelion than it does in the same amount of time near aphelion. That is, the planet travels faster when it is nearer to the Sun and slower when it is farther from the Sun.

Kepler's third law: More distant planets orbit the Sun at slower average speeds, obeying the precise mathematical relationship $p^2 = a^3$.

Kepler's third law tells us that more distant planets orbit the Sun at slower average speeds, obeying a precise mathematical relationship (Figure 3.16). The relationship is written $p^2 = a^3$, where p is the planet's orbital period in years and a is its average distance from the Sun in astronomical units. Figure 3.16a shows the $p^2 = a^3$ law graphically. Notice that the square of each planet's orbital period (p^2) is indeed equal to the cube of its average distance from the Sun (a^3). Because Kepler's third law relates a planet's orbital distance to its orbital time (period), we can use the law to calculate a planet's average orbital speed. Figure 3.16b shows the result, confirming that more distant planets orbit the Sun more slowly.

think about it Suppose a comet has an orbit that brings it quite close to the Sun at its perihelion and beyond Mars at its aphelion, but with an average distance (semimajor axis) of 1 AU. According to Kepler's laws, how long would the comet take to complete each orbit of the Sun? Would it spend most of its time close to the Sun, far from the Sun, or somewhere in between? Explain.

The fact that more distant planets move more slowly led Kepler to suggest that planetary motion might be the result of a force from the Sun. He did not know the nature of the force, but others worked to discover it. The mystery was finally solved by Isaac Newton, who explained planetary motion and Kepler's laws as consequences of gravity [Section 4.4].

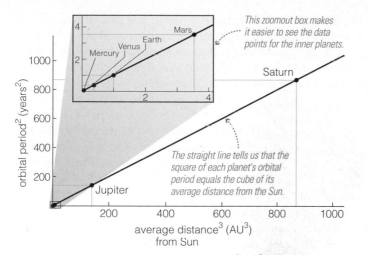

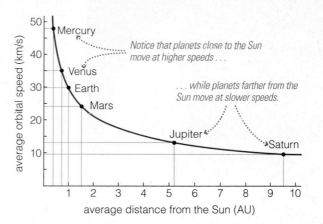

This zoomout box makes it easier to see the data points for the inner planets.

The straight line tells us that the square of each planet's orbital period equals the cube of its average distance from the Sun.

Notice that planets close to the Sun move at higher speeds . . .

. . . while planets farther from the Sun move at slower speeds.

a This graph shows that Kepler's third law ($p^2 = a^3$) does indeed hold true; for simplicity, the graph shows only the planets known in Kepler's time.

b This graph shows how orbital speed depends on distance from the Sun. (Kepler knew the form of this relationship but not the actual speeds, because the numerical value of the astronomical unit was not yet known.)

Figure 3.16

Graphs based on Kepler's third law.

How did Galileo solidify the Copernican revolution?

The success of Kepler's laws in matching Tycho's data provided strong evidence in favor of Copernicus's placement of the Sun at the center of the solar system. Nevertheless, many scientists still voiced reasonable objections to the Copernican view. There were three basic objections, all rooted in the 2000-year-old beliefs of Aristotle and other ancient Greeks.

- First, Aristotle had held that Earth could not be moving because, if it were, objects such as birds, falling stones, and clouds would be left behind as Earth moved along its way.

- Second, the idea of noncircular orbits contradicted Aristotle's claim that the heavens—the realm of the Sun, Moon, planets, and stars— must be perfect and unchanging.

- Third, no one had detected the stellar parallax that should occur if Earth orbits the Sun.

Galileo Galilei (1564–1642), usually known by his first name, answered all three objections.

Galileo defused the first objection with experiments that almost single-handedly overturned the Aristotelian view of physics. In particular, he used experiments with rolling balls to demonstrate that a moving object remains in motion *unless* a force acts to stop it (an idea now codified in Newton's first law of motion [Section 4.2]). This insight explained why objects that share Earth's motion through space—such as birds, falling stones, and clouds—should *stay* with Earth rather than falling behind as Aristotle had argued. This same idea explains why passengers stay with a moving airplane even when they leave their seats.

Tycho's supernova and comet observations already had challenged the validity of the second objection by showing that the heavens could change. Galileo shattered the idea of heavenly perfection after he built a telescope in late 1609. (The telescope was invented in 1608 by Hans Lippershey, but Galileo's was much more powerful.) Through his telescope, Galileo saw sunspots on the Sun, which were considered "imperfections" at the time.

Galileo (1564–1642)

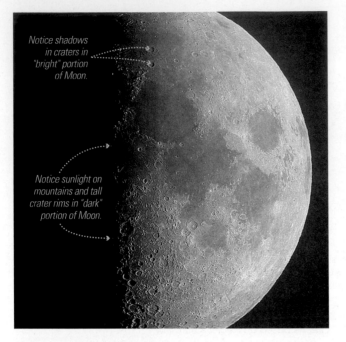

Notice shadows in craters in "bright" portion of Moon.

Notice sunlight on mountains and tall crater rims in "dark" portion of Moon.

Figure 3.17 interactive figure

The shadows cast by mountains and crater rims near the dividing line between the light and dark portions of the lunar face prove that the Moon's surface is not perfectly smooth.

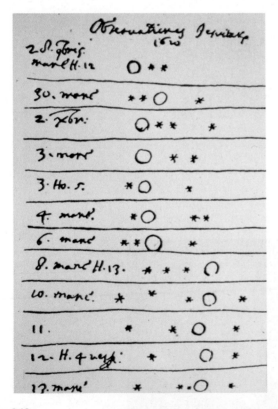

Figure 3.18

A page from Galileo's notebook written in 1610. His sketches show four "stars" near Jupiter (the circle) but in different positions at different times (and sometimes hidden from view). Galileo soon realized that the "stars" were actually moons orbiting Jupiter.

He also used his telescope to prove that the Moon has mountains and valleys like the "imperfect" Earth by noticing the shadows cast near the dividing line between the light and dark portions of the lunar face (Figure 3.17). If the heavens were in fact not perfect, then the idea of elliptical orbits (as opposed to "perfect" circles) was not so objectionable.

Galileo's experiments and telescopic observations overcame remaining scientific objections to the Copernican idea, sealing the case for the Sun-centered solar system.

The third objection—the absence of observable stellar parallax—had been of particular concern to Tycho. Based on his estimates of the distances of stars, Tycho believed that his naked-eye observations were sufficiently precise to detect stellar parallax if Earth did in fact orbit the Sun. Refuting Tycho's argument required showing that the stars were more distant than Tycho had thought and therefore too distant for him to have observed stellar parallax. Although Galileo didn't actually prove this fact, he provided strong evidence in its favor. For example, he saw with his telescope that the Milky Way resolved into countless individual stars. This discovery helped him argue that the stars were far more numerous and more distant than Tycho had believed.

In hindsight, the final nails in the coffin of the Earth-centered model came with two of Galileo's earliest discoveries through the telescope. First, he observed four moons clearly orbiting Jupiter, *not* Earth (Figure 3.18). Soon thereafter, he observed that Venus goes through phases in a way that proved that it must orbit the Sun and not Earth (Figure 3.19).

Although we now recognize that Galileo won the day, the story was more complex in his own time, when Catholic Church doctrine still held Earth to be the center of the universe. On June 22, 1633, Galileo was brought before a Church inquisition in Rome and ordered to recant his claim that Earth orbits the Sun. Nearly 70 years old and fearing for his life, Galileo did as ordered. His life was spared. However, legend has it that as he rose from his knees he whispered under his breath, *Eppur si muove*—Italian for "And yet it moves." (Given the likely consequences if Church officials had heard him say this, most historians doubt the legend.)

The Church did not formally vindicate Galileo until 1992, but Church officials gave up the argument long before that: In 1757, all works backing the idea of a Sun-centered solar system were removed from the Church's Index of banned books. Today, Catholic scientists are at the forefront of much astronomical research, and official Church teachings are compatible not only with Earth's planetary status but also with the theories of the Big Bang and the subsequent evolution of the cosmos and of life.

3.4 The Nature of Science

The story of how our ancestors gradually figured out the basic architecture of the cosmos exhibits many features of what we now consider "good science." For example, we have seen how models were formulated and tested against observations and were modified or replaced when they failed those tests. The story also illustrates some classic mistakes, such as the apparent failure of anyone before Kepler to question the belief that orbits must be circles. The ultimate success of the Copernican revolution led scientists, philosophers, and theologians to reassess the various modes of thinking that played a role in the 2000-year process of discovering Earth's place in the universe. Let's examine how the principles of modern science emerged from the lessons learned in the Copernican revolution.

Ptolemaic View of Venus

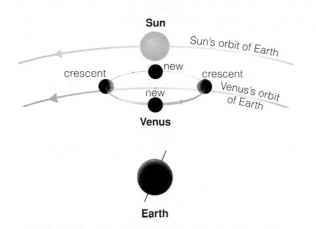

a In the Ptolemaic model, Venus orbits Earth, moving around a smaller circle on its larger orbital circle; the center of the smaller circle lies on the Earth–Sun line. If this view were correct, Venus's phases would range only from new to crescent.

Copernican View of Venus

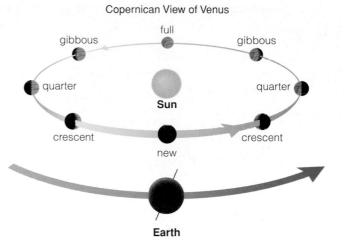

b In reality, Venus orbits the Sun, so from Earth we can see it in many different phases. This is just what Galileo observed, allowing him to prove that Venus orbits the Sun.

Figure 3.19 interactive figure ↖

Galileo's telescopic observations of Venus proved that it orbits the Sun rather than Earth.

• How can we distinguish science from nonscience?

It's surprisingly difficult to define the term *science* precisely. The word comes from the Latin *scientia,* meaning "knowledge," but not all knowledge is science. For example, you may know what music you like best, but your musical taste is not a result of scientific study.

Approaches to Science One reason science is difficult to define is that not all science works in the same way. For example, you've probably heard it said that science is supposed to proceed according to something called the "scientific method." As an idealized illustration of this method, consider what you would do if your flashlight suddenly stopped working. In hopes of fixing the flashlight, you might *hypothesize* that its batteries have died. This type of tentative explanation, or **hypothesis**, is sometimes called an *educated guess*—in this case, it is "educated" because you already know that flashlights need batteries. Your hypothesis allows you to make a simple prediction: If you replace the batteries with new ones, the flashlight should work. You can test this prediction by replacing the batteries. If the flashlight now works, you've confirmed your hypothesis. If it doesn't, you must revise or discard your hypothesis, perhaps in favor of some other one that you can also test (such as that the bulb is burned out). Figure 3.20 illustrates the basic flow of this process.

> The scientific method is a useful idealization of scientific thinking, but science rarely progresses in such an orderly way.

The scientific method can be a useful idealization, but real science rarely progresses in such an orderly way. Scientific progress often begins with someone going out and looking at nature in a general way, rather than by conducting a careful set of experiments. For example, Galileo wasn't looking for anything in particular when he pointed his telescope at the sky and made his first startling discoveries. Furthermore, scientists are human beings, and their intuition and personal beliefs inevitably influence their work. Copernicus, for example, adopted the idea that Earth orbits the Sun not because he had carefully tested it but because he believed it made more sense

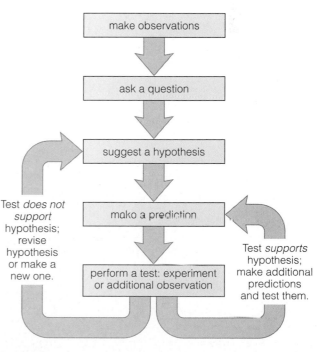

Figure 3.20

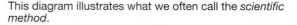

This diagram illustrates what we often call the *scientific method.*

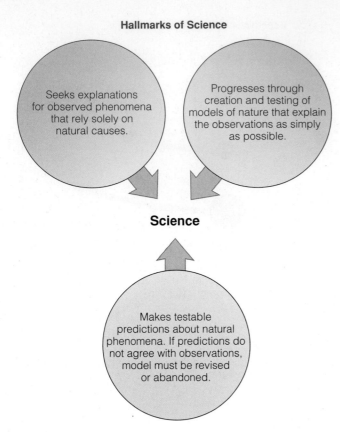

Hallmarks of Science

Seeks explanations for observed phenomena that rely solely on natural causes.

Progresses through creation and testing of models of nature that explain the observations as simply as possible.

Science

Makes testable predictions about natural phenomena. If predictions do not agree with observations, model must be revised or abandoned.

Figure 3.21

Hallmarks of science.

than the prevailing view of an Earth-centered universe. While his intuition guided him to the right general idea, he erred in the specifics because he still held Plato's ancient belief that heavenly motion must be in perfect circles.

Given that the idealized scientific method is an overly simplistic characterization of science, how can we tell what is science and what is not? To answer this question, we must look a little deeper at the distinguishing characteristics of scientific thinking.

Hallmarks of Science   One way to define scientific thinking is to list the criteria that scientists use when they judge competing models of nature. Historians and philosophers of science have examined (and continue to examine) this issue in great depth, and different experts express different viewpoints on the details. Nevertheless, everything we now consider to be science shares the following three basic characteristics, which we will refer to as the "hallmarks" of science (Figure 3.21):

- Modern science seeks explanations for observed phenomena that rely solely on natural causes.

- Science progresses through the creation and testing of models of nature that explain the observations as simply as possible.

- A scientific model must make testable predictions about natural phenomena that would force us to revise or abandon the model if the predictions do not agree with observations.

> Science seeks to explain observed phenomena using testable models of nature that explain the observations as simply as possible.

Each of these hallmarks is evident in the story of the Copernican revolution. The first shows up in the way Tycho's careful measurements of planetary motion motivated Kepler to come up with a better explanation for those motions. The second is evident in the way several competing models were compared and tested, most notably those of Ptolemy, Copernicus, and Kepler. We see the third in the fact that each model could make precise predictions about the future motions of the Sun, Moon, planets, and stars in our sky. When a model's predictions failed, the model was modified or ultimately discarded. Kepler's model gained acceptance in large part because its predictions were so much better than those of the Ptolemaic model in matching Tycho's observations. Figure 3.22 (pp. 74–75) summarizes the Copernican revolution and how it illustrates the hallmarks of science.

Occam's Razor   The criterion of simplicity in the second hallmark deserves further explanation. Remember that the original model of Copernicus did *not* match the data noticeably better than Ptolemy's model. If scientists had judged Copernicus's model solely on the accuracy of its predictions, they might have rejected it immediately. However, many scientists found elements of the Copernican model appealing, such as its simple explanation for apparent retrograde motion. They therefore kept the model alive until Kepler found a way to make it work.

In fact, if agreement with data were the sole criterion for judgment, we could imagine a modern-day Ptolemy adding millions or billions of additional circles to the geocentric model in an effort to improve its agreement with observations. A sufficiently complex geocentric model could in principle reproduce the observations with almost perfect accuracy—but it still would not convince us that Earth is the center of the universe. We would still choose the Copernican view over the geocentric view because its predictions would be just as accurate while arising from a much simpler model of nature. The idea that scientists should prefer the simpler of two models that agree equally well with observations is called *Occam's razor,* after the medieval scholar William of Occam (1285–1349).

Verifiable Observations The third hallmark of science forces us to face the question of what counts as an "observation" against which a prediction can be tested. Consider the claim that aliens are visiting Earth in UFOs. Proponents of this claim say that thousands of eyewitness observations of UFO encounters provide evidence that it is true. But do these personal testimonials count as *scientific* evidence? On the surface, the answer isn't obvious, because all scientific studies involve eyewitness accounts on some level. For example, only a handful of scientists have personally made detailed tests of Einstein's theory of relativity, and it is their personal reports of the results that have convinced other scientists of the theory's validity. However, there's an important difference between personal testimony about a scientific test and an observation of a UFO: The first can be verified by anyone, at least in principle, while the second cannot.

Understanding this difference is crucial to understanding what counts as science and what does not. Even though you may never have conducted a test of Einstein's theory of relativity yourself, there's nothing stopping you from doing so. It might require several years of study before you have the necessary background to conduct the test, but you could then confirm the results reported by other scientists. In other words, while you may currently be trusting the eyewitness testimony of scientists, you always have the option of verifying their testimony for yourself.

In contrast, there is no way for you to verify someone's eyewitness account of a UFO. Moreover, scientific studies of eyewitness testimony show it to be notoriously unreliable, because different eyewitnesses often disagree on what they saw even immediately after an event has occurred. As time passes, memories of the event may change further. In some cases in which memory has been checked against reality, people have reported vivid memories of events that never happened at all. This explains something that virtually all of us have experienced: disagreements with a friend about who did what and when. Since both people cannot be right in such cases, at least one person must have a memory that differs from reality.

Because of its demonstrated unreliability, eyewitness testimony alone should *never* be used as evidence in science, no matter who reports it or how many people offer similar testimony. It can be used in support of a scientific model only when it is backed up by independently verifiable evidence that anyone could in principle check. (For much the same reason, eyewitness testimony is usually insufficient for a conviction in criminal court; other evidence, such as motive, is required.)

Objectivity in Science It's important to realize that science is not the only valid way of seeking knowledge. For example, suppose you are shopping for a car, learning to play drums, or pondering the meaning of life. In each case, you might make observations, exercise logic, and test hypotheses. Yet these pursuits clearly are not science, because they are not directed at developing testable explanations for observed natural phenomena. As long as nonscientific searches for knowledge make no claims about how the natural world works, they do not conflict with science.

The boundaries between science and nonscience are sometimes blurry. We generally think of science as being objective, meaning that all

Ancient Earth-centered models of the universe easily explained the simple motions of the Sun and Moon through our sky, but had difficulty explaining the more complicated motions of the planets. The quest to understand planetary motions ultimately led to a revolution in our thinking about Earth's place in the universe that illustrates the process of science. This figure summarizes the major steps in that process.

1 Night-by-night, planets usually move from west to east relative to the stars. However, during periods of *apparent retrograde motion,* they reverse direction for a few weeks to months [Section 2.4]. The ancient Greeks knew that any credible model of the solar system had to explain these observations.

Jupiter's position at 1-month intervals.

Jan. 2008

Feb. 2009

Sagittarius

Not to scale.

2 Most ancient Greek thinkers assumed that Earth remained fixed at the center of the solar system. To explain retrograde motion, they therefore added a complicated scheme of circles moving upon circles to their Earth-centered model. However, at least some Greeks, such as Aristarchus, preferred a Sun-centered model, which offered a simpler explanation for retrograde motion.

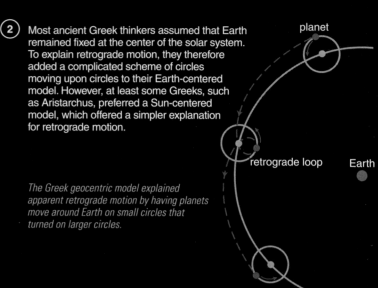

planet

retrograde loop

Earth

The Greek geocentric model explained apparent retrograde motion by having planets move around Earth on small circles that turned on larger circles.

HALLMARK OF SCIENCE **A scientific model must seek explanations for observed phenomena that rely solely on natural causes.** The ancient Greeks used geometry to explain their observations of planetary motion.

(Left page)
A schematic map of the universe from 1539 with Earth at the center and the Sun (Solis) orbiting it between Venus and Mars (Martis).

(Right page)
A page from Copernicus's De Revolutionibus, published in 1543, showing the Sun (Sol) at the center and Earth (Terra) orbiting between Venus and Mars.

LIBRI COSMO.   Fo.V.

Schema huius præmiffæ diuifionis Sphærarum.

3 By the time of Copernicus (1473–1543), predictions based on the Earth-centered model had become noticeably inaccurate. Hoping for improvement, Copernicus revived the Sun-centered idea. He did not succeed in making substantially better predictions because he retained the ancient belief that planets must move in perfect circles, but he inspired a revolution continued over the next century by Tycho, Kepler, and Galileo.

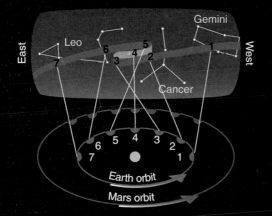

Apparent retrograde motion is simply explained in a Sun-centered system. Notice how Mars appears to change direction as Earth moves past it.

HALLMARK OF SCIENCE **Science progresses through creation and testing of models of nature that explain the observations as simply as possible.** Copernicus developed a simple Sun-centered model in hopes of explaining observations better than the more complicated Earth-centered model.

4 Tycho exposed flaws in both the ancient Greek and Copernican models by observing planetary motions with unprecedented accuracy. His observations led to Kepler's breakthrough insight that planetary orbits are elliptical, not circular, and enabled Kepler to develop his three laws of planetary motion.

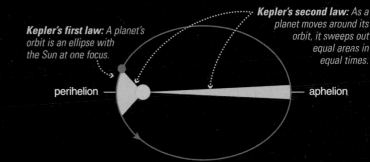

Kepler's first law: *A planet's orbit is an ellipse with the Sun at one focus.*

Kepler's second law: *As a planet moves around its orbit, it sweeps out equal areas in equal times.*

perihelion — aphelion

Kepler's third law: *More distant planets orbit at slower average speeds, obeying $p^2 = a^3$.*

HALLMARK OF SCIENCE **A scientific model makes testable predictions about natural phenomena. If predictions do not agree with observations, the model must be revised or abandoned.** Kepler could not make his model agree with observations until he abandoned the belief that planets move in perfect circles.

5 Galileo's experiments and telescopic observations overcame remaining scientific objections to the Sun-centered model. Together, Galileo's discoveries and the success of Kepler's laws in predicting planetary motion overthrew the Earth-centered model once and for all.

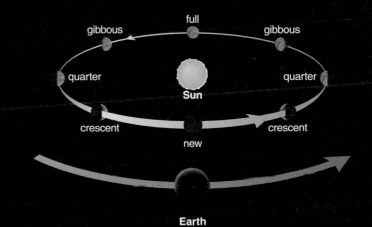

With his telescope, Galileo saw phases of Venus that are consistent only with the idea that Venus orbits the Sun rather than Earth.

people should be able to find the same answers to scientific questions. However, there is a difference between the overall objectivity of science and the objectivity of individual scientists.

Science is practiced by human beings, and individual scientists may bring their personal biases and beliefs to their scientific work. For example, most scientists choose their research projects based on personal interests rather than on some objective formula. In extreme cases, scientists have been known to cheat—either deliberately or subconsciously—to obtain a result they desire. In one famous case that occurred a little over a century ago, astronomer Percival Lowell claimed to see a network of artificial canals in blurry telescopic images of Mars, leading him to conclude that there was a great Martian civilization [Section 9.4]. But no such canals actually exist, so Lowell must have allowed his beliefs about extraterrestrial life to influence the way he interpreted what he saw—in essence, a form of cheating, though probably not intentional.

Bias can sometimes show up even in the thinking of the scientific community as a whole. Some valid ideas may not be considered by any scientist because they fall too far outside the general patterns of thought, or **paradigm**, of the time. Einstein's theory of relativity is an example. Many scientists in the decades before Einstein had gleaned hints of the theory but did not investigate them, at least in part because they seemed too outlandish.

> Individual scientists inevitably carry personal biases into their work, but the collective action of many scientists should ultimately make science objective.

The beauty of science is that it encourages continued testing by many people. Even if personal biases affect some results, tests by others should eventually uncover the mistakes. Similarly, if a new idea is correct but falls outside the accepted paradigm, sufficient testing and verification of the idea should eventually force a paradigm shift. In that sense, *science ultimately provides a means of bringing people to agreement,* at least on topics that can be subjected to scientific study.

• What is a scientific theory?

The most successful scientific models explain a wide variety of observations in terms of just a few general principles. When a powerful yet simple model makes predictions that survive repeated and varied testing, scientists elevate its status and call it a **theory**. Some famous examples are Isaac Newton's theory of gravity, Charles Darwin's theory of evolution, and Albert Einstein's theory of relativity.

Note that the scientific meaning of the word *theory* is quite different from its everyday meaning, in which we equate a theory more closely with speculation or a hypothesis. For example, someone might get a new idea and say, "I have a new theory about why people enjoy the beach." Without the support of a broad range of evidence that others have tested and confirmed, this "theory" is really only a guess. In contrast, Newton's theory of gravity qualifies as a scientific theory because it uses simple physical principles to explain many observations and experiments.

> A scientific theory is a simple yet powerful model whose predictions have been borne out by repeated and varied testing.

Despite its success in explaining observed phenomena, a scientific theory can never be proved true beyond all doubt, because future observations may disagree with its

common misconceptions

Eggs on the Equinox

One of the hallmarks of science holds that you needn't take scientific claims on faith. In principle, at least, you can always test them for yourself. Consider the claim, repeated in news reports every year, that the spring equinox is the only day on which you can balance an egg on its end. Many people believe this claim, but you'll be immediately skeptical if you think about the nature of the spring equinox. The equinox is merely a point in time at which sunlight strikes both hemispheres equally (see Figure 2.13). It's difficult to see how sunlight could affect an attempt to balance eggs (especially if the eggs are indoors), and there is no difference in the strength of either Earth's gravity or the Sun's gravity on that day compared to any other day.

More important, you can test this claim directly. It's not easy to balance an egg on its end, but with practice you'll find that you can do it on any day of the year, not just on the spring equinox. Not all scientific claims are so easy to test for yourself, but the basic lesson should be clear: Before you accept any scientific claim, you should demand at least a reasonable explanation of the evidence that backs it up.

predictions. However, anything that qualifies as a scientific theory must be supported by a large, compelling body of evidence.

In this sense, a scientific theory is not at all like a hypothesis or any other type of guess. We are free to change a hypothesis at any time, because it has not yet been carefully tested. In contrast, we can discard or replace a scientific theory only if we have an alternate way of explaining the evidence that supports it.

Again, the theories of Newton and Einstein offer good examples. A vast body of evidence supports Newton's theory of gravity, but by the late 1800s scientists had begun to discover cases where its predictions did not perfectly match observations. These discrepancies were explained only when Einstein developed his general theory of relativity, which was able to match the observations. Still, the many successes of Newton's theory could not be ignored, and Einstein's theory would not have gained acceptance if it had not been able to explain these successes equally well. It did, and that is why we now view Einstein's theory as a broader theory of gravity than Newton's theory. Some scientists today are seeking a theory of gravity that will go beyond Einstein's. If any new theory ever gains acceptance, it will have to match all the successes of Einstein's theory as well as work in new realms where Einstein's theory does not.

think about it When people claim that something is "only a theory," what do you think they mean? Does this meaning of "theory" agree with the definition of a theory in science? Do scientists always use the word *theory* in its "scientific" sense? Explain.

special topic:

Astrology

ALTHOUGH THE TERMS *astrology* and *astronomy* sound very similar, today they describe very different practices. In ancient times, however, astrology and astronomy often went hand in hand, and astrology played an important role in the historical development of astronomy. Indeed, astronomers and astrologers were usually one and the same.

The basic tenet of astrology is that human events are influenced by the apparent positions of the Sun, Moon, and planets among the stars in our sky. The origins of this idea are easy to understand. The position of the Sun in the sky clearly influences our lives—it determines the seasons and hence the times of planting and harvesting, of warmth and cold, and of daylight and darkness. Similarly, the Moon determines the tides, and the cycle of lunar phases coincides with many biological cycles. Because the planets also appear to move among the stars, it seemed reasonable to imagine that planets also influence our lives, even if these influences were much more difficult to discover.

Ancient astrologers hoped that they might learn *how* the positions of the Sun, Moon, and planets influence our lives. They charted the skies, seeking correlations with events on Earth. For example, if an earthquake occurred when Saturn was entering the constellation of Leo, might Saturn's position have caused the earthquake? If the king became ill when Mars was in Gemini and the first-quarter moon was in Scorpio, might it mean another tragedy for the king when this particular alignment of the Moon and Mars next recurred? Ancient astrologers thought that the patterns of influence eventually would become clear and they would then be able to forecast human events with the same reliability with which observations of the Sun could forecast the coming of spring.

This hope was never realized. Although many astrologers still attempt to predict future events, scientific tests have shown that their predictions come true no more often than would be expected by pure chance. Moreover, in light of our current understanding of the universe, the original ideas behind astrology no longer make sense. For example, today we use ideas of gravity and energy to explain the influences of the Sun and the Moon, and these same ideas tell us that the planets are too far from Earth to have a similar influence.

Of course, many people continue to practice astrology, perhaps because of its ancient and rich traditions. Scientifically, we cannot say anything about such traditions, because traditions are not testable predictions. But if you want to understand the latest discoveries about the cosmos, you'll need a science that can be tested and refined—and astrology fails to meet these requirements.

the big picture
Putting Chapter 3 into Context

In this chapter, we focused on the scientific principles through which we have learned so much about the universe. Key "big picture" concepts from this chapter include the following:

- The basic ingredients of scientific thinking—careful observation and trial-and-error testing—are a part of everyone's experience. Modern science simply provides a way of organizing this everyday thinking to facilitate the learning and sharing of new knowledge.

- Although our understanding of the universe is growing rapidly today, each new piece of knowledge rests on ideas that came before.

- The Copernican revolution, which overthrew the ancient Greek belief in an Earth-centered universe, did not occur instantaneously. It unfolded over a period of more than a century, during which many of the characteristics of modern science first appeared.

- Science exhibits several key features that distinguish it from nonscience and that in principle allow anyone to come to the same conclusions when studying a scientific question.

summary of key concepts

3.1 The Ancient Roots of Science

• In what ways do all humans use scientific thinking?
Scientific thinking relies on the same type of trial-and-error thinking that we use in our everyday lives, but done in a carefully organized way.

• How did astronomical observations benefit ancient societies?

Ancient cultures used astronomical observations to help them keep track of time and the seasons, crucial skills for people who depended on agriculture for survival, as well as to aid them in navigation.

• What did ancient civilizations achieve in astronomy?
Ancient astronomers were accomplished observers who learned to tell the time of day and the time of year, to track cycles of the Moon, and to observe planets and stars. Many ancient structures aided in astronomical observations.

3.2 Ancient Greek Science

• Why does modern science trace its roots to the Greeks?
The Greeks developed **models** of nature and emphasized the importance of having the predictions of those models agree with observations of nature.

• How did the Greeks explain planetary motion?

retrograde loop

The Greek **geocentric model** reached its culmination with the **Ptolemaic model,** which explained apparent retrograde motion by having each planet move on a small circle whose center moves around Earth on a larger circle.

• How did Islamic scientists preserve and extend Greek science?
While Europe was in its Dark Ages, Islamic scholars preserved and extended ancient Greek knowledge. After the fall of Constantinople, some of these scholars moved west to Europe, where their knowledge helped ignite the European Renaissance.

3.3 The Copernican Revolution

• How did Copernicus, Tycho, and Kepler challenge the Earth-centered model?
Copernicus created a Sun-centered model of the solar system designed to replace the Ptolemaic model, but it was no more accurate than Ptolemy's because Copernicus still used perfect circles. Tycho's accurate, naked-eye observations provided the data needed to improve on Copernicus's model. Kepler developed a model of planetary motion that fit Tycho's data.

- **What are Kepler's three laws of planetary motion?**

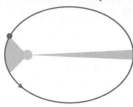

(1) The orbit of each planet is an ellipse with the Sun at one focus. (2) As a planet moves around its orbit, it sweeps out equal areas in equal times. (3) More distant planets orbit the Sun at slower average speeds, obeying the precise mathematical relationship $p^2 = a^3$.

- **How did Galileo solidify the Copernican revolution?**

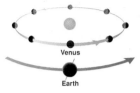

Galileo's experiments and telescopic observations overcame remaining objections to the Copernican idea of Earth as a planet orbiting the Sun. Although not everyone accepted his results immediately, in hindsight we see that Galileo sealed the case for the Sun-centered solar system.

3.4 The Nature of Science

- **How can we distinguish science from nonscience?**
Science generally exhibits these three hallmarks: (1) Modern science seeks explanations for observed phenomena that rely solely on natural causes. (2) Science progresses through the creation and testing of models of nature that explain the observations as simply as possible. (3) A scientific model must make testable predictions about natural phenomena that would force us to revise or abandon the model if the predictions do not agree with observations.

- **What is a scientific theory?**
A scientific **theory** is a simple yet powerful model that explains a wide variety of observations in terms of just a few general principles and has attained the status of a theory by surviving repeated and varied testing.

exercises and problems

For instructor-assigned homework go to **www.masteringastronomy.com.**

Review Questions

Short-Answer Questions Based on the Reading

1. In what way is scientific thinking natural to all of us? How does modern science differ from this everyday type of thinking?
2. Why did ancient peoples study astronomy? Describe an astronomical achievement of at least three ancient cultures.
3. How are the names of the seven days of the week related to astronomical objects?
4. What is a lunar calendar? Are lunar calendars still used today?
5. What do we mean by a *model* in science? Briefly summarize the Greek *geocentric model.*
6. What do we mean by the *Ptolemaic model?* How did this model account for the apparent retrograde motion of planets in our sky?
7. What was the *Copernican revolution,* and how did it change the human view of the universe?
8. Why wasn't the Copernican model immediately accepted? Describe the roles of Tycho, Kepler, and Galileo in the eventual triumph of the Sun-centered model.
9. What is an *ellipse?* Define the *focus* and the *eccentricity* of an ellipse. Why are ellipses important in astronomy?
10. State each of *Kepler's laws of planetary motion.* Describe the meaning of each law in a way that anyone could understand.
11. Describe the three hallmarks of science and explain how we can see them in the Copernican revolution. What is *Occam's razor?* Why doesn't science accept personal testimony as evidence?
12. What is the difference between a *hypothesis* and a *theory* in science?

Test Your Understanding

Science or Nonscience?

Each of the following statements makes some type of claim. Decide in each case whether the claim could be evaluated scientifically or whether it falls into the realm of nonscience. Explain clearly; not all of these have definitive answers, so your explanation is more important than your chosen answer.

13. The Yankees are the best baseball team of all time.
14. Several kilometers below its surface, Jupiter's moon Europa has an ocean of liquid water.
15. My house is haunted by ghosts who make the creaking noises I hear each night.
16. There is no liquid water on the surface of Mars today.
17. Dogs are smarter than cats.
18. Children born when Jupiter is in the constellation Taurus are more likely to be musicians than other children.
19. Aliens can manipulate time so that they can abduct and perform experiments on people who never realize they were taken.
20. Newton's law of gravity works as well for explaining orbits of planets around other stars as it does for explaining the planets in our own solar system.
21. God created the laws of motion that were discovered by Newton.
22. A huge fleet of alien spacecraft will land on Earth and introduce an era of peace and prosperity on January 1, 2020.

Quick Quiz

Choose the best answer to each of the following. Explain your reasoning with one or more complete sentences.

23. In the Greek geocentric model, the retrograde motion of a planet occurs when (a) Earth is about to pass the planet in its orbit around the Sun. (b) The planet actually goes backward in its orbit around Earth. (c) The planet is aligned with the Moon in our sky.
24. Which of the following was *not* a major advantage of Copernicus's Sun-centered model over the Ptolemaic model? (a) It made significantly better predictions of planetary positions in our sky. (b) It offered a more natural explanation for the apparent retrograde motion of planets in our sky. (c) It allowed calculation of the orbital periods and distances of the planets.
25. When we say that a planet has a highly eccentric orbit, we mean that (a) it is spiraling in toward the Sun. (b) its orbit is an ellipse with the Sun at one focus. (c) in some parts of its orbit it is much closer to the Sun than in other parts.
26. Earth is closer to the Sun in January than in July. Therefore, in accord with Kepler's second law, (a) Earth travels faster in its

orbit around the Sun in July than in January. (b) Earth travels faster in its orbit around the Sun in January than in July. (c) It is summer in January and winter in July.

27. According to Kepler's third law, (a) Mercury travels fastest in the part of its orbit in which it is closest to the Sun. (b) Jupiter orbits the Sun at a faster speed than Saturn. (c) All the planets have nearly circular orbits.

28. Tycho Brahe's contribution to astronomy included (a) inventing the telescope. (b) proving that Earth orbits the Sun. (c) collecting data that enabled Kepler to discover the laws of planetary motion.

29. Galileo's contribution to astronomy included (a) discovering the laws of planetary motion. (b) discovering the law of gravity. (c) making observations and conducting experiments that dispelled scientific objections to the Sun-centered model.

30. Which of the following is *not* true about scientific progress? (a) Science progresses through the creation and testing of models of nature. (b) Science advances only through the scientific method. (c) Science avoids explanations that invoke the supernatural.

31. Which of the following is *not* true about a scientific theory? (a) A theory must explain a wide range of observations or experiments. (b) Even the strongest theories can never be proved true beyond all doubt. (c) A theory is essentially an educated guess.

32. When Einstein's theory of gravity (general relativity) gained acceptance, it demonstrated that Newton's theory had been (a) wrong. (b) incomplete. (c) really only a guess.

Process of Science

Examining How Science Works

33. *What Makes It Science?* Choose a single idea in the modern view of the cosmos discussed in Chapter 1, such as "The universe is expanding," or "We are made from elements manufactured by stars," or "The Sun orbits the center of the Milky Way Galaxy once every 230 million years."
    a. Describe how this idea reflects each of the three hallmarks of science, discussing how it is based on observations, how our understanding of it depends on a model, and how that model is testable.
    b. No matter how strongly the evidence may support a scientific idea, we can never be certain beyond all doubt that the idea is true. Describe an observation that might cause us to call the idea you chose into question. Then briefly discuss whether you think that, overall, the idea is likely or unlikely to hold up to future observations. Defend your opinion.

34. *Earth's Shape.* It took thousands of years for humans to deduce that Earth is spherical. For each of the following alternative models of Earth's shape, identify one or more observations that you could make for yourself and that would invalidate the model.
    a. A flat Earth
    b. A cylindrical Earth (which was actually proposed by the Greek philosopher Anaximander (c. 610–546 B.C.))
    c. A football-shaped Earth

35. *Your Own Astrological Test.* Devise your own scientific test of astrology. Clearly define your methods and how you will evaluate the results. Carry out the test, and write a short report about it.

Investigate Further

In-Depth Questions to Increase Your Understanding

Short-Answer/Essay Questions

36. *Copernican Players.* Using a bulleted list format, make a one-page "executive summary" of the major roles that Copernicus, Tycho, Kepler, and Galileo played in overturning the ancient belief in an Earth-centered universe.

37. *Influence on History.* Based on what you have learned about the Copernican revolution, write a one- to two-page essay about how you believe it altered the course of human history.

38. *Cultural Astronomy.* Choose a particular culture of interest to you, and research the astronomical knowledge and accomplishments of that culture. Write a two- to three-page summary of your findings.

Quantitative Problems

Be sure to show all calculations clearly and state your final answers in complete sentences.

39. *Method of Eratosthenes.* You are an astronomer on planet Nearth, which orbits a distant star. It has recently been accepted that Nearth is spherical in shape, though no one knows its size. One day, while studying in the library of Alectown, you learn that on the equinox your sun is directly overhead in the city of Nyene, located 1000 km due north of you. On the equinox, you go outside in Alectown and observe that the altitude of your sun is 80°. What is the circumference of Nearth? (*Hint:* Apply the technique used by Eratosthenes to measure Earth's circumference.)

40. *Eris Orbit.* The recently discovered Eris orbits the Sun every 560 years. What is its average distance (semimajor axis) from the Sun? How does its average distance compare to that of Pluto?

41. *Halley Orbit.* Halley's comet orbits the Sun every 76.0 years and has an orbital eccentricity of 0.97.
    a. Find its average distance (semimajor axis).
    b. Halley's orbit is a very eccentric (stretched-out) ellipse, so that at perihelion it is only about 90 million km from the Sun, compared to more than 5 billion km at aphelion. Does Halley's comet spend most of its time near its perihelion distance, its aphelion distance, or halfway in between? Explain.

Discussion Questions

42. *The Impact of Science.* The modern world is filled with ideas, knowledge, and technology that developed through science and application of the scientific method. Discuss some of these things and how they affect our lives. Which of these impacts do you think are positive? Which are negative? Overall, do you think science has benefited the human race? Defend your opinion.

43. *The Importance of Ancient Astronomy.* Why was astronomy important to people in ancient times? Discuss both the practical importance of astronomy and the importance it may have had for religious or other traditions. Which do you think was more important in the development of ancient astronomy, its practical or its philosophical role? Defend your opinion.

44. *Astronomy and Astrology.* Why do you think astrology remains so popular around the world even though it has failed all scientific tests of its validity? Do you think this popularity has any social consequences? Defend your opinions.

Web Projects

45. *The Ptolemaic Model.* This chapter gives only a very brief description of Ptolemy's model of the universe. Investigate this model in greater depth. Using diagrams and text as needed, create a two-to three-page description of the model.

46. *The Galileo Affair.* In recent years, the Roman Catholic Church has devoted a lot of resources to learning more about the trial of Galileo and to understanding past actions of the Church in the Galilean case. Learn more about these studies, and write a two-to three-page report about the current Vatican view of the case.

47. *Science or Pseudoscience.* Choose a pseudoscientific claim related to astronomy; learn more about it and about how scientists have debunked it. (A good starting point is the Bad Astronomy Web site: www.badastronomy.com.) Write a short summary of your findings.

visual skills check

Use the following questions to check your understanding of some of the many types of visual information used in astronomy. Answers are provided in Appendix K. For additional practice, try the Chapter 3 Visual Quiz at **www.masteringastronomy.com**.

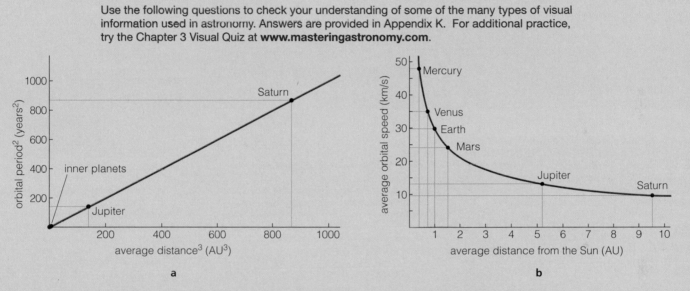

a

b

Study the two graphs above, based on Figure 3.16. Use the information in the graphs to answer the following questions.

1. Approximately how fast is Jupiter orbiting the Sun?
   a. cannot be determined from the information provided
   b. 20 km/s
   c. 10 km/s
   d. a little less than 15 km/s

2. An asteroid with an average orbital distance of 2 AU will orbit the Sun at an average speed that is _____.
   a. a little slower than the orbital speed of Mars
   b. a little faster than the orbital speed of Mars
   c. the same as the orbital speed of Mars

3. Uranus, not shown on the graph, orbits about 19 AU from the Sun. Based on the graph, its approximate orbital speed is between about _____.
   a. 20 and 25 km/s
   b. 15 and 20 km/s
   c. 10 and 15 km/s
   d. 5 and 10 km/s

4. Kepler's third law is often stated as $p^2 = a^3$. The value a^3 for a planet is shown on _____.
   a. the horizontal axis of Figure a
   b. the vertical axis of Figure a
   c. the horizontal axis of Figure b
   d. the vertical axis of Figure b

5. On Figure a, you can see Kepler's third law ($p^2 = a^3$) from the fact that _____.
   a. the data fall on a straight line
   b. the axes are labeled with values for p^2 and a^3
   c. the planet names are labeled on the graph

6. Suppose Figure a showed a planet on the red line directly above a value of 1000 AU^3 along the horizontal axis. On the vertical axis, this planet would be at _____.
   a. 1000 years^2
   b. 1000^2 years^2
   c. $\sqrt{1000}$ years^2
   d. 100 years

7. How far does the planet in question 6 orbit from the Sun?
   a. 10 AU
   b. 100 AU
   c. 1000 AU
   d. $\sqrt{1000}$ AU

Our perspective on the universe has changed dramatically throughout human history. This timeline summarizes some of the key discoveries that have shaped our modern perspective.

Stonehenge

Earth-centered model of the universe

Galileo's telescope

< 2500 B.C.	400 B.C. –170 A.D.	1543–1648 A.D.

(1) Ancient civilizations recognized patterns in the motion of the Sun, Moon, planets, and stars through our sky. They also noticed connections between what they saw in the sky and our lives on Earth, such as the cycles of seasons and of tides [Section 3.1].

(2) The ancient Greeks tried to explain observed motions of the Sun, Moon, and planets using a model with Earth at the center, surrounded by spheres in the heavens. The model explained many phenomena well, but could explain the apparent retrograde motion of the planets only with the addition of many complex features— and even then, its predictions were not especially accurate [Section 3.2].

(3) Copernicus suggested that Earth is a planet orbiting the Sun. The Sun-centered model explained apparent retrograde motion simply, though it made accurate predictions only after Kepler discovered his three laws of planetary motion. Galileo's telescopic observations confirmed the Sun-centered model, and revealed that the universe contains far more stars than had been previously imagined [Section 3.3].

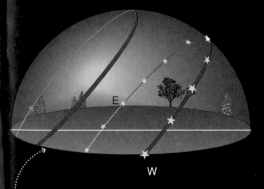

Earth's rotation around its axis leads to the daily east-to-west motions of objects in the sky.

The tilt of Earth's rotation axis leads to seasons as Earth orbits the Sun.

Planets are much smaller than the Sun. At a scale of 1-to-10 billion, the Sun is the size of a grapefruit, Earth is the size of a ball point of a pen, and the distance between them is about 15 meters.

Yerkes Observatory

Edwin Hubble at the Mt. Wilson telescope

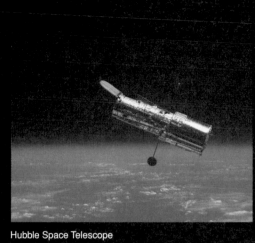

Hubble Space Telescope

1838–1920 A.D.

1924–1929 A.D.

1990 A.D.–present

(4) Larger telescopes and photography made it possible to measure the parallax of stars, offering direct proof that Earth really does orbit the Sun and showing that even the nearest stars are light-years away. We learned that our Sun is a fairly ordinary star in the Milky Way [Section 2.4, 11.1].

(5) Edwin Hubble measured the distances of galaxies, showing that they lay far beyond the bounds of the Milky Way and proving that the universe is far larger than our own galaxy. He also discovered that more distant galaxies are moving away from us faster, telling us that the entire universe is expanding and suggesting that it began in an event we call the Big Bang [Section 1.3, 15.2].

(6) Improved measurements of galactic distances and the rate of expansion have shown that the universe is about 14 billion years old. These measurements have also revealed still unexplained surprises, including evidence for the existence of mysterious dark matter and dark energy [Section 1.3, 16.1].

Distances between stars are enormous. At a scale of 1-to-10 billion, you can hold the Sun in your hand, but the nearest stars are thousands of kilometers away.

Our solar system is located about 28,000 light-years from the center of the Milky Way Galaxy.

The Milky Way Galaxy contains over 100 billion stars.

The observable universe contains over 100 billion galaxies.

4

making sense of the universe:
understanding motion, energy, and gravity

The history of the universe is essentially a story about the interplay between matter and energy since the beginning of time. Interactions between matter and energy began in the Big Bang and continue today in everything from the microscopic jiggling of atoms to gargantuan collisions of galaxies. Understanding the universe therefore depends on becoming familiar with how matter responds to the ebb and flow of energy.

You might guess that it would be difficult to understand the many interactions that shape the universe, because they occur on so many different size scales. However, we now know that just a few physical laws govern the movements of everything from atoms to galaxies. The Copernican revolution spurred the discovery of these laws, and Galileo deduced some of them from his experiments. But it was Sir Isaac Newton who put all of the pieces together into a simple system of laws describing both motion and gravity.

In this chapter, we'll discuss the laws that govern motion and energy, including Newton's laws of motion, the laws of conservation of angular momentum and of energy, and the universal law of gravitation. Understanding these laws will enable you to make sense of many of the wide-ranging phenomena you will encounter as you study astronomy.

4.1 Describing Motion: Examples from Daily Life

We all have experience with motion and a natural intuition as to what motion is, but in science we need to define our ideas and terms precisely. In this section, we'll use examples from everyday life to explore some of the fundamental ideas of motion.

• How do we describe motion?

You are probably familiar with the terms used to describe motion in science—terms such as *velocity, acceleration,* and *momentum.* However, their scientific definitions may differ subtly from those you use in casual conversation. Let's investigate the precise meanings of these terms.

Speed, Velocity, and Acceleration   A car provides a good illustration of the three basic terms that we use to describe motion:

- The **speed** of the car tells us how far it will go in a certain amount of time. For example, 100 kilometers per hour (about 60 miles per hour) is a speed, and it tells us that the car will cover a distance of 100 kilometers if it is driven at this speed for an hour.

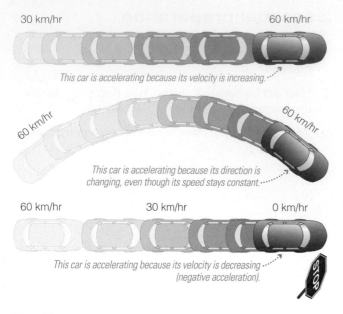

30 km/hr   60 km/hr

This car is accelerating because its velocity is increasing.

60 km/hr

60 km/hr

This car is accelerating because its direction is changing, even though its speed stays constant.

60 km/hr   30 km/hr   0 km/hr

This car is accelerating because its velocity is decreasing (negative acceleration).

Figure 4.1

Speeding up, turning, and slowing down are all examples of acceleration.

- The **velocity** of the car tells us both its speed and direction. For example, "100 kilometers per hour going due north" describes a velocity.

- The car has an **acceleration** if its velocity is changing in any way, whether in speed or direction or both.

An object is accelerating if either its speed or its direction is changing.

You are undoubtedly familiar with the term *acceleration* as it applies to increasing speed. In science, we also say that you are accelerating when you slow down or turn (Figure 4.1). Slowing occurs when acceleration is in a direction opposite to the motion. In this case, we say that your acceleration is negative, causing your velocity to decrease. Turning changes your velocity because it changes the direction in which you are moving, so turning is a form of acceleration even if your speed remains constant.

You can often feel the effects of acceleration. For example, as you speed up in a car, you feel yourself being pushed back into your seat. As you slow down, you feel yourself being pulled forward. As you drive around a curve, you feel yourself being pushed away from the direction of your turn. In contrast, you don't feel such effects when moving at *constant velocity*. That is why you don't feel any sensation of motion when you're traveling in an airplane on a smooth flight.

The Acceleration of Gravity One of the most important types of acceleration is the acceleration caused by gravity. In a legendary experiment in which he supposedly dropped weights from the Leaning Tower of Pisa, Galileo demonstrated that gravity accelerates all objects by the same amount, regardless of their mass. This fact may be surprising because it seems to contradict everyday experience: A feather floats gently to the ground, while a rock plummets. However, air resistance causes this difference in acceleration. If you dropped a feather and a rock on the Moon, where there is no air, both would fall at exactly the same rate.

see it for yourself Find a piece of paper and a small rock. Hold both at the same height, one in each hand, and let them go at the same instant. The rock, of course, hits the ground first. Next, crumple the paper into a small ball and repeat the experiment. What happens? Explain how this experiment suggests that gravity accelerates all objects by the same amount.

The acceleration of a falling object is called the **acceleration of gravity**, abbreviated g. On Earth, the acceleration of gravity causes falling objects to fall faster by 9.8 meters per second (m/s), or about 10 m/s, with each passing second. For example, suppose you drop a rock from a tall building. At the moment you let it go, its speed is 0 m/s. After 1 second, the rock will be falling downward at about 10 m/s. After 2 seconds, it will be falling at about 20 m/s. In the absence of air resistance, its speed will continue to increase by about 10 m/s each second until it hits the ground (Figure 4.2). We therefore say that the acceleration of gravity is about 10 *meters per second per second*, or 10 *meters per second squared*, which we write as 10 m/s^2 (more precisely, $g = 9.8$ m/s^2).

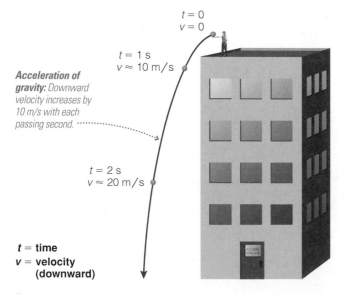

$t = 0$
$v = 0$

$t = 1$ s
$v \approx 10$ m/s

Acceleration of gravity: Downward velocity increases by 10 m/s with each passing second.

$t = 2$ s
$v \approx 20$ m/s

$t = $ time
$v = $ velocity (downward)

Figure 4.2

On Earth, gravity causes an unsupported object to accelerate downward at about 10 m/s^2, which means its downward velocity increases by about 10 m/s with each passing second. (Gravity does not affect horizontal velocity.)

Momentum and Force The concepts of speed, velocity, and acceleration describe how an individual object moves, but most of the interesting phenomena we see in the universe result from interactions between objects. We need two additional concepts to describe these interactions:

- An object's **momentum** is the product of its mass and its velocity; that is, momentum = mass × velocity.

- The only way to change an object's momentum is to apply a **force** to it.

We can understand these concepts by considering the effects of collisions. Imagine that you're stopped in your car at a red light when a bug flying at a velocity of 30 km/hr due south slams into your windshield. What will happen to your car? Not much, except perhaps a bit of a mess on your windshield. Next, imagine that a 2-ton truck runs the red light and hits you head-on with the same velocity as the bug. Clearly, the truck will cause far more damage. We can understand why by considering the momentum and force in each collision.

Before the collisions, the truck's much greater mass means it has far more momentum than the bug, even though both the truck and the bug are moving with the same velocity. During the collisions, the bug and the truck each transfer some of their momentum to your car. The bug has very little momentum to give to your car, so it does not exert much of a force. In contrast, the truck imparts enough of its momentum to cause a dramatic and sudden change in your car's momentum. You feel this sudden change in momentum as a force, and it can do great damage to you and your car.

The mere presence of a force does not always cause a change in momentum. For example, a moving car is always affected by forces of air resistance and friction with the road—forces that will slow your car if you take your foot off the gas pedal. However, you can maintain a constant velocity, and hence constant momentum, if you step on the gas pedal hard enough to overcome the slowing effects of these forces.

In fact, forces of some kind are always present, such as the force of gravity or the electromagnetic forces acting between atoms. The **net force** (or *overall force*) acting on an object represents the combined effect of all the individual forces put together. There is no net force on your car when you are driving at constant velocity, because the force generated by the engine to turn the wheels precisely offsets the forces of air resistance and road friction. A change in momentum occurs only when the net force is not zero.

An object must accelerate whenever a net force acts on it.

Changing an object's momentum means changing its velocity, as long as its mass remains constant. A net force that is not zero therefore causes an object to accelerate. Conversely, whenever an object accelerates, a net force must be causing the acceleration. That is why you feel forces (pushing you forward, backward, or to the side) when you accelerate in your car. We can use the same ideas to understand many astronomical processes. For example, planets are always accelerating as they orbit the Sun, because their direction of travel constantly changes as they go around their orbits. We can therefore conclude that some force must be causing this acceleration. As we'll discuss shortly, Isaac Newton identified this force as gravity.

• How is mass different from weight?

In daily life, we usually think of *mass* as something you can measure with a bathroom scale, but technically the scale measures your *weight*, not your mass. The distinction between mass and weight rarely matters when we are talking about objects on Earth, but it is very important in astronomy:

Figure 4.3 interactive figure

Mass is not the same as weight. The man's mass never changes, but his weight is different when the elevator accelerates.

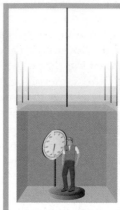

When the elevator moves at constant velocity (or is stationary)...

...your weight is normal.

*When the elevator **accelerates** upward...*

...you weigh more.

*When the elevator **accelerates** downward...*

...you weigh less.

*If the cable breaks so that you are in **free-fall**...*

...you are weightless.

- Your **mass** is the amount of matter in your body.

- Your **weight** (or **apparent weight***) is the *force* that a scale measures when you stand on it; that is, weight depends both on your mass and on the forces (including gravity) acting on your mass.

To understand the difference between mass and weight, imagine standing on a scale in an elevator (Figure 4.3). Your mass will be the same no matter how the elevator moves, but your weight can vary. When the elevator is stationary or moving at constant velocity, the scale reads your "normal" weight. When the elevator accelerates upward, the floor exerts a greater force than it does when you are at rest. You feel heavier, and the scale verifies your greater weight. When the elevator accelerates downward, the floor and the scale exert a weaker force on you, so the scale registers less weight. Note that the scale shows a weight different from your "normal" weight only when the elevator is *accelerating,* not when it is going up or down at constant speed.

see it for yourself Find a small bathroom scale and take it with you on an elevator ride. How does your weight change when the elevator accelerates upward or downward? Does it change when the elevator is moving at constant speed? Explain your observations.

Your mass is the same no matter where you are, but your weight can vary. Your mass therefore depends only on the amount of matter in your body and is the same anywhere, but your weight can vary because the forces acting on you can vary. For example, your mass would be the same on the Moon as on Earth, but you would weigh less on the Moon because of its weaker gravity.

---

*Some physics texts distinguish between "true weight," which is due only to gravity, and "apparent weight," which also depends on other forces (as in an elevator). In this book, the word *weight* means "apparent weight."

Free-Fall and Weightlessness Now consider what happens if the elevator cable breaks (see the last frame in Figure 4.3). The elevator and you are suddenly in **free-fall**—falling without any resistance to slow you down. The floor drops away at the same rate that you fall, allowing you to "float" freely above it, and the scale reads zero because you are no longer held to it. In other words, your free-fall has made you **weightless**.

In fact, you are in free-fall whenever there's nothing to *prevent* you from falling. For example, you are in free-fall when you jump off a chair or spring from a diving board or trampoline. Surprising as it may seem, you have therefore experienced weightlessness many times in your life. You can experience it right now simply by jumping off your chair—though your weightlessness lasts for only a very short time until you hit the ground.

Weightlessness in Space You've probably seen videos of astronauts floating weightlessly in the Space Shuttle or the Space Station. But why are they weightless? Many people guess that there's no gravity in space, but that's not true. After all, it is gravity that makes the Space Shuttle and the Space Station orbit Earth. Astronauts are weightless for the same reason you are weightless when you jump off a chair: They are in free-fall.

> People or objects are weightless whenever they are falling freely, and astronauts in orbit are weightless because they are in a constant state of free-fall.

Astronauts are weightless the entire time they orbit Earth because they are in a *constant state of free-fall*. To understand this idea, imagine a tower that reaches all the way to the Space Station's orbit, about 350 kilometers above Earth (Figure 4.4). If you stepped off the tower, you would fall downward, remaining weightless until you hit the ground (or until air resistance had a noticeable effect on you). Now, imagine that instead of stepping off the tower, you ran and jumped out of the tower. You'd still fall to the ground, but because of your forward motion you'd land a short distance away from the base of the tower.

The faster you ran out of the tower, the farther you'd go before landing. If you could somehow run fast enough—about 28,000 km/hr (17,000 mi/hr) at the orbital altitude of the Space Station—a very interesting thing would happen: By the time gravity had pulled you downward as far as the length of the tower, you'd already have moved far enough around Earth that you'd no longer be going down at all. Instead, you'd be just as high above Earth as you'd been all along, but a good portion of the way around the world. In other words, you'd be orbiting Earth.

The Space Shuttle, the Space Station, and all other orbiting objects stay in orbit because they are constantly "falling around" Earth. Their constant state of free-fall makes these spacecraft and everything in them weightless.

 Motion and Gravity Tutorial, Lesson 1

4.2 Newton's Laws of Motion

The complexity of motion in daily life might lead you to guess that the laws governing motion would also be complex. For example, if you watch a falling piece of paper waft lazily to the ground, you'll see it rock back and forth in a seemingly unpredictable pattern. However, the complexity

The faster you run from the tower, the farther you go before falling to Earth.

Using a rocket to gain enough speed, you could continually "fall" around Earth; that is, you'd be in orbit.

Not to scale!

Figure 4.4 interactive figure

This figure explains why astronauts are weightless and float freely in space. It shows that if you could leap from a tall tower with enough speed (with the aid of a rocket), you could travel forward so fast that you'd orbit Earth. You'd then be in a constant state of free-fall, which means you'd be weightless. Note: On the scale shown here, the tower extends far higher than the Space Station's orbit; the rocket orientation assumes that it rotates once with each orbit, as is the case for the Space Shuttle. (Adapted from *Space Station Science* by Marianne Dyson.)

 **common misconceptions**

No Gravity in Space?

If you ask people why astronauts are weightless in space, one of the most common answers is "There is no gravity in space." But you can usually convince people that this answer must be wrong by following up with another simple question: Why does the Moon orbit Earth? Most people know that the Moon orbits Earth because of gravity, proving that there is gravity in space. In fact, at the altitude of the Space Station's orbit, the acceleration of gravity is scarcely less than it is on Earth's surface.

The real reason astronauts are weightless is that they are in a constant state of free-fall. Imagine being an astronaut. You'd have the sensation of free-fall—just as when you are falling from a diving board—the entire time you were in orbit. This constant falling sensation makes most astronauts sick to their stomachs when they first experience weightlessness. Fortunately, they quickly get used to the sensation, which allows them to work hard and enjoy the view.

Sir Isaac Newton (1642–1727)

of this motion arises because the paper is affected by a variety of forces, including gravity and the changing forces caused by air currents. If you could analyze the forces individually, you'd find that each force affects the paper's motion in a simple, predictable way. Sir Isaac Newton (1642–1727) discovered the remarkably simple laws that govern motion.

• How did Newton change our view of the universe?

Newton was born in Lincolnshire, England, on Christmas Day in 1642. He had a difficult childhood and showed few signs of unusual talent. He attended Trinity College at Cambridge, where he earned his keep by performing menial labor, such as cleaning the boots and bathrooms of wealthier students and waiting on their tables.

The plague hit Cambridge shortly after Newton graduated, and he returned home. By his own account, he experienced a moment of inspiration in 1666 when he saw an apple fall to the ground. He suddenly realized that the gravity making the apple fall was the same force that held the Moon in orbit around Earth. In that moment, Newton shattered the remaining vestiges of the Aristotelian view of the world, which for centuries in Europe had been taken as near-gospel truth.

Aristotle had made many claims about the physics of motion, using his ideas to support his belief in an Earth-centered cosmos. He had also maintained that the heavens were totally distinct from Earth, so that physical laws on Earth did not apply to heavenly motion. By the time Newton saw the apple fall, the Copernican revolution had displaced Earth from a central position, and Galileo's experiments had shown that the laws of physics were not what Aristotle had believed.

Newton showed that the same physical laws that operate on Earth also operate in the heavens.

Newton's sudden insight delivered the final blow to Aristotle's physics. When Newton realized that gravity operated in the heavens as well as on Earth, he eliminated Aristotle's distinction between the two realms. For the first time in history, the heavens and Earth were brought together as one *universe*. Newton's insight also heralded the birth of the modern science of *astrophysics* (although the term wasn't coined until much later). Astrophysics applies physical laws discovered on Earth to phenomena throughout the cosmos.

Over the next 20 years, Newton's work completely revolutionized mathematics and science. He quantified the laws of motion and gravity, conducted crucial experiments regarding the nature of light, built the first reflecting telescopes, and invented the mathematics of calculus. We'll discuss his laws of motion in the rest of this section, and later in the chapter we'll turn our attention to Newton's discoveries about gravity.

• What are Newton's three laws of motion?

Newton published the laws of motion and gravity in 1687, in his book *Philosophiae Naturalis Principia Mathematica* ("Mathematical Principles of Natural Philosophy"), usually called *Principia*. He enumerated three laws that apply to all motion, what we now call **Newton's laws of motion**. These laws govern the motion of everything from our daily movements

Newton's first law of motion:
An object moves at constant velocity unless a net force acts to change its speed or direction.

Example: A spaceship needs no fuel to keep moving in space.

Newton's second law of motion:
Force = mass × acceleration

Example: A baseball accelerates as the pitcher applies a force by moving his arm. (Once the ball is released, the force from the pitcher's arm ceases, and the ball's path changes only because of the forces of gravity and air resistance.)

Newton's third law of motion:
For any force, there is always an equal and opposite reaction force.

Example: A rocket is propelled upward by a force equal and opposite to the force with which gas is expelled out its back.

Figure 4.5

Newton's three laws of motion.

here on Earth to the movements of planets, stars, and galaxies throughout the universe. Figure 4.5 summarizes the three laws.

Newton's First Law Newton's first law of motion states that in the absence of a net force, an object will move with constant velocity. Objects at rest (velocity = 0) tend to remain at rest, and objects in motion tend to remain in motion with no change in either their speed or their direction.

> Newton's first law: An object moves at constant velocity if there is no net force acting upon it.

The idea that an object at rest should remain at rest is rather obvious: A car parked on a flat street won't suddenly start moving for no reason. But what if the car is traveling along a flat, straight road? Newton's first law says that the car should keep going at the same speed forever *unless* a force acts to slow it down. You know that the car eventually will come to a stop if you take your foot off the gas pedal, so we must conclude that one or more forces are stopping the car—in this case, forces arising from friction and air resistance. If the car were in space, and therefore unaffected by friction or air, it would keep moving forever (though gravity would eventually alter its speed and direction). That is why interplanetary spacecraft need no fuel to keep going after they are launched into space, and why astronomical objects don't need fuel to travel through the universe.

Newton's first law also explains why you don't feel any sensation of motion when you're traveling in an airplane on a smooth flight. As long as the plane is traveling at constant velocity, no net force is acting on it or on you. Therefore, you feel no different from the way you would feel at rest. You can walk around the cabin, play catch with someone, or relax and go to sleep just as though you were "at rest" on the ground.

Newton's Second Law Newton's second law of motion tells us what happens to an object when a net force *is* present. We have already seen that a net force will change an object's momentum, accelerating it in the direction of the force. Newton's second law quantifies this relationship, telling us that the amount of the acceleration depends on the object's mass and the strength of the net force. We usually write this law as an equation: force = mass × acceleration, or $F = ma$ for short.

Newton's second law:
Force = mass × acceleration ($F = ma$).

This law explains why you can throw a baseball farther than you can throw a shot-put. The force your arm delivers to both the baseball and the shot-put equals the product of mass and acceleration. Because the mass of the shot-put is greater than that of the baseball, the same force from your arm gives the shot-put a smaller acceleration. Because of its smaller acceleration, the shot-put leaves your hand with less speed than the baseball and therefore travels a shorter distance before hitting the ground.

Newton's second law also explains why large planets such as Jupiter have a greater effect on asteroids and comets than small planets such as Earth [Section 9.4]. Because Jupiter is much more massive than Earth, it exerts a stronger gravitational force on passing asteroids and therefore sends them scattering with a greater acceleration.

Newton's Third Law   Think for a moment about standing still on the ground. Your weight exerts a downward force; if this force were acting alone, Newton's second law would demand that you accelerate downward. The fact that you are not falling means there must be no *net* force acting on you, which is possible only if the ground is exerting an upward force on you that precisely offsets the downward force you exert on the ground. The fact that the downward force you exert on the ground is offset by an equal and opposite force that pushes upward on you is one example of the more general rule embodied in Newton's third law of motion: Every force is always paired with an equal and opposite reaction force.

Newton's third law: For any force, there is always an equal and opposite reaction force.

This law is very important in astronomy, because it tells us that objects always attract *each other* through gravity. For example, your body always exerts a gravitational force on Earth identical to the force that Earth exerts on you, except that it acts in the opposite direction. Of course, the same force means a much greater acceleration for you than for Earth (because your mass is so much smaller than Earth's), which is why you fall toward Earth when you jump off a chair, rather than Earth falling toward you.

Newton's third law also explains how a rocket works: A rocket engine generates a force that drives hot gas out the back, which creates an equal and opposite force that propels the rocket forward.

common misconceptions

What Makes a Rocket Launch?

If you've ever watched a rocket launch, it's easy to see why many people believe that the rocket "pushes off" the ground. In fact, the ground has nothing to do with the rocket launch. The rocket's launch is explained by Newton's third law of motion. To balance the force driving gas out the back of the rocket, an equal and opposite force must propel the rocket forward. Rockets can be launched horizontally as well as vertically, and a rocket can be "launched" in space (for example, from a space station) with no need for a solid surface to push off from.

4.3   Conservation Laws in Astronomy

Newton's laws of motion are easy to state, but they may seem a bit arbitrary. Why, for example, should every force be opposed by an equal and opposite reaction force? In the centuries since Newton first stated his laws, we have learned that they are not arbitrary at all, but instead reflect deeper aspects of nature known as *conservation laws*.

Consider what happens when two objects collide. Newton's second law tells us that object 1 exerts a force that will change the momentum of object 2. At the same time, Newton's third law tells us that object 2 exerts an equal and opposite force on object 1—which means that object 1's momentum changes by precisely the same amount as object 2's momentum, but in the opposite direction. The total combined momentum of objects 1 and 2 remains the same both before and after the collision. We say that the total momentum of the colliding objects is conserved, reflecting a principle that we call *conservation of momentum.* In essence, the law of conservation of momentum tells us that the total momentum of all interacting objects always stays the same. An individual object can gain or lose momentum only when a force causes it to exchange momentum with another object.

Conservation of momentum is one of several important conservation laws that underlie Newton's laws of motion and other physical laws in the universe. Two other conservation laws are especially important in astronomy. They go by the names *conservation of angular momentum* and *conservation of energy.* Let's see how these important laws work.

• What keeps a planet rotating and orbiting the Sun?

Perhaps you've wondered how Earth manages to keep rotating and going around the Sun day after day and year after year. The answer relies on a special type of momentum that we use to describe objects turning in circles or going around curves. This special type of "circling momentum" is called **angular momentum**. (The term *angular* arises because a circle turns through an *angle* of 360°.)

Conservation of angular momentum: An object's angular momentum cannot change unless it transfers angular momentum to or from another object.

The **law of conservation of angular momentum** tells us that total angular momentum can never change. An individual object can change its angular momentum only by transferring some angular momentum to or from another object.

Consider Earth's orbit around the Sun. A simple formula tells us Earth's angular momentum at any point in its orbit:

$$\text{angular momentum} = m \times v \times r$$

where m is Earth's mass, v is its speed (or velocity) around the orbit, and r is the "radius" of the orbit, by which we mean Earth's distance from the Sun (Figure 4.6). Because there are no objects around to give or take angular momentum from Earth as it orbits the Sun, Earth's orbital angular momentum must always stay the same. This explains two key facts about Earth's orbit:

1. Earth needs no fuel or push of any kind to keep orbiting the Sun—it will keep orbiting as long as nothing comes along to take angular momentum away.
2. Because Earth's angular momentum at any point in its orbit depends on the product of its speed and orbital radius (distance from the Sun), Earth's orbital speed must be faster when it is nearer to the Sun (and the radius is shorter) and slower when it is farther from the Sun (and the radius is longer).

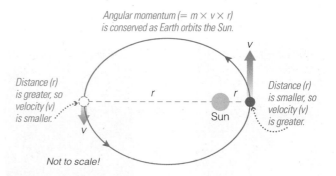

Angular momentum ($= m \times v \times r$) is conserved as Earth orbits the Sun.

Distance (r) is greater, so velocity (v) is smaller.

Distance (r) is smaller, so velocity (v) is greater.

Sun

Not to scale!

Figure 4.6

Earth's angular momentum always stays the same as it orbits the Sun, so it moves faster when it is closer to the Sun and slower when it is farther from the Sun. It needs no fuel to keep orbiting because no forces are acting in a way that could change its angular momentum.

In the product m × v × r, extended arms mean larger radius and smaller velocity of rotation.

Bringing in her arms decreases her radius and therefore increases her rotational velocity.

Figure 4.7

A spinning skater conserves angular momentum.

Energy can be converted from one form to another.

kinetic energy
(energy of motion)

radiative energy
(energy of light)

potential energy
(stored energy)

Figure 4.8

The three basic categories of energy. Energy can be converted from one form to another, but it can never be created or destroyed, an idea embodied in the law of conservation of energy.

The second fact is just what Kepler's second law of planetary motion states [Section 3.3]. That is, the law of conservation of angular momentum tells us *why* Kepler's law is true.

The same idea explains why Earth keeps rotating. As long as Earth isn't transferring any of the angular momentum of its rotation to another object, it keeps rotating at the same rate. (In fact, Earth is very gradually transferring some of its rotational angular momentum to the Moon, and as a result Earth's rotation is gradually slowing down; see Special Topic, p. 103.)

Earth is not exchanging substantial angular momentum with any other object, so its rotation rate and orbit must stay about the same.

Conservation of angular momentum also explains why we see so many spinning disks in the universe, such as the disks of galaxies like the Milky Way and disks of material orbiting young stars. The idea is easy to illustrate with an ice skater spinning in place (Figure 4.7). Because there is so little friction on ice, the angular momentum of the ice skater remains essentially constant. When she pulls in her extended arms, she decreases her radius—which means her velocity of rotation must increase. Stars and galaxies are both born from clouds of gas that start out much larger in size. These clouds almost inevitably have some small net rotation, though it may be imperceptible. But like the spinning skater as she pulls in her arms, these clouds must spin faster as gravity makes them shrink in size. (We'll discuss why the clouds also flatten into disks in Chapter 6.)

think about it How does conservation of angular momentum explain the spiraling of water going down a drain?

(MA)™ **Energy Tutorial, Lesson 1**

• Where do objects get their energy?

The **law of conservation of energy** tells us that, like momentum and angular momentum, energy cannot appear out of nowhere or disappear into nothingness. Objects can gain or lose energy only by exchanging energy with other objects. Because of this law, the story of the universe is a story of the interplay of energy and matter: All actions involve exchanges of energy or the conversion of energy from one form to another.

Conservation of energy: Energy can be transferred from one object to another or transformed from one type to another, but the total amount of energy is always conserved.

Throughout the rest of this book, we'll see numerous cases in which we can understand astronomical processes simply by studying how energy is transformed and exchanged. For example, we'll see that planetary interiors cool with time only because they radiate energy into space, and that the Sun became hot because of energy released by the gas that formed it. By applying the laws of conservation of angular momentum and conservation of energy, we can understand almost every major process that occurs in the universe.

Basic Types of Energy Before we can fully understand the law of conservation of energy, we need to know exactly what energy is. In essence, energy is what makes matter move. Because this statement is so broad, we often distinguish between many different types of energy. For example, we talk about the energy we get from the food we eat, the energy that makes our cars go, and the energy put out by a light bulb. Fortunately, scientists have found a way to classify all these various types of energy into just three major categories (Figure 4.8):

- Energy of motion, or **kinetic energy** (*kinetic* comes from a Greek word meaning "motion"). Falling rocks, orbiting planets, and the molecules moving in the air around us are all examples of objects with kinetic energy.

- Energy carried by light, or **radiative energy** (the word *radiation* is often used as a synonym for *light*). All light carries energy, which is why light can cause changes in matter. For example, light can alter molecules in our eyes—thereby allowing us to see—or warm the surface of a planet.

- Stored energy, or **potential energy**, which might later be converted into kinetic or radiative energy. For example, a rock perched on a ledge has *gravitational* potential energy because it will fall if it slips off the edge, and gasoline contains *chemical* potential energy that can be converted into the kinetic energy of a moving car.

There are three basic categories of energy: energy of motion (kinetic), energy of light (radiative), and stored energy (potential).

Regardless of which type of energy we are dealing with, we can measure the amount of energy with the same standard units. For Americans, the most familiar units of energy are *Calories*, which are shown on food labels to tell us how much energy our bodies can draw from the food. A typical adult needs about 2500 Calories of energy from food each day. In science, the standard unit of energy is the **joule**. One food Calorie is equivalent to about 4184 joules, so the 2500 Calories used daily by a typical adult is equivalent to about 10 million joules. Table 4.1 compares various energies in joules.

Thermal Energy—The Kinetic Energy of Many Particles Although there are only three major categories of energy, we sometimes divide them into various subcategories. In astronomy, the most important subcategory of kinetic energy is **thermal energy**, which represents the collective kinetic energy of the many individual particles (atoms and molecules) moving randomly within a substance like a rock or the air or the gas within a distant star. In such cases, it is much easier to talk about the thermal energy of the object rather than about the kinetic energies of its billions upon billions of individual particles.

Thermal energy gets its name because it is related to temperature, but temperature and thermal energy are not quite the same thing. Thermal energy measures the *total* kinetic energy of all the randomly moving particles in a substance, while **temperature** measures the *average* kinetic energy of the particles. For a particular object, a higher temperature simply means that the particles on average have more kinetic energy and hence are moving faster (Figure 4.9). You're probably familiar with temperatures measured on the *Fahrenheit* or *Celsius* scale, but in science we often use the **Kelvin** temperature scale (Figure 4.10). The Kelvin scale does not have negative temperatures, because it starts from the coldest possible temperature, known as *absolute zero* (0 K), at which there are no random motions at all.

TABLE 4.1 *Energy Comparisons*

Item	Energy (joules)
Energy of sunlight at Earth (per square meter per second)	1.3×10^3
Energy from metabolism of a candy bar	1×10^6
Energy needed to walk for 1 hour	1×10^6
Kinetic energy of a car going 60 mi/hr	1×10^6
Daily food energy need of average adult	1×10^7
Energy released by burning 1 liter of oil	1.2×10^7
Thermal energy of parked car	1×10^8
Energy released by fission of 1 kilogram of uranium-235	5.6×10^{13}
Energy released by fusion of hydrogen in 1 liter of water	7×10^{13}
Energy released by 1-megaton H-bomb	5×10^{15}
Energy released by major earthquake (magnitude 8.0)	2.5×10^{16}
Annual U.S. energy consumption	10^{20}
Annual energy generation of Sun	10^{34}
Energy released by a supernova	$10^{44}-10^{46}$

lower temperature

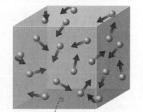

These particles are moving relatively slowly, which means lower temperature . . .

higher temperature

. . . and now the same particles are moving faster, which means higher temperature.

Figure 4.9

Temperature is a measure of the average kinetic energy of the particles (atoms and molecules) in a substance. (Longer arrows represent faster speeds.)

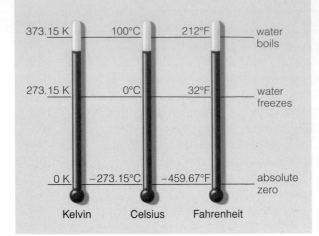

Figure 4.10

Three common temperature scales: Kelvin, Celsius, and Fahrenheit. Scientists generally prefer the Kelvin scale. (The degree symbol ° is not usually used with the Kelvin scale.)

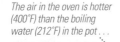

The air in the oven is hotter (400°F) than the boiling water (212°F) in the pot . . .

. . . but the water in the pot contains more thermal energy because of its much higher density.

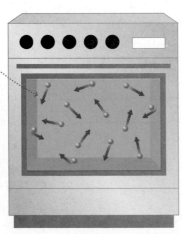

212°F   400°F

Figure 4.11

Thermal energy depends on both the temperature and the density of a substance.

Thermal energy is the total kinetic energy of many individual particles.

Thermal energy depends on temperature, because a higher average kinetic energy for the particles in a substance must also lead to a higher total energy. But thermal energy also depends on the number and density of the particles, as you can see by imagining that you quickly thrust your arm in and out of a hot oven and a pot of boiling water. The air in a hot oven is much hotter in temperature than the water boiling in a pot (typically 400°F for the oven versus 212°F for boiling water). However, the boiling water would scald your arm almost instantly, while you can safely put your arm into the oven air for a few seconds. The reason for this difference is density (Figure 4.11). If air or water is hotter than your body, molecules striking your skin transfer thermal energy to molecules in your arm. The higher temperature in the oven means that the air molecules strike your skin harder, on average, than the molecules in the boiling water. However, because the *density* of water is so much higher than the density of air (meaning water has far more molecules in the same amount of space), many more molecules strike your skin each second in the water. While each individual molecule that strikes your skin transfers a little less energy in the boiling water than in the oven, the sheer number of molecules hitting you in the water means that more thermal energy is transferred to your arm. That is why boiling water causes a burn almost instantly.

think about it   In air or water that is colder than your body temperature, thermal energy is transferred from you to the surrounding cold air or water. Use this fact to explain why falling into a 32°F (0°C) lake is much more dangerous than standing naked outside on a 32°F day.

Potential Energy in Astronomy   Many types of potential energy are important in astronomy, but two are particularly important: *gravitational potential energy* and the potential energy of mass itself, or *mass-energy*.

An object's gravitational potential energy increases when it moves higher and decreases when it moves lower.

An object's **gravitational potential energy** depends on its mass and how far it can fall as a result of gravity. An object has more gravitational potential energy when it is higher and less when it is lower. For example, if you throw a ball up into the air, it has more potential energy when it is high up than it does near the ground. Because energy must be conserved during the ball's flight, the ball's kinetic energy increases when its gravitational potential energy decreases, and vice versa (Figure 4.12a). That is why the ball travels fastest (has the most kinetic energy) when it is closest to the ground, where it has the least gravitational potential energy. The higher it is, the more gravitational potential energy it has and the slower the ball travels (less kinetic energy).

The same general idea explains how stars become hot (Figure 4.12b). Before a star forms, its matter is spread out in a large, cold cloud of gas. Most of the individual gas particles are far from the center of this large cloud and therefore have a lot of gravitational potential energy. The particles lose gravitational potential energy as the cloud contracts under its own gravity, and this "lost" potential energy ultimately gets converted into thermal energy, making the center of the cloud hot.

Einstein discovered that mass itself is a form of potential energy, often called **mass-energy**. The amount of potential energy contained in mass is described by Einstein's famous equation

$$E = mc^2$$

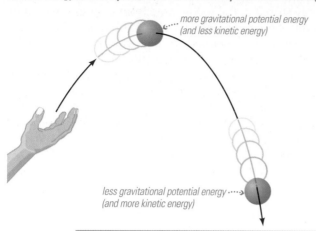

The total energy (kinetic + potential) is the same at all points in the ball's flight.

more gravitational potential energy
(and less kinetic energy)

less gravitational potential energy
(and more kinetic energy)

a The ball has more gravitational potential energy when it is high up than when it is near the ground.

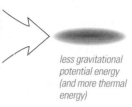

Energy is conserved: As the cloud contracts, gravitational potential energy is converted to thermal energy (and some of this energy is converted to radiation).

less gravitational potential energy (and more thermal energy)

more gravitational potential energy (and less thermal energy)

b A cloud of interstellar gas can contract due to its own gravity. It has more gravitational potential energy when it is spread out than when it shrinks in size.

Figure 4.12

Two examples of gravitational potential energy.

Mass itself is a form of potential energy, as described by Einstein's equation $E = mc^2$. where E is the amount of potential energy, m is the mass of the object, and c is the speed of light. This equation tells us that a small amount of mass contains a huge amount of energy. For example, the energy released by a 1-megaton H-bomb comes from converting only about 0.1 kilogram of mass (about 3 ounces—a quarter of a can of soda) into energy (Figure 4.13). The Sun generates energy by converting a tiny fraction of its mass into energy through a similar process of nuclear fusion [Section 10.2].

Just as Einstein's formula tells us that mass can be converted into other forms of energy, it also tells us that energy can be transformed into mass. This process is especially important in understanding what we think happened during the early moments in the history of the universe, when some of the energy of the Big Bang turned into the mass from which all objects, including us, are made [Section 17.1]. Scientists also use this idea to search for undiscovered particles of matter, by using large machines called *particle accelerators* to create subatomic particles from energy.

Conservation of Energy We have seen that energy comes in three basic categories—kinetic, radiative, and potential—and explored several subcategories that are especially important in astronomy: thermal energy, gravitational potential energy, and mass-energy. Now we are ready to return to the question of where objects get their energy. Because energy cannot be created or destroyed, objects always get their energy from other objects. Ultimately, we can always trace an object's energy back to the Big Bang [Section 1.1], the beginning of the universe in which all matter and energy is thought to have come into existence.

The energy of any object can be traced back to the origin of the universe in the Big Bang. For example, imagine that you've thrown a baseball. It is moving, so it has kinetic energy. Where did this kinetic energy come from? The baseball got its kinetic energy from the motion of your arm as you threw it. Your arm, in turn, got its kinetic energy from the release of chemical potential energy stored in your muscle tissues. Your muscles got

Figure 4.13

The energy released by this H-bomb comes from converting only about 0.1 kilogram of mass into energy in accordance with the formula $E = mc^2$.

this energy from the chemical potential energy stored in the foods you ate. The energy stored in the foods came from sunlight, which plants convert into chemical potential energy through photosynthesis. The radiative energy of the Sun was generated through the process of nuclear fusion, which releases some of the mass-energy stored in the Sun's supply of hydrogen. The mass-energy stored in the hydrogen came from the birth of the universe in the Big Bang. After you throw the ball, its kinetic energy will ultimately be transferred to molecules in the air or ground. According to present understanding, the total energy content of the universe was determined in the Big Bang. It remains the same today and will stay the same in the future.

4.4 The Force of Gravity

Newton's three laws of motion describe how objects in the universe move in response to forces. The laws of conservation of momentum, angular momentum, and energy offer an alternative and often simpler way of thinking about what happens when a force causes some change in the motion of one or more objects. However, we cannot fully understand motion unless we also understand the forces that lead to changes in motion. In astronomy, the most important force is gravity, which governs virtually all large-scale motion in the universe.

(MA) **Motion and Gravity Tutorial, Lesson 2**

• What determines the strength of gravity?

Isaac Newton discovered the basic law that describes how gravity works. Newton expressed the force of gravity mathematically with his **universal law of gravitation**. Three simple statements summarize this law:

- Every mass attracts every other mass through the force called *gravity*.

- The strength of the gravitational force attracting any two objects is *directly proportional* to the product of their masses. For example, doubling the mass of *one* object doubles the force of gravity between the two objects.

- The strength of gravity between two objects decreases with the *square* of the distance between their centers. We therefore say that the gravitational force follows an **inverse square law**. For example, doubling the distance between two objects weakens the force of gravity by a factor of 2^2, or 4.

Doubling the distance between two objects weakens the force of gravity by a factor of 2^2, or 4.

These three statements tell us everything we need to know about Newton's universal law of gravitation. Mathematically, all three statements can be combined into a single equation, usually written like this:

$$F_g = G \frac{M_1 M_2}{d^2}$$

where F_g is the force of gravitational attraction, M_1 and M_2 are the masses of the two objects, and d is the distance between their centers (Figure 4.14). The symbol G is a constant called the **gravitational**

The ***universal law of gravitation*** *tells us the strength of the gravitational attraction between the two objects.*

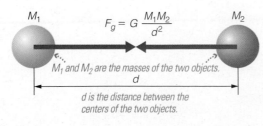

M_1 $F_g = G \dfrac{M_1 M_2}{d^2}$ M_2

M_1 and M_2 are the masses of the two objects.

d

d is the distance between the centers of the two objects.

Figure 4.14

The universal law of gravitation is an *inverse square law*, which means that the force of gravity declines with the *square* of the distance *d* between two objects.

constant, and its numerical value has been measured to be $G = 6.67 \times 10^{-11} \text{m}^3/(\text{kg} \times \text{s}^2)$.

> **think about it** How does the gravitational force between two objects change if the distance between them triples? If the distance between them drops by half?

MA **Orbits and Kepler's Law Tutorial, Lessons 1–4**

• How does Newton's law of gravity extend Kepler's laws?

By the time Newton published *Principia* in 1687, Kepler's three laws of planetary motion [Section 3.3] had already been known and tested for some 70 years. Kepler's laws had proven so successful that there was little doubt about their validity. However, there was great debate among scientists about *why* Kepler's laws hold true, a debate resolved only when Newton showed mathematically that Kepler's laws are consequences of the laws of motion and the universal law of gravitation. In doing so, Newton discovered that he could generalize Kepler's laws in several ways, three of which are particularly important for our purposes.

First, Newton discovered that Kepler's first two laws apply to all orbiting objects, not just to planets going around the Sun. For example, the orbits of a satellite around Earth, of a moon around a planet, and of an asteroid around the Sun are all ellipses in which the orbiting object moves faster at the nearer points in its orbit and slower at the farther points.

Second, Newton found that ellipses are not the only possible orbital paths (Figure 4.15). Kepler was right when he found that ellipses (which include circles) are the only possible shapes for **bound orbits**—orbits in which an object goes around another object over and over again. (The term *bound orbit* comes from the idea that gravity creates a *bond* that holds the objects together.) However, Newton discovered that objects can also follow **unbound orbits**—paths that bring an object close to another object just once. For example, some comets that enter the inner solar system follow unbound orbits. They come in from afar just once, loop around the Sun, and never return.

Newton's version of Kepler's third law allows us to calculate the masses of distant objects.

Third, and perhaps most important, Newton generalized Kepler's third law in a way that allows us to calculate the masses of distant objects. Recall that the precise statement of Kepler's third law is $p^2 = a^3$, where p is a planet's orbital period in years and a is the planet's average distance from the Sun in AU. Newton found that this statement is actually a special case of a more general equation that we call **Newton's version of Kepler's third law** (see Cosmic Calculations 4.1). This more general equation allows us to calculate the mass of a distant object if we can observe another object orbiting it and measure the orbiting object's orbital period and distance. For example, it allows us to calculate the mass of the Sun from Earth's orbital period (1 year) and its average distance (1 AU) from the Sun; it allows us to calculate Jupiter's mass by measuring the orbital period and average distance of one of Jupiter's

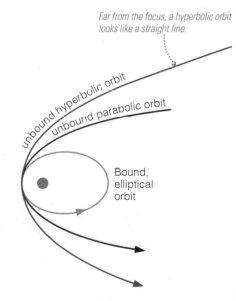

Far from the focus, a hyperbolic orbit looks like a straight line.

unbound hyperbolic orbit

unbound parabolic orbit

Bound, elliptical orbit

Figure 4.15

Newton showed that ellipses are not the only possible orbital paths. Orbits can also be unbound, taking the mathematical shapes of either parabolas or hyperbolas.

cosmic calculations 4.1

Newton's Version of Kepler's Third Law

For an object of mass M_1 orbiting another object of mass M_2, Newton's version of Kepler's third law states

$$p^2 = \frac{4\pi^2}{G(M_1 + M_2)} a^3$$

($G = 6.67 \times 10^{-11} \frac{m^3}{kg \times s^2}$ is the gravitational constant.)

This equation allows us to calculate the sum $M_1 + M_2$ if we know the orbital period p and average (semimajor axis) distance a. The equation is especially useful when one object is much more massive than the other.

Example: Use the fact that Earth orbits the Sun in 1 year at an average distance of 1 AU to calculate the Sun's mass.

Solution: Newton's version of Kepler's third law becomes

$$p_{Earth}^2 = \frac{4\pi^2}{G(M_{Sun} + M_{Earth})} a_{Earth}^3$$

Because the Sun is much more massive than Earth, the sum of their masses is nearly the mass of the Sun alone: $M_{Sun} + M_{Earth} \approx M_{Sun}$. Using this approximation, we find

$$p_{Earth}^2 \approx \frac{4\pi^2}{GM_{Sun}} a_{Earth}^3$$

We now solve for the mass of the Sun and plug in Earth's orbital period ($p_{Earth} = 1$ year $\approx 3.15 \times 10^7$ seconds) and average orbital distance ($a_{Earth} = 1$ AU $\approx 1.5 \times 10^{11}$ m):

$$M_{Sun} \approx \frac{4\pi^2 a_{Earth}^3}{G p_{Earth}^2} \approx \frac{4\pi^2 (1.5 \times 10^{11} \text{ m})^3}{\left(6.67 \times 10^{-11} \frac{m^3}{kg \times s^2}\right)(3.15 \times 10^7 \text{ s})^2}$$

$$= 2.0 \times 10^{30} \text{ kg}$$

The Sun's mass is about 2×10^{30} kilograms.

moons; and it allows us to determine the masses of distant stars if they are members of binary star systems, in which two stars orbit one another. In fact, Newton's version of Kepler's third law is the primary means by which we determine masses throughout the universe.

• How do gravity and energy allow us to understand orbits?

We've seen that Newton's law of universal gravitation explains Kepler's laws of planetary motion, which describe the simple and stable orbits of the planets. By extending Kepler's laws, Newton also explained many other stable orbits, such as the orbit of a satellite around Earth or of a moon around a planet. But orbits do not always stay the same. For example, you've probably heard of satellites crashing to Earth from orbit, proving that orbits can sometimes change dramatically. To understand how and why orbits sometimes change, we need to consider the role of energy in orbits.

Orbital Energy   Consider the orbit of a planet around the Sun. An orbiting planet has both kinetic energy (because it is moving around the Sun) and gravitational potential energy (because it would fall toward the Sun if it stopped orbiting). The planet's kinetic energy depends on its orbital speed, and its gravitational potential energy depends on its distance from the Sun. Because the planet's distance and speed both vary as it orbits the Sun, its gravitational potential energy and kinetic energy also vary (Figure 4.16). However, the planet's total **orbital energy**—the sum of its kinetic and gravitational potential energies—always stays the same. This fact is a consequence of the law of conservation of energy. As long as no other object causes the planet to gain or lose orbital energy, its orbital energy cannot change and its orbit must remain the same.

Orbits cannot change spontaneously—an object's orbit can change only if it gains or loses orbital energy.

Generalizing from planets to other objects leads us to a very important idea about motion throughout the cosmos: *Orbits cannot change spontaneously.* Left undisturbed, planets would forever keep the same orbits around the Sun, moons would keep the same orbits around planets, and stars would keep the same orbits in their galaxies.

Gravitational Encounters   Although orbits cannot change spontaneously, they can change through exchanges of energy. One way that two objects can exchange orbital energy is through a **gravitational**

Figure 4.16

The total orbital energy of a planet stays the same throughout its orbit, because its gravitational potential energy increases when its kinetic energy decreases, and vice versa.

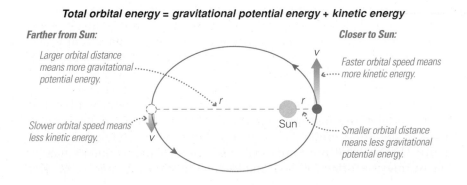

Total orbital energy = gravitational potential energy + kinetic energy

Farther from Sun:
Larger orbital distance means more gravitational potential energy.
Slower orbital speed means less kinetic energy.

Closer to Sun:
Faster orbital speed means more kinetic energy.
Smaller orbital distance means less gravitational potential energy.

Sun

encounter, in which they pass near enough so that each can feel the effects of the other's gravity. For example, in the rare cases in which a comet happens to pass near a planet, the comet's orbit can change dramatically. Figure 4.17 shows a comet headed toward the Sun on an unbound orbit. The comet's close passage by Jupiter allows the comet and Jupiter to exchange energy. In this case, the comet loses so much orbital energy that its orbit changes from unbound to bound and elliptical. Jupiter gains exactly as much energy as the comet loses, but the effect on Jupiter is unnoticeable because of its much greater mass.

Spacecraft engineers can use the same basic idea in reverse. For example, the *New Horizons* spacecraft now en route to Pluto was deliberately sent past Jupiter on a path that allowed it to gain orbital energy at Jupiter's expense. This extra orbital energy sped up the spacecraft so that the trip to Pluto will take four years less than it would have taken otherwise. Of course, the effect of the tiny spacecraft on Jupiter was unnoticeable.

A similar dynamic sometimes occurs naturally and may explain why most comets orbit so far from the Sun. Astronomers think that most comets once orbited in the same region of the solar system as the large outer planets [Section 9.2]. Gravitational encounters with Jupiter or the other large planets then caused some of these comets to be "kicked out" into much more distant orbits around the Sun, or ejected from the solar system completely.

Atmospheric Drag Friction can cause objects to lose orbital energy. For example, consider a satellite orbiting Earth. If the orbit is fairly low—say, just a few hundred kilometers above Earth's surface—the satellite experiences a bit of drag from Earth's thin upper atmosphere. This drag gradually causes the satellite to lose orbital energy until it finally plummets to Earth. The satellite's lost orbital energy is converted to thermal energy in the atmosphere, which is why a falling satellite usually burns up.

Friction may also have played a role in shaping the current orbits of some moons and planets. Some of the small moons of Jupiter and the other outer planets may once have orbited the Sun independently. Their orbits could not have changed spontaneously. However, the outer planets probably once were surrounded by clouds of gas [Section 6.4], and friction would have slowed objects passing through this gas. Some of these small objects may have lost just enough energy to friction to allow them to be "captured" as moons.

Escape Velocity An object that gains orbital energy moves into an orbit with a higher average altitude. For example, if we want to boost the orbital altitude of a spacecraft, we can give it more orbital energy by firing a rocket. The chemical potential energy released by the rocket fuel is converted to orbital energy for the spacecraft.

A spacecraft that achieves escape velocity can escape Earth completely.

If we give a spacecraft enough orbital energy, it may end up in an unbound orbit that allows it to *escape* Earth completely (Figure 4.18). For example, when we send a space probe to Mars, we must use a large rocket that gives the probe enough energy to leave Earth orbit. Although it would probably make more sense to say that the probe achieves "escape energy," we instead say that it achieves **escape velocity**.

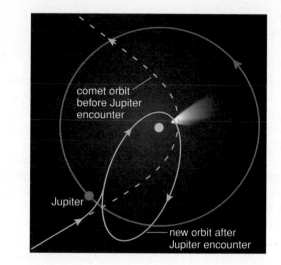

Figure 4.17

This diagram shows a comet in an unbound orbit of the Sun that happens to pass near Jupiter. The comet loses orbital energy to Jupiter, changing its unbound orbit to a bound orbit around the Sun.

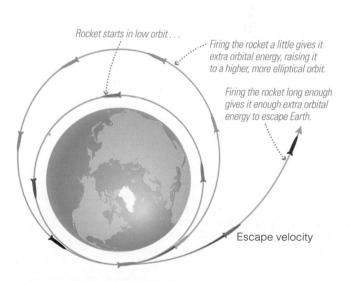

Rocket starts in low orbit . . .

. . . Firing the rocket a little gives it extra orbital energy, raising it to a higher, more elliptical orbit.

Firing the rocket long enough gives it enough extra orbital energy to escape Earth.

Escape velocity

Figure 4.18 interactive figure

If an object orbiting Earth gains orbital energy, it moves to a higher or more elliptical orbit. With enough extra orbital energy, it may achieve escape velocity. Escape velocity depends on how high the object is when it starts. From Earth's surface, escape velocity is about 11 km/s.

The Origin of Tides

Many people believe that tides arise because the Moon pulls Earth's oceans toward it. But if that were the whole story, there would be a bulge only on the side of Earth facing the Moon, and hence only one high tide each day. The correct explanation for tides must account for why Earth has two tidal bulges.

Only one explanation works: Earth must be stretching from its center in both directions (toward and away from the Moon). This stretching force, or tidal force, arises from the difference between the force of gravity attracting different parts of Earth to the Moon. In fact, stretching due to tides affects many objects, not just Earth. Many moons are stretched into slightly oblong shapes by tidal forces caused by their parent planets, and mutual tidal forces stretch close binary stars into teardrop shapes. In regions where gravity is extremely strong, such as near a black hole, tides can have even more dramatic effects (see Chapter 13).

The escape velocity from Earth's surface is about 40,000 km/hr, or 11 km/s, meaning that this is the minimum velocity required to escape Earth's gravity for a spacecraft that starts near the surface.

Notice that the escape velocity does not depend on the mass of the escaping object—*any* object must travel at a velocity of 11 km/s to escape from Earth, whether it is an individual atom or molecule escaping from the atmosphere, a spacecraft being launched into deep space, or a rock blasted into the sky by a large impact. Escape velocity *does* depend on whether you start from the surface or from someplace high above the surface. Because gravity weakens with distance, it takes less energy—and hence a lower escape velocity—to escape from a point high above Earth than from Earth's surface.

• How does gravity cause tides?

Newton's universal law of gravitation has applications that go far beyond explaining Kepler's laws and orbits. For our purposes, however, there is just one more topic we need to cover: how gravity causes tides.

If you've spent time near an ocean, you've probably observed the rising and falling of the tides. In most places, tides rise and fall twice each day. Tides arise because gravity attracts Earth and the Moon toward each other (with the Moon staying in orbit as it "falls around" Earth), but it affects different parts of Earth slightly differently: Because the strength of gravity declines with distance, the gravitational attraction of each part of Earth to the Moon becomes weaker as we go from the side of Earth facing the Moon to the side facing away from the Moon. This difference in attraction creates a "stretching force," or **tidal force**, that stretches the entire Earth to create two tidal bulges—one facing the Moon and one opposite the Moon (Figure 4.19). If you are still unclear about why there are *two* tidal bulges, think about a rubber band: If you pull on a rubber band it will stretch in both directions relative to its center, even if you pull on only one side. In the same way, Earth stretches on both sides even though the Moon is tugging harder on only one side.

Tides affect both land and ocean, but we generally notice only the ocean tides because water flows much more readily than land. Earth's rotation carries any location through each of the two bulges each day, creating two high tides. Low tides occur when the location is at the points

Figure 4.19

Tides are created by the difference in the force of attraction between different parts of Earth and the Moon. There are two daily high tides as any location on Earth rotates through the two tidal bulges. (The diagram highly exaggerates the tidal bulges, which raise the oceans only about 2 meters and the land only about a centimeter.)

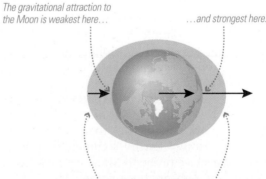

The gravitational attraction to the Moon is weakest here...   ...and strongest here.

The difference in gravitational attraction tries to pull Earth apart, raising tidal bulges both toward and away from the Moon.

Not to scale!

Why Does the Moon Always Show the Same Face to Earth?

YOU ARE PROBABLY aware that we always see (nearly) the same face of the Moon. This happens because the Moon rotates on its axis in exactly the same time period that it takes to orbit Earth, a trait called **synchronous rotation**. A simple demonstration shows this idea (Figure 1). Place a ball on a table to represent Earth while you represent the Moon. The only way you can face the ball at all times is by completing exactly one rotation while you complete one orbit. But *why* does the Moon have this synchronous rotation? We can trace the answer directly to tides.

a If you do not rotate while walking around the model, you will not always face it.

b You will face the model at all times only if you rotate exactly once during each orbit.

Figure 1

The fact that we always see the same face of the Moon means that the Moon must rotate once in the same amount of time that it takes to orbit Earth once. You can see why by walking around a model of Earth while imagining that you are the Moon.

It's easiest to start by considering the effects of tides on Earth. So far, we have talked as if Earth rotates smoothly through the tidal bulges. But because tidal forces stretch Earth itself, the process causes some friction, called *tidal friction*. Figure 2 shows the effects of this friction. In essence, the Moon's gravity tries to keep the tidal bulges on the Earth–Moon line, while Earth's rotation tries to pull the bulges around with it. The resulting "compromise" keeps the bulges just ahead of the Earth–Moon line at all times, which causes two important effects. First, the Moon's gravity always pulls back on the bulges, slowing Earth's rotation. Second, the gravity of the bulges pulls the Moon slightly ahead in its orbit, causing the Moon to move farther from Earth. These effects are barely noticeable on human time scales, but they add up over bil-

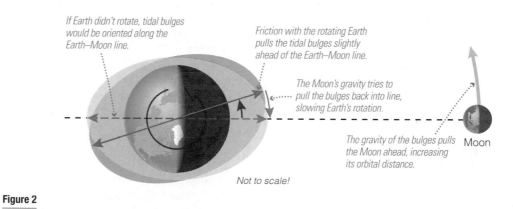

If Earth didn't rotate, tidal bulges would be oriented along the Earth–Moon line.

Friction with the rotating Earth pulls the tidal bulges slightly ahead of the Earth–Moon line.

The Moon's gravity tries to pull the bulges back into line, slowing Earth's rotation.

The gravity of the bulges pulls the Moon ahead, increasing its orbital distance. Moon

Not to scale!

Figure 2

Earth's rotation pulls its tidal bulges slightly ahead of the Earth–Moon line, leading to gravitational effects that gradually slow Earth's rotation and increase the Moon's orbital distance.

lions of years. Early in Earth's history, a day may have been only 5 or 6 hours long and the Moon may have been one-tenth or less of its current distance from Earth. These changes also provide a great example of conservation of angular momentum: The Moon's growing orbit gains the angular momentum that Earth loses as its rotation slows.

Now, let's turn the situation around to see how tides affect the Moon. Because Earth is more massive than the Moon, Earth exerts a greater tidal force on the Moon than the Moon does on Earth. This tidal force gives the Moon two tidal bulges along the Earth–Moon line, much like the two tidal bulges that the Moon creates on Earth. (The Moon does not have visible tidal bulges, but it does indeed have excess mass along the Earth–Moon line.) As a result, if the Moon were rotating through its tidal bulges in the same way that Earth rotates through its tidal bulges, the resulting friction would cause the Moon's rotation to slow down. This is exactly what we think happened long ago.

The Moon probably once rotated much faster than it does today. As a result, it *did* rotate through its tidal bulges, and its rotation gradually slowed. Once the Moon's rotation slowed to the point at which the Moon and its bulges rotated at the same rate—that is, synchronously with the orbital period—there was no further source for tidal friction. The Moon's synchronous rotation therefore was a natural outcome of Earth's tidal effects on the Moon.

Similar tidal friction has led to synchronous rotation in many other cases. For example, Jupiter's four large moons (Io, Europa, Ganymede, and Callisto) keep nearly the same face toward Jupiter at all times, as do many other moons. Pluto and its moon Charon *both* rotate synchronously: Like two dancers, they always keep the same face toward each other. Many binary star systems also rotate in this way. Tidal forces may be most familiar because of their effects on our oceans, but they are important throughout the universe.

Figure 4.20

Photographs of high and low tide at the abbey of Mont-Saint-Michel, France. Here the tide rushes in much faster than a person can swim. Before a causeway was built (visible to the left), the Mont was accessible by land only at low tide. At high tide, it became an island.

Spring tides occur at new moon and full moon:

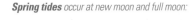

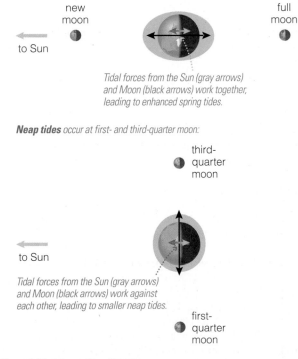

Tidal forces from the Sun (gray arrows) and Moon (black arrows) work together, leading to enhanced spring tides.

Neap tides occur at first- and third-quarter moon:

Tidal forces from the Sun (gray arrows) and Moon (black arrows) work against each other, leading to smaller neap tides.

Figure 4.21 interactive figure ↖

The Sun exerts a tidal force on Earth about one-third as strong as that from the Moon. When the tidal forces from the Sun and Moon work together at new moon and full moon, we get enhanced *spring tides*. When they work against each other, at first- and third-quarter moons, we get smaller *neap tides*.

halfway between the two tidal bulges. The height and timing of ocean tides can vary considerably from place to place on Earth. For example, while the tide rises gradually in most locations, the incoming tide near the famous abbey on Mont-Saint-Michel, France, moves much faster than a person can swim (Figure 4.20). In centuries past, the Mont was an island twice a day at high tide but was connected to the mainland at low tide. Many pilgrims drowned when they were caught unprepared by the tide rushing in. Another unusual tidal pattern occurs in coastal states along the northern shore of the Gulf of Mexico, where topography and other factors combine to make only one noticeable high tide and low tide each day.

> Tidal forces cause the entire Earth to stretch along the Earth–Moon line, creating two tidal bulges.

The Sun also affects the tides. Although the Sun is much more massive than the Moon, its tidal effect on Earth is smaller because its much greater distance means that the *difference* in the Sun's pull on the near and far sides of Earth is relatively small. The overall tidal force caused by the Sun is about one-third that caused by the Moon (Figure 4.21). When the tidal forces of the Sun and the Moon work together, as is the case at both new moon and full moon, we get the especially pronounced *spring tides* (so named because the water tends to "spring up" from the Earth). When the tidal forces of the Sun and the Moon counteract each other, as is the case at first- and third-quarter moon, we get the relatively small tides known as *neap tides*.

Tidal forces affect not only Earth, but also many other objects. Earth exerts tidal forces on the Moon that explain why the Moon always shows the same face to Earth (see Special Topic on page 103), and in Chapter 9 we'll see how tidal forces have led to the astonishing volcanic activity of Jupiter's moon Io and the possibility of a subsurface ocean on its moon Europa.

think about it → Explain why any tidal effects on Earth caused by the other planets would be unnoticeably small.

the big picture

Putting Chapter 4 into Context

We've covered a lot of ground in this chapter, from the scientific terminology of motion to the overarching principles that govern motion throughout the universe. Be sure you understand the following "big picture" ideas:

- Understanding the universe requires understanding motion. Motion may seem complex, but it can be described simply using Newton's three laws of motion.

- Today, we know that Newton's laws of motion stem from deeper physical principles, including the laws of conservation of angular momentum and of energy. These principles enable us to understand a wide range of astronomical phenomena.

- Newton also discovered the universal law of gravitation, which explains how gravity holds planets in their orbits and much more—including how satellites can reach and stay in orbit, the nature of tides, and why the Moon rotates synchronously around Earth.

- Perhaps most important, Newton's discoveries showed that the same physical laws we observe on Earth apply throughout the universe.

summary of key concepts

4.1 Describing Motion: Examples from Daily Life

• How do we describe motion?
Speed is the rate at which an object is moving. **Velocity** is speed in a certain direction. **Acceleration** is a change in velocity, meaning a change in either speed or direction. **Momentum** is mass $\times$ velocity. A **force** can change an object's momentum, causing it to accelerate.

• How is mass different from weight?

An object's **mass** is the same no matter where it is located, but its **weight** varies with the strength of gravity or other forces acting on the object. An object becomes **weightless** when it is in **free-fall**, even though its mass is unchanged.

4.2 Newton's Laws of Motion

• How did Newton change our view of the universe?
Newton showed that the same physical laws that operate on Earth also operate in the heavens, making it possible to learn about the universe by studying physical laws on Earth.

• What are Newton's three laws of motion?
(1) An object moves at constant velocity if there is no net force acting upon it. (2) Force = mass $\times$ acceleration ($F = ma$). (3) For any force, there is always an equal and opposite reaction force.

4.3 Conservation Laws in Astronomy

• What keeps a planet rotating and orbiting the Sun?

Conservation of angular momentum means that a planet's rotation and orbit cannot change unless it transfers angular momentum to another object. The planets in our solar system do not exchange substantial angular momentum with each other or anything else, so their orbits and rotation rates remain quite steady.

• Where do objects get their energy?

kinetic energy

radiative energy   potential energy

Energy is always conserved—it can be neither created nor destroyed. Objects received whatever energy they now have from exchanges of energy with other objects. Energy comes in three basic categories—**kinetic**, **radiative**, and **potential**—though sometimes we use subcategories for convenience.

4.4 The Force of Gravity

• What determines the strength of gravity?

According to the **universal law of gravitation**, every object attracts every other object with a gravitational force that is directly proportional to the product of the objects' masses and declines with the square of the distance between their centers:

$$F_g = G \frac{M_1 M_2}{d^2}$$

• How does Newton's law of gravity extend Kepler's laws?

(1) Newton showed that Kepler's first two laws apply to all orbiting objects, not just planets. (2) He showed that elliptical **bound orbits** are not the only possible orbital shape—orbits can also be **unbound** (taking the shape of a parabola or a hyperbola). (3) **Newton's version of Kepler's third law** allows us to calculate the masses of orbiting objects from their orbital periods and distances.

• How do gravity and energy allow us to understand orbits?

Gravity determines orbits, and an object cannot change its orbit unless it gains or loses **orbital energy**—the sum of its kinetic and gravitational potential energy—through energy transfer with other objects. If an object gains enough orbital energy, it may achieve **escape velocity** and leave the gravitational influence of the object it was orbiting.

• How does gravity cause tides?

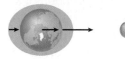

The Moon's gravity creates a **tidal force** that stretches Earth along the Earth–Moon line, causing Earth to bulge both toward and away from the Moon. Earth's rotation carries us through each of the two bulges each day, giving us two daily high tides and two daily low tides.

exercises and problems

For instructor-assigned homework go to **www.masteringastronomy.com.**

Review Questions

Short-Answer Questions Based on the Reading

1. How does *speed* differ from *velocity?* Give an example in which you can be traveling at constant speed but not at constant velocity.
2. What do we mean by *acceleration?* What is the *acceleration of gravity?* Explain what we mean when we state an acceleration in units of meters per second squared (m/s²).
3. What is *momentum?* How can momentum be affected by a *force?* What do we mean when we say that momentum can be changed only by a *net force?*
4. What is *free-fall,* and why does it make you *weightless?* Briefly describe why astronauts are weightless in the Space Station.
5. State *Newton's three laws of motion.* For each law, give an example of its application.
6. What are the laws of *conservation of momentum, conservation of angular momentum,* and *conservation of energy?* For each, give an example of how it is important in astronomy.
7. Define *kinetic energy, radiative energy,* and *potential energy.* For each type of energy, give at least two examples of objects that either have it or use it.
8. Define *temperature* and *thermal energy.* How are they related? How are they different?
9. Which has more gravitational potential energy: a rock on the ground or a rock that you hold out the window of a 10-story building? Explain.
10. What do we mean by *mass-energy?* Is it a form of kinetic, radiative, or potential energy? How is the idea of mass-energy related to the formula $E = mc^2$?
11. Summarize the *universal law of gravitation* in words. Then state the law mathematically, explaining the meaning of each symbol in the equation.
12. What is the difference between *bound orbits* and *unbound orbits?*
13. What do we need to know if we want to measure an object's mass with *Newton's version of Kepler's third law?* Explain.
14. Explain why orbits cannot change spontaneously. How can atmospheric drag affect an orbit? How can a *gravitational encounter* cause an orbit to change? How can an object achieve *escape velocity?*
15. Explain how the Moon creates tides on Earth. Why do we have two high and low tides each day?
16. How do the tides vary with the phase of the Moon? Why?

Test Your Understanding

Does It Make Sense?

Decide whether the statement makes sense (or is clearly true) or does not make sense (or is clearly false). Explain clearly; not all these have definitive answers, so your explanation is more important than your chosen answer.

17. If you could buy a pound of chocolate on the Moon, you'd get a lot more chocolate than if you bought a pound on Earth. (*Hint:* Pounds are a unit of weight, not mass.)

18. Suppose you could enter a vacuum chamber (on Earth), that is, a chamber with no air in it. Inside this chamber, if you dropped a hammer and a feather from the same height at the same time, both would hit the bottom at the same time.

19. When an astronaut goes on a space walk outside the Space Station, she will quickly float away from the station unless she has a tether holding her to the station or constantly fires thrusters on her space suit.

20. I used Newton's version of Kepler's third law to calculate Saturn's mass from orbital characteristics of its moon Titan.

21. If the Sun was magically replaced with a giant rock that had precisely the same mass, Earth's orbit would not change.

22. The fact that the Moon rotates once in precisely the time it takes to orbit Earth once is such an astonishing coincidence that scientists probably never will be able to explain it.

23. Venus has no oceans, so it could not have tides even if it had a moon (which it doesn't).

24. If an asteroid passed by Earth at just the right distance, Earth's gravity would capture it and make it our second moon.

25. When I drive my car at 30 miles per hour, it has more kinetic energy than it does at 10 miles per hour.

26. Someday soon, scientists are likely to build an engine that produces more energy than it consumes.

Quick Quiz

Choose the best answer to each of the following. Explain your reasoning with one or more complete sentences.

27. Which one of the following describes an object that is accelerating? (a) A car traveling on a straight, flat road at 50 miles per hour. (b) A car traveling on a straight uphill road at 30 miles per hour. (c) A car going around a circular track at a steady 100 miles per hour.

28. Suppose you visit another planet: (a) Your mass and weight would be the same as they are on Earth. (b) Your mass would be the same as on Earth, but your weight would be different. (c) Your weight would be the same as on Earth, but your mass would be different.

29. Which person is weightless? (a) a child in the air as she plays on a trampoline (b) a scuba diver exploring a deep-sea wreck (c) an astronaut on the Moon

30. Consider the statement "There's no gravity in space." This statement is (a) completely false. (b) false if you are close to a planet or moon, but true in between the planets. (c) completely true.

31. To make a rocket turn left, you need to (a) fire an engine that shoots out gas to the left. (b) fire an engine that shoots out gas to the right. (c) spin the rocket clockwise.

32. Compared to its angular momentum when it is farthest from the Sun, Earth's angular momentum when it is nearest to the Sun is (a) greater. (b) less. (c) the same.

33. The gravitational potential energy of a contracting interstellar cloud (a) stays the same at all times. (b) gradually transforms into other forms of energy. (c) gradually grows larger.

34. If Earth were twice as far from the Sun, the force of gravity attracting Earth to the Sun would be (a) twice as strong. (b) half as strong. (c) one-quarter as strong.

35. According to the law of universal gravitation, what would happen to Earth if the Sun were somehow replaced by a black hole of the same mass? (a) Earth would be quickly sucked into the black hole. (b) Earth would slowly spiral into the black hole. (c) Earth's orbit would not change.

36. If the Moon were closer to Earth, high tides would (a) be higher than they are now. (b) be lower than they are now. (c) occur three or more times a day rather than twice a day.

Process of Science

Examining How Science Works

37. *Testing Gravity.* Scientists are constantly trying to learn whether our current understanding of gravity is complete or must be modified. Describe how the observed motion of spacecraft headed out of the solar system (such as the *Voyager* spacecraft) can be used to test the accuracy of our current theory of gravity.

38. *How Does the Table Know?* Thinking deeply about seemingly simple observations sometimes reveals underlying truths that we might otherwise miss. For example, think about holding a golf ball in one hand and a bowling ball in the other. To keep them motionless, you must actively adjust the tension in your arm muscles so that each arm exerts a different upward force that exactly balances the weight of each ball. Now, think about what happens when you set the balls on a table. Somehow, the table exerts exactly the right amount of upward force to keep the balls motionless, even though their weights are very different. How does a table "know" to make the same type of adjustment that you make consciously when you hold the balls motionless in your hands? (*Hint:* Think about the origin of the force pushing upward on the objects.)

Investigate Further

In-Depth Questions to Increase Your Understanding

Short-Answer/Essay Questions

39. *Weightlessness.* Astronauts are weightless when in orbit in the Space Shuttle. Are they also weightless during the Shuttle's launch? How about during its return to Earth? Explain.

40. *Einstein's Famous Formula.*
    a. What is the meaning of the formula $E = mc^2$? Be sure to define each variable.
    b. How does this formula explain the generation of energy by the Sun?
    c. How does this formula explain the destructive power of nuclear bombs?

41. *The Gravitational Law.*
    a. How does quadrupling the distance between two objects affect the gravitational force between them?
    b. Suppose the Sun were somehow replaced by a star with twice as much mass. What would happen to the gravitational force between Earth and the Sun?
    c. Suppose Earth were moved to one-third of its current distance from the Sun. What would happen to the gravitational force between Earth and the Sun?

42. *Allowable Orbits?*
    a. Suppose the Sun were replaced by a star with twice as much mass. Could Earth's orbit stay the same? Why or why not?
    b. Suppose Earth doubled in mass (but the Sun stayed the same as it is now). Could Earth's orbit stay the same? Why or why not?

43. *Head-to-Foot Tides.* You and Earth attract each other gravitationally, so you should also be subject to a tidal force resulting from the difference between the gravitational attraction felt by your feet and that felt by your head (at least when you are standing). Explain why you can't feel this tidal force.

Quantitative Problems

Be sure to show all calculations clearly and state your final answers in complete sentences.

44. *Energy Comparisons.* Use the data in Table 4.1 to answer each of the following questions.
    a. Compare the energy of a 1-megaton H-bomb to the energy released by a major earthquake.
    b. If the United States obtained all its energy from oil, how much oil would be needed each year?
    c. Compare the Sun's annual energy output to the energy released by a supernova.

45. *Fusion Power.* No one has yet succeeded in creating a commercially viable way to produce energy through nuclear fusion. However, suppose we could build fusion power plants using the hydrogen in water as a fuel. Based on the data in Table 4.1, how much water would we need each minute to meet U.S. energy needs? Could such a reactor power the entire United States with the water flowing from your kitchen sink? Explain. (*Hint:* Use the annual U.S. energy consumption to find the energy consumption per minute, and then divide by the energy yield from fusing 1 liter of water to figure out how many liters would be needed each minute.)

46. *Understanding Newton's Version of Kepler's Third Law I.* Imagine another solar system, with a star of the same mass as the Sun. Suppose there is a planet in that solar system with a mass twice that of Earth orbiting at a distance of 1 AU from the star. What is the orbital period of this planet? Explain. (*Hint:* The calculations for this problem are so simple that you will not need a calculator.)

47. *Understanding Newton's Version of Kepler's Third Law II.* Suppose a solar system has a star that is four times as massive as our Sun. If that solar system has a planet the same size as Earth orbiting at a distance of 1 AU, what is the orbital period of the planet? Explain. (*Hint:* The calculations for this problem are so simple that you will not need a calculator.)

48. *Using Newton's Version of Kepler's Third Law I.*
    a. The Moon orbits Earth in an average time of 27.3 days at an average distance of 384,000 km. Use these facts to determine the mass of Earth. (*Hint:* You may neglect the mass of the Moon, since its mass is only about $\frac{1}{80}$ of Earth's.)
    b. Jupiter's moon Io orbits Jupiter every 42.5 hours at an average distance of 422,000 km from the center of Jupiter. Calculate the mass of Jupiter. (*Hint:* Io's mass is very small compared to Jupiter's.)
    c. You discover a planet orbiting a distant star that has about the same mass as the Sun. Your observations show that the planet orbits the star every 63 days. What is its orbital distance?

49. *Using Newton's Version of Kepler's Third Law II.*
    a. Pluto's moon Charon orbits Pluto every 6.4 days with a semimajor axis of 19,700 kilometers. Calculate the *combined* mass of Pluto and Charon. Compare this combined mass to the mass of Earth, which is about 6×10^{24} kg.
    b. Calculate the orbital period of the Space Shuttle in an orbit 300 kilometers above Earth's surface.
    c. The Sun orbits the center of the Milky Way Galaxy every 230 million years at a distance of 28,000 light-years. Use these facts to determine the mass of the galaxy. (As we'll discuss in Chapter 14, this calculation actually tells us only the mass of the galaxy *within* the Sun's orbit.)

Discussion Questions

50. *Knowledge of Mass-Energy.* Einstein's discovery that energy and mass are equivalent has led to technological developments that are both beneficial and dangerous. Discuss some of these developments. Overall, do you think the human race would be better or worse off if we had never discovered that mass is a form of energy? Defend your opinion.

51. *Perpetual Motion Machines.* Every so often, someone claims to have built a machine that can generate energy perpetually from nothing. Why isn't this possible according to the known laws of nature? Why do you think claims of perpetual motion machines sometimes receive substantial media attention?

Web Projects

52. *Space Station.* Visit a NASA site with pictures from the Space Station. Choose two photos that illustrate some facet of Newton's laws. Explain how Newton's laws apply to each photo.

53. *Energy Comparisons.* Using information from the U.S. Energy Information Administration Web site, choose some aspect of national or international energy use that interests you. Write a short report on the topic.

54. *Nuclear Power.* There are two basic ways to generate energy from atomic nuclei: through nuclear fission (splitting nuclei) and through nuclear fusion (combining nuclei). All current nuclear reactors are based on fission, but using fusion would have many advantages if we could develop the technology. Research some of the advantages of fusion and some of the obstacles to developing fusion power. Do you think fusion power will be a reality in your lifetime? Explain.

Use the following questions to check your understanding of some of the many types of visual information used in astronomy. Answers are provided in Appendix K. For additional practice, try the Chapter 4 Visual Quiz at **www.masteringastronomy.com**.

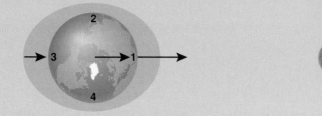

The figure above, based on Figure 4.19, shows how the Moon causes tides on Earth. Note that the North Pole is in the center of the diagram, so the numbers 1 through 4 label points along Earth's equator.

1. What do the three black arrows represent?
   a. the tidal force Earth exerts on the Moon
   b. the Moon's gravitational force at different points on Earth
   c. the direction in which Earth's water is flowing
   d. Earth's orbital motion
2. Where is it high tide?
   a. Point 1 only
   b. Point 2 only
   c. Points 1 and 3
   d. Points 2 and 4
3. Where is it low tide?
   a. Point 1 only
   b. Point 2 only
   c. Points 1 and 3
   d. Points 2 and 4

4. What time is it at Point 1?
   a. noon
   b. midnight
   c. 6 a.m.
   d. cannot be determined from the information in the figure
5. The light blue region represents tidal bulges. In what way are these bulges drawn inaccurately?
   a. There should be only one bulge rather than two.
   b. They should be aligned with the Sun rather than the Moon.
   c. They should be much smaller compared to Earth.
   d. They should be more pointy in shape.

5

light:
the cosmic messenger

learning goals

Ancient observers could discern only the most basic features of the light that they saw, such as color and brightness. Over the past several hundred years, we have discovered that light carries far more information. Today, we can analyze the light of distant objects to learn what they are made of, how hot they are, how fast they are moving, and much more. Light is truly the cosmic messenger, bringing the stories of distant objects to Earth.

Understanding the messages carried by light requires familiarity with the way light and matter interact. In this chapter, we'll explore the basic properties of light and matter that allow us to learn so much about the universe by studying light from distant objects. We'll also discuss how telescopes are used to collect light, and the technologies that make telescopes so much more powerful than our eyes.

essential preparation

1. How did Galileo solidify the Copernican revolution? [Section 3.3]

2. What are Newton's three laws of motion? [Section 4.2]

3. Where do objects get their energy? [Section 4.3]

 Light and Spectroscopy Tutorial, Lesson 1

5.1 Basic Properties of Light and Matter

The photograph that opens this chapter shows a detailed view of the Sun's **spectrum**—the light from the Sun as it appears when we pass it through a prism or similar device. The rainbow of color, which stretches in horizontal rows from the upper left to the lower right of the photograph, probably reminds you of what we see whenever we pass white light through a prism (Figure 5.1). However, notice that the Sun's spectrum is not a pure rainbow. Instead, its spectrum shows hundreds of dark lines, representing places where a small piece of the rainbow is missing from the sunlight. All the features of the spectrum, including the rainbow and the dark lines, are created by interactions between light and matter in the Sun. Careful study of these features can tell us the Sun's chemical composition, its temperature, the motions of its atmosphere, and more.

We see similar dark or bright lines when we look at almost any spectrum in detail, whether it is the spectrum of the flame from a backyard gas grill or the spectrum of a distant galaxy whose light we collect with a gigantic telescope. As long as we collect enough light to see details in the spectrum, we can learn many fundamental properties of the object we are viewing, no matter how far away it is located.

Our primary goal in this chapter is to understand how we can learn about distant objects from their spectra. But before we discuss the information encoded in spectra, we must first clarify the nature of light and matter.

• What is light?

Light is familiar to all of us, but its nature remained a mystery until quite recently in human history. Experiments performed by Isaac Newton in the 1660s provided the first real insights into the nature of light. It was already known that passing white light through a prism produced a rainbow of color, but many people thought the colors came from the prism rather than from the light itself. Newton proved that the colors came from the light by placing a second prism in front of the light of just one color,

Figure 5.1

When we pass white light through a prism, it disperses into a rainbow of color that we call a *spectrum*.

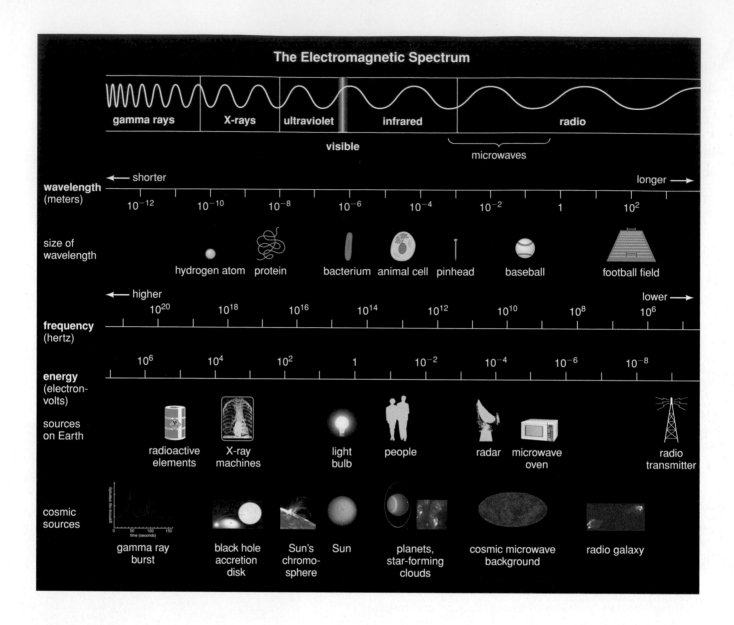

The Electromagnetic Spectrum

gamma rays | X-rays | ultraviolet | infrared | radio

visible

microwaves

wavelength (meters)

← shorter   longer →

10^{-12} 10^{-10} 10^{-8} 10^{-6} 10^{-4} 10^{-2} 1 10^{2}

size of wavelength

hydrogen atom   protein   bacterium   animal cell   pinhead   baseball   football field

frequency (hertz)

← higher   lower →

10^{20} 10^{18} 10^{16} 10^{14} 10^{12} 10^{10} 10^{8} 10^{6}

energy (electron-volts)

10^{6} 10^{4} 10^{2} 1 10^{-2} 10^{-4} 10^{-6} 10^{-8}

sources on Earth

radioactive elements   X-ray machines   light bulb   people   radar   microwave oven   radio transmitter

cosmic sources

gamma ray burst   black hole accretion disk   Sun's chromosphere   Sun   planets, star-forming clouds   cosmic microwave background   radio galaxy

Figure 5.2 interactive figure

The electromagnetic spectrum. Notice that wavelength increases as we go from gamma rays to radio waves, while frequency and energy increase in the opposite direction.

such as red, from the first prism. If the rainbow of color came from the prism itself, the second prism would have produced a rainbow just like the first. But it did not: When only red light entered the second prism, only red light emerged, proving that the color was a property of the light and not of the prism.

Light is also known as electromagnetic radiation.

Newton's experiment proved that *white* light is actually a mix of all the colors in the rainbow. Later scientists found that there is light "beyond the rainbow" as well. Just as there are sounds that our ears cannot hear (such as the sound of a dog whistle), there is light that our eyes cannot see. In fact, the **visible light** that splits into the rainbow of color is only a tiny part of the complete spectrum of light. Figure 5.2 shows this complete spectrum, usually called the **electromagnetic spectrum**. Light itself is often called **electromagnetic radiation**. Let's investigate why.

Wave Properties of Light You've probably heard that light is a wave, but what exactly does that mean? In general, a wave is something that can transmit energy without carrying material along with it. For example, you can make waves move along a rope by shaking one end of it up and down (Figure 5.3a). The shaking creates a series of peaks and troughs that move along the rope, making every piece of the rope bob up and down as the peaks and troughs go by. We define the **wavelength** as the distance between adjacent peaks and the **frequency** as the number of times that any piece of the rope moves up and down each second. For example, if a piece of the rope moves up and down three times each second, we say the wave has a frequency of three cycles per second, or three *hertz* for short. Notice that the rope itself stays intact as the wave moves along it, showing that it is energy and not material that is moving with the wave.

Light is different from waves on a rope because we cannot see anything moving up and down as it travels. However, we can tell that light is a wave from its effect on matter. If you could set up a row of electrically charged particles such as electrons, it would wriggle like a snake as a wave of light passed by (Figure 5.3b). The distance between adjacent peaks in this row of electrons would tell us the wavelength of the light wave, while the number of times each electron bobs up and down would tell us the frequency (Figure 5.3c). Because light can affect both electrically charged particles and magnets, we say that light is an **electromagnetic wave**—which is why light is called *electromagnetic radiation* and the spectrum of light is called the *electromagnetic spectrum*.

The longer the wavelength of light, the lower its frequency and energy.

All light travels through empty space at the same speed—the **speed of light**—which is about 300,000 kilometers per second. Because the speed of any wave is its wavelength times its frequency, we find an important relationship between wavelength and frequency for light: *The longer the wavelength, the lower the frequency, and vice versa* (Figure 5.4). For example, gamma rays have the shortest wavelengths and the highest frequencies of any form of light (see Figure 5.2).

Particle Properties of Light In everyday life, waves seem to be quite different from particles. A wave exists only as a pattern of motion with a wavelength and a frequency, while a particle is a "thing" such as a marble, a baseball, or an individual atom. However, experiments show that light can behave *both* as a wave and as a particle.

Light comes in "pieces" called photons, each with a precise wavelength, frequency, and energy.

The idea that light can be both a wave and a particle may seem quite strange, but it is fundamental to our modern understanding of physics. We think of light as consisting of many individual "pieces," or **photons**. Like baseballs, photons of light can be counted individually and can hit a wall one at a time. Like waves, each photon travels at the speed of light and is characterized by a wavelength and frequency. Moreover, each photon carries a particular amount of energy that depends on its frequency: the higher the frequency of the photon, the more energy it carries. That is why energy increases in the same direction as frequency in Figure 5.2.

think about it How does the energy of a photon depend on its wavelength? Briefly explain why.

The Many Forms of Light Figure 5.2 also shows that we give special names to different portions of the electromagnetic spectrum. Visible light has wavelengths ranging from about 400 nm at the blue or violet end of the rainbow to about 700 nm at the red end. (A nanometer [nm] is a

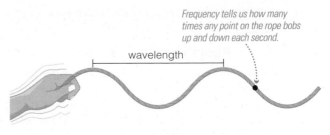
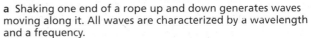

a Shaking one end of a rope up and down generates waves moving along it. All waves are characterized by a wavelength and a frequency.

b If you could line up electrons, they would wriggle up and down as light passes by, demonstrating that light is a wave.

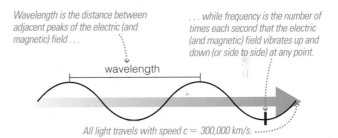

c Light can affect both electrically charged particles and magnets, so we say that light is an *electromagnetic wave*.

Figure 5.3 interactive figure

These diagrams explain the wave properties of light.

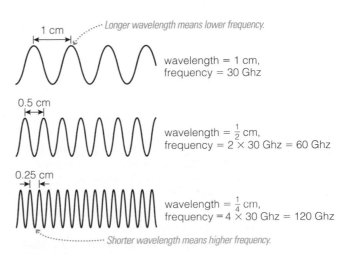

wavelength = 1 cm, frequency = 30 Ghz

wavelength = $\frac{1}{2}$ cm, frequency = 2 × 30 Ghz = 60 Ghz

wavelength = $\frac{1}{4}$ cm, frequency = 4 × 30 Ghz = 120 Ghz

Figure 5.4

Because all light travels through space at the same speed, light of longer wavelength must have lower frequency, and vice versa. (GHz stands for gigahertz, or 10^9 hertz.)

⌒ **common misconceptions**
◉

Is Radiation Dangerous?

Many people associate the word *radiation* with danger. However, the word *radiate* simply means "to spread out from a center" (note the similarity between *radiation* and *radius* [of a circle]). Radiation is energy being carried through space. If energy is being carried by particles of matter, such as protons or neutrons, we call it *particle radiation*. If energy is being carried by light, we call it *electromagnetic radiation*.

High-energy forms of radiation, such as particles from radioactive materials or X rays, are dangerous because they can penetrate body tissues and cause cell damage. Low-energy forms of radiation, such as radio waves, are usually harmless. The visible-light radiation from the Sun sustains life on Earth. Thus, while some forms of radiation are dangerous, others are harmless or beneficial.

⌒ **common misconceptions**
◉

Can You Hear Radio or See an X Ray?

Most people associate the term *radio* with sound, but radio waves are a form of *light* with long wavelengths—too long for our eyes to see. Radio stations encode sounds (such as voices and music) as electrical signals, which they broadcast as radio waves. What we call "a radio" in daily life is an electronic device that receives these radio waves and decodes them to re-create the sounds played at the radio station. Televisions, cell phones, and other wireless devices also work by encoding and decoding information in the form of light called radio waves.

X rays are also a form of light, with wavelengths far too short for our eyes to see. In a doctor's or dentist's office, a special machine works somewhat like the flash on an ordinary camera but emits X rays instead of visible light. This machine flashes the X rays at you, and a piece of photographic film or an electronic detector records the X rays that are transmitted through your body. You never see the X rays themselves—you see only the image recorded by the film or detector.

billionth of a meter.) Light with wavelengths somewhat longer than red light is called **infrared**, because it lies beyond the red end of the rainbow. **Radio waves** are the longest-wavelength light. That is, radio waves are a form of light, *not* a form of sound. The region near the border between infrared and radio waves, where wavelengths range from micrometers to millimeters, is sometimes given the name **microwaves**.

On the other side of the spectrum, light with wavelengths somewhat shorter than blue light is called **ultraviolet**, because it lies beyond the blue (or violet) end of the rainbow. Light with even shorter wavelengths is called **X rays**, and the shortest-wavelength light is called **gamma rays**. Notice that visible light is an extremely small part of the entire electromagnetic spectrum: The reddest red that our eyes can see has only about twice the wavelength of the bluest blue, but the radio waves from your favorite radio station are a billion times as long as the X rays used in a doctor's office.

Radio waves, microwaves, infrared, visible light, ultraviolet, X rays, and gamma rays are all forms of light.

The different energies of different forms of light explain many familiar effects in everyday life. Radio waves carry so little energy that they have no noticeable effect on our bodies. However, radio waves can make electrons move up and down in an antenna, which is how your car radio receives the radio waves coming from a radio station. Molecules moving around in a warm object emit infrared light, which is why we sometimes associate infrared light with heat. Receptors in our eyes respond to visible-light photons, making vision possible. Ultraviolet photons carry enough energy to harm cells in our skin, causing sunburn or skin cancer. X-ray photons have enough energy to penetrate through skin and muscle but can be blocked by bones or teeth. That is why doctors and dentists can see our bone and tooth structures on photographs taken with X-ray light.

• What is matter?

Light carries information about matter across the universe, but we are usually more interested in the matter the light is coming from than we are in the light itself. Planets, stars, and galaxies are made of matter, and we must understand the nature of matter if we are to decode the messages we receive in light.

Like the nature of light, the nature of matter remained mysterious for most of human history. The ancient Greeks imagined that all material was made of four elements: fire, water, earth, and air. Some Greeks, beginning with the philosopher Democritus (c. 470–380 B.C.), further imagined that these four elements came in the form of tiny particles they called *atoms,* a Greek term meaning "indivisible." Our modern ideas of atoms differ in many details from the ideas of the ancient Greeks. For example, we now know of more than 100 types of atoms, or chemical **elements**, and fire, water, earth, and air are *not* among them. Some of the most familiar elements are hydrogen, helium, carbon, oxygen, silicon, iron, gold, silver, lead, and uranium. (See Appendix D for a complete list.)

Atomic Structure Each chemical element represents a different type of **atom**, and atoms are in turn made of particles that we call **protons**, **neutrons**, and **electrons** (Figure 5.5). Protons and neutrons are found in the tiny **nucleus** at the center of the atom. The rest of the atom's volume contains the electrons that surround the nucleus. Although the nucleus is very small compared to the atom as a whole, it contains most of the atom's mass, because protons and neutrons are each about 2000 times as massive as an electron. Note that atoms are incredibly small: Millions could fit end to end across the period at the end of this sentence. The number of atoms in a single drop of water (typically, 10^{22} to 10^{23} atoms) may exceed the number of stars in the observable universe.

The chemical elements are made of atoms, which in turn are made of protons, neutrons, and electrons.

The properties of an atom depend mainly on the **electrical charge** in its nucleus. Electrical charge is a fundamental physical property that is always conserved, just as energy is always conserved. We define the electrical charge of a proton as the basic unit of positive charge, which we write as $+1$. An electron has an electrical charge that is precisely opposite that of a proton, so we say it has negative charge (-1). Neutrons are electrically neutral, meaning that they have no charge.

Oppositely charged particles attract one another, and similarly charged particles repel one another. The attraction between the positively charged protons in the nucleus and the negatively charged electrons that surround it is what holds an atom together. Ordinary atoms have identical numbers of electrons and protons, making them electrically neutral overall. (You may wonder why electrical repulsion doesn't cause the positively charged protons in a nucleus to fly apart from one another. The answer is that an even stronger force, called the *strong force,* overcomes electrical repulsion and holds the nucleus together **[Section 10.2]**.)

Although we can think of electrons as tiny particles, they are not quite like tiny grains of sand and they don't orbit the nucleus the way planets orbit the Sun. Instead, the electrons in an atom form a kind of "smeared out" cloud that surrounds the nucleus and gives the atom its apparent size. The electrons aren't really cloudy, but it is impossible to pinpoint their positions in the atom. In Figure 5.5, you can see that the electrons give the atom a size far larger than its nucleus even though they represent only a tiny portion of the atom's mass. If we imagine an atom on a scale that makes its nucleus the size of your fist, its electron cloud would be many kilometers wide.

Atomic Terminology
You've probably learned the basic terminology of atoms in past science classes, but let's review it just to be sure. Figure 5.6 summarizes the key terminology we will use in this book.

Atoms of different chemical elements have different numbers of protons.

Each different chemical element contains a different number of protons in its nucleus. This number is its **atomic number**. For example, a hydrogen nucleus contains just one proton, so its atomic number is 1. A helium nucleus contains two protons, so its atomic number is 2. The *combined* number of protons and neutrons in an atom is called its **atomic mass number**. The atomic mass number of ordinary hydrogen is 1 because its nucleus is just a single proton. Helium usually has two neutrons in addition to its two protons, giving it an atomic mass number of 4. Carbon usually has six protons and six neutrons, giving it an atomic mass number of 12.

Isotopes of a particular chemical element all have the same number of protons but different numbers of neutrons.

Every atom of a given element contains exactly the same number of protons, but the number of neutrons can vary. For example, all carbon atoms have six protons, but they may have six, seven, or eight neutrons. Versions of an element with different numbers of neutrons are called **isotopes** of that element. Isotopes are named by listing their element name and atomic mass number.

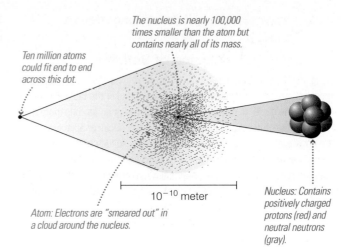

The nucleus is nearly 100,000 times smaller than the atom but contains nearly all of its mass.

Ten million atoms could fit end to end across this dot.

10^{-10} meter

Atom: Electrons are "smeared out" in a cloud around the nucleus.

Nucleus: Contains positively charged protons (red) and neutral neutrons (gray).

Figure 5.5

The structure of a typical atom. Notice that atoms are extremely tiny: The atom shown in the middle is magnified to about 1 billion times its actual size, and the nucleus on the right is magnified to about 100 trillion times its actual size.

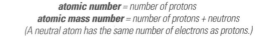

atomic number = number of protons
atomic mass number = number of protons + neutrons
(A neutral atom has the same number of electrons as protons.)

Hydrogen (^{1}H)	Helium (^{4}He)	Carbon (^{12}C)
atomic number = 1	atomic number = 2	atomic number = 6
atomic mass number = 1	atomic mass number = 4	atomic mass number = 12
(1 electron)	(2 electrons)	(6 electrons)

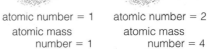

Different **isotopes** of a given element contain the same number of protons, but different numbers of neutrons.

Isotopes of Carbon

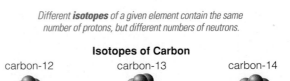

carbon-12	carbon-13	carbon-14
^{12}C	^{13}C	^{14}C
(6 protons + 6 neutrons)	(6 protons + 7 neutrons)	(6 protons + 8 neutrons)

Figure 5.6

Terminology of atoms.

For example, the most common isotope of carbon has 6 protons and 6 neutrons, giving it atomic mass number $6 + 6 = 12$, so we call it carbon-12. The other isotopes of carbon are carbon-13 (six protons and seven neutrons) and carbon-14 (six protons and eight neutrons). We can also write the atomic mass number of an isotope as a superscript to the left of the element symbol: ^{12}C, ^{13}C, ^{14}C. We read ^{12}C as "carbon-12."

think about it The symbol ^4He represents helium with an atomic mass number of 4. ^4He is the most common form of helium, containing two protons and two neutrons. What does the symbol ^3He represent?

The number of different material substances is far greater than the number of chemical elements because atoms can combine to form **molecules**. Some molecules consist of two or more atoms of the same ele-ment. For example, we breathe O_2, oxygen molecules made of two oxygen atoms. Other molecules, such as water, are made up of atoms of two or more different elements. The symbol H_2O tells us that a water molecule contains two hydrogen atoms and one oxygen atom. The chemical proper-ties of a molecule are different from those of its individual atoms. For ex-ample, water behaves very differently than pure hydrogen or pure oxygen.

• How do light and matter interact?

Now that we have discussed the nature of light and of matter individu-ally, we are ready to explore how light and matter interact. Energy car-ried by light can interact with matter in four general ways:

- **Emission**: When you turn on a lamp, electric current heats the fila-ment of the light bulb to a point at which it *emits* visible light.

- **Absorption**: When you place your hand near a lit light bulb, your hand *absorbs* some of the light, and this absorbed energy warms your hand.

- **Transmission**: Some forms of matter, such as glass or air, *transmit* light, which means allowing it to pass through.

- **Reflection/scattering**: Light can bounce off matter, leading to what we call *reflection* (when the bouncing is all in the same general direction) or *scattering* (when the bouncing is more random).

Matter can emit, absorb, transmit, or reflect light. Materials that transmit light are said to be *transparent*, and mate-rials that absorb light are called *opaque*. Many materials are neither perfectly transparent nor perfectly opaque. For example, dark sunglasses and clear eyeglasses are both par-tially transparent, but the dark glasses absorb more light and transmit less. Materials can also affect different colors of light differently. For ex-ample, red glass transmits red light but absorbs other colors, while a green lawn reflects (scatters) green light but absorbs all other colors.

Let's put these ideas together to understand what happens when you walk into a room and turn on the light switch (Figure 5.7). The light bulb begins to emit white light, which is a mix of all the colors in the visible spectrum. Some of this light exits the room, transmitted through the windows. The rest of the light strikes the surfaces of objects inside the room, and the material properties of each object determine the colors it absorbs or reflects. The light coming from each object therefore carries an enormous amount of information about the object's location, shape and

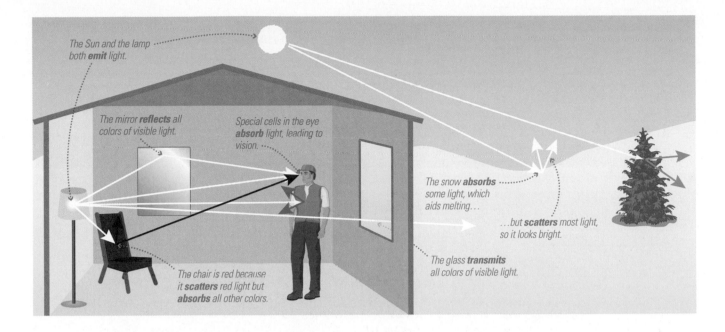

Figure 5.7 interactive figure ↖

When light strikes any piece of matter in the universe, that matter reacts in one or a combination of four ways: emission, absorption, transmission, and reflection (or scattering).

structure, and composition. You acquire this information when light enters your eyes, where special cells in your retina absorb it and send signals to your brain. Your brain interprets the messages that light carries, recognizing materials and objects in the process we call *vision*.

5.2 Learning from Light

Light carries much more information than our naked eyes can recognize. Modern instruments can reveal otherwise hidden details in the spectrum of light, and specially equipped telescopes can record forms of light that are invisible to our eyes. In this section, we'll learn how detailed studies of light help us unlock the secrets of the universe. The key to unlocking those secrets lies in learning how to read astronomical spectra. Let's start with the basic types of spectra and how they are produced, and then we will be ready to see what we can learn from them.

 Light and Spectroscopy Tutorial, Lessons 2–4

• What are the three basic types of spectra?

Laboratory studies show that spectra come in three basic types,* summarized in Figure 5.8:

1. The spectrum of an ordinary (incandescent) light bulb is a rainbow of color. Because the rainbow spans a broad range of wavelengths without interruption, we call it a **continuous spectrum**.
2. A thin or low-density cloud of gas does not produce a continuous spectrum. Instead, it emits light only at specific wavelengths that depend on its composition and temperature. The spectrum therefore consists of bright **emission lines** against a black background and is called an **emission line spectrum**.

---

*The rules that specify the conditions producing each type are often called *Kirchhoff's laws.*

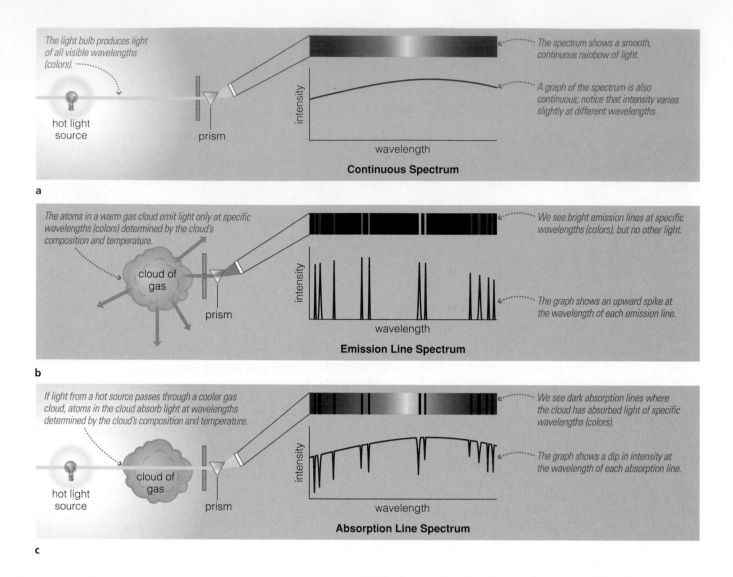

The light bulb produces light of all visible wavelengths (colors).

hot light source

prism

The spectrum shows a smooth, continuous rainbow of light.

A graph of the spectrum is also continuous; notice that intensity varies slightly at different wavelengths.

intensity

wavelength

Continuous Spectrum

a

The atoms in a warm gas cloud emit light only at specific wavelengths (colors) determined by the cloud's composition and temperature.

cloud of gas

prism

We see bright emission lines at specific wavelengths (colors), but no other light.

The graph shows an upward spike at the wavelength of each emission line.

intensity

wavelength

Emission Line Spectrum

b

If light from a hot source passes through a cooler gas cloud, atoms in the cloud absorb light at wavelengths determined by the cloud's composition and temperature.

hot light source

cloud of gas

prism

We see dark absorption lines where the cloud has absorbed light of specific wavelengths (colors).

The graph shows a dip in intensity at the wavelength of each absorption line.

intensity

wavelength

Absorption Line Spectrum

c

Figure 5.8 interactive figure

These diagrams show examples of the conditions under which we see the three basic types of spectra.

3. If the cloud of gas lies between us and a light bulb, we still see most of the continuous light emitted by the light bulb. However, the cloud absorbs light of specific wavelengths, so that the spectrum shows dark **absorption lines** over the background rainbow from the light bulb.* We call this an **absorption line spectrum**.

There are three basic types of spectra: continuous, emission line, and absorption line.

As you study Figure 5.8, notice that each of the spectra is shown both as a band of light and as a graph. The band of light is essentially what you would see if you projected the light that passes through the prism onto a wall. The graph shows the amount, or **intensity**, of the light at each wavelength in the spectrum. The intensity is high at wavelengths where there is a lot of light and low where there is little light. For example, notice how the graph of the absorption line spectrum shows dips in intensity at the wavelengths where the band of light shows dark lines. Astronomers usually display spectra as graphs because they make it easier to tell how the precise intensity of the light varies across the spectrum.

---

*More technically, we'll see an absorption line spectrum as long as the cloud is cooler in temperature than the source of background light (which is the light bulb filament in this case).

We can apply the ideas of Figure 5.8 to the solar spectrum that opens this chapter. Notice that it shows numerous absorption lines over a background rainbow of colors. This tells us that we are essentially looking at a hot light source through gas that is absorbing some of the colors, much as we see when looking through the cloud of gas to the light bulb in Figure 5.8c. For the solar spectrum, the hot light source is the hot interior of the Sun, while the "cloud" is the relatively cool and low-density layer of gas that makes up the Sun's visible surface, or *photosphere* [Section 10.1].

• How does light tell us what things are made of?

We have just seen *how* different viewing conditions lead to different types of spectra, so we are now ready to discuss *why*. Let's start with absorption and emission line spectra. As we'll see, the positions of the lines in these spectra can tell us what distant objects are made of.

Energy Levels in Atoms   To understand why we sometimes see emission and absorption lines, we must first discuss a strange fact about electrons in atoms: The electrons can have only particular amounts of energy, and not other energies in between. As an analogy, suppose you're washing windows on a building. If you use an adjustable platform to reach high windows, you can stop the platform at any height above the ground. But if you use a ladder, you can stand only at *particular* heights—the heights of the rungs of the ladder—and not at any height in between. The possible energies of electrons in atoms are like the possible heights on a ladder. Only a few particular energies are possible, and energies between these special few are not possible. The possible energies are known as the **energy levels** of an atom.

Electrons in atoms can have only particular amounts of energy, and not other energies in between.

Figure 5.9 shows the energy levels in hydrogen, the simplest of all elements. The energy levels are labeled on the left in numerical order and on the right with energies in units of *electron-volts*, or *eV* for short. (1 eV = 1.60 × 10⁻¹⁹ joule.) The lowest possible energy level—called level 1 or the *ground state*—is defined as an energy of 0 eV. Each of the higher energy levels (sometimes called *excited states*) is labeled with the "extra" energy of an electron in that level compared to the ground state.

An electron can rise from a low energy level to a higher one or fall from a high level to a lower one. However, because energy must be conserved [Section 4.3], these *energy level transitions* can occur only when an electron gains or loses the specific amount of energy separating two levels. For example, an electron in level 1 can rise to level 2 only if it gains 10.2 eV of energy. If you try to give the electron 5 eV of energy, it won't accept it because that is not enough energy to reach level 2. Similarly, if you try to give it 11 eV, it won't accept it because it is too much for level 2 but not enough to reach level 3. Once in level 2, the electron can return to level 1 by giving up 10.2 eV of energy.

Notice that the amount of energy separating the various levels gets smaller at higher levels. For example, it takes more energy to raise the electron from level 1 to level 2 than from level 2 to level 3, which in turn takes more energy than the transition from level 3 to level 4. If the electron gains enough energy to reach the *ionization level*, it escapes the atom completely. Because the escaping electron carries away negative electrical charge, the atom is left with positive electrical charge. Electrically charged atoms are called **ions**, so we say that the escape of the electron *ionizes* the atom.

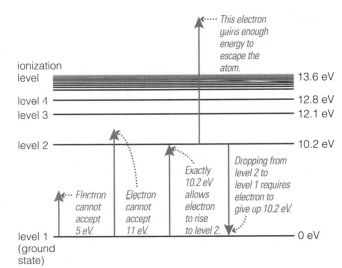

Figure 5.9

Energy levels for the electron in a hydrogen atom. The electron can change energy levels only if it gains or loses the amount of energy separating the levels. If the electron gains enough energy to reach the ionization level, it can escape from the atom, leaving behind a positively charged ion.

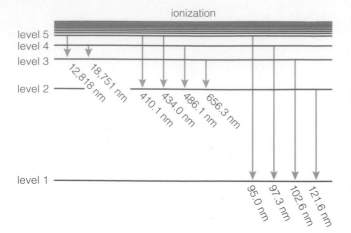

a Energy level transitions in hydrogen correspond to photons with specific wavelengths. Only a few of the many possible transitions are labeled.

410.1 434.0   486.1   656.3
 nm   nm   nm   nm

b This spectrum shows emission lines produced by downward transitions between higher levels and level 2 in hydrogen.

410.1 434.0   486.1   656.3
 nm   nm   nm   nm

c This spectrum shows absorption lines produced by upward transitions between level 2 and higher levels in hydrogen.

Figure 5.10 interactive figure

An atom emits or absorbs light only at specific wavelengths that correspond to changes in the atom's energy as an electron undergoes transitions between its allowed energy levels.

think about it Are there any circumstances under which an electron in a hydrogen atom can gain 2.6 eV of energy? Explain.

Other atoms also have distinct energy levels, but the levels correspond to different amounts of energy than those of hydrogen. Every type of ion and every type of molecule also has a distinct set of energy levels.

Emission and Absorption Lines The fact that each type of atom, ion, or molecule possesses a unique set of energy levels is what causes emission and absorption lines to appear at specific wavelengths in spectra. It is also what allows us to learn the compositions of distant objects in the universe. To see how, let's consider what happens in a cloud of gas consisting solely of hydrogen atoms.

The atoms in any cloud of gas are constantly colliding with one another, exchanging energy in each collision. Most of the collisions simply send the atoms careening off in new directions. However, a few of the collisions transfer the right amount of energy to bump an electron from a low energy level to a higher energy level. Electrons can't stay in higher energy levels for long. They always fall back down to level 1, usually in a tiny fraction of a second. The energy the electron loses when it falls to a lower energy level must go somewhere, and often it goes to *emitting* a photon of light. The emitted photon must have the same amount of energy that the electron loses, which means that it has a specific wavelength and frequency. Figure 5.10a again shows the energy levels in hydrogen that we saw in Figure 5.9, but it is also labeled with the wavelengths of the photons emitted by various downward transitions of an electron from a higher energy level to a lower one. For example, the transition from level 2 to level 1 emits an ultraviolet photon of wavelength 121.6 nm, and the transition from level 3 to level 2 emits a red visible-light photon of wavelength 656.3 nm.

The photons that produce emission lines are created when electrons fall to lower energy levels.

Although electrons that rise to higher energy levels in a gas quickly return to level 1, new collisions can raise other electrons into higher levels. As long as the gas remains moderately warm, collisions are always bumping some electrons into higher levels from which they fall back down and emit photons with some of the wavelengths shown in Figure 5.10a. The gas therefore emits light with these specific wavelengths. That is why a warm gas cloud produces an emission line spectrum, as shown in Figure 5.10b. The bright emission lines appear at the wavelengths that correspond to downward transitions of electrons, and the rest of the spectrum is dark (black). The specific set of lines that we see depends on the cloud's temperature as well as its composition: At higher temperatures, electrons are more likely to be bumped to higher energy levels.

think about it If nothing continues to heat the hydrogen gas, all the electrons eventually will end up in the lowest energy level (the ground state, or level 1). Use this fact to explain why we should not expect to see an emission line spectrum from a very cold cloud of hydrogen gas.

Now, suppose a light bulb illuminates the hydrogen gas from behind (as in Figure 5.8c). The light bulb emits light of all wavelengths, producing a spectrum that looks like a rainbow of color. However, the hydrogen atoms can absorb those photons that have the right amount of energy

Absorption lines occur when photons cause electrons to rise to higher energy levels.

needed to raise an electron from a low energy level to a higher one. Figure 5.10c shows the result. It is an absorption line spectrum, because the light bulb produces a continuous rainbow of color while the hydrogen atoms absorb light at specific wavelengths.*

You can now see why the dark absorption lines in Figure 5.10c occur at the same wavelengths as the emission lines in Figure 5.10b: Both types of lines represent the same energy level transitions, except in opposite directions. For example, electrons moving downward from level 3 to level 2 in hydrogen can emit photons of wavelength 656.3 nm (producing an emission line at this wavelength), while electrons absorbing photons with this wavelength can jump up from level 2 to level 3 (producing an absorption line at this wavelength).

Chemical Fingerprints The fact that hydrogen emits and absorbs at specific wavelengths makes it possible to detect its presence in distant objects. For example, imagine that you look through a telescope at an interstellar gas cloud, and its spectrum looks like that shown in Figure 5.10b. Because this particular set of lines is produced only by hydrogen, you can conclude that the cloud is made of hydrogen. In essence, the spectrum contains a "fingerprint" left by hydrogen atoms.

Every kind of atom, ion, and molecule produces a unique spectral "fingerprint."

Real interstellar clouds are not made solely of hydrogen. However, the other chemical constituents in the cloud leave fingerprints on the spectrum in much the same way. Every type of atom, ion, and molecule has its own unique spectral fingerprint, because it has its own unique set of energy levels. Over the past century, scientists have done laboratory experiments to identify the spectral lines of every chemical element and many ions and molecules. When we see any of those lines in the spectrum of a distant object, we can determine what chemicals produced them. For example, if we see spectral lines of hydrogen, helium, and carbon in the spectrum of a distant star, we know that all three elements are present in the star. With more detailed analysis, we can determine the relative proportions of the various elements. That is how we have learned the chemical compositions of objects throughout the universe.

• How does light tell us the temperatures of planets and stars?

We have seen how emission and absorption line spectra form, and how we can use them to determine the composition of a cloud of gas. Now we are ready to turn our attention to continuous spectra. Although continuous spectra can be produced in more than one way, light bulbs, planets, and stars produce a particular kind of continuous spectrum that can help us determine their temperatures.

Thermal Radiation: Every Body Does It In a cloud of gas that produces a simple emission or absorption line spectrum, the individual atoms or molecules are essentially independent of one another. Most photons

---

*You might wonder what happens to the electrons after they absorb photons and jump to a higher energy level: The electrons quickly fall back down, emitting photons of the same energy in random directions. We therefore see absorption lines because most of the emitted photons are not sent along our line of sight.

pass easily through such a gas, except those that cause energy level transitions in the atoms or molecules of the gas. However, the atoms and molecules within most of the objects we encounter in everyday life—such as rocks, light bulb filaments, and people—cannot be considered independent and therefore have much more complex sets of energy levels. These objects tend to absorb light across a broad range of wavelengths, which means light that strikes them cannot easily pass through and light emitted inside them cannot easily escape. The same is true of almost any large or dense object, including planets and stars.

In order to understand the spectra of such objects, let's consider an idealized case, in which an object absorbs all photons that strike it and does not allow photons inside it to escape easily. Photons tend to bounce randomly around inside such an object, constantly exchanging energy with its atoms or molecules. By the time the photons finally escape the object, their radiative energies have become randomized so that they are spread over a wide range of wavelengths. The wide wavelength range of the photons explains why the spectrum of light from such an object is smooth, or *continuous*, like a pure rainbow without any absorption or emission lines.

Most important, the spectrum from such an object depends on only one thing: the object's *temperature*. To understand why, remember that temperature represents the average kinetic energy of the atoms or molecules in an object [Section 4.3]. Because the randomly bouncing photons interact so many times with those atoms or molecules, they end up with energies that match the kinetic energies of the object's atoms or molecules—which means the photon energies depend only on the object's temperature, regardless of what the object is made of. The temperature dependence of this light explains why we call it **thermal radiation** (sometimes known as *blackbody* radiation) and why its spectrum is called a **thermal radiation spectrum**.

Planets, stars, rocks, and people emit thermal radiation that depends only on temperature.

No real object emits a perfect thermal radiation spectrum, but almost all familiar objects—including the Sun, the planets, rocks, and even you—emit light that approximates thermal radiation. Figure 5.11 shows a graph of the idealized thermal radiation spectra of three stars and a human, each with its temperature given on the Kelvin scale (see Figure 4.10). Be sure to notice that these spectra show the intensity of light *per unit surface area*, not the total amount of light emitted by the object. For example, a very large 3000 K star can emit more total light than a small 15,000 K star, even though the hotter star emits much more light per unit area of its surface.

The Two Laws of Thermal Radiation   If you compare the spectra in Figure 5.11, you'll see that temperature affects them according to the two laws of thermal radiation:

- Law 1 (Stefan-Boltzmann law): *Each square meter of a hotter object's surface emits more light at all wavelengths.* For example, each square meter on the surface of the 15,000 K star emits a lot more light at every wavelength than each square meter of the 3000 K star, and the hotter star emits light at some ultraviolet wavelengths that the cooler star does not emit at all.

- Law 2 (Wien's law ["Wien" is pronounced *veen*]): *Hotter objects emit photons with a higher average energy,* which means a shorter average wavelength. That is why the "humps" of the spectra are at shorter

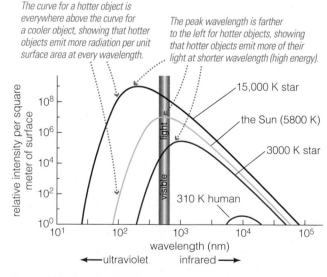

The curve for a hotter object is everywhere above the curve for a cooler object, showing that hotter objects emit more radiation per unit surface area at every wavelength.

The peak wavelength is farther to the left for hotter objects, showing that hotter objects emit more of their light at shorter wavelength (high energy).

Figure 5.11 interactive figure

This graph of idealized thermal radiation spectra demonstrates the two laws of thermal radiation: (1) Each square meter of a hotter object's surface emits more light at all wavelengths; (2) hotter objects emit photons with a higher average energy. Notice that the graph uses power-of-10 scales on both axes, so that we can see all the curves even though the differences between them are quite large.

wavelengths for hotter objects. For example, the hump for the 15,000 K star is in ultraviolet light, the hump for the 5800 K Sun is in visible light, and the hump for the 3000 K star is in the infrared.

You can see these laws in action with a fireplace poker (Figure 5.12). While the poker is still relatively cool, it emits only infrared light, which we cannot see. As it gets hot (above about 1500 K), it begins to glow with visible light, and it glows more brightly as it gets hotter, demonstrating the first law. Its color demonstrates the second law. At first it glows "red hot," because red light has the longest wavelengths of visible light. As it gets even hotter, the average wavelength of the emitted photons moves toward the blue (short wavelength) end of the visible spectrum. The mix of colors emitted at this higher temperature makes the poker look white to your eyes, which is why "white hot" is hotter than "red hot."

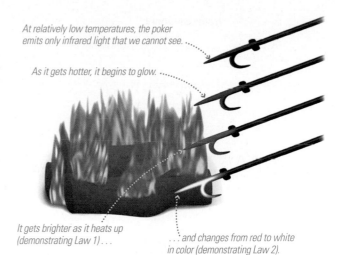

At relatively low temperatures, the poker emits only infrared light that we cannot see.

As it gets hotter, it begins to glow.

It gets brighter as it heats up (demonstrating Law 1)...

...and changes from red to white in color (demonstrating Law 2).

Figure 5.12

A fireplace poker shows the two laws of thermal radiation in action.

 **see it for yourself** Find a light that has a dimmer switch. What happens to the bulb temperature (which you can check by placing your hand near it) as you turn the switch up? How does the light change color? Explain how these observations demonstrate the two laws of thermal radiation.

Hotter objects emit more total light per unit surface area and emit photons with a higher average energy. Because thermal radiation spectra depend only on temperature, we can use them to measure the temperatures of distant objects. In many cases we can estimate temperatures simply from the object's colors. Notice that while hotter objects emit more light at *all* wavelengths, the biggest difference appears at the shortest wavelengths. At human body temperature of about 310 K, people emit mostly in the infrared and emit no visible light at all—which explains why we don't glow in the dark! A relatively cool star, with a 3000 K surface temperature, emits mostly red light. That is why some bright stars in our sky, such as Betelgeuse (in Orion) and Antares (in Scorpius), appear reddish in color. The Sun's 5800 K surface emits most strongly in green light (around 500 nm), but the Sun looks yellow or white to our eyes because it also emits other colors throughout the visible spectrum. Hotter stars emit mostly in the ultraviolet but appear blue-white in color because our eyes cannot see their ultraviolet light. If an object were heated to a temperature of millions of degrees, it would radiate mostly X rays. Some astronomical objects are indeed hot enough to emit X rays, such as disks of gas encircling exotic objects like neutron stars and black holes (see Chapter 13.)

 **The Doppler Effect Tutorial, Lessons 1–2**

• How does light tell us the speed of a distant object?

There is still more that we can learn from light: We can use light to learn about the motion of distant objects (relative to us) from changes in their spectra caused by the **Doppler effect**.

The Doppler Effect You've probably noticed the Doppler effect on the *sound* of a train whistle near train tracks. If the train is stationary, the pitch of its whistle sounds the same no matter where you stand (Figure 5.13a). But if the train is moving, the pitch sounds higher when the train is coming toward you and lower when it's moving away from you.

cosmic calculations 5.1

Laws of Thermal Radiation

The two laws of thermal radiation have simple mathematical formulas. Law 1 (the *Stefan-Boltzmann law*) is expressed

Law 1: emitted power (per square meter of surface) $= \sigma T^4$

T is temperature (in Kelvin) and $\sigma = 5.7 \times 10^{-8} \frac{\text{watt}}{(\text{m}^2 \times \text{K}^4)}$ is a constant. (A *watt* is a unit of power, equivalent to 1 joule per second.) Law 2 (*Wien's law*) is expressed

$$\text{Law 2: } \lambda_{\max}(\text{in nanometers}) \approx \frac{2{,}900{,}000}{T\,(\text{in Kelvin})}$$

where $\lambda_{\max}$ (read as "lambda max") is the wavelength (in nanometers) of maximum intensity, which is the peak of the hump in a thermal radiation spectrum.

Example: Consider a 15,000 K object that emits thermal radiation. How much power does it emit per square meter? What is its wavelength of maximum intensity?

Solution: We use the first law to calculate the emitted power per square meter for an object with $T = 15{,}000$ K:

$$\sigma T^4 = 5.7 \times 10^{-8} \frac{\text{watt}}{\text{m}^2 \times \text{K}^4} \times (15{,}000 \text{ K})^4$$
$$= 2.9 \times 10^9 \text{ watt/m}^2$$

The second law gives the wavelength of maximum intensity:

$$\lambda_{\max} \approx \frac{2{,}900{,}000}{15{,}000 \text{ K}} \text{ nm} \approx 190 \text{ nm}$$

A 15,000 K object emits a total power 2.9 billion watts per square meter of surface. Its wavelength of maximum intensity is about 190 nm, which is in the ultraviolet portion of the electromagnetic spectrum.

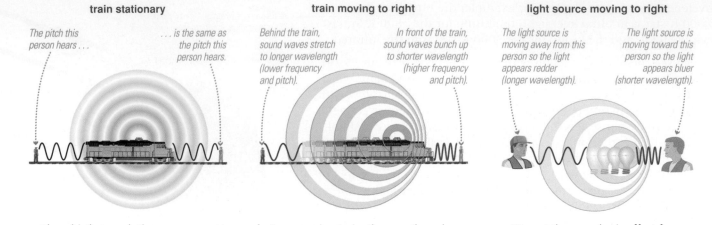

train stationary

The pitch this person hears . . .   *. . . is the same as the pitch this person hears.*

train moving to right

Behind the train, sound waves stretch to longer wavelength (lower frequency and pitch).   *In front of the train, sound waves bunch up to shorter wavelength (higher frequency and pitch).*

light source moving to right

The light source is moving away from this person so the light appears redder (longer wavelength).   *The light source is moving toward this person so the light appears bluer (shorter wavelength).*

a The whistle sounds the same no matter where you stand near a stationary train.

b For a moving train, the sound you hear depends on whether the train is moving toward you or away from you.

c We get the same basic effect from a moving light source (although the shifts are usually too small to notice by eye).

Figure 5.13

The Doppler effect. Each circle represents the crests of sound (or light) waves going in all directions from the source. For example, the circles from the train might represent waves emitted 0.001 second apart.

cosmic calculations 5.2

The Doppler Shift

We can calculate an object's radial velocity from its Doppler shift. For velocities that are small compared to the speed of light (less than a few percent of c), the formula is

$$\frac{v_{rad}}{c} = \frac{\lambda_{shift} - \lambda_{rest}}{\lambda_{rest}}$$

where v_{rad} is the object's radial velocity, λ_{rest} is the rest wavelength of a particular spectral line, and λ_{shift} is the shifted wavelength of the same line. (As always, c is the speed of light.) A positive answer means the object is redshifted and moving away from us; a negative answer means it is blueshifted and moving toward us.

Example: One of the visible lines of hydrogen has a rest wavelength of 656.285 nm, but it appears in the spectrum of the star Vega at 656.255 nm. How is Vega moving relative to us?

Solution: We use the rest wavelength $\lambda_{rest} = 656.285$ nm and the shifted wavelength $\lambda_{shift} = 656.255$ nm:

$$\frac{v_{rad}}{c} = \frac{\lambda_{shift} - \lambda_{rest}}{\lambda_{rest}}$$
$$= \frac{656.255 \text{ nm} - 656.285 \text{ nm}}{656.285 \text{ nm}}$$
$$= -4.5712 \times 10^{-5}$$

The negative answer tells us that Vega is moving *toward* us. Its speed is 4.5712×10^{-5} of the speed of light c. Because $c = 300,000$ km/s, this is equivalent to $4.5712 \times 10^{-5} \times \left(3 \times 10^5 \frac{km}{s}\right) \approx 13.7$ km/s.

Just as the train passes by, you can hear the dramatic change from high to low pitch—a sort of "weeeeeeee–ooooooooooh" sound. To understand why, we have to think about what happens to the sound waves coming from the train (Figure 5.13b). When the train is moving toward you, each pulse of a sound wave is emitted a little closer to you. The result is that waves are bunched up between you and the train, giving them a shorter wavelength and higher frequency (pitch). After the train passes you by, each pulse comes from farther away, stretching out the wavelengths and giving the sound a lower frequency.

Spectral lines shift to shorter wavelengths when an object is moving toward us, and to longer wavelengths when an object is moving away from us.

The Doppler effect causes similar shifts in the wavelengths of light (Figure 5.13c). If an object is moving toward us, the light waves bunch up between us and the object, so that its entire spectrum is shifted to shorter wavelengths. Because shorter wavelengths of visible light are bluer, the Doppler shift of an object coming toward us is called a **blueshift**. If an object is moving away from us, its light is shifted to longer wavelengths. We call this a **redshift** because longer wavelengths of visible light are redder. For convenience, astronomers use the terms *blueshift* and *redshift* even when they aren't talking about visible light.

Spectral lines provide the reference points we use to identify and measure Doppler shifts (Figure 5.14). For example, suppose we recognize the pattern of hydrogen lines in the spectrum of a distant object. We know the **rest wavelengths** of the hydrogen lines—that is, their wavelengths in stationary clouds of hydrogen gas—from laboratory experiments in which a tube of hydrogen gas is heated so that the wavelengths of the spectral lines can be measured. If the hydrogen lines from the object appear at longer wavelengths, then we know they are redshifted and the object is moving away from us. The larger the shift, the faster the object is moving. If the lines appear at shorter wavelengths, then we know they are blueshifted and the object is moving toward us.

think about it Suppose the hydrogen emission line with a rest wavelength of 121.6 nm (the transition from level 2 to level 1) appears at a wavelength of 120.5 nm in the spectrum of a particular star. Given that these wavelengths are in the ultraviolet, is the shifted wavelength closer to or farther from blue visible light? Why, then, do we say that this spectral line is *blueshifted*?

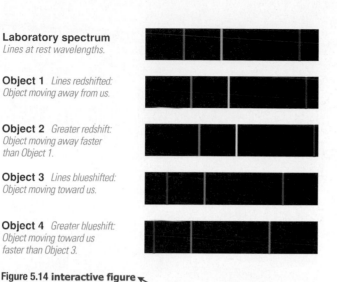

Laboratory spectrum
Lines at rest wavelengths.

Object 1 *Lines redshifted:*
Object moving away from us.

Object 2 *Greater redshift:*
Object moving away faster
than Object 1.

Object 3 *Lines blueshifted:*
Object moving toward us.

Object 4 *Greater blueshift:*
Object moving toward us
faster than Object 3.

Figure 5.14 interactive figure

Spectral lines provide the crucial reference points for measuring Doppler shifts.

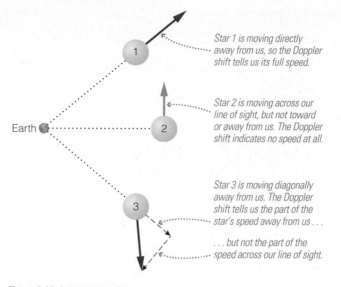

Star 1 is moving directly away from us, so the Doppler shift tells us its full speed.

Star 2 is moving across our line of sight, but not toward or away from us. The Doppler shift indicates no speed at all.

Star 3 is moving diagonally away from us. The Doppler shift tells us the part of the star's speed away from us . . .

. . . but not the part of the speed across our line of sight.

Earth

Figure 5.15 interactive figure

The Doppler shift tells us only the portion of an object's speed that is directed toward or away from us. It does not give us any information about how fast an object is moving across our line of sight.

Notice that the Doppler shift tells us only the part of an object's full motion that is directed toward or away from us (the object's *radial* component of motion). Doppler shifts do not give us any information about how fast an object is moving across our line of sight (the object's *tangential* component of motion). For example, consider three stars all moving at the same speed, with one moving directly away from us, one moving across our line of sight, and one moving diagonally away from us (Figure 5.15). The Doppler shift will tell us the full speed only of the first star. It will not indicate any speed for the second star, because none of this star's motion is directed toward or away from us. For the third star, the Doppler shift will tell us only the part of the star's speed that is directed away from us. To measure how fast an object is moving across our line of sight, we must observe it long enough to notice how its position gradually shifts across our sky.

Spectral Summary We've covered the major ways in which we can learn from an object's spectrum, discussing how we learn about an object's composition, temperature, and motion. Figure 5.16 (on pages 126–127) summarizes the ways in which we learn from spectra.

(MA) ™ **Telescopes Tutorial, Lessons 1–2**

5.3 Collecting Light with Telescopes

We've seen that light carries a great deal of information, but only a little of that information can be obtained when we look at the sky with our naked eyes. Most of the great advances that have taken place in astronomy in the past three centuries have been made possible through the use of telescopes. In this section, we'll briefly explore how telescopes work and how they help us learn about the universe.

An astronomical spectrum carries an enormous amount of information. This figure illustrates some of what we can learn from a spectrum, using a schematic spectrum of Mars as an example.

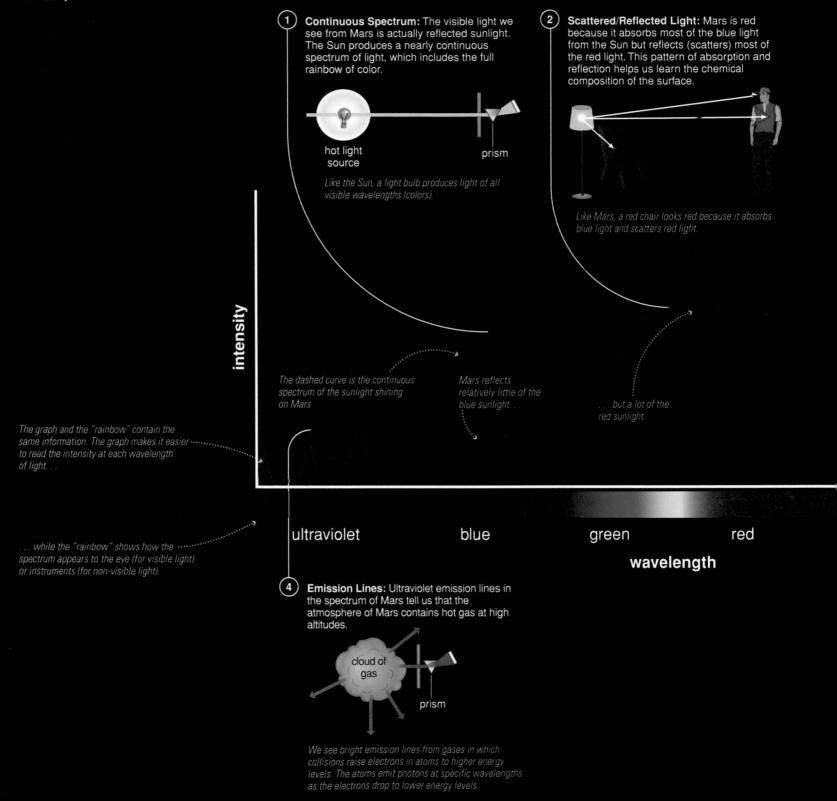

① Continuous Spectrum: The visible light we see from Mars is actually reflected sunlight. The Sun produces a nearly continuous spectrum of light, which includes the full rainbow of color.

hot light source

prism

Like the Sun, a light bulb produces light of all visible wavelengths (colors).

② Scattered/Reflected Light: Mars is red because it absorbs most of the blue light from the Sun but reflects (scatters) most of the red light. This pattern of absorption and reflection helps us learn the chemical composition of the surface.

Like Mars, a red chair looks red because it absorbs blue light and scatters red light.

intensity

The dashed curve is the continuous spectrum of the sunlight shining on Mars

Mars reflects relatively little of the blue sunlight. . .

. . . but a lot of the red sunlight.

The graph and the "rainbow" contain the same information. The graph makes it easier to read the intensity at each wavelength of light. . .

. . . while the "rainbow" shows how the spectrum appears to the eye (for visible light) or instruments (for non-visible light).

ultraviolet   blue   green   red

wavelength

④ Emission Lines: Ultraviolet emission lines in the spectrum of Mars tell us that the atmosphere of Mars contains hot gas at high altitudes.

cloud of gas

prism

We see bright emission lines from gases in which collisions raise electrons in atoms to higher energy levels. The atoms emit photons at specific wavelengths as the electrons drop to lower energy levels.

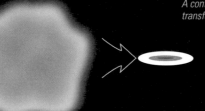

A contracting gas cloud in space heats up because it transforms gravitational potential energy into thermal energy.

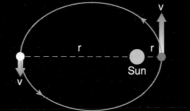

Conservation of angular momentum also explains why a planet's orbital speed increases when it is closer to the Sun.

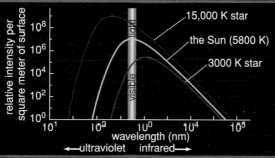

Gravity also operates in space—its attractive force can act across great distances to pull objects closer together or to hold them in orbit.

$$F_g = G \frac{M_1 M_2}{d^2}$$

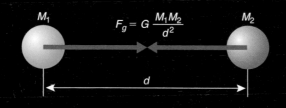

Sunlight is also a visible form of thermal radiation. The Sun is much brighter and whiter than a fireplace poker because its surface is much hotter.

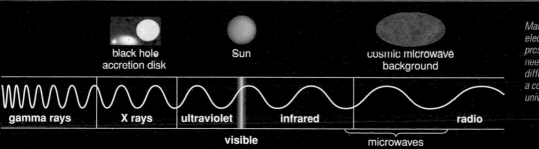

Many different forms of electromagnetic radiation are present in space. We therefore need to observe light of many different wavelengths to get a complete picture of the universe.

6

formation of planetary systems:
our solar system and beyond

Now that we have discussed some of the key laws that govern nature, we can apply these laws to the study of objects throughout our universe. We will begin with our solar system in this and the next three chapters, and later study stars, galaxies, and the universe.

In this chapter, we'll explore the nature of our solar system and current scientific ideas about its birth. After a brief overview of the solar system and its individual worlds, we'll focus on characteristics of the solar system that offer key clues about how it formed. Finally, we'll learn how astronomers have discovered planets around other stars, and how these other planetary systems may help us understand our own.

essential preparation

1. What is our place in the universe? [Section 1.1]

2. How did we come to be? [Section 1.1]

3. How big is Earth compared to our solar system? [Section 1.2]

4. What keeps a planet rotating and orbiting the Sun? [Section 4.3]

5. How does light tell us the speed of a distant object? [Section 5.2]

 Scale of the Universe Tutorial, Lesson 1

6.1 A Brief Tour of the Solar System

Our ancestors long ago recognized the motions of the planets through the sky, but it has been only a few hundred years since we learned that Earth is also a planet that orbits the Sun. Even then, we knew little about the other planets until the development of large telescopes. More recently, space exploration has brought us far greater understanding of other worlds. We've lived in this solar system all along, but only now are we getting to know it. Let's begin with a quick tour of our planetary system, which will provide context for the more detailed study that will follow.

• What does the solar system look like?

The first step in getting to know our solar system is to visualize what it looks like as a whole. Imagine viewing the solar system from beyond the orbits of the planets. What would we see?

Without a telescope, the answer would be "not much." Remember that the Sun and planets are all quite small compared to the distances between them [Section 1.2]—so small that if we viewed them from the outskirts of our solar system, the planets would be only pinpoints of light, and even the Sun would be just a small bright dot in the sky. But if we magnify the sizes of the planets by about a million times compared to their distances from the Sun and show their orbital paths, we get the central picture in Figure 6.1 (pp. 144–145).

The ten pages that follow Figure 6.1 offer a brief tour through our solar system, beginning at the Sun and continuing to each of the planets. The tour highlights a few of the most important features of each world we visit—just enough information so that you'll be ready for the comparative study we'll undertake in later chapters. The side of each page shows the planets to scale, using the 1-to-10-billion scale introduced in Chapter 1. The map along the bottom of each page shows the locations of the Sun and each of the planets in the Voyage scale model solar system (see Figures 1.5 and 1.6) so that you can see relative distances from the Sun. Table 6.1, which follows the tour, summarizes key planetary data.

As you study Figure 6.1, the tour pages, and Table 6.1, you'll quickly see that our solar system is *not* a random collection of worlds. Figure 6.1 shows that all the planets orbit the Sun in the same direction and in nearly the same plane, while the figure and tour show that the four inner planets are quite different in character from the next four planets.

(main text continues on page 157)

The solar system's layout and composition offer four major clues to how it formed. The main illustration below shows the orbits of planets in the solar system from a perspective beyond Neptune, with the planets themselves magnified by about a million times relative to their orbits.

① **Large bodies in the solar system have orderly motions.** All planets have nearly circular orbits going in the same direction in nearly the same plane. Most large moons orbit their planets in this same direction, which is also the direction of the Sun's rotation.

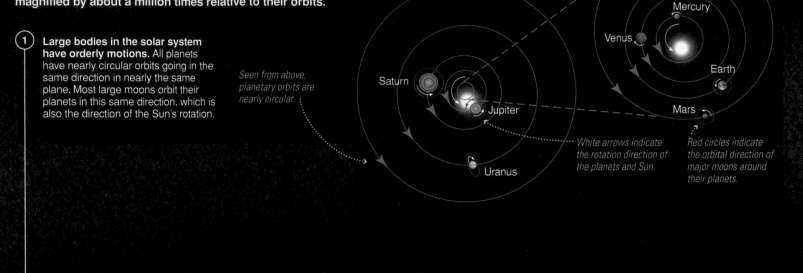

Seen from above, planetary orbits are nearly circular.

White arrows indicate the rotation direction of the planets and Sun.

Red circles indicate the orbital direction of major moons around their planets.

Each planet's axis tilt is shown, with small circling arrows to indicate the direction of the planet's rotation.

Orbits are shown to scale, but planet sizes are exaggerated about 1 million times relative to orbits. The Sun is not shown to scale.

Jupiter

Mercury

Sun

Venus

Earth

Mars

Asteroid belt

Neptune

Orange arrows indicate the direction of orbital motion.

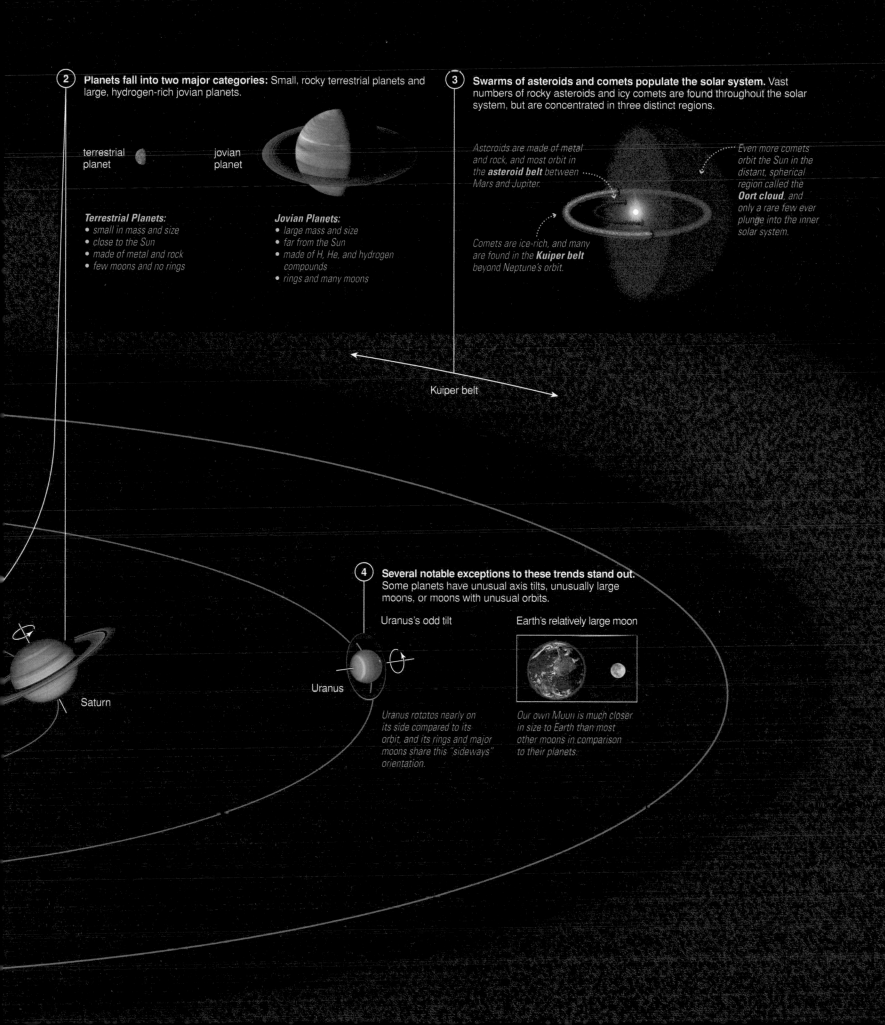

2 **Planets fall into two major categories:** Small, rocky terrestrial planets and large, hydrogen-rich jovian planets.

terrestrial planet

jovian planet

Terrestrial Planets:
- small in mass and size
- close to the Sun
- made of metal and rock
- few moons and no rings

Jovian Planets:
- large mass and size
- far from the Sun
- made of H, He, and hydrogen compounds
- rings and many moons

3 **Swarms of asteroids and comets populate the solar system.** Vast numbers of rocky asteroids and icy comets are found throughout the solar system, but are concentrated in three distinct regions.

Asteroids are made of metal and rock, and most orbit in the **asteroid belt** *between Mars and Jupiter.*

Even more comets orbit the Sun in the distant, spherical region called the **Oort cloud***, and only a rare few ever plunge into the inner solar system.*

Comets are ice-rich, and many are found in the **Kuiper belt** *beyond Neptune's orbit.*

Kuiper belt

4 **Several notable exceptions to these trends stand out.** Some planets have unusual axis tilts, unusually large moons, or moons with unusual orbits.

Uranus's odd tilt

Earth's relatively large moon

Uranus

Saturn

Uranus rotates nearly on its side compared to its orbit, and its rings and major moons share this "sideways" orientation.

Our own Moon is much closer in size to Earth than most other moons in comparison to their planets.

Figure 6.2

The Sun contains more than 99.9% of the total mass in our solar system.

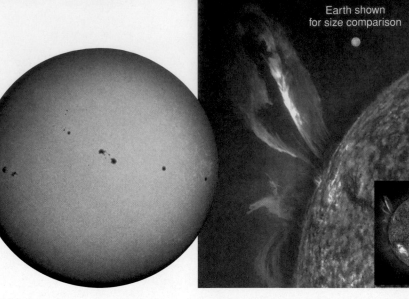

Earth shown
for size comparison

a A visible-light photograph of the Sun's surface. The dark splotches are sunspots—each large enough to swallow several Earths.

b This ultraviolet photograph, from the *SOHO* spacecraft, shows a huge streamer of hot gas on the Sun. The image of Earth was added for size comparison.

• The Sun

- Radius: 695,000 km = $108R_{Earth}$
- Mass: $333,000M_{Earth}$
- Composition (by mass): 98% hydrogen and helium, 2% other elements

The Sun is by far the largest and brightest object in our solar system. It contains more than 99.9% of the solar system's total mass, making it more than a thousand times as massive as everything else in the solar system combined.

The Sun's surface looks solid in photographs (Figure 6.2), but it is actually a roiling sea of extremely hot (about 5800 K, or 6100°C or 11,000°F) hydrogen and helium gas. The surface is speckled with sunspots that appear dark in photographs only because they are slightly cooler than their surroundings. Solar storms sometimes send streamers of hot gas soaring far above the surface.

The Sun's interior is also gaseous. If you could plunge into the Sun, you'd find the gas temperature and pressure getting ever higher as you went deeper. The source of the Sun's energy lies deep in its core, where the temperatures

and pressures are so high that the Sun is a nuclear fusion power plant. Each second, fusion transforms about 600 million tons of the Sun's hydrogen into 596 million tons of helium. The "missing" 4 million tons becomes energy in accord with Einstein's famous formula, $E = mc^2$ **[Section 4.3]**. Despite losing 4 million tons of mass each second, the Sun contains so much hydrogen that it has already shone steadily for almost 5 billion years and will continue to shine for another 5 billion years.

The Sun is the most influential object in our solar system. Its gravity governs the orbits of the planets. Its heat is the primary influence on the temperatures of planetary surfaces and atmospheres. It is the source of virtually all visible light in our solar system—the Moon and planets shine only by virtue of the sunlight they reflect. In addition, charged particles flowing outward from the Sun (the *solar wind*) help shape planetary magnetic fields and can influence planetary atmospheres. Nevertheless, we can understand almost all the present characteristics of the planets without knowing much more about the Sun than what we have just discussed. We'll save more detailed discussion of the Sun for Chapter 10, where we will study it as our prototype for understanding other stars.

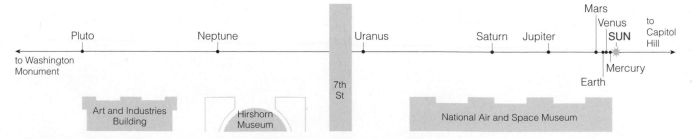

The Voyage scale model solar system represents sizes and distances in our solar system at one ten-billionth of their actual values (see Figure 1.6). The strip along the side of the page shows the sizes of the Sun and planets on this scale, and the map above shows their locations in the Voyage model on the National Mall in Washington, D.C. The Sun is about the size of a large grapefruit on this scale.

Figure 6.3

This image of Mercury's battered crust was taken by the *MESSENGER* spacecraft during its January 2008 flyby. In addition to abundant craters, Mercury has smooth plains, probably of volcanic origin, and wrinkles caused by global shrinking. The inset shows a global map of Mercury obtained by the *Mariner 10* mission more than 30 years ago.

Mercury

- Average distance from the Sun: 0.39 AU
- Radius: 2440 km = $0.38R_{Earth}$
- Mass: $0.055M_{Earth}$
- Average density: 5.43 g/cm^3
- Composition: rocks, metals
- Average surface temperature: 700 K (day), 100 K (night)
- Moons: 0

Mercury is the innermost planet of our solar system and the smallest of the eight official planets. It is a desolate, cratered world with no active volcanoes, no wind, no rain, and no life. Because there is virtually no air to scatter sunlight or color the sky, you could see stars even in the daytime if you stood on Mercury with your back toward the Sun.

You might expect Mercury to be very hot because of its closeness to the Sun, but in fact it is a world of both hot and cold extremes. Tidal forces from the Sun have forced Mercury into an unusual rotation pattern: Its 58.6-day rotation period means it rotates exactly three times for every two of its 87.9-day orbits of the Sun. This combination of rotation and orbit gives Mercury days and nights that last about 3 Earth months each. Daytime temperatures reach 425°C—nearly the temperature of hot coals. At night or in shadow, the temperature falls below −150°C—far colder than Antarctica in winter.

Mercury is the least studied of the inner planets. Its proximity to the Sun makes it difficult to observe through telescopes. Until 2008, it had been visited by just one spacecraft: *Mariner 10,* which obtained images of one hemisphere (Figure 6.3) during three rapid flybys in 1974–1975. The images show vast numbers of craters, ancient lava flows, and tall, steep cliffs that run hundreds of kilometers in length. As we'll discuss in Chapter 7, the cliffs may be wrinkles from an episode of "planetary shrinking" early in Mercury's history. Mercury's mass and density (learned from its gravitational effects on *Mariner 10*'s orbit) indicate that it is made mostly of iron, making it the most metal-rich of the planets.

We will soon learn much more about Mercury. NASA's *MESSENGER* spacecraft flew past Mercury in 2008 and will return for two more flybys before entering orbit around the planet in 2011.

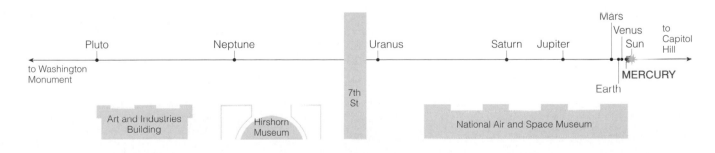

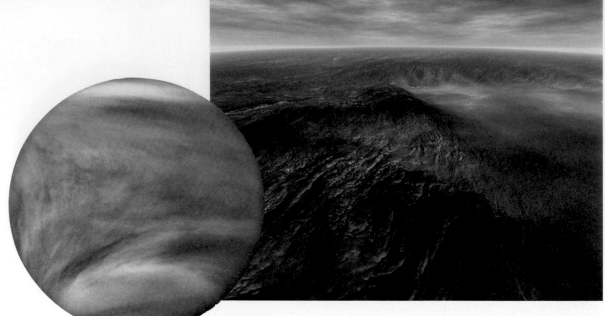

Figure 6.4

The image above shows an artistic rendition of the surface of Venus as scientists think it would appear to our eyes. The surface topography is based on data from NASA's *Magellan* spacecraft. The inset (left) shows the full disk of Venus photographed by NASA's *Pioneer* Venus orbiter with cameras sensitive to ultraviolet light. With visible light, cloud features cannot be distinguished from the general haze. (Image above from the Voyage scale model solar system, developed by the Challenger Center for Space Science Education, the Smithsonian Institution, and NASA. Image by David P. Anderson, Southern Methodist University © 2001.)

• Venus

- Average distance from the Sun: 0.72 AU
- Radius: 6051 km = $0.95R_{Earth}$
- Mass: $0.82M_{Earth}$
- Average density: 5.24 g/cm^3
- Composition: rocks, metals
- Average surface temperature: 740 K
- Moons: 0

Venus, the second planet from the Sun, is nearly identical in size to Earth. Before the era of spacecraft visits, Venus stood out largely for its strange rotation: It rotates on its axis very slowly and in the opposite direction from Earth, so its days and nights are very long and the Sun rises in the west and sets in the east instead of rising in the east and setting in the west. Its surface is completely hidden from view by dense clouds, so we knew little about it until a few decades ago, when spacecraft began to map Venus with cloud-penetrating radar (Figure 6.4).

Because we knew so little about it, some science fiction writers used its Earth-like size, thick atmosphere, and closer distance to the Sun to speculate that it might be a lush, tropical paradise—a "sister planet" to Earth.

The reality is far different. Venus's atmosphere contains no oxygen to breathe, and there is no liquid water. Worse still for the possibility of life, an extreme *greenhouse effect* bakes the surface to an incredible 470°C (about 880°F), trapping heat so effectively that nighttime offers no relief. Day and night, Venus is hotter than a pizza oven, and the thick atmosphere bears down on the surface with a pressure equivalent to that felt nearly a kilometer (0.6 mile) beneath the surface of Earth's oceans. Far from being a beautiful sister planet to Earth, Venus resembles a traditional view of hell.

The geology of Venus features mountains, valleys, and craters, along with many signs of past or present volcanic activity. But Venus also has geological features unlike any on Earth, and we see no evidence of Earth-like plate tectonics. We are rapidly learning how and why Venus became so different from Earth, in part through ongoing studies by the European Space Agency's *Venus Express* spacecraft, which has been orbiting Venus since 2006.

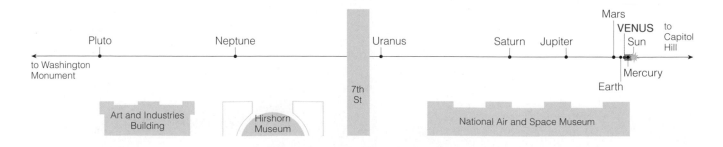

a This image (left), computer generated from satellite data, shows the striking contrast between the daylight and nighttime hemispheres of Earth. The day side reveals little evidence of human presence, but at night our presence is revealed by the lights of human activity. (From the Voyage scale model solar system, developed by the Challenger Center for Space Science Education, the Smithsonian Institution, and NASA. Image created by ARC Science Simulations © 2001.)

b Earth and the Moon, shown to scale. The Moon's diameter is about one-fourth of Earth's diameter, and its mass is about 1/80 of Earth's mass. If you wanted to show the distance between Earth and Moon on the same scale, you'd need to hold these two photographs about 1 meter (3 feet) apart.

Figure 6.5

Earth, our home planet.

• Earth

- Average distance from the Sun: 1.00 AU
- Radius: 6378 km $= 1R_{Earth}$
- Mass: $1.00M_{Earth}$
- Average density: 5.52 g/cm^3
- Composition: rocks, metals
- Average surface temperature: 290 K
- Moons: 1

Beyond Venus, we next encounter our home planet, Earth, the only known oasis of life in our solar system. Earth is also the only world on which humans could survive without a space suit or protective enclosure. It is the only planet in our solar system with oxygen for us to breathe and ozone to shield us from deadly solar radiation, and the only planet with abundant surface water to nurture life. Temperatures are pleasant for us because Earth's atmosphere contains just enough carbon dioxide and water vapor to maintain a moderate greenhouse effect.

Despite Earth's small size, its beauty is striking (Figure 6.5a). Blue oceans cover nearly three-fourths of the surface, broken by the continental land masses and scattered islands. The polar caps are white with snow and ice, and white clouds are scattered above the surface. At night, the glow of artificial lights clearly reveals the presence of an intelligent civilization.

Earth is the first planet on our tour with a moon. Moreover, our Moon is surprisingly large compared to Earth (Figure 6.5b); although it is not the largest moon in the solar system, almost all other moons are much smaller relative to the planets they orbit. Why Earth has such a large moon has long been a mystery, but as we'll discuss later in this chapter, it is now thought to have been due to a *giant impact* in which a Mars-size object slammed into the young Earth.

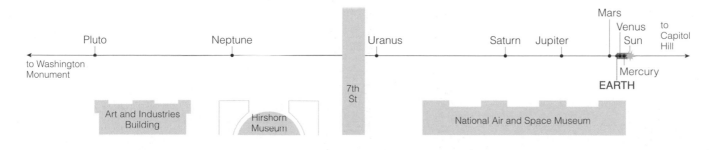

Figure 6.6

The image below shows the surface walls of a Martian crater photographed by NASA's *Opportunity* rover in 2004; and a simulated image of the rover at the appropriate scale is included. The globe shows the full disk of Mars; the horizontal "gash" across the center is the giant canyon Valles Marineris.

• Mars

- Average distance from the Sun: 1.52 AU
- Radius: 3397 km = $0.53R_{Earth}$
- Mass: $0.11M_{Earth}$
- Average density: 3.93 g/cm³
- Composition: rocks, metals
- Average surface temperature: 225 K
- Moons: 2 (very small)

The next planet on our tour is Mars, the last of the four inner planets of our solar system (Figure 6.6). Mars is larger than Mercury and the Moon but smaller than Venus and Earth. About half Earth's size in diameter, Mars has a mass about 10% that of Earth. Mars has two tiny moons, Phobos and Deimos, that look like asteroids and may once have roamed freely in the asteroid belt.

Mars is a world of wonders, with ancient volcanoes that dwarf the largest mountains on Earth, a great canyon that runs nearly one-fifth of the way around the planet, and polar ice caps made of frozen carbon dioxide ("dry ice") and water. Although Mars is frozen today, the presence of dried-up riverbeds, rock-strewn floodplains, and minerals that form in water offers clear evidence that Mars had at least some warm and wet periods in the distant past. Major flows of liquid water probably ceased 2–3 billion years ago, but the past wet periods offer some possibility that Mars may once have been hospitable to life.

Mars looks almost Earth-like in photographs taken by spacecraft on its surface, but you wouldn't want to visit without a space suit. The air pressure is far less than that on top of Mount Everest, the temperature is usually well below freezing, the trace amounts of oxygen would not be enough to breathe, and the lack of atmospheric ozone would leave you exposed to deadly ultraviolet radiation from the Sun.

Mars is the most studied planet besides Earth. More than a dozen spacecraft have flown past, orbited, or landed on Mars. The *Phoenix* lander touched down near Mars's north polar cap in May 2008, and plans are in the works for many more missions to the planet. We may even send humans to Mars within the next few decades. By overturning rocks in ancient riverbeds or chipping away at ice in the polar caps, explorers will help us learn whether Mars has ever been home to life.

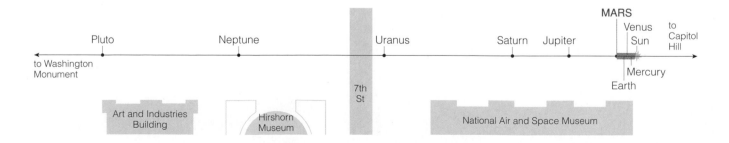

Figure 6.7

This image shows what it would look like to orbit near Jupiter's moon Io as Jupiter comes into view. Notice the Great Red Spot to the left of Jupiter's center. The extraordinarily dark rings discovered during the *Voyager* missions are exaggerated to make them visible. This computer visualization was created using data from NASA's *Voyager* and *Galileo* missions. (From the Voyage scale model solar system, developed by the Challenger Center for Space Science Education, the Smithsonian Institution, and NASA. Image created by ARC Science Simulations © 2001.)

• Jupiter

- Average distance from the Sun: 5.20 AU
- Radius: 71,492 km = $11.2 R_{Earth}$
- Mass: $318 M_{Earth}$
- Average density: 1.33 g/cm^3
- Composition: mostly hydrogen and helium
- Cloud-top temperature: 125 K
- Moons: at least 63

To reach the orbit of Jupiter from Mars, we must traverse a distance that is more than double the distance from the Sun to Mars, passing through the asteroid belt along the way. Upon our arrival, we find a planet much larger than any we have encountered so far (Figure 6.7).

Jupiter is so different from the planets of the inner solar system that we must adopt an entirely new mental image of the term *planet*. Its mass is more than 300 times that of Earth, and its volume is more than 1,000 times that of Earth. Its most famous feature—a long-lived storm called the Great Red Spot—is itself large enough to swallow two or three Earths.

Like the Sun, Jupiter is made primarily of hydrogen and helium and has no solid surface. If we plunged deep into Jupiter, the increasing gas pressure would crush us long before we ever reached its core.

Jupiter reigns over dozens of moons and a thin set of rings (too faint to be seen in most photographs). Most of the moons are very small, but four are large enough that we'd consider them planets if they orbited the Sun. These four moons—Io, Europa, Ganymede, and Callisto—are called the *Galilean moons*, because Galileo discovered them shortly after he first turned his telescope toward the heavens [Section 3.3]. They are also planetlike in having varied and interesting geology. Io is the most volcanically active body in the solar system. Europa has an icy crust that may hide a subsurface ocean of liquid water, making it a promising place to search for life. Ganymede and Callisto also have icy surfaces that may hide subsurface oceans. We still have much to learn about all these moons.

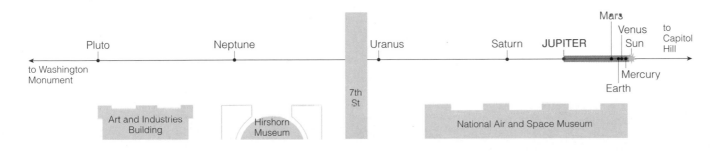

Figure 6.8

Cassini's view of Saturn. We see the shadow of the rings on Saturn's sun-lit face; the rings become lost in Saturn's shadow on the night side. The inset shows Saturn's largest moon, Titan, which is shrouded in a thick, cloudy atmosphere.

• Saturn

- Average distance from the Sun: 9.54 AU
- Radius: 60,268 km $= 9.4R_{Earth}$
- Mass: $95.2M_{Earth}$
- Average density: 0.70 g/cm^3
- Composition: mostly hydrogen and helium
- Cloud-top temperature: 95 K
- Moons: at least 50

The journey from Jupiter to Saturn is a long one: Saturn orbits nearly twice as far from the Sun as Jupiter. Saturn, the second-largest planet in our solar system, is only slightly smaller than Jupiter in diameter. However, its lower density makes it considerably less massive (about one-third Jupiter's mass). Like Jupiter, Saturn is made mostly of hydrogen and helium and has no solid surface.

Saturn is famous for its spectacular rings (Figure 6.8). Although all four of the giant outer planets have rings, only Saturn's rings can be seen easily through a small telescope. The rings may look solid from a distance, but in reality they are made of countless small particles, each of which orbits Saturn like a tiny moon. If you could fly through the rings, you'd find yourself surrounded by chunks of rock and ice that range in size from dust grains to city blocks.

Like Jupiter, Saturn is orbited by many moons. Most are the size of small asteroids or comets, but a few are much larger. Perhaps the most fascinating moon is Titan, which is larger than the planet Mercury and blanketed by a thick atmosphere. On Titan's surface, you'd find an atmospheric pressure even greater than Earth's, and you could inhale air with roughly the same nitrogen content as air on Earth. However, you'd need to bring your own oxygen and a warm space suit for protection against the frigid outside temperatures. NASA's *Cassini* spacecraft, designed to explore Saturn and its rings and moons, carried a probe called *Huygens* that landed on Titan in January 2005. From its data and *Cassini*'s radar mapping, we now see Titan as a world with lakes of liquid methane or ethane and an erosion-carved landscape—oddly similar to Earth, despite Titan's −180°C temperature.

think about it Find the current status of the *Cassini* mission (from news reports or the mission Web site). Is the spacecraft still operating? What is its next target of study? What recent discoveries has it helped us make?

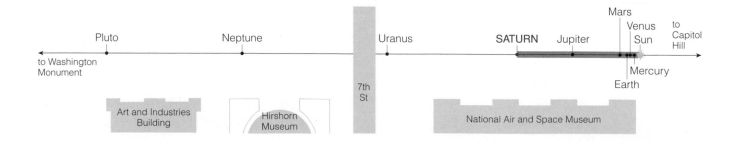

Figure 6.9

This image shows a view of Uranus from high above its moon Ariel. The ring system is shown, although it would actually be too dark to see from this vantage point. This computer simulation is based on data from NASA's *Voyager 2* mission. (From the Voyage scale model solar system, developed by the Challenger Center for Space Science Education, the Smithsonian Institution, and NASA. Image created by ARC Science Simulations © 2001.)

• Uranus

- Average distance from the Sun: 19.2 AU
- Radius: 25,559 km $= 4.0R_{Earth}$
- Mass: $14.5M_{Earth}$
- Average density: 1.32 g/cm^3
- Composition: hydrogen, helium, hydrogen compounds
- Cloud-top temperature: 60 K
- Moons: at least 27

It's another long journey to our next stop on the tour, as Uranus lies twice as far from the Sun as Saturn. Uranus (normally pronounced *YUR-uh-nus*) is much smaller than either Jupiter or Saturn but still is much larger than Earth. It is made largely of hydrogen, helium, and *hydrogen compounds* such as water (H_2O), ammonia (NH_3), and methane (CH_4). Methane gas gives Uranus its pale blue-green color (Figure 6.9). Like the other giants of the outer solar system, Uranus lacks a solid surface. More than two dozen moons orbit Uranus, along with a set of rings somewhat similar to Saturn's but much darker and more difficult to see.

The entire Uranus system—planet, rings, and moon orbits—is tipped on its side compared to the rest of the planets. This sideways axis tilt makes seasonal patterns of daylight and darkness more extreme on Uranus than on any other planet. If you lived on a platform floating in Uranus's atmosphere near its north pole, you'd have continuous daylight for half of each orbit, or 42 years. Then, after a very gradual sunset, you'd enter into a 42-year-long night.

Only one spacecraft has visited Uranus: *Voyager 2*, which flew past all four of the giant outer planets before heading out of the solar system. Much of our current understanding of Uranus comes from that mission, though powerful new telescopes are also capable of studying this planet. Scientists would love to study Uranus and its rings and moons in much greater detail, but no missions to the planet are currently under development.

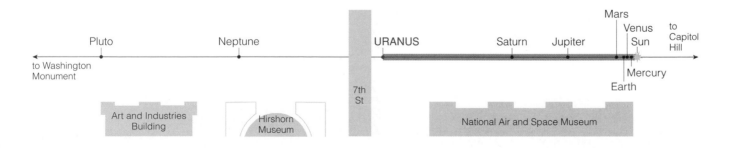

Figure 6.10

This image shows what it would look like to be orbiting Neptune's moon Triton as Neptune itself comes into view. The dark rings are exaggerated to make them visible in this computer simulation using data from NASA's *Voyager 2* mission. (From the Voyage scale model solar system, developed by the Challenger Center for Space Science Education, the Smithsonian Institution, and NASA. Image created by ARC Science Simulations © 2001.)

Neptune

- Average distance from the Sun: 30.1 AU
- Radius: 24,764 km = $3.9R_{Earth}$
- Mass: $17.1M_{Earth}$
- Average density: 1.64 g/cm^3
- Composition: hydrogen, helium, hydrogen compounds
- Cloud-top temperature: 60 K
- Moons: at least 13

The journey from the orbit of Uranus to the orbit of Neptune is the longest yet in our tour, calling attention to the vast emptiness of the outer solar system. Neptune looks nearly like a twin of Uranus, although it is more strikingly blue (Figure 6.10). It is slightly smaller than Uranus in size, but a higher density makes it slightly more massive even though the two planets share very similar compositions. Like Uranus, Neptune has been visited only by the *Voyager 2* spacecraft. No further missions are currently planned.

Neptune has rings and numerous moons. Its largest moon, Triton, is larger than Pluto and is one of the most fascinating moons in the solar system. Its icy surface has features that appear to be somewhat like geysers, although they spew nitrogen gas rather than water into the sky [Section 8.2]. Even more surprisingly, Triton is the only large moon in the solar system that orbits its planet "backward"—that is, in a direction opposite to the direction in which Neptune rotates. This backward orbit indicates that Triton almost certainly orbited the Sun independently before somehow being captured into Neptune's orbit.

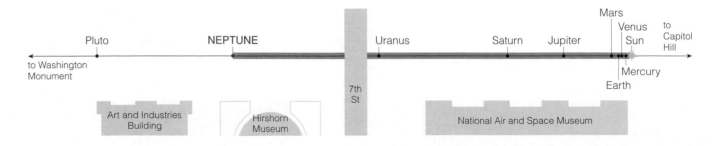

Figure 6.11

Pluto and its moons, as photographed by the Hubble Space Telescope.

Hydra

Nix

Charon

Pluto

Pluto (and Other Dwarf Planets)

- Average distance from the Sun: 39.5 AU
- Radius: 1160 km = $0.18R_{Earth}$
- Mass: $0.0022M_{Earth}$
- Average density: 2.0 g/cm^3
- Composition: ices, rock
- Average surface temperature: 40 K
- Moons: 3

We conclude our tour at Pluto, which reigned for 75 years as the ninth and last planet in our solar system. Pluto's average distance from the Sun lies as far beyond Neptune as Neptune lies beyond Uranus. Its great distance makes Pluto cold and dark. From Pluto, the Sun would be little more than a bright light among the stars. Pluto's largest moon, Charon, is locked together with it in synchronous rotation [Section 4.4], so that Charon dominates the sky on one side of Pluto but is never seen from the other side.

We've known for decades that Pluto is much smaller and less massive than any of the first eight planets, and its orbit is much more eccentric and inclined to the ecliptic plane. Its composition of ice and rock is also quite different from that of any of those planets, although it is virtually identical to that of many known comets. Moreover, since the early 1990s, astronomers have discovered more than 1000 objects of similar composition orbiting in Pluto's general neighborhood, which is the region of the solar system known as the *Kuiper belt*. Pluto is not even the largest of these Kuiper belt objects: Eris, discovered in 2005, is slightly larger than Pluto.

Scientifically, these facts leave no room for doubt that both Pluto and Eris belong to a different class of objects than the first eight planets. Rather than seeing Pluto as a "misfit" planet, we now see it and Eris as just the two largest among hundreds of large iceballs—essentially large comets—located in the Kuiper belt. The only question has been one of words: Should Pluto and Eris be called "planets" or something else? In 2006, the International Astronomical Union (IAU) voted to classify Pluto and Eris as *dwarf planets* (see Special Topic, p.11), thereby trimming the official list of planets to eight. More technically, the IAU defined a dwarf planet as any object that orbits the Sun and is large enough for its own gravity to have made it nearly round in shape, but that has not cleared its orbital neighborhood of other objects. The largest asteroid in the asteroid belt, Ceres, thereby also made the list of dwarf planets, and it is possible that dozens of other objects (mostly in the Kuiper belt) will also qualify.

The great distances and small sizes of Pluto and other dwarf planets make them difficult to study, regardless of whether they are located in the asteroid belt or the Kuiper belt. As you can see in Figure 6.11, even the best telescopic views of Pluto and its moons reveal little detail. Better information should be coming soon. A spacecraft called *New Horizons*, launched in 2006, will fly past Pluto in mid-2015 and may then visit other objects of the Kuiper belt. Meanwhile, the *Dawn* spacecraft, launched in 2007, should give us our first good views of large asteroids in the asteroid belt beginning in about 2011, with a pass by Ceres four years later that will nearly coincide with *New Horizons*'s pass by Pluto.

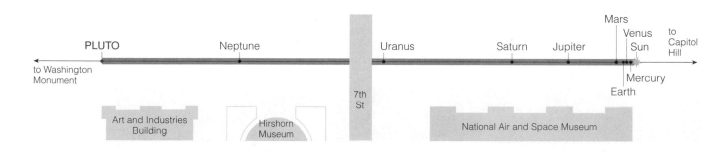

TABLE 6.1 *Planetary Data**

Photo	Planet	Relative Size	Average Distance from Sun (AU)	Average Equatorial Radius (km)	Mass (Earth = 1)	Average Density (g/m³)	Orbital Period	Rotation Period	Axis Tilt	Average Surface (or Cloud Top) Temperature†	Composition	Known Moons (2008)	Rings?
	Mercury	·	0.387	2440	0.055	5.43	87.9 days	58.6 days	0.0°	700 K (day) 100 K (night)	Rocks, metals	0	No
	Venus	•	0.723	6051	0.82	5.24	225 days	243 days	177.3°	740 K	Rocks, metals	0	No
	Earth	•	1.00	6378	1.00	5.52	1.00 year	23.93 hours	23.5°	290 K	Rocks, metals	1	No
	Mars	·	1.52	3397	0.11	3.93	1.88 years	24.6 hours	25.2°	220 K	Rocks, metals	2	No
	Jupiter	●	5.20	71,492	318	1.33	11.9 years	9.93 hours	3.1°	125 K	H, He, hydrogen compounds§	63	Yes
	Saturn	●	9.54	60,268	95.2	0.70	29.4 years	10.6 hours	26.7°	95 K	H, He, hydrogen compounds§	60	Yes
	Uranus	●	19.2	25,559	14.5	1.32	83.8 years	17.2 hours	97.9°	60 K	H, He, hydrogen compounds§	27	Yes
	Neptune	●	30.1	24,764	17.1	1.64	165 years	16.1 hours	29.6°	60 K	H, He, hydrogen compounds§	13	Yes
	Pluto	·	39.5	1160	0.0022	2.0	248 years	6.39 days	112.5°	40 K	Ices, rock	3	No
	Eris	·	67.7	1200	0.0028	2.3	557 years	?	?	?	Ices, rock	1	?

*Including the dwarf planets Pluto and Eris; Appendix E gives a more complete list of planetary properties.

†Surface temperatures for all objects except Jupiter, Saturn, Uranus, and Neptune, for which cloud-top temperatures are listed.

§Includes water (H_2O), methane (CH_4), and ammonia (NH_3).

(continued from page 143)

The planets are tiny compared to the distances between them, but they exhibit clear patterns of composition and motion.

In science, we always seek explanations for the existence of patterns like those evident in Figure 6.1 and the planetary tour. We will therefore devote most of this chapter to learning how our modern theory of solar system formation explains these and other features of the solar system. We will then see how recent discoveries of other planetary systems fit in with this theory, even as they have led us to refine some of its details.

 Orbits and Kepler's Laws Tutorial, Lessons 2–4

6.2 Clues to the Formation of Our Solar System

Let's begin by taking a more in-depth look at the general features of our solar system that must be explained by any successful theory of its origin. We can then discuss what theory best describes the major characteristics of our solar system and accounts for how it formed.

• What features of our solar system provide clues to how it formed?

We have already seen that our solar system is not a random collection of worlds but rather a family of worlds exhibiting many traits that would be difficult to attribute to coincidence. A valid theory of our solar system's formation must successfully account for these common traits. We could make a long list of such traits, but it is easier to develop a scientific theory by focusing on the more general structure of our solar system. For our purposes, four major features stand out:

1. **Patterns of motion among large bodies.** The Sun, planets, and large moons generally orbit and rotate in a very organized way.
2. **Two major types of planets.** The eight official planets divide clearly into two groups: the small, rocky planets that are close together and close to the Sun, and the large, gas-rich planets that are farther apart and farther from the Sun.
3. **Asteroids and comets.** Between and beyond the planets, vast numbers of asteroids and comets orbit the Sun; some are large enough to qualify as dwarf planets. The locations, orbits, and compositions of these asteroids and comets follow distinct patterns.
4. **Exceptions to the rules.** The generally orderly solar system also has some notable exceptions. For example, only Earth has a large moon among the inner planets, and Uranus is tipped on its side. A successful theory must make allowances for exceptions even as it explains the general rules.

Because these four features are so important to our study of the solar system, let's investigate each of them in a little more detail.

Feature 1: Patterns of Motion Among Large Bodies If you look back at Figure 6.1, you'll notice several clear patterns of motion among the large bodies of our solar system. (In this context, a "body" is simply an individual object such as the Sun, a planet, or a moon.) For example:

- All planetary orbits are nearly circular and lie nearly in the same plane.

- All planets orbit the Sun in the same direction: counterclockwise as viewed from high above Earth's North Pole.

- Most planets rotate in the same direction in which they orbit, with fairly small axis tilts. The Sun also rotates in this direction.

- Most of the solar system's large moons exhibit similar properties in their orbits around their planets, such as orbiting in their planet's equatorial plane in the same direction that the planet rotates.

The Sun, planets, and large moons orbit and rotate in an organized way. We consider these orderly patterns together as the first major feature of our solar system. As we'll see shortly, our theory of solar system formation explains these patterns as consequences of processes that occurred during the early stages of the birth of our solar system.

Feature 2: The Existence of Two Types of Planets

Our brief planetary tour showed that the four inner planets are quite different from the four large outer planets. We say that these two groups represent two distinct planetary classes: *terrestrial* and *jovian*.

Terrestrial planets are small, rocky, and close to the Sun. Jovian planets are large, gas-rich, and far from the Sun. The **terrestrial planets** are the four planets of the inner solar system: Mercury, Venus, Earth, and Mars. (*Terrestrial* means "Earth-like.") These planets are relatively small and dense, with rocky surfaces and an abundance of metals deep in their interiors. They have few moons, if any, and no rings. We often count our Moon as a fifth terrestrial world, because its history has been shaped by the same processes that have shaped the terrestrial planets.

The **jovian planets** are the four large planets of the outer solar system: Jupiter, Saturn, Uranus, and Neptune. (*Jovian* means "Jupiter-like.") The jovian planets are much larger in size and lower in average density than the terrestrial planets, and they have rings and many moons. They lack solid surfaces and are made mostly of hydrogen, helium, and **hydrogen compounds**—compounds containing hydrogen, such as water (H_2O), ammonia (NH_3), and methane (CH_4). Because these substances are gases under earthly conditions, the jovian planets are sometimes called "gas giants." However, throughout most of the jovian planet interiors, the pressures and temperatures are too high for these "gases" to remain in gaseous form. Instead, they become liquid or enter phases unlike anything we ordinarily see on Earth. Table 6.2 contrasts the general traits of the terrestrial and jovian planets.

Notice that Pluto and Eris are left out in the cold, both literally and figuratively. They are much smaller and farther from the Sun (on average) than any of the eight official planets. As we discussed in our tour (p. 155) and will discuss further in Chapter 9, Pluto and Eris are just the two largest of many similar objects orbiting the Sun beyond Neptune—which is a major reason why both are now classified as dwarf planets.

Feature 3: Asteroids and Comets

The third major feature of the solar system is the existence of vast numbers of small objects orbiting the Sun. These objects fall into two major groups: asteroids and comets.

TABLE 6.2 *Comparison of Terrestrial and Jovian Planets*

Terrestrial Planets	Jovian Planets
Smaller size and mass	Larger size and mass
Higher density	Lower density
Made mostly of rock and metal	Made mostly of hydrogen, helium, and hydrogen compounds
Solid surface	No solid surface
Few (if any) moons and no rings	Rings and many moons
Closer to the Sun (and closer together), with warmer surfaces	Farther from the Sun (and farther apart), with cool temperatures at cloud tops

Rocky asteroids and icy comets far outnumber the planets and their moons.

Asteroids are rocky bodies that orbit the Sun much like planets, but they are much smaller (Figure 6.12). Even Ceres, an asteroid large enough to qualify as a dwarf planet, is considerably smaller than our Moon. Most known asteroids are found within the **asteroid belt** between the orbits of Mars and Jupiter (see Figure 6.1).

Comets are also small objects that orbit the Sun, but they are made largely of ices (such as water ice, ammonia ice, and methane ice) mixed with rock. You are probably familiar with the occasional appearance of comets in the inner solar system, where they may become visible to the naked eye with long, beautiful tails (Figure 6.13). These visitors, which may delight sky watchers for a few weeks or months, are actually quite rare among comets.

The vast majority of comets never visit the inner solar system. Instead, they orbit the Sun in one of two distinct regions in the outer reaches of the solar system. The first is a donut-shaped region beyond the orbit of Neptune that we call the **Kuiper belt** (*Kuiper* rhymes with *piper*). Pluto and Eris are the largest known objects in the Kuiper belt, which contains at least 100,000 icy objects. The second cometary region is much farther from the Sun and roughly spherical in shape, and it is called the **Oort cloud** (*Oort* rhymes with *court*). The Oort cloud may contain a trillion comets. You can see the general regions of the Kuiper belt and Oort cloud in feature 3 of Figure 6.1.

Feature 4: Exceptions to the Rules The fourth key feature of our solar system is that there are a few notable exceptions to the general rules. Two such exceptions are the rotations of Uranus and Venus: While most of the planets rotate in the same direction as they orbit, Uranus rotates nearly on its side and Venus rotates "backward" (clockwise as viewed from high above Earth's North Pole). Similarly, while most large moons orbit their planets in the same direction as their planets rotate, many small moons have much more unusual orbits.

A successful theory of solar system formation must allow for exceptions to the general rules.

One of the most interesting exceptions concerns our own Moon. While the other terrestrial planets have either no moons (Mercury and Venus) or very tiny moons (Mars), Earth has one of the largest moons in the solar system.

We are now ready to turn our attention to the task of finding a theory that can explain all four features. First, however, you might wish to look back at Figure 6.1, which summarizes them.

• What theory best explains the features of our solar system?

After the Copernican revolution, many scientists speculated about the origin of the solar system. However, we generally credit two 18th-century scientists with proposing the hypothesis that ultimately blossomed into our modern scientific theory of the origin of the solar system. Around 1755, German philosopher Immanuel Kant proposed that our solar system formed from the gravitational collapse of an interstellar cloud of gas. About 40 years later, French mathematician Pierre-Simon

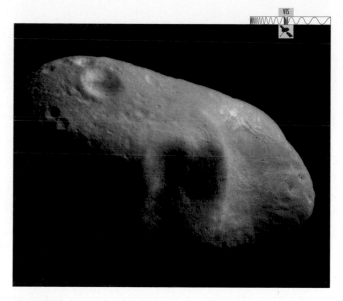

Figure 6.12

The asteroid Eros (photographed from the *NEAR* spacecraft). Its appearance is probably typical of most asteroids. Eros is about 40 kilometers in length, and like other small objects in the solar system, it is not spherical.

Figure 6.13

Comet Hale–Bopp, photographed over Boulder, Colorado, during its appearance in 1997.

Laplace put forth the same idea independently. Because an interstellar cloud is usually called a *nebula* (Latin for "cloud"), their idea became known as the *nebular hypothesis*.

The nebular hypothesis remained popular throughout the 19th century. By the early 20th century, however, scientists had found a few aspects of our solar system that the nebular hypothesis did not seem to explain well—at least in its original form as described by Kant and Laplace. While some scientists sought to modify the nebular hypothesis, others looked for entirely different explanations for how the solar system might have formed.

During the first half of the 20th century, the nebular hypothesis faced stiff competition from a hypothesis proposing that the planets represent debris from a near-collision between the Sun and another star. According to this *close encounter hypothesis,* the planets formed from blobs of gas that had been gravitationally pulled out of the Sun during the near-collision.

Today, the close encounter hypothesis has been discarded. It began to lose favor when calculations showed that it could not account for either the observed orbital motions of the planets or the neat division of the planets into two major categories (terrestrial and jovian). Moreover, the close encounter hypothesis required a highly improbable event: a near-collision between our Sun and another star. Given the vast separation between star systems in our region of the galaxy, the chance of such an encounter is so small that it would be difficult to imagine it happening even in the one case needed to make our own solar system. It certainly could not account for the many other planetary systems that we have discovered in recent years.

The nebular theory holds that our solar system formed from the gravitational collapse of a great cloud of gas, and it explains all the general features of our solar system.

While the close encounter hypothesis was losing favor, new discoveries about the physics of planet formation led to modifications of the nebular hypothesis. Using more sophisticated models of the processes that occur in a collapsing cloud of gas, scientists realized that the nebular hypothesis offered natural explanations for all four general features of our solar system. Indeed, so much evidence has accumulated in favor of the nebular hypothesis that it has achieved the status of a scientific *theory* [Section 3.4]—the **nebular theory** of our solar system's birth. In the rest of this chapter, we will discuss how the nebular theory explains the major features of our solar system and learn how it has been further improved through the study of other planetary systems.

(MA) Formation of the Solar System Tutorial, Lessons 1–2

6.3 The Birth of the Solar System

According to the nebular theory, the first general feature of our solar system—orderly patterns of motion—arose during the early stages of the birth of our solar system. In this section, we'll see how these patterns arose, and why similar patterns should be expected in other planetary systems.

• Where did the solar system come from?

The nebular theory begins with the idea that our solar system was born from a cloud of gas, called the **solar nebula,** that collapsed under its own gravity. But where did this gas come from?

According to current understanding, the gas that gave birth to our solar system was the product of billions of years of galactic recycling that occurred before the Sun and planets were born. Recall that the universe as a whole is thought to have been born in the Big Bang [Section 1.1], which essentially produced only two chemical elements [Section 1.1]: hydrogen and helium. Heavier elements were produced later by massive stars, through nuclear fusion in their cores and the explosions that accompany their deaths. Once they are released into space, these elements can be recycled into new generations of stars (Figure 6.14).

> The gas that made up the solar nebula contained hydrogen and helium from the Big Bang and heavier elements produced by stars.

The heavy element content of the galaxy rose with each generation of massive stars, so the gas cloud that gave birth to our Sun should have contained more heavy elements than those that gave birth to more ancient stars. Nevertheless, the chemical composition of the galaxy has remained predominantly hydrogen and helium. By studying the composition of the Sun, of other stars of the same age, and of interstellar gas clouds, we have learned that the gas that made up the solar nebula contained (by mass) about 98% hydrogen and helium and 2% all other elements combined. The Sun still has this basic composition, while the planets tend to have higher proportions of heavy elements (for reasons we will discuss in Section 6.4). In particular, Earth and the other terrestrial worlds ended up being made primarily from the heavier elements mixed within the solar nebula.

> **think about it** Could a solar system like ours have formed with the first generation of stars after the Big Bang? Explain.

Strong observational evidence supports this scenario. With telescopes we can witness stars in the process of formation today, and these forming stars are always found within interstellar clouds [Section 12.1]. Moreover, careful studies of gas in the Milky Way Galaxy have allowed us to put together a clear picture of the entire galactic recycling process [Section 14.2], leaving little doubt that new stars are born from gas that has been recycled from prior generations of stars. As we saw in Chapter 1, we are "star stuff," because we and our planet are made of elements forged in stars that lived and died long ago.

• What caused the orderly patterns of motion in our solar system?

The solar nebula probably began as a large and roughly spherical cloud of very cold, low-density gas. Initially, this gas was so spread out—perhaps over a region a few light-years in diameter—that gravity alone may not have been strong enough to pull it together and start its collapse. Instead, the collapse may have been triggered by a cataclysmic event, such as the impact of a shock wave from the explosion of a nearby star (a supernova).

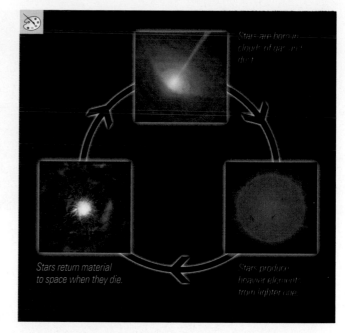

Figure 6.14 interactive figure

This figure (which is a portion of Figure 1.2) summarizes the galactic recycling process.

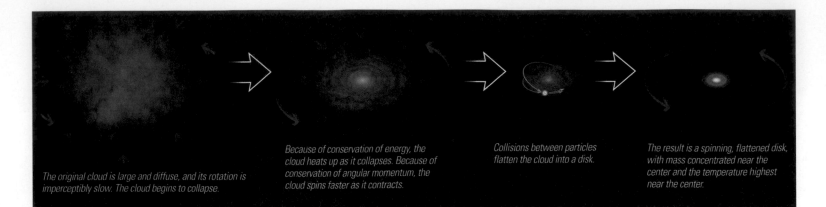

The original cloud is large and diffuse, and its rotation is imperceptibly slow. The cloud begins to collapse.

Because of conservation of energy, the cloud heats up as it collapses. Because of conservation of angular momentum, the cloud spins faster as it contracts.

Collisions between particles flatten the cloud into a disk.

The result is a spinning, flattened disk, with mass concentrated near the center and the temperature highest near the center.

Figure 6.15 interactive figure

This sequence of illustrations shows how the gravitational collapse of a large cloud of gas causes it to become a spinning disk of matter. The hot, dense central bulge becomes a star, while planets can form in the surrounding disk.

Once the collapse started, the law of gravity ensured that it would continue. Remember that the strength of gravity follows an inverse square law with distance [Section 4.4]. Because the mass of the cloud remained the same as it shrank, the strength of gravity increased as the diameter of the cloud decreased. For example, when the diameter decreased by half, the force of gravity increased by a factor of four.

Because gravity pulls inward in all directions, you might at first guess that the solar nebula would have remained spherical as it shrank. Indeed, the idea that gravity pulls in all directions explains why the Sun and the planets are spherical. However, we must also consider other physical laws that apply to a collapsing gas cloud in order to understand how orderly motions arose in the solar nebula.

Heating, Spinning, and Flattening As the solar nebula shrank in size, three important processes altered its density, temperature, and shape, changing it from a large, diffuse (spread-out) cloud to a much smaller spinning disk (Figure 6.15):

- *Heating.* The temperature of the solar nebula increased as it collapsed. Such heating represents energy conservation in action [Section 4.3]. As the cloud shrank, its gravitational potential energy was converted to the kinetic energy of individual gas particles falling inward. These particles crashed into one another, converting the kinetic energy of their inward fall to the random motions of thermal energy (see Figure 4.12b). The Sun formed in the center, where temperatures and densities were highest.

- *Spinning.* Like an ice skater pulling in her arms as she spins, the solar nebula rotated faster and faster as it shrank in radius. This increase in rotation rate represents conservation of angular momentum in action [Section 4.3]. The rotation of the cloud may have been imperceptibly slow before its collapse began, but the cloud's shrinkage made fast rotation inevitable. The rapid rotation helped ensure that not all the material in the solar nebula collapsed into the center: The greater the angular momentum of a rotating cloud, the more spread out it will be.

- *Flattening.* The solar nebula flattened into a disk. This flattening is a natural consequence of collisions between particles in a spinning cloud. A cloud may start with any size or shape, and different clumps of gas within the cloud may be moving in random directions at random speeds. These clumps collide and merge as the cloud collapses, and each new clump has the average velocity of the clumps that formed it. In this way, the random motions of the original cloud become more orderly as the cloud collapses, changing the cloud's original lumpy shape into a rotating, flattened disk. Similarly, collisions between clumps of material in highly elliptical orbits reduce their eccentricities, making their orbits more circular.

The orderly motions of our solar system today are a direct result of the solar system's birth in a spinning, flattened cloud of gas.

The formation of the spinning disk explains the orderly motions of our solar system today. The planets all orbit the Sun in nearly the same plane because they formed in the flat disk. The direction in which the disk was spinning became the direction of the Sun's rotation and the orbits of the planets. Computer models show that planets would have tended to rotate in the same direction as they formed—which is why most planets rotate the same way—though the small sizes of planets compared to the entire disk allowed some exceptions to arise. The fact that collisions in the disk tended to make orbits more circular explains why most planets have nearly circular orbits.

see it for yourself You can demonstrate the development of orderly motion, much as it occurred in the solar system, by sprinkling pepper into a bowl of water and stirring it quickly in random directions. The water molecules constantly collide with one another, so the motion of the pepper grains will tend to settle into a slow rotation representing the average of the original, random velocities. Try the experiment several times, stirring the water differently each time. Do the random motions ever cancel out exactly, resulting in no rotation at all? Describe what occurs, and explain how this is similar to what took place in the solar nebula.

Testing the Model Because the same processes should affect other collapsing gas clouds, we can test our model by searching for disks around other forming stars. Observational evidence does indeed support our model of spinning, heating, and flattening.

The heating that occurs in a collapsing cloud of gas means the gas should emit thermal radiation [Section 5.2], primarily in the infrared. We've detected infrared radiation from many nebulae where star systems appear to be forming. More direct evidence comes from flattened, spinning disks around other stars (Figure 6.16), some of which appear to be ejecting jets of

Figure 6.16

These Hubble Space Telescope photos show flattened, spinning disks of material around other stars.

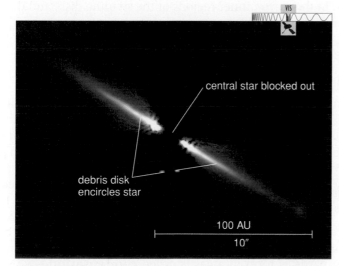

a We see this disk edge-on around the star AU Microscopii, confirming its flattened shape.

central star blocked out

debris disk encircles star

100 AU
10"

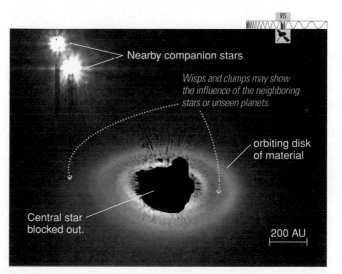

Nearby companion stars

Wisps and clumps may show the influence of the neighboring stars or unseen planets.

orbiting disk of material

Central star blocked out.

200 AU

b This photo shows a disk around the star HD 141569A. The colors are not real; a black-and-white image has been tinted red to bring out faint detail.

material perpendicular to their disks [Section 12.1]. These jets are thought to result from the flow of material from the disk onto the forming star.

Observations of disks around other stars support the idea that our own solar system was once a spinning disk of gas.

Other support for the model comes from computer simulations of the formation process. A simulation begins with a set of data representing the conditions we observe in interstellar clouds. Then, with the aid of a computer, we apply the laws of physics to predict the changes that should occur over time. These computer simulations successfully reproduce most of the general characteristics of motion in our solar system, suggesting that the nebular theory is on the right track.

Additional evidence that our ideas about the formation of flattened disks are correct comes from many other structures in the universe. We expect flattening to occur anywhere that orbiting particles can collide, which explains why we find so many cases of flat disks, including the disks of spiral galaxies like the Milky Way, the disks of planetary rings, and the *accretion disks* surrounding many neutron stars and black holes [Sections 13.3 and 13.4].

MA Formation of the Solar System Tutorial, Lesson 3

6.4 The Formation of Planets

The planets began to form after the solar nebula had collapsed into a flattened disk of perhaps 200 AU in diameter (about twice the present-day diameter of Pluto's orbit). In this section, we'll discuss planetary formation and address three major features of our solar system that we have not yet explained: the existence of two types of planets, the existence of asteroids and comets, and the exceptions to the rules.

• Why are there two major types of planets?

The churning and mixing of gas in the solar nebula should have ensured that the nebula had the same composition throughout, so how did the terrestrial planets end up being so different in composition from the jovian planets? The key clue comes from their locations: The terrestrial planets formed in the warm, inner regions of the swirling disk, and the jovian planets formed in the colder, outer regions.

Condensation: Sowing the Seeds of Planets In the center of the collapsing solar nebula, gravity drew together enough material to form the Sun. In the surrounding disk, however, the gaseous material was too spread out for gravity alone to clump it together. Instead, material had to begin clumping in some other way and to grow in size until gravity could start pulling it together into planets. In essence, planet formation required the presence of "seeds"—solid bits of matter around which gravity could ultimately build planets.

Planet formation began around tiny "seeds" of solid metal, rock, or ice.

The basic process of seed formation was probably much like the formation of snowflakes in clouds on Earth: When the temperature is low enough, some atoms or molecules in a gas may bond and solidify. The general process in which solid (or liquid) particles form in a gas is called **condensation**—we say that the particles *condense* out of the gas. These particles start out microscopic in size, but they can grow larger with time.

TABLE 6.3 *Materials in the Solar Nebula*

A summary of the four types of materials present in the solar nebula. The squares represent the relative proportions of each type (by mass).

	Examples	Typical condensation temperature	Relative abundance (by mass)
Hydrogen and Helium Gas	hydrogen, helium	do not condense in nebula	98%
Hydrogen Compounds	water (H_2O) methane (CH_4) ammonia (NH_3)	<150 K	1.4%
Rock	various minerals	500–1300 K	0.4%
Metals	iron, nickel, aluminum	1000–1600 K	0.2%

Different materials condense at different temperatures. As summarized in Table 6.3, the ingredients of the solar nebula fell into four major categories:

- **Hydrogen and helium gas (98% of the solar nebula).** These gases never condense under the conditions present in a nebula.

- **Hydrogen compounds (1.4% of the solar nebula).** Materials such as water (H_2O), methane (CH_4), and ammonia (NH_3) can solidify into **ices** at low temperatures (below about 150 K under the low pressure of the solar nebula).

- **Rock (0.4% of the solar nebula).** Rocky material is gaseous at high temperatures but condenses into solid bits of mineral at temperatures between about 500 K and 1300 K, depending on the type of rock. (A *mineral* is a piece of rock with a particular chemical composition and structure.)

- **Metals (0.2% of the solar nebula).** Metals such as iron, nickel, and aluminum are also gaseous at very high temperatures but condense into solid form at temperatures higher than rock—typically in the range of 1000 K to 1600 K.

Because hydrogen and helium gas made up 98% of the solar nebula's mass and did not condense, the vast majority of the nebula remained gaseous at all times. However, other materials could condense wherever the temperature allowed (Figure 6.17). Close to the forming Sun, where the temperature was above 1600 K, it was too hot for any material to condense. Near what is now Mercury's orbit, the temperature was low enough for metals and some types of rock to condense into tiny solid particles, but other types of rock and all the hydrogen compounds remained gaseous. More types of rock could condense, along with the metals, at the distances from the Sun where Venus, Earth, and Mars would form. In the region where the asteroid belt would eventually be located, temperatures were low enough to allow dark, carbon-rich minerals to condense, along with minerals containing small amounts of water. Hydrogen compounds could condense into ices only beyond the **frost line**—the minimum distance at which it was cold enough for ice to condense—which lay between the present-day orbits of Mars and Jupiter.

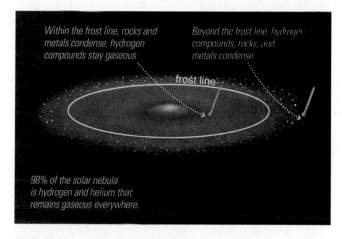

Within the frost line, rocks and metals condense, hydrogen compounds stay gaseous

Beyond the frost line, hydrogen compounds, rocks, and metals condense

frost line

98% of the solar nebula is hydrogen and helium that remains gaseous everywhere.

Figure 6.17 interactive figure

Temperature differences in the solar nebula led to different kinds of condensed materials, sowing the seeds for two different kinds of planets.

think about it Consider a region of the solar nebula in which the temperature was about 1300 K. Based on the data in Table 6.3, what fraction of the material in this region was gaseous? What were the solid particles in this region made of? Answer the same questions for a region with a temperature of 100 K. Would the 100 K region be closer to or farther from the Sun? Explain.

The solid seeds in the inner solar system were made only of metal and rock, but in the outer solar system they included the more abundant ices.

The frost line marked the key transition between the warm inner regions of the solar system where terrestrial planets formed and the cool outer regions where jovian planets formed. Inside the frost line, only metal and rock could condense into solid "seeds," which is why the terrestrial planets ended up being made of metal and rock. Beyond the frost line, where it was cold enough for hydrogen compounds to condense into ices, the solid seeds were built of ice along with metal and rock. Moreover, because hydrogen compounds were nearly three times as abundant in the nebula as metal and

common misconceptions

Solar Gravity and the Density of Planets

You might think that it was the Sun's gravity that pulled the dense rocky and metallic materials to the inner part of the solar nebula, or that gases escaped from the inner nebula because gravity couldn't hold them. But this is not the case—all the ingredients were orbiting the Sun together under the influence of the Sun's gravity. The orbit of a particle or a planet does not depend on its size or density, so the Sun's gravity cannot be the cause of the different kinds of planets. Rather, the different temperatures in the solar nebula are the cause.

rock combined (see Table 6.3), the total amount of solid material was far greater beyond the frost line than within it. The stage was set for the birth of two types of planets: planets born from seeds of metal and rock in the inner solar system and planets born from seeds of ice (as well as metal and rock) in the outer solar system.

Building the Terrestrial Planets From this point, the story of the inner solar system seems fairly clear: The solid seeds of metal and rock in the inner solar system ultimately grew into the terrestrial planets we see today, but these planets ended up relatively small in size because rock and metal made up such a small amount of the material in the solar nebula.

The process by which small "seeds" grew into planets is called **accretion** (Figure 6.18). Accretion began with the microscopic solid particles that condensed from the gas of the solar nebula. These particles orbited the forming Sun with the same orderly, circular paths as the gas from which they condensed. Individual particles therefore moved at nearly the same speed as neighboring particles, so "collisions" were more like gentle touches. Although the particles were far too small to attract each other gravitationally at this point, they were able to stick together through electrostatic forces—the same "static electricity" that makes hair stick to a comb. Small particles thereby began to combine into larger ones. As the particles grew in mass, gravity began to aid in the accretion process, accelerating their growth into boulders large enough to count as **planetesimals**, which means "pieces of planets."

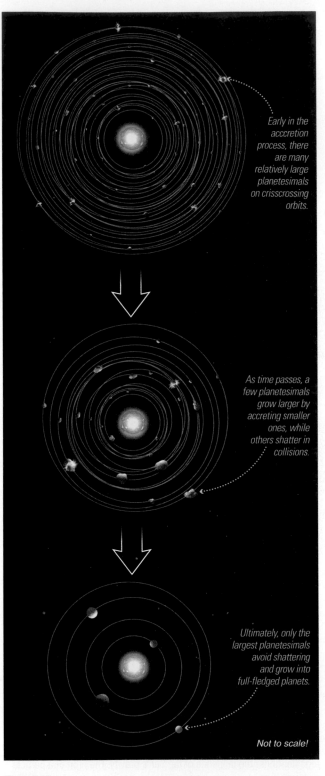

Early in the accretion process, there are many relatively large planetesimals on crisscrossing orbits.

As time passes, a few planetesimals grow larger by accreting smaller ones, while others shatter in collisions.

Ultimately, only the largest planetesimals avoid shattering and grow into full-fledged planets.

Not to scale!

Figure 6.18

These diagrams show how planetesimals gradually accrete into terrestrial planets.

The terrestrial planets were made from the solid bits of metal and rock that condensed in the inner solar system.

The planetesimals grew rapidly at first. As they grew large, they had both more surface area to make contact with other planetesimals and more gravity to attract them. Some planetesimals probably grew to hundreds of kilometers in size in only a few million years—a long time in human terms, but only about $\frac{1}{1000}$ the present age of the solar system. However, once the planetesimals reached these relatively large sizes, further growth became more difficult.

Gravitational encounters [Section 4.4] between planetesimals tended to alter their orbits, particularly those of the smaller planetesimals. With different orbits crossing each other, collisions between planetesimals tended to occur at higher speeds and hence became more destructive. Such collisions tended to shatter planetesimals rather than help them grow. Only the largest planetesimals avoided being shattered and could grow into full-fledged planets.

Theoretical evidence in support of this model comes from computer simulations of the accretion process. Observational evidence comes from meteorites that appear to be surviving fragments from the period of condensation [Section 9.1]. These meteorites contain metallic grains embedded in rocky minerals (Figure 6.19), just as we would expect if metal and rock condensed in the inner solar system. Meteorites thought to come from the outskirts of the asteroid belt contain abundant carbon-rich materials, and some contain water—again, as we would expect for material that condensed in that region.

Making the Jovian Planets Accretion should have occurred similarly in the outer solar system, but condensation of ices meant both that there was more solid material and that this material contained ice in addition to metal and rock. The solid objects that reside in the outer solar system today, such as comets and the moons of the jovian planets, still show this ice-rich composition. However, the growth of icy planetesimals cannot be the whole story of jovian planet formation, because the jovian planets themselves are *not* made mostly of ice. Instead, the leading model for jovian planet formation holds that these planets formed as gravity drew gas around large, icy planetesimals.

According to this model, the abundance of ices in the outer solar system should have allowed some icy planetesimals to grow to masses many times that of Earth. With these large masses, their gravity became strong enough to capture and hold some of the hydrogen and helium gas that made up the vast majority of the surrounding solar nebula. As the growing planets accumulated gas, their gravity grew stronger still, allowing them to capture even more gas. Ultimately, the jovian planets accreted so much gas that they bore little resemblance to the icy seeds from which they started.

The jovian planets began as large, icy planetesimals, which then captured hydrogen and helium gas from the solar nebula.

This model also explains most of the large moons of the jovian planets. The same processes of heating, spinning, and flattening that made the disk of the solar nebula should also have affected the gas drawn by gravity to the young jovian planets. Each jovian planet came to be surrounded by its own disk of gas, spinning in the same direction as the planet rotated (Figure 6.20). Moons that accreted from icy planetesimals within these disks therefore ended up with nearly circular orbits going in the same direction as their planet's rotation and lying close to their planet's equatorial plane.

This model of accretion followed by gas capture explains the observed features of the jovian planets quite well. Nevertheless, some scientists have advanced a competing model in which disturbances in the disk of the solar nebula could have led clumps of gas to collapse and form jovian planets (without first forming icy planetesimals). Most planetary scientists still favor the accretion and gas capture model, but good practice of science demands that we fully consider this alternative model; planetary scientists are therefore exploring it to see if it can work.

Clearing the Nebula The vast majority of the hydrogen and helium gas in the solar nebula never became part of any planet. So what happened to it? Apparently, it was cleared away by a combination of radiation from the young Sun and the *solar wind*—a stream of charged particles continually blown outward in all directions from the Sun [Section 10.1]. Although the solar wind is fairly weak today, observations show that stars tend to have much stronger winds when they are young. The young Sun therefore should have had a strong solar wind—strong enough to have swept huge quantities of gas out of the solar system.

Remaining gas in the solar nebula was cleared away into space, ending the era of planet formation.

The clearing of the gas sealed the compositional fate of the planets. If the gas had remained longer, it might have continued to cool until hydrogen compounds could have condensed into ices even in the inner solar system. In that case, the terrestrial planets might have accreted abundant ice, and perhaps hydrogen and helium gas as well, changing their basic nature. At the other extreme, if the gas had been blown out much earlier, the raw materials of the planets might have been swept away before the planets could fully form. Although these extreme scenarios did not occur in our solar system, they may sometimes occur around other stars. Planet formation may also sometimes be interrupted when radiation from hot, neighboring stars drives away material in a solar nebula.

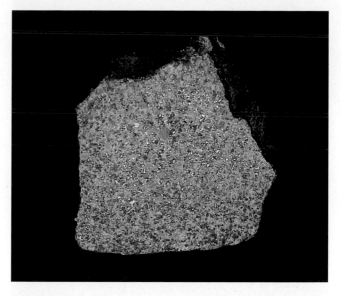

Figure 6.19

Shiny flakes of metal are clearly visible in this slice through a meteorite (a few centimeters across), mixed in among the rocky material. Such metallic flakes are just what we would expect to find if condensation really occurred in the solar nebula as described by the nebular theory.

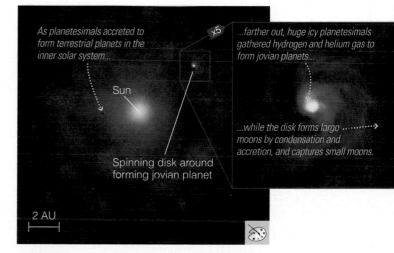

Figure 6.20

The forming jovian planets were surrounded by disks of gas, much like the disk of the entire solar nebula but smaller in size. According to the leading model, the planets grew as large, icy planetesimals that captured hydrogen and helium gas from the solar nebula. This painting shows the gas and planetesimals surrounding one jovian planet in the larger solar nebula.

Figure 6.21

Around 4 billion years ago, Earth, its Moon, and the other planets were heavily bombarded by leftover planetesimals. This painting shows the young Earth and Moon, with an impact in progress on Earth.

• Where did asteroids and comets come from?

The process of planet formation also explains the origin of the many asteroids and comets that populate our solar system (including those large enough to qualify as dwarf planets): They are "leftovers" from the era of planet formation. Asteroids are the rocky leftover planetesimals of the inner solar system, while comets are the icy leftover planetesimals of the outer solar system. We'll see in Chapter 9 why most asteroids ended up grouped in the asteroid belt while most comets ended up split between two regions (the Kuiper belt and the Oort cloud).

Rocky asteroids and icy comets are leftover planetesimals from the era of planet formation.

Evidence that asteroids and comets are leftover planetesimals comes from analysis of meteorites, spacecraft visits to comets and asteroids, and theoretical models of solar system formation. In fact, the nebular theory allowed scientists to make predictions about the locations of comets that weren't verified until decades later, when we discovered large comets orbiting in the vicinity of Neptune and Pluto.

The asteroids and comets that exist today probably represent only a small fraction of the leftover planetesimals that roamed the young solar system. The rest are now gone. Some of these "lost" planetesimals may have been flung into deep space by gravitational encounters, but many others must have collided with the planets. When impacts occur on solid worlds, they leave behind *impact craters* as scars. These impacts have transformed planetary landscapes, and, in the case of Earth, they have altered the course of evolution. For example, an impact is thought to have been responsible for the death of the dinosaurs [Section 9.4].

Leftover planetesimals battered the planets during the solar system's first few hundred million years.

Although impacts occasionally still occur, the vast majority of these collisions occurred in the first few hundred million years of our solar system's history, during the period we call the **heavy bombardment.** Every world in our solar system must have been pelted by impacts during the heavy bombardment (Figure 6.21), and most of the craters we see on the Moon and other worlds date from this period. These impacts did more than just batter the planets. They also brought materials from other regions of the solar system—a fact that is critical to our existence on Earth today.

Remember that the terrestrial planets were built from planetesimals made of metal and rock. These planetesimals probably contained no water or other hydrogen compounds at all, because it was too hot for these compounds to condense in our region of the solar nebula. How, then, did Earth come to have the water that makes up our oceans and the gases that first formed our atmosphere? The likely answer is that water, along with other hydrogen compounds, was brought to Earth and other terrestrial planets by the impacts of water-bearing planetesimals that formed farther from the Sun. Remarkably, the water we drink and the air we breathe probably once were part of planetesimals that accreted beyond the orbit of Mars.

• How do we explain the existence of our Moon and other exceptions to the rules?

We have now explained all the major features of our solar system except for the exceptions to the rules, including our surprisingly large Moon. Today, we think that most of these exceptions arose from collisions or close gravitational encounters.

Captured Moons We have explained the orbits of most large jovian planet moons by their formation in a disk that swirled around the forming planet. But how do we explain moons with less orderly orbits, such as those that go in the "wrong" direction (opposite their planet's rotation) or that have large inclinations to their planet's equator? These moons are probably leftover planetesimals that originally orbited the Sun but were then captured into planetary orbit.

It's not easy for a planet to capture a moon. An object cannot switch from an unbound orbit (for example, an asteroid whizzing by Jupiter) to a bound orbit (for example, a moon orbiting Jupiter) unless it somehow loses orbital energy [Section 4.4]. For the jovian planets, captures probably occurred when passing planetesimals lost energy to drag in the extended and relatively dense gas that surrounded these planets as they formed. The planetesimals would have been slowed by friction with the gas, just as artificial satellites are slowed by drag in encounters with Earth's atmosphere. If friction reduced a passing planetesimal's orbital energy enough, it could have become an orbiting moon. Because of the random nature of the capture process, captured moons would not necessarily orbit in the same direction as their planet or in its equatorial plane. Most of the small moons of the jovian planets are a few kilometers across, supporting the idea that they were captured in this way. Mars may have similarly captured its two small moons, Phobos and Deimos, at a time when the planet had a much more extended atmosphere than it does today (Figure 6.22).

The Giant Impact Formation of Our Moon

Capture processes cannot explain our own Moon, because it is much too large to have been captured by a small planet like Earth. We can also rule out the possibility that our Moon formed simultaneously with Earth, because if both had formed together, they would have accreted from planetesimals of the same type and should therefore have approximately the same composition and density. But this is not the case: The Moon's density is considerably lower than Earth's, indicating that it has a very different average composition. So how did we get our Moon? Today, the leading hypothesis suggests that it formed as the result of a **giant impact** between Earth and a huge planetesimal.

According to models, a few leftover planetesimals may have been as large as Mars. If one of these Mars-size objects struck a young planet, the blow might have tilted the planet's axis, changed the planet's rotation rate, or completely shattered the planet. The giant impact hypothesis holds that a Mars-size object hit Earth at a speed and angle that blasted Earth's outer layers into space. According to computer simulations, this material could have collected into orbit around our planet, and accretion within this ring of debris could have formed the Moon (Figure 6.23).

Our Moon is probably the result of a giant impact that blasted Earth's outer layers into orbit, where the material accreted to form the Moon.

Strong support for the giant impact hypothesis comes from two features of the Moon's composition. First, the Moon's overall composition is quite similar to that of Earth's outer layers—just as we should expect if it were made from material blasted away from those layers. Second, the Moon has a much smaller proportion of easily vaporized ingredients (such as water) than Earth. This fact supports the hypothesis because the heat of the impact would have vaporized these ingredients. As gases, they would not have participated in the process of accretion that formed the Moon.

Other Exceptions

Giant impacts may also explain other exceptions to the general trends. For example, Pluto's moon Charon shows signs of

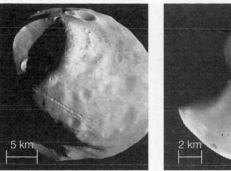
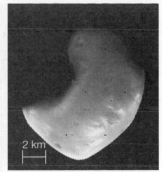

a Phobos   **b** Deimos

Figure 6.22

The two moons of Mars are probably captured asteroids. Phobos (imaged by *Mars Express*) is only about 13 kilometers across, and Deimos (imaged by the *Viking* orbiter) is only about 8 kilometers across—making each of these two moons small enough to fit within the boundaries of a typical large city.

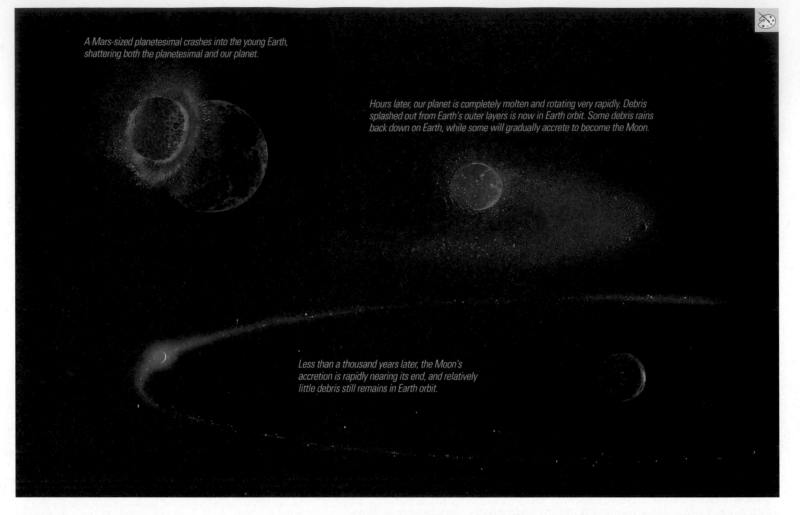

A Mars-sized planetesimal crashes into the young Earth, shattering both the planetesimal and our planet.

Hours later, our planet is completely molten and rotating very rapidly. Debris splashed out from Earth's outer layers is now in Earth orbit. Some debris rains back down on Earth, while some will gradually accrete to become the Moon.

Less than a thousand years later, the Moon's accretion is rapidly nearing its end, and relatively little debris still remains in Earth orbit.

Figure 6.23

Artist's conception of the giant impact hypothesis for the formation of our Moon. The fact that ejected material came mostly from Earth's outer rocky layers explains why the Moon contains very little metal. The impact must have occurred more than 4.4 billion years ago, since that is the age of the oldest Moon rocks. As shown, the Moon formed quite close to a rapidly rotating Earth, but over billions of years, tidal forces have slowed Earth's rotation and moved the Moon's orbit outward [Section 4.4].

having formed in a giant impact similar to the one thought to have formed our Moon, and Mercury's surprisingly high density may be the result of a giant impact that blasted away its outer, lower-density layers. Giant impacts could have also been responsible for tilting the axes of many planets (including Earth) and perhaps for tipping Uranus on its side. Venus's slow and backward rotation could also be the result of a giant impact, though some scientists suspect it is a consequence of processes attributable to Venus's thick atmosphere.

Although we cannot definitively explain these exceptions to the general rules, the overall lesson is clear: The chaotic processes that accompanied planet formation, including the many collisions that surely occurred, are *expected* to have led to at least a few exceptions. We therefore conclude that nebular theory can account for all four of the major features of our solar system. Figure 6.24 summarizes what we have discussed.

• When did the planets form?

Computer models of planetary formation suggest that the entire process took no more than about 50 million years, and perhaps significantly less. But when did it all occur, and how do we know? The answer is that the planets began to form through accretion just over $4\frac{1}{2}$ billion years ago, a fact we learn by determining the age of the oldest rocks in the solar system.

Dating Rocks The most reliable method for measuring the age of a rock is **radiometric dating**, which relies on careful measurement of the

Figure 6.24

A summary of the process by which our solar system formed, according to the nebular theory.

A large, diffuse interstellar gas cloud (solar nebula) contracts due to gravity.

Contraction of Solar Nebula: *As it contracts, the cloud heats, flattens, and spins faster, becoming a spinning disk of dust and gas.*

The Sun will be born in the center.

Planets will form in the disk.

Warm temperatures allow only metal/rock "seeds" to condense in inner solar system.

Condensation of Solid Particles: *Hydrogen and helium remain gaseous, but other materials can condense into solid "seeds" for building planets.*

Cold temperatures allow "seeds" to contain abundant ice in the outer solar system.

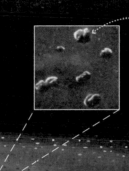

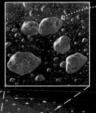

Terrestrial planets are built from metal and rock.

Accretion of Planetesimals: *Solid "seeds" collide and stick together. Larger ones attract others with their gravity, growing bigger still.*

The seeds of jovian planets grow large enough to attract hydrogen and helium gas, making them into giant, mostly gaseous planets; moons form in disks of dust and gas that surround the planets.

Clearing the Nebula: *The solar wind blows remaining gas into interstellar space.*

Terrestrial planets remain in the inner solar system.

Jovian planets remain in the outer solar system.

"Leftovers" from the formation process become asteroids (metal/rock) and comets (mostly ice).

Not to scale

proportions of various atoms and isotopes in the rock. The method works because some atoms undergo changes with time that allow us to determine how long they have been held in place within the rock's solid structure. By analyzing these changes we learn the amount of time that has passed since the atoms became locked together in their present arrangement, which in most cases means the time *since the rock last solidified*.

Remember that each chemical element is uniquely characterized by the number of protons in its nucleus. Different *isotopes* of the same element differ only in their number of neutrons [Section 5.1]. A **radioactive** isotope has a nucleus prone to spontaneous change, or *decay*, such as breaking apart or having one of its protons turn into a neutron. This decay always occurs at the same rate for any particular radioactive isotope, and scientists can measure these rates in the laboratory. We generally characterize decay rates by stating a **half-life**—the length of time it would take for half the nuclei in the collection to decay.

For example, potassium-40 is a radioactive isotope with nuclei that decay when a proton turns into a neutron, changing the potassium-40 into argon-40. The half-life for this decay process is 1.25 billion years. (Note that it takes only a few months to years of laboratory measurements to pin down the decay rate and determine the half-life.) Now, consider a small piece of rock that contained 1 microgram of potassium-40 and no argon-40 when it formed (solidified) long ago. The half-life of 1.25 billion years means that half the original potassium-40 would have decayed into argon-40 by the time the rock was 1.25 billion years old, so at that time the rock would have contained $\frac{1}{2}$ microgram of potassium-40 and $\frac{1}{2}$ microgram of argon-40. Half of this remaining potassium-40 would then decay by the end of the next 1.25 billion years, so after 2.5 billion years the rock contained $\frac{1}{4}$ microgram of potassium-40 and $\frac{3}{4}$ microgram of argon-40. After three half-lives, or 3.75 billion years, only $\frac{1}{8}$ microgram of potassium-40 remained, while $\frac{7}{8}$ microgram had become argon-40. Figure 6.25 summarizes the gradual decrease in the amount of potassium-40 and the corresponding rise in the amount of argon-40.

We can determine the age of a rock through careful analysis of the proportions of various atoms and isotopes within it.

We can now see the essence of radiometric dating. Suppose you find a rock that contains equal numbers of atoms of potassium-40 and argon-40. If you assume that all the argon came from potassium decay (and if the rock shows no evidence of subsequent heating that could have allowed any argon to escape), then it must have taken precisely one half-life for the rock to end up with equal amounts of the two isotopes. You could therefore conclude that the rock is 1.25 billion years old. The only question is whether you are right in assuming that the rock lacked argon-40 when it formed. In this case, knowing a bit of "rock chemistry" helps. Potassium-40 is a natural ingredient of many minerals in rocks, but argon-40 is a gas that does not combine with other elements and did not condense in the solar nebula. If you find argon-40 gas trapped inside minerals, it must have come from radioactive decay of potassium-40.

Radiometric dating is possible with many other radioactive isotopes as well. In many cases, we can date a rock that contains more than one radioactive isotope, so agreement between the ages calculated from the different isotopes gives us confidence that we have dated the rock correctly. We can also check results from radiometric dating against those from other methods of measuring or estimating ages. For example, some fairly recent archaeological artifacts have original dates printed on them, and the dates agree with ages found by radiometric dating. We can validate

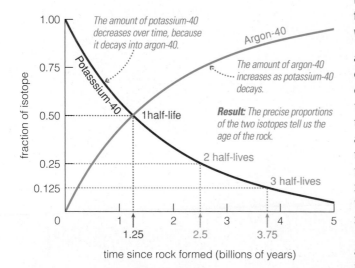

Figure 6.25

Potassium-40 is radioactive, decaying into argon-40 with a half-life of 1.25 billion years. The red curve shows the decreasing amount of potassium-40, and the blue curve shows the increasing amount of argon-40. The remaining amount of potassium-40 drops in half with each successive half-life.

the $4\frac{1}{2}$-billion-year radiometric age for the solar system as a whole by comparing it to an age based on detailed study of the Sun. Theoretical models of the Sun, along with observations of other stars, show that stars slowly expand and brighten as they age. The model ages are not nearly as precise as radiometric ages, but they confirm that the Sun is between about 4 and 5 billion years old. Overall, the technique of radiometric dating has been checked in so many ways and relies on such basic scientific principles that there is no longer any serious scientific debate about its validity.

Earth Rocks, Moon Rocks, and Meteorites Radiometric dating tells us how long it has been since a rock solidified, which is not the same as the age of a planet as a whole. For example, we find rocks of many different ages on Earth. Some rocks are quite young because they formed recently from molten lava; others are much older. The oldest Earth rocks are about 4 billion years old, and some small mineral grains are about 4.4 billion years old, but even these are not as old as Earth itself.

Moon rocks brought back by the *Apollo* astronauts date as far back as 4.4 billion years ago. Although they are older than Earth rocks, these Moon rocks must still be younger than the Moon itself. The ages of these rocks also tell us that the giant impact thought to have created the Moon must have occurred more than 4.4 billion years ago.

Age dating of meteorites that are unchanged since they condensed and accreted tells us that the solar system is about $4\frac{1}{2}$ billion years old.

To go all the way back to the origin of the solar system, we must find rocks that have not melted or vaporized since they first condensed in the solar nebula. Meteorites that have fallen to Earth are our source of such rocks. Many meteorites appear to have remained unchanged since they condensed and accreted in the early solar system. Careful analysis of radioactive isotopes in these meteorites shows that the oldest ones formed about 4.55 billion years ago, so this time must mark the beginning of accretion in the solar nebula. Because the planets apparently accreted within about 50 million (0.05 billion) years, Earth and the other planets formed about 4.5 billion years ago.

 Detecting Extrasolar Planets Tutorial, Lessons 1–3

6.5 Other Planetary Systems

When you were born, the complete list of known planets in the universe consisted only of those in our own solar system. The nebular theory made it seem likely that planets existed around other stars, but technology was not yet at the point where we could test the idea. As we discussed in Chapter 1, seeing planets around other stars is equivalent to looking for dim ball points or marbles from a distance of thousands of kilometers away—with the star typically a billion times brighter than the planet. Remarkably, we can now detect some of these planets, and this fact has ushered in a new era in astronomy: For the first time, we can engage in comparative study of planetary *systems*, which allows us to test and refine our ideas about the formation of stars and planets.

• How do we detect planets around other stars?

The first clear-cut discovery of a planet around another Sun-like star—a star called 51 Pegasi—came in 1995. Hundreds of additional extrasolar

Radiometric Dating

From the fact that the amount of a radioactive substance decays by half with each half-life, it is possible to derive a simple formula for the age of a rock. If you have measured the current amount of a radioactive substance and determined the original amount (by measuring the abundance of its decay products), and if you know its half-life t_{half}, then the time t since the rock formed is

$$t = t_{half} \times \frac{\log_{10}\left(\frac{\text{current amount}}{\text{original amount}}\right)}{\log_{10}\left(\frac{1}{2}\right)}$$

Even if you are unfamiliar with logarithms, you can work with this formula by using the "log" button on your calculator.

Example: You chemically analyze a small sample of a meteorite. Potassium-40 and argon-40 are present in a ratio of approximately 0.85 unit of potassium-40 atoms to 9.15 units of gaseous argon-40 atoms. (The units are unimportant, because only the relative amounts of the parent and daughter materials matter.) How old is the meteorite?

Solution: Because no argon gas could have been present in the meteorite when it formed, the 9.15 units of argon-40 must originally have been potassium-40 that has decayed with a half-life of 1.25 billion years. The sample must therefore have started with 0.85 + 9.15 = 10 units of potassium-40 (the original amount), of which 0.85 unit remains (the current amount). The formula now reads

$$t = 1.25 \text{ billion yr} \times \frac{\log_{10}\left(\frac{0.85}{10}\right)}{\log_{10}\left(\frac{1}{2}\right)} = 4.45 \text{ billion yr}$$

This meteorite solidified about 4.45 billion years ago.

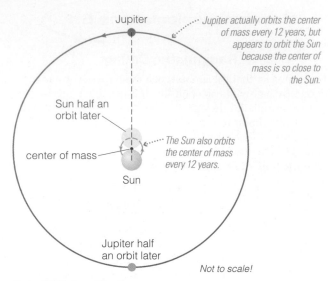

Figure 6.26 interactive figure

This diagram shows how both the Sun and Jupiter actually orbit around their mutual center of mass, which lies very close to the Sun. The diagram is not to scale; the sizes of the Sun and its orbit are exaggerated about 100 times compared to the size shown for Jupiter's orbit (and Jupiter's size is exaggerated even more).

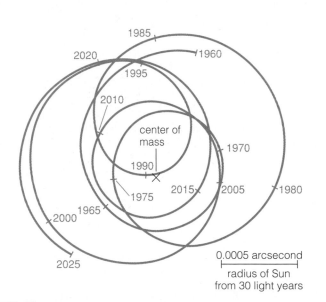

Figure 6.27

This diagram shows the orbital path of the Sun from 1960 to 2025 around the center of mass of our solar system, as it would appear if viewed face-on from a distance of 30 light-years away. The complex motion reveals the gravitational effects of the planets (primarily Jupiter and Saturn). The *astrometric technique* for detecting extrasolar planets works by looking for similar changes in the position of other stars. Notice that the entire range of motion during this period is only about 0.0015 arcsecond, which is almost 100 times smaller than the angular resolution of the Hubble Space Telescope.

planets have been discovered since that time, using several planet-finding strategies. If we strip away the details, however, there are really only two basic ways to search for extrasolar planets:

1. **Directly:** Pictures or spectra of the planets themselves constitute direct evidence of their existence.
2. **Indirectly:** Precise measurements of stellar properties such as position, brightness, or spectra may indirectly reveal the effects of orbiting planets.

Almost all extrasolar planets detected to date have been found indirectly rather than through direct imaging.

Direct detection is preferable because it can tell us far more about the planet's properties, but to date nearly all detections have been indirect.

think about it Do a quick Web search on "extrasolar planets" to find the current number of known extrasolar planets. How many have been found in the past year alone?

Gravitational Tugs Two indirect techniques—the *astrometric* and *Doppler* techniques—rely on observing stars in search of motion that we can attribute to gravitational tugs from orbiting planets. Although we usually think of a star as remaining still while planets orbit around it, that is only approximately correct. In reality, all the objects in a star system, including the star itself, orbit the system's *center of mass*, which is in essence the balance point for all the mass of the solar system. Because the Sun is far more massive than all the planets combined, the center of mass of our solar system lies close to the Sun—but not exactly at the Sun's center.

We can see how this fact allows us to discover extrasolar planets by imagining the viewpoint of extraterrestrial astronomers observing our solar system from afar. Let's start by considering only the influence of Jupiter, the most massive planet in our solar system (Figure 6.26). The center of mass between the Sun and Jupiter lies just outside the Sun's visible surface, so what we usually think of as Jupiter's 12-year orbit around the Sun is really a 12-year orbit around this center of mass. Because the Sun and Jupiter are always on opposite sides of the center of mass (otherwise it wouldn't be a "center"), the Sun must orbit this point with the same 12-year period. The Sun's orbit traces out a very small ellipse with each 12-year period, because the Sun's average orbital distance is barely larger than its own radius. Nevertheless, with sufficiently precise measurements, extraterrestrial astronomers could detect this orbital movement of the Sun and thereby deduce the existence of Jupiter—without having ever seen the planet. They could even determine Jupiter's mass from the orbital characteristics of the Sun as it goes around the center of mass. A more massive planet located at the same distance would pull the center of mass farther from the Sun's center, thereby giving the Sun a larger orbit and a faster orbital speed around the center of mass.

see it for yourself To see how a small planet can make a big star wobble, find a pencil and tape a heavier object (such as a set of keys) to one end and a lighter object (perhaps a small stack of coins) to the other end. Tie a string (or piece of floss) at the balance point—the center of mass—so that the pencil is horizontal; then tap the lighter object into "orbit" around the heavier object. What does the heavier object do, and why? How does your model correspond to a planet orbiting a star? You can experiment further with objects of different weights or shorter pencils; explain the differences you see.

Orbiting planets exert gravitational tugs on their star, so we can detect the planets by observing the star's resulting "wobble" around its average position in the sky.

The other planets also exert gravitational tugs on the Sun, each adding a small additional effect to the effects of Jupiter. In principle, with sufficiently precise measurements of the Sun's orbital motion made over many decades, an extraterrestrial astronomer could deduce the existence of all the planets of our solar system (Figure 6.27). This is the essence of the **astrometric technique**, in which we make very precise measurements of stellar positions in the sky (*astrometric* means "measurement of the stars"). If a star "wobbles" gradually around its average position (the center of mass), we must be observing the influence of unseen planets. The primary difficulty with the astrometric technique is that we are looking for changes to position that are very small even for nearby stars, and these changes become smaller for more distant stars. In addition, the stellar motions are largest for massive planets orbiting *far* from their star, but the long orbital periods of such planets mean that it can take decades to notice the motion. As a result, the astrometric technique has been of only limited use to date, but astronomers hope it will prove successful with future space-based telescopes.

The **Doppler technique** searches for a star's orbital movement around the center of mass by looking for changing Doppler shifts in a star's spectrum [Section 5.2]. As long as a planet's orbit is *not* face-on to us, its gravitational influence will cause its star to move alternately with an orbital toward and away from us—motions that cause spectral lines to shift alternately toward the blue and red ends of the spectrum (Figure 6.28a). The 1995 discovery of a planet orbiting 51 Pegasi came when this star was found to have alternating blueshifts and redshifts within a period of four days (Figure 6.28b). The 4-day period of the star's motion must be the orbital period of its planet. We can then use this period with the star's mass and Newton's version of Kepler's third law [Section 4.4] to calculate the planet's orbital distance. (In Chapter 11, we'll see how the star's mass and other properties can be known.) The Doppler technique even allows us to estimate the planet's mass from the measured change in the star's velocity.* The data in Figure 6.28b thereby enabled us to learn that the planet orbiting 51 Pegasi is similar to Jupiter in mass but orbits only about 0.05 AU from its star—so close that its surface temperature is probably over 1000 K. It is therefore an example of what we call a "hot Jupiter," because it has a Jupiter-like mass but a much higher surface temperature.

Alternating Doppler shifts in a star's spectrum, indicating back-and-forth motion, can also reveal the influence of orbiting planets.

The Doppler technique has been used for the vast majority of planet discoveries to date. In some cases, Doppler data are good enough to tell us whether the star has more than one planet. Remember that if two or more planets exert a noticeable gravitational tug on their star, the Doppler data will show the combined effect of these tugs. As of 2008, at least 25 multiple-planet systems had been identified, including one with five planets. Keep in mind, however, that the Doppler technique is best suited to identifying massive planets that orbit relatively close to their star, because the star's orbital speed depends on the strength of the gravitational tug, and gravity is strongest for massive planets with small orbital distances [Section 4.4].

*The Doppler shift tells us the star's full orbital velocity only if we are viewing its planetary system edge-on; in all other cases, it gives us a lower limit on the star's velocity and therefore a lower limit to the planet's mass. However, statistical arguments show that in two out of three cases, the planet's true mass will be no more than double this lower limit.

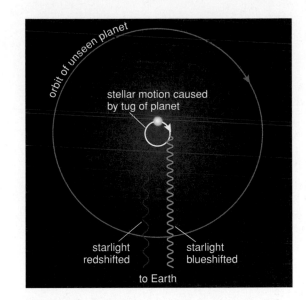

a Doppler shifts allow us to detect the slight motion of a star caused by an orbiting planet.

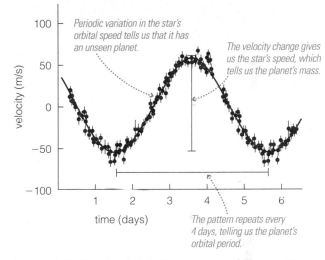

b A periodic Doppler shift in the spectrum of the star 51 Pegasi shows the presence of a large planet with an orbital period of about 4 days. Dots are actual data points; bars through dots represent measurement uncertainty.

Figure 6.28 interactive figure

The Doppler technique for discovering extrasolar planets.

This fact probably explains why most of the extrasolar planets discovered to date orbit relatively close to their stars—these planets are easier to find than planets orbiting far from their stars, which would have weaker gravitational effects and such long orbital periods that it might take decades of observations to detect them. It also explains why we have yet to detect any planets with Earth-like masses: These planets would have such weak gravitational effects on their stars that we could not apply the Doppler technique to find them using current technology.

Transits and Eclipses   A third indirect way of detecting distant planets relies on searching for slight changes in a star's brightness that occur when a planet passes in front of or behind it. If we were to examine a large sample of stars with planets, a small number of them (<1%) will by chance be aligned in such a way that one or more of the star's planets pass directly between us and the star during each orbit. The result is a **transit**, in which the planet appears to move across the face of the star, causing a small, temporary dip in the star's brightness. Because a star's brightness can also vary for other reasons, we can assume that a transiting planet is the cause only if the dimming repeats with a regular period.

think about it   What kind of planet is most likely to cause a transit across its star that we could observe from Earth: (a) a large planet close to its star? (b) a large planet far from its star? (c) a small planet close to its star? or (d) a small planet far from its star? Explain.

If a planet happens to orbit edge-on as seen from Earth, it will periodically pass in front of its star, causing a dip in the star's brightness.

Figure 6.29 shows transit data for a planet orbiting the star HD209458. Transits occur every $3\frac{1}{2}$ days, telling us the planet's orbital period. We can then use Newton's version of Kepler's third law to calculate the planet's orbital distance, and the 1.7% dips in the star's brightness tell us how the planet's radius compares to its star's radius.

Half an orbit after a transit, a planet may pass behind its star. This event is usually called an **eclipse**. Observing an eclipse is much like observing a transit: In both cases, we actually measure the *combined* light from the star and planet, so in principle there can be a dip in brightness whenever either object blocks light from the other. However, because planets generally emit in the infrared, not in the visible [Section 5.2], the dips that occur during eclipses are usually measurable only at infrared wavelengths. For example, during eclipses in the HD209458 system, the infrared brightness drops by about 0.25%, telling us that the planet emits

Figure 6.29 interactive figure

This diagram shows the planet orbiting the star HD209458. The graphs show how the star's brightness changes during transits and eclipses, which each occur once during every $3\frac{1}{2}$-day orbit. During a transit, the star's brightness drops for about 2 hours by 1.7%, which tells us how the planet's radius compares to the radius of its star. During an eclipse, the infrared signal drops by 0.25%, which tells us about the planet's thermal emission.

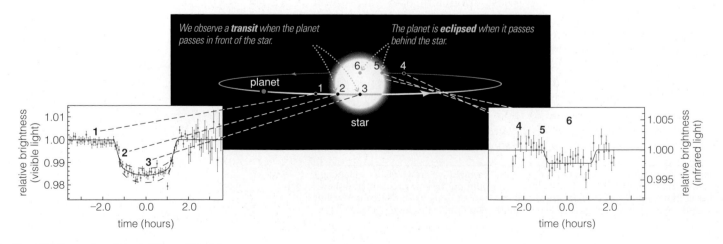

0.25% as much infrared radiation as the star (see Figure 6.29). By combining this fact with the planet's radius measured during the transits, astronomers calculate the planet's temperature to be more than 1100 K.

The primary limitation of the transit and eclipse methods is that they work only for the small fraction of planets whose orbits are nearly edge-on. But the method also has advantages, including the ability to take a spectrum of starlight transmitted through a planet's atmosphere. So far, astronomers have confirmed the existence of hydrogen, water, methane, and even a hint of sodium in the atmospheres of extrasolar planets. The transit method can also be used to search simultaneously for planets around vast numbers of stars and to detect much smaller planets than is possible with the Doppler technique. NASA's *Kepler* mission, scheduled for launch in 2009, is designed to monitor some 100,000 stars for transits, and if Earth-size planets are common, it should be able to detect dozens of them. A European Space Agency (ESA) spacecraft called *COROT* has already detected four transiting planets but may not be able to detect planets as small as Earth. In addition, telescopes as small as 4 inches in diameter have been used to discover transiting planets, and it's relatively easy to confirm for yourself some of the transits already detected. What was once considered impossible can now be assigned as homework (see Problem 54 at the end of the chapter).

Direct Detection The indirect planet-hunting techniques we have discussed so far have started a revolution in planetary science by demonstrating that our solar system is just one of many planetary systems. However, these indirect techniques tell us relatively little about the planets themselves, aside from their orbital properties and their masses or radii. To learn more about their nature, we need to be able to observe the planets themselves. For example, even low-resolution images might reveal important surface features, and spectra could tell us about their compositions and properties of their atmospheres.

Direct images and spectra could allow us to learn much more about the nature of extrasolar planets. For the most part, direct detection of extrasolar planets remains beyond our current technology, primarily because of the incredible glare of the stars themselves. Even the best telescopes blur the light from stars at least a little, so the glare of scattered starlight tends to overwhelm the small blips of planetary light. However, the situation is somewhat better if we observe in infrared light instead of with visible light—primarily because stars emit less infrared light than visible light—and astronomers are beginning to achieve success. The infrared image in Figure 6.30 shows a candidate planet orbiting a very dim "star"—technically a substellar object known as a *brown dwarf* [Section 12.1]—whose dimness made it easier to detect the infrared emission from the planet. The two-page Cosmic Context spread (Figure 6.31) summarizes the major planet detection techniques.

Other Planet-Hunting Strategies The astonishing success of recent efforts to find extrasolar planets has led astronomers to think of many other possible ways of enhancing the search. One example is the Optical Gravitational Lensing Experiment (OGLE), a large survey of thousands of distant stars. Although it was not originally designed for planet detection, OGLE has already detected several planets by observing transits. It has also succeeded in detecting three using *gravitational lensing,* an effect predicted by Einstein's general theory of relativity that occurs when one object's gravity bends or brightens the light of a more distant object [Section 16.2]. This method has led to the detection of the smallest planet found so far, and to

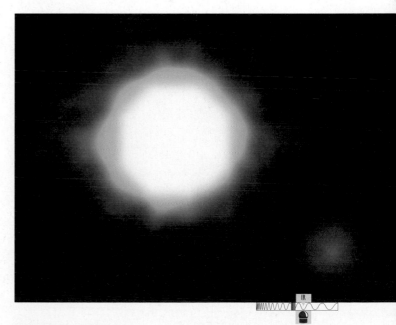

Figure 6.30

This infrared image from the European Southern Observatory's Very Large Telescope shows a brown dwarf called 2M1207 (blue) and what is probably a jovian planet in orbit around it (red).

The search for planets around other stars is one of the fastest growing and most exciting areas of astronomy. Although it has been only a little more than a decade since the first discoveries, known extrasolar planets already number well above 250. This figure summarizes major techniques that astronomers use to search for and study extrasolar planets.

① Gravitational Tugs: We can detect a planet by observing the small orbital motion of its star as both the star and its planet orbit their mutual center of mass. The star's orbital period is the same as that of its planet, and the star's orbital speed depends on the planet's distance and mass. Any additional planets around the star will produce additional features in the star's orbital motion.

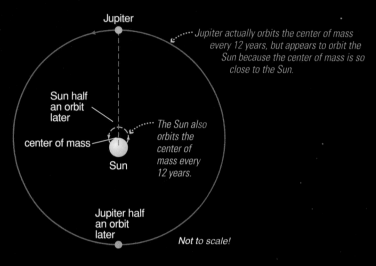

Jupiter

······ Jupiter actually orbits the center of mass every 12 years, but appears to orbit the Sun because the center of mass is so close to the Sun.

Sun half an orbit later

center of mass

······ The Sun also orbits the center of mass every 12 years.

Sun

Jupiter half an orbit later

Not to scale!

①a The Doppler Technique: As a star moves alternately toward and away from us around the center of mass, we can detect its motion by observing alternating Doppler shifts in the star's spectrum: a blueshift as the star approaches and a redshift as it recedes. This technique has revealed the vast majority of known extrasolar planets.

①b The Astrometric Technique: A star's orbit around the center of mass leads to tiny changes in the star's position in the sky. As we improve our ability to measure these tiny changes, we should discover many more extrasolar planets.

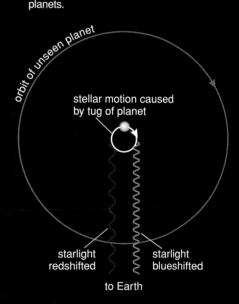

orbit of unseen planet

stellar motion caused by tug of planet

starlight redshifted   starlight blueshifted

to Earth

Current Doppler-shift measurements can detect an orbital velocity as small as 1 meter per second—walking speed.

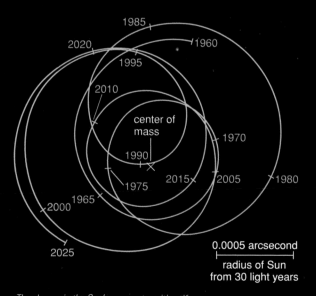

1985
2020   1960
1995
2010
center of mass
1990   1970
1975   2015 2005 1980
2000 1965
2025

0.0005 arcsecond
radius of Sun from 30 light years

The change in the Sun's apparent position, if seen from a distance of 10 light years, would be similar to the angular width of a human hair at a distance of 5 kilometers.

Artist's conception of another planetary system, viewed near a ringed jovian planet.

(2) **Transits and Eclipses:** If a planet's orbital plane happens to lie along our line of sight, the planet will transit in front of its star once each orbit, while being eclipsed behind its star half an orbit later. The amount of starlight blocked by the transiting planet can tell us the planet's size, and changes in the spectrum can tell us about the planet's atmosphere.

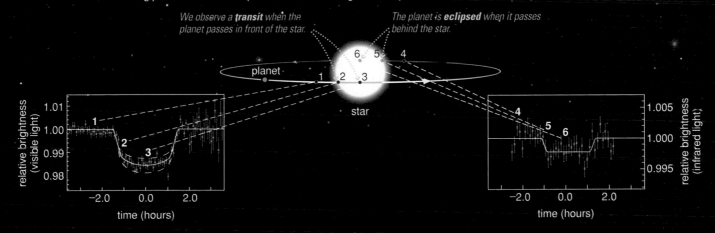

We observe a **transit** when the planet passes in front of the star.

The planet is **eclipsed** when it passes behind the star.

planet

star

(3) **Direct Detection:** In principle, the best way to learn about an extrasolar planet is to observe directly either the visible starlight it reflects or the infrared light that it emits. Our technology is only beginning to reach the point where direct detection is possible, but someday we will be able to study both images and spectra of distant planets.

This infrared image shows a brown dwarf called 2M1207 (blue) . . .

. . . and what is probably a jovian planet (red) in orbit around it.

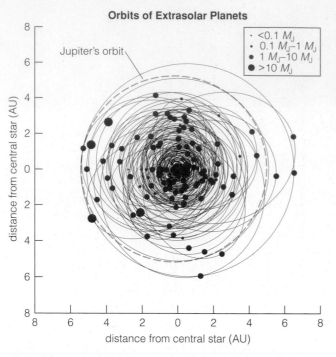

Orbits of Extrasolar Planets

Jupiter's orbit

· <0.1 M_J
· 0.1 M_J–1 M_J
● 1 M_J–10 M_J
● >10 M_J

distance from central star (AU)

distance from central star (AU)

Figure 6.32

Orbital properties of the first 170 known extrasolar planets. This diagram shows all the orbits superimposed on each other, as if all the planets orbited a single star. The sizes of the dots indicate approximate masses for the planets; the dots are located at the farthest point for each orbit. Notice that many extrasolar planets orbit quite close to their stars, and most have larger orbital eccentricities than the planets of our solar system.

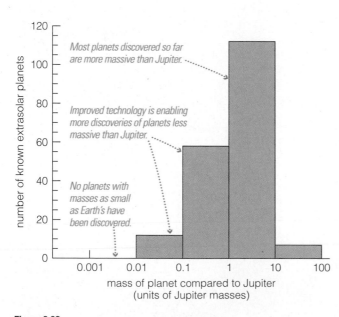

number of known extrasolar planets

Most planets discovered so far are more massive than Jupiter.

Improved technology is enabling more discoveries of planets less massive than Jupiter.

No planets with masses as small as Earth's have been discovered.

mass of planet compared to Jupiter
(units of Jupiter masses)

Figure 6.33

This bar chart shows the number of planets of different masses among the first 170 extrasolar planets discovered. The masses are approximate and are based on minimums that can be found with the Doppler technique. Notice that the horizontal axis uses an exponential scale so that the wide range of masses can all fit on the graph.

the discovery of a system with two jovian planets as far from their star as Jupiter and Saturn are from the Sun. While gravitational lensing is a useful technique, the geometry required for its application never repeats, giving no opportunity for follow-up observations. A different strategy looks for the gravitational effects of unseen planets on the disks of dust that surround many stars, while another method searches for the thermal emission from the impacts of accreting planetesimals. As we learn more about extrasolar planets, new search methods are sure to arise.

• How do extrasolar planets compare with planets in our solar system?

We have discovered a large enough number of extrasolar planets that we can begin to search for patterns, trends, and groupings that might give us insight into how these planets compare to the planets of our own solar system and how they formed. Even though we generally lack direct images or spectra, we can still learn a lot from orbital data and the mass and size estimates that we can make for extrasolar planets.

Much as Kepler first appreciated the true layout of our own solar system [Section 3.3], we can now step back and see the layout of many other solar systems. Figure 6.32 shows the orbits of the first 170 known extrasolar planets all superimposed on each other. Despite the crowding of the orbits when viewed this way, at least two important facts should jump out at you. First, notice that only a handful of these planets have orbits that take them beyond about 5 AU, which is Jupiter's distance from our Sun. Most of them orbit very close to their host star. Second, notice that many of the orbits are clearly elliptical, rather than nearly circular like the orbits of planets in our own solar system. As we'll see shortly, both facts give us important clues about the formation of these extrasolar planets.

Figure 6.33 presents the estimated masses of these same 170 planets as a bar chart. Notice that most of the planets are more massive than Jupiter, and even the least massive ones are almost as massive as Uranus and Neptune. As we discussed earlier, the lack of detections of lower-mass planets is due to the limits of current detection technology.

The masses of the known extrasolar planets suggest they are jovian in nature, but mass alone cannot rule out the possibility of "supersize" terrestrial planets—that is, very massive planets made of metal or rock. To distinguish between these possibilities, we need to know the sizes (radii) of the planets, which we can use along with their masses to calculate their densities. We expect jovian planets to have large sizes and low densities, and terrestrial planets to have small sizes and higher densities. Unfortunately, the vast majority of known extrasolar planets have been detected by the Doppler technique, which gives us reasonable mass estimates but no information about size. However, in the cases where we also have size data from transits, the planets turn out to have sizes and densities consistent with what we expect for jovian planets. Some recent discoveries have also allowed us to determine whether individual planets are more like Jupiter or like Neptune.

The extrasolar planets discovered so far resemble our jovian planets in size and mass, though some orbit very close to their stars.

We have found one surprising characteristic among these extrasolar planets: Many of them orbit quite close to their stars or have highly elliptical orbits. Those on close-in orbits must be much higher in temperature than Jupiter, so these planets have been nicknamed *hot Jupiters*. These planets would probably have clouds much like the real Jupiter but of a different type (Figure 6.34). The temperatures on hot Jupiters would be far too high for gases

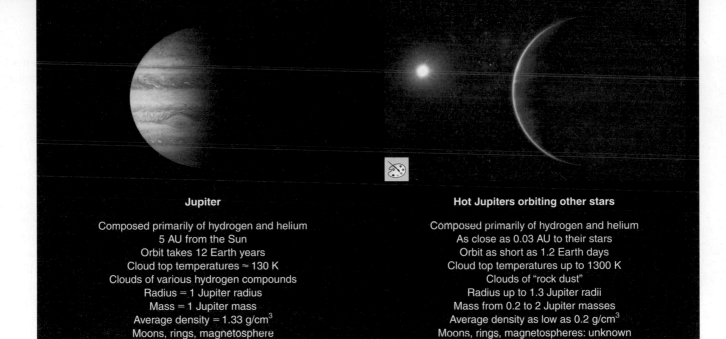

Jupiter	Hot Jupiters orbiting other stars
Composed primarily of hydrogen and helium	Composed primarily of hydrogen and helium
5 AU from the Sun	As close as 0.03 AU to their stars
Orbit takes 12 Earth years	Orbit as short as 1.2 Earth days
Cloud top temperatures ≈ 130 K	Cloud top temperatures up to 1300 K
Clouds of various hydrogen compounds	Clouds of "rock dust"
Radius = 1 Jupiter radius	Radius up to 1.3 Jupiter radii
Mass = 1 Jupiter mass	Mass from 0.2 to 2 Jupiter masses
Average density = 1.33 g/cm^3	Average density as low as 0.2 g/cm^3
Moons, rings, magnetosphere	Moons, rings, magnetospheres: unknown

Figure 6.34

A summary of the expected similarities and differences between the real Jupiter and extrasolar hot Jupiters orbiting Sun-like stars.

such as ammonia or water vapor to condense into liquid droplets or ice flakes, as they do on our Jupiter. Instead, models suggest that a hot Jupiter with a temperature above about 1000 K would have clouds of "rock dust" containing common minerals.

Can we yet say anything about how common planetary systems are overall? Among the thousands of Sun-like stars that astronomers have so far examined in search of extrasolar planets, more than 1 in 10 show evidence of planets around them. While this could imply that planetary systems are relatively rare, a more likely hypothesis is that planets are present but more difficult to detect in the other 9 in 10 systems. In essence, we have been hunting for planets with "elephant traps"—and we have been catching elephants. The more common systems with smaller planets may simply be beyond the grasp of our current traps. In that case, hot Jupiters might actually be relatively rare, and known in large numbers only because they are easier to detect. Indeed, as technology has improved, we have begun to find systems that more closely resemble our own. The idea that terrestrial planets are common is supported by observations that reveal a correlation between the fraction of elements heavier than helium in a star and the chance that it has planets orbiting it. The more rocks, metals, and hydrogen compounds present in a solar nebula, the more likely the star is to have planets—just as we'd expect from the nebular theory.

see it for yourself It's impossible to see planets orbiting other stars with your naked eye, but you can see some of the stars known to have planets. As of 2008, the brightest star known to have a planet was Pollux, located in the constellation Gemini. Its planet has a mass three times that of Jupiter and orbits Pollux every $1\frac{1}{2}$ years. Use the star charts in Appendix J to find out if, when, and where you can observe Pollux tonight, and look for it if you can. Does knowing that Pollux has its own planetary system alter your perspective when you look at the night sky? Why or why not?

● Do we need to modify our theory of solar system formation?

The discovery of extrasolar planets presents us with an opportunity to test our theory of solar system formation, and it has already presented challenges. For example, the nebular theory clearly predicts that jovian

The orbiting planet nudges particles in the disk . . .

. . . causing material to bunch up: These dense regions in turn tug on the planet, causing it to migrate inward.

Figure 6.35

This figure shows a simulation of waves created by a planet embedded in a dusty disk of material surrounding its star; these waves may cause the planet to migrate inward.

planets should form only in the cold outer regions of star systems and should have nearly circular orbits, so how can our theory account for hot Jupiters or planets with highly elliptical orbits?

One possibility that scientists must always consider is that something is fundamentally wrong with our model of solar system formation, and scientists have indeed considered this possibility. However, a decade of re-examination has not turned up any obvious flaws in the basic theory. As a result, scientists now suspect that the hot Jupiters were indeed born with circular orbits far from their stars and that those that now have close-in or highly elliptical orbits underwent some sort of "planetary migration" or suffered gravitational interactions with other massive objects.

Hot Jupiters probably were born in their outer solar systems as the nebular theory predicts, but later migrated inward.

How might planetary migration occur? Our best guess is that it can be caused by waves passing through a gaseous disk (Figure 6.35). A planet's gravity and motion tend to disturb the otherwise evenly distributed disk material, generating waves that travel through the disk. The waves cause material to bunch up as they pass by, and these clumps exert their own gravitational pull on the planet, robbing it of energy and causing it to move inward.

Computer models confirm that waves in the nebula can cause young planets to spiral slowly toward their star. In our own solar system, this migration did not play a major role because the solar wind probably cleared out the gas before it could have much effect. But planets may form earlier in other solar systems, allowing time for jovian planets to migrate substantially inward. In some cases, the planets may form so early that they end up spiraling into their stars.

Another way to account for some of the observed extrasolar planet orbits invokes close encounters between young jovian planets. Such an encounter might send one planet out of the star system entirely while the other is flung inward into a highly elliptical orbit. Other possibilities include the idea that a jovian planet could migrate inward as a result of multiple close encounters with much smaller planetesimals (there is evidence that the jovian planets in our solar system have migrated a little bit by this mechanism) or that jovian planets might periodically line up with one another in a way that would cause their orbits to become more elliptical.

The bottom line is that discoveries of extrasolar planets have shown us that the nebular theory is incomplete. It explains the formation of planets and the simple layout of a solar system such as ours, but it needs new features—such as planetary migration and gravitational encounters—to explain the differing layouts of other solar systems. A much wider range of solar system arrangements now seems possible than we had guessed before the discovery of extrasolar planets.

Planetary scientists are anxious to learn more, and over the past few years NASA and the European Space Agency have developed a series of plans for ambitious missions to try to find many more planets—including Earth-like planets, if they exist—and to study those planets through imaging and spectroscopy. However, budgetary pressures have placed all of those plans on hold for now.

think about it Look back at the discussion of the nature of science in Chapter 3, especially the definition of a scientific theory. Should the nebular theory qualify as a scientific theory even though we know that it needs modification to account for the orbits of planets in other solar systems? Does this mean that the theory was "wrong" as we understood it before? Explain.

the big picture
Putting Chapter 6 into Context

In this chapter, we've introduced the major features of our solar system and described the current scientific theory of its formation. We've seen how this theory explains the major features we observe and how it can be extended to other planetary systems. As you continue your study of the solar system, keep in mind the following "big picture" ideas:

- Our solar system is not a random collection of objects moving in random directions. Rather, it is highly organized, with clear patterns of motion and common traits among families of objects.

- We can explain the major features of our solar system with a theory that holds that the solar system formed from the gravitational collapse of an interstellar gas cloud.

- Most of the general features of the solar system were determined by processes that occurred very early in the solar system's history, which began some $4\frac{1}{2}$ billion years ago.

- Planet-forming processes are universal. Discoveries of planets around other stars have begun an exciting new era in planetary science.

summary of key concepts

6.1 A Brief Tour of the Solar System

• What does the solar system look like?
The planets are tiny compared to the distances between them. Our solar system consists of the Sun, the planets and their moons, and vast numbers of asteroids and comets. Each world has its own unique character, but there are many clear patterns among the worlds.

6.2 Clues to the Formation of Our Solar System

• What features of our solar system provide clues to how it formed?
Four major features provide clues: (1) The Sun, planets, and large moons generally rotate and orbit in a very organized way. (2) The eight official planets divide clearly into two groups: **terrestrial** and **jovian**. (3) The solar system contains vast numbers of asteroids and comets, some large enough to qualify as dwarf planets. (4) There are some notable exceptions to these general patterns.

• What theory best explains the features of our solar system?
The **nebular theory**, which holds that the solar system formed from the gravitational collapse of a great cloud of gas and dust, successfully explains all the major features of our solar system.

6.3 The Birth of the Solar System

• Where did the solar system come from?

The cloud of gas that gave birth to our solar system was the product of recycling of gas through many generations of stars within our galaxy. This gas consisted of 98% hydrogen and helium and 2% all other elements.

• What caused the orderly patterns of motion in our solar system?

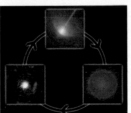

A collapsing gas cloud tends to heat up, spin faster, and flatten out as it shrinks in size. Our solar system began as a spinning disk of gas and dust, so the orderly motions we observe today came from the orderly motion of this spinning disk.

6.4 The Formation of Planets

• Why are there two major types of planets?

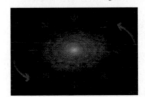

Planets formed around solid "seeds" that condensed from gas and then grew through accretion. In the inner solar system, temperatures were so high that only metal and rock could

Chapter 6 Formation of Planetary Systems **183**

condense, which explains why terrestrial worlds are made of metal and rock. In the outer solar system, cold temperatures allowed more abundant ices to condense along with metal and rock. Icy planetesimals grew large enough for their gravity to draw in hydrogen and helium gas, forming the massive jovian planets.

• **Where did asteroids and comets come from?**
Asteroids are the rocky leftover planetesimals of the inner solar system, and comets are the icy leftover planetesimals of the outer solar system.

• **How do we explain the existence of our Moon and other exceptions to the rules?**

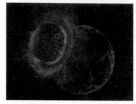

Most of the exceptions probably arose from collisions or close encounters with leftover planetesimals. Our Moon is most likely the result of a **giant impact** between a Mars-size planetesimal and the young Earth.

• **When did the planets form?**
The planets began to accrete in the solar nebula about 4.55 billion years ago, a fact we determine from radiometric dating of the oldest meteorites.

6.5 Other Planetary Systems

• **How do we detect planets around other stars?**

So far, we are best able to detect extrasolar planets indirectly by observing the planet's effects on the star it orbits. Most discoveries to date have been made with the **Doppler technique**, in which Doppler shifts reveal the gravitational tug of a planet (or planets) on a star. We can also search for **transits** and **eclipses** in which a system becomes slightly dimmer as a planet passes in front of or behind its star.

• **How do extrasolar planets compare with planets in our solar system?**
The known extrasolar planets are all much more massive than Earth. Many of them orbit surprisingly close to their stars or have large orbital eccentricities. We have limited information about sizes and compositions, but these data are consistent with the idea that most of these planets are jovian in nature.

• **Do we need to modify our theory of solar system formation?**
Our basic theory of solar system formation seems to be sound, but we have had to modify it to allow for planetary migration and gravitational encounters.

exercises and problems

For instructor-assigned homework go to **www.masteringastronomy.com**.

Review Questions

Short-Answer Questions Based on the Reading

1. Briefly describe the layout of the solar system as it would appear from beyond the orbit of Neptune.
2. For the Sun and each of the planets in our solar system, describe at least two features that you find interesting.
3. What are the four major features of our solar system that provide clues to how it formed? Describe each one briefly.
4. What are the basic differences between *terrestrial* and *jovian* planets? Which planets in our solar system fall into each group? How does Pluto fit in?
5. What is the *nebular theory,* and why is it widely accepted by scientists today?
6. What do we mean by the *solar nebula?* What was it made of, and where did it come from?
7. Describe each of the three key processes that led the solar nebula to take the form of a spinning disk. What observational evidence supports this scenario?
8. List the four categories of materials in the solar nebula by their condensation properties and abundance. Which ingredients are present in terrestrial planets? In jovian planets? Explain why.
9. What was the *frost line* in the solar nebula? Explain how temperature differences led to the formation of two distinct types of planets.
10. Briefly describe the process by which terrestrial planets are thought to have formed. How was the formation of jovian planets similar? How was it different? Why did the jovian planets end up with so many moons?
11. What are asteroids and comets? How and why are they different?
12. What was the heavy bombardment? When did it occur?
13. How do we think the Moon formed, and what evidence supports this hypothesis?
14. Briefly explain the technique of radiometric dating, and describe how we use it to determine the age of the solar system.
15. Describe three major methods used to detect extrasolar planets indirectly. What does each method tell you about the planet?
16. Why is direct detection of extrasolar planets so difficult? What could we learn from direct detection?
17. What data suggest that the known extrasolar planets are similar in nature to the jovian planets in our solar system? In what ways are some of the orbits of these planets surprising?
18. Based on the known extrasolar planets, should we conclude that Earth-like planets are rare? Why or why not?
19. What properties of extrasolar planets and their orbits have forced a re-examination of the nebular theory? How have we modified the theory to explain these properties?
20. *True or False.* Decide whether each statement is true or false, and explain why.
    a. On average, Venus has the hottest surface temperature of any planet in the solar system—even hotter than Mercury.
    b. Our Moon is about the same size as moons of the other terrestrial planets.

c. The weather conditions on Mars today are much different than they were at some times in the distant past.

d. Moons cannot have atmospheres, active volcanoes, or liquid water.

e. Saturn is the only planet in the solar system with rings.

f. Neptune orbits the Sun in the opposite direction of all the other planets.

g. If Pluto were as large as the planet Mercury, we would classify it as a terrestrial planet.

h. Asteroids are made of essentially the same materials as the terrestrial planets.

i. When scientists say that our solar system is about $4\frac{1}{2}$ billion years old, they are making a rough estimate based on guesswork about how long it should have taken planets to form.

Test Your Understanding

Surprising Discoveries?

Suppose we found a solar system with the property described (these are not real discoveries). Decide whether the discovery should be considered reasonable or surprising. Explain clearly; not all these have definitive answers, so your explanation is more important than your chosen answer.

21. A solar system is discovered with four large jovian planets in its inner region and seven small terrestrial planets in its outer reaches.

22. A solar system has ten planets that all orbit the star in approximately the same plane. However, five planets orbit in one direction (e.g., counterclockwise), while the other five orbit in the opposite direction (e.g., clockwise).

23. A solar system has four Earth-size terrestrial planets. Each of the four planets has a single moon that is nearly identical in size to Earth's Moon.

24. A solar system has many rocky asteroids and many icy comets. However, most of the comets orbit in the inner solar system, while the asteroids orbit in far-flung regions much like the Kuiper belt and Oort cloud of our solar system.

25. A solar system has several planets similar in composition to the jovian planets of our solar system but similar in mass to the terrestrial planets of our solar system.

26. Radiometric dating of meteorites from another solar system shows that they are a billion years younger than rocks from the terrestrial planets of the same system.

27. An extrasolar planet is discovered with a year that lasts only 3 days.

28. Within the next few years, astronomers confirm all the planet detections made with the Doppler technique by observing transits of these same planets.

29. The number of known extrasolar planets increases from around 300 in 2008 to more than 1000 by the year 2010.

30. The fact that we have not yet discovered an Earth-size extrasolar planet tells us that such planets must be very rare.

Quick Quiz

Choose the best answer to each of the following. Explain your reasoning with one or more complete sentences.

31. The largest terrestrial planet and jovian planet are, respectively, (a) Venus and Jupiter. (b) Earth and Jupiter. (c) Earth and Pluto.

32. Which of the following three kinds of objects resides closer to the Sun on average? (a) comets (b) asteroids (c) jovian planets

33. Planetary orbits are (a) very eccentric (stretched-out) ellipses and in the same plane. (b) fairly circular and in the same plane. (c) fairly circular but oriented in every direction.

34. The composition of the solar nebula was 98% (a) rock and metal. (b) hydrogen compounds. (c) hydrogen and helium.

35. What's the leading theory for the origin of the Moon? (a) It formed from the solar nebula along with the Earth. (b) It formed from the material ejected in a giant impact. (c) It split out of a rapidly rotating Earth.

36. About how old is the solar system? (a) 4.5 million years (b) 4.5 billion years (c) 4.5 trillion years

37. The extrasolar planets discovered so far most resemble (a) terrestrial planets. (b) jovian planets. (c) large icy worlds.

38. How many extrasolar planets have been detected as of 2008? (a) between 10 and 100 (b) between 100 and 1000 (c) more than 1000

39. Which technique could detect a planet in an orbit that is face-on to the Earth? (a) Doppler technique (b) transit technique (c) astrometric technique

40. Observations to date suggest that Earth-size planets orbiting Sun-like stars (a) do not exist at all. (b) are extremely rare. (c) may be common, though we cannot yet detect them.

Process of Science

Examining How Science Works

41. *Explaining the Past.* Is it really possible for science to inform us about things that may have happened billions of years ago? To address this question, test the nebular theory against each of the three hallmarks of science discussed in Chapter 3. Be as detailed as possible in explaining whether the theory does or does not satisfy these hallmarks. Use your explanations to decide whether the theory can really tell us about how our solar system formed. Defend your opinion.

42. *Dating the Past.* The method of radiometric dating that tells us the age of our solar system is also used to determine when many other past events occurred. For example, it is used to determine ages of fossils that tell us when humans first evolved and ages of relics that teach us about the rise of civilization. Research one key aspect of human history for which radiometric dating has helped us piece the story together. Write two or three paragraphs explaining how radiometric dating was used in this case (such as what materials were dated and what radioactive elements were used) and what the study or studies concluded. Does your understanding of the method lead you to accept the results? Why or why not?

43. *Confirming Observations.* After the first few discoveries of shifts in stars' spectra using the Doppler technique, some astronomers hypothesized that the stars' companions were brown dwarves in nearly face-on orbits, instead of planets with a random distribution of orbits. How did later observations refute this hypothesis? Discuss both later discoveries with the Doppler technique and observations with other techniques.

44. *Refuting the Theory.* Consider the following three hypothetical observations: (1) the discovery of a lone extrasolar planet that is small and dense like a terrestrial planet but has a Jupiter-like orbit; (2) the discovery of a planetary system in which three terrestrial planets have orbits outside those of two jovian planets; (3) the discovery that in a majority of planetary systems the jovian planets are nearer to their star than 1 AU and the terrestrial planets are beyond 5 AU. Each of these observations

would challenge our current theory of solar system formation, but would any of them shake the very foundations of the theory? Which one(s) would do so, and why? Also explain why the other(s), while posing a challenge, would not necessarily cause major problems.

Investigate Further

In-Depth Questions to Increase Your Understanding

Short-Answer/Essay Questions

45. *Planetary Tour.* Based on the brief planetary tour in this chapter, which planet besides Earth do you think is the most interesting, and why? Defend your opinion clearly in two or three paragraphs.

46. *Patterns of Motion.* In one or two paragraphs, summarize the orderly patterns of motion in our solar system and explain why their existence should suggest that the Sun and the planets all formed at one time from one cloud of gas, rather than as individual objects at different times.

47. *Solar System Trends.* Study the planetary data in Table 6.1 to answer each of the following.
    a. Notice the relationship between distance from the Sun and surface temperature. Describe the trend, explain why it exists, and explain any notable exceptions to the trend.
    b. The text says that planets can be classified as either terrestrial or jovian. Describe in general how the columns for density, composition, and distance from the Sun support this classification.
    c. Describe the trend you see in orbital periods and explain the trend in terms of Kepler's third law.
    d. Which column of data would you use to find out which planet has the shortest days? Do you see any notable differences in the length of a day for the different types of planets? Explain.
    e. Which planets would you expect not to have seasons? Why?

48. *Two Kinds of Planets.* The jovian planets differ from the terrestrial planets in a variety of ways. Using phrases or sentences that members of your family would understand, explain why the jovian planets differ from the terrestrial planets in each of the following: composition, size, density, distance from the Sun, and number of satellites.

49. *An Early Solar Wind.* Suppose the solar wind had cleared away the solar nebula before the seeds of the jovian planets could gravitationally draw in hydrogen and helium gas. How would the planets of the outer solar system be different? Would they still have many moons? Explain your answer in a few sentences.

50. *History of the Elements.* Our bodies (and most living things) are made mostly of water (H_2O). Summarize the "history" of a typical hydrogen atom from its creation to Earth's formation. Do the same for a typical oxygen atom. (*Hint:* Which elements were created in the Big Bang, and where were the others created?)

51. *Understanding Radiometric Dating.* Imagine you had the good fortune to find a rocky meteorite in your backyard. Qualitatively, how would you expect its ratio of potassium-40 and argon-40 to be different from other rocks in your yard? Explain why, in a few sentences.

52. *No Hot Jupiters Here.* How do we think hot Jupiters formed? Why didn't one form in our solar system?

53. *Comparing Methods.* What are the advantages and disadvantages of the Doppler and transit techniques? What kinds of planets are easiest to detect with each method? Are there planets that each method cannot detect, even if the planets are very large? Explain. What are the advantages of being able to detect a planet by both methods?

54. *Detect an Extrasolar Planet for Yourself.* Most colleges and many amateur astronomers have the equipment necessary to detect known extrasolar planets using the transit method (Figure 6.29). All that's required is a telescope 10 or more inches in diameter, a CCD camera system, and a computer system for data analysis. The basic method is to take exposures of a few minutes' duration over a period of several hours around the times of predicted transit and to compare the brightness of the star being transited relative to other stars in the same CCD frame. For complete instructions, see www.masteringastronomy.com.

Quantitative Problems

Be sure to show all calculations clearly and state your final answers in complete sentences.

55. *Dating Lunar Rocks.* You are analyzing Moon rocks that contain small amounts of uranium-238, which decays into lead with a half-life of about 4.5 billion years.
    a. In a rock from the lunar highlands, you determine that 55% of the original uranium-238 remains, while the other 45% has decayed into lead. How old is the rock?
    b. In a rock from the lunar maria, you find that 63% of the original uranium-238 remains, while the other 37% has decayed into lead. Is this rock older or younger than the highlands rock? By how much?

56. *Size Comparisons.* How many Earths could fit inside Jupiter (assuming you could fill up all the volume)? How many Jupiters could fit inside the Sun? The equation for the volume of a sphere is $V = \left(\frac{4}{3}\right)\pi r^3$.

57. *Monster Iceballs.* The ice-rich planetesimals that formed the cores of the jovian planets were about 10 times more massive than Earth. Assuming such an iceball has a density of about 2 g/cm^3, what would its radius be? How does this compare to Earth's radius?

58. *Transit of TrES-1.* The planet orbiting the star TrES-1 has been detected by both the transit and Doppler methods, so we can calculate its density and get an idea of what kind of planet it is.
    a. Using the method described in the text, calculate the radius of the transiting planet. The planetary transits block 2% of the star's light. The star TrES-1 has a radius of about 85% of our Sun's radius.
    b. The mass of the planet is approximately 0.75 times the mass of Jupiter, and Jupiter's mass is about 1.9×10^{27} kilograms. Calculate the average density of the planet. Give your answer in grams per cubic centimeter. Compare this density to the average densities of Saturn (0.7 g/cm^3) and Earth (5.5 g/cm^3). Is the planet terrestrial or jovian in nature? (*Hint:* To find the volume of the planet, use the formula for the volume of a sphere: $V = \left(\frac{4}{3}\right)\pi r^3$. Be careful with unit conversions.)

59. *Planet Around 51 Pegasi.* The star 51 Pegasi has about the same mass as our Sun. A planet discovered orbiting around it has an orbital period of 4.23 days. The mass of the planet is estimated to be 0.6 times the mass of Jupiter. Use Kepler's third law to find the planet's average distance (semimajor axis) from its star. (*Hint:* Because the mass of 51 Pegasi is about the same as the mass of our Sun, you can use Kepler's third law in its original form, $p^2 = a^3$ (Section 3.3). Be sure to convert the period into years before using this equation.)

Discussion Questions

60. *Planetary Priorities.* Suppose you were in charge of developing and prioritizing future planetary missions for NASA. What would you choose as your first priority for a new mission, and why?

61. *Lucky to Be Here?* Considering the overall process of solar system formation, do you think it was likely for a planet like Earth to have formed? Could random events in the early history of the solar system have prevented our being here today? What implications do your answers have for the possibility of Earth-like planets around other stars? Defend your opinions.

62. *So What?* What is the significance of the discovery of extrasolar planets, if any? Justify your answer in the context of this book's discussion of the history of astronomy. Should NASA fund missions to search for more extrasolar planets? Defend your opinion.

Web Projects

63. *Current Planetary Mission.* Find out what missions to the planets of our solar system are currently underway. Visit the Web page for one of these missions. Write a one- to two-page summary of the mission's basic design, goals, and status.

64. *Spitzer Space Telescope.* The Spitzer Space Telescope operates at the infrared wavelengths that are especially useful for studying star and planet formation. Visit the Spitzer Web site to see if recent discoveries are confirming the nebular theory of solar system formation or are requiring us to broaden our understanding of the process. Summarize your findings in a one- to two-page report.

65. *New Planets.* Find the latest information on discoveries of extrasolar planets. Create a personal planet journal, complete with illustrations, with a page for each of at least three recent discoveries of new planets. On each journal page, note the technique that was used to find the planet, give any known information about the nature of the planet, and discuss how the planet does or does not fit in with our current understanding of planetary systems.

66. *The Kepler Mission.* The *Kepler* mission was designed expressly to look for Earth-size planets around other stars. Go to the *Kepler* Web site and learn more about the mission. Write a one- to two-page summary of the mission's goals and its current status.

visual skills check

Use the following questions to check your understanding of some of the many types of visual information used in astronomy. Answers are provided in Appendix K. For additional practice, try the Chapter 6 Visual Quiz at **www.masteringastronomy.com**.

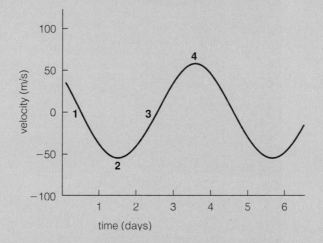

This plot, based on Figure 6.28b, shows the periodic variations in the Doppler shift of a star caused by a planet orbiting around it. Positive velocities mean the star is moving away from Earth, and negative velocities mean the star is moving toward Earth. (You can assume that the orbit appears edge-on from Earth.) Answer the following questions based on the information in the graph.

1. How long does it take the star and planet to complete one orbit around their center of mass?
2. What maximum velocity does the star attain?
3. Match the *star's* position at points 1, 2, 3, and 4 in the plot with the descriptions below.
   a. headed straight toward Earth
   b. headed straight away from Earth
   c. closest to Earth
   d. farthest from Earth

4. Match the *planet's* position at points 1, 2, 3, and 4 in the plot with the descriptions in question 3.
5. How would the plot change if the planet were more massive?
   a. It would not change, because it describes the motion of the star, not the planet.
   b. The peaks and valleys would get larger (greater positive and negative velocities) because of larger gravitational tugs.
   c. The peaks and valleys would get closer together (shorter period) because of larger gravitational tugs.

7

earth and the terrestrial worlds

It's easy to take for granted the qualities that make Earth so suitable for human life: a temperature neither boiling nor freezing, abundant water, a protective atmosphere, and a relatively stable environment. But we need look only as far as our neighboring terrestrial worlds to see how fortunate we are. The Moon is airless and barren, and Mercury is much the same. Venus is a searing hothouse, while Mars has an atmosphere so thin and cold that liquid water cannot last on its surface today.

How did the terrestrial worlds come to be so different, when all were made from metal and rock that had condensed in the solar nebula? Why did Earth alone develop conditions that permit abundant life? We'll begin to answer these questions by exploring key processes that have shaped Earth and the other terrestrial worlds over time, and then we'll consider the history of each world individually. We will see that the histories of the worlds are not random accidents, but consequences of properties endowed at their births. Once we understand what has happened on other worlds, we'll be ready to return to Earth at the end of the chapter, seeing it in an entirely different way than we could have before the era of planetary exploration.

(MA) Formation of the Solar System Tutorial, Lesson 1

7.1 Earth as a Planet

Earth's surface seems solid and steady, but every so often it offers us a reminder that nothing about it is permanent. If you live in Alaska or California, you've probably felt the ground shift beneath you in an earthquake. In Washington State, you may have witnessed the rumblings of Mount St. Helens. In Hawaii, a visit to the still-active Kilauea volcano will remind you that you are standing on mountains of volcanic rock protruding from the ocean floor.

Volcanoes and earthquakes are not the only processes acting to reshape Earth's surface. They are not even the most dramatic: Far greater change can occur on the rare occasions when an asteroid or a comet slams into Earth. More gradual processes can also have spectacular effects. The Colorado River causes only small changes in the landscape from year to year, but its unrelenting flow over the past few million years carved the Grand Canyon. The Rocky Mountains were once twice as tall as they are today; they have been cut down in size through tens of millions of years of erosion by wind, rain, and ice. Entire continents move slowly about, completely rearranging the map of Earth every few hundred million years.

Earth is not alone in having undergone tremendous change since its birth. The surfaces of all five terrestrial worlds—Mercury, Venus, Earth, the Moon, and Mars—must have looked quite similar when they were young. All five were made of rocky material that condensed in the solar nebula, and all five were subjected early on to the impacts of the heavy bombardment [Section 6.4]. The great differences in their present-day appearance must therefore be the result of changes that have occurred through time. Ultimately, these changes can be traced to fundamental properties of the planets.

essential preparation

1. How do light and matter interact? [Section 5.1]

2. What does the solar system look like? [Section 6.1]

3. Why are there two major types of planets? [Section 6.4]

4. Where did asteroids and comets come from? [Section 6.4]

Figure 7.1 shows global views of the terrestrial worlds to scale, along with sample surface views from orbit. Profound differences between these worlds are immediately obvious. Mercury and the Moon show the scars of their battering during the heavy bombardment: They are densely covered by craters except in areas that appear to be volcanic plains. Venus is covered by a thick atmosphere with clouds that hide its surface from view, but radar mapping reveals a surface dotted with volcanoes and other features indicating active geology. Mars, despite its middling size, has the solar system's largest volcanoes and a huge canyon cutting across its surface, along with many features that appear to have been shaped by running water. Earth has surface features similar to all those on the other terrestrial worlds, and more—including a unique layer of living organisms that covers almost the entire surface of the planet.

Our primary goal in this chapter is to gain a deeper understanding of our own planet Earth by investigating how the terrestrial worlds came to be so different. We'll begin by examining the basic nature of our planet.

• Why is Earth geologically active?

All the terrestrial worlds have changed since their birth, but Earth is unique in the degree to which it continues to change today. We say that Earth is *geologically active,* meaning that its surface is continually being reshaped by volcanic eruptions, earthquakes, erosion, and other geological processes. Most of this geological activity is the result of what goes on deep inside our planet. Consequently, to understand why Earth is so much more geologically active than other worlds, we must examine what the terrestrial worlds are like inside.

Interior Structure   Studies of internal structure (see Special Topic, p. 193) show that all the terrestrial worlds have layered interiors. We often divide these layers by density into three major categories:

- **Core:** The highest-density material, consisting primarily of metals such as nickel and iron, resides in the central core.

Figure 7.1 VIS

The terrestrial worlds, shown to scale, along with sample surface close-ups from orbiting spacecraft. All the photos were taken with visible light except the Venus close-up.

Mercury

Venus

Earth

Earth's moon

Mars

Heavily cratered Mercury has long, steep cliffs—one is the long curve going from upper right to lower left.

Cloud-penetrating radar obtained this view of a twin-peaked volcano on Venus. (The colors are not real.)

We see land, ocean, and snow-capped peaks in this view of Earth from orbit.

The Moon's surface is heavily cratered in most places.

Mars has features that look like dry riverbeds, along with impact craters like those near the upper right.

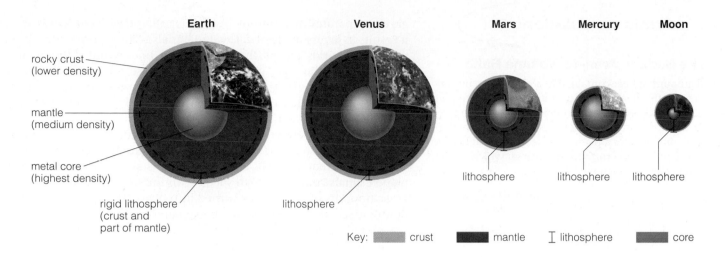

Earth   Venus   Mars   Mercury   Moon

rocky crust
(lower density)

mantle
(medium density)

metal core
(highest density)

rigid lithosphere
(crust and
part of mantle)

lithosphere

lithosphere   lithosphere   lithosphere

Key: ▉ crust   ▉ mantle   ⊥ lithosphere   ▉ core

- **Mantle**: Rocky material of moderate density—mostly minerals that contain silicon, oxygen, and other elements—forms the thick mantle that surrounds the core.

- **Crust**: The lowest-density rock, such as granite and basalt (a common form of volcanic rock) forms the thin crust, essentially representing the world's outer skin.

Figure 7.2 shows these layers for the five terrestrial worlds. Although not shown in the figure, Earth's metallic core actually consists of two distinct regions: a solid *inner core* and a molten (liquid) *outer core*.

In geology, it's often more useful to categorize interior layers by rock strength instead of density. The idea that rock can vary in strength may seem surprising, but like all matter built of atoms, rock is mostly empty space [Section 5.1]. The solidity of rock comes from electrical bonds between its atoms and molecules, and while these bonds are strong, they can still break and re-form when subjected to heat or sustained stress. Over millions and billions of years, even "solid" rock can slowly deform and flow. The long-term behavior of rock is much like that of the popular toy Silly Putty, which breaks like a brittle solid when you pull it sharply but deforms and stretches when you pull it slowly (Figure 7.3). Also like Silly Putty, rock becomes softer and easier to deform when it is warmer.

see it for yourself Roll some room-temperature Silly Putty into a ball and measure its diameter. Place the ball on a table and gently place a heavy book on top. After 5 seconds, measure the height of the squashed ball. Repeat the experiment, but warm the Silly Putty in hot water before you start; repeat again but cool the Silly Putty in ice water before you start. How does temperature affect the rate of "squashing"? How does the experiment relate to planetary geology?

A planet's *lithosphere* is its outer layer of cool, rigid rock.

In terms of rock strength, Earth's outer layer consists of relatively cool and rigid rock, called the **lithosphere** (*lithos* is Greek for "stone"), that "floats" on warmer, softer rock beneath. The lithosphere encompasses the crust and part of the upper mantle on Earth and extends deeper into the mantle on smaller worlds.

Differentiation and Internal Heat The distinct layering of the terrestrial worlds occurs for the same reason that liquids separate by density. For example, in a mixture of oil and water, gravity pulls the denser water to the bottom, driving the less dense oil to the top. The process by which

Figure 7.2

Interior structures of the terrestrial worlds, shown to scale and in order of decreasing size. Color coding shows the core-mantle-crust layering by density; a dashed circle represents the inner boundary of the lithosphere, defined by strength of rock rather than by density. The thicknesses of the crust and the lithosphere of Venus and Earth are exaggerated to make them visible in this figure.

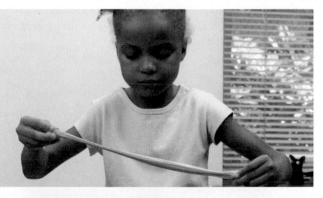

Figure 7.3

Silly Putty stretches when pulled slowly but breaks cleanly when pulled rapidly. Rock behaves just the same, but on a longer time scale.

cosmic calculations 7.1

The Surface Area–to–Volume Ratio

The total amount of heat contained in a planet depends on its volume, but this heat can escape into space only from its *surface*. As heat escapes, more heat flows upward from the interior to replace it until the interior is no hotter than the surface. The time it takes a planet to lose its internal heat is related to the ratio of the *surface area* through which it loses heat to the *volume* that contains heat:

$$\text{surface area–to–volume ratio} = \frac{\text{surface area}}{\text{volume}}$$

A spherical planet (radius r) has surface area $4\pi r^2$ and volume $\frac{4}{3}\pi r^3$, so the ratio becomes:

$$\underset{\text{(for a sphere)}}{\text{surface area–to–volume ratio}} = \frac{4\pi r^2}{\frac{4}{3}\pi r^3} = \frac{3}{r}$$

Because r appears in the denominator, we conclude that *larger objects have smaller surface area–to–volume ratios*.

Example: Compare the surface area–to–volume ratios of the Moon and Earth. Data:

$$r_{\text{Moon}} = 1738 \text{ km}; \ r_{\text{Earth}} = 6378 \text{ km}.$$

Solution: Dividing the surface area–to–volume ratios for the Moon and Earth, we find:

$$\frac{\text{surface area–to–volume ratio (Moon)}}{\text{surface area–to–volume ratio (Earth)}} = \frac{^3/_{r_{\text{Moon}}}}{^3/_{r_{\text{Earth}}}} = \frac{r_{\text{Earth}}}{r_{\text{Moon}}}$$

$$= \frac{6378 \text{ km}}{1738 \text{ km}}$$

$$= 3.7$$

The Moon's surface area–to–volume ratio is about four times that of Earth, which means the Moon would cool four times as fast if both worlds started with the same temperature and gained no additional heat.

gravity separates materials by density is called **differentiation** (because it results in layers made of *different* materials). The layered interiors of the terrestrial worlds tell us that all of them underwent differentiation at some time in the past, which means all these worlds must once have been hot enough inside for their interior rock and metal to melt and separate by density. Dense metals like iron sank toward the center, driving less dense rocky material toward the surface.

When they were young, Earth and the other terrestrial worlds were hot inside for two major reasons. First, the planets gained heat from the process of formation itself. During the later stages of accretion, incoming planetesimals collided at high speed with the forming planets, depositing large amounts of energy that turned into heat. The process of differentiation released additional heat as dense materials sank to the core. Second, the metal and rock that make up the terrestrial planets include small but important amounts of radioactive elements. As these radioactive materials decay, they release heat directly into the planetary interiors. Radioactive decay still supplies heat to the terrestrial interiors, though at a lower level than it did when the planets were young (because some of the radioactive material has already decayed).

> Earth and the other terrestrial worlds were once hot enough inside for their interiors to melt, allowing material to settle into layers of differing density.

None of the terrestrial worlds are still hot enough to remain liquid throughout their interiors. However, they differ considerably in the amount of heat they have retained. Size is the most important factor in planetary cooling (Cosmic Calculations 7.1): Just as a hot potato remains hot inside much longer than a hot pea, a large planet stays hot inside much longer than a small one. You can see why size is the critical factor by picturing a large planet as a smaller planet wrapped in extra layers of rock. The extra rock acts as insulation, so it takes much longer for interior heat to reach the surface and escape.

 **see it for yourself** The fact that large objects stay warmer longer than small objects is easy to observe with food and drink. The next time you eat something large and hot, cut off a small piece; notice how much more quickly the small piece cools compared to the rest of it. A similar experiment demonstrates the time it takes a cold object to warm up: Find two ice cubes of the same size; crack one into small pieces with a spoon, and then compare how fast each melts. Explain your observations in terms that your friends would understand.

Internal Heat and Geological Activity Interior heat is the primary driver of geological activity, because this heat supplies the energy needed to move rock and reshape the surface. Inside a planet, temperature increases with depth. If the interior is hot enough, hot rock can gradually rise within the mantle, slowly cooling as it rises. Cooler rock at the top of the mantle gradually falls (Figure 7.4). The process by which hot material expands and rises while cooler material contracts and falls is called **convection**. Keep in mind that mantle convection primarily involves solid rock, not molten rock. Because solid rock flows quite slowly, mantle convection is a very slow process. At the typical rate of mantle convection on Earth—a few centimeters per year—it would take 100 million years for a piece of rock to be carried from the base of the mantle to the top.

Mantle convection: hot rock rises and cooler rock falls.

Figure 7.4

Earth's hot interior allows the mantle to undergo convection, in which hot rock gradually rises upward while cool rock gradually falls. Arrows indicate the direction of flow in a portion of the mantle.

> Larger planets retain internal heat much longer than smaller ones, and this heat drives geological activity.

Just as planetary size determines how long a planet stays hot, it is also the primary factor in the strength of mantle convection and lithospheric thickness.

As a planet's interior cools, the rigid lithosphere grows thicker and convection occurs only deeper inside the planet. A thick lithosphere inhibits volcanic and tectonic activity, because any molten rock is too deeply buried to erupt to the surface and the strong lithosphere resists distortion by tectonic stresses. If the interior cools enough, convection may stop entirely, leaving the planet geologically dead, with no eruptions or crustal movement.

We can now understand the differences in lithospheric thickness shown in Figure 7.2, which go along with differences in geological activity. Earth, the largest of the terrestrial planets, remains quite hot inside and therefore has a thin lithosphere. Venus is probably similar to Earth in its internal heat, though it may have a thicker lithosphere (for reasons we will discuss later). With their small sizes, Mercury and the Moon have very thick lithospheres and no geological activity. Mars, intermediate in size, has cooled significantly but probably retains some internal heat.

special topic:

How Do We Know What's Inside Earth?

OUR DEEPEST DRILLS have barely pricked Earth's surface, penetrating less than 1% of the way into the interior. How, then, can we claim to know what our planet is like on the inside?

For Earth, much of our information about the interior comes from *seismic waves*, vibrations created by earthquakes. Seismic waves come in two basic types that are analogous to the two ways that you can generate waves in a Slinky (Figure 1). Pushing and pulling on one end of a Slinky (while someone holds the other end still) generates a wave in which the Slinky is bunched up in some places and stretched out in others. Waves like this in rock are called *P waves*. The P stands for *primary*, because these waves travel fastest and are the first to arrive after an earthquake, but it is easier to think of P for *pressure* or *pushing*. P waves can travel through almost any material—whether solid, liquid, or gas—because molecules can always push on their neighbors no matter how weakly they are bound together. (Sound also travels as a pressure wave quite similar to P waves.)

Shaking a Slinky slightly up and down generates an up-and-down motion all along its length. Such up-and-down (or side-to-side) waves in rock are called *S waves*. The S stands for *secondary* but is easier to remember as meaning *shear* or *side-to-side*. S waves travel only through solids, because the bonds between neighboring molecules in a liquid or gas are too weak to transmit up-and-down or sideways forces.

The speeds and directions of seismic waves traveling through Earth depend on the composition, density, pressure, temperature, and phase (solid or liquid) of the material they pass through. For example, P waves reach the side of the world opposite an earthquake, but S waves do not. This tells us that a liquid layer has stopped the S waves, which is how we know that Earth has a liquid outer core (Figure 2). More careful analysis of seismic waves has allowed geologists to develop a detailed picture of Earth's interior structure.

We have also used seismic waves to study the Moon's interior, thanks to monitoring stations left behind by the *Apollo* astronauts. We use less direct clues to learn about the interiors of other worlds. For example, knowing that the density of surface rock is much less than a planet's overall average density tells us that the planet must contain denser rock or metal inside. We can also learn about a planet's interior from precise measurements of its gravity, which tell us how mass is distributed within the planet; from studies of its magnetic field, which is generated deep inside the planet; and from observations of surface rocks that appear to have emerged from deep within the interior.

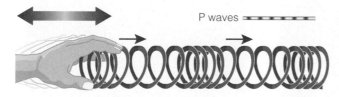

P waves

P waves result from compression and stretching in the direction of travel.

S waves

S waves vibrate up and down or side to side perpendicular to the direction of travel.

Figure 1

Slinky examples demonstrating P and S waves.

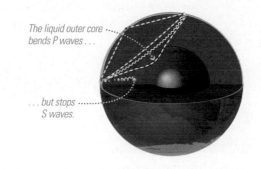

The liquid outer core bends P waves . . .

. . . but stops S waves.

Figure 2

Because S waves do not reach the side of Earth opposite an earthquake, we infer that part of Earth's core is liquid.

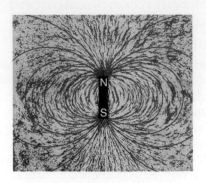

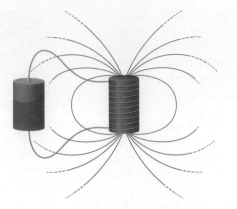

a This photo shows how a bar magnet influences iron filings (small black specks) around it. The *magnetic field lines* (red) represent this influence graphically.

b A similar magnetic field is created by an electromagnet, which is essentially a wire wrapped around a metal bar and attached to a battery. The field is created by the battery-forced motion of charged particles (electrons) along the wire.

c Earth's magnetic field also arises from the motion of charged particles. The charged particles move within Earth's liquid outer core, which is made of electrically conducting, convecting molten metals.

Figure 7.5

Sources of magnetic fields.

Figure 7.6 interactive figure ↖

Earth's magnetosphere acts like a protective bubble, shielding our planet from charged particles coming from the solar wind.

The Magnetic Field Interior heat is also responsible for Earth's global **magnetic field**. You are probably familiar with the general pattern of the magnetic field created by an iron bar (Figure 7.5a). Earth's magnetic field is generated by a process more similar to that of an *electromagnet,* in which the magnetic field arises as a battery forces charged particles to move along a coiled wire (Figure 7.5b).

Earth's magnetic field is generated by the motions of molten metal in its liquid outer core.

Earth does not contain a battery, but charged particles move with the molten metal in its liquid outer core (Figure 7.5c). Internal heat causes the liquid metal to rise and fall (convection), while Earth's rotation twists and distorts the convection pattern. The result is that electrons in the molten metal move within the outer core in much the same way they move in an electromagnet, generating Earth's magnetic field.

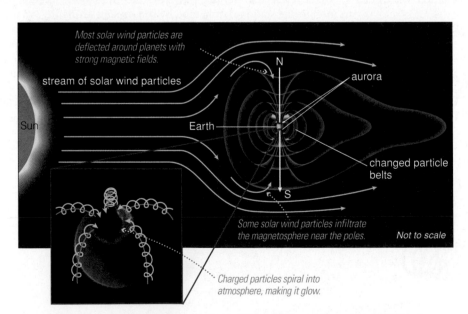

a This diagram shows how Earth's magnetosphere deflects solar wind particles. Some particles accumulate in *charged particle belts* encircling our planet. The inset shows a photo of ring auroras (the crater-like ridge) around the North Pole; the bright and dark regions below the ring are day and night regions of the Earth.

b This photograph shows the aurora near Yellowknife, Northwest Territories, Canada. In a video, you would see these lights dancing about in the sky.

Earth's magnetic field helps protect the surface from energetic particles that continually flow outward from the Sun with the *solar wind* [Section 6.4]. These particles could strip away atmospheric gas and cause genetic damage to living organisms. The magnetic field shields us by creating a **magnetosphere**—a kind of protective bubble that surrounds our planet (Figure 7.6a). The magnetosphere deflects most of the charged particles from the Sun around our planet. The relatively few particles that make it through the magnetosphere tend to be channeled toward the poles, where they collide with atoms and molecules in our atmosphere and produce the beautiful lights of the **aurora** (Figure 7.6b). The aurora is strongest when solar winds buffet the magnetosphere, energizing the charged particles trapped within it.

None of the other terrestrial worlds have magnetic fields as strong as Earth's. As a result, they lack protective magnetospheres—a fact that, as we'll discuss later, has had a profound effect on the planetary histories of Venus and Mars.

(MA) Shaping Planetary Surfaces Tutorial, Lessons 1–3

• What processes shape Earth's surface?

We are now ready to turn to planetary surfaces. Earth offers a huge variety of geological surface features, and the variety only increases when we survey other worlds. Nevertheless, almost all surface features can be explained by just four major geological processes:

- **Impact cratering**: the excavation of bowl-shaped *impact craters* by asteroids or comets crashing into a planet's surface.

- **Volcanism**: the eruption of molten rock, or *lava*, from a planet's interior onto its surface.

- **Tectonics**: the disruption of a planet's surface by internal stresses.

- **Erosion**: the wearing down or building up of geological features by wind, water, ice, and other phenomena of planetary weather.

Virtually all geological features originate from impact cratering, volcanism, tectonics, and/or erosion.

Before we examine these processes in greater detail, notice that impact cratering is the only one of the four processes with an external cause—impacts of objects from space. The other three processes are attributable to the planet itself and represent what we usually define as *geological activity*.

Impact Cratering   An impact crater forms when an asteroid or comet slams into a solid surface (Figure 7.7). Impacting objects typically hit the surface at a speed between about 40,000 and 250,000 kilometers per hour. At such a tremendous speed, the impact releases enough energy to vaporize solid rock and blast out a crater (the Greek word for "cup"). Craters are usually circular because an impact blasts out material in all directions, regardless of the incoming object's direction. Laboratory experiments show that craters are typically about 10 times as wide as the objects that create them and about 10–20% as deep as they are wide. For example, an asteroid 1 kilometer in diameter will blast out a crater about 10 kilometers wide and 1–2 kilometers deep.

We have never witnessed a major impact on Earth (though we have witnessed one on Jupiter [Section 9.4]), but we have studied the results of past impacts (Figure 7.8). We also see numerous impact craters on other worlds, especially the Moon and Mercury (see Figure 7.1).

Figure 7.7 interactive figure

Artist's conception of the impact process.

Figure 7.8

Meteor Crater in Arizona is more than a kilometer across and nearly 200 meters deep. It was created around 50,000 years ago by the impact of a metallic asteroid about 50 meters across.

molten rock in upper mantle

Figure 7.9 interactive figure

Volcanism. The photo shows the eruption of an active volcano on the flanks of Kilauea on the Big Island of Hawaii. The inset shows the underlying process: Molten rock collects in a "magma chamber" and can erupt upward.

Figure 7.10

This photo shows the eruption of Mount St. Helens (Washington State) on May 18, 1980. Note the tremendous outgassing that accompanied the eruption.

Comparing the number of impact craters on the Moon and Earth leads us to an important insight. Throughout its history, Earth must have had at least as many impacts as the Moon, since we occupy the same region of the solar system. Why, then, are there so many more impact craters on the Moon? The answer is that most of Earth's impact craters have been erased with time by geological activity such as volcanic eruptions and erosion.

Like all terrestrial worlds, Earth was bombarded by impacts when it was young, but most ancient craters have been erased by other geological processes.

The idea that craters can be erased with time offers us a way to estimate the age of a world's surface—that is, how long it has been since craters were last erased on the surface. Remember that all the planets were battered by impacts during the heavy bombardment that occurred early in our solar system's history [Section 6.4]. Most impact craters were made during that time, and relatively few impacts have occurred since. In places where we see numerous craters, such as on much of the Moon's surface, we must be looking at a surface that has stayed virtually unchanged for billions of years. In contrast, when we see very few craters, as we do on Earth, we must be looking at a surface that has undergone recent change. Careful studies of the Moon, where different surface regions have both different numbers of craters and different ages (determined by radiometric dating of Moon rocks), have allowed planetary scientists to determine the rate at which craters were made during much of the solar system's history. Knowing this rate allows scientists to estimate the age of a planetary surface just by photographing it from orbit and counting its craters.

Volcanism Volcanism occurs when underground molten rock finds a path through the lithosphere to the surface (Figure 7.9). Molten rock tends to rise for three main reasons. First, molten rock is generally less dense than solid rock, and lower-density materials tend to rise when surrounded by higher-density materials. Second, because most of Earth's interior is not molten, the solid rock surrounding a chamber of molten rock can squeeze the molten rock, driving it upward under pressure. Third, molten rock often contains trapped gases that expand as it rises, which can make it rise much faster and lead to dramatic eruptions. Erupting lava can make tall, steep volcanoes if the lava is very thick, or vast, flat lava plains if the lava is very runny.

Earth's atmosphere and oceans were made from gases released from the interior by volcanic outgassing.

Volcanic mountains are the most obvious result of volcanism, but volcanism has had a much more profound effect on our planet: It explains the existence of our atmosphere and oceans. Recall that Earth accreted from rocky and metallic planetesimals, while water and other ices were brought in by planetesimals from more distant reaches of the solar system [Section 6.4]. Water and gases became trapped beneath the surface in much the same way the gas in a carbonated beverage is trapped in a pressurized bottle. Volcanic eruptions later released some of this gas into the atmosphere in a process known as **outgassing** (Figure 7.10).

Measurements show that the most common gases released by outgassing are water vapor (H_2O), carbon dioxide (CO_2), nitrogen (N_2), and sulfur-bearing gases (H_2S or SO_2). The outgassed water vapor rained down to form our oceans, while the other gases helped make our atmosphere. Much of the nitrogen remains in Earth's atmosphere to this day, where it is now the dominant ingredient (77%). We'll discuss how oxygen came to make up most of the rest of our atmosphere in Section 7.5.

Internal stresses can cause compression in the crust . . .

. . . creating mountains like the Himalayas.

Internal stresses can also pull the crust apart . . .

. . . creating cracks and seas.

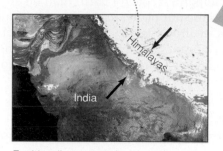

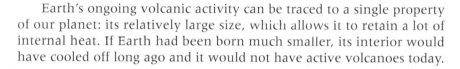

Earth's tallest mountain range, the Himalayas, created as India pushes into the rest of Asia.

The Red Sea, created as the Arabian Peninsula was torn away from Africa.

Earth's ongoing volcanic activity can be traced to a single property of our planet: its relatively large size, which allows it to retain a lot of internal heat. If Earth had been born much smaller, its interior would have cooled off long ago and it would not have active volcanoes today.

Tectonics   The third major geological process, tectonics, refers to any surface reshaping that results from stretching, compression, or other forces acting on the lithosphere. Figure 7.11 shows two examples of tectonic features on Earth, one created by surface compression (the Himalayas) and one created by surface stretching (the Red Sea).

Much of the tectonic activity on any planet is a direct or indirect result of mantle convection. Tectonics is particularly important on Earth, because the underlying mantle convection fractured Earth's lithosphere into more than a dozen pieces, or *plates*. These plates move over, under, and around each other, leading to a special type of tectonics that we call **plate tectonics**. While some type of tectonics has affected every terrestrial world, plate tectonics appears to be unique to Earth. Moreover, as we'll discuss in Section 7.5, plate tectonics may be crucial to explaining life's abundance on Earth.

Tectonics and volcanism generally occur together because both require internal heat and therefore depend on a planet's size.

Tectonic activity usually goes hand in hand with volcanism, because both require internal heat. Like volcanism, Earth's ongoing tectonics is possible only because of our planet's relatively large size, which has allowed it to retain plenty of internal heat.

Erosion   The last of our four major geological processes is erosion. *Erosion* is a blanket term for a variety of processes that break down or transport rock through the action of ice, liquid, or gas. The shaping of valleys by glaciers (ice), the carving of canyons by rivers (liquid), and the shifting of sand dunes by wind (gas) are all examples of erosion (Figure 7.12).

Erosion can both break down and build up geological features.

We often associate erosion with the breakdown of existing features, but erosion also builds things. Sand dunes, river deltas, and lake bed deposits are all examples of features built by erosion. Indeed, much of the surface rock on Earth was built by erosion. Over long periods of time, erosion has piled sediments into layers on the floors of oceans and seas, forming what we call **sedimentary**

Figure 7.11 interactive figure

Tectonic forces can produce a wide variety of features. Mountains created by tectonic compression and valleys or seas created by tectonic stretching are among the most common. Both images are satellite photos.

a The Colorado River has carved the Grand Canyon over the past few million years.

b Glaciers created Yosemite Valley.

c Wind erosion wears away rocks and builds up these desert sand dunes.

d This river delta consists of sediments worn away by wind and rain and then carried downstream.

Figure 7.12

A few examples of erosion on Earth.

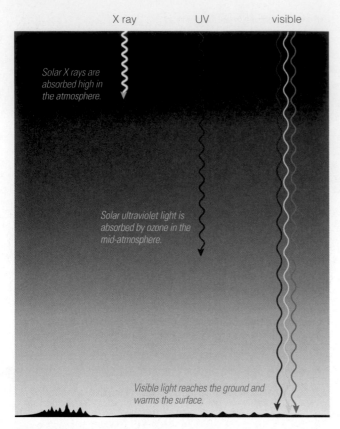

X ray   UV   visible

Solar X rays are absorbed high in the atmosphere.

Solar ultraviolet light is absorbed by ozone in the mid-atmosphere.

Visible light reaches the ground and warms the surface.

Figure 7.13

This diagram summarizes how different forms of light from the Sun are affected by Earth's atmosphere.

common misconceptions

Why Is the Sky Blue?

If you ask around, you'll find a variety of misconceptions about why the sky is blue. Some people think the sky is blue because of light reflecting from the oceans, but anyone who lives in Nebraska knows that this must be wrong—Nebraska is too far from any ocean for its sky to be influenced by such reflection. Others claim that "air is blue." It's not obvious what someone might mean by this statement, but it's clearly wrong: If air molecules emitted blue light, then air would glow blue even in the dark; if they were blue because they reflected blue light and absorbed red light, then no red light could reach us at sunset. The real explanation for the blue sky is light scattering, as shown in Figure 7.14. This explanation not only makes sense for our blue sky but also explains our red sunsets.

rock. The layered rock of the Grand Canyon is an example of sedimentary rock, built up by erosion long before the canyon formed.

Erosion plays a far more important role on Earth than on any other terrestrial world, primarily because our planet has both strong winds and plenty of water. Strong winds are driven largely by our planet's relatively rapid rotation. Water exists as a result of outgassing by volcanism, and it causes erosion because our planet's temperature is just right to allow water to exist in both liquid and solid form on the surface.

(MA) Surface Temperatures of Terrestrial Planets Tutorial, Lessons 1–4

• How does Earth's atmosphere affect the planet?

Our atmosphere consists of about 77% nitrogen (N_2), 21% oxygen (O_2), and small amounts of other gases. This atmosphere provides the air we breathe and supplies the pressure that allows liquid water to flow, and as we've seen, it explains the extensive erosion on Earth. But its importance goes far deeper: Without the atmosphere, Earth's surface would be rendered lifeless by dangerous solar radiation and would be so cold that all water would be perpetually frozen.

Remarkably, our atmosphere plays all these roles despite being very thin compared to the planet. About two-thirds of the air in Earth's atmosphere lies within 10 kilometers of the surface. You could represent this air on a standard globe with a layer only as thick as a dollar bill. Let's look more closely at the fundamental roles of our thin atmosphere.

Surface Protection   The Sun emits the visible light that allows us to see, but it also emits dangerous ultraviolet and X-ray radiation. In space, astronauts need thick spacesuits to protect them from the hazards of this radiation. On Earth, we are protected by our atmosphere (Figure 7.13).

X-ray photons carry enough energy to knock electrons free from almost any atom or molecule. That is, they *ionize* [Section 5.2] the atoms or molecules they strike, which is also why they can damage living tissue. Fortunately, X rays are absorbed so easily by atoms and molecules that all the X rays from the Sun are absorbed high in our atmosphere, leaving none to reach the ground. That is why X-ray telescopes must be placed in space (see Figure 5.21).

Ozone absorbs the Sun's dangerous ultraviolet radiation, while the X rays are absorbed by atoms and molecules higher up in Earth's atmosphere.

Ultraviolet photons from the Sun are not so easily absorbed. Most gases are transparent to ultraviolet light, allowing it to pass through unhindered. We owe our protection from ultraviolet light to a relatively rare gas called **ozone** (O_3). Ozone resides primarily in a middle layer of Earth's atmosphere (the *stratosphere*), where it absorbs most of the dangerous ultraviolet radiation from the Sun. Without ozone, we could not survive.

The Sun emits most of its radiation in the form of visible light. Visible light passes easily through the atmosphere and reaches the surface, allowing us to see and providing energy for photosynthesis. More important, visible light heats the ground when it is absorbed, making it the primary source of heat for Earth's surface.

Although most visible-light photons pass straight through Earth's atmosphere, a few are scattered randomly around the sky. This scattering is the reason our sky is bright rather than dark—which is why we cannot see stars in the daytime. Without scattering, our sky would look like the lunar sky does to an astronaut, with the Sun just a very bright circle set against a black, star-studded background. Scattering also explains why the daytime sky is blue (Figure 7.14). Visible light consists of all the colors of the rainbow, but the colors are not all scattered equally. Gas molecules scatter blue light (higher energy) much more effectively than red light (lower energy). The difference in scattering is so great that, for practical purposes, we can imagine that only the blue light gets scattered. When the Sun is overhead, this scattered blue light reaches our eyes from all directions and the sky appears blue. At sunset or sunrise, the sunlight must pass through a greater amount of atmosphere on its way to us. Most of the blue light is scattered away, leaving only red light to color the sky.

The Greenhouse Effect

Visible light warms Earth's surface, but not as much as you might guess. Calculations based on Earth's distance from the Sun (and the percentages of visible light absorbed and reflected by the ground) show that, by itself, visible light would give Earth an average surface temperature of only −16°C (+3°F)—well below the freezing point of water. In fact, Earth's global average temperature is about 15°C (59°F), plenty warm enough for liquid water to flow and life to thrive. Why is Earth so much warmer than it would be from visible-light warming alone? The answer is that our atmosphere traps additional heat through what we call the **greenhouse effect**.

> The greenhouse effect keeps Earth's surface much warmer than it would be otherwise, allowing water to stay liquid over most of the surface.

Figure 7.15 shows the basic idea behind the greenhouse effect. Some of the visible light that reaches the ground is reflected and some is absorbed. The absorbed energy must be returned to space—otherwise the ground would rapidly heat up—but planetary surfaces are too cool to emit visible light. (Remember that the type of thermal radiation an object emits depends on its temperature [Section 5.2].) Instead, planetary temperatures are in the range in which they emit energy primarily in the form of infrared rather than visible light. The greenhouse effect occurs when the atmosphere temporarily "traps" some of the infrared light that the ground emits, slowing its return to space.

The greenhouse effect therefore occurs only when the atmosphere contains gases that can absorb infrared light. Gases that are particularly good at absorbing infrared light are called **greenhouse gases**, and they include water vapor (H_2O), carbon dioxide (CO_2), and methane (CH_4). These gases absorb infrared light effectively because their molecular structures begin rotating or vibrating when they absorb a photon of infrared light. A molecule then reemits a photon of infrared light in some random direction. This photon can then be absorbed by another gas molecule, which does the same thing. The net result is that greenhouse gases tend to slow the escape of infrared radiation from the lower atmosphere, while their molecular motions heat the surrounding air. In this way, the greenhouse effect makes the surface and the lower atmosphere warmer than they would be from sunlight alone. The more greenhouse gases present, the greater the degree of surface warming.

think about it Molecules that consist of two atoms of the same type—such as the N_2 and O_2 molecules that together make up 98% of Earth's atmosphere—are poor infrared absorbers. But imagine this were not the case. How would Earth be different if nitrogen and oxygen absorbed infrared light effectively?

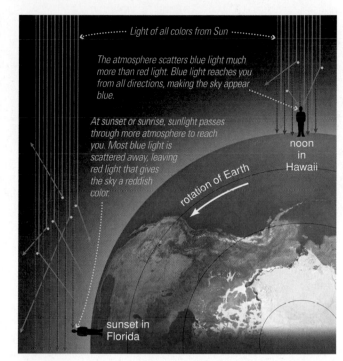

Figure 7.14

This diagram summarizes why the sky is blue and sunsets (and sunrises) are red.

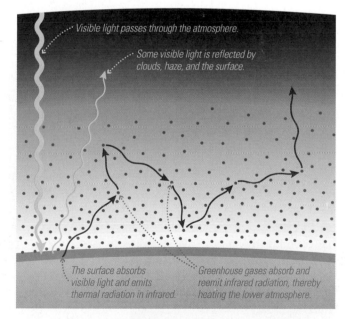

Figure 7.15 interactive figure

The greenhouse effect. The lower atmosphere becomes warmer than it would be if it had no greenhouse gases such as water vapor, carbon dioxide, and methane.

Moon   Mercury

Figure 7.16

Similar views of the Moon and Mercury, shown to scale. The Mercury photo was obtained during the January 2008 *MESSENGER* flyby.

7.2 The Moon and Mercury: Geologically Dead

In the rest of this chapter, we will investigate the histories of the terrestrial worlds, with the ultimate goal of learning how and why Earth became unique. We'll start in this section with the two worlds that have the simplest histories: the Moon and Mercury (Figure 7.16).

The simple histories of the Moon and Mercury are a direct consequence of their small sizes. Both these worlds are considerably smaller than Venus, Earth, or Mars, so they long ago lost most of their internal heat, leaving them without any energy source to power ongoing geological activity. Small size also explains their lack of significant atmospheres: Their gravity is too weak to hold gas for long periods of time, and without ongoing volcanism they lack the outgassing needed to replenish gas lost in the past.

• Was there ever geological activity on the Moon or Mercury?

Impact cratering has been by far the most important geological process on both worlds. However, closer examination shows that a few features are volcanic or tectonic in origin and indicate that the Moon and Mercury had geological activity in the past. Long ago, before they had a chance to cool, these worlds were hot enough inside for some volcanism and tectonics to occur.

Geological Features of the Moon The familiar face of the full moon shows that not all regions of the surface look the same (Figure 7.17). Some regions are heavily cratered. Other regions, known as the **lunar maria**, look smoother and darker. Indeed, the maria got their name because they look much like oceans when seen from afar. *Maria* (singular, *mare*) is Latin for "seas." The smooth and dark appearance of the lunar maria suggests that they were made by a flood of molten lava, and studies of moon rocks confirm this suggestion.

Figure 7.18 shows how we think the maria formed. During the heavy bombardment, craters covered the Moon's entire surface. The largest impacts violently fractured the Moon's lithosphere beneath the huge craters they created. However, the Moon's interior had already cooled since its formation, and there was no molten rock to flood these craters immediately. Instead, the lava floods came hundreds of millions of years later, thanks to heat released by the decay of radioactive elements in the Moon's interior. This heat gradually built up during the Moon's early history, until mantle material melted about 3 to 4 billion years ago. The molten rock then welled up through the cracks in the lithosphere, flooding the largest impact craters with lava. The maria are generally circular because they are essentially flooded craters (and craters are almost always round). Their dark color comes from the dense, iron-rich rock (basalt) that rose up from the lunar mantle as molten lava.

200 km

The Moon's dark, smooth maria were made by floods of molten lava billions of years ago, when the Moon's interior was heated by radioactive decay.

The Moon's interior cooled quickly after that, and there was never again enough radioactive heat to cause further melting. Because the lava floods occurred after the heavy bombardment subsided, the maria have remained much as they were when they first formed. The relatively few craters that we see within them today were made by impacts that occurred after the maria formed.

Figure 7.17

The familiar face of the full Moon shows numerous dark, smooth maria.

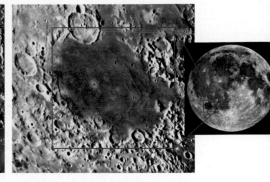

a This illustration shows the Mare Humorum region as it probably looked about 4 billion years ago, when it would have been completely covered in craters.

b Around that time, a huge impact excavated the crater that would later become Mare Humorum. The impact fractured the Moon's lithosphere and erased the many craters that existed earlier.

c A few hundred million years later, heat from radioactive decay built up enough to melt the Moon's upper mantle. Molten lava welled up through the lithospheric cracks, flooding the impact crater.

d This photo shows Mare Humorum as it appears today, and the inset shows its location on the Moon.

The Moon's era of geological activity is long gone. Today, the Moon is a desolate and nearly unchanging place. Rare impacts may occur in the future, but we are unlikely ever to witness a major one. Little happens on the Moon, aside from the occasional visit of robotic spacecraft or astronauts from Earth (Figure 7.19a).

The only ongoing geological change on the Moon is a very slow "sandblasting" of the surface by *micrometeorites,* sand-size particles from space. These tiny particles burn up as meteors in the atmospheres of Earth, Venus, and Mars but rain directly onto the surface of the airless Moon. The micrometeorites gradually pulverize the surface rock, which explains why the lunar surface is covered by a thin layer of powdery "soil." The *Apollo* astronauts left their footprints in this powdery surface (Figure 7.19b). Pulverization by micrometeorites is a very slow process, and the astronauts' footprints will last millions of years before they are finally erased.

Figure 7.18

The lunar maria formed between 3 and 4 billion years ago, when molten lava flooded large craters that had formed hundreds of millions of years earlier. This sequence of diagrams represents the formation of Mare Humorum.

Figure 7.19

The Moon today is geologically dead, but it can still tell us a lot about the history of our solar system.

a Astronaut Gene Cernan takes the Lunar Roving Vehicle for a spin during the final *Apollo* mission to the Moon (*Apollo 17*, December 1972).

b The *Apollo* astronauts left clear footprints, like this one, in the Moon's powdery "soil." The powder is the result of gradual pulverization of surface rock by micrometeorites. Micrometeorites will eventually erase the astronauts' footprints, but not for millions of years.

Although the Moon has been "geologically dead" since the maria formed more than 3 billion years ago, it remains a prime target of exploration, with missions in progress or planned not only by NASA but also by nations including Japan, China, and India. These missions are helping us understand both the Moon's history and the history of our solar system. NASA and other nations also hope to establish a human outpost on the Moon within the next decade or so. There, astronauts will explore the potential value of lunar resources and learn about the challenges of staying for extended periods on other worlds.

Geological Features of Mercury Mercury looks so much like the Moon that it's often difficult at first to tell which world you are looking at in surface photos. The similarities extend to all the geological processes, though there are also a few differences. Mercury's closeness to the Sun and slow rotation make it a world of extremes. The combination of its 88-day orbit and its 59-day rotation gives Mercury days and nights that last about 3 Earth months each [Section 6.1]. As the Sun rises above the horizon for 3 months of daylight, the equatorial surface temperature soars to 425°C. At night or in shadow, the temperature falls below −150°C.

Impact craters are visible almost everywhere on Mercury, indicating an ancient surface. However, Mercury's craters are less crowded together than the craters in the most ancient regions of the Moon, suggesting that molten lava later covered up some of the craters that formed on Mercury during the heavy bombardment (Figure 7.20a). As on the Moon, these lava flows probably occurred when heat from radioactive decay accumulated enough to melt part of the mantle. Although we have not found evidence of lava flows as large as those that created the lunar maria, the lesser crater crowding and the many smaller lava plains suggest that Mercury had at least as much volcanism as the Moon.

The largest single surface feature on Mercury is a huge impact crater called the *Caloris Basin* (Figure 7.20b and c). The Caloris Basin spans more than half of Mercury's radius, and its multiple rings bear witness to the violent impact that created it. The Caloris Basin has few craters within it, indicating that it must have formed at a time when the heavy bombardment was already subsiding.

Figure 7.20

Features of impact cratering and volcanism on Mercury.

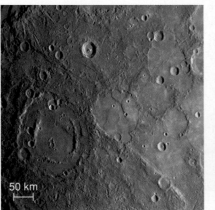 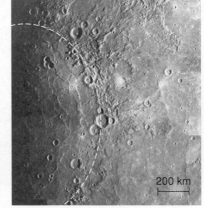 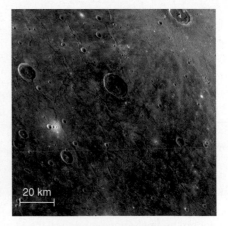

a *MESSENGER* close-up view of Mercury's surface, showing impact craters and smooth regions where lava apparently covered up craters.

b Part of the Caloris Basin (outlined by the dashed ring), a large impact crater on Mercury (*Mariner 10* image).

c This *MESSENGER* close-up of the floor of the Caloris Basin shows stretch marks that were probably created as the surface gradually rebounded after the impact.

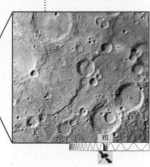

Mercury's core and mantle shrank . . .

Some portions of the crust were forced to slide under others. ·

Today we see long, steep cliffs created by this crustal movement.

. . . causing Mercury's crust to contract.

Shrinkage not to scale!

VIS

Figure 7.21

Long cliffs on Mercury offer evidence that the entire planet shrank early in its history, perhaps by as much as 20 kilometers in radius.

a Mercury's long cliffs probably formed when its core cooled and contracted, causing the mantle and lithosphere to shrink. This diagram shows how the cliffs probably formed as the surface crumpled.

b This cliff extends about 100 kilometers in length, and its vertical face is as much as 2 kilometers tall. (Photo from *Mariner 10*.)

The most surprising features of Mercury are its many tremendous cliffs—evidence of a type of past tectonics quite different from anything we have found on any other terrestrial world (Figure 7.21). Mercury's cliffs have vertical faces up to 3 or more kilometers high and typically run for hundreds of kilometers across the surface. They probably formed when tectonic forces compressed the crust, causing the surface to crumple. Because crumpling would have shrunk the portions of the surface it affected, Mercury as a whole could not have stayed the same size unless other parts of the surface expanded. However, we find no evidence of "stretch marks" on Mercury. Can it be that the whole planet simply shrank?

The planet Mercury appears to have shrunk long ago, leaving behind long, steep cliffs.

Apparently so. Early in its history, Mercury's larger size and greater iron content allowed it to gain more internal heat from accretion and differentiation than did the Moon. This heat caused the large iron core to swell. Later, as the core cooled, it contracted by perhaps as much as 20 kilometers in radius. The mantle and lithosphere must have contracted along with the core, generating the tectonic stresses that created the great cliffs. The contraction probably also closed off any remaining volcanic vents, ending Mercury's period of volcanism. Today, Mercury seems just as geologically dead as the Moon. We'll learn much more about this planet after the *MESSENGER* spacecraft begins orbiting it in 2011.

7.3 Mars: A Victim of Planetary Freeze-Drying

Based on its relative size (see Figure 7.1), we expect Mars to be more geologically interesting and varied than the Moon or Mercury but geologically less active than Earth or Venus. Observations confirm this basic picture, though Mars's greater distance from the Sun—about 50% farther than Earth is from the Sun—also plays a role in its geology. Mars's size and distance from the Sun have dictated much of its geological history.

The present-day surface of Mars looks much like some deserts or volcanic plains on Earth (see Figure 6.6), and it offers several other similarities to Earth. A Martian day is less than an hour longer than an Earth

Seasons on Mars

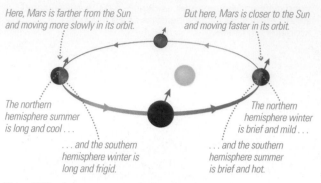

Here, Mars is farther from the Sun and moving more slowly in its orbit.

But here, Mars is closer to the Sun and moving faster in its orbit.

The northern hemisphere summer is long and cool . . .

The northern hemisphere winter is brief and mild . . .

. . . and the southern hemisphere winter is long and frigid.

. . . and the southern hemisphere summer is brief and hot.

Figure 7.22

The ellipticity of Mars's orbit makes seasons more extreme (hotter summers and colder winters) in the southern hemisphere than in the northern hemisphere.

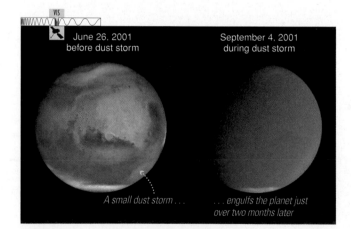

June 26, 2001
before dust storm

September 4, 2001
during dust storm

A small dust storm . . .

. . . engulfs the planet just over two months later

Figure 7.23

These two Hubble Space Telescope photos contrast the appearance of the same face of Mars in the absence (left) and presence (right) of a global dust storm.

day, and Mars has polar caps that resemble Earth's, although they contain frozen carbon dioxide in addition to water ice. Mars's rotation axis is tilted about the same amount as Earth's, and as a result it has seasons much like those on Earth. However, unlike Earth's, Mars's seasons are affected by its orbit as well as its tilt (Figure 7.22). Mars's more elliptical orbit puts it significantly closer to the Sun during the southern hemisphere summer and farther from the Sun during the southern hemisphere winter. Mars therefore has more extreme seasons in its southern hemisphere—that is, shorter, hotter summers and longer, colder winters—than in its northern hemisphere.

The superficial similarities between Earth and Mars have made the idea of life on Mars a staple of science fiction for more than a century. However, Mars in reality is quite different from Earth. The atmosphere is so thin that it creates only a weak greenhouse effect despite being made mostly of the greenhouse gas carbon dioxide. The temperature is usually well below freezing, with a global average of about −50°C (−58°F), and the atmospheric pressure is less than 1% of that on the surface of Earth. The lack of oxygen means that Mars lacks an ozone layer, so much of the Sun's damaging ultraviolet radiation passes unhindered to the surface.

Even Martian winds are very different from those on Earth. Winds on Earth are driven primarily by effects of Earth's rotation and by heat flow from the equator to the poles. In contrast, winds on Mars are strongly affected by its extreme seasonal changes. Temperatures at the winter pole drop so low (to about −130°C) that carbon dioxide condenses into "dry ice" at the polar cap. At the same time, frozen carbon dioxide at the summer pole sublimates into carbon dioxide gas. (*Sublimation* is the process by which an ice turns to a gas without first melting into liquid.) During the peak of summer, nearly all the carbon dioxide may sublime from the summer pole, leaving only a residual polar cap of water ice. The atmospheric pressure therefore increases at the summer pole and decreases at the winter pole, driving strong pole-to-pole winds. As much as one-third of the total carbon dioxide of the Martian atmosphere moves seasonally between the north and south polar caps. Sometimes these winds initiate huge dust storms, particularly when the more extreme summer approaches in the southern hemisphere (Figure 7.23).

All in all, surface conditions on Mars today make it seem utterly inhospitable to life. However, careful study of Martian geology offers evidence that Mars had a warmer and wetter past. If so, it might have had conditions under which life could have arisen, and it's conceivable that such life could still survive under the surface. The search for past or present life is a major reason why we are sending more spacecraft to Mars than to any other planet.

• What geological features tell us that water once flowed on Mars?

No liquid water exists on the surface of Mars today. We know this not only because we've studied most of the surface in reasonable detail but also because the surface conditions do not allow it. In most places and at most times, Mars is so cold that any liquid water would immediately freeze into ice. Even when the temperature rises above freezing, as it often does at midday near the equator, the air pressure is so low that liquid water would quickly evaporate. If you donned a spacesuit and took a cup of water outside your pressurized spaceship, the water would rapidly freeze or boil away (or do a combination of both).

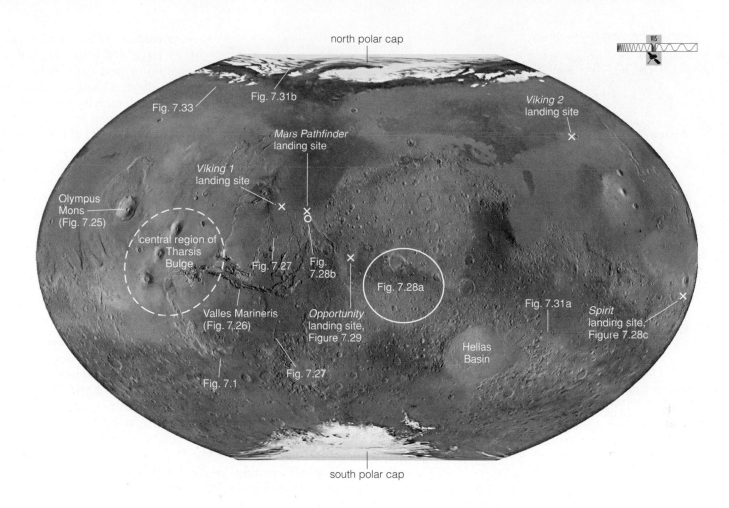

north polar cap

Fig. 7.33

Fig. 7.31b

Mars Pathfinder
landing site

Viking 1
landing site

Viking 2
landing site

Olympus
Mons
(Fig. 7.25)

central region of
Tharsis
Bulge

Fig. 7.27

Fig.
7.28b

Fig. 7.28a

Valles Marineris
(Fig. 7.26)

Opportunity
landing site,
Figure 7.29

Fig. 7.31a

Spirit
landing site,
Figure 7.28c

Hellas
Basin

Fig. 7.1

Fig. 7.27

south polar cap

Figure 7.24 interactive figure

This image showing the full surface of Mars is a composite
made by combining more than 1000 images with more than
200 million altitude measurements from the *Mars Global
Surveyor* mission. Several key geological features are labeled,
and the locations of features shown in close-up photos
elsewhere in this book are marked.

Nevertheless, Mars offers ample evidence of past water flows. Because water could not have flowed for long unless the Martian atmosphere were thicker and warmer, it appears that Mars must have had at least some warm and wet periods in the distant past. Let's investigate the geological evidence for this past, more hospitable world.

The Geology of Mars   To recognize features we can attribute to past water flows, we first need a global picture of Martian geology. Figure 7.24 shows the full surface of Mars. Aside from the polar caps, the most striking feature is the dramatic difference in terrain around different parts of the planet. Much of the southern hemisphere has relatively high elevation and is scarred by numerous large impact craters, including the large crater known as the Hellas Basin. In contrast, the northern plains show few impact craters and tend to be below the average Martian surface level. The differences in crater crowding tell us that the southern highlands are a much older surface than the northern plains, which must have had their early craters erased by other geological processes. Further study suggests that volcanism was the most important of these processes.

More dramatic evidence of volcanism on Mars comes from several towering volcanoes. One of these, Olympus Mons, is the tallest known volcano in the solar system (Figure 7.25). Its base is some 600 kilometers across, large enough to cover an area the size of Arizona. Its peak stands about 26 kilometers above the average Martian surface level, about three times as high as Mount Everest stands above sea level on Earth. Much of Olympus Mons is rimmed by a cliff that in places is 6 kilometers high.

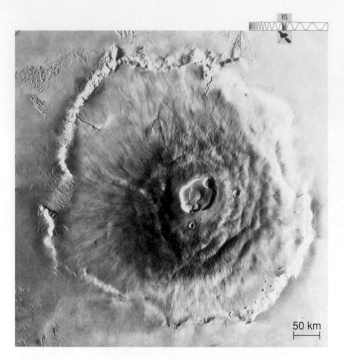

Figure 7.25

Olympus Mons, the largest volcano in the solar system, covers an area the size of Arizona and rises higher than Mount Everest on Earth. Note the tall cliff around its rim and the central volcanic crater from which lava erupted.

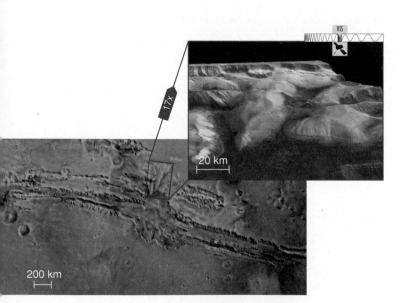

Figure 7.26

Valles Marineris is a huge valley on Mars created in part by tectonic stresses. It extends nearly a fifth of the way around the planet (see Figure 7.24), and in some places is 10 kilometers deep. The inset shows a perspective view looking north across the center of the canyon, obtained by *Mars Express*.

Mars has had extremely active volcanism in the past, and its surface is dotted with numerous large volcanoes.

Olympus Mons and several other large volcanoes are concentrated on or near the continent-size *Tharsis Bulge*. Tharsis is some 4000 kilometers across, and most of it rises several kilometers above the average Martian surface level. It was probably created by a long-lived plume of rising mantle material that bulged the surface upward and provided the molten rock for the eruptions that built the giant volcanoes.

Mars also has tectonic features, though none on a global scale like the plate tectonics of Earth. The most prominent tectonic feature is the long, deep system of valleys called *Valles Marineris* (Figure 7.26). Valles Marineris extends almost a fifth of the way along the planet's equator. It is as long as the United States is wide and almost four times as deep as Earth's Grand Canyon. No one knows exactly how Valles Marineris formed, but its location suggests a link to the Tharsis Bulge. Perhaps it formed through tectonic stresses accompanying the uplift of material that created Tharsis, cracking the surface and leaving the tall cliff walls of the valleys.

Is there any ongoing volcanic or tectonic activity on Mars? Until recently, we didn't think so. We expect Mars to be much less geologically active than Earth, because its smaller size has allowed its interior to cool faster. However, radiometric dating of meteorites that appear to have come from Mars (so-called *Martian meteorites* [Section 9.1]) shows some of them to be made of volcanic rock that solidified from molten lava as little as 180 million years ago—quite recent in the $4\frac{1}{2}$-billion-year history of the solar system. In that case, it is likely that Martian volcanoes will erupt again someday. Nevertheless, the Martian interior is presumably cooling and its lithosphere thickening. Within a few billion years, Mars will become as geologically dead as the Moon and Mercury.

Ancient Water Flows   Impacts, volcanism, and tectonics explain most of the major geological features of Mars, but close examination also shows features of water erosion. For example, Figure 7.27 looks much like a dry riverbed on Earth seen from above. These channels appear to have been carved by running water, though no one knows whether the water came from runoff after rainfall, from erosion by water-rich debris flows, or from an underground source. Regardless of the specific mechanism, water was almost certainly responsible, because it is the only substance that could have been liquid under past Martian conditions and that is sufficiently abundant to have created such extensive erosion features. But the water apparently stopped flowing long ago. Notice that a few impact craters lie on top of the channels. From counts of the craters in and near them, it appears that these channels are at least 2–3 billion years old, meaning that water has not flowed through them since that time.

Dried up riverbeds and other signs of erosion show that water flowed on Mars in the distant past.

Other orbital evidence also argues that Mars had rain and surface water in the distant past. Figure 7.28a shows a broad region of the ancient, heavily cratered southern highlands. Notice the indistinct rims of many large craters and the relative lack of small craters. Both facts argue for ancient rainfall, which would have eroded crater rims and erased small craters altogether. Figure 7.28b shows a close-up of a crater floor with sculpted patterns that suggest it was once the site of a lake. Ancient rains may have filled the crater, allowing sediments to build up from material that settled to the bottom. The sculpted patterns in the crater bottom may

Figure 7.27

This photo, taken by the *Mars Reconnaissance Orbiter*, shows what appears to be a dried-up meandering riverbed, now filled with dunes of windblown dust. Notice the numerous small craters, indicating that the riverbed dried out billions of years ago.

have been created by erosion as wind exposed layer upon layer of sedimentary rock, much as erosion by the water in the Colorado River exposed the layers visible in the walls of the Grand Canyon on Earth. Figure 7.28c shows a three-dimensional perspective of the surface that suggests water once flowed between two ancient crater lakes.

Surface studies provide additional evidence for past water. In 2004, the robotic rovers *Spirit* and *Opportunity* landed on nearly opposite sides of Mars. The twin rovers carried cameras, instruments to identify rock composition, and a grinder to expose fresh rock for analysis. The rovers, designed for just 3 months of operation, were still going strong as this book was being written, more than 4 years after their arrival.

Both rovers found compelling evidence that liquid water was once plentiful on Mars. Rocks at the *Opportunity* landing site contain tiny spheres—nicknamed "blueberries," although they're neither blue nor as large as berries—and odd indentations suggesting that they formed in standing water, or possibly by groundwater percolating through

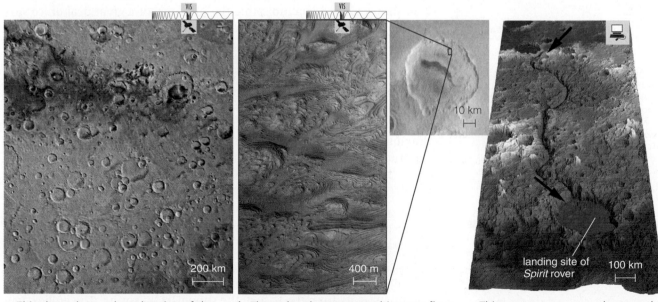

Figure 7.28

More evidence of past water on Mars.

a This photo shows a broad region of the southern highlands on Mars. The eroded rims of large craters and the lack of many small craters suggest erosion by rainfall.

b The sculpted patterns on this crater floor may be layers of sedimentary rock that were laid down at a time when the crater was filled with water.

c This computer-generated perspective view shows how a Martian valley forms a natural passage between two possible ancient lakes (shaded blue). Vertical relief is exaggerated 14 times to reveal the topography.

a This sequence zooms in on a knee-high rock outcropping near the rover's landing site. The close-up shows a piece of the rock about 3 centimeters across. The layered structure, odd indentations, and small sphere all support the idea that the rock formed from sediments in standing water.

Mars   **Earth**

b "Blueberries" on two planets. In both cases, the foreground shows hematite "blueberries," which formed within sedimentary rock layers like those in the background, then eroded out and rolled downhill; the varying tilts of the rock layers hint at changing winds or waves during formation. The background rocks are about twice as far away from the camera in the Earth photo as in the Mars photo (taken by the *Opportunity* rover).

Figure 7.29

Surface studies by the *Mars Exploration* rovers have provided evidence of a warmer, wetter Mars long ago.

rocks (Figure 7.29). Compositional analysis shows that the "blueberries" contain the iron-rich mineral hematite, and other rocks contain the sulfur-rich mineral jarosite. Both minerals form in water, and chemical analysis supports the case for formation in a salty environment such as a sea or ocean, apparently made acidic by dissolving in some atmospheric SO_2. Moreover, a close look at the layering of the sedimentary rocks suggests a changing environment of waves and/or wind. Taken together, the orbital and surface studies provide convincing evidence for abundant water in Mars's past.

Martian Water Today If water once flowed over the surface of Mars, where did it all go? As we'll discuss shortly, much of the water was probably lost to space forever. However, significant amounts of water still remain, frozen at the polar caps and in the top meter or so of the surface soil around much of the rest of the planet. Recent research has shown that the polar caps are made mostly of water ice, overlaid with a thin layer (at most a few meters thick) of CO_2 ice. More water ice is found in vast layers of dusty ice surrounding the poles. The extent of this water ice is only beginning to become clear; scientists were surprised to find surface ice sitting right under the *Phoenix* lander when it arrived in 2008. Figure 7.30 shows orbital data indicating the presence of a great deal of ice surrounding the south polar ice cap. If all this ice melted, it could make an ocean 11 meters deep over the planet.

If water ice exists on Mars this close to the surface, even more water probably lies deeper underground. It is even possible that some liquid water exists underground near sources of volcanic heat, providing a potential home to microscopic life. Moreover, if there is enough volcanic heat, ice might occasionally melt and flow along the surface for the short time until it freezes or evaporates. Although we have found no geological evidence to suggest that any large-scale water flows have occurred on Mars in the past billion years, orbital photographs offer evidence of smaller-scale water flows in much more recent times.

Gullies on crater walls suggest that water might still occasionally flow on Mars today.

The strongest evidence comes from photos of gullies on crater and channel walls (Figure 7.31a). These gullies look strikingly similar to the gullies we see on eroded slopes on Earth. One hypothesis suggests that the gullies form when snow accumulates on the crater walls in winter and melts away from the base of the snowpack in spring. The gullies may form when subsurface ice melts and gushes out from the crater walls before rapidly freezing or evaporating in the thin atmosphere. Spacecraft images show clear changes in the gullies over just a few years, so the process that creates them is still happening. If they're not formed by liquid water, the gullies may result from landslides, seen to occur elsewhere on Mars with the change of seasons (Figure 7.31b).

• Why did Mars change?

There seems little doubt that Mars had wetter and possibly warmer periods, probably with rainfall, before about 3 billion years ago. The full extent of these periods is a topic of considerable scientific debate. Some scientists suspect that Mars was continuously warm and wet for much of its first billion years of existence. Others think that Mars had only intermittent periods of rainfall, perhaps triggered by the heat of large impacts, and that ancient lakes, ponds, or oceans may have been completely ice-covered. Either way, Mars apparently underwent major and permanent climate change, turning a world that could sustain liquid water into a frozen wasteland.

The idea that Mars once had a thicker atmosphere makes sense, because we would expect that its many volcanoes outgassed plenty of atmospheric gas. Much of this gas should have been water vapor and carbon dioxide, and these greenhouse gases would have warmed the planet. Ancient sulfur dioxide levels (inferred from study of sulfurous minerals on the surface) should have also contributed to a stronger greenhouse effect. If Martian volcanoes outgassed greenhouse gases in the same proportions as do volcanoes on Earth, Mars would have had enough water to fill oceans tens or even hundreds of meters deep. The heat of impacts may also have released water vapor into the atmosphere, enhancing the greenhouse effect until the water rained out.

Early in its history, Mars probably had a dense atmosphere from volcanic outgassing, with a stronger greenhouse effect than it has today.

The bigger question is not whether Mars once had a denser atmosphere, but where all the atmospheric gas went. Mars must have somehow lost most of its carbon dioxide gas. This loss would have weakened the greenhouse effect until the planet essentially froze over. Some of the carbon dioxide condensed to make the polar caps, and some may be chemically bound to surface rock. However, the bulk of the gas was probably lost to space.

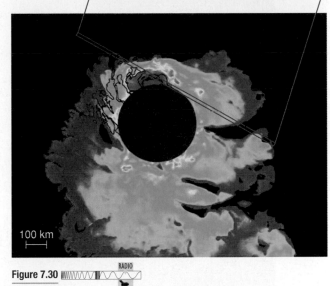

radar reflection from the base of the ice cap

2 km

100 km

Figure 7.30 RADIO

The MARSIS radar on the European Space Agency's *Mars Express* spacecraft measured the thickness of layers of dusty water ice surrounding the south polar ice cap. In the top panel, the upper line shows the radar reflection from the surface of the polar cap, and the line below is the reflection of the base of the ice layers 3 kilometers below. In the bottom panel, the thickness of the ice layers is shown around the whole polar region. The black lines outline the permanent polar cap.

September 2005

new deposit formed between 1999 and 2005

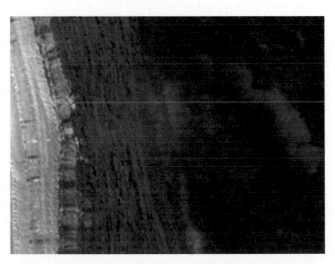

Figure 7.31 VIS

Recent geological changes on Mars.

a New gully features appear on Martian crater walls every few years. Some scientists believe they're caused by occasional meltwater, while others think they're caused by landslides of dry material.

b The *Reconnaissance* orbiter captured this landslide in Mars's north polar regions. In a cliff over 700 meters high, layers of ice mixed with dust thawed during the northern spring, causing the landslide.

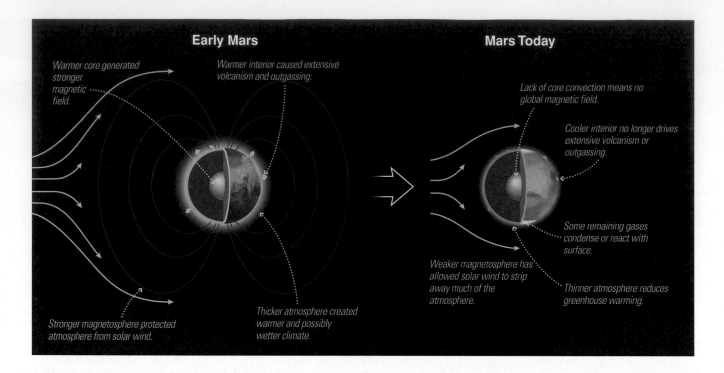

Early Mars

Warmer core generated stronger magnetic field.

Warmer interior caused extensive volcanism and outgassing.

Stronger magnetosphere protected atmosphere from solar wind.

Thicker atmosphere created warmer and possibly wetter climate.

Mars Today

Lack of core convection means no global magnetic field.

Cooler interior no longer drives extensive volcanism or outgassing.

Some remaining gases condense or react with surface.

Weaker magnetosphere has allowed solar wind to strip away much of the atmosphere.

Thinner atmosphere reduces greenhouse warming.

Figure 7.32

Some 3 billion years ago, Mars underwent dramatic climate change, ensuring that rain could never fall again.

The precise way in which Mars lost its carbon dioxide gas is not clear, but recent data suggest a close link to a change in Mars's magnetic field (Figure 7.32). Early in its history, Mars probably had molten convecting metals in its core, much like Earth today. Given Mars's rapid rotation, these convecting metals should have produced a magnetic field and a protective magnetosphere around Mars. The magnetic field would have weakened as Mars cooled and the core ceased to convect, leaving the atmosphere vulnerable to solar wind particles. These solar wind particles could have stripped gases out of the Martian atmosphere and into space.

Mars underwent permanent climate change about 3 billion years ago, when it lost much of its atmospheric carbon dioxide and water to space.

Much of the water once present on Mars is also probably gone for good. Like the carbon dioxide, some water vapor may have been stripped away by the solar wind. However, Mars also lost water in another way. Because the Martian atmosphere lacks ultraviolet-absorbing gases, atmospheric water molecules would have been easily broken apart by ultraviolet photons. The hydrogen atoms that broke away from the water molecules would have been lost rapidly to space. With these hydrogen atoms gone, the water molecules could not be made whole again. Initially, oxygen from the water molecules would have remained in the atmosphere, but over time this oxygen was lost too. Some was probably stripped away by the solar wind, and the rest was drawn out of the atmosphere through chemical reactions with surface rock. This process literally rusted the Martian rocks, giving the "red planet" its distinctive tint.

We should learn more about Martian history with the next wave of spacecraft missions. The *Mars Reconnaissance Orbiter* (*MRO*), which entered Martian orbit in 2006, is snapping photos with unprecedented resolution, and it carries a radar instrument that should tell us much more about subsurface ice. On the surface, the *Phoenix* lander touched down near the north polar ice cap in May 2008 (Figure 7.33), and it is scheduled to be followed in 2010 by the *Mars Science Laboratory*—a nuclear-powered lab on wheels, designed to determine whether Mars has ever been suitable for life.

Still, we already know enough to draw a general conclusion: Mars's fate was shaped primarily by its relatively small size. It was big enough for volcanism and outgassing to release water and atmospheric gas early in its history, but too small to maintain the internal heat needed to keep this water and gas. As its interior cooled, its volcanoes quieted and it lost its magnetic field, allowing gas to be stripped away to space. If Mars had been as large as Earth, so that it could still have outgassing and a global magnetic field, it might still have a moderate climate today.

think about it Could Mars still have flowing water today if it had formed closer to the Sun? Use your answer to explain how Mars's fate has been shaped not just by its size, but also by its distance from the Sun.

7.4 Venus: A Hothouse World

We have seen that planetary size is a major factor in explaining why the histories of the Moon, Mercury, and Mars have been so different from that of Earth. On the basis of size alone, we would expect Venus and Earth to be quite similar: Venus is only about 5% smaller than Earth in radius (see Figure 7.1), and its overall composition is about the same as that of Earth. However, as we saw in our planetary tour in Section 6.1, the surface of Venus is a searing hothouse, quite unlike the surface of Earth. In this section, we'll investigate how a planet so similar in size and composition to Earth ended up so different in almost every other respect.

• Is Venus geologically active?

As we did for Mars, let's begin our study of Venus by looking at what its surface geology tells us about its history. Venus's thick cloud cover prevents us from seeing through to its surface, but we can study its geological features with radar (because radio waves can pass through clouds). *Radar mapping* bounces radio waves off the surface and uses the reflections to create three-dimensional images of the surface. From 1990 to 1993, the *Magellan* spacecraft used radar to map the surface of Venus, discerning features as small as 100 meters across. Careful study suggests that Venus is indeed geologically active, just as we would expect for a planet almost as large as Earth.

Geological Features of Venus Figure 7.34 shows a global map and selected close-ups of Venus based on the *Magellan* radar observations. We see many geological features similar to Earth's, including occasional impact craters, volcanoes, and a lithosphere that has been contorted by tectonic forces. Venus also has some unique features, such as the large, circular *coronae* (Latin for "crowns") that were probably made by hot, rising plumes of mantle rock. These plumes probably also forced lava to the surface, explaining the volcanoes found near coronae.

Venus shows features of volcanism and tectonics, just as we expect for a planet of similar size to Earth. Venus almost certainly remains geologically active today, since it should still retain nearly as much internal heat as Earth. Its relatively few impact craters confirm that its surface is geologically young. In addition, the composition of Venus's clouds suggests that volcanoes must still be active on geological time scales (erupting within the past 100 million years). The clouds contain sulfuric acid, which is made from sulfur dioxide (SO_2) and water. Sulfur dioxide enters the atmosphere through volcanic outgassing, but once in the atmosphere it is gradually removed by

Figure 7.33

The *Phoenix* lander has a robotic arm for scooping up Martian soil and on-board instruments for analyzing it. Results are helping scientists determine whether and when Mars's polar regions may have been habitable.

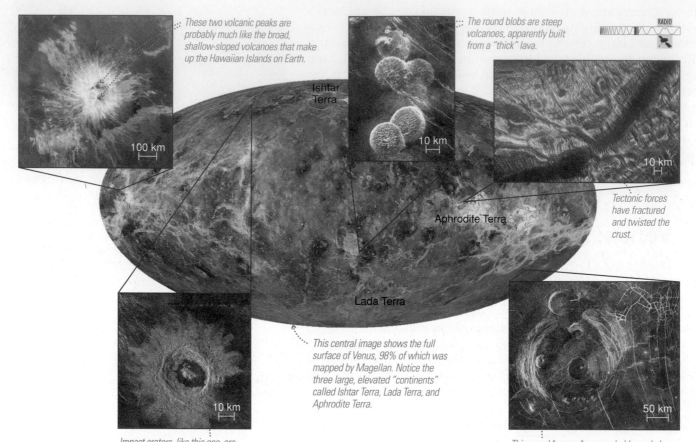

These two volcanic peaks are probably much like the broad, shallow-sloped volcanoes that make up the Hawaiian Islands on Earth.

100 km

The round blobs are steep volcanoes, apparently built from a "thick" lava.

10 km

RADIO

Tectonic forces have fractured and twisted the crust.

10 km

Ishtar Terra

Aphrodite Terra

Lada Terra

This central image shows the full surface of Venus, 98% of which was mapped by Magellan. Notice the three large, elevated "continents" called Ishtar Terra, Lada Terra, and Aphrodite Terra.

10 km

Impact craters, like this one, are relatively rare on Venus and are distributed uniformly over the surface.

50 km

This round "corona" was probably made by a mantle plume. It is dotted with small volcanoes (the round dots) and surrounded by tectonic stress marks.

Figure 7.34

These images show the surface of Venus as revealed by radar observations from the *Magellan* spacecraft. Bright regions represent rough areas or higher altitudes.

chemical reactions with surface rocks. The fact that sulfuric acid clouds still exist means that outgassing must continue to supply sulfur dioxide to the atmosphere. The European Space Agency's *Venus Express* spacecraft, orbiting Venus since 2006, is currently studying Venus's atmosphere (Figure 7.35). It has detected rapid changes in the amount of atmospheric sulfur dioxide, which suggest recent volcanic activity.

The biggest difference between the geology of Venus and that of Earth is the lack of erosion on Venus. We might naively expect Venus's thick atmosphere to produce strong erosion, but the view both from orbit and on the surface suggests that erosion is only a minor process. The former Soviet Union sent several landers to Venus in the 1970s and early 1980s. Before the intense surface heat destroyed them, the probes returned images of a bleak, volcanic landscape with little evidence of erosion (Figure 7.36). We can trace the lack of erosion on Venus to two facts. First, Venus is far too hot for any type of rain or snow on its surface. Second, Venus has virtually no wind or weather because of its slow rotation. Venus rotates so slowly—once every 243 days—that its atmosphere barely stirs the surface. Without any glaciers, rivers, rain, or strong winds, there is very little erosion on Venus.

The Absence of Plate Tectonics   We can easily explain the lack of erosion on Venus, but another "missing feature" of its geology is more surprising: Venus shows no evidence of Earth-like plate tectonics. Plate tectonics shapes nearly all of Earth's major geological features, including mid-ocean ridges, deep ocean trenches, and long mountain ranges like the Rockies and Himalayas.

Venus lacks any similar features. Instead, Venus shows evidence of a very different type of global geological change. On Earth, plate tectonics resculpts the surface gradually, so that different regions have different ages. On Venus, the relatively few impact craters are distributed fairly uniformly over the entire planet, suggesting that the surface is about the same age everywhere. Crater counts suggest that the surface is about 750 million years old. Apparently, the entire surface of Venus was somehow "repaved" at that time.

We do not know how much of the repaving was due to tectonic processes and how much was due to volcanism, but both probably were important. It is even possible that plate tectonics played a role before and during the repaving, only to stop after the repaving episode was over. Either way, the absence of present-day plate tectonics on Venus poses a major mystery.

Earth's lithosphere was broken into plates by forces due to the underlying mantle convection. The lack of plate tectonics on Venus therefore suggests either that it has weaker mantle convection or that its lithosphere somehow resists fracturing. The first possibility seems unlikely: Venus's similar size to Earth means it should have a similar level of mantle convection. Most scientists therefore suspect that Venus's lithosphere resists fracturing into plates because it is thicker and stronger than Earth's lithosphere, though we have no direct evidence to support this hypothesis.

Venus's lack of Earth-like plate tectonics poses a scientific mystery, but may arise because Venus has a thicker and stronger lithosphere than Earth. Even if a thicker and stronger lithosphere explains the lack of plate tectonics on Venus, we are still left with the question of why the lithospheres of Venus and Earth should differ. One possible answer invokes Venus's high surface temperature. Venus is so hot that any water in its crust and mantle has probably been baked out over time. Water tends to soften and lubricate rock, so its loss would have thickened and strengthened Venus's lithosphere. If this idea is correct, then Venus might have had plate tectonics if it had not become so hot in the first place.

• Why is Venus so hot?

It's tempting to attribute Venus's high surface temperature solely to the fact that it is closer than Earth to the Sun, but Venus would actually be quite cold without its strong greenhouse effect. Venus absorbs less sunlight than Earth, despite being closer to the Sun, because its clouds reflect so much sunlight back to space. Calculations show that Venus's average surface temperature would be a frigid −40°C (−40°F) without the greenhouse effect, rather than its actual temperature of about 470°C (880°F). The real question, then, is why Venus has such a strong greenhouse effect.

Venus's thick carbon dioxide atmosphere creates the extremely strong greenhouse effect that makes Venus so hot. On a simple level, the answer is the huge amount of carbon dioxide in Venus's atmosphere. Venus has a far thicker atmosphere than Earth—its surface pressure is about 90 times that on Earth—and this atmosphere is about 96% carbon dioxide. The total amount of carbon dioxide in Venus's atmosphere is nearly 200,000 times that in Earth's atmosphere, and it creates an extremely strong greenhouse effect.

However, a deeper question still remains. Given their similar sizes and compositions, we expect Venus and Earth to have had similar levels of volcanic outgassing—and the released gas ought to have had about the same composition on both worlds. Why, then, is Venus's atmosphere so different from Earth's?

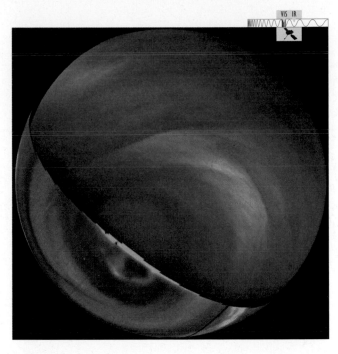

Figure 7.35

This composite from *Venus Express* combines a visible-wavelength image of the dayside (upper right) and an infrared image of the nightside (lower left). Venus's south pole lies midway along the boundary between the two images.

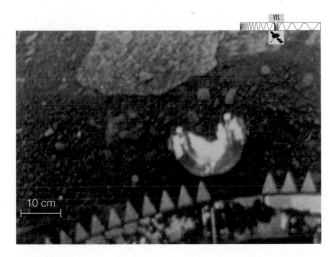

Figure 7.36

This photo from one of the former Soviet Union's *Venera* landers shows a small patch of Venus; part of the lander is in the foreground. Many volcanic rocks are visible, hardly affected by erosion despite their age—this region of the planet is covered in volcanic plains that formed some 750 million years ago.

Atmospheric Composition We expect that huge amounts of water and carbon dioxide should have been outgassed into the atmospheres of both Venus and Earth. Venus's atmosphere does indeed have huge amounts of carbon dioxide, but it has virtually no water. Earth's atmosphere has very little of either gas. What happened to all the outgassed water on Venus and to the outgassed water and carbon dioxide on Earth?

Earth has as much carbon dioxide as Venus, but it is mostly locked away in rocks rather than in our atmosphere.

We can easily account for both missing gases on Earth. The huge amounts of water vapor released into our atmosphere condensed into rain, forming our oceans. In other words, the water is still here, but mostly in liquid rather than gaseous form. The huge amount of carbon dioxide released into our atmosphere is also still here, but in solid form: Carbon dioxide dissolves in water, where it can undergo chemical reactions to make **carbonate** rocks (rocks rich in carbon and oxygen) such as limestone. Earth has about 170,000 times as much carbon dioxide locked up in rocks as in its atmosphere—which means that Earth does indeed have about as much total carbon dioxide as Venus. Of course, the fact that Earth's carbon dioxide is mostly in rocks rather than in the atmosphere makes all the difference in the world: If this carbon dioxide were in our atmosphere, our planet would be nearly as hot as Venus and certainly uninhabitable.

We are left with the question of what happened to Venus's water. Venus has no oceans and little atmospheric water, and as noted earlier, any water in its crust and mantle was probably baked out long ago. This absence of water explains why Venus retains so much carbon dioxide in its atmosphere: Without oceans, carbon dioxide cannot dissolve or become locked away in carbonate rocks. If it is true that a huge amount of water was outgassed on Venus, it has somehow disappeared.

Venus retains carbon dioxide in its atmosphere because it lacks oceans to dissolve the carbon dioxide and lock it away in rock.

The leading hypothesis for the disappearance of Venus's water invokes one of the same processes thought to have removed water from Mars. Ultraviolet light from the Sun breaks apart water molecules in Venus's atmosphere. The hydrogen atoms then escape to space, ensuring that the water molecules can never re-form. The oxygen from the water molecules is lost to a combination of chemical reactions with surface rocks and stripping by the solar wind; Venus's lack of magnetic field leaves its atmosphere vulnerable to the solar wind.

Acting over billions of years, the breakdown of water molecules and the escape of hydrogen can easily explain the loss of an ocean's worth of water from Venus—as long as the water was in the atmosphere rather than in liquid oceans. Our quest to understand Venus's high temperature thereby leads to one more question: Why didn't Venus get oceans like Earth, which would have prevented its water from being lost to space?

The Runaway Greenhouse Effect To understand why Venus does not have oceans, let's consider what would happen if we could magically move Earth to the orbit of Venus (Figure 7.37).

The greater intensity of sunlight would almost immediately raise Earth's global average temperature by about 30°C, from its current 15°C to about 45°C (113°F). Although this is still well below the boiling point of water, the higher temperature would lead to increased evaporation of water from the oceans. The higher temperature would also allow the atmosphere to hold more water vapor before the vapor condensed to make rain. The combination of more evaporation and greater atmospheric capacity for water vapor would substantially increase the total amount of

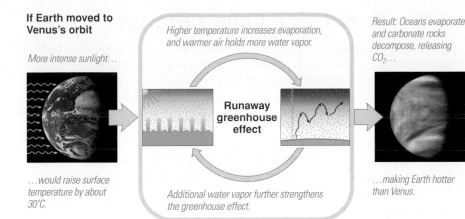

If Earth moved to Venus's orbit

More intense sunlight…

…would raise surface temperature by about 30°C.

Higher temperature increases evaporation, and warmer air holds more water vapor.

Runaway greenhouse effect

Additional water vapor further strengthens the greenhouse effect.

Result: Oceans evaporate and carbonate rocks decompose, releasing CO_2…

…making Earth hotter than Venus.

Figure 7.37

This diagram shows how, if Earth were placed at Venus's distance from the Sun, the runaway greenhouse effect would cause the oceans to evaporate completely.

water vapor in Earth's atmosphere. Now, remember that water vapor, like carbon dioxide, is a greenhouse gas. The added water vapor would therefore strengthen the greenhouse effect, driving temperatures a little higher. The higher temperatures, in turn, would lead to even more ocean evaporation and more water vapor in the atmosphere—strengthening the greenhouse effect even further. In other words, we'd have a positive feedback loop in which each little bit of additional water vapor in the atmosphere would mean higher temperature and even more water vapor. The process would careen rapidly out of control as a **runaway greenhouse effect.**

The runaway process would cause Earth to heat up until the oceans were completely evaporated and the carbonate rocks had released all their carbon dioxide back into the atmosphere. By the time the process was complete, temperatures on our "moved Earth" would be even higher than they are on Venus today, thanks to the combined greenhouse effects of carbon dioxide and water vapor in the atmosphere. The water vapor would then gradually disappear, as ultraviolet light broke water molecules apart and the hydrogen escaped to space. In short, moving Earth to Venus's orbit would essentially turn our planet into another Venus.

Venus is too close to the Sun to have liquid water oceans. Without water to dissolve carbon dioxide gas, Venus was doomed to its runaway greenhouse effect.

We have arrived at a simple explanation of why Venus is so much hotter than Earth. Even though Venus is only about 30% closer to the Sun than Earth, this difference was enough to be critical. On Earth, it was cool enough for water to rain down to make oceans. The oceans then dissolved carbon dioxide and chemical reactions locked it away in carbonate rocks, leaving our atmosphere with only enough greenhouse gases to make our planet pleasantly warm. On Venus, the greater intensity of sunlight made it just enough warmer that oceans either never formed or soon evaporated. Without oceans to dissolve carbon dioxide and make carbonate rock, carbon dioxide accumulated in the atmosphere, leading to a runaway greenhouse effect.

The next time you see Venus shining brightly as the morning or evening "star," consider the radically different path it has taken from that taken by Earth—and thank your lucky star. If Earth had formed a bit closer to the Sun or if the Sun had been slightly hotter, our planet might have suffered the same greenhouse-baked fate.

think about it We've seen that moving Earth to Venus's orbit would cause our planet to become Venus-like. If we could somehow move Venus to Earth's orbit, would it become Earth-like? Why or why not?

7.5 Earth as a Living Planet

We began this chapter by discussing Earth as a planet, looking at features and processes that it shares in common with some of our planetary neighbors. We then explored the histories of the other terrestrial worlds, finding that we could understand them by thinking about fundamental planetary properties (such as size and distance from the Sun) and processes familiar to us from Earth. However, we have not yet discussed the feature that makes Earth truly unique: its abundance of life, including human life. It is time for us to turn our attention back to Earth, to see how and why our planet is such a pleasant place for us to live.

• What unique features of Earth are important for life?

If you think about what we've learned about the terrestrial worlds, you can probably identify a number of features that are unique to Earth. Four unique features turn out to be particularly important to life on Earth:

- **Surface liquid water**: Earth is the only planet on which temperature and pressure conditions allow surface water to be stable as a liquid.

- **Atmospheric oxygen**: Earth is the only planet with significant oxygen in its atmosphere and an ozone layer.

- **Plate tectonics**: Earth is the only planet with a surface shaped largely by this distinctive type of tectonics.

- **Climate stability**: Earth differs from the other terrestrial worlds with significant atmospheres (Venus and Mars) in having a climate that has remained relatively stable throughout its history.

Our Unique Oceans and Atmosphere The first and second items in our list—abundant liquid water and atmospheric oxygen—are clearly important to our existence. Life as we know it requires water [Section 18.1], and animal life requires oxygen. We have already explained the origin of Earth's water: Water vapor outgassed from volcanoes rained down on the surface to make the oceans and neither froze nor evaporated thanks to our moderate greenhouse effect and distance from the Sun. But where did the oxygen in our atmosphere come from?

Oxygen (O_2) is not a product of volcanic outgassing. In fact, no geological process can explain how oxygen came to make up such a large fraction (21%) of Earth's atmosphere. Moreover, oxygen is a highly reactive gas that would disappear from the atmosphere in just a few million years if it were not continuously resupplied. Fire, rust, and the discoloration of freshly cut fruits and vegetables are everyday examples of chemical reactions that remove oxygen from the atmosphere. Similar reactions between oxygen and surface materials give rise to the reddish appearance of much of Earth's rock and clay, including the beautiful reds of Arizona's Grand Canyon. So we must explain not only how oxygen got into Earth's atmosphere in the first place, but also how the amount of oxygen remains relatively steady even though chemical reactions remove it rapidly from the atmosphere.

Without life, there would be no oxygen in Earth's atmosphere.

The answer to the oxygen mystery is life itself. Plants and many microorganisms release oxygen through photosynthesis. Photosynthesis takes in CO_2 and releases O_2. The carbon becomes incorporated into living tissues. Virtually all Earth's oxygen was originally released into the atmosphere by photosynthetic life. Today,

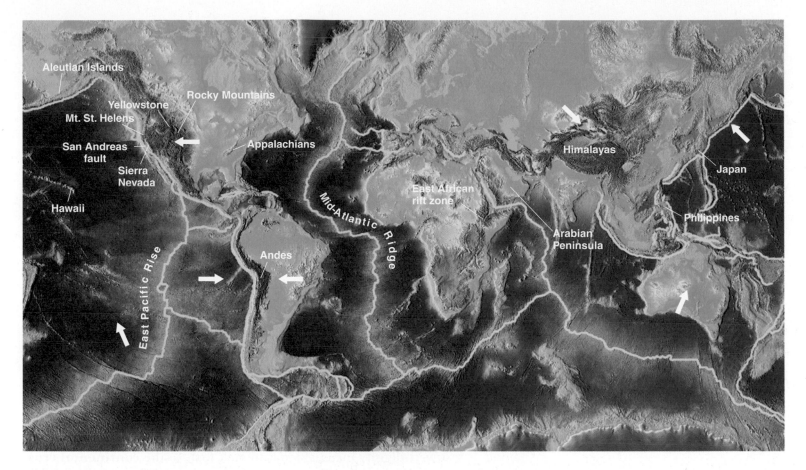

Aleutian Islands
Rocky Mountains
Yellowstone
Mt. St. Helens
San Andreas fault
Sierra Nevada
Hawaii
Appalachians
East Pacific Rise
Andes
Mid-Atlantic Ridge
East African rift zone
Himalayas
Arabian Peninsula
Japan
Philippines

photosynthetic organisms return oxygen to the atmosphere in approximate balance with the rate at which animals and chemical reactions consume it, which is why the oxygen concentration remains relatively steady. This oxygen is also what makes possible Earth's protective ozone layer, since ozone (O_3) is produced from ordinary oxygen (O_2).

Figure 7.38

This relief map shows plate boundaries (solid yellow lines), with arrows to represent directions of plate motion. Color represents elevation, progressing from blue (lowest) to red (highest).

think about it Suppose that, somehow, all photosynthetic life (such as plants) died out. What would happen to the oxygen in our atmosphere? Could animals, including us, still survive?

Plate Tectonics The third and fourth items on our list—plate tectonics and climate stability—turn out to be closely linked. We will therefore begin by exploring how plate tectonics has created the ocean basins and the continents on which we live. Recall that Earth's lithosphere is broken into more than a dozen plates that slowly move about through the action we call *plate tectonics* (Figure 7.38). The plate motions are barely noticeable on human time scales. On average, plates move at speeds of only a few centimeters per year—about the same speed at which your fingernails grow. Nevertheless, over millions of years, these motions rearrange the locations of continents, open ocean basins between continents, build mountain ranges, and much more.

The movements of plate tectonics act like a giant conveyor belt for Earth's lithosphere, creating new crust and recycling old crust back into the mantle (Figure 7.39). Mantle material rises upward and erupts to the surface along mid-ocean ridges, becoming new crust for the seafloor, or **seafloor crust** for short. This newly emerging material causes the seafloor to spread away from the ridge, which is why the ridges are found in the middle of the ocean. Over tens of millions of years, any piece of

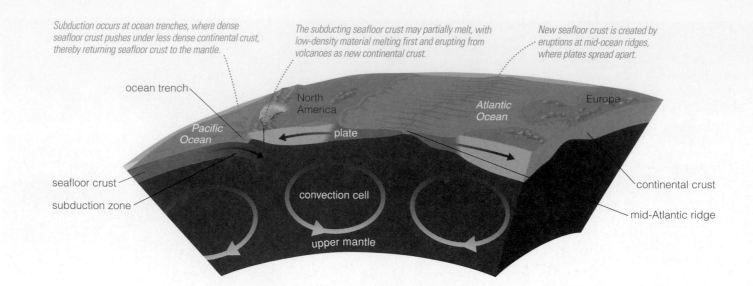

Subduction occurs at ocean trenches, where dense seafloor crust pushes under less dense continental crust, thereby returning seafloor crust to the mantle.

The subducting seafloor crust may partially melt, with low-density material melting first and erupting from volcanoes as new continental crust.

New seafloor crust is created by eruptions at mid-ocean ridges, where plates spread apart.

ocean trench

North America

Europe

Pacific Ocean

Atlantic Ocean

plate

seafloor crust

subduction zone

convection cell

continental crust

mid-Atlantic ridge

upper mantle

Figure 7.39

Plate tectonics acts like a giant conveyor belt for Earth's lithosphere.

Plate tectonics acts like a giant conveyor belt for Earth's lithosphere, continually recycling seafloor crust and building up the continents.

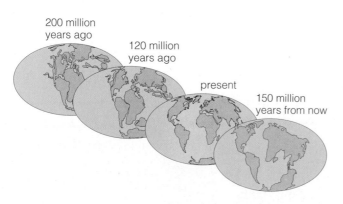

200 million years ago

120 million years ago

present

150 million years from now

Figure 7.40

Past, present, and future arrangements of Earth's continents. Notice how the present shapes of South America and Africa reflect the way they fit together in the past. The continents are always in motion, so the arrangement looked different at earlier times and will look different in the future.

seafloor crust gradually makes its way across the ocean bottom, then finally gets recycled into the mantle in the process we call **subduction**.

Subduction occurs where seafloor plates run into continental plates. As seafloor crust descends into the mantle, it is heated and may partially melt. This molten rock then erupts from volcanoes over the subduction zones, which is why so many active volcanoes tend to be found along the edges of continents. Moreover, the lowest-density material tends to melt first, so the **continental crust** emerging from these landlocked volcanoes is much lower in density than seafloor crust. This lower density explains why continents rise above the seafloor. Radiometric dating confirms this picture of plate tectonics. Seafloor crust is never more than a couple of hundred million years old (because it is recycled at the subduction zones) and is younger near the mid-ocean ridges (where it first emerges and solidifies).

In fact, almost all Earth's active geology is tied to plate tectonics. For example, compression of the crust and mountain building occurs where continental plates are pushed together, and valleys or seas can form where continental plates are pulling apart (see Figure 7.11). Earthquakes tend to occur when two plates get "stuck" against one another and then lurch violently when the pressure builds to the breaking point. Even the present arrangement of the continents is due to plate tectonics, since the continents are slowly pushed around as seafloors spread and subduct (Figure 7.40).

think about it By studying the plate boundaries in Figure 7.38, explain why the west coast states of California, Oregon, and Washington are prone to more earthquakes and volcanoes than other parts of the United States. Find the locations of recent earthquakes and volcanic eruptions worldwide. Do the locations fit the pattern you expect?

Climate Stability Earth's long-term climate stability has clearly been important to the ongoing evolution of life—and hence to our own relatively recent arrival as a species (see Figure 1.10). Had our planet undergone a runaway greenhouse effect like Venus, life would certainly have been extinguished. If Earth had suffered loss of atmosphere and a global freezing like Mars, any surviving life would have been driven to hide in underground pockets of liquid water.

Earth's climate is not perfectly stable—our planet has endured numerous ice ages and warm periods in the past. Nevertheless, even in the deepest ice ages and warmest warm periods, Earth's temperature has remained in a range in which some liquid water could still exist and harbor life. On the surface, the key to this climate stability is simple: Earth has always kept just enough carbon dioxide in its atmosphere to keep the temperature in a range suitable for liquid water and life. This long-term climate stability is even more remarkable, because models suggest that the Sun has brightened substantially (about 30%) over the past 4 billion years, yet Earth's temperature has managed to stay in the same range throughout this time. Apparently, the strength of the greenhouse effect somehow self-adjusts to keep the climate stable.

The mechanism by which Earth self-regulates its temperature is called the **carbon dioxide cycle**, or the **CO_2 cycle** for short. Let's follow the cycle as illustrated in Figure 7.41, starting at the top center:

- Atmospheric carbon dioxide dissolves in rainwater, creating a mild acid.

- The mildly acidic rainfall erodes rocks on Earth's continents, and rivers carry the broken-down minerals to the oceans.

- In the oceans, the eroded minerals combine with dissolved carbon dioxide and fall to the ocean floor, making carbonate rocks such as limestone.

- Over millions of years, the conveyor belt of plate tectonics carries the carbonate rocks to subduction zones, and subduction carries them down into the mantle.

- As they are pushed deeper into the mantle, some of the subducted carbonate rock melts and releases its carbon dioxide, which then outgasses back into the atmosphere through volcanoes.

The CO_2 cycle acts as a long-term thermostat for Earth, because the overall rate at which carbon dioxide is pulled from the atmosphere is very sensitive to temperature: the higher the temperature, the higher the rate at which carbon dioxide is removed. As a result, a small change in Earth's temperature will be offset by a change in the CO_2 cycle.

Earth has remained habitable for billions of years because its climate is kept stable by the natural action of the carbon dioxide cycle.

Consider first what happens if Earth warms up a bit. The warmer temperature means more evaporation and rainfall, pulling more CO_2 out of the atmosphere. The reduced atmospheric CO_2 concentration leads to a weakened greenhouse effect that counteracts the initial warming and cools the planet back down. Similarly, if Earth cools a bit, precipitation decreases and less CO_2 is dissolved in rainwater, allowing the CO_2 released by volcanism to build back up in the atmosphere. The increased CO_2 concentration strengthens the greenhouse effect and warms the planet back up. Overall, the natural thermostat of the carbon dioxide cycle has allowed the greenhouse effect to strengthen or weaken just enough to keep Earth's climate fairly stable, regardless of what other changes have occurred on our planet. Note, however, that this cycle operates on a time scale of a few hundred thousand years, which means it has no effect on short-term changes, such as those we will discuss next.

We can now see why plate tectonics is so intimately connected to our existence. Plate tectonics is a crucial part of the CO_2 cycle; without plate tectonics, CO_2 would remain locked up in seafloor rocks rather than being recycled through outgassing. Earth's climate might then have undergone changes as dramatic as those that occurred on Venus and Mars.

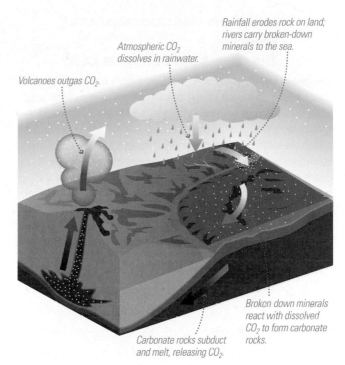

Atmospheric CO_2 dissolves in rainwater.

Rainfall erodes rock on land; rivers carry broken-down minerals to the sea.

Volcanoes outgas CO_2.

Broken down minerals react with dissolved CO_2 to form carbonate rocks.

Carbonate rocks subduct and melt, releasing CO_2.

Figure 7.41

This diagram shows how the CO_2 cycle continually moves carbon dioxide from the atmosphere to the ocean to rock and back to the atmosphere. Notice that plate tectonics (subduction in particular) plays a crucial role in the cycle.

The Greenhouse Effect Is Bad

The greenhouse effect is often in the news, usually in discussions about environmental problems, but in itself the greenhouse effect is not a bad thing. In fact, we could not exist without it, since it is responsible for keeping our planet warm enough for liquid water to flow in the oceans and on the surface. The "no greenhouse" temperature of Earth is well below freezing. Why, then, is the greenhouse effect discussed as an environmental problem? The reason is that human activity is adding more greenhouse gases to the atmosphere—and scientists agree that the additional gases are changing Earth's climate. While the greenhouse effect makes Earth livable, it is also responsible for the searing 470°C temperature of Venus—proving that it's possible to have too much of a good thing.

• How is human activity changing our planet?

We humans are well adapted to the unique, present-day conditions on our planet. The amount of oxygen in our atmosphere, the average temperature of our planet, and the ultraviolet-absorbing ozone layer are just what we need to survive. We have seen that these "ideal" conditions are no accident—they are consequences of our planet's unique geology and biology.

Nevertheless, the stories of the dramatic and permanent climate changes that occurred on Venus and Mars should teach us to take nothing for granted. Our planet may regulate its own climate quite effectively over long time scales, but fossil and geological evidence tells us that substantial and rapid swings in global climate can occur on shorter ones. For example, Earth cycles in and out of ice ages on geologically short time scales of tens of thousands of years, and in some cases the climate seems to have warmed several degrees Celsius in just decades. Evidence also shows that these past climate changes have had dramatic effects on local climates by raising or lowering sea level, altering ocean currents that keep coastlines warm, and transforming rainforests into deserts.

These past climate changes have been due to "natural" causes. For example, cycles of ice ages have been linked to small, cyclical changes in Earth's axis tilt and other characteristics of Earth's rotation and orbit (called *Milankovitch cycles*). More rapid changes may be linked to major volcanic eruptions, sudden releases of trapped carbon dioxide from the oceans, or a variety of other geological processes. Today, however, Earth is undergoing climate change for a new reason: human activity that is rapidly increasing the atmospheric concentration of carbon dioxide and other greenhouse gases (such as methane). This increase in greenhouse gas concentration is apparent already: Global average temperatures have risen by about 0.8°C (1.4°F) in the past century (Figure 7.42). This **global warming** poses a potentially serious threat to our civilization.

Global Warming   The potential threat of global warming is a hot political issue because alleviating the threat would require finding new energy sources and making other changes that would dramatically affect the world's economy. However, a major research effort has gradually added to our understanding of the potential threat, particularly in the past decade. Although some controversy still exists about how we should respond to the problem of global warming, at least three facts are clear:

1. The burning of fossil fuels and other human activity is clearly increasing the amounts of greenhouse gases in the atmosphere. The current atmospheric concentration of carbon dioxide is signifi-

Figure 7.42

Average global temperatures from 1860 through 2007. Notice the clear global warming trend of the past few decades. (Data from the National Climate Data Center.)

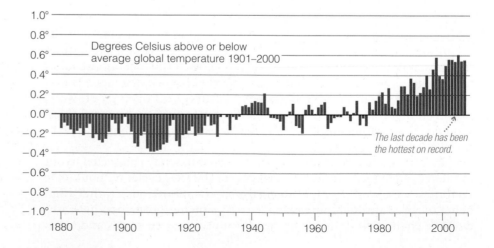

The last decade has been the hottest on record.

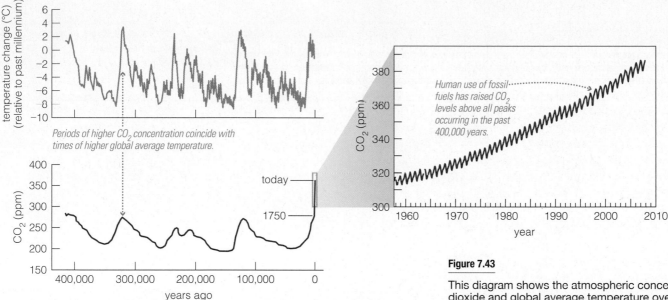

Periods of higher CO_2 concentration coincide with times of higher global average temperature.

Human use of fossil fuels has raised CO_2 levels above all peaks occurring in the past 400,000 years.

Figure 7.43

This diagram shows the atmospheric concentration of carbon dioxide and global average temperature over the past 400,000 years. Notice that both vary together: More carbon dioxide goes with higher temperature, and vice versa. The CO_2 data for the past half century come from direct measurements; most of the earlier data come from studies of air bubbles trapped in Antarctic ice. (Additional data, not shown, extend the record back to about a million years.) The concentration is measured in parts per million (ppm), which is the number of CO_2 molecules among every 1 million air molecules.

cantly higher (by about 30%) than it has been at any time during the past million years, and it is rising rapidly (Figure 7.43).

2. We understand the basic mechanism of the greenhouse effect (see Figure 7.15) and have seen its clear effect on other planets. There is no doubt that a continually rising concentration of greenhouse gases would eventually make our planet warm up; the only debate is about how soon and how much.

3. Scientists have created sophisticated computer models of the climate, and these models match actual climate data quite well. While the models still contain some uncertainties, they indicate that global temperatures are increasing in a manner consistent with what we would expect if human activity is the cause (Figure 7.44).

Human activity is rapidly increasing the atmospheric concentrations of greenhouse gases, causing the global average temperature to rise.

These facts, summarized in Figure 7.45, offer convincing evidence that we humans are now tinkering with the climate in a way that may cause major changes not just in the distant future, but in our own lifetimes. The same models that convince scientists of the reality of human-induced global warming tell us that if current trends in greenhouse gas concentrations continue—that is, if we do nothing to slow our emissions of carbon dioxide and other greenhouse gases—the warming trend will continue to accelerate. By the end of this century, the global average temperature will be 3°–5°C (6°–10°F) higher than it is now, giving our children and grandchildren the warmest climate that any generation of *Homo sapiens* has ever experienced.

Consequences of Global Warming
An increase in temperature of a few degrees might not sound so bad, but small changes in *average* temperature can lead to much more dramatic changes in climate patterns. These changes will cause some regions to warm much more than the average, while other regions may actually cool. Some regions might experience more rainfall or might become deserts.

Polar regions will warm the most, causing polar ice to melt. This is clearly threatening to the species of these regions (polar bears, which depend on an abundance of ice floes, are already under pressure), but it also warms the oceans everywhere and changes their salt content as

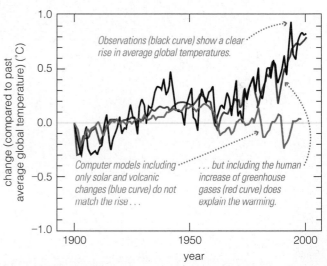

Observations (black curve) show a clear rise in average global temperatures.

Computer models including only solar and volcanic changes (blue curve) do not match the rise . . .

. . . but including the human increase of greenhouse gases (red curve) does explain the warming.

Figure 7.44

This graph compares observed temperature changes (black curve) with the predictions of climate models. The blue curve represents model predictions that include only natural factors, such as changes in the brightness of the Sun and effects of volcanoes. The red curve represents model predictions that include the human contribution due to increasing greenhouse gas concentrations along with the natural factors. Notice that only the red curve matches the observations well, especially for recent decades, providing very strong evidence that global warming is a result of human activity. (The red and blue model curves are each averages of many scientists' independent models of global warming, which generally agree with each other within 0.1–0.2°C.)

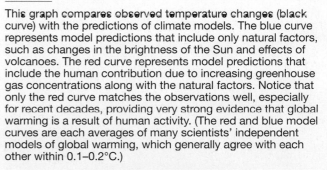

Chapter 7 Earth and the Terrestrial Worlds **221**

Scientific studies of global warming apply the same basic approach used in all areas of science: We create models of nature, compare the predictions of those models with observations, and use our comparisons to improve the models. We have found that climate models agree more closely with observations if they include human production of greenhouse gases like carbon dioxide, making scientists confident that human activity is indeed causing global warming.

1 The greenhouse effect makes a planetary surface warmer than it would be otherwise because greenhouse gases such as carbon dioxide, methane, and water vapor slow the escape of infrared light radiated by the planet. Scientists have great confidence in models of the greenhouse effect because they successfully predict the surface temperatures of Venus, Earth, and Mars.

2 Human activity is adding carbon dioxide and other greenhouse gases to the atmosphere. While the carbon dioxide concentration also varies naturally, its concentration is now much higher than it has been at any time in the previous million years, and it is continuing to rise rapidly.

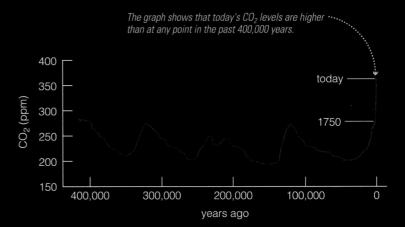

The graph shows that today's CO_2 levels are higher than at any point in the past 400,000 years.

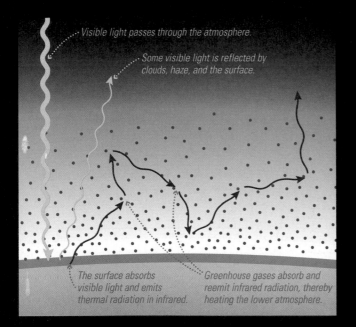

Visible light passes through the atmosphere.

Some visible light is reflected by clouds, haze, and the surface.

The surface absorbs visible light and emits thermal radiation in infrared.

Greenhouse gases absorb and reemit infrared radiation, thereby heating the lower atmosphere.

Global Average Surface Temperature

Planet	Temperature Without Greenhouse Effect	Temperature With Greenhouse Effect
Venus	−40°C	470°C
Earth	−16°C	15°C
Mars	−56°C	−50°C

This table shows planetary temperatures as they would be without the greenhouse effect and as they actually are with it. The greenhouse effect makes Earth warm enough for liquid water and Venus hotter than a pizza oven.

③ Observations show that Earth's average surface temperature has risen during the last several decades. Computer models of Earth's climate show that an increased greenhouse effect triggered by CO_2 from human activities can explain the observed temperature increase.

④ Models can also be used to predict the consequences of a continued rise in greenhouse gas concentrations. These models show that, without significant reductions in greenhouse gas emissions, we should expect further increases in global average temperature, rising sea levels, and more intense and destructive weather patterns.

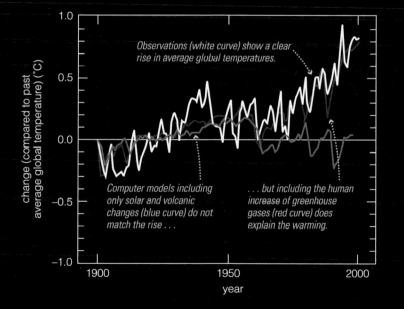

Observations (white curve) show a clear rise in average global temperatures.

Computer models including only solar and volcanic changes (blue curve) do not match the rise . . .

. . . but including the human increase of greenhouse gases (red curve) does explain the warming.

This diagram shows the change in Florida's coastline that would occur if sea levels rose by 1 meter. Some models predict that this rise could occur within a century. The light blue regions show portions of the existing coastline that would be flooded.

HALLMARK OF SCIENCE **Science progresses through creation and testing of models of nature that explain the observations as simply as possible.** Observations showing a rise in Earth's temperature demand a scientific explanation. Models that include an increased greenhouse effect due to human activity explain those observations better than models without human activity.

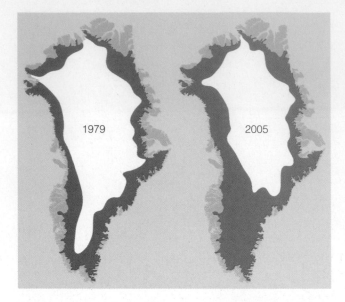

Figure 7.46

These maps contrast the extent of the year-round Greenland ice sheet, shown in white, in 1979 and 2005. The orange area indicates the region that melts during the warm season. Notice that the melt region has expanded significantly, extending both further inland (to higher elevations) and further north.

melting ice pours more fresh water into the sea. The fact that the waters of the Gulf of Mexico are at their warmest level in at least a century may be contributing to the greater strength of hurricanes that have recently blown out of the Caribbean, though it is difficult to cite causes for specific storms. More generally, the greater overall warmth of the atmosphere will increase evaporation from the oceans, leading to more numerous and more intense storms; ironically, this fact means that global warming could mean more severe winter blizzards. Some researchers also worry that the influx of large quantities of fresh water into the oceans may alter major ocean currents, such as the Gulf Stream—a "river" within the ocean that regulates the climate of western Europe and parts of the United States.

Melting polar ice may significantly increase sea level in the future, but global warming has already caused a rise in sea level for a different reason. Water expands very slightly as it warms—so slightly that we don't notice the change in a glass of water, but enough that sea level has risen some 20 centimeters in the past hundred years. This effect alone could cause sea level to rise as much as another meter during this century, with potentially devastating effect to coastal communities and low-lying countries such as Bangladesh. The added effect of melting ice could increase sea level much more. While the melting of ice in the Arctic Ocean does not affect sea level—it is already floating—melting of land-locked ice does. Such melting appears to be occurring already. For example, the famous "snows" (glaciers) of Mount Kilimanjaro are rapidly retreating and may be gone within the next decade or so. More ominously, recent data suggest that the Greenland ice sheet is melting much more rapidly than models have predicted (Figure 7.46). If this trend continues, sea level could rise as much as several *meters*—enough to flood much of Florida by the end of this century. Looking further ahead, complete melting of the polar ice caps would increase sea level by some 70 meters (more than 200 feet). Although such melting would probably take centuries or millennia, it suggests the disconcerting possibility that future generations will have to send deep-sea divers to explore the underwater ruins of many of our major cities.

Fortunately, most scientists believe that we still have time to avert the most serious consequences of global warming, provided we dramatically and rapidly curtail our greenhouse gas emissions. The most obvious way to cut back on these emissions is to improve energy efficiency. Doubling the average gas mileage of cars—which we could do easily with current technology—would immediately cut automobile-related carbon dioxide emissions in half. Other tactics include replacing fossil fuels with alternative energy sources such as solar, wind, nuclear, and biofuels, and perhaps even finding ways to bury the carbon dioxide by-products of the fossil fuels that we still use. The key idea to keep in mind is that global warming is a global problem, and it will therefore require significant international cooperation if we hope to solve it. But there is precedent for success: In the 1980s and 1990s, as we learned that human-produced chemicals (known as CFCs) were causing great damage to the ozone layer, the nations of the world agreed on a series of treaties that ultimately phased those chemicals out of production. As a result, the ozone layer is beginning to recover from its earlier damage, and we learned that people can indeed be moved to act in the face of a threat to the environment on which we depend for survival.

think about it If you were a political leader, how would you deal with the threat of global warming?

What makes a planet habitable?

We have discussed the features of our planet that have made our world habitable for a great variety of life, including us. But why is Earth the only terrestrial world that has these features?

Our comparative study of the terrestrial worlds tells us there are two primary answers. First, Earth is habitable because it is large enough to have remained geologically active since its birth, so that outgassing could release the water and gases that formed our atmosphere and oceans. In addition, the core has remained hot enough so that Earth has retained a global magnetic field, which generates a magnetosphere that protects our atmosphere from the solar wind. Second, we are located at a distance from the Sun where outgassed water vapor was able to condense and rain down to form oceans, making possible the carbon dioxide cycle that regulates our climate.

Earth is habitable because it is large enough to remain geologically active and located at a distance from the Sun where oceans were able to form.

Figure 7.47 summarizes the lessons we have learned about how a planet's size and distance from the Sun determine its fate.

Figure 7.47

This illustration shows how a terrestrial world's size and distance from the Sun help determine its geological history and whether it has conditions suitable for life. Earth is habitable because it is large enough and at a suitably moderate distance from the Sun.

The Role of Planetary Size

Small Terrestrial Planets

Interior cools rapidly...

...so that tectonic and volcanic activity cease after a billion years or so. Many ancient craters therefore remain.

Lack of volcanism means little outgassing, and low gravity allows gas to escape more easily; no atmosphere means no erosion.

Large Terrestrial Planets

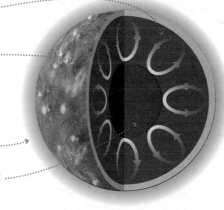

Warm interior causes mantle convection...

...leading to ongoing tectonic and volcanic activity; most ancient craters have been erased.

Outgassing produces an atmosphere and strong gravity holds it, so that erosion is possible.

Core may be molten, producing a magnetic field if rotation is fast enough, and a magnetosphere that can shield an atmosphere from the solar wind.

The Role of Distance from the Sun

Sun

Planets Close to the Sun

Surface is too hot for rain, snow, or ice, so little erosion occurs.

High atmospheric temperature allows gas to escape more easily.

Planets at Intermediate Distances from the Sun

Moderate surface temperatures can allow for oceans, rain, snow, and ice, leading to substantial erosion.

Gravity can more easily hold atmospheric gases.

Planets Far from the Sun

Low surface temperatures can allow for ice and snow, but no rain or oceans, limiting erosion.

Atmosphere may exist, but gases can more easily condense to make surface ice.

Figure 7.48

This diagram summarizes the geological histories of the terrestrial worlds. The brackets along the top indicate that impact cratering has affected all worlds similarly. The arrows represent volcanic and tectonic activity. A thicker and darker arrow means more volcanic/tectonic activity, and the arrow length tells us how long this activity persisted. Notice that the trend follows the order of planetary size: Earth remains active to this day. Venus has also been quite active, though we are uncertain whether it remains so. Mars has had an intermediate level of activity and might still have low-level volcanism. Mercury and the Moon have had very little volcanic/tectonic activity. Erosion is not shown, because it has played a significant role only on Earth (ongoing) and on Mars (where it was quite significant in the past and continues at low levels today).

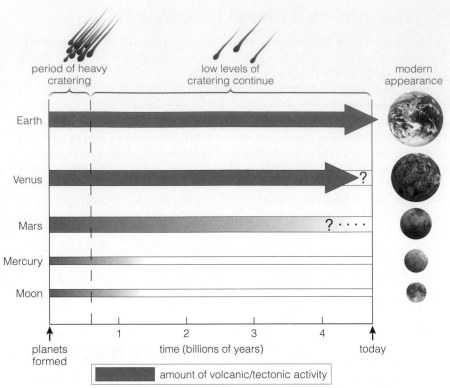

Figure 7.48 shows the trends we've seen for the terrestrial planets in our solar system, which should make sense as you examine the role of planetary size. In principle, these lessons mean we can now predict the geological and atmospheric properties of terrestrial worlds that we may someday find around other stars. Only a suitably large terrestrial planet located at an intermediate distance from its star is likely to have conditions under which life could thrive. Of course, we do not yet know whether simply having conditions suitable for life means that life will actually arise. We will consider current understanding of this fascinating question in Chapter 18, when we consider the prospects of finding life beyond Earth.

the big picture
Putting Chapter 7 into Context

In this chapter, we have explored the histories of the terrestrial worlds. As you think about the details you have learned, keep the following "big picture" ideas in mind:

* The terrestrial worlds all looked much the same when they were born, so their present-day differences are a result of geological processes that occurred in the ensuing $4\frac{1}{2}$ billion years.

* The primary factor in determining a terrestrial world's geological history is its size, because only a relatively large world can retain internal heat long enough for ongoing geological activity. However, the

differences between Venus and Earth show that distance from the Sun also plays an important role.

- A planet's distance from the Sun is important to its surface temperature, but the cases of Venus and Earth show that the strength of the greenhouse effect can play an even bigger role. Humans are currently altering the balance of greenhouse gases in Earth's atmosphere, with potentially dire consequences.

- The histories of Venus and Mars show that a stable climate like Earth's is more the exception than the rule. The stable climate that makes our existence possible is a direct consequence of our planet's unique geology, including its plate tectonics and carbon dioxide cycle.

summary of key concepts

7.1 Earth as a Planet

• Why is Earth geologically active?

 Internal heat drives geological activity, and Earth retains internal heat because of its relatively large size for a terrestrial world. This heat causes mantle **convection** and keeps Earth's lithosphere thin, ensuring active surface geology. It also keeps part of Earth's core melted, and circulation of this molten metal creates Earth's magnetic field.

• What processes shape Earth's surface?

The four major geological processes are **impact cratering**, **volcanism**, **tectonics**, and **erosion**. Earth has experienced many impacts, but most craters have been erased by other processes. We owe the existence of our atmosphere and oceans to volcanic **outgassing**. A special type of tectonics—**plate tectonics**—shapes much of Earth's surface. Ice, water, and wind drive rampant erosion on our planet.

• How does Earth's atmosphere affect the planet?

Two crucial effects are (1) protecting the surface from dangerous solar radiation—ultraviolet is absorbed by ozone and X rays are absorbed high in the atmosphere—and (2) the **greenhouse effect**, without which the surface temperature would be below freezing.

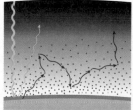

7.2 The Moon and Mercury: Geologically Dead

• Was there ever geological activity on the Moon or Mercury?

Both the Moon and Mercury had some volcanism and tectonics when they were young. However, because of their small sizes, their interiors long ago cooled too much for ongoing geological activity.

7.3 Mars: A Victim of Planetary Freeze-Drying

• What geological features tell us that water once flowed on Mars?

 Dry riverbeds, eroded craters, and chemical analysis of Martian rocks all show that water once flowed on Mars, though any periods of rainfall seem to have ended at least 3 billion years ago. Mars today still has water ice underground and in its polar caps and could possibly have pockets of underground liquid water.

• Why did Mars change?

Mars's atmosphere must once have been thicker with a stronger greenhouse effect, so change must have occurred due to loss of atmospheric gas. Much of the lost gas probably was stripped away by the solar wind, after Mars lost its magnetic field and protective magnetosphere. Mars also lost water, because solar ultraviolet light split water molecules apart and the hydrogen escaped to space.

7.4   Venus: A Hothouse World

• Is Venus geologically active?

Venus almost certainly remains geologically active today. Its surface shows evidence of major volcanic or tectonic activity in the past billion years, and it should retain nearly as much internal heat as Earth. However, geological activity on Venus differs from that on Earth in at least two key ways: lack of erosion and lack of plate tectonics.

• Why is Venus so hot?

Venus's extreme surface heat is a result of its thick, carbon dioxide atmosphere, which creates a very strong greenhouse effect. The reason Venus has such a thick atmosphere is its distance from the Sun: It was too close to develop liquid oceans like those on Earth, where most of the outgassed carbon dioxide dissolved in water and became locked away in **carbonate** rock. Carbon dioxide remained in Venus's atmosphere, creating a **runaway greenhouse effect**.

7.5   Earth as a Living Planet

• What unique features of Earth are important for life?

Unique features of Earth on which we depend for survival are (1) surface liquid water, made possible by Earth's moderate temperature; (2) atmospheric oxygen, a product of photosynthetic life; (3) plate tectonics, driven by internal heat; and (4) climate stability, a result of the **carbon dioxide cycle**, which in turn requires plate tectonics.

• How is human activity changing our planet?

CO$_2$ concentration over the past 400,000 years

The global average temperature has risen about 0.8°C over the past hundred years, accompanied by an even larger rise in the atmospheric CO$_2$ concentration—a result of fossil fuel burning and other human activity. The current CO$_2$ concentration is higher than at any time in the past million years, and climate models indicate that this higher concentration is indeed the cause of **global warming**.

• What makes a planet habitable?

We can trace Earth's habitability to its relatively large size and its distance from the Sun. Its size keeps the internal heat that allowed volcanic outgassing to lead to our oceans and atmosphere, and also drives the plate tectonics that helps regulate our climate through the carbon dioxide cycle. Its distance from the Sun is neither too close nor too far, thereby allowing liquid water to exist on Earth's surface.

exercises and problems

Mastering ASTRONOMY™

For instructor-assigned homework go to **www.masteringastronomy.com.**

Review Questions

Short-Answer Questions Based on the Reading

1. What are Earth's basic layers by composition? What do we mean by the *lithosphere,* and why isn't it listed as one of the three layers by composition?

2. What is differentiation, and how did it affect the internal structures of the terrestrial worlds?

3. Why do large planets retain internal heat longer than smaller planets? Briefly explain how internal heat is related to mantle convection and lithospheric thickness.

4. Why does Earth have a global *magnetic field?* What is the *magnetosphere?*

5. Define each of the four major geological processes, and give examples of features on Earth shaped by each process.

6. What is *outgassing,* and how did it lead to the existence of Earth's atmosphere and oceans?

7. Describe how Earth's atmosphere protects the surface from harmful radiation. What is the role of *ozone?*

8. What does the *greenhouse effect* do to a planet? Explain the role of greenhouse gases and describe the basic mechanism of the greenhouse effect.

9. How do crater counts tell us the age of a planetary surface? Briefly explain why the Moon is so much more heavily cratered than Earth.

10. Briefly summarize the geological history of the Moon. How did the lunar maria form?

11. Briefly summarize the geological history of Mercury. How are Mercury's great cliffs thought to have formed?

12. Describe at least three similarities and three differences between Earth and Mars.

13. Choose five features on the global map of Mars (Figure 7.24), and explain the nature and likely origin of each.

14. Explain why liquid water is not stable on Mars today, but why we nonetheless think it flowed in the distant past on Mars. Could there still be liquid water anywhere on Mars today? Explain.

15. Briefly summarize how and why Mars lost much of its atmosphere some 3 billion years ago.

16. Describe at least three major geological features of Venus. Why is it surprising that Venus lacks plate tectonics? What might explain this lack?

17. What do we mean by a *runaway greenhouse effect?* Explain why this process occurred on Venus but not on Earth.

18. List four unique features of Earth in comparison to other terrestrial worlds, and briefly explain what we mean by each one.

19. What is *plate tectonics?* How does it change the arrangement of the continents with time?

20. What is the *carbon dioxide cycle,* and why is it so crucial to life on Earth?

21. Briefly summarize the problem of global warming and its potential consequences if we do not act to stop it.
22. Based on Figure 7.47, write a paragraph each on the role of planetary size and the role of distance from the Sun in explaining the current nature of the terrestrial worlds.

Test Your Understanding

Surprising Discoveries?

Suppose we were to make the following discoveries. (These are not real discoveries.) In light of your understanding of planetary geology, decide whether the discovery should be considered reasonable or surprising. Explain your reasoning clearly, if possible tracing your logic back to basic planetary properties of size or distance from the Sun; because not all of these have definitive answers, your explanation is more important than your chosen answer.

23. The *MESSENGER* mission to Mercury photographs part of the surface never seen before and detects vast fields of sand dunes.
24. New observations show that several of the volcanoes on Venus have erupted within the past few million years.
25. A Venus radar mapper discovers extensive regions of layered sedimentary rocks, similar to those found on Earth.
26. Radiometric dating of rocks brought back from one lunar crater shows that the crater was formed only a few tens of millions of years ago.
27. New, high-resolution orbital photographs of Mars show many crater bottoms filled with pools of liquid.
28. Clear-cutting in the Amazon rain forest on Earth exposes vast regions of ancient terrain that is as heavily cratered as the lunar highlands.
29. Drilling into the Martian surface, a robotic spacecraft discovers liquid water a few meters beneath the slopes of a Martian volcano.
30. We find a planet in another solar system that has an Earth-like atmosphere with plentiful oxygen but no life of any kind.
31. We find a planet in another solar system that has Earth-like plate tectonics; the planet is the size of the Moon and orbits 1 AU from its star.
32. We find evidence that the early Earth had more carbon dioxide in its atmosphere than Earth does today.

Quick Quiz

Choose the best answer to each of the following. Explain your reasoning with one or more complete sentences.

33. Which heat source continues to contribute to Earth's *internal* heat? (a) accretion (b) radioactive decay (c) sunlight
34. In general, what kind of terrestrial planet would you expect to have the thickest lithosphere? (a) a large planet (b) a small planet (c) a planet located far from the Sun
35. Which of a planet's fundamental properties has the greatest effect on its level of volcanic and tectonic activity? (a) size (b) distance from the Sun (c) rotation rate
36. Which describes our understanding of flowing water on Mars? (a) It was never important. (b) It was important once, but no longer. (c) It is a major process on the Martian surface today.
37. What do we conclude if a planet has few impact craters of any size? (a) The planet was never bombarded by asteroids or comets. (b) Its atmosphere stopped impactors of all sizes. (c) Other geological processes have erased craters.

38. How many of the five terrestrial worlds are considered "geologically dead"? (a) none (b) two (c) four (Be sure to explain *why* these worlds became geologically dead.)
39. Which terrestrial world has the most atmospheric gas? (a) Venus (b) Earth (c) Mars
40. Which of the following is a strong greenhouse gas? (a) nitrogen (b) water vapor (c) oxygen
41. The oxygen in Earth's atmosphere was released by (a) volcanic outgassing. (b) the CO_2 cycle. (c) life.
42. Where is most of the CO_2 that has outgassed from Earth's volcanoes? (a) in the atmosphere (b) escaped to space (c) locked up in rocks

Process of Science

Examining How Science Works

43. *What Is Predictable?* We've found that much of a planet's geological history is destined from its birth. Briefly explain why, and discuss the level of detail that is predictable. For example, was Mars's general level of volcanism predictable? Could we have predicted a mountain as tall as Olympus Mons or a canyon as long as Valles Marineris? Explain.
44. *Science with Consequences.* Some people are still skeptical of global warming. Research the opinions of such skeptics in newspapers, in magazines, and on the Internet. Do they disagree with the data, saying that Earth is not getting warmer? Do they disagree with the conclusion that humans are the primary cause? Do they disagree with the idea that action must be taken? Defend or refute their findings based on your own studies and your understanding of the hallmarks of science discussed in Chapter 3.
45. *Unanswered Questions.* As discussed in this chapter, our exploration of Mars suggests that it may have been habitable in the past. Choose one important but unanswered question about Mars's past, and write two or three paragraphs discussing how we might answer this question in the future. Be as specific as possible, focusing on the type of evidence necessary to answer the question and the method(s) that should be used to gather the evidence. What would be the benefits of finding answers to this question?

Investigate Further

In-Depth Questions to Increase Your Understanding

Short-Answer/Essay Questions

46. *Miniature Mars.* Suppose Mars had turned out to be significantly smaller than its current size—say, the size of our Moon. How would this have affected the number of geological features due to each of the four major geological processes? Do you think Mars would still be a good candidate for harboring extraterrestrial life? Summarize your answers in two or three paragraphs.
47. *Two Paths Diverged.* By looking back to fundamental properties such as size and distance from the Sun, explain why Earth has oceans and very little atmospheric carbon dioxide, while similar-size Venus has a thick, carbon dioxide atmosphere.
48. *Change in Formation Properties.* Consider Earth's size and distance from the Sun. Choose one property and suppose that it had been different (for example, smaller size or greater distance). Describe how this change might have affected Earth's subsequent history and the possibility of life on Earth.

49. *Experiment: Planetary Cooling in a Freezer.* Fill two small plastic containers of similar shape but different size with cold water and put both into the freezer at the same time. Every hour or so, record the time and your estimate of the thickness of the "lithosphere" (the frozen layer) in each container. How long does it take the water in each container to freeze completely? Describe in a few sentences the relevance of your experiment to planetary geology. Extra credit: Plot your results on a graph with time on the *x*-axis and lithospheric thickness on the *y*-axis. What is the ratio of the two freezing times?

50. *Amateur Astronomy: Observing the Moon.* Any amateur telescope has a resolution adequate to identify geological features on the Moon. The light highlands and dark maria should be evident, and shadowing is visible near the line between night and day. Try to observe the Moon near the first- or third-quarter phase. Sketch or photograph the Moon at low magnification, and then zoom in on a region of interest. Again sketch or photograph your field of view, label its features, and identify the geological process that created them. Look for craters, volcanic plains, and tectonic features. Estimate the size of each feature by comparing it to the size of the whole Moon (radius = 1738 kilometers).

51. *Global Warming.* What, if anything, should we be doing to alleviate the threat of global warming that we are not doing already? Write a one-page editorial summarizing and defending your opinion.

Quantitative Problems

Be sure to show all calculations clearly and state your final answers in complete sentences.

52. *Moon and Mars.* Compare the surface area–to–volume ratios of the Moon and Mars. Based on your answer, what should you expect about the cooling time of Mars compared to that of the Moon?

53. *Earth and Venus.* Compare the surface area–to–volume ratios of Venus and Earth. Based on your answer, how should you expect the cooling time to compare for these two worlds?

54. *Doubling Your Size.* Just as the surface area–to–volume ratio depends on size, so can other properties. To see how, suppose that your size suddenly doubled; that is, your height, width, and depth all doubled. (For example, if you were 5 feet tall before, you now are 10 feet tall.)
    a. By what factor has your waist size increased?
    b. How much more material will be required for your clothes? (*Hint:* Clothes cover the *surface area* of your body.)
    c. By what factor has your weight increased? (*Hint:* Weight depends on the *volume* of your body.)
    d. The pressure on your weight-bearing joints depends on how much *weight* is supported by the *surface area* of each joint. How has the pressure on your weight-bearing joints changed?

55. *Impact Energies.* An asteroid 1 kilometer in diameter will make a crater of about 10 kilometers in diameter. How much kinetic energy does the asteroid have if it strikes the surface at 20 kilometers per second? (The amount of kinetic energy is given by the formula $\frac{1}{2}mv^2$.) Convert your answer to megatons of TNT: 1 megaton is about 4×10^{15} joules. Compare the impactor to the largest nuclear weapons, currently less than 100 megatons of energy.

56. *Internal vs. External Heating.* In daylight, the Earth's surface absorbs about 400 watts per square meter. Earth's internal radioactivity produces a total of 3 trillion watts that leak out through our planet's entire surface. Calculate the amount of heat from radioactive decay that flows outward through each square meter of Earth's surface (your answer should have units of watts per square meter). Compare this quantitatively to solar heating, and comment on why internal heating drives geological activity.

57. *Plate Tectonics.* Typical motions of one plate relative to another are 1 centimeter per year. At this rate, how long would it take for two continents 3000 kilometers apart to collide? What are the global consequences of motions like this?

58. *Planet Berth.* Imagine a planet, which we'll call *Berth*, orbiting a star identical to the Sun at a distance of 1 AU. Assume that Berth has eight times as much mass as Earth and is twice as large as Earth in diameter.
    a. How does Berth's density compare to Earth's?
    b. How does Berth's surface area compare to Earth's?
    c. Based on your answers to (a) and (b), discuss how Berth's geological history is likely to have differed from Earth's.

Discussion Questions

59. *Worth the Effort?* Politicians often argue over whether planetary missions are worth the expense involved. Based on what we have learned by comparing the geologies of the terrestrial worlds, do you think the missions that have given us this knowledge have been worth their expense? Defend your opinion.

60. *Lucky Earth.* The climate histories of Venus and Mars make it clear that it's not "easy" to get a pleasant climate like that of Earth. How does this affect your opinion about whether Earth-like planets might exist around other stars? Explain.

61. *Terraforming Mars.* Some people have suggested that we might be able to carry out planetwide engineering of Mars that would cause its climate to warm and its atmosphere to thicken. This type of planet engineering is called *terraforming*, because its objective is to make a planet more Earth-like and therefore easier for humans to live on. Discuss possible ways to terraform Mars, at least in principle. Do any of these ideas seem practical? Does it seem like a good idea? Defend your opinions.

Web Projects

62. *"Coolest" Surface Photo.* Visit the Astronomy Picture of the Day Web site, and search for past images of the terrestrial worlds. After looking at many of the images, choose the one you think is the "coolest." Make a printout, write a short description of what it shows, and explain what you like about it.

63. *Water on Mars.* Go to the Web site for NASA's Mars Exploration Program, and look for the latest evidence concerning recent water flows on Mars. Write a few paragraphs describing the new evidence and what it tells us.

64. *Mars Colonization.* Visit the Web site of a group that advocates human colonization of Mars, such as the Mars Society. Learn about the challenges of human survival on Mars and about prospects for terraforming Mars. Do you think colonization of Mars is a good idea? Write a short essay describing what you've learned and defend your opinions.

65. *Human Threats to Earth.* Write a three- to five-page research report about current understanding of and controversy over one of the following issues: global warming, ozone depletion, or the loss of species on Earth due to human activity. Be sure to address both the latest knowledge about the issue and proposals for alleviating any dangers associated with it. End your report by making your own recommendations about what, if anything, needs to be done to prevent further damage.

visual skills check

Use the following questions to check your understanding of some of the many types of visual information used in astronomy. Answers are provided in Appendix K. For additional practice, try the Chapter 7 Visual Quiz at **www.masteringastronomy.com.**

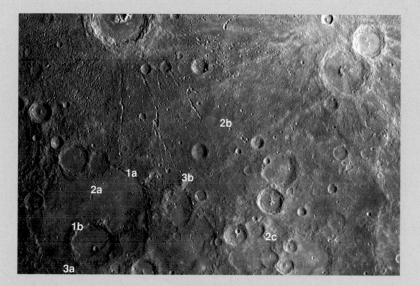

This image from the *MESSENGER* flyby shows evidence of impact cratering, volcanism, and tectonic activity on Mercury. Answer the following questions based on the image. Remember that craters are bowl-shaped and rough-floored when they form, and wipe out any pre-existing features in the area. Lava on Mercury appears to be fairly runny and makes flat, smooth plains as it spreads out.

1. Label 1a lies on the rim of a large crater, and 1b lies on the rim of a smaller one. Which crater must have formed first?
   a. crater 1a
   b. crater 1b
   c. cannot be determined
2. The region around 2b has far fewer craters than the region around 2c. The crater floor at 2a is also flat and smooth, without many smaller craters on it. Why are regions 2a and 2b so smooth?
   a. Few small craters ever formed in these regions.
   b. Erosion erased craters that once existed in these regions.
   c. Lava flows covered craters that once existed in these regions.
3. A tectonic ridge appears to connect points 3a and 3b, crossing several craters. From its appearance, we can conclude that it must have formed
   a. before the area was cratered.
   b. after the area was cratered.
   c. at the same time the area was cratered.
4. Using your answers from questions 1–3, list the following features in order from oldest to youngest:
   a. the tectonic ridge from 3a to 3b
   b. crater 1a
   c. the smooth floor of crater 1b

8

jovian planet systems

232

In Roman mythology, the namesakes of the jovian planets are rulers among gods: Jupiter is the king of the gods, Saturn is Jupiter's father, Uranus is the lord of the sky, and Neptune rules the sea. However, our ancestors could not have foreseen the true majesty of the four jovian planets. The smallest, Neptune, is large enough to contain the volume of more than 50 Earths. The largest, Jupiter, has a volume some 1400 times that of Earth. These worlds are totally unlike the terrestrial planets. They are essentially giant balls of gas, with no solid surface on which to stand.

Why should we care about a set of worlds so different from our own? Apart from satisfying natural curiosity, studies of the jovian planets and their moons help us understand the birth and evolution of our solar system—which in turn helps us understand our own planet Earth. In addition, the jovian planets provide stepping stones to understanding the hundreds of planets so far discovered around other stars, because most of these planets are probably jovian in nature. In this chapter, we'll explore the jovian planet systems, first focusing on the planets themselves, then on their many moons, and finally on their beautifully complex rings.

(MA) **Formation of the Solar System Tutorial, Lesson 1**

8.1 A Different Kind of Planet

The jovian planets are radically different from the terrestrial planets. They are far larger in size and very different in composition. They are orbited by rings and numerous moons. They even rotate much faster than the terrestrial planets.

We toured the jovian planets briefly in Section 6.1. Now we are ready to explore these planets in a little more depth. As you'll see from both the discussion and the selection of photos in this chapter, much of our present knowledge has come from spacecraft visits, especially from the *Voyager 1* and *2* missions that flew past these planets in the late 1970s and 1980s and the *Galileo* spacecraft that orbited Jupiter from 1995 until 2003. Currently, scientists are learning much more from the *Cassini* spacecraft now orbiting Saturn and *New Horizons*, which flew past Jupiter in 2007 on its way to Pluto.

• What are jovian planets made of?

The jovian planets are often called "gas giants," making it sound as if they are entirely gaseous like air on Earth. While this idea is not entirely wrong, the reality is somewhat more complex.

Figure 8.1 shows a montage of the jovian planets compiled by the *Voyager* spacecraft, along with basic data and Earth included for scale. The immense sizes of the jovian worlds are apparent. But while all four are enormous, there are important differences between them. In particular, note that they differ substantially in mass, density, and overall composition.

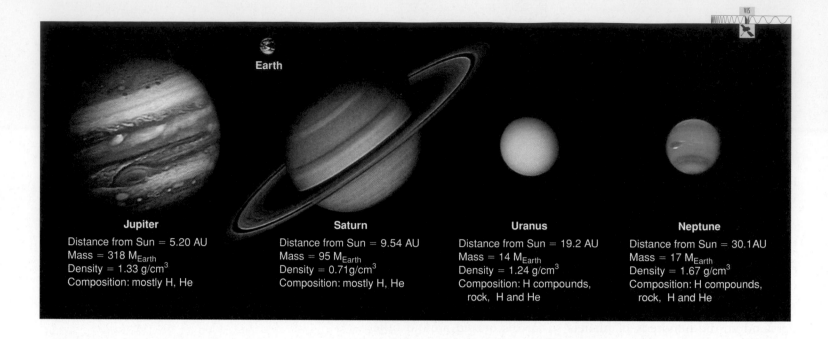

Earth

Jupiter
Distance from Sun = 5.20 AU
Mass = 318 M_{Earth}
Density = 1.33 g/cm^3
Composition: mostly H, He

Saturn
Distance from Sun = 9.54 AU
Mass = 95 M_{Earth}
Density = 0.71g/cm^3
Composition: mostly H, He

Uranus
Distance from Sun = 19.2 AU
Mass = 14 M_{Earth}
Density = 1.24 g/cm^3
Composition: H compounds,
     rock, H and He

Neptune
Distance from Sun = 30.1AU
Mass = 17 M_{Earth}
Density = 1.67 g/cm^3
Composition: H compounds,
     rock, H and He

Figure 8.1

Jupiter, Saturn, Uranus, and Neptune, shown to scale with Earth for comparison.

General Composition of Jovian Planets Jupiter and Saturn are made almost entirely of hydrogen and helium, with just a few percent of their masses coming from hydrogen compounds and even smaller amounts of rock and metal. In fact, their overall compositions are much more similar to the composition of the Sun than to the compositions of the terrestrial planets. Some people even call Jupiter a "failed star" because it has a starlike composition but lacks the nuclear fusion needed to make it shine. This is a consequence of its size: Although Jupiter is large for a planet, it is much less massive than any star. As a result, its gravity is too weak to compress its interior to the extreme temperatures and densities needed for nuclear fusion. (Jupiter would have needed to grow to about 80 times its current mass to have become a star.) Of course, where some people see a failed star, others see an extremely successful planet.

The jovian planets are made mostly of hydrogen, helium, and hydrogen compounds, making them very different in composition from terrestrial worlds.

Uranus and Neptune are much smaller than Jupiter and Saturn overall, and also contain proportionally much smaller amounts of hydrogen and helium. These gases make up much less than half their total masses. Rather than being made mostly of hydrogen and helium, Uranus and Neptune are made primarily of hydrogen compounds such as water (H_2O), methane (CH_4), and ammonia (NH_3), along with smaller amounts of metal and rock. We can understand the differences in composition and size among the jovian planets by looking at the way in which we think they formed.

Gas Capture in the Solar Nebula Recall that the jovian planets formed in a very different way from the terrestrial planets [Section 6.4]. The terrestrial planets grew only through accretion of solid planetesimals containing rock and metal. Because rock and metal were quite rare in the solar nebula, the terrestrial planets never grew massive enough for their gravity to hold any of the abundant hydrogen and helium gas that made up most of the nebula.

The jovian planets formed in the outer solar system (beyond the frost line), where it was cold enough for hydrogen compounds to condense into ices (see Figure 6.17). Because hydrogen compounds were so much more abundant than metal and rock, some of the ice-rich planetesimals of the outer solar system grew to great size. Once these planetesimals became sufficiently massive, their gravity allowed them to draw in the hydrogen and helium gas that surrounded them. All four jovian planets are thought to have grown from ice-rich planetesimals of about the same mass—roughly 10 times the mass of Earth. The differences in their composition stem from the amounts of hydrogen and helium gas that they captured.

Jupiter and Saturn captured so much hydrogen and helium gas that these gases now make up the vast majority of their masses. The ice-rich planetesimals from which they grew now represent only a small fraction of their overall masses—about 3% in Jupiter's case and about 10% in Saturn's case.

Uranus and Neptune pulled in much less hydrogen and helium gas. For example, Uranus has about 14 times the mass of Earth. Assuming that it began as an ice-rich planetesimal with about 10 times Earth's mass, Uranus obtained only about a third of its total mass from drawn-in hydrogen and helium gas. The bulk of its mass consists of material from the original ice-rich planetesimal: hydrogen compounds mixed with smaller amounts of rock and metal. The same is true for Neptune, though its higher density suggests that it may have formed around a slightly more massive ice-rich planetesimal.

The jovian planets nearer to the Sun captured more hydrogen and helium gas, making them larger and leaving them with smaller proportions of hydrogen compounds, rock, and metal.

We can trace these differences in the amounts of captured gas to differences in distance from the Sun. As the solar system formed, the solid particles that condensed far from the Sun must have been much more widely spread out than particles that condensed nearer to the Sun. At greater distances from the Sun, it took longer for small particles to accrete into large, icy planetesimals with gravity strong enough to pull in gas from the surrounding nebula. Jupiter would have been the first jovian planet whose icy planetesimal grew large enough to start drawing in gas, followed by Saturn, Uranus, and Neptune. Because all the planets stopped accreting gas at the same time—when the solar wind blew the remaining gas into interstellar space—the more distant planets had less time to capture gas and ended up smaller in size.

Density Differences Notice in Figure 8.1 that Saturn is considerably less dense than Uranus or Neptune. This should make sense when you compare compositions, because the hydrogen compounds, rock, and metal that make up Uranus and Neptune are normally much more dense than hydrogen or helium gas. By the same logic, we'd expect Jupiter to be even less dense than Saturn—but it's not. To understand Jupiter's surprisingly high density, we need to think about how massive planets are affected by their own gravity.

Building a planet of hydrogen and helium is a bit like making one out of fluffy pillows. Imagine assembling a planet pillow by pillow. As each new pillow is added, those on the bottom are compressed more by those above. As the lower pillows are forced closer together, their mutual gravitational attraction increases, compressing them even further. At first the stack grows substantially with each pillow, but eventually the growth slows until adding pillows barely increases the height of the stack (Figure 8.2a).

Measure the thickness of your pillow; then put it at the bottom of a stack of other pillows, folded blankets, or clothing. How much has the pillow been

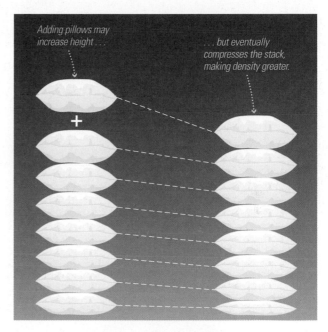

a Adding pillows to a stack may increase its height at first but eventually compresses the stack, making its density greater.

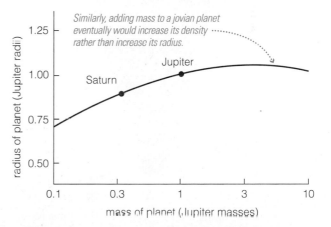

b This graph shows how radius depends on mass for a hydrogen/helium planet. Notice that Jupiter is only slightly larger in radius than Saturn, despite being three times as massive. Gravitational compression of a planet much more massive than Jupiter would actually make it smaller in size.

Figure 8.2

The relationship between mass and radius for a planet made of hydrogen and helium.

compressed by the stack above it? Insert your hand between the different layers to feel the pressure differences—and imagine the kind of pressures and compression you'd find in a stack tens of thousands of kilometers tall!

This analogy explains why Jupiter is only slightly larger than Saturn in radius even though it is more than three times as massive. The extra mass of Jupiter compresses its interior to a much higher density. More precise calculations show that Jupiter's radius is almost the maximum possible radius for a jovian planet. If much more gas were added to Jupiter, its weight would actually compress the interior enough to make the planet *smaller* rather than larger (Figure 8.2b). Indeed, some extrasolar planets that are larger in mass than Jupiter are probably smaller in size.

think about it Saturn's average density of 0.71 g/cm^3 is less than that of water. As a result, it is sometimes said that Saturn could float on a giant ocean. Suppose there really were a gigantic planet with a gigantic ocean and we put Saturn on the ocean's surface. Would it float? If not, what would happen?

• What are jovian planets like on the inside?

Jupiter and Saturn might at first seem to deserve the name "gas giants," since they are made primarily of hydrogen and helium—substances that are gaseous on Earth and were gaseous in the solar nebula. The name may seem less fitting for Uranus and Neptune, since they are made mostly of materials other than pure hydrogen and helium. However, closer inspection shows that the name is a little misleading even for Jupiter and Saturn, because their strong gravity compresses most of the "gas" into forms of matter quite unlike anything we are familiar with in everyday life on Earth. Let's begin by considering what Jupiter is like on the inside, then extend these ideas to the other jovian planets.

Inside Jupiter Jupiter's lack of a solid surface makes it tempting to think of the planet as "all atmosphere," but you could not fly through Jupiter's interior in the way airplanes fly through air. A spacecraft plunging into Jupiter would find increasingly higher temperatures and pressures as it descended. The *Galileo* spacecraft dropped a scientific probe into Jupiter in 1995 that collected measurements for about an hour before the ever-increasing pressures and temperatures destroyed it. The probe provided valuable data about Jupiter's atmosphere but didn't last long enough to sample the interior: It survived to a depth of only about 200 kilometers, or about 0.3% of Jupiter's radius.

If you plunged below Jupiter's clouds, you'd never encounter a solid surface—just ever-denser and hotter hydrogen/helium compressed into bizarre liquid and metallic phases.

While Jupiter has no solid surface, computer models tell us that it still has fairly distinct interior layers. The layers do not differ much in composition—all except the core are mostly hydrogen and helium. Instead, they differ in the phase of their hydrogen. From the tops of its clouds to its core, Jupiter's interior layers are (Figure 8.3):

• Gaseous hydrogen: In the outer layer, conditions are moderate enough for hydrogen to remain in its familiar, gaseous form. This layer extends about 10% of the way from the cloudtops toward the center. In the outer portions of this layer, the gas would seem much like air on Earth (but with a different composition), so we usually think of this region as Jupiter's atmosphere.

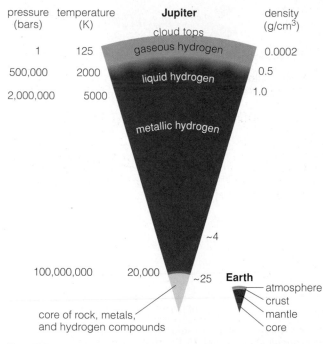

Figure 8.3

Jupiter's interior structure, labeled with the pressure, temperature, and density at various depths. Earth's interior structure is shown to scale for comparison. (Notes on units: 1 bar is approximately the atmospheric pressure at sea level on Earth; the density of liquid water is 1 g/cm³.)

- Liquid hydrogen: This layer occupies about the next 10% of Jupiter's interior. The temperature exceeds 2000 K and the pressure exceeds 500,000 times the pressure on Earth's surface. Laboratory experiments show that hydrogen acts more like a liquid than a gas under these conditions. Notice that the density in this layer (ranging from 0.5 to almost 1.0 g/cm^3) is only slightly less than the density of water.

- Metallic hydrogen: In most of the rest of Jupiter, the temperatures and pressures are so extreme that hydrogen is forced into a compact, metallic form (which still flows like a liquid). Just as is the case with everyday metals, electrons are free to move around in metallic hydrogen, so it conducts electricity quite effectively. As we'll see shortly, Jupiter's magnetic field is generated in this layer of metallic hydrogen.

- Core: The core is a mix of hydrogen compounds, rock, and metal. However, the high temperature and extreme pressure ensure that this mix bears little resemblance to familiar solids or liquids. The core contains about 10 times as much mass as the entire Earth, but it is only about the same size as Earth because it is compressed to such high density.

Comparing Jovian Interiors Because all four jovian planets have cores of about the same mass, their interiors differ mainly in the hydrogen/helium layers that surround their cores. Figure 8.4 contrasts the four jovian interiors. Remember that while the outer layers are named for the phase of their hydrogen, they also contain helium and hydrogen compounds.

Saturn is the most similar to Jupiter, just as we should expect given its similar size and composition. Its four interior layers differ from those of Jupiter only because of its lower mass and weaker gravity. The lower mass makes the weight of the overlying layers less on Saturn than on Jupiter, so you must travel deeper into Saturn to find the layer where pressure changes hydrogen from one form to another. That is why Saturn has a thicker layer of gaseous hydrogen and a much thinner and more deeply buried layer of metallic hydrogen.

Pressures within Uranus and Neptune are not high enough to form liquid or metallic hydrogen at all. Each of these two planets has only a thick layer of gaseous hydrogen surrounding its core of hydrogen compounds, rock, and metal. This core material may be liquid, making for very odd "oceans" buried deep inside Uranus and Neptune. The less extreme interior conditions make Uranus's and Neptune's cores less compressed, and allow them to differentiate so that hydrogen compounds reside in a layer above a central layer of rock and metal.

Figure 8.4

These diagrams compare the interior structures of the jovian planets, shown approximately to scale. All four planets have cores of rock, metal, and hydrogen compounds, with masses about 10 times the mass of Earth's core. They differ primarily in the depth of the hydrogen/helium layers that surround their cores. The cores of Uranus and Neptune are differentiated into separate layers of rock/metal and hydrogen compounds.

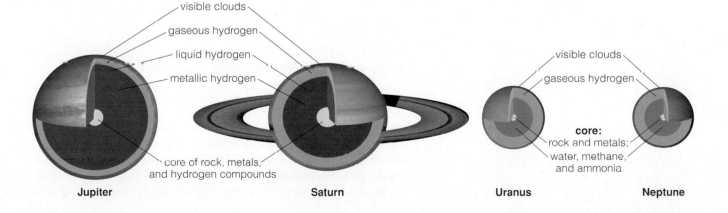

visible clouds

gaseous hydrogen

liquid hydrogen

metallic hydrogen

visible clouds

gaseous hydrogen

core:
rock and metals;
water, methane, and ammonia

core of rock, metals, and hydrogen compounds

Jupiter   **Saturn**   **Uranus**   **Neptune**

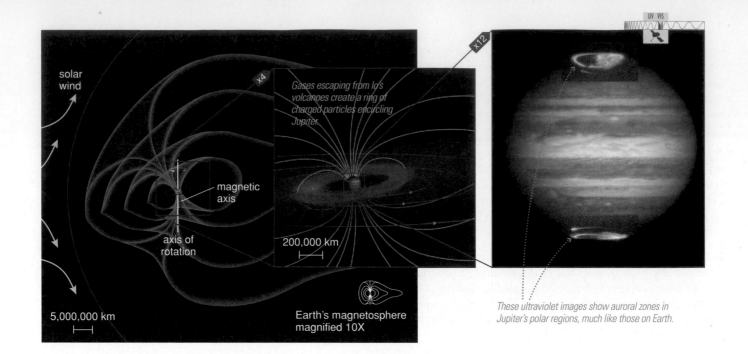

solar
wind

magnetic
axis

axis of
rotation

Gases escaping from Io's volcanoes create a ring of charged particles encircling Jupiter.

200,000 km

5,000,000 km

Earth's magnetosphere
magnified 10X

These ultraviolet images show auroral zones in Jupiter's polar regions, much like those on Earth.

Figure 8.5

Jupiter's strong magnetic field gives it an enormous magneto-sphere. Gases escaping from Io feed the donut-shaped Io torus, and particles entering Jupiter's atmosphere near its magnetic poles contribute to auroras on Jupiter. The image at the right is a composite of ultraviolet images of the polar re-gions overlaid on a visible image of the whole planet, all taken by the Hubble Space Telescope.

Magnetic Fields   Recall that Earth has a global magnetic field generated by the movements of charged particles in our planet's outer core of molten metal (see Figure 7.5). The jovian planets also have global magnetic fields generated by motions of charged particles deep in their interiors.

Jupiter's magnetic field is by far the strongest—some 20,000 times as strong as Earth's. This strong field is generated in Jupiter's thick layer of metallic hydrogen. Just as on Earth, Jupiter's magnetic field creates a *magnetosphere* that surrounds the planet and shields it from the solar wind. But like almost everything else about Jupiter, its magnetosphere is enormous. It begins to deflect the solar wind some 3 million kilometers (about 40 Jupiter radii) before the solar wind even reaches the planet (Figure 8.5). If our eyes could see this part of Jupiter's magnetosphere, it would be larger than the full moon in our sky.

Jupiter's magnetosphere traps far more charged particles than Earth's magnetosphere. These particles contribute to auroras on Jupiter. They also create belts of very intense radiation around Jupiter, which can cause damage to orbiting spacecraft. The main source of the many charged par-ticles is Jupiter's volcanically active moon Io. Gases escaping from Io's vol-canoes become ionized and feed a donut-shaped charged particle belt (called the *Io torus*) that approximately traces Io's orbit.

The other jovian planets also have magnetic fields and magnetos-pheres, but theirs are much weaker than Jupiter's (although still much stronger than Earth's). Saturn's magnetic field is weaker than Jupiter's because it has a thinner layer of electrically conducting metallic hydro-gen. Uranus and Neptune, smaller still, have no metallic hydrogen at all. Their relatively weak magnetic fields must be generated in their core "oceans" of hydrogen compounds, rock, and metal.

• What is the weather like on jovian planets?

Jovian atmospheres have dynamic winds and weather, with colorful clouds and enormous storms readily visible to telescopes and spacecraft. Weather on these planets is driven not only by energy from the Sun (as on the ter-restrial planets), but also by heat generated within the planets themselves. All but Uranus generate a great deal of internal heat. No one knows the precise source of the internal heat on jovian planets, but it probably comes from the conversion of gravitational potential energy to thermal energy inside them. The best guesses are that this conversion comes from a slow but imperceptible contraction in overall size, or from ongoing differentiation as heavier materials continue to sink toward the core.

As we did for the jovian planetary interiors, let's examine different aspects of their atmospheres by starting with Jupiter as the prototype for each feature. We'll then use the general differences between the jovian planets to understand differences in their weather.

Clouds and Colors   The spectacular colors of the jovian planets are probably the first thing that jumps out at you when you look at the photos in Figure 8.1. Many mysteries remain about precisely why the jovian planets are so colorful, but at least some major color features are caused by clouds. Earth's clouds look white from space because they are made of water that reflects the white light of the Sun. The jovian planets have clouds of several different types, and some of these reflect light of other colors.

Clouds form when a gas condenses to make tiny liquid droplets or solid flakes. Water vapor is the only gas that can condense in Earth's atmosphere, which is why clouds on Earth are made of water droplets or flakes that can produce rain or snow. In contrast, Jupiter's atmosphere has several gases that can condense to form clouds. Each of these gases condenses at a different temperature, leading to distinctive cloud layers at different altitudes.

Jupiter has three primary cloud layers, which we can understand by considering temperatures at different altitudes (Figure 8.6). Just as the temperature tends to fall as you climb up a mountain on Earth, Jupiter's atmosphere is colder at higher altitudes. About 100 kilometers below the highest cloudtops, the temperatures are nearly Earth-like and water can condense to form clouds. Higher up, at about 50 kilometers above the water clouds, it is cold enough for a gas called ammonium hydrosulfide (NH_4SH) to condense into clouds. These ammonium hydrosulfide clouds reflect brown and red light (though no one knows why), producing many of the dark colors of Jupiter. Higher still, the temperature is so cold that ammonia (NH_3) condenses to make an upper layer of white clouds.

Saturn has the same set of three cloud layers as Jupiter, but these layers occur deeper in Saturn's atmosphere. The reason is that Saturn's outer atmosphere is colder than Jupiter's, both because Saturn is farther from the Sun and because it has weaker gravity. For example, to find the relatively warm temperatures at which water vapor can condense to form water clouds, we must look about 200 kilometers deeper into Saturn than into Jupiter. The fact that Saturn's clouds lie deeper in its atmosphere than Jupiter's probably explains Saturn's more subdued colors: Less light penetrates to the depths at which Saturn's clouds are found, and the light they reflect is more obscured by the atmosphere above them.

All the jovian planets have cloudy skies, but clouds of different kinds form at different altitudes in each planet's atmosphere.

Uranus and Neptune are so cold that any cloud layers similar to those of Jupiter or Saturn would be buried too deep in their atmospheres for us to see. Instead, the colors of Uranus and Neptune come mainly from methane gas and clouds (Figure 8.7). These two planets have much more methane gas than Jupiter or Saturn, and the cold temperatures allow some of this methane to condense into clouds. Methane gas absorbs red light, allowing only blue light to penetrate to the level at which the methane clouds form. The methane clouds reflect this blue light upward, giving the planets their blue colors.

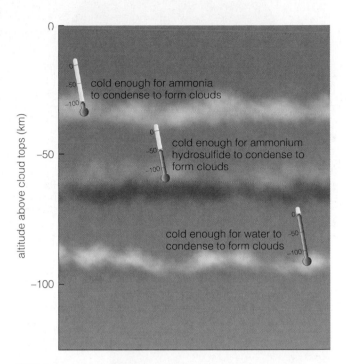

Figure 8.6

This diagram shows the temperature structure of Jupiter's atmosphere. Jupiter has at least three distinct cloud layers because different atmospheric gases condense at different temperatures and hence at different altitudes. The tops of the ammonia clouds are usually considered the zero altitude for Jupiter, which is why lower altitudes are negative.

Labels in figure: altitude above cloud tops (km); 0; −50; −100; cold enough for ammonia to condense to form clouds; cold enough for ammonium hydrosulfide to condense to form clouds; cold enough for water to condense to form clouds

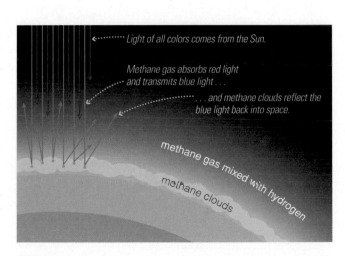

Figure 8.7

Neptune and Uranus look blue because methane gas absorbs red light but transmits blue light. Clouds of methane snowflakes reflect the transmitted blue light back to space.

Labels in figure: Light of all colors comes from the Sun. Methane gas absorbs red light and transmits blue light . . . . . . and methane clouds reflect the blue light back into space. methane gas mixed with hydrogen; methane clouds

a This photograph shows how storms circulate around low pressure regions (L) on Earth. Earth's rotation causes this circulation, which is in opposite directions in the two hemispheres.

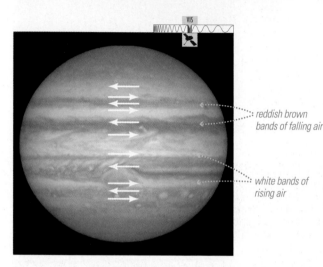

reddish brown
bands of falling air

white bands of
rising air

b Jupiter's faster rotation and larger size essentially stretch out the circulation patterns that occur on Earth into planet-wide bands of fast moving air.

Figure 8.8 interactive figure

Wind patterns on both Earth and Jupiter arise from the way planetary rotation affects rising and falling air.

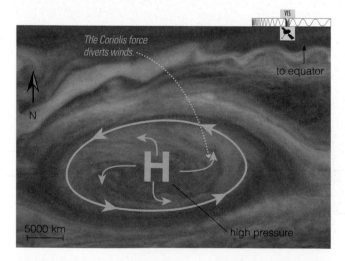

The Coriolis force
diverts winds.

to equator

N

H

high pressure

5000 km

Figure 8.9 interactive figure

This photograph shows Jupiter's Great Red Spot, a huge, high-pressure storm that is large enough to swallow two or three Earths. The overlaid diagram shows a weather map of the region.

Global Winds and Storms

In photographs, Jupiter shows alternating east-west stripes of white and reddish-brown clouds. The stripes represent alternating bands of rising and falling air, and they stretch around the planet because of Jupiter's rapid rotation.

You are probably familiar with circulation of storms on Earth (Figure 8.8a). Storms around low-pressure regions tend to circulate counterclockwise in the Northern Hemisphere and clockwise in the Southern Hemisphere. Earth's rotation causes these circulation patterns by diverting north- or south-flowing air. (More technically, Earth's rotation produces something called the *Coriolis effect*, which diverts the paths of missiles or rockets as well as winds.) The same effect occurs on Jupiter, but Jupiter's faster rotation and larger size make the effect much stronger. In essence, the circular patterns we see in Earth's atmosphere become stretched out to the east and west on Jupiter, to such an extent that they end up going all the way around the planet (Figure 8.8b). This leads to very high east-west wind speeds—sometimes more than 400 kilometers per hour (250 miles per hour).

The colors of the stripes come from the cloud layers. The entire planet is blanketed with the reddish-brown clouds that make up the middle layer of ammonium hydrosulfide clouds (see Figure 8.6). However, we see these clouds only in places where none of the white ammonia clouds lie above them. The white clouds form in bands of rising air, because this air rises to the high, cold altitudes at which white ammonia snowflakes can condense. Ammonia "snow" falls from these clouds, depleting the rising air of ammonia. When the rising air reaches its highest point, it spills north or south and descends in the bands of falling air. Because this air now lacks ammonia, no ammonia clouds can form and we can see down to the reddish-brown ammonium hydrosulfide clouds that lie deeper in the atmosphere. That is why the bands of falling air have their dark color.

The rapid rotation of the jovian planets helps drive strong winds, creating their banded appearances and sometimes giving rise to huge storms.

Just as unusually large hurricanes occasionally arise on Earth, Jupiter also has its share of powerful storms. Of course, because it is Jupiter, its storms dwarf those on Earth. White and brown ovals that frequently appear in Jupiter's atmosphere are low-pressure storms, but many are as large as the entire Earth. Jupiter's most famous feature—its **Great Red Spot**—is also a gigantic storm.

The Great Red Spot is more than twice as wide as Earth. It is somewhat like a hurricane on Earth, except that its winds circulate around a high-pressure region rather than a low-pressure region (Figure 8.9). It is also extremely long-lived compared to storms on Earth: Astronomers have seen it throughout the three centuries during which telescopes have been powerful enough to detect it. No one knows why the Great Red Spot has lasted so long. However, storms on Earth tend to lose their strength when they pass over land. Perhaps Jupiter's biggest storms last for centuries simply because there's no solid surface effect to sap their energy. In 2006, an amateur astronomer from the Philippines was the first person to notice that a storm known for years had changed color and become a new red spot, nicknamed "Red Jr." (Figure 8.10a). Scientists hope to learn more about Jupiter's weather by studying this new storm.

The other jovian planets also have dramatic weather patterns (Figure 8.10b–d). As on Jupiter, Saturn's rapid rotation creates alternating bands of rising and falling air, along with rapid east-west winds. In fact, Saturn's winds are even faster than Jupiter's—a surprise that scientists have yet to explain. Neptune's atmosphere is also banded, and we have seen a high-pressure storm, called the Great Dark Spot, similar to

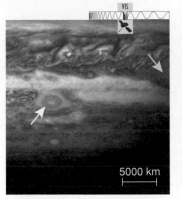

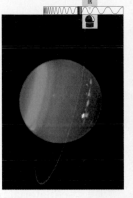

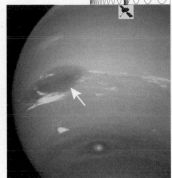

Figure 8.10

Selected views of weather patterns on the four jovian planets.

a This Hubble Space Telescope image shows Jupiter's southern hemisphere with the storm "Red Jr." (white arrow); a piece of the Great Red Spot is visible at the right (blue arrow).

b Saturn's atmosphere, photographed by *Voyager 1*. Its banded appearance is very similar to that of Jupiter, but it has even faster winds.

c This infrared photograph from the Keck telescope shows several storms (the bright blotches) brewing on Uranus. Uranus's rings show up in red.

d Neptune's atmosphere, viewed from *Voyager 2*, shows bands and occasional strong storms. The large storm (white arrow) was called the Great Dark Spot.

Jupiter's Great Red Spot. However, the Great Dark Spot did not last as long; it disappeared from view just 6 years after its discovery.

The greatest surprise in jovian weather comes from Uranus. When *Voyager 2* flew past Uranus in 1986, photographs revealed virtually no clouds and no banded structure like those found on the other jovian planets. Scientists attributed the lack of weather to the lower internal heat of Uranus. However, more recent observations from the Hubble Space Telescope and ground-based adaptive optics telescopes show storms raging in Uranus's atmosphere. The storms may be brewing because of the changing seasons: Thanks to Uranus's extreme axis tilt and 84-year orbit of the Sun, its northern hemisphere is just beginning to see sunlight for the first time in decades.

8.2   A Wealth of Worlds: Satellites of Ice and Rock

The jovian planets are majestic and fascinating, but they are only the beginning of our exploration of jovian planet *systems*. Each of the four jovian systems includes numerous moons and a set of rings. The total mass of all the moons and rings put together is minuscule compared to any one of the jovian planets, but the remarkable diversity of these satellites makes up for their lack of size. In this section, we'll explore a few of the most interesting aspects of the jovian moons.

• What kinds of moons orbit the jovian planets?

We now know of 150 moons orbiting the jovian planets. Jupiter has the most, with more than 60 moons known to date. It's helpful to organize these moons into three groups by size: small moons less than about 300 kilometers in diameter, medium-size moons ranging from about 300 to 1500 kilometers in diameter, and large moons more than 1500 kilometers in diameter. These categories are useful because size relates to geological activity. In general, larger moons are more likely to show evidence of past or present geological activity.

Figure 8.11 shows a montage of all the medium-size and large moons. These moons resemble the terrestrial planets in many ways. Each

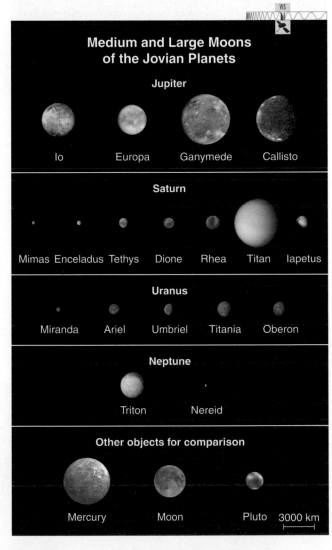

Medium and Large Moons of the Jovian Planets

Jupiter

Io   Europa   Ganymede   Callisto

Saturn

Mimas Enceladus Tethys Dione Rhea Titan Iapetus

Uranus

Miranda   Ariel   Umbriel   Titania   Oberon

Neptune

Triton   Nereid

Other objects for comparison

Mercury   Moon   Pluto   3000 km

Figure 8.11

The medium-size and large moons of the jovian planets, with sizes (but not distances) shown to scale. Mercury, the Moon, and Pluto are included for comparison.

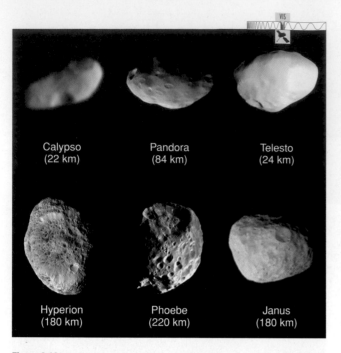

Figure 8.12

These photos from the *Cassini* spacecraft show six of Saturn's smaller moons. All are much smaller than the smallest moons shown in Figure 8.11. Their irregular shapes are due to their small size, which makes their gravities too weak to force them into spheres. The sizes in parentheses represent approximate lengths along their longest axes.

Calypso (22 km) Pandora (84 km) Telesto (24 km)
Hyperion (180 km) Phoebe (220 km) Janus (180 km)

Figure 8.13

This set of photos, taken by the *Galileo* spacecraft orbiting Jupiter, shows global views of the four Galilean moons as we know them today. Sizes are shown to scale. (Io is about the size of Earth's Moon.)

is spherical with a solid surface and its own unique geology. Some possess atmospheres, hot interiors, and even magnetic fields. The two largest—Jupiter's moon Ganymede and Saturn's moon Titan—are larger than the planet Mercury. Four others are larger than the largest known dwarf planets, Pluto and Eris: These are Jupiter's moons Io, Europa, and Callisto and Neptune's moon Triton. However, they differ from terrestrial worlds in their compositions: Because they formed in the cold outer solar system, they contain substantial amounts of ice in addition to metal and rock.

Most of the medium-size and large moons probably formed by accretion within the disks of gas surrounding individual jovian planets [Section 6.4]. That explains why their orbits are almost circular and lie close to the equatorial plane of their parent planet, and also why these moons orbit in the same direction in which their planet rotates. In contrast, many of the numerous small moons are probably captured asteroids or comets, and as a result they do not follow any particular orbital patterns. Dozens of the smallest moons have been discovered only within the past few years (see Appendix E.3 for a current list), and many more may yet be discovered.

> A few jovian moons rival the smallest planets in size and geological interest, while vast numbers of smaller moons are captured asteroids and comets.

The small moons' shapes generally resemble potatoes (Figure 8.12), because their gravities are too weak to force their rigid material into spheres. We have not studied these moons in depth, but we expect their small sizes to allow for little if any geological activity. For the most part, the small moons are just chunks of ice and rock held captive by the gravity of their parent planet.

• Why are Jupiter's Galilean moons geologically active?

We are now ready to embark on a brief tour of the most interesting moons of the jovian planets. Our first stop is Jupiter. Jupiter's four largest moons, known as the *Galilean moons* because Galileo discovered them [Section 3.3], are all large enough that they would count as planets or dwarf planets if they orbited the Sun (Figure 8.13).

Io: The Most Volcanically Active World in the Solar System For anyone who thinks of moons as barren, geologically dead places like our own Moon, Io shatters the stereotype. When the *Voyager* spacecraft first photographed Io up close about three decades ago, we discovered a world with a surface so young that not a single impact crater has survived from

1000 km

Io   Europa   Ganymede   Callisto

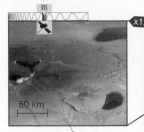

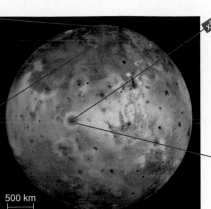

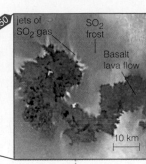

50 km

This close-up shows the glow of intensely hot lava from a volcanic eruption.

500 km

jets of SO$_2$ gas

SO$_2$ frost

Basalt lava flow

10 km

This 80-km-high gas plume was created when hot lava flowed over sulfur dioxide frost, causing it to sublimate explosively into gas.

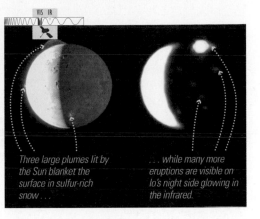

Three large plumes lit by the Sun blanket the surface in sulfur-rich snow . . .

. . . while many more eruptions are visible on Io's night side glowing in the infrared.

a Most of the black, brown, and red spots on Io's surface are recently active volcanic features. White and yellow areas are sulfur dioxide (SO$_2$) and sulfur deposits, respectively, from volcanic gases. (Photographs from the *Galileo* spacecraft; some colors slightly enhanced or altered.)

b Two views of Io's volcanoes taken by *New Horizons* on its way to Pluto.

past impacts. Moreover, *Voyager* cameras recorded volcanic eruptions in progress as the spacecraft passed by. We now know that Io is by far the most volcanically active world in our solar system. Large volcanoes pockmark its entire surface (Figure 8.14), and eruptions are so frequent that they have buried virtually every impact crater. Io probably also has tectonic activity, because tectonics and volcanism generally go hand in hand. However, debris from volcanic eruptions has probably buried most tectonic features.

Io's active volcanoes tell us that it must be quite hot inside. However, Io is only about the size of our geologically dead Moon, so it should have long ago lost any heat from its birth and is too small for radioactivity to provide much ongoing heat. How, then, can Io be so hot inside? The only possible answer is that some other ongoing process must be heating Io's interior. Scientists have identified this process and call it **tidal heating**, because it arises from effects of tidal forces exerted by Jupiter.

Just as Earth exerts a tidal force that causes the Moon to keep the same face toward us at all times [Section 4.4], a tidal force from Jupiter makes Io keep the same face toward Jupiter as it orbits. But Jupiter's mass makes this tidal force far larger than the tidal force that Earth exerts on the Moon. Moreover, Io's orbit is slightly elliptical, so its orbital speed and distance from Jupiter vary. This variation means that the strength and direction of the tidal force change slightly as Io moves through each orbit, which in turn changes the size and orientation of Io's tidal bulges (Figure 8.15a). The result is that Io is continuously being flexed in different directions, which generates friction inside it. The flexing heats the interior in the same way that flexing warms Silly Putty. Tidal heating generates tremendous heat on Io—precise calculations show that it can indeed explain Io's incredible volcanic activity.

see it for yourself Pry apart the overlapping ends of a paper clip, and hold one end in each hand. Flex the ends apart and together until the paper clip breaks. Lightly touch the broken end to your finger or lips—can you feel the the warmth produced by flexing? How is this heating similar to the tidal heating of Io?

Orbital resonances among the Galilean moons make Io's orbit slightly elliptical, leading to the tidal heating that makes Io the most volcanically active place in the solar system.

But we are still left with a deeper question: Why is Io's orbit slightly elliptical, when almost all other large satellites' orbits are virtually circular? The answer lies in an interesting dance executed by Io

Figure 8.14 interactive figure

Io is the most volcanically active body in the solar system.

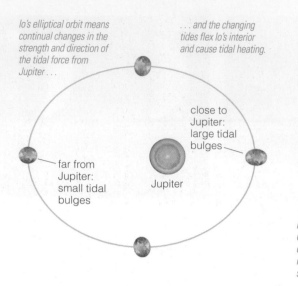

Io's elliptical orbit means continual changes in the strength and direction of the tidal force from Jupiter . . .

. . . and the changing tides flex Io's interior and cause tidal heating.

close to Jupiter: large tidal bulges

far from Jupiter: small tidal bulges

Jupiter

a Tidal heating arises because Io's elliptical orbit (exaggerated in this diagram) causes varying tides.

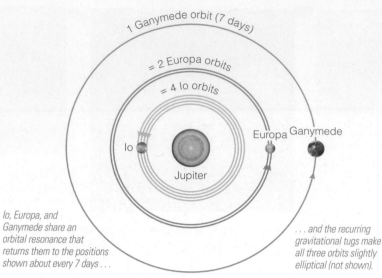

1 Ganymede orbit (7 days)

= 2 Europa orbits

= 4 Io orbits

Io

Europa Ganymede

Jupiter

Io, Europa, and Ganymede share an orbital resonance that returns them to the positions shown about every 7 days . . .

. . . and the recurring gravitational tugs make all three orbits slightly elliptical (not shown).

b Io's orbit is elliptical because of the orbital resonance it shares with Europa and Ganymede.

Figure 8.15 interactive figure ↖

These diagrams explain the cause of tidal heating on Io. Tidal heating has a weaker effect on Europa and Ganymede, because they are farther from Jupiter and tidal forces weaken with distance.

Figure 8.16

Europa's icy crust may hide a deep, liquid water ocean beneath its surface. These photos are from the *Galileo* spacecraft; colors are enhanced in the global view.

and its neighboring moons (Figure 8.15b). During the time Ganymede takes to complete one orbit of Jupiter, Europa completes exactly two orbits and Io completes exactly four orbits. The three moons therefore line up periodically, and the gravitational tugs they exert on one another add up over time. Because the tugs are always in the same direction with each alignment, they tend to stretch out the orbits, making them slightly elliptical. The effect is much like that of pushing a child on a swing. If timed properly, a series of small pushes can add up to a *resonance* that causes the child to swing quite high. For the three moons, the **orbital resonance** that makes their orbits elliptical comes from the small gravitational tugs that repeat at each alignment.

Europa: The Water World? Europa offers a stark contrast to Io. Instead of active volcanoes dotting its surface, Europa is covered by water ice (Figure 8.16). Nevertheless, its fractured, frozen surface must hide an interior made hot by the same type of tidal heating that powers Io's volcanoes—but tidal heating is weaker on Europa because it lies farther from Jupiter. Europa has only a handful of impact craters, which means that ongoing geological activity must have erased the evidence of nearly all past impacts. Scientists suspect that this geological activity is driven either by ice that is soft enough to undergo convection or by liquid water beneath the icy crust.

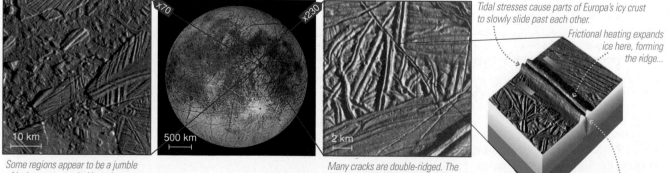

Some regions appear to be a jumble of icebergs suspended in a place where liquid or slushy water froze.

Many cracks are double-ridged. The diagram (right) shows how tidal forces may create them.

Tidal stresses cause parts of Europa's icy crust to slowly slide past each other.

Frictional heating expands ice here, forming the ridge...

...and may melt ice here, collapsing the ridge center.

In fact, we have good reason to suspect that a deep water ocean lies beneath Europa's icy skin. Data collected by the *Galileo* spacecraft suggest that Europa has a metallic core and rocky mantle surrounded by a thick layer of water (H_2O). While the top of this layer is frozen as brittle ice, calculations suggest that tidal heating could supply enough heat to make most of this layer into an ocean of liquid water (Figure 8.17). Several pieces of observational evidence support the existence of a liquid water ocean, including close-up photos of the surface and careful studies of Europa's magnetic field.

Tidal heating may create a deep ocean of liquid water beneath Europa's icy crust.

If it really exists, Europa's liquid ocean may be more than 100 kilometers deep. If so, it contains more than twice as much liquid water as all of Earth's oceans combined. Perhaps as on Earth's seafloor, lava erupts from vents on Europa's seafloor, sometimes violently enough to jumble the icy crust above. And, knowing that primitive life thrives near seafloor vents on Earth, we can wonder whether Europa might also be a home to life—a possibility we will explore further in Chapter 18.

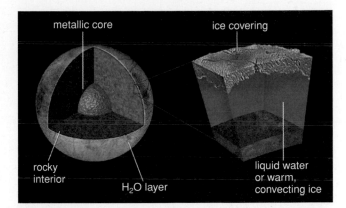

Figure 8.17

This diagram shows one model of Europa's interior structure. There is little doubt that the H_2O layer is real, but questions remain about whether the material beneath the icy crust is actually liquid water or just relatively warm, convecting ice. The latest model, shown at right, suggests both are present.

Ganymede and Callisto Jupiter's two other large moons, Ganymede and Callisto, also show intriguing geology. Like Europa, both have surfaces of water ice.

The surface of Ganymede, the largest moon in the solar system, appears to have a dual personality (Figure 8.18). Some regions are dark and densely cratered, suggesting that they look much the same today as they did billions of years ago. Other regions are light-colored with very few craters, suggesting that liquid water has recently erupted and refrozen. Moreover, magnetic field data indicate that Ganymede, like Europa, could have a subsurface ocean of liquid water. If so, we'd need to explain the source of the heat that melts Ganymede's subsurface ice. Ganymede has some tidal heating, but calculations suggest that it is not strong enough to account for an ocean. Perhaps ongoing radioactive decay supplies enough additional heat to make an ocean. Or perhaps not—no one yet knows what secrets Ganymede hides.

Tidal heating is weak on Ganymede and absent on Callisto, yet both moons show some evidence of subsurface oceans.

The outermost Galilean moon, Callisto, looks most like what scientists originally expected for an outer solar system satellite: a heavily cratered iceball (Figure 8.19). The bright patches on its surface are impact craters. However, the surface still holds some surprises. Close-up images show a dark, powdery substance concentrated in low-lying areas, leaving ridges and crests bright white. The nature of this material and how it got there are unknown. Even more surprising, magnetic field data suggest that Callisto, too, could hide a subsurface ocean. No one knows what might heat the interior of Callisto, since it does not participate in the orbital resonances of the other Galilean moons and therefore has no tidal heating at all. Nevertheless, the potential for an ocean raises the intriguing possibility that there could be three "water worlds" orbiting Jupiter—with far more total ocean than we find here on Earth.

• What geological activity do we see on Titan and other moons?

Aside from the four Galilean moons, the rest of Jupiter's moons fall in the small category, for which we expect no geological activity. However, if

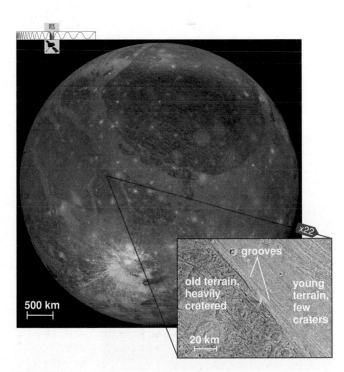

Figure 8.18

Ganymede, the largest moon in the solar system, has both old and young regions on its surface of water ice. The dark regions are heavily cratered and must be billions of years old, while the light regions are younger landscapes where eruptions of water have presumably erased ancient craters; the long grooves in the light regions were probably formed by water erupting along surface cracks. Notice that the boundary between the two types of terrain can be quite sharp.

Callisto is heavily cratered, indicating an old surface that nonetheless may hide a deeply buried ocean.

2 km

Close-up photo shows a dark powder overlaying the low areas of the surface.

Figure 8.19

Callisto, the outermost of the four Galilean moons, has a heavily cratered icy surface.

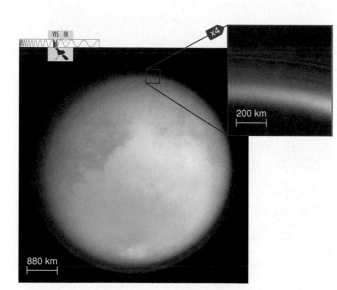

x4

200 km

880 km

Figure 8.20

Titan, as photographed by the *Cassini* spacecraft, is enshrouded by a thick atmosphere of nitrogen and hydrogen compounds. The inset shows the haze layers extending a hundred kilometers above a portion of the moon. *Cassini* was outfitted with filters designed to look at the specific near-infrared wavelengths of light that are least affected by the atmosphere or haze.

you look back at Figure 8.11, you'll see that the remaining jovian planets have 14 more medium- or large-size moons among them: seven for Saturn, five for Uranus, and two for Neptune. Thanks especially to the *Cassini* spacecraft, orbiting Saturn since 2004, we are learning that even some of the smaller of these moons can have surprisingly active geology.

Titan Saturn's moon Titan is the second-largest moon in the solar system (after Ganymede). It is also unique among the moons of our solar system in having a thick atmosphere—so thick that it hides the surface from view, except at a few specific wavelengths of light (Figure 8.20). Titan's reddish color comes from chemicals in its atmosphere much like those that make smog over cities on Earth. The atmosphere is about 90% nitrogen, not that different from the 77% nitrogen content of Earth's atmosphere. However, on Earth the rest of the atmosphere is mostly oxygen, while the rest of Titan's atmosphere consists of argon, methane (CH_4), ethane (C_2H_6), and other hydrogen compounds.

Titan has a thick atmosphere, leading to methane rain and surprising erosional geology.

Titan's atmospheric composition is different from that of any other world in our solar system. Titan's icy composition supplies methane and ammonia gas through evaporation, sublimation, and possibly volcanic eruptions. Solar ultraviolet light breaks down some of those molecules, releasing hydrogen atoms and leaving highly reactive compounds containing carbon and nitrogen. The light hydrogen atoms can completely escape from Titan, while the remaining molecular fragments react to make the other ingredients of Titan's atmosphere. For example, the abundant molecular nitrogen is made after ultraviolet light breaks down ammonia (NH_3) molecules, and ethane is made from methane.

The methane and ethane in Titan's atmosphere are both greenhouse gases, and therefore give Titan an appreciable greenhouse effect [Section 7.1] that makes it warmer than it would be otherwise. Still, because of its great distance from the Sun, its surface temperature is a frigid 93 K (−180°C). The surface pressure on Titan is about 1.5 times the sea level pressure on Earth, which would be fairly comfortable if not for the lack of oxygen and the cold temperatures.

A moon with a thick atmosphere is intriguing enough, but we have at least two other reasons for our special interest in Titan. First, its complex atmospheric chemistry probably produces numerous organic chemicals—the chemicals that are the basis of life. Few scientists think there could be actual life on Titan, due to the cold temperatures, but many hope that further study of Titan will teach us about the chemistry that may have occurred on Earth before life actually arose. Second, although it is far too cold for liquid water to exist on Titan, conditions are right for methane or ethane rain, which may create rivers feeding into lakes or oceans.

NASA and the European Space Agency (ESA) combined forces to explore Titan, with NASA's *Cassini* "mother ship" releasing the ESA-built probe called *Huygens* (pronounced "Hoy-guns") to parachute to the surface in 2005 (Figure 8.21). During its descent, the probe photographed river valleys merging together, flowing down to what looks like a shoreline. On impact, instruments on the probe discovered that the landing was not so hard: The surface has a hard crust but is a bit squishy below, perhaps like sand with liquid mixed in. The view from the surface shows "ice boulders" rounded by erosion. All these results support the idea of a wet climate—but wet with liquid methane rather than liquid water.

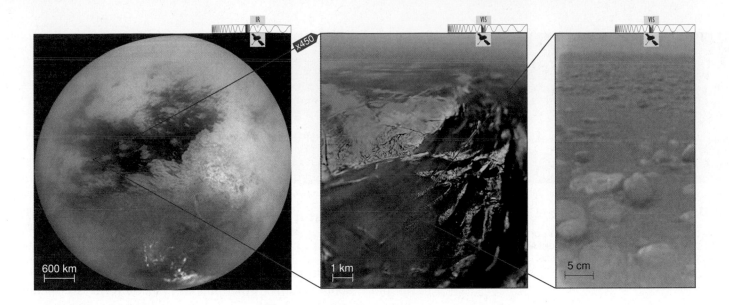

The *Cassini* observations have taught us a great deal about Titan's surface. The brighter regions in the photographs in Figure 8.21 are icy hills that may have been made by ice volcanoes. The dark valleys are probably created when methane rain carries down "smog particles" that concentrate on river bottoms. The vast plains into which the valleys appear to empty are indeed the low regions, but they do not appear to be liquid. They too are probably made dark from smog particles carried there by the rivers. All in all, conditions in Titan's equatorial regions appear to be analogous to the desert southwest of the United States, where infrequent rainfall carves valleys and creates vast dry lakes called "playas" where the water evaporates or soaks into the ground.

The polar regions of Titan, revealed by *Cassini* radar, contain numerous lakes of liquid methane or ethane (Figure 8.22). Almost all the lakes lie at high northern latitudes, where it was winter when they were discovered; the cooler winter temperatures may favor condensation. Images also reveal polar storm clouds (see Figure 8.21) and rivers flowing into the lakes, suggesting that Titan has a methane/ethane cycle resembling the water cycle on Earth.

One of the most astonishing results from the *Huygens* mission is how familiar the landscape looks in this alien environment with unfamiliar materials. Instead of liquid water, Titan has liquid methane and ethane. Instead of rock, Titan has ice. Instead of molten lava, Titan has a slush of water ice mixed with ammonia. Instead of surface dirt, Titan's surface has smog-like particles that rain out of the sky and accumulate on the ground. The similarities between the physical processes that occur on Titan and Earth appear to be far more important in shaping the landscapes than the fact that the two worlds have very different compositions and temperatures.

Huygens data recently revealed another surprise: Titan appears to have an ocean below its icy crust. Scientists reached this conclusion by watching surface features slowly drift from their expected locations from one month to the next. The icy crust is not quite in synchronous rotation [Section 4.4] with the interior, which is possible only if a liquid ocean separates them. Even stranger, the rate of the drift is changing, possibly because of Titan's winds. This hypothesis will be put to the test as the seasonal winds change.

The *Cassini* spacecraft should continue to operate in orbit around Saturn until at least 2010, and scientists hope that additional radar mapping of Titan will teach us more about its surface.

Figure 8.21 interactive figure

This sequence zooms in on the *Huygens* landing site on Titan. Left: a global view taken by the orbiting *Cassini* spacecraft. Center: an aerial view from the descending probe. Right: a surface view taken by the probe after landing; the "rocks," which are 10–20 centimeters across, are presumably made of ice.

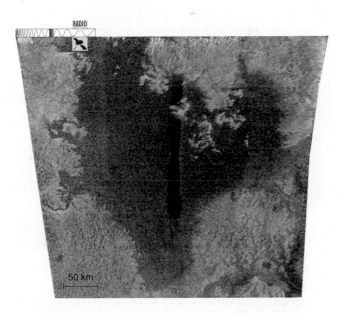

Figure 8.22

Radar image of Titan's north polar region. The blue regions may be lakes of liquid methane or ethane at a temperature of −180°C.

Mimas   Tethys   Dione   Rhea   Iapetus

Figure 8.23

Portraits taken by the *Cassini* spacecraft of five of Saturn's medium-size moons (not to scale). All but Mimas show evidence of past volcanism and/or tectonics.

think about it ➔ What other geological features might you expect on Titan, given its similarities to Earth? How might those features be different, given its differences?

Saturn's Medium-Size Moons The *Cassini* mission is also teaching us more about Saturn's other moons, especially its six medium-size moons (Figures 8.23 and 8.24). The photographs suggest that these moons have had a complex history.

Only Mimas, the smallest of these six moons, shows little evidence of past volcanism or tectonics. It is essentially a heavily cratered iceball, with one huge crater nicknamed "Darth Crater" because of Mimas's resemblance to the Death Star in the *Star Wars* movies. Most of Saturn's other medium-size moons also have heavily cratered surfaces, confirming that they lack global geological activity today. However, we find abundant evidence of volcanism and/or tectonics that must have occurred more recently. For example, the light regions visible in the photos may be places where icy lava once flowed. Iapetus is particularly bizarre, with an astonishing ridge that spans more than a quarter of its circumference, curiously aligned along the equator. No one knows its origin, but some kind of tectonic activity is a likely explanation.

> Enceladus is the smallest moon in the solar system known to be geologically active today.

Enceladus provided an even bigger surprise: This moon is barely 500 kilometers across—small enough to fit inside the borders of Colorado— and yet it shows clear evidence of *ongoing* geological activity (Figure 8.24). Scientists knew that Enceladus undergoes some tidal heating through resonances, but were surprised to learn that the heating is enough to make this moon active today. Its surface has very few impact craters—and some regions have none at all—telling us that recent geological activity has erased older craters. Moreover, the strange grooves near its south pole are measurably warmer than the surrounding terrain, and photographs show this region venting huge clouds of water vapor and ice crystals. These fountains must have some subsurface source, which could potentially mean the existence of an underground reservoir of liquid water. In that case, there would be at least a slim possibility that Enceladus could harbor life [Section 18.2].

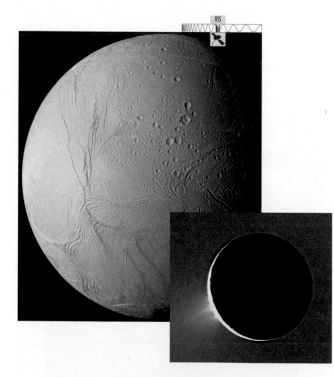

Figure 8.24

Cassini photo of Enceladus. The blue "tiger stripes" near the bottom of the main photo are regions of fresh ice that must have recently emerged from below. The colors are exaggerated; the image is a composite made at near-ultraviolet, visible, and near-infrared wavelengths. The inset shows Enceladus backlit by the Sun, with fountains of ice particles (and water vapor) clearly visible as they spray out to the lower left.

Moons of Uranus and Neptune We know far less about the moons of Uranus and Neptune, because they have been photographed close-up only once each, during the *Voyager 2* flybys of the 1980s. Nevertheless, we again see evidence of surprising geological activity.

Uranus has five medium-size moons (and no large moons), and at least three of them show evidence of past volcanism or tectonics. Miranda, the smallest of the five, is the most surprising (Figure 8.25). Despite its small size, it shows tremendous tectonic features and relatively few craters. Apparently, it underwent geological activity well after the heavy bombardment ended [Section 6.4], erasing its early craters.

The surprises continue with Neptune's moon Triton (Figure 8.26). Triton is a strange moon to begin with: It is a large moon, but it does not follow the orbital patterns of all other large moons in the outer solar system. Instead, it orbits Neptune "backward" (opposite to Neptune's rotation) and at a high inclination to Neptune's equator. These are telltale signs of a moon that was captured rather than having formed in the disk of gas around its planet. No one knows how a moon as large as Triton could have been captured, but it still seems almost certain that Triton once orbited the Sun rather than Neptune. Recent research suggests that Triton might have been one member of a binary Kuiper belt object that passed too close to Neptune. Triton may have been captured while its companion could have been flung off at high speed.

Triton orbits Neptune "backward" and shows evidence of relatively recent geological activity.

Triton's geology is just as surprising as its origin. It is smaller than our own Moon, yet its surface shows evidence of relatively recent geological activity. Some regions show signs of past volcanism, while others show wrinkly ridges (nicknamed "cantaloupe terrain") that appear tectonic in nature. Triton even has a very thin atmosphere that has left some wind streaks on its surface. It's likely that Triton was originally captured into an elliptical orbit, which may have led to enough tidal heating to explain its geological activity.

• Why are jovian moons more geologically active than small rocky planets?

Based on what we learned when studying the geology of the terrestrial worlds, the active geology of the jovian moons seems out of character with their sizes. Numerous jovian moons remained geologically active far longer than Mercury or our Moon, yet they are no bigger and in many cases much smaller in size. However, there is a crucial difference between the jovian moons and the terrestrial worlds: composition.

Most of the jovian moons contain ices that can melt or deform at far lower temperatures than rock. As a result, they can experience geological activity even when their interiors have cooled to temperatures far below what they were at their births. Indeed, except on Io, most of the volcanism that has occurred in the outer solar system probably did not produce

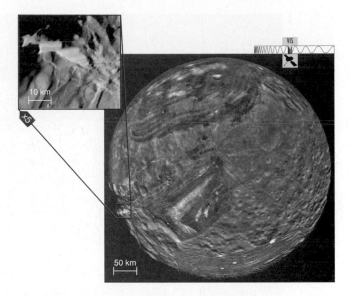

Figure 8.25

The surface of Miranda shows astonishing tectonic activity despite its small size. The cliff walls seen in the inset are higher than those of the Grand Canyon on Earth.

Figure 8.26

Neptune's moon Triton shows evidence of a surprising level of past geological activity.

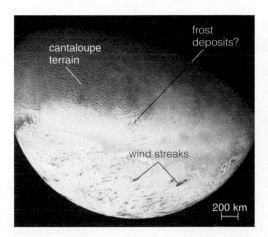

Triton's southern hemisphere as seen by *Voyager 2*.

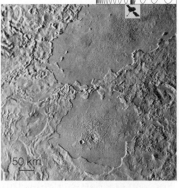

This close-up shows lava-filled impact basins similar to the lunar maria, but the lava was water or slush rather than molten rock.

Terrestrial Planet Geology

- Internal heat, primarily from radioactive decay, can cause volcanic and tectonic activity.
- Only large planets retain enough internal heat to stay geologically active today.
- Example: Mars (photo above) probably retains some internal heat. If it had been smaller, like Mercury, it would be geologically "dead" today. If it had been larger, like Earth, it would probably have much more active and ongoing tectonics and volcanism.

Jovian Moon Geology

- Tidal heating can cause tremendous geological activity on moons with elliptical orbits around massive planets.
- Even without tidal heating, icy materials can melt and deform at lower temperatures than rock, increasing the likelihood of geological activity.
- Together, these effects explain why icy moons are much more likely to have ongoing geological activity than rocky terrestrial worlds of the same size.
- Example: Ganymede (photo above) shows evidence of recent geological activity, even though it is similar in size to the geologically dead terrestrial planet Mercury.

Figure 8.27

Jovian moons can be much more geologically active than terrestrial worlds of similar size due to their icy compositions and a heat source (tidal heating) that is not an important factor on the terrestrial worlds.

any hot lava at all. Instead, it produced icy lava that was essentially liquid water, perhaps mixed with methane and ammonia.

Tidal heating plus the easy melting and deformation of ices mean that even small icy moons can sustain geological activity.

The major lesson, then, is that "ice geology" is possible at far lower temperatures than "rock geology." This fact, combined with the occasional extra heat source of tidal heating, explains how the jovian moons have had such interesting geological histories despite their small sizes. Figure 8.27 summarizes the differences between the geology of jovian moons and that of the terrestrial worlds.

8.3   Jovian Planet Rings

The jovian planet systems have three major components: the planets themselves, the moons, and the rings that encircle the planets. We have already studied the planets and their moons, so we now turn our attention to their amazing rings. We'll begin by exploring the rings of Saturn, since they are by far the most spectacular.

• What are Saturn's rings like?

You can see Saturn's rings through a backyard telescope, but learning their true nature requires higher resolution (Figure 8.28). From Earth, the rings appear to be continuous, concentric sheets of material separated by a large gap (called the *Cassini division*). Spacecraft images reveal these "sheets" to

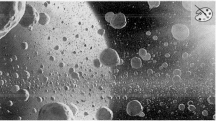

a This Earth-based telescopic view of Saturn makes the rings look like large, concentric sheets. The dark gap within the rings is called the Cassini division.

b This image of Saturn's rings from the *Cassini* spacecraft reveals many individual rings separated by narrow gaps.

c Artist's conception of particles in a ring system. All the particles are moving slowly relative to one another and occasionally collide.

Figure 8.28 interactive figure

Zooming in on Saturn's rings.

be made of many individual rings, each separated from the next by a narrow gap. But even these appearances are somewhat deceiving. If we could wander into Saturn's rings, we'd find that they are made of countless icy particles ranging in size from dust grains to large boulders. All are far too small to be photographed even from spacecraft passing nearby.

Ring Particle Characteristics Spectroscopy reveals that Saturn's ring particles are made of relatively reflective water ice. The rings look bright where they contain enough particles to intercept sunlight and scatter it back toward us. We see gaps in places where there are few particles to reflect sunlight.

> Saturn's rings are made of vast numbers of icy particles ranging in size from dust grains to boulders, each circling Saturn according to Kepler's laws.

Each individual ring particle orbits Saturn independently in accord with Kepler's laws, so the rings are much like myriad tiny moons. The individual ring particles are so close together that they collide frequently. In the densest parts of the rings, each particle collides with another every few hours. However, the collisions are fairly gentle: Despite the high orbital speeds of the ring particles, nearby ring particles are moving at nearly the same speed and touch only gently when they collide.

think about it Which ring particles travel faster: those closer to Saturn or those farther away? Explain why. (*Hint:* Review Kepler's third law.)

The frequent collisions explain why Saturn's rings are one of the thinnest known astronomical structures. They span more than 270,000 kilometers in diameter but are only a few tens of *meters* thick. To understand how collisions keep the rings thin, imagine what would happen to a ring particle on an orbit slightly inclined to the central ring plane. The particle would collide with other particles every time its orbit intersected the ring plane, and its orbital tilt would be reduced with every collision. Before long, these collisions would force the particle to conform to the orbital pattern of the other particles, and any particle that moved away from the narrow ring plane would soon be brought back within it.

Rings and Gaps Close-up photographs show an astonishing number of rings, gaps, ripples, and other features—as many as 100,000 altogether (Figure 8.29). Scientists are still struggling to explain all the features, but some general ideas are now clear.

Rings and gaps are caused by particles bunching up at some orbital distances and being forced out at others. This bunching happens when

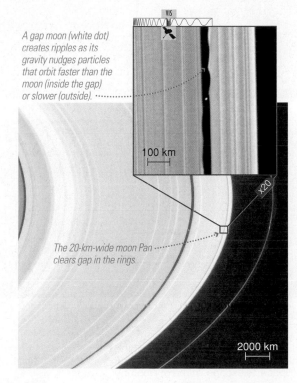

A gap moon (white dot) creates ripples as its gravity nudges particles that orbit faster than the moon (inside the gap) or slower (outside).

100 km

x20

The 20-km-wide moon Pan clears gap in the rings.

2000 km

Figure 8.29

Small moons within the rings have important effects on ring structure (*Cassini* photos).

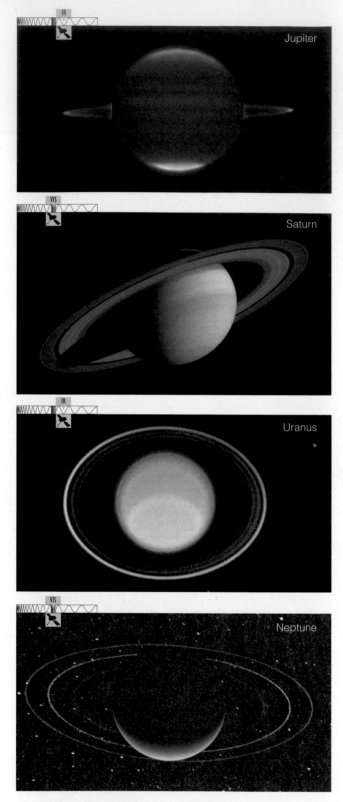

Figure 8.30

Four ring systems (not to scale). The rings differ in appearance and in the composition and sizes of the ring particles. (Jupiter: Keck telescope, infrared; Saturn: *Cassini*, visible; Uranus: Hubble Space Telescope, infrared; Neptune: *Voyager 2*, visible.)

gravity nudges the orbits of ring particles in some particular way. One source of nudging comes from small moons located within the gaps in the rings themselves, sometimes called *gap moons*. The gravity of a gap moon can effectively keep the gap clear of smaller ring particles. In some cases, two nearby gap moons can force particles between them into a very narrow ring. (The gap moons are often called *shepherd moons* in those cases, because they shepherd particles into line.)

Ring particles also may be nudged by the gravity from larger, more distant moons. For example, a ring particle orbiting about 120,000 kilometers from Saturn's center will circle the planet in exactly half the time it takes the moon Mimas to orbit. Every time Mimas returns to a certain location, the ring particle will also be at its original location and therefore will experience the same gravitational nudge from Mimas. The periodic nudges reinforce one another and clear a gap in the rings—in this case, the Cassini division. This type of reinforcement due to repeated gravitational tugs is another example of an *orbital resonance*, much like the orbital resonance that makes Io's orbit elliptical (see Figure 8.15b). Other orbital resonances, caused by moons both within the rings and farther out from Saturn, probably explain most of the intricate structures we see.

• Why do the jovian planets have rings?

Saturn's rings were once thought to be unique in the solar system, leading scientists to assume they were formed by some kind of rare event, such as a moon wandering too close to Saturn and being torn apart by tidal forces. However, we now know that all four jovian planets have rings (Figure 8.30). Although Saturn's rings have more numerous and more reflective particles than the other ring systems, we can no longer think that rings are rare. We therefore need an explanation for rings that doesn't require rare events to have happened for all four planets.

Some scientists once guessed that the ring particles might be leftover chunks of rock and ice that condensed in the disks of gas that orbited each jovian planet when it was young. This would explain why all four jovian planets have rings, because tidal forces near each planet would have prevented these chunks from accreting into a full-fledged moon. However, we now know that the ring particles cannot be leftovers from the birth of the planets, because particles of the size we find in the rings today could not survive for billions of years. Ring particles are continually being ground down in size, primarily by the impacts of the countless sand-size particles that orbit the Sun—the same types of particles that become meteors in Earth's atmosphere and cause micrometeorite impacts on the Moon [Section 7.2]. Millions of years of such tiny impacts would have ground the existing ring particles to dust long ago.

Ring particles cannot last for billions of years, so the rings we see today must be made of particles created recently.

We are left with only one reasonable possibility: New particles must be continually supplied to the rings to replace those that are destroyed. These new particles must come from a source that lies in each planet's equatorial plane. The most likely source is numerous small "moonlets"—moons the size of gap moons (see Figure 8.29)—that formed in the disks of material orbiting the young jovian planets. Like the ring particles themselves, tiny impacts are gradually grinding away these small moons, but they are large enough to still exist despite $4\frac{1}{2}$ billion years of such sandblasting.

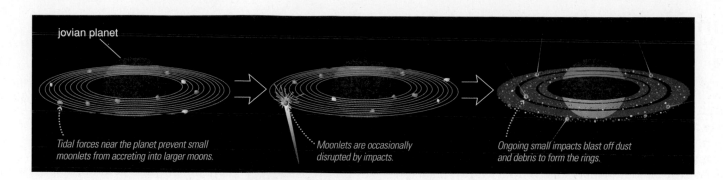

jovian planet

Tidal forces near the planet prevent small moonlets from accreting into larger moons.

Moonlets are occasionally disrupted by impacts.

Ongoing small impacts blast off dust and debris to form the rings.

New ring particles are released by impacts on small moons within the rings.

The small moons contribute ring particles in two ways. First, each tiny impact releases particles from a small moon's surface, and these released particles become new, dust-size ring particles. Ongoing impacts ensure that some ring particles are present at all times. Second, occasional larger impacts can shatter a small moon completely, creating a supply of boulder-size ring particles. The frequent tiny impacts then slowly grind these boulders into smaller ring particles. Some of these particles are "recycled" by forming into small clumps, only to come apart again later on; others are ground down to dust and slowly spiral onto their planet. In summary, all ring particles ultimately come from the gradual dismantling of small moons that formed during the birth of the solar system (Figure 8.31).

Figure 8.31

This illustration summarizes the origin of rings around the jovian planets.

---

the big picture

Putting Chapter 8 into Context

In this chapter, we saw that the jovian planets really are a different kind of planet and, indeed, a different kind of planetary system. The jovian planets dwarf the terrestrial planets. Even some of their moons are as large as terrestrial worlds. As you continue your study of the solar system, keep in mind the following "big picture" ideas:

- The jovian planets may lack solid surfaces on which geology can occur, but they are interesting and dynamic worlds with rapid winds, huge storms, strong magnetic fields, and interiors in which common materials behave in unfamiliar ways.

- Despite their relatively small sizes and frigid temperatures, many jovian moons are geologically active by virtue of their icy compositions—a result of their formation in the outer regions of the solar nebula—and tidal heating.

- Ring systems probably owe their existence to small moons formed in the disks of gas that produced the jovian planets billions of years ago. The rings we see today are composed of particles liberated from those moons quite recently.

- Understanding jovian planet systems forced us to modify many of our earlier ideas about the solar system, in particular by adding the concepts of ice geology, tidal heating, and orbital resonance. Each new set of circumstances we discover offers new opportunities to learn how our universe works.

summary of key concepts

8.1 A Different Kind of Planet

• What are jovian planets made of?

Jupiter and Saturn are made almost entirely of hydrogen and helium, while Uranus and Neptune are made mostly of hydrogen compounds mixed with metal and rock. These differences arose because all four planets started from ice-rich planetesimals of about the same size but captured different amounts of hydrogen and helium gas from the solar nebula.

• What are jovian planets like on the inside?

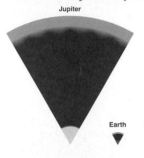

The jovian planets have layered interiors with very high internal temperatures and pressures. All have a core about 10 times as massive as Earth, consisting of hydrogen compounds, metals, and rock. They differ mainly in their surrounding layers of hydrogen and helium, which can take on unusual forms under the extreme internal conditions of the planets.

• What is the weather like on jovian planets?

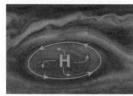

The jovian planets all have multiple cloud layers that help determine the colors of the planets, fast winds, and large storms. Some storms, such as the Great Red Spot, can apparently rage for centuries or longer.

8.2 A Wealth of Worlds: Satellites of Ice and Rock

• What kinds of moons orbit the jovian planets?

We can categorize the sizes of the many known moons as small, medium, or large. Most of the medium and large moons probably formed with their planet in the disks of gas that surrounded the jovian planets when they were young. Smaller moons are often captured asteroids or comets.

• Why are Jupiter's Galilean moons geologically active?

Io is the most volcanically active object in the solar system, thanks to an interior kept hot by **tidal heating**—which occurs because Io's close orbit is made elliptical by **orbital resonance** with other moons. Europa (and possibly

Ganymede) may have a deep, liquid water ocean under its icy crust, also due to tidal heating. Callisto is the least geologically active, since it has no orbital resonance or tidal heating.

• What geological activity do we see on Titan and other moons?

Titan, the only moon in our solar system with a thick atmosphere, shows evidence of active surface geology, including erosion caused by methane rain. Other medium-size moons of Saturn and Uranus show evidence of past geology. Saturn's moon Enceladus is geologically active today, as evidenced by fountains of ice and water vapor that shoot out from its surface. Neptune's large moon Triton is almost certainly a captured object and also shows evidence of recent geological activity.

• Why are jovian moons more geologically active than small rocky planets?

Ices deform and melt at much lower temperatures than rock, allowing icy volcanism and tectonics at surprisingly low temperatures. In addition, some jovian moons have a heat source—tidal heating—that is not important for the terrestrial worlds.

8.3 Jovian Planet Rings

• What are Saturn's rings like?

Saturn's rings are made up of countless individual particles, each orbiting Saturn independently like a tiny moon. The rings lie in Saturn's equatorial plane, and they are extremely thin.

• Why do the jovian planets have rings?

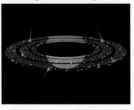

Ring particles probably come from the dismantling of small moons formed in the disks of gas that surrounded the jovian planets billions of years ago. Small ring particles come from countless tiny impacts on the surfaces of these moons, while larger ones come from impacts that shatter the moons.

exercises and problems

Review Questions

Short-Answer Questions Based on the Reading

1. Briefly describe how the differences in composition among the jovian planets can be traced to their formation in the solar nebula.

2. Why is Jupiter so much more dense than Saturn? Could a planet be smaller in size than Jupiter but greater in mass? Explain.

3. Briefly describe the interior structure of Jupiter and why it is layered in this way. How do the interiors of the other jovian planets compare to that of Jupiter?

4. Why does Jupiter have such a strong magnetic field? Describe a few features of Jupiter's magnetosphere.

5. Briefly describe Jupiter's cloud layers. How do the cloud layers help explain Jupiter's colors? Why are Saturn's colors more subdued? Why are Uranus and Neptune blue?

6. Briefly describe Jupiter's weather patterns and contrast them with those on the other jovian planets. What is the *Great Red Spot?*

7. Briefly describe how we categorize jovian moons by size. What is the origin of most of the medium and large moons? What is the origin of many of the small moons?

8. What are the key features of Jupiter's four Galilean moons? Explain the role of tidal heating and orbital resonance in explaining these features.

9. Describe the atmosphere of Titan. What did the *Cassini/Huygens* mission learn about Titan's surface?

10. Briefly describe Triton and explain why we think it is a captured moon.

11. How do we know that Enceladus has active geology? Briefly explain why active geology is seen on icy moons that are much smaller than rocky worlds with no active geology.

12. What are planetary rings made of, and how do they differ among the four jovian planets? Briefly describe the effects of gap moons and orbital resonance on ring systems.

13. Explain why we think that ring particles must be replenished over time. Will the jovian planet rings always look the same?

Test Your Understanding

Surprising Discoveries?

Suppose someone claimed to make the discoveries described below. (These are not real discoveries.) Decide whether each discovery should be considered reasonable or surprising. Explain clearly; not all these have definitive answers, so your explanation is more important than your chosen answer.

14. Saturn's core is pockmarked with impact craters and dotted with volcanoes erupting basaltic lava.

15. Neptune's deep blue color is not due to methane, as previously thought, but instead is due to its surface being covered with an ocean of liquid water.

16. A jovian planet in another star system has a moon as big as Mars.

17. An extrasolar planet is discovered that is made primarily of hydrogen and helium. It has approximately the same mass as Jupiter but is the same size as Neptune.

18. A new small moon orbits Jupiter outside the orbits of other known moons. It is smaller than Jupiter's other moons but has several large, active volcanoes.

19. A new moon orbits Neptune in the planet's equatorial plane and in the same direction that Neptune rotates, but it is made almost entirely of metals such as iron and nickel.

20. An icy, medium-size moon orbits a jovian planet in a star system that is only a few hundred million years old. The moon shows evidence of active tectonics.

21. A jovian planet is discovered in a star system that is much older than our solar system. The planet has no moons at all, but it has a system of rings as spectacular as the rings of Saturn.

22. Radar measurements of Titan indicate that most of the moon is heavily cratered.

23. During a future mission to Uranus, scientists discover it is orbited by another 20 previously unknown moons.

Quick Quiz

Choose the best answer to each of the following. Explain your reasoning with one or more complete sentences.

24. Which lists the jovian planets in order of increasing distance from the Sun? (a) Jupiter, Saturn, Uranus, Pluto (b) Saturn, Jupiter, Uranus, Neptune (c) Jupiter, Saturn, Uranus, Neptune

25. Why does Neptune appear blue and Jupiter red? (a) Neptune is hotter, which gives bluer thermal emission. (b) Methane in Neptune's atmosphere absorbs red light. (c) Neptune's air molecules scatter blue light, much as Earth's atmosphere does.

26. Why is Jupiter denser than Saturn? (a) It has a larger proportion of rock and metal. (b) It is more compressed by the Sun's gravity. (c) Its higher mass and gravity compress its interior.

27. Some jovian planets give off more energy than they receive because (a) fusion is taking place in their cores. (b) tidal heating is occurring. (c) they are still slowly contracting after formation.

28. The main ingredients of most moons of the jovian planets are (a) rock and metal. (b) hydrogen compound ices. (c) hydrogen and helium.

29. Why is Io more volcanically active than our moon? (a) Io is much larger. (b) Io has a higher concentration of radioactive elements. (c) Io has a different internal heat source.

30. What is unusual about Triton? (a) It orbits its planet backward. (b) It does not keep the same face toward its planet. (c) It is the only moon with its own rings.

31. Which moon shows evidence of rainfall and erosion by some liquid substance? (a) Europa (b) Titan (c) Ganymede

32. Saturn's many moons affect the rings through (a) tidal forces. (b) orbital resonances. (c) magnetic field interactions.

33. Saturn's rings (a) have looked basically the same since they formed along with Saturn. (b) were created long ago when tidal forces tore apart a large moon. (c) are continually supplied by impacts with small moons.

Process of Science

Examining How Science Works

34. *Europan Ocean.* Scientists strongly suspect that Europa has a subsurface ocean, even though we cannot see through the surface ice. Briefly explain why scientists think this ocean exists. Is the "belief" in a Europan ocean scientific? Why or why not?

35. *Breaking the Rules.* As discussed in Chapter 7, the geological "rules" for the terrestrial worlds tell us that a world as small as Io should not have any geological activity. However, the *Voyager* images of Io's volcanoes proved that the old "rules" had been wrong. Based on your understanding of the nature of science [Section 3.4], should this be seen as a failure in the process of the science? Defend your opinion.

36. *Unanswered Question.* NASA may turn off the fully functional *Cassini* spacecraft to divert its budget to new missions. Perform an Internet search on *Cassini* to identify one important, still-unanswered question about the Saturn system. Write two or three paragraphs in which you discuss how *Cassini* might answer this question in the future. Be as specific as possible, focusing on the type of evidence necessary to answer the question and the method(s) that could be used to gather it. In your opinion, should NASA keep *Cassini* going or spend its money on other missions?

Investigate Further

In-Depth Questions to Increase Your Understanding

Short-Answer/Essay Questions

37. *The Importance of Rotation.* Suppose the material that formed Jupiter came together without any rotation so that no "jovian nebula" formed and the planet today wasn't spinning. How else would the jovian system be different? Think of as many effects as you can, and explain each in a sentence.

38. *Comparing Jovian Planets.* You can do comparative planetology armed only with telescopes and an understanding of gravity.
    a. The small moon Amalthea orbits Jupiter at about the same distance in kilometers at which Mimas orbits Saturn, yet Mimas takes almost twice as long to orbit. From this observation, what can you conclude about how Jupiter and Saturn differ? Explain.
    b. Jupiter and Saturn are not very different in radius. When you combine this information with your answer to part (a), what can you conclude? Explain.

39. *Minor Ingredients Matter.* Suppose the jovian planets' atmospheres were composed only of hydrogen and helium, with no hydrogen compounds at all. How would the atmospheres be different in terms of clouds, color, and weather? Explain.

40. *Hot Jupiters.* Many of the newly discovered planets orbiting other stars are more massive than Jupiter but orbit much closer to their stars. Assuming that they would be Jupiter-like if they orbited at a greater distance from their stars, how would you expect these new planets to differ from the jovian planets of our solar system? How would you expect their moons to differ? Explain.

41. *The New View of Titan.* What planet or moon in the solar system does Titan most resemble, in your opinion? Summarize the similarities and differences in a few sentences.

42. *Observing Project: Jupiter's Moons.* Using binoculars or a small telescope, view the moons of Jupiter. Make a sketch of what you see, or take a photograph. Repeat your observations several times (nightly, if possible) over a period of a couple of weeks. Can you determine which moon is which? Can you measure the moons' orbital periods? Can you determine their approximate distances from Jupiter? Explain.

43. *Observing Project: Saturn's Rings.* Using binoculars or a small telescope, view the rings of Saturn. Make a sketch of what you see, or take a photograph. What season is it in Saturn's northern hemisphere? How far do the rings extend above Saturn's atmosphere? Can you identify any gaps in the rings? Describe any other features you notice.

Quantitative Problems

Be sure to show all calculations clearly and state your final answers in complete sentences.

44. *Disappearing Moon.* Io loses about a ton (1000 kilograms) of sulfur dioxide per second to Jupiter's magnetosphere.
    a. At this rate, what fraction of its mass would Io lose in $4\frac{1}{2}$ billion years?
    b. Suppose sulfur dioxide currently makes up 1% of Io's mass. When will Io run out of this gas at the current loss rate?

45. *Ring Particle Collisions.* Each ring particle in the densest part of Saturn's rings collides with another about every 5 hours. If a ring particle survived for the age of the solar system, how many collisions would it undergo?

46. *Prometheus and Pandora.* These two moons orbit Saturn at 139,350 and 141,700 kilometers, respectively.
    a. Using Newton's version of Kepler's third law, find their two orbital periods. Find the percent difference in their distances and in their orbital periods.
    b. Consider the two in a race around Saturn: In one Prometheus orbit, how far behind is Pandora (in units of time)? In how many Prometheus orbits will Pandora have fallen behind by one of its own orbital periods? Convert this number of periods back into units of time. This is how often the satellites pass by each other.

47. *Orbital Resonances.* Using the data in Appendix E, identify the orbital resonance relationship between Titan and Hyperion. (*Hint:* If the orbital period of one were 1.5 times that of the other, we would say that they are in a 3:2 resonance.) Which medium-size moon is in a 2:1 resonance with Enceladus?

48. *Titanic Titan.* What is the ratio of Titan's mass to all the other satellites of Saturn whose mass is listed in Appendix E? Calculate the strength of gravity on Titan compared to that on Mimas. Comment on how this affects the possibility of atmospheres on each.

49. *Saturn's Thin Rings.* Saturn's ring system is over 270,000 kilometers wide and approximately 50 meters thick. Assuming the rings could be shrunk down so that their diameter is the width of a dollar bill (6.6 centimeters), how thick would the rings be? Compare your answer to the actual thickness of a dollar bill (0.01 centimeter).

Discussion Questions

50. *Jovian Planet Mission.* We can study terrestrial planets up close by landing on them, but jovian planets have no surfaces to land on. Suppose that you are in charge of planning a long-term mission to "float" in the atmosphere of a jovian planet. Describe the technology you would use and how you would ensure survival for any people assigned to this mission.

51. *Pick a Moon.* Suppose you could choose any one moon to visit in the solar system. Which one would you pick, and why? What dangers would you face in your visit to this moon? What kinds of scientific instruments would you want to bring along for studies?

Web Projects

52. *News from Cassini.* Find the latest news about the *Cassini* mission to Saturn. What is the current mission status? Write a short report about the mission's status and results too recent to be in the textbooks.

53. *Oceans of Europa.* The possibility of a subsurface ocean on Europa holds great scientific interest. Investigate plans for future study of Europa, either from Earth or with spacecraft. Write a short summary of the plans and how they might help us learn whether Europa really has an ocean and, if so, what it might contain.

visual skills check

Use the following questions to check your understanding of some of the many types of visual information used in astronomy. Answers are provided in Appendix K. For additional practice, try the Chapter 8 Visual Quiz at **www.masteringastronomy.com.**

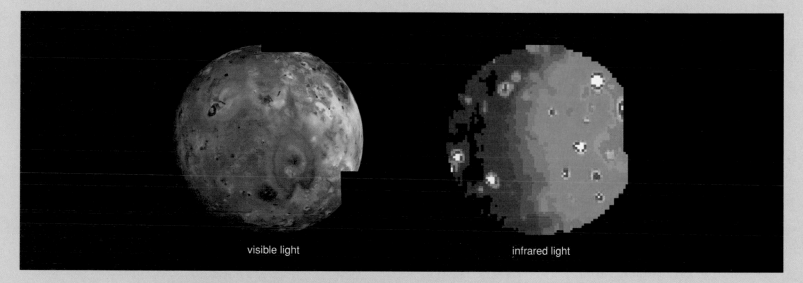

visible light   infrared light

The left image shows the approximate colors of Io in visible light; the black spots are volcanoes that are still active or have recently gone inactive. The right image shows infrared thermal emission from Io; the bright spots are active volcanoes. (Both images are from *Galileo* data, but taken at different times.) Answer the following questions by comparing the two images.

1. What do the colors represent in the right image?
   a. the actual colors of Io's surface
   b. the colors we would see if we had infrared eyes
   c. the intensity of the infrared light
   d. regions of different chemical composition on the surface
2. Which color in the right image represents regions with the highest temperature?
   a. blue
   b. green
   c. orange
   d. red
   e. white
3. The image on the right was obtained when only part of Io was in sunlight. Based on the colors shown over Io's surface as a whole, which part of the image was in sunlight?
   a. the left side
   b. the right side
   c. only the peaks of the volcanoes

4. The bright spots in the infrared image are active volcanoes, and the black spots in the visible image are volcanoes that may or may not be active now. By comparing the two images, what can you conclude about Io's volcanoes?
   a. Every black spot in the visible image has a bright spot in the infrared image, so all of Io's volcanoes were active when the photos were taken.
   b. There are more black spots in the visible image than bright spots in the infrared image, so many of Io's volcanoes were inactive when the photos were taken.
   c. There are more bright spots in the infrared image than black spots in the visible image, so new eruptions must have started after the visible photo was taken.

9

asteroids, comets, and dwarf planets:
their nature, orbits, and impacts

learning goals

Comet McNaught and the Milky Way Galaxy over Patagonia, Argentina (2007). The fuzzy patch above the comet tail is the Small Magellanic Cloud, a satellite galaxy of the Milky Way.

Asteroids and comets might at first seem insignificant compared to the planets and moons we've discussed so far, but there is strength in numbers. The trillions of smaller bodies orbiting our Sun are far more important than their sizes might suggest.

The appearance of comets has more than once altered the course of human history when our ancestors acted on superstitions related to comet sightings. More profoundly, asteroids or comets falling to Earth have scarred our planet with impact craters and have altered the course of biological evolution. Asteroids and comets are also important scientifically: As remnants from the birth of our solar system, they have taught us a great deal about how the solar system formed.

In this chapter, we will explore asteroids, comets, and the rocks that fall to Earth as meteorites. Along the way, we'll see that Pluto, until recently considered the ninth planet, is actually just one of many similar objects orbiting the Sun beyond Neptune—and not even the largest of these objects. Finally, we will explore the dramatic effects of the occasional collisions between small bodies and the planets, including collisions that have altered the course of Earth's history.

essential preparation

1. What does the solar system look like? [Section 6.1]

2. Where did asteroids and comets come from? [Section 6.4]

3. What processes shape Earth's surface? [Section 7.1]

9.1 Asteroids and Meteorites

We begin our study of small bodies by focusing on asteroids and meteorites. Recall that asteroids are rocky leftover planetesimals—chunks of rock that still orbit the Sun because they never managed to become part of a planet [Section 6.4]. Meteorites are pieces of rock that have fallen to the ground from space. Asteroids and meteorites are therefore closely related: Most meteorites are pieces of asteroids that orbited the Sun for billions of years before colliding with Earth.

Asteroids are virtually undetectable to the naked eye and remained unknown for almost two centuries after the invention of the telescope. The first asteroids were discovered only about 200 years ago, and it took 50 years to discover the first 10 asteroids. Today, advanced telescopes can discover far more than that in a single night, and more than 400,000 asteroids have been catalogued. Asteroids can be recognized in telescopic images because they move relative to the stars (Figure 9.1).

Asteroids come in a wide range of sizes. The largest, Ceres, is just under 1000 kilometers in diameter, or a little less than a third the diameter of our Moon. At this size, it is large enough for its own gravity to have made it round, so Ceres qualifies as a *dwarf planet*. (Remember that a dwarf planet is large enough to be nearly round but not large enough to have cleared its orbital neighborhood; see Special Topic, p. 11.) About a dozen other asteroids are large enough that we would call them medium-size moons if they orbited a planet, and a few of these may yet prove to be round enough to qualify as dwarf planets. Scientists hope to learn more about large asteroids with the *Dawn* mission, which launched in 2007. *Dawn* will visit the third-largest asteroid, Vesta, in 2011 and Ceres in 2015.

Smaller asteroids are far more numerous. There are probably more than a million asteroids with diameters greater than 1 kilometer and

Figure 9.1 interactive photo

Because asteroids orbit the Sun, they move through our sky relative to the stars. In this long-exposure image, stars show up as distinct dots, while the motion of an asteroid relative to the stars makes it show up as a short streak.

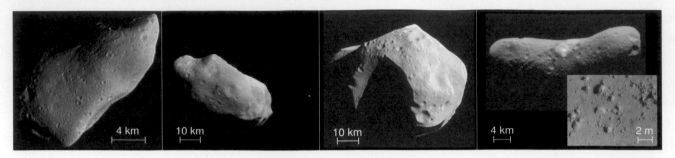

a Gaspra, photographed by the *Galileo* spacecraft.

b Ida, photographed by the *Galileo* spacecraft. The small dot to the right is Dactyl, a tiny moon orbiting Ida.

c Mathilde, photographed by the *Near-Earth Asteroid Rendezvous* (*NEAR*) spacecraft on its way to Eros.

d Eros, photographed by the *NEAR* spacecraft, which orbited Eros for a year before ending its mission with a soft landing on the asteroid's surface. The inset photo was taken by *NEAR* just before it landed.

Figure 9.2

Close-up views of asteroids studied by spacecraft.

many more even smaller in size. A few small asteroids have already been photographed up close by spacecraft (Figure 9.2). They are not spherical, because they are too small for their gravity to have reshaped their rocky material. The images also reveal numerous impact craters, telling us that asteroids, like planets and moons, have been battered by impacts. Indeed, many asteroids have odd shapes because they are fragments of larger asteroids that were shattered in collisions.

Despite their large numbers, asteroids don't add up to much in total mass. If we could put all the asteroids together (including Ceres and the other large asteroids) and allow gravity to compress them into a sphere, they'd make an object less than 2000 kilometers in diameter, just over half the diameter of our Moon.

• Why is there an asteroid belt?

The asteroid belt between Mars and Jupiter gets its name because it is where we find the majority of asteroids (Figure 9.3). But why are asteroids located mainly in this region, rather than being spread throughout the inner solar system?

The answer is that the asteroid belt was the only place where rocky planetesimals could survive for billions of years. During the birth of the solar system, planetesimals formed throughout the inner solar system. However, most of those within Mars's orbit ultimately accreted into one of the four inner planets. The relatively few asteroids that orbit in the inner solar system today are almost certainly "impacts waiting to happen." Some of these asteroids pass near Earth's orbit and may pose a potential threat to our planet—a threat we will discuss in Section 9.4.

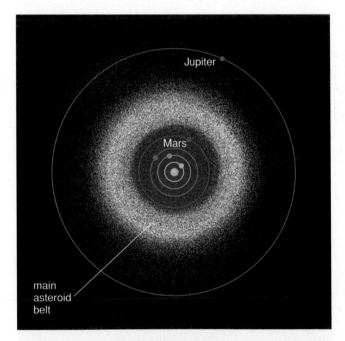

Figure 9.3

Calculated positions of 152,942 asteroids for midnight, January 1, 2004. The asteroids that share Jupiter's orbit, found 60° ahead of and behind Jupiter, are called *Trojan asteroids*. To scale, the asteroids themselves would be much smaller than the dots shown.

Rocky planetesimals survived in the asteroid belt between Mars and Jupiter because they did not accrete into a planet.

In contrast, the asteroids in the asteroid belt stay clear of any planet and can therefore survive in their current orbits for billions of years. But this leaves us with a deeper question: Why didn't a fifth terrestrial planet form in this region beyond the orbit of Mars, sweeping up asteroids as it grew in size?

think about it Recall that the frost line in the solar nebula lay between the present-day orbits of Mars and Jupiter (see Figure 6.17), just outside the region where we find the asteroid belt today. Use this fact to explain why we generally do not find asteroids in the outer solar system.

Resonances with Jupiter A key to understanding the asteroid belt comes from examining asteroid orbits more closely. Figure 9.4 shows the number of asteroids with various orbital periods (listed along the lower axis) that correspond to particular average orbital distances (listed along the top of the graph). Notice that most asteroids tend to share a few particular orbital periods and distances, leaving gaps between them in which there are very few asteroids. (The gaps are often called *Kirkwood gaps*, after their discoverer.)

The gaps in the asteroid belt are not random. They occur at orbital periods that bear special and simple relationships to Jupiter's nearly 12-year orbital period. For example, the arrow labeled $\frac{1}{4}$ in Figure 9.4 points to a gap at an orbital period that is exactly one-quarter the length of Jupiter's orbital period. Similarly, the arrow labeled $\frac{1}{2}$ points to a gap at an orbital period that is exactly half as long as Jupiter's.

The gaps occur at these special places because of *orbital resonances* with Jupiter. We've encountered orbital resonances twice before: first, in explaining the elliptical orbits of Jupiter's moons Io, Europa, and Ganymede [Section 8.2]; second, in explaining the gaps in Saturn's rings [Section 8.3]. Recall that resonances arise whenever objects periodically line up with each other so that gravity affects them over and over again in the same direction. Asteroids with orbital periods that are simple fractions of Jupiter's orbital period are repeatedly nudged by Jupiter's gravity. For example, any asteroid in an orbit that takes 6 years to circle the Sun—half of Jupiter's 12-year orbital period—would receive the same gravitational nudge from Jupiter every 12 years. These repeated nudges would soon push the asteroid out of this 6-year orbit, which is why we do not find any asteroids with orbital periods exactly half that of Jupiter. Other resonances explain the other gaps identified in Figure 9.4.

Jupiter's Role in the Asteroid Belt Jupiter's gravity does much more than just explain the locations of the gaps. It also explains why no planet ever formed in the asteroid belt.

> Jupiter's gravity, through the influence of orbital resonances, stirred up asteroid orbits and thereby prevented their accretion into a planet.

When the solar system was forming, the region now occupied by the asteroid belt probably contained enough rocky material to form another planet as large as Earth or Mars. However, resonances with the young planet Jupiter disrupted the orbits of this region's planetesimals, preventing them from accreting into a full-fledged terrestrial planet. Over the next $4\frac{1}{2}$ billion years, ongoing orbital disruptions gradually kicked pieces of this "unformed planet" out of the asteroid belt altogether. Once booted from the asteroid belt, these objects either crashed into a planet or moon or were flung out of the solar system. The asteroid belt thereby lost most of its original mass, which explains why the total mass of all its asteroids is now less than that of any terrestrial planet.

The asteroid belt is still undergoing slow change today. Jupiter's gravity continues to nudge asteroid orbits, sending asteroids on collision courses with each other and occasionally the planets. A major collision occurs somewhere in the asteroid belt every 100,000 years or so. Over long periods of time, larger asteroids continue to be broken into smaller ones, with each collision also creating numerous dust-size particles. The asteroid belt has been grinding itself down for more than 4 billion years and will continue to do so for as long as the solar system exists.

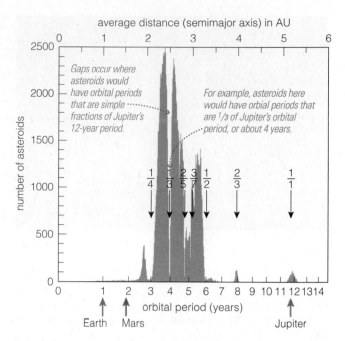

Figure 9.4

This graph shows the number of asteroids with various orbital periods, which correspond to different average distances from the Sun (labeled along the top). Notice the gaps created by orbital resonances with Jupiter. (Some orbital resonances, such as $\frac{1}{1}$ [meaning asteroids with the same orbital period as Jupiter] are stable and therefore have many asteroids rather than few.)

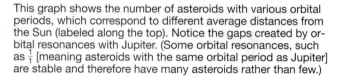

common misconceptions

Dodge Those Asteroids!

Science fiction movies often show brave spacecraft pilots navigating through crowded fields of asteroids, dodging this way and that as they heroically pass through with only a few bumps and bruises. It's great drama, but not very realistic. The asteroid belt looks crowded when we draw it on paper as in Figure 9.3, but in reality it is an enormous region of space. Despite their large numbers, asteroids are thousands to millions of kilometers apart on average—so far apart that it would take incredibly bad luck to crash into one by accident. Indeed, spacecraft must be carefully guided to fly close enough to an asteroid to take a decent photograph. Future space travelers will have plenty of dangers to worry about, but dodging asteroids is not likely to be one of them.

Figure 9.5

This large meteorite, called the *Ahnighito Meteorite*, is located at the American Museum of Natural History in New York. Its dark, pitted surface comes from its fiery passage through Earth's atmosphere. Meteorites enter the atmosphere at speeds of up to 250,000 kilometers per hour (150,000 miles per hour).

• How are meteorites related to asteroids?

Because asteroids are leftovers from the birth of our solar system, studying them should teach us a lot about how Earth and the other planets formed. We can study asteroids with telescopes and spacecraft, but it would be far better if we could study an asteroid sample in a laboratory. No spacecraft has yet returned an asteroid sample (though the damaged Japanese spacecraft *Hayabusa* may return a sample to Earth in a few years). Nonetheless, we already have pieces of asteroids—the rocks called *meteorites* that fall from the sky.

The Difference Between Meteors and Meteorites   Before we discuss what we can learn from meteorites, let's be clear about what they are. In everyday language, people often use the terms *meteors* and *meteorites* interchangeably. Technically, however, a **meteor** is only a flash of light caused by a particle entering our atmosphere at high speed, not the particle itself. The vast majority of the particles that make meteors are no larger than peas and burn up completely before ever reaching the ground.

Only in rare cases are meteors caused by something large enough to survive the plunge through our atmosphere and leave a **meteorite** on the ground. Those cases make unusually bright meteors, called *fireballs*. Observers find a few meteorites each year by following the trajectories of fireballs.

Unless you actually see a meteorite fall, it can be difficult to distinguish meteorites from terrestrial rocks. Fortunately, a few clues can help. Meteorites are usually covered with a dark, pitted crust resulting from their fiery passage through the atmosphere (Figure 9.5). Some have an unusually high metal content, enough to attract a magnet hanging on a string. The ultimate judge of extraterrestrial origin is laboratory analysis. Meteorites often contain elements such as iridium that are rare in Earth rocks, and even common elements in meteorites tend to have different ratios among their isotopes [Section 5.1] than rocks from Earth. Many museums will analyze a small chip of a suspected meteorite free of charge.

Types of Meteorites   The origin of meteorites was long a mystery, but in recent decades we've been able to determine where in our solar system they come from. The most direct evidence comes from the relatively few meteorites whose trajectories have been observed or filmed as they fell to the ground. In every case so far, these meteorites clearly originated in the asteroid belt. Detailed analysis of thousands of meteorites shows that they come in two basic types:

- **Primitive meteorites** (Figure 9.6a) are simple mixtures of rock and metal, sometimes also containing carbon compounds and small amounts of water. Radiometric dating [Section 6.4] shows them to be just under 4.6 billion years old, making them remnants from the birth of our solar system, essentially unchanged since they first accreted in the solar nebula.

- **Processed meteorites** (Figure 9.6b) appear to be pieces of large asteroids that, like the terrestrial worlds, underwent differentiation into a core-mantle-crust structure [Section 7.1]. Some are made mostly of iron, suggesting that they came from the core of a shattered asteroid. Others are rocky and either came from the mantle or crust of a shattered asteroid or were blasted off the surface of a large asteroid by an impact. Radiometric dating shows that processed meteorites are slightly younger than primitive meteorites.

Most meteorites are pieces of asteroids, and they teach us much about the early history of our solar system.

Both types of meteorite teach us important lessons about our solar system. Primitive meteorites represent samples of some of the first material to condense and accrete in the solar nebula. They therefore provide information about the composition of the solidified material from which the planets originally formed. In addition, their ages tell us when the process of accretion first began, which is why we use them to determine the age of the solar system [Section 6.4]. Differences in composition among primitive meteorites reflect where they condensed: Those made only of metal and rock must have condensed in the inner regions of the asteroid belt, while those with carbon compounds must have condensed farther out, where it was cool enough for such compounds and even some water to condense. Indeed, asteroids in the outer regions of the asteroid belt are much darker in color, as expected for their carbon-rich composition.

Processed meteorites represent direct samples of shattered worlds. Those that come from the surfaces of asteroids are often so close in composition to volcanic rocks on Earth that we conclude that some asteroids had active volcanoes when they were young. Processed meteorites that come from the cores or mantles of shattered asteroids tell us what those asteroids were like on the inside. They also represent direct evidence of the fact that large enough worlds really did undergo differentiation (confirming what we infer from seismic studies of Earth [Section 7.1]). Indeed, processed meteorites show that some large asteroids must be geologically quite similar to small terrestrial worlds, one reason why some are now considered to be dwarf planets.

Meteorites from the Moon and Mars

In a few cases, the compositions of processed meteorites do not appear to match any known asteroids. Instead they appear to match either the Moon or Mars.

Careful analysis of these meteorite compositions makes us very confident that we really do have a few meteorites that were once part of the Moon or Mars. Moderately large impacts can blast surface material from terrestrial worlds into interplanetary space. Once they are blasted into space, the rocks orbit the Sun until they come crashing down on another world. Calculations show that it is not surprising that we should have found a few meteorites chipped off the Moon and Mars in this way. Study of these *lunar meteorites* and *Martian meteorites* is providing new insights into conditions on the Moon and Mars. In at least one case, a Martian meteorite may offer clues about whether life ever existed on Mars [Section 18.2].

9.2 Comets

Asteroids are one of the two major categories of small bodies in the solar system. The other is comets, to which we now turn our attention. Asteroids and comets have much in common. Both are leftover planetesimals from the birth of our solar system. Both come in a wide range of sizes. The primary difference between them is in composition, which reflects where they formed. Asteroids are rocky because they formed in the inner solar system where metal and rock condensed. Comets are ice-rich because they formed beyond the frost line, where abundant hydrogen compounds condensed into ice [Section 6.4]. Note that we refer to any icy, leftover planetesimal orbiting the Sun as a comet, regardless of its size, whether it has a tail, or where it resides or comes from.

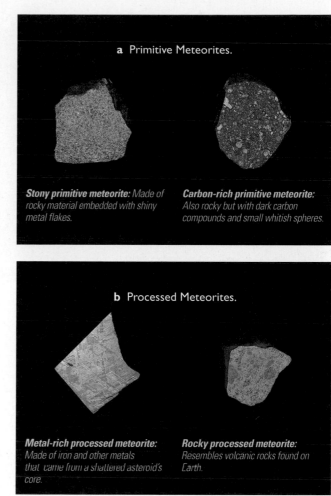

a Primitive Meteorites.

Stony primitive meteorite: *Made of rocky material embedded with shiny metal flakes.*

Carbon-rich primitive meteorite: *Also rocky but with dark carbon compounds and small whitish spheres.*

b Processed Meteorites.

Metal-rich processed meteorite: *Made of iron and other metals that came from a shattered asteroid's core.*

Rocky processed meteorite: *Resembles volcanic rocks found on Earth.*

Figure 9.6

There are two basic types of meteorites: (a) primitive and (b) processed. Each also has two subtypes. They are shown slightly smaller than actual size. (The meteorites have flat faces because they have been sliced with rock saws.)

a Comet Hyakutake.

b Comet Hale–Bopp, photographed at Mono Lake, CA.

Figure 9.7 interactive photo

Brilliant comets can appear at almost any time, as demonstrated by the back-to-back appearances of Comet Hyakutake in 1996 and Comet Hale–Bopp in 1997.

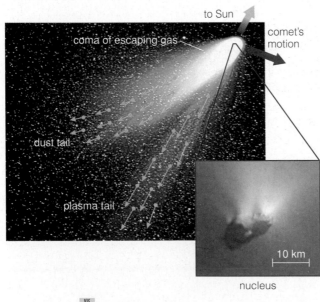

Figure 9.8

Anatomy of a comet. The inset photo is the nucleus of Halley's Comet, photographed by the *Giotto* spacecraft; the coma and tails shown are those of Comet Hale–Bopp in a ground-based photo. A comet grows a coma and tail around its nucleus only if it happens to come close to the Sun. Most comets never do this, instead remaining perpetually frozen in the far outer solar system.

• How do comets get their tails?

For most of human history, comets were familiar only from their occasional presence in the night sky. Every few years, a comet becomes visible to the naked eye, appearing as a fuzzy ball with a long tail (Figure 9.7; see also the photo on page 258). In photographs, the tails make it look as if comets are racing across the sky, but they are not. If you watch a comet for minutes or hours, you'll see it staying nearly stationary relative to the stars around it in the sky. Over many days it will rise and set just like the stars, while gradually moving relative to the constellations. You'll be able to see it night after night for several weeks or more, until it finally fades from view.

Today, we know that the vast majority of comets do not have tails and are never visible in our skies. Most comets never venture anywhere close to Earth, instead remaining perpetually in the far outer reaches of our solar system. There, they slowly orbit the Sun forever unless their orbits are changed by the gravitational influence of a planet, another comet, or stars passing by in the distance.

Most comets remain perpetually frozen in the outer solar system. Only a few enter the inner solar system, where they can grow tails.

The comets that appear with long tails in the night sky are the rare ones that have had their orbits changed, causing them to venture into the inner solar system. Most of these comets will not return to the inner solar system for thousands of years, if ever. A few happen to pass near enough to a planet to have their orbits changed further, and some end up on elliptical orbits that periodically bring them close to the Sun. The most famous example is Halley's Comet, which orbits the Sun every 76 years. It last passed through the inner solar system in 1986, and it will return in 2061.

see it for yourself As you can see from the chapter-opener photo and Figure 9.7, bright comets can be quite photogenic. Go to the Astronomy Picture of the Day Web site, search for comet photos taken from Earth, and examine a few. Which is your favorite? Which has the best combination of beauty and scientifically interesting detail?

Comet Structure and Composition Comets grow tails only as they enter the solar system, where they are heated by the warmth of the Sun.

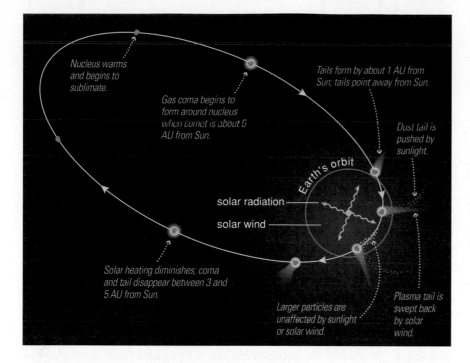

Figure 9.9

This diagram shows the changes that occur when a comet's orbit takes it on a passage into the inner solar system. (Not to scale.)

Within the diagram:

Nucleus warms and begins to sublimate.

Gas coma begins to form around nucleus when comet is about 5 AU from Sun.

Tails form by about 1 AU from Sun; tails point away from Sun.

Dust tail is pushed by sunlight.

Earth's orbit

solar radiation

solar wind

Solar heating diminishes; coma and tail disappear between 3 and 5 AU from Sun.

Larger particles are unaffected by sunlight or solar wind.

Plasma tail is swept back by solar wind.

Figure 9.8 shows the anatomy of a comet when it is in the inner solar system, and Figure 9.9 shows how this anatomy takes shape along the comet's orbital path. Far from the Sun, the comet is completely frozen. If you could see it, the comet would look like a large "dirty snowball"—a chunk of ice mixed with rocky dust and some more complex chemicals. We call this chunk of ice the **nucleus** of the comet. Comet nuclei are typically a few kilometers across.

As a comet accelerates toward the Sun, its surface temperature increases, and ices begin to sublimate into gas that easily escapes the comet's weak gravity. Some of the escaping gas drags away dust particles from the nucleus, and the gas and dust create a huge, dusty atmosphere called a **coma**. The coma is far larger than the nucleus it surrounds. The coma grows as the comet continues into the inner solar system, and some of the gas and dust is pushed away from the Sun, forming the comet's tails.

When a comet nears the Sun, its ices can sublimate into gas and carry off dust, creating a coma and long tails.

Comets have two visible tails, and each can be hundreds of millions of kilometers in length. The **plasma tail** consists of gas escaping from the coma. Ultraviolet light from the Sun ionizes the gas, and the solar wind then carries this gas straight outward from the Sun at speeds of hundreds of kilometers per second. That is why the plasma tail extends almost directly away from the Sun at all times. The **dust tail** is made of dust-size particles escaping from the coma. These particles are not affected by the solar wind and instead are pushed away from the Sun by the much weaker pressure of sunlight itself (*radiation pressure*). The dust tail therefore points generally away from the Sun, but has a slight curve back in the direction the comet came from.

After the comet loops around the Sun and begins to head back outward, sublimation declines, the coma dissipates, and the tails disappear. Nothing happens until the comet again comes sunward—in a century, a millennium, a million years, or perhaps never. Comets that do return eventually use up their ices, some crumbling before our eyes (Figure 9.10) and others becoming inactive nuclei. In 2007, the inactive Comet Holmes

500 km

Figure 9.10

One possible fate for a comet is disintegration, as the ice binding the comet together is used up. This photo shows how a comet fragment (one of dozens of fragments that made up Comet Schwassmann-Wachmann 3) crumbled as it passed Earth in 2006.

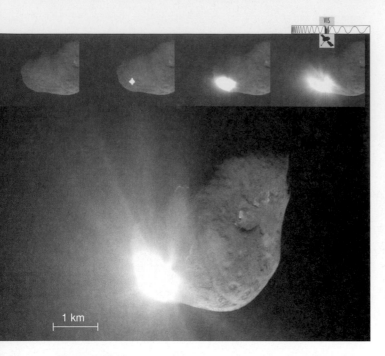

Figure 9.11 interactive photo ↖

These photos were taken by the *Deep Impact* spacecraft as its 370-kilogram impactor crashed into Comet Tempel 1. The entire sequence, which starts just before impact at the upper left, unfolded over just 67 seconds.

came briefly back to life, perhaps because of a comet-quake that occurred when the comet cooled and contracted as it receded from the Sun.

In the past, all our data about comets came from images and spectroscopy of their tails and comas. Recently, however, spacecraft missions to comets have begun to provide us with much more detailed information. We now have our first sample of actual comet material for laboratory study. In 2004, NASA's *Stardust* spacecraft used a material called aerogel to capture dust particles from Comet Wild 2 (pronounced "vilt two"). The aerogel was then sealed in a re-entry capsule, which the main spacecraft ejected on a return trajectory to Earth. Scientists are still puzzling over the composition of some of the comet dust, which indicates that it must have formed in the inner solar system and somehow became mixed in with other cometary materials that formed in the outer solar system.

NASA's *Deep Impact* mission offered an even closer look at a comet. On July 4, 2005, a 370-kilogram impactor slammed into Comet Tempel 1 at 37,000 kilometers per hour as the main spacecraft followed behind to record the event (Figure 9.11). The impactor penetrated many meters into a relatively soft and porous surface, vaporized from the heat of the collision, and ejected a huge plume of dust and gas. In an example of space age recycling, NASA has redirected the *Deep Impact* spacecraft on an orbit to revisit Comet Tempel 1 in 2011 to study the interior exposed by the impact. Scientists are also looking forward to data from the European Space Agency's *Rosetta* mission, on track for a 2014 landing on a comet that is now approaching the Sun.

Comet Tails and Meteor Showers   Comets also eject sand- to pebble-size particles that are too big to be affected by either the solar wind or sunlight. These particles essentially form a third, invisible tail that follows the comet around its orbit. They are also the particles responsible for most meteors and meteor showers.

We see a meteor light up the sky when one of these small particles (or a similar-size particle from an asteroid) burns up in our atmosphere. The sand- to pebble-size particles are much too small to be seen themselves, but they enter the atmosphere at such high speeds that they make the surrounding air glow with heat. It is this glow that we see as the brief but brilliant flash of a meteor. The small particles are vaporized by the heat and never reach the ground. You can typically see a few meteors each hour on any clear night.

A comet ejects small particles that follow it around in its orbit and cause meteor showers when Earth crosses the comet's orbit.

Comet dust is sprinkled throughout the inner solar system, but the "third tails" of ejected particles make the dust much more concentrated along the orbits of comets. This dust rains down on our planet whenever we cross a comet's orbit, producing a **meteor shower**. Meteor showers recur at about the same time each year because the orbiting Earth passes through a particular comet's orbit at the same time each year. For example, the meteor shower known as the *Perseids* occurs on about August 12 each year, when Earth passes through the orbit of a comet called Swift–Tuttle. Table 9.1 lists major annual meteor showers and their parent comet, if known.

If you go outside during one of the annual meteor showers, you may see dozens of meteors per hour. The meteors generally appear to radiate from a particular direction in the sky, for essentially the same reason that snow or heavy rain seems to come from a particular direction in front of a moving car (Figure 9.12). Because more meteors hit Earth from the front than from behind (just as more snow hits the front windshield of a

TABLE 9.1 *Major Annual Meteor Showers*

Shower Name	Approximate Date	Associated Comet
Quadrantids	January 3	?
Lyrids	April 22	Thatcher
Eta Aquarids	May 5	Halley
Delta Aquarids	July 28	?
Perseids	August 12	Swift–Tuttle
Orionids	October 22	Halley
Taurids	November 3	Encke
Leonids	November 17	Tempel–Tuttle
Geminids	December 14	Phaeton
Ursids	December 23	Tuttle

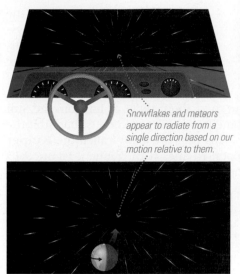

Snowflakes and meteors appear to radiate from a single direction based on our motion relative to them.

a Meteors appear to radiate from a particular point in the sky for the same reason that we see snow or heavy rain come from a single point in front of a moving car.

b This digital composite photo, taken near Uluru (also known as Ayers Rock) in Australia during the 2001 Leonids meteor shower, shows meteors as streaks of light.

moving car), meteor showers are best observed in the predawn sky, when part of the sky faces in the direction of Earth's motion.

Figure 9.12

The geometry of meteor showers.

see it for yourself Try to observe the next meteor shower (see Table 9.1); be prepared with an expendable star chart, a marker, and a dim (preferably red) flashlight. Each time you see a meteor, record its path on your star chart. Record at least a dozen meteors, and try to determine the "radiant" of the shower—that is, the constellation from which the meteors appear to radiate. Does the meteor shower live up to its name?

• Where do comets come from?

Comets that repeatedly visit the inner solar system, like Halley's Comet, cannot last long on the time scale of our solar system. A comet loses a small fraction of its material on every pass around the Sun, so there is nothing left after just a few hundred passages. The comets that humans have seen in the skies cannot be the same ones that passed overhead when dinosaurs or earlier life forms inhabited Earth. So where do comets come from?

We've already stated that comets come from the outer solar system, but we can be much more specific. By analyzing the orbits of comets that pass close to the Sun, scientists learned that there are two major "reservoirs" of comets in the outer solar system.

Most comets that visit the inner solar system follow orbits that seem almost random. They do not orbit the Sun in the same direction as the planets, and their elliptical orbits have random orientations. Moreover, their orbits show that they come from far beyond the orbits of the planets—sometimes nearly a quarter of the distance to the nearest star. These comets must come plunging sunward from a vast, spherical region of space that scientists call the **Oort cloud** (after astronomer Jan Oort; rhymes with "court"). Be sure to note that the Oort cloud is not a cloud of gas but rather a collection of many individual comets. Based on the number of Oort cloud comets that enter the inner solar system each year, we conclude that the Oort cloud must contain about a trillion (10^{12}) comets.

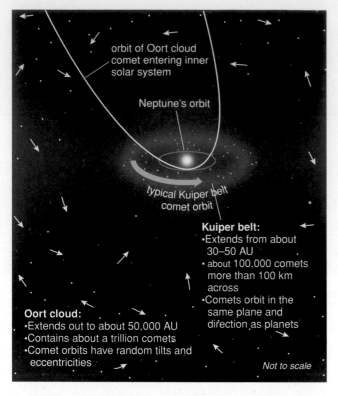

Figure 9.13

The comets we occasionally see in the inner solar system come from two major reservoirs of comets in the outer solar system: the Kuiper belt and the Oort cloud.

The comets we occasionally see in the inner solar system come from two major reservoirs of comets in the outer solar system: the *Kuiper belt* and the *Oort cloud*.

A smaller number of the comets that visit the inner solar system have a pattern to their orbits. They travel around the Sun in the same direction and in nearly the same plane as the planets, and their elliptical orbits carry them no more than about twice as far from the Sun as Neptune. These comets must come from a ring of comets that orbit the Sun beyond the orbit of Neptune. This ring is usually called the **Kuiper belt** (after astronomer Gerald Kuiper; rhymes with "piper"). Figure 9.13 contrasts the general features of the Kuiper belt and the Oort cloud.

The Origin of Comets How did comets end up in these far-flung regions of the solar system? The only answer that makes scientific sense comes from thinking about what happened to the icy, leftover planetesimals that roamed the region in which the jovian planets formed.

The leftover planetesimals that cruised the spaces between Jupiter, Saturn, Uranus, and Neptune were doomed to suffer either a collision or a close gravitational encounter with one of the young jovian planets. The planetesimals that escaped being swallowed up tended to be flung off in all directions. (Recall that when a small object passes near a large planet, the planet is hardly affected but the small object may be flung off at high speed [Section 4.4].) Some may have been cast away at such high speeds that they completely escaped the solar system and now drift through interstellar space. The rest ended up with orbits at very large average distances from the Sun, becoming the comets of the Oort cloud. The random directions in which these comets were flung explain why the Oort cloud is roughly spherical in shape. Oort cloud comets are so far from the Sun that they can be nudged by the gravity of nearby stars, and even by the mass of the galaxy as a whole. These nudges further randomize their orbits, preventing some of them from ever returning to the solar system and sending others plummeting toward the Sun.

Kuiper belt comets orbit in the region in which they formed, just beyond Neptune's orbit. The more distant Oort cloud contains comets that once orbited among the jovian planets.

Beyond the orbit of Neptune, the icy planetesimals were much less likely to be cast off by gravitational encounters. Instead, they remained in orbits going in the same directions as planetary orbits and concentrated relatively near the ecliptic plane. These are the comets of the Kuiper belt. In other words, the comets of the Kuiper belt seem to have originated farther from the Sun than the comets of the Oort cloud, even though the Kuiper belt comets now reside much closer. Kuiper belt comets can be nudged by the gravity of the jovian planets through orbital resonances, sending some on close passes through the inner solar system.

9.3 Pluto: Lone Dog No More

Pluto seemed a misfit among the planets almost since the day it was discovered in 1930. Its 248-year orbit is more elliptical and more inclined to the ecliptic plane than that of the first eight planets (Figure 9.14). In fact, Pluto sometimes comes closer to the Sun than Neptune, although there is no danger of collision: Neptune orbits the Sun precisely three times for every two Pluto orbits, and this stable orbital resonance means that Neptune is always a safe distance away whenever Pluto approaches its orbit. Neptune and Pluto will probably continue their dance of avoidance until the end of the solar system.

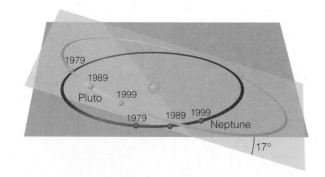

Figure 9.14 interactive figure

Pluto's orbit is significantly elliptical and tilted relative to the ecliptic. It comes closer to the Sun than Neptune for 20 years in each 248-year orbit, as was the case between 1979 and 1999. There's no danger of a collision, however, thanks to an orbital resonance in which Neptune completes three orbits for every two of Pluto's orbits.

Pluto's composition and orbit indicate that it is essentially a large comet of the Kuiper belt.

As we learned more about Pluto, it proved to be a misfit among the planets in size and composition as well: It is far smaller than even the terrestrial planets, and its ice-rich composition fits neither the terrestrial nor jovian categories. These seeming oddities long made Pluto seem like a lone dog of the outer solar system. But it is not: We now know that Pluto is part of a vast pack of similar objects—essentially large comets—that orbit in its region of the solar system. In fact, Pluto is not even the largest member of this pack.

• How big can a comet be?

We've known since the 1950s that many of the comets that visit the inner solar system come from the reservoir that we call the Kuiper belt. We've also known that Pluto orbits the Sun near the middle of this reservoir. Scientists now realize that Pluto's location is no mere coincidence, but rather evidence of an astonishing fact: Some of the objects that populate the Kuiper belt are far larger than any of the comets that we ever see in the inner solar system.

All the comets we've observed in the inner solar system are fairly small, with nuclei no larger than about 20 kilometers across. This shouldn't be surprising based on our understanding of comet origins, because only relatively small objects are likely to have their orbits changed enough to send them from the Kuiper belt or Oort cloud into the inner solar system. But it made scientists wonder: Given that the comets we see are rare and relatively small visitors from distant reaches of the solar system, could these outer realms hold much larger objects as well?

According to our understanding of solar system formation, the icy planetesimals of the Kuiper belt should have been able to continue their accretion as long as other icy particles were nearby. In principle, one of these planetesimals could have grown large enough to become the seed of a fifth jovian planet, beyond Neptune, but that did not occur—probably because the density of material was too low at this great distance from the Sun. Nevertheless, it's reasonable to think that many of these planetesimals grew to hundreds or even thousands of kilometers in diameter.

Discovering Large Iceballs Finding such large, icy objects requires powerful telescope technology. Trying to see a 1000-kilometer iceball at Pluto's distance from Earth is equivalent to looking for a snowball the size of your fist that is 600 kilometers away (and in dim light, too). Aside from Pluto, the first large objects in the Kuiper belt were discovered in the early 1990s. The discoveries have come at a rapid pace ever since.

As astronomers surveyed more and more of the sky in search of large, icy objects, the record size seemed to rise with each passing year. In 2002, the record was set by an object named *Quaoar* (pronounced "kwa-o-whar"), which is more than half the diameter of Pluto. Two objects discovered in 2004—Sedna and Orcus—are between two-thirds and three-quarters of Pluto's size. It seemed only a matter of time until scientists would find an object larger than Pluto, and the time came in July 2005, when Caltech astronomer Michael Brown announced the discovery of the object now named Eris (Figure 9.15). Eris is named for a Greek goddess who caused strife among humans (an allusion to the arguments its discovery caused about the definition of *planet*). Eris even has a moon of its own, named Dysnomia, the mythological daughter of Eris and a goddess of lawlessness.

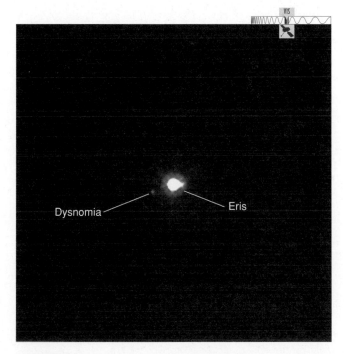

Dysnomia — Eris

Figure 9.15

Eris and its moon, photographed by the Hubble Space Telescope.

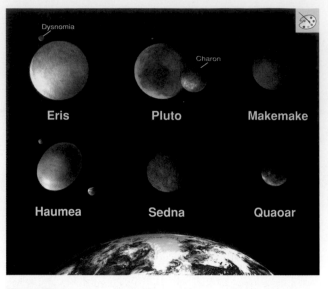

Figure 9.16

Artist paintings of the largest known objects of the Kuiper belt (as of 2008), along with their known moons, shown to scale with Earth for comparison. Pluto and Eris officially qualify as dwarf planets, and others may yet join the list. (The paintings are guesses about appearance, since the Kuiper belt objects have not been photographed with high resolution.)

Eris, discovered in 2005, is about 5% larger than Pluto, making it the largest known member of the Kuiper belt.

Eris currently lies about twice as far as Pluto from the Sun. Its 557-year orbit is inclined to the ecliptic plane more than twice as much as Pluto's (shown in Figure 9.14), and it is so eccentric that it must sometimes come closer to the Sun than Pluto. Hubble Space Telescope observations indicate that Eris is about 5% larger and 27% more massive than Pluto. Pluto has therefore fallen to second place in the rankings of large objects in the Kuiper belt, and it may yet fall further—as of 2008, just over half the sky has been carefully searched for large Kuiper belt objects.

Classifying Pluto and Its Siblings What should we call Pluto, Eris, and the other big iceballs (Figure 9.16)? The terminology has been a topic of much debate, but the science behind it seems clear. By 2008, nearly 1100 icy objects had been directly observed in the Kuiper belt, allowing scientists to infer that the region contains at least 100,000 objects more than 100 kilometers across and many more that are smaller. The smaller ones are easy to refer to as comets, but the larger ones pose a classification challenge, as evidenced by the fact that Pluto was generally considered a planet from the time of its discovery in 1930 until 2006, when the International Astronomical Union (IAU) demoted Pluto to *dwarf planet*.

Under the new classification scheme (see Special Topic, p. 11), Pluto, Eris, Makemake, and Haumea all qualify as *dwarf planets* because they are big enough to be round but have not cleared their orbital neighborhoods. (In the case of Haumea, it is actually oblong, rather than round, only because of its high rotation rate.) Indeed, although the distances between individual objects in the Kuiper belt are quite large on average (tens of millions of kilometers), many of these objects share the same orbital periods and distances. Dozens of other objects of the Kuiper belt may also qualify as dwarf planets, though further observations will be needed to establish whether these objects are really round.

In terms of composition, Pluto, Eris, and all the other large objects of the Kuiper belt are essentially large comets.

Regardless of what we call them, all the objects of the Kuiper belt—from the smallest boulders to the largest dwarf planets—probably share the same basic composition of ice and rock. In other words, they are all essentially comets of different sizes. That is why we often refer to all of them as *comets* of the Kuiper belt. However, some astronomers object to calling objects "comets" if they never venture into the inner solar system and show tails; therefore, you may also hear these objects referred to as *Kuiper Belt Objects* (*KBOs*) or *Trans-Neptunian Objects* (*TNOs*).

• What are Pluto and other large objects of the Kuiper belt like?

Although Pluto, Eris, and other similar objects are large compared to most comets of the Kuiper belt, they are quite small compared to any of the terrestrial or jovian planets. These small sizes, combined with their great distances, make them difficult to study. Even our best photographs show little more than fuzzy blobs (see Figures 9.15 and 9.17). As a result, we generally know little about these objects besides their orbits and the fact that their reflectivities and spectra suggest that they have ice-rich compositions. To understand these objects a little better, let's first consider Pluto and then turn our attention to the rest.

Pluto We've had more time to study Pluto than any other object of the Kuiper belt, but the most important reason we've been able to learn a lot about Pluto is its relatively large moon, Charon. Pluto also has two much smaller moons, discovered in 2005 (Figure 9.17a).

Remember that we can determine an object's mass only if we can observe its gravitational effect on some other object. As a result, we could only estimate Pluto's mass until Charon's discovery in 1978. Charon's orbital characteristics then allowed us to find the combined Pluto–Charon mass using Newton's version of Kepler's third law [Section 4.4].

More precise data came from some good luck in timing: From 1985 to 1990, Pluto and Charon happened to be aligned in a way that made them eclipse each other every few days as seen from Earth—something that won't happen again for more than 100 years. Detailed analysis of brightness variations during these eclipses allowed the calculation of accurate sizes, masses, and densities for both Pluto and Charon. These measurements confirmed that Pluto and Charon are both made of ice mixed with rock—just like comets. The eclipse data even allowed astronomers to construct rough maps of Pluto's surface markings, which have not yet been explained (Figure 9.17b).

Curiously, Pluto is slightly higher in density than Charon. This has led astronomers to guess that Charon might have been created by a *giant impact* similar to the one thought to have formed our Moon [Section 6.4]. A large comet may have crashed into Pluto and blasted away its low-density outer layers. The debris could have then formed a ring of debris around Pluto, with the material eventually accreting to make its three moons. The impact may also explain why Pluto's rotation axis is tipped almost on its side.

Pluto is very cold, with an average temperature of only 40 K, as we would expect at its great distance from the Sun. Nevertheless, Pluto currently has a thin atmosphere of nitrogen and other gases formed by sublimation of surface ices. The atmosphere should gradually refreeze onto the surface as Pluto's 248-year orbit carries it farther from the Sun, although recent results have puzzled astronomers by showing the opposite effect.

Despite the cold, the view from Pluto would be stunning. Charon would dominate the sky, appearing almost 10 times larger in angular size than our Moon appears from Earth. Pluto and Charon's mutual tidal pulls long ago made them rotate synchronously with each other [Section 4.4], so that Charon is visible from only one side of Pluto and always shows the same face to Pluto. Moreover, the synchronous rotation means that Pluto's "day" is the same length as Charon's "month" (orbital period) of 6.4 Earth days. Viewed from Pluto's surface, Charon neither rises nor sets but hangs motionless as it cycles through its phases every 6.4 days. The Sun would appear more than a thousand times fainter than it appears here on Earth and would be no larger in angular size than Jupiter appears in our skies.

see it for yourself The data show that Charon is both relatively large compared to Pluto and relatively close to it. But this doesn't do justice to the visual impact Charon would have in Pluto's sky. Let your head stand for Pluto, and find a round object about 5 centimeters across to represent Charon. Place it 2 meters away. How does this compare to the angular size of our Moon, which is about as large as your pinky fingernail held at arm's length?

Other Kuiper Belt Objects We don't know much about other large comets of the Kuiper belt, but because they all probably formed in the same region of the solar system, we expect them to be similar in nature and composition to Pluto and Charon. Evidence of similarities comes from careful study of orbits. Like Pluto, many Kuiper belt comets have stable orbital resonances with Neptune. In fact, hundreds of Kuiper belt

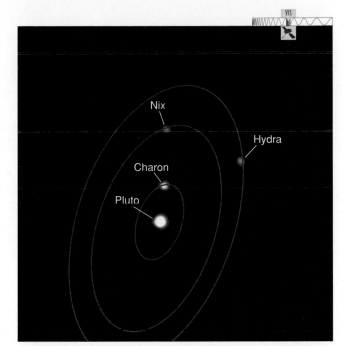

a Pluto and its moons, photographed by the Hubble Space Telescope with orbital paths drawn in.

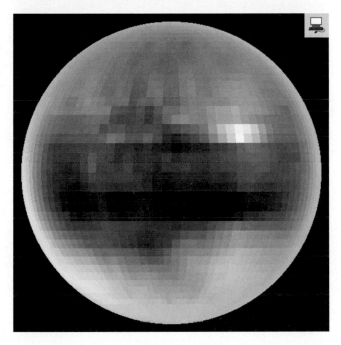

b The surface of Pluto in approximately true color, as derived through computer processing from brightness measurements made during mutual eclipses of Pluto and Charon. The many squares in the image represent the smallest regions for which color could be estimated accurately.

Figure 9.17

Pluto's great distance and small size make it difficult to study.

comets have the same orbital period and average distance from the Sun as Pluto itself. Besides Pluto and Eris, several other Kuiper belt comets are also known to have moons, and in those cases we can use Newton's version of Kepler's third law [Section 4.4] to calculate their masses and then infer their densities. The results confirm that these objects have comet-like compositions of ice and rock.

Pluto and other larger Kuiper belt objects are smaller, icier, and more distant than any of the planets. They can have moons, atmospheres, and possibly geological activity.

We should learn much more about the objects of the Kuiper belt beginning in 2015, when the *New Horizons* spacecraft will fly by Pluto and hopefully continue on to fly past one or two other Kuiper belt comets. Meanwhile, if you think back to what we learned about jovian moons in Chapter 8, you'll realize that we probably already have close-up photos of at least one *former* member of the Kuiper belt: Neptune's moon Triton (see Figure 8.26). Like Triton, it's possible that Pluto and other large Kuiper belt comets are geologically active.

think about it What is your opinion of the new definitions that make Pluto and Eris *dwarf planets* rather than "real" planets? How do you think an object like Triton (which presumably qualified as a dwarf planet until it was captured by Neptune) should be classified? What about moons such as Titan or Ganymede, which are larger than the planet Mercury? Consider several possible ways of classifying solar system objects, and discuss the pros and cons of each one.

9.4 Cosmic Collisions: Small Bodies versus the Planets

The hordes of small bodies orbiting the Sun are slowly shrinking in number through collisions with the planets and ejection from the solar system. Many more roamed the solar system in the days of the heavy bombardment, when most impact craters formed [Section 6.4]. Plenty of small bodies still remain, however, and cosmic collisions still occur on occasion. These collisions have had important effects on Earth and other planets.

• Have we ever witnessed a major impact?

Modern scientists have never witnessed a major impact on a solid world, but in 1994 we were privileged to witness one on Jupiter. The dramatic event involved a comet named *Shoemaker–Levy 9,* or *SL9* for short.

Rather than having a single nucleus, SL9 consisted of a string of comet nuclei lined up in a row (Figure 9.18a). Apparently, tidal forces from Jupiter ripped apart a single comet nucleus during a previous close pass near the planet. Crater chains on Jupiter's moons are evidence that similar breakups of comets near Jupiter have occurred in the past.

Comet SL9 caused a string of violent impacts on Jupiter in 1994, reminding us that catastrophic collisions still happen.

Comet SL9 was discovered more than a year before its impact on Jupiter, and orbital calculations told astronomers precisely when the impact would occur. When the impacts began, they were observed with nearly every major telescope on Earth and from well-positioned spacecraft. Each of the individual nuclei in comet SL9 crashed into Jupiter with an energy equivalent to that of a million hydrogen bombs (Figure 9.18b–d). Comet nuclei barely a kilometer across left scars larger than the entire Earth. The scars lasted for months before Jupiter's strong winds finally made them fade from view.

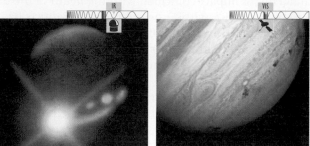

a Jupiter's tidal forces ripped apart a single comet nucleus into a chain of smaller nuclei.

b This infrared photo shows the brilliant glow of a rising fireball from the impact of one nucleus. Jupiter is the round disk, with the impact occurring near the lower left.

c Each of the black dots in this Hubble Space Telescope photo is a scar from the impact of one of the SL9 nuclei.

d This painting shows how the impacts might have looked from the surface of Io. The impacts occurred on Jupiter's night side.

Figure 9.18

The impacts of Comet Shoemaker–Levy 9 on Jupiter allowed astronomers their first direct view of a cosmic collision.

The SL9 impacts allowed us to study both the impact process and material splashed out from deep inside Jupiter. They also provided two important sociological lessons. First, "Comet Crash Week" proved to be one of the best examples of international collaboration in history. With the aid of the Internet, scientists quickly and effectively shared data from observatories around the world. Second, extensive media coverage helped the event capture the public imagination, leading to awareness that impacts are not merely relegated to ancient geological history. If such violent impacts can happen on other planets in our lifetime, could they also happen on Earth?

• Did an impact kill the dinosaurs?

There's no doubt that major impacts have occurred on Earth in the past: Geologists have identified more than 150 impact craters on our planet. So before we consider whether an impact might occur in our lifetimes, it's worth examining the potential consequences if it did. Clearly, an impact could cause widespread physical damage. But a growing body of evidence, accumulated over the past three decades, suggests that an impact can do much more—in some cases, large impacts may have altered the entire course of evolution.

In 1978, while analyzing geological samples collected in Italy, a scientific team led by father-and-son team Luis and Walter Alvarez made a startling discovery. They found that a thin layer of dark sediments deposited about 65 million years ago—about the time the dinosaurs went extinct—was unusually rich in the element iridium. Iridium is a metal that is rare on Earth's surface (because it sank to Earth's core with other metals during differentiation) but common in meteorites. Subsequent studies found the same iridium-rich layer in 65-million-year-old sediments around the world (Figure 9.19). The Alvarez team suggested a stunning hypothesis: The extinction of the dinosaurs was caused by the impact of an asteroid or comet.

A layer rich in iridium and soot tells us a huge impact occurred at this point in geological (and biological) history.

Figure 9.19

Around the world, sedimentary rock layers dating to 65 million years ago share the evidence of the impact of a comet or asteroid. Fossils of dinosaurs and many other species appear only in rocks below the iridium-rich layer.

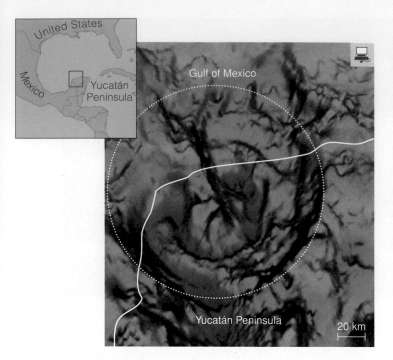

Figure 9.20

This computer-generated image, based on measurements of small local variations in the strength of gravity, shows an impact crater about 200 kilometers across (dashed circle). The crater straddles the coast of Mexico's Yucatán Peninsula (see inset).

Figure 9.21

This painting shows an asteroid or comet moments before its impact on Earth, some 65 million years ago. The impact probably caused the extinction of the dinosaurs, and if it hadn't occurred, the dinosaurs might still rule Earth today.

In fact, the death of the dinosaurs was only a small part of the biological devastation that seems to have occurred 65 million years ago. The fossil record suggests that up to 99% of all living organisms died around that time and that up to 75% of all existing *species* were driven to extinction. This event was clearly a **mass extinction**—the rapid extinction of a large fraction of all living species. Could it really have been caused by an impact?

There's still some scientific controversy about whether the impact was the sole cause of the mass extinction or just one of many causes, but there's little doubt that a major impact coincided with the death of the dinosaurs. In addition to the evidence within the sediments, scientists have identified a 65-million-year-old crater (Figure 9.20), apparently created by the impact of an asteroid or a comet measuring about 10 kilometers across. Astronomers have also found evidence that the breakup of a 170-kilometer asteroid 160 million years ago may have created a fragment that struck Earth a hundred million years later.

An iridium-rich sediment layer and an impact crater on the Mexican coast show that a large impact occurred at the time the dinosaurs died out, 65 million years ago.

If an impact was indeed the cause of the mass extinction, here's how it probably happened. On that fateful day some 65 million years ago, the asteroid or comet slammed into Mexico with the force of a hundred million hydrogen bombs (Figure 9.21). North America may have been devastated immediately. Hot debris from the impact rained around the rest of the world, igniting fires that killed many more living organisms.

The longer-term effects were even more severe. Dust and smoke remained in the atmosphere for weeks to months, blocking sunlight and causing temperatures to fall as if Earth were experiencing a harsh global winter. The reduced sunlight would have stopped photosynthesis for up to a year, killing large numbers of species throughout the food chain. Acid rain may have been another by-product, killing vegetation and acidifying lakes around the world. Chemical reactions in the atmosphere

probably produced nitrous oxides and other compounds that dissolved in the oceans and killed marine organisms.

Perhaps the most astonishing fact is not that 75% of all species died but that 25% *survived*. Among the survivors were a few small mammals. These mammals may have survived in part because they lived in underground burrows and managed to store enough food to outlast the global winter that immediately followed the impact.

The evolutionary impact of the extinctions was profound. For 180 million years, dinosaurs had diversified into a wide variety of species, while most mammals (which had arisen at almost the same time as the dinosaurs) had generally remained small and rodent-like. With the dinosaurs gone, mammals became the new kings of the planet. Over the next 65 million years, small mammals rapidly evolved into an assortment of much larger mammals—ultimately including us.

• Is the impact threat a real danger or just media hype?

The dinosaur extinction is only one of several known mass extinctions in Earth's past, and some of the others also seem to coincide with past impacts. Given that impacts have altered the course of life on Earth in the past, could a future impact endanger our own survival? Hollywood seems to take the threat seriously and has made several feature movies about it. Should we take it seriously, too?

Space is filled with plenty of objects that could hit our planet. As of 2008, astronomers had identified more than 5000 near-Earth asteroids, of which nearly 1000 were classified as potentially hazardous. The threat from comets is lower, but their high speeds would give us less time to prepare for an impact.

The threat of a major impact is undoubtedly real, but the chance of a large impact in our lifetime is quite small. Geological data show that impacts large enough to cause mass extinctions happen many tens of millions of years apart on average. We're far more likely to do ourselves in than to be done in by a large asteroid or comet.

Smaller impacts can be expected more frequently. While such impacts would not wipe out our civilization, they could kill thousands or millions of people. We know of one close call in modern times. In 1908, a tremendous explosion occurred over Tunguska, Siberia (Figure 9.22). The explosion, estimated to have released energy equivalent to that of several atomic bombs, is thought to have been caused by a small asteroid less than 40 meters across. If the asteroid had exploded over a major city instead of Siberia, it would have been the worst natural disaster in human history. A less devastating impact occurred in 2007, when eye-witnesses saw a bright fireball streak across the mid-day sky near Carancas, Peru. A stony meteorite about 1–2 meters across slammed into the ground and excavated a crater 15 meters across (Figure 9.23). The impact spewed debris more than 300 meters and shattered windows as far away as a kilometer.

Impacts will certainly occur in the future, and while the chance of a major impact in our lifetimes is small, the effects could be devastating.

Figure 9.24 shows how often, on average, we expect Earth to be hit by objects of different sizes. Notice that objects large enough to cause mass extinctions come tens of millions of years apart. Objects of the size that caused the Tunguska event probably strike our planet every few hundred years or so. That's a level of threat that we cannot discount. Nevertheless, these rare events are

Figure 9.22

This photo shows forests burned and flattened by the 1908 impact over Tunguska, Siberia. Atmospheric friction caused the small asteroid to explode completely before it hit the ground, so it left no impact crater.

Figure 9.23

The impact crater made by a 1–2 meter stony meteorite near Carancas, Peru, in 2007. The crater filled with groundwater soon after the impact.

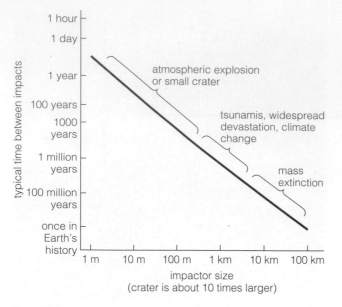

Figure 9.24

This graph shows that larger objects (asteroids or comets) hit the Earth less frequently than smaller ones. The labels describe the effects of impacts of different sizes.

unlikely to strike populated areas. After all, most of Earth's surface is ocean, and even on land humans are concentrated in relatively small urban areas.

If we were to find an asteroid or a comet on a collision course with Earth, could we do anything about it? Many people have proposed schemes to save Earth by using nuclear weapons or other means to demolish or divert an incoming asteroid, but no one knows whether current technology is up to the task. We can only hope that the threat doesn't become a reality before we're ready.

think about it Study Figure 9.24. Based on the frequency of impacts large enough to cause serious damage, do you think we should be spending time and resources to counter the impact threat? Or should we focus resources on other threats first? Defend your opinion.

• How do other planets affect impact rates and life on Earth?

Ancient people imagined that the mere movement of planets relative to the visible stars in our sky could somehow have an astrological influence on our lives. Although scientists no longer give credence to this ancient superstition, we now know that planets can have a real effect on life on Earth. By catapulting asteroids and comets in Earth's direction, other planets have caused cosmic collisions that have helped shape Earth's destiny.

Jupiter and the other jovian planets have had the greatest effects because of their influence on the small bodies of the solar system. As we saw earlier in this chapter, Jupiter disturbed the orbits of rocky planetesimals outside Mars's orbit, preventing a planet from forming and instead creating the asteroid belt. The jovian planets ejected icy planetesimals to create the distant Oort cloud of comets, and orbital resonances with Neptune shape the orbits of many comets in the Kuiper belt. Ultimately, every asteroid or comet that has impacted Earth since the end of the heavy bombardment was in some sense sent our way by the influence of Jupiter or one of the other jovian planets. Figure 9.25 summarizes the ways the jovian planets have controlled the motions of asteroids and comets.

Nearly every asteroid or comet that has ever struck Earth was in some sense sent our way by the influence of the jovian planets.

We thereby find a deep connection between the jovian planets and the survival of life on Earth. If Jupiter did not exist, the threat from asteroids might be much smaller, since the objects that make up the asteroid belt might instead have become part of a planet. On the other hand, the threat from comets might be much greater: Jupiter probably ejected more comets to the Oort cloud than any other jovian planet, and without Jupiter, those comets might have remained dangerously close to Earth. Of course, even if Jupiter has protected us from impacts, it's not clear whether that has been good or bad for life overall. The dinosaurs appear to have suffered from an impact, but the same impact may have paved the way for our existence. So while some scientists argue that more impacts would have damaged life on Earth, others argue that more impacts might have sped up evolution.

The role of Jupiter has led some scientists to wonder whether we could exist if our solar system had been laid out differently. Could it be that civilizations can arise only in solar systems that happen to have a Jupiter-like planet in a Jupiter-like orbit? No one yet knows the answer to this question. What we do know is that Jupiter has had profound effects on life on Earth and will continue to have effects in the future.

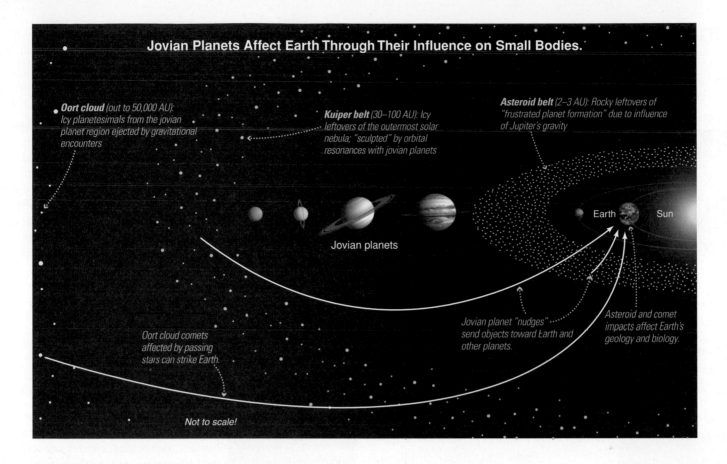

Jovian Planets Affect Earth Through Their Influence on Small Bodies.

Oort cloud (out to 50,000 AU): Icy planetesimals from the jovian planet region ejected by gravitational encounters

Kuiper belt (30–100 AU): Icy leftovers of the outermost solar nebula; "sculpted" by orbital resonances with jovian planets

Asteroid belt (2–3 AU): Rocky leftovers of "frustrated planet formation" due to influence of Jupiter's gravity

Jovian planets

Earth   Sun

Oort cloud comets affected by passing stars can strike Earth.

Jovian planet "nudges" send objects toward Earth and other planets.

Asteroid and comet impacts affect Earth's geology and biology.

Not to scale!

Figure 9.25

The connections between the jovian planets, small bodies, and Earth. The gravity of the jovian planets helped shape both the asteroid belt and the Kuiper belt, and the Oort cloud consists of comets ejected from the jovian planet region by gravitational encounters with these large planets. Ongoing gravitational influences can send asteroids or comets heading toward Earth.

the big picture

Putting Chapter 9 into Context

In this chapter we concluded our study of the solar system by focusing on its smallest objects, finding that these objects can have big consequences. Keep in mind the following "big picture" ideas:

- Asteroids, comets, and meteorites may be small compared to planets, but they provide much of the evidence that has helped us understand how the solar system formed.

- The small bodies are subject to the gravitational influences of the largest. The jovian planets shaped the asteroid belt, the Kuiper belt, and the Oort cloud, and they continue to nudge objects onto collision courses with the planets.

- Pluto, once considered a "misfit" among the planets, is now viewed as just one of many moderately sized objects in the Kuiper belt. In terms of composition, the dwarf planets Pluto and Eris—along with other similar objects—are essentially comets of unusually large size.

- Collisions not only bring meteorites and leave impact craters but also can profoundly affect life on Earth. An impact probably wiped out the dinosaurs, and future impacts pose a threat that we cannot ignore.

summary of key concepts

9.1 Asteroids and Meteorites

• Why is there an asteroid belt?

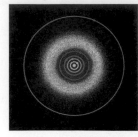

Orbital resonances with Jupiter disrupted the orbits of planetesimals in the asteroid belt, preventing them from accreting into a terrestrial planet. Many were ejected, but some remained and make up the asteroid belt today. Most asteroids in other regions of the inner solar system crashed into one of the planets.

• How are meteorites related to asteroids?

Most **meteorites** are pieces of asteroids. **Primitive meteorites** are essentially unchanged since the birth of the solar system. **Processed meteorites** are fragments of larger asteroids that underwent differentiation.

9.2 Comets

• How do comets get their tails?

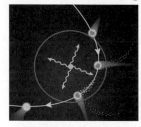

Comets are icy leftovers from the era of planet formation, and most orbit far from the Sun. If a comet approaches the Sun, its **nucleus** heats up and its ice sublimates into gas. The escaping gases carry along some dust, forming a **coma** and two tails: a **plasma tail** of ionized gas and a **dust tail**. Larger particles can also escape, becoming the particles that cause **meteors** and **meteor showers** on Earth.

• Where do comets come from?

Comets come from two reservoirs: the **Kuiper belt** and the **Oort cloud**. The Kuiper belt comets still reside in the region beyond Neptune in which they formed. The Oort cloud comets formed between the jovian planets and were kicked out to a great distance by gravitational encounters with these planets.

9.3 Pluto: Lone Dog No More

• How big can a comet be?

In the Kuiper belt, icy planetesimals were able to grow to hundreds or thousands of kilometers in size. The recently discovered Eris is the largest known of these objects and Pluto is the second largest.

• What are Pluto and other large objects of the Kuiper belt like?

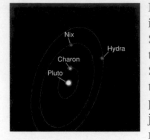

Like smaller comets, these objects are ice-rich in composition. They orbit the Sun roughly between the orbit of Neptune and twice that distance from the Sun. Their orbits tend to be more elliptical and more inclined to the ecliptic plane than those of the terrestrial and jovian planets. Many share orbital resonances with Neptune. A few have moons, including Pluto.

9.4 Cosmic Collisions: Small Bodies Versus the Planets

• Have we ever witnessed a major impact?

In 1994, we observed the fragmented Comet Shoemaker–Levy 9 impacting Jupiter, scarring its atmosphere for months.

• Did an impact kill the dinosaurs?

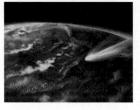

It may not have been the sole cause, but a major impact clearly coincided with the **mass extinction** in which the dinosaurs died out, about 65 million years ago. Sediments from this era contain iridium and other evidence of an impact, and an impact crater of the same age lies along the coast of Mexico.

• Is the impact threat a real danger or just media hype?

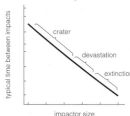

Impacts certainly pose a threat, though the probability of a major impact in our lifetimes is fairly low. Impacts like the Tunguska event may occur every few hundred years and would be catastrophic if they struck populated areas.

• How do other planets affect impact rates and life on Earth?

Impacts are always linked in at least some way to the gravitational influences of Jupiter and the other jovian planets. These influences have shaped the asteroid belt, the Kuiper belt, and the Oort cloud and continue to determine when an object is flung our way.

exercises and problems

For instructor-assigned homework go to **www.masteringastronomy.com.**

Review Questions

Short-Answer Questions Based on the Reading

1. Briefly explain why comets, asteroids, and meteorites are so useful in helping us understand the history of the solar system.

2. How does the largest asteroid compare in size to the planets? How does the total mass of all asteroids compare to the mass of a terrestrial world?

3. Where is the *asteroid belt* located, and why? Briefly explain how orbital resonances with Jupiter have affected the asteroid belt.

4. What is the difference between a *meteor* and a *meteorite?* How can we distinguish a meteorite from a terrestrial rock?

5. Distinguish between *primitive meteorites* and *processed meteorites* in terms of both composition and origin.

6. What does a comet look like when it is far from the Sun? How does its appearance change when it is near the Sun? What happens to comets that make many passes near the Sun?

7. What produces the *coma* and tails of a comet? What is the *nucleus?* Why do tails point away from the Sun?

8. Explain how meteor showers are linked to comets. Why do meteor showers recur at about the same time each year?

9. Describe the *Kuiper belt* and *Oort cloud* in terms of their locations and the orbits of comets within them. How did comets come to exist in these two regions?

10. How large are the largest objects in the Kuiper belt? How many qualify—or may qualify—as *dwarf planets?* In what sense are all these objects really just large comets?

11. Briefly describe Pluto and Charon. How do Eris and other large Kuiper belt objects compare to Pluto and Charon?

12. Briefly describe the impact of comet Shoemaker–Levy 9 on Jupiter.

13. Briefly describe the evidence suggesting that an impact caused the mass extinction that killed off the dinosaurs. How might the impact have led to the mass extinction?

14. How often should we expect impacts of various sizes on Earth? How serious a threat do we face from these impacts?

Test Your Understanding

Surprising Discoveries?

Consider the following hypothetical discoveries. (These are *not* real discoveries.) In light of what you've learned about the formation of our solar system, decide whether each discovery should be considered reasonable or surprising. Explain clearly; not all these have definitive answers, so your explanation is more important than your chosen answer.

15. A small asteroid that orbits within the asteroid belt has an active volcano.

16. Scientists discover a meteorite that, based on radiometric dating, is 7.9 billion years old.

17. An object that resembles a comet in size and composition is discovered orbiting in the inner solar system.

18. Studies of a large object in the Kuiper belt reveal that it is made almost entirely of rocky (as opposed to icy) material.

19. Astronomers discover a previously unknown comet that will be brightly visible in our night sky about 2 years from now.

20. A mission to Pluto finds that it has lakes of liquid water on its surface.

21. Geologists discover a crater from a 5-kilometer object that impacted Earth more than 100 million years ago.

22. Archaeologists learn that the fall of ancient Rome was caused in large part by an asteroid impact in Asia.

23. Astronomers discover three objects with the same average distance from the Sun (and the same orbital period) as Pluto.

24. Astronomers discover an asteroid with an orbit suggesting that it will impact the Earth in the year 2064.

Quick Quiz

Choose the best answer to each of the following. Explain your reasoning with one or more complete sentences.

25. The asteroid belt lies between the orbits of (a) Earth and Mars. (b) Mars and Jupiter. (c) Jupiter and Saturn.

26. Jupiter nudges the asteroids through the influence of (a) tidal forces. (b) orbital resonances. (c) magnetic fields.

27. Can an asteroid be pure metal? (a) No, all asteroids contain rock. (b) Yes, it must have formed where only metal could condense in the solar nebula. (c) Yes, it must have been the core of a shattered asteroid.

28. Did a large terrestrial planet ever form in the region of the asteroid belt? (a) No, because there was never enough mass there. (b) No, because Jupiter prevented one from accreting. (c) Yes, but it was shattered by a giant impact.

29. What does Pluto most resemble? (a) a terrestrial planet (b) a jovian planet (c) a comet

30. How big an object causes a typical shooting star? (a) a grain of sand or a small pebble (b) a boulder (c) an object the size of a car

31. Which have the most elliptical and tilted orbits? (a) asteroids (b) Kuiper belt comets (c) Oort cloud comets

32. Which are thought to have formed farthest from the Sun? (a) asteroids (b) Kuiper belt comets (c) Oort cloud comets

33. About how often does a 1-kilometer object strike Earth? (a) every year (b) every million years (c) every billion years

34. What would happen if a 1-kilometer object struck Earth? (a) It would break up in the atmosphere without causing widespread damage. (b) It would cause widespread devastation and climate change. (c) It would cause a mass extinction.

Process of Science

Examining How Science Works

35. *The Pluto Debate.* Research the decision to reclassify Pluto as a dwarf planet. In your opinion, is this a good example of the application of the scientific process? Does it exhibit the hallmarks of science described in Chapter 3? Compare your conclusions to opinions you find about the debate, and describe how you think astronomers should handle similar debates in the future.

36. *Life or Death Astronomy.* In most cases, the study of the solar system has little direct effect on our lives. But the discovery of an asteroid or comet on a collision course with Earth is another matter. How should the standards for verifiable observations described in Chapter 3 apply in this case? Is the potential danger

so great that any astronomer with evidence of an impending impact should spread the word as soon as possible? Or is the potential for panic so great that even higher standards of verification ought to be applied? What kind of review process, if any, would you set in place? Who should be informed of an impact threat, and when?

37. *Unanswered Questions.* NASA has two missions en route to dwarf planets: *New Horizons* to Pluto, and the *Dawn* mission to Ceres and Vesta. Do a Web search to identify one important but still unanswered question about these destinations, and write two or three paragraphs discussing how one of these missions might answer this question. Be as specific as possible, focusing on the type of evidence necessary to answer the question and how the evidence could be gathered. What are the benefits of finding an answer to this question?

Investigate Further

In-Depth Questions to Increase Your Understanding

Short-Answer/Essay Questions

38. *The Role of Jupiter.* Suppose that Jupiter had never existed. Describe at least three ways in which our solar system would be different, and clearly explain why.

39. *Life Story of an Iron Atom.* Imagine that you are an iron atom in a processed meteorite made mostly of iron. Tell the story of how you got to Earth, beginning from the time you were part of the gas in the solar nebula 4.6 billion years ago. Include as much detail as possible. Your story should be scientifically accurate but also creative and interesting

40. *Asteroids vs. Comets.* Contrast the compositions and locations of comets and asteroids, and explain in your own words why they have turned out differently.

41. *Comet Tails.* Describe in your own words why comets have tails. Why do most comets have two distinct visible tails, and why do the tails go in different directions? Why is the third, invisible tail of small pebbles of interest to us on Earth?

42. *Oort Cloud vs. Kuiper Belt.* Explain in your own words how and why there are two different reservoirs of comets. Be sure to discuss where the two groups of comets formed, and what kinds of orbits they travel on.

43. *Project: Dirty Snowballs.* If there is snow where you live or study, make a dirty snowball. (The ice chunks that form behind tires work well.) How much dirt does it take to darken snow? Find out by allowing your dirty snowball to melt in a container and measuring the approximate proportions of water and dirt afterward. What do your results tell you about comet composition?

Quantitative Problems

Be sure to show all calculations clearly and state your final answers in complete sentences.

44. *Adding Up Asteroids.* It's estimated that there are a million asteroids 1 kilometer across or larger. If a million asteroids 1 kilometer across were all combined into one object, how big would it be? How many 1-kilometer asteroids would it take to make an object as large as the Earth? (*Hint:* You can assume they'e spherical. The equation for the volume of a sphere is $\left(\frac{4}{3}\right)\pi r^3$, where r is the radius—not the diameter.)

45. *Impact Energies.* A relatively small impact crater 20 kilometers in diameter could be made by a comet 2 kilometers in diameter traveling at 30 kilometers (30,000 meters) per second.

a. Assume that the comet has a total mass of 4.2×10^{12} kilograms. What is its total kinetic energy? (*Hint:* The kinetic energy is equal to $\frac{1}{2}mv^2$, where m is the comet's mass and v is its speed. If you use mass in kilograms and velocity in meters per second, the answer for kinetic energy will have units of joules.)

b. Convert your answer from part (a) to an equivalent in megatons of TNT, the unit used for nuclear bombs. Comment on the degree of devastation the impact of such a comet could cause if it struck a populated region on Earth. (*Hint:* One megaton of TNT releases 4.2×10^{15} joules of energy.)

46. *The "Near Miss" of Toutatis.* The 5-kilometer asteroid Toutatis passed a mere 1.5 million kilometers from Earth in 2004. Suppose Toutatis were destined to pass *somewhere* within 3 million kilometers of Earth. Calculate the probability that this "somewhere" would have meant that it slammed into Earth. Based on your result, do you think it is fair to call the 2004 passage a "near miss"? Explain. (*Hint:* You can calculate the probability by imagining a dartboard of radius 3 million kilometers in which the bull's-eye has Earth's radius, 6378 kilometers.)

47. *Room to Roam.* It's estimated that there are a trillion comets in the Oort cloud, which extends out to about 50,000 AU. What is the total volume of the Oort cloud in cubic AU? How much space does each comet have in cubic AU, on average? Take the cube root of the average volume per comet to find their typical spacing in AU. (*Hints:* For this calculation, you can assume the Oort cloud fills the whole sphere out to 50,000 AU. The equation for the volume of a sphere is $\left(\frac{4}{3}\right)\pi r^3$, where r is the radius.)

48. *Dust Accumulation.* A few hundred tons of comet and asteroid dust are added to Earth daily from the millions of meteors that enter our atmosphere. Estimate the time it would take for Earth to get 0.1% heavier at this rate. Is this mass accumulation significant for Earth as a planet? Explain.

Discussion Questions

49. *Rise of the Mammals.* Suppose the impact 65 million years ago had not occurred. How do you think our planet would be different? Do you think that mammals still would eventually have come to dominate Earth? Would we be here? Defend your opinions.

50. *How Should Kids Count Planets?* The new definitions that have officially demoted Pluto from planet to dwarf planet have many educational implications. For example, many children learn songs in school that refer to "nine planets" and Pluto. How would you recommend that school teachers deal with the new definitions? Be sure to consider the fact that while these definitions have "official" status today (from the International Astronomical Union), it is possible that they may change again in the future.

51. *Reducing the Impact Threat.* Based on your own opinion of how the impact threat compares to other threats, how much money and resources do you think should be used to alleviate it? What types of programs would you support? (Examples include programs to search for objects that could strike Earth, to develop defenses against impacts, or to build actual defenses.) Defend your opinions.

Web Projects

52. *Recent Asteroid and Comet Missions.* Learn about the goals and status of the *NEAR, Stardust,* or *Hayabusa* mission. What did it (or will it) accomplish? Write a one- to two-page summary of your findings.

53. *Future Asteroid and Comet Missions.* Learn about another proposed space mission to study asteroids or comets, such as *Dawn* or *Rosetta.* For the mission you choose, write a short report about its plans, goals, and prospects for success.

54. *The New Horizons Mission to Pluto.* The *New Horizons* mission is on its way! Find out the current status of the mission. What are its goals? What has it done so far? Summarize your findings in a few paragraphs.

55. *Impact Hazards.* Many groups are searching for near-Earth asteroids that might impact the Earth. They use the *Torino Scale* to evaluate the possible danger posed by an asteroid, based on our knowledge of its orbit. What is this scale? What object has reached the highest level on this scale? What were the estimated chances of impact, and when?

visual skills check

Use the following questions to check your understanding of some of the many types of visual information used in astronomy. Answers are provided in Appendix K. For additional practice, try the Chapter 9 Visual Quiz at **www.masteringastronomy.com**.

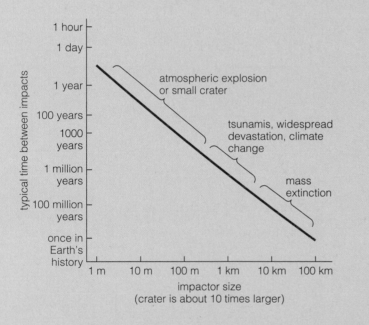

1 km

The graph above (from Figure 9.24) shows how often impacts occur for objects of different sizes. The photo above shows comet Tempel 1 moments before the *Deep Impact* spacecraft crashed into it.

1. Estimate comet Tempel 1's diameter, using the scale bar in the photo above.
2. According to the graph above, how frequently do objects the size of comet Tempel 1 strike Earth?
   a. once in Earth's history
   b. about once every hundred million years
   c. about once every million years
   d. about once every thousand years
3. According to the graph, what kind of damage would this object cause if it hit our planet?
   a. mass extinction
   b. widespread devastation and climate change
   c. atmospheric explosion or a small crater

4. The Meteor Crater in Arizona is about 1.2 kilometers across. According to the graph, about how big was the object that made this crater? (*Note*: Be sure to read the axis labels carefully.)
   a. 1 meter
   b. 10 meters
   c. 100 meters
   d. 1 kilometer
5. How often do objects big enough to create craters like the Meteor Crater impact Earth?
   a. once in Earth's history
   b. about once every few tens of thousands of years
   c. about once every few million years
   d. about once every day, but most burn up in the atmosphere or land in the ocean

Comparing the worlds in the solar system has taught us important lessons about Earth and why it is so suitable for life. This illustration summarizes some of the major lessons we've learned by studying other worlds both in our own solar system and beyond it.

(1) Comparing the terrestrial worlds shows that a planet's size and distance from the Sun are the primary factors that determine how it evolves through time [Chapter 7].

Venus demonstrates the importance of distance from the Sun: If Earth were moved to the orbit of Venus, it would suffer a runaway greenhouse effect and become too hot for life.

Mars shows why size is important: A planet smaller than Earth loses interior heat faster, which can lead to a decline in geological activity and loss of atmospheric gas.

The smallest terrestrial worlds, Mercury and the Moon, became geologically dead long ago. They therefore retain ancient impact craters, which provide a record of how impacts must have affected Earth and other worlds.

2 Jovian planets are gas-rich and far more massive than Earth. They and their ice-rich moons have opened our eyes to the diversity of processes that shape worlds [Chapter 8].

Our Moon led us to expect all small objects to be geologically dead . . .

Earth and the Moon

The strong gravity of the jovian planets has shaped the asteroid and Kuiper belts, and flung comets into the distant Oort cloud, ultimately determining how frequently asteroids and comets strike Earth.

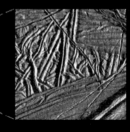

Jupiter and Europa

. . . but Europa—along with Io, Titan and other moons—proved that tidal heating or icy composition can lead to geological activity, in some cases with subsurface oceans and perhaps even life.

3 Asteroids and comets may be small bodies in the solar system, but they have played major roles in the development of life on Earth [Chapter 9].

Comets or water-rich asteroids from the outer asteroid belt brought Earth the ingredients of its oceans and atmosphere.

Impacts of comets and asteroids have altered the course of life on Earth and may do so again.

4 The discovery of planets around other stars has shown that our solar system is not unique. Studies of other solar systems are teaching us new lessons about how planets form and about the likelihood of finding other Earth-like worlds [Section 6.5]

Current detection techniques are best at finding extrasolar planets similar in mass to Jupiter, but improving technology will soon enable us to detect planets as small as Earth.

10

our star

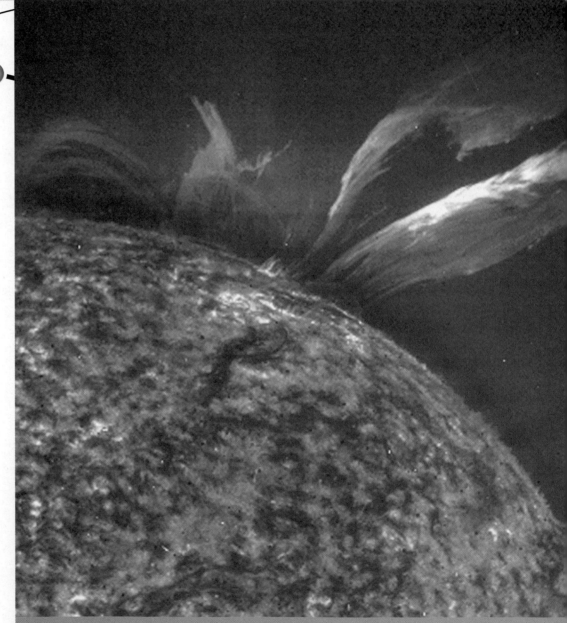

Astronomy today encompasses the study of the entire universe, but the root of the word *astronomy* comes from the Greek word for "star." Although we have learned a lot about the universe up to this point in the book, only now do we turn our attention to the study of the stars, the namesakes of astronomy.

When we think of stars, we usually think of the beautiful points of light visible on a clear night. But the nearest and most easily studied star—our Sun—is visible only in the daytime. Of course, the Sun is important to us in many more ways than just as an object for astronomical study. The Sun is the source of virtually all light, heat, and energy reaching Earth, and life on Earth's surface could not survive without it.

In this chapter, we will study the Sun in some depth. We will learn how the Sun generates the energy that supports life on Earth. Equally important, we will study our Sun as a star so that in subsequent chapters we can more easily understand stars throughout the universe.

essential preparation

1. Where do objects get their energy? [Section 4.3]

2. What is matter? [Section 5.1]

3. How does light tell us the temperatures of planets and stars? [Section 5.2]

(MA)™ The Sun Tutorial, Lesson 1

10.1 A Closer Look at the Sun

We discussed the general features of the Sun in our tour of the solar system in Chapter 6 (see p. 146). Now it's time to get better acquainted with our nearest star.

Ancient peoples recognized the vital role of the Sun in their lives. Some worshipped the Sun as a god. Others created mythologies to explain its daily rise and set. But no one who lived before the 20th century knew how the Sun provides us with light and heat.

Most ancient thinkers imagined the Sun to be some type of fire, perhaps a lump of burning coal or wood. It was a reasonable suggestion for the times, since science had not yet advanced to the point where the idea could be put to the test. Ancient people did not know the size or distance of the Sun, so they could not imagine how incredible its energy output really is. And they did not know how long Earth had existed, so they had no way of knowing that the Sun has provided light and heat for a very long time.

Scientists began to address the question of how the Sun shines around the middle of the 19th century, by which time the Sun's size and distance had been measured with reasonable accuracy. The ancient idea that the Sun was composed of burning coal or wood was quickly ruled out: Calculations showed that such burning could not possibly account for the Sun's huge output of energy. Other ideas based on chemical processes were likewise ruled out.

In the late 19th century, astronomers came up with an idea that seemed more plausible, at least at first. They suggested that the Sun generates energy by slowly contracting in size, a process called **gravitational contraction** (or *Kelvin-Helmholtz contraction,* after the scientists who proposed the mechanism). Recall that a shrinking gas cloud heats up because some of the gravitational potential energy of gas particles far from the cloud center is converted into thermal energy as the gas moves inward (see Figure 4.12b). A gradually shrinking Sun would always have some gas moving inward, converting gravitational potential energy into thermal energy. This thermal energy would keep the inside of the Sun hot. Because of its large mass, the Sun would need to contract only very slightly each year to maintain its temperature—so slightly that the contraction would have been unnoticeable to 19th century astronomers. Calculations showed that gravitational contraction could have kept the Sun shining steadily for up to 25 million years. For a while, some astronomers thought that this idea had solved the ancient mystery of how the Sun shines. However, geologists pointed out a fatal flaw: Studies of rocks and fossils had already shown Earth to be far older than 25 million years, which meant that gravitational contraction could not be the mechanism by which the Sun generates its energy.

Figure 10.1

An acrobat stack is in gravitational equilibrium: The lowest person supports the most weight and feels the greatest pressure, and the overlying weight and underlying pressure decrease for those higher up.

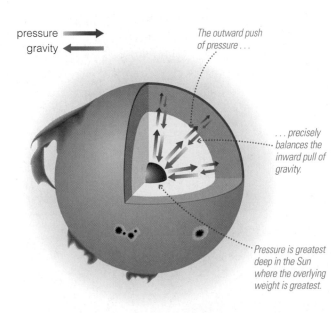

pressure ⟶
gravity ⟵

The outward push of pressure . . .

. . . precisely balances the inward pull of gravity.

Pressure is greatest deep in the Sun where the overlying weight is greatest.

Figure 10.2

Gravitational equilibrium in the Sun: At each point inside, the pressure pushing outward balances the weight of the overlying layers.

• Why does the Sun shine?

With both chemical processes and gravitational contraction ruled out as the explanation for why the Sun shines, scientists of the late 19th century were at a loss. There was no known way that an object the size of the Sun could generate so much energy for billions of years. A completely new type of explanation was needed, and it came with Einstein's publication of his special theory of relativity in 1905.

Einstein's theory included his famous discovery of $E = mc^2$ [Section 4.3]. This equation shows that mass itself contains an enormous amount of potential energy. Calculations demonstrated that the Sun's mass contained more than enough energy to account for billions of years of sunshine, if only there were some way for the Sun to convert the energy of mass into thermal energy. It took a few decades for scientists to work out the details, but by the end of the 1930s we had learned that the Sun converts mass into energy through the process of nuclear fusion.

The Stable Sun Nuclear fusion requires extremely high temperatures and densities (for reasons we will discuss in the next section). In the Sun, these conditions are found deep in the core. For the Sun to shine steadily, it must have a way of keeping the core hot and dense. It maintains these internal conditions through a natural balance between two competing forces: gravity pulling inward and pressure pushing outward. This balance is called **gravitational equilibrium** (or *hydrostatic equilibrium*).

A stack of acrobats provides a simple example of gravitational equilibrium (Figure 10.1). The bottom person supports the weight of everybody above him, so his arms must push upward with enough pressure to support all this weight. At each higher level, the overlying weight is less, so it's a little easier for each additional person to hold up the rest of the stack.

Everywhere inside the Sun, the outward push of pressure balances the inward pull of gravity.

Gravitational equilibrium works much the same in the Sun, except the outward push against gravity comes from internal gas pressure rather than an acrobat's arms. The Sun's internal pressure precisely balances gravity at every point within it, thereby keeping the Sun stable in size (Figure 10.2). Because the weight of overlying layers is greater as we look deeper into the Sun, the pressure must increase with depth. Deep in the Sun's core, the pressure makes the gas hot and dense enough to sustain nuclear fusion. The energy released by fusion, in turn, heats the gas and maintains the pressure that keeps the Sun in balance against the inward pull of gravity.

think about it Earth's atmosphere is also in gravitational equilibrium, with the weight of upper layers supported by the pressure in lower layers. Use this idea to explain why the air gets thinner at higher altitudes.

How Fusion Started The steadiness of our Sun's gravitational equilibrium is easy to understand: Gravity pushes inward while the energy released by fusion maintains the pressure that pushes outward. But how did the Sun become hot enough for fusion to begin in the first place?

The answer invokes the mechanism of gravitational contraction that astronomers of the late 19th century mistakenly thought might be responsible for the Sun's heat today. Recall that our Sun was born about $4\frac{1}{2}$ billion years ago from a collapsing cloud of interstellar gas [Section 6.2]. The contraction of the cloud released gravitational potential energy, raising the interior temperature and pressure. When the central temperature finally grew high enough to sustain nuclear fusion, energy generation from the Sun's interior finally came into balance with the energy lost from the surface in the form of radiation. This energy balance stabilized the size of the Sun, bringing it into a state of gravitational equilibrium that has persisted to this day.

Compression of gas in the Sun's core made it hot enough for fusion, which provides the energy that now keeps pressure and gravity in balance.

In summary, the answer to the question "Why does the Sun shine?" is that about $4\frac{1}{2}$ billion years ago *gravitational contraction* made the Sun hot enough to sustain nuclear fusion in its core. Ever since, energy liberated by fusion has maintained the Sun's *gravitational equilibrium* and kept the Sun shining steadily, supplying the light and heat that sustain life on Earth.

• What is the Sun's structure?

The Sun is essentially a giant ball of hot gas or, more technically, *plasma*—a gas in which atoms are ionized [Section 5.2] because of the high temperature. A plasma behaves much like an ordinary gas, but its many positively charged ions and freely moving electrons enable it to create and respond to magnetic fields.

The differing temperatures and densities of the plasma at different depths give the Sun the layered structure shown in Figure 10.3. To make sense of what you see in the figure, imagine that you have a spaceship that can somehow withstand the immense heat and pressure of the Sun and take an imaginary journey from Earth to the center of the Sun. This journey will acquaint you with the basic properties of the Sun, which we'll discuss in greater detail in the rest of this chapter.

Basic Properties of the Sun As you begin your journey from Earth, the Sun appears as a whitish ball of glowing gas. Just as astronomers have done in real life, you can use simple observations to determine basic properties of the Sun. Spectroscopy [Section 5.2] tells you that the Sun is made almost entirely of hydrogen and helium. From the Sun's angular size and distance, you can determine that its radius is just under 700,000 km, or more than 100 times the radius of Earth. Even **sunspots**, which appear as dark splotches on the Sun's surface, can be larger in size than Earth. You can measure the Sun's mass using Newton's version of Kepler's third law. It is about 2×10^{30} kilograms, which is some 300,000 times the mass of Earth and more than 1000 times the mass of all the planets in our solar system put together. You can observe the Sun's rotation by tracking the

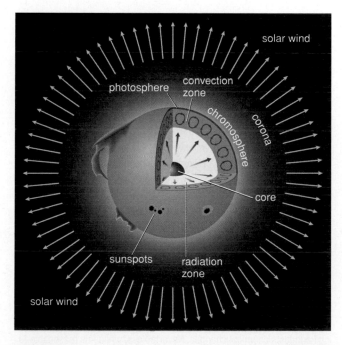

Figure 10.3

The basic structure of the Sun.

TABLE 10.1 *Basic Properties of the Sun*

Radius (R_{Sun})	696,000 km (about 109 times the radius of Earth)
Mass (M_{Sun})	2×10^{30} kg (about 300,000 times the mass of Earth)
Luminosity (L_{Sun})	3.8×10^{26} watts
Composition (by percentage of mass)	70% hydrogen, 28% helium, 2% heavier elements
Rotation rate	25 days (equator) to 30 days (poles)
Surface temperature	5800 K (average); 4000 K (sunspots)
Core temperature	15 million K

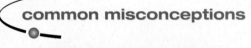

common misconceptions

The Sun Is Not on Fire

We often say that the Sun is "burning," a term that conjures up images of a giant bonfire in the sky. However, the Sun does not burn in the same sense as a fire burns on Earth. Fires generate light through chemical changes that consume oxygen and produce a flame. The glow of the Sun has more in common with the glowing embers left over after the flames have burned out. Much like hot embers, the Sun's surface shines because it is hot enough to emit thermal radiation that includes visible light [Section 5.2].

Hot embers quickly stop glowing as they cool, but the Sun keeps shining because its surface is kept hot by the energy rising from its core. Because this energy is generated by nuclear fusion, we sometimes say that it is the result of "nuclear burning"—a term intended to suggest nuclear changes in much the same way that "chemical burning" suggests chemical changes. Nevertheless, while it is reasonable to say that the Sun undergoes nuclear burning in its core, it is not accurate to speak of any kind of burning on the Sun's surface, where light is produced primarily by thermal radiation.

motion of sunspots or by measuring Doppler shifts [Section 5.2] on opposite sides of the Sun. Unlike a spinning ball, the entire Sun does *not* rotate at the same rate. Instead, the solar equator completes one rotation in about 25 days, and the rotation period increases with latitude to about 30 days near the solar poles.

think about it As a brief review, describe how astronomers use Newton's version of Kepler's third law to determine the mass of the Sun. What two properties of Earth's orbit do we need to know in order to apply this law? (*Hint:* See Section 4.4.)

The Sun releases an enormous amount of radiative energy into space. In science, we measure energy in units of *joules* [Section 4.3]. We define **power** as the *rate* at which energy is used or released. The standard unit of power is the **watt**, defined as 1 joule of energy per second; that is, 1 watt = 1 joule/s. For example, a 100-watt light bulb requires 100 joules of energy for every second it is left turned on. The Sun's total power output, or **luminosity**, is an incredible 3.8×10^{26} watts. If we could somehow capture and store just 1 second's worth of the Sun's luminosity, it would be enough to meet current human energy demands for roughly the next 500,000 years. Table 10.1 summarizes the basic properties of the Sun.

The Sun's Atmosphere Even at great distances from the Sun, you and your spacecraft can feel slight effects from the **solar wind**—the stream of charged particles continually blown outward in all directions from the Sun. Recall that the solar wind helps shape the magnetospheres of planets (see Figure 7.6) and blows back the material that forms the plasma tails of comets [Section 9.2].

As you approach the Sun more closely, you'll begin to encounter the low-density gas that represents what we usually think of as the Sun's atmosphere. The outermost layer of this atmosphere, called the **corona**, extends several million kilometers above the visible surface of the Sun. The temperature of the corona is astonishingly high—about 1 million K—explaining why this region emits most of the Sun's X rays. However, the corona's density is so low that your spaceship feels relatively little heat, despite the million-degree temperature [Section 4.3].

The Sun's upper atmosphere is much hotter than the visible surface, or *photosphere*, but its density is much lower.

Nearer the surface, the temperature suddenly drops to about 10,000 K in the **chromosphere**, the middle layer of the solar atmosphere and the region that radiates most of the Sun's ultraviolet radiation. Then you plunge through the lowest layer of the atmosphere, or **photosphere**, which is the visible surface of the Sun. Although the photosphere looks like a well-defined surface from Earth, it consists of gas far less dense than Earth's atmosphere. The temperature of the photosphere averages just under 6000 K, and its surface seethes and churns like a pot of boiling water. The photosphere is also where you'll find sunspots, regions of intense magnetic fields that would cause your compass needle to swing wildly about.

The Sun's Interior Up to this point in your journey, you may have seen Earth and the stars when you looked back. But blazing light engulfs you as you slip beneath the photosphere. You are inside the Sun, and incredible turbulence tosses your spacecraft about. If you can hold steady long enough to see what is going on around you, you'll notice spouts of hot gas rising upward, surrounded by cooler gas cascading down from above. You are in the **convection zone**, where energy generated in the solar core travels upward, transported by the rising of hot gas and falling of cool gas called *convection* **[Section 7.1]**. With some quick thinking, you may realize that the photosphere above you is the top of the convection zone and that convection is the cause of the Sun's seething, churning appearance.

About a third of the way down to the Sun's center, the turbulence of the convection zone gives way to the calmer plasma of the **radiation zone**, where energy moves outward primarily in the form of photons of light. The temperature rises to almost 10 million K, and your spacecraft is bathed in X rays trillions of times more intense than the visible light at the solar surface.

Inside the Sun, temperature rises with depth, reaching 15 million K in the core.

No real spacecraft could survive, but your imaginary one keeps plunging straight down to the solar **core**. There you finally find the source of the Sun's energy: nuclear fusion transforming hydrogen into helium. At the Sun's center, the temperature is about 15 million K, the density is more than 100 times that of water, and the pressure is 200 billion times that on the surface of Earth. The energy produced in the core today will take about a few hundred thousand years to reach the surface.

With your journey complete, it's time to turn around and head back home. We'll continue this chapter by studying fusion in the solar core and then tracing the flow of the energy generated by fusion as it moves outward through the Sun.

 **The Sun Tutorial, Lessons 2–3**

10.2 Nuclear Fusion in the Sun

We've seen that the Sun shines because of energy generated by nuclear fusion and that this fusion occurs under the extreme temperatures and densities found deep in the Sun's core. But exactly how does fusion occur and release energy? And how can we claim to know about something taking place out of sight in the Sun's hidden interior?

Before we begin to answer these questions, it's important to realize that the nuclear reactions that generate energy in the Sun are very different from those used to generate energy in human-built nuclear reactors on Earth. Our nuclear power plants generate energy by splitting large nuclei—such as those of uranium or plutonium—into smaller ones. The process of splitting an atomic nucleus is called *nuclear fission*. In contrast, the Sun makes energy by combining, or fusing, two or more small nuclei into a larger one. That is why we call the process *nuclear fusion*. Figure 10.4 summarizes the difference between fission and fusion.

• How does nuclear fusion occur in the Sun?

Fusion occurs within the Sun because the 15 million K plasma in the solar core is like a "soup" of hot gas, with bare, positively charged atomic nuclei (and negatively charged electrons) whizzing about at extremely high speeds. At any time, some of these nuclei are on high-speed collision courses with each other. In most cases, electromagnetic forces deflect the nuclei, preventing actual collisions, because positive charges repel one

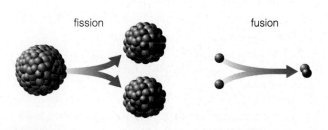

Figure 10.4

Nuclear fission splits a nucleus into smaller nuclei (not usually of equal size), while nuclear fusion combines smaller nuclei into a larger nucleus.

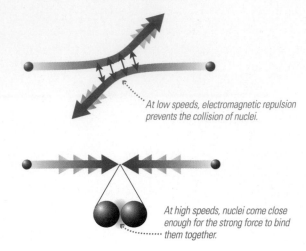

At low speeds, electromagnetic repulsion prevents the collision of nuclei.

At high speeds, nuclei come close enough for the strong force to bind them together.

Figure 10.5

Positively charged nuclei can fuse only if a high-speed collision brings them close enough for the strong force to come into play.

cosmic calculations 10.1

The Ideal Gas Law

The pressure of a gas depends on its temperature, because particles in a hot gas move more quickly and collide with more force than those in a cooler gas. But pressure also depends on *number density*—the number of gas particles contained in each cubic centimeter—because higher density means more particles that can potentially collide. The *ideal gas law* mathematically expresses the relationship between temperature T, number density n, and pressure P:

$$P = nkT$$

where $k = 1.38 \times 10^{-23}$ joules/K is *Boltzmann's constant*.

Example: The Sun's core density is about 10^{26} particles per cubic centimeter (which we write as 10^{26} cm^{-3}) and its temperature is about 15 million K (1.5×10^7 K). Compare the Sun's core pressure to that of Earth's atmosphere at sea level, where the density is about 2.4×10^{19} particles per cubic centimeter and the temperature is about 300 K.

Solution: The pressure of the Sun's core is

$$P_{core} = (10^{26} \text{ cm}^{-3})(1.38 \times 10^{-23} \text{ joules/K})(1.5 \times 10^7 \text{ K})$$
$$= 2.1 \times 10^{10} \text{ joules cm}^{-3}$$

The pressure of Earth's atmosphere is

$$P_{atmos} = (2.4 \times 10^{19} \text{ cm}^{-3})(1.38 \times 10^{-23} \text{ joules/K})(300 \text{ K})$$
$$= 0.10 \text{ joules cm}^{-3}$$

Dividing the two pressures above shows that the Sun's core pressure is about 200 billion (2×10^{11}) times as great as Earth's atmospheric pressure.

another. If nuclei collide with sufficient energy, however, they can stick together to form a heavier nucleus.

Sticking positively charged nuclei together is not easy (Figure 10.5). The **strong force**, which binds protons and neutrons together in atomic nuclei, is the only force in nature that can overcome the electromagnetic repulsion between two positively charged nuclei. In contrast to gravitational and electromagnetic forces, which drop off gradually as the distances between particles increase (by an inverse square law **[Section 4.4]**), the strong force is more like glue or Velcro: It overpowers the electromagnetic force over very small distances but is insignificant when the distances between particles exceed the typical sizes of atomic nuclei. The key to nuclear fusion is therefore to push the positively charged nuclei close enough together for the strong force to outmuscle electromagnetic repulsion.

Positively charged nuclei fuse together if they pass close enough for the strong force to overpower electromagnetic repulsion.

The high pressures and temperatures in the solar core are just right for fusion of hydrogen nuclei into helium nuclei. The high temperature is important because the nuclei must collide at very high speeds if they are to come close enough together to fuse.* The higher the temperature, the harder the collisions, making fusion reactions more likely at higher temperatures. The high pressure of the overlying layers is also necessary; without it, the hot plasma of the solar core would simply explode into space, shutting off the nuclear reactions.

think about it The Sun generates energy by fusing hydrogen into helium, but as we'll see in later chapters, some stars fuse helium or even heavier elements. Do temperatures need to be higher or lower than the Sun's core temperature for the fusion of heavier elements? Why? (*Hint*: How does the positive charge of a nucleus affect its ability to fuse to another nucleus?)

The Proton–Proton Chain Let's investigate the fusion process in the Sun in a little more detail. Recall that hydrogen nuclei are simply individual protons, while the most common form of helium consists of two protons and two neutrons (see Figure 5.6). The overall hydrogen fusion reaction therefore transforms four individual protons into a helium nucleus containing two protons and two neutrons:

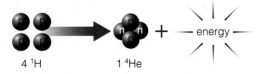

4 ^{1}H   1 ^{4}He

Nuclear fusion in the Sun combines four hydrogen nuclei into one helium nucleus, releasing energy in the process.

This overall reaction actually proceeds through several steps involving just two nuclei at a time. The sequence of steps that occurs in the Sun is called the **proton–proton chain** because it begins with collisions between individual protons (hydrogen nuclei). Figure 10.6 illustrates the steps in the proton–proton chain. Notice that the overall reaction is just as described above, with four protons combining to make one helium nucleus. Energy is carried off by

*The fusion reaction happens only if the protons come close enough for an effect known as *quantum tunneling* to operate.

Hydrogen Fusion by the Proton–Proton Chain

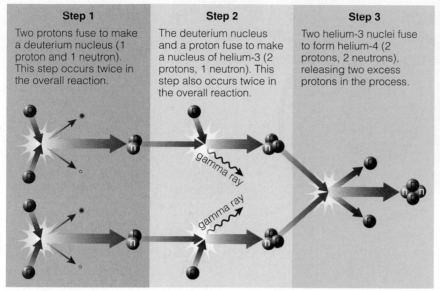

Step 1
Two protons fuse to make a deuterium nucleus (1 proton and 1 neutron). This step occurs twice in the overall reaction.

Step 2
The deuterium nucleus and a proton fuse to make a nucleus of helium-3 (2 protons, 1 neutron). This step also occurs twice in the overall reaction.

Step 3
Two helium-3 nuclei fuse to form helium-4 (2 protons, 2 neutrons), releasing two excess protons in the process.

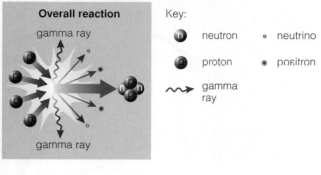

Overall reaction

Key:
- neutron
- proton
- gamma ray
- neutrino
- positron

Figure 10.6 interactive figure

In the Sun, four hydrogen nuclei (protons) fuse into one helium-4 nucleus by way of the proton–proton chain. Gamma rays and subatomic particles known as neutrinos and positrons carry off the energy released in the reaction.

the gamma rays and subatomic particles (neutrinos and positrons) released in the process.

Fusion of hydrogen into helium generates energy because a helium nucleus has a mass slightly less (by about 0.7%) than the combined mass of four hydrogen nuclei. Thus, when four hydrogen nuclei fuse into a helium nucleus, a little bit of mass disappears. This disappearing mass becomes energy in accord with Einstein's formula $E = mc^2$. Overall, fusion in the Sun converts about 600 million tons of hydrogen into 596 million tons of helium every second, which means that 4 million tons of matter is turned into energy each second. Although this sounds like a lot, it is a minuscule fraction of the Sun's total mass and does not affect the overall mass of the Sun in any measurable way.

The Solar Thermostat Nuclear fusion is the source of all the energy the Sun releases into space. If the fusion rate varied, so would the Sun's energy output, and large variations in the Sun's luminosity would almost surely be lethal to life on Earth. Fortunately, the Sun fuses hydrogen at a steady rate, thanks to a natural feedback process that acts as a thermostat for the Sun's interior. We have encountered a natural thermostat once before, when we discussed how Earth's carbon dioxide cycle keeps our planet's surface temperature relatively steady [Section 7.5]. The solar thermostat is even more efficient.

Solar energy production remains steady because the rate of nuclear fusion is very sensitive to temperature. A slight increase in the Sun's core temperature would mean a much higher fusion rate, and a slight decrease in temperature would mean a much lower fusion rate. Either kind of change in the fusion rate would alter the Sun's gravitational equilibrium—the balance between the pull of gravity and the push of internal pressure—in a way that would quickly restore the original temperature and fusion rate. To see how, let's examine what would happen if there were a small change in the core temperature (Figure 10.7).

Suppose the Sun's core temperature rose very slightly. Protons in the core would collide with more energy, bringing them closer together and

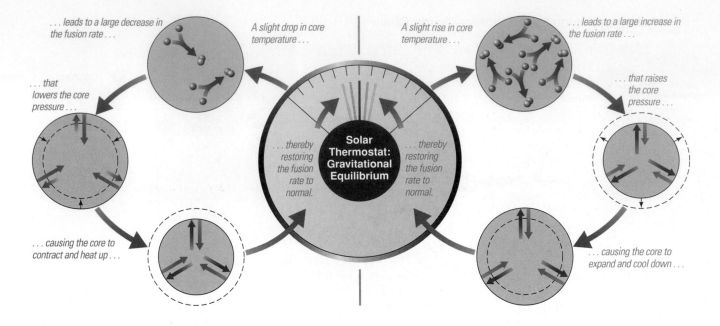

... leads to a large decrease in the fusion rate ...

A slight drop in core temperature ...

A slight rise in core temperature ...

... leads to a large increase in the fusion rate ...

... that lowers the core pressure ...

... that raises the core pressure ...

... thereby restoring the fusion rate to normal.

Solar Thermostat: Gravitational Equilibrium

... thereby restoring the fusion rate to normal.

... causing the core to contract and heat up ...

... causing the core to expand and cool down ...

Figure 10.7

The solar thermostat. Gravitational equilibrium regulates the Sun's core temperature. Everything is in balance if the amount of energy leaving the core equals the amount of energy produced by fusion. A rise in core temperature triggers a chain of events that causes the core to expand, lowering its temperature to the original value. A decrease in core temperature triggers an opposite chain of events, also restoring the original core temperature.

thereby causing more fusion reactions to happen. The rate of nuclear fusion would then soar, generating lots of extra energy. Because energy moves so slowly through the Sun, this extra energy would be bottled up in the core, causing an increase in the core pressure. The push of this pressure would temporarily exceed the pull of gravity, causing the core to expand and cool. This cooling, in turn, would cause the fusion rate to drop back down until the core was restored to its original size and temperature, returning the fusion rate back to its normal value.

Gravitational equilibrium acts as a thermostat that keeps the Sun's core temperature and fusion rate steady.

A slight drop in the Sun's core temperature would trigger an opposite chain of events. The reduced core temperature would lead to a decrease in the rate of nuclear fusion, causing a drop in pressure and contraction of the core. As the core shrank, its temperature would rise until the fusion rate returned to normal and restored the core to its original size and temperature.

• How does the energy from fusion get out of the Sun?

The solar thermostat balances the Sun's fusion rate so that the amount of nuclear energy generated in the core equals the amount of energy radiated from the surface as sunlight. However, the journey of solar energy from the core to the photosphere takes hundreds of thousands of years.

Most of the energy released by fusion starts its journey out of the solar core in the form of photons. Although photons travel at the speed of light, the paths they take through the Sun's interior zigzag so much that it takes them a very long time to make any outward progress. Deep in the solar interior, the plasma is so dense that a photon can travel only a fraction of a millimeter in any one direction before it interacts with an electron. Each time a photon "collides" with an electron, the photon gets deflected into a new and random direction. Thus, the photon bounces around the dense interior in a haphazard way (sometimes called a *random walk*) and only very gradually works its way outward from the Sun's center (Figure 10.8).

Randomly bouncing photons carry energy through the deepest layers of the Sun, and convection carries energy through the upper layers to the surface.

Energy released by fusion moves outward through the Sun's radiation zone (see Figure 10.3) primarily by way of these randomly bouncing photons. At the top of the radiation zone, where the temperature drops to about 2 million K, the solar plasma absorbs photons more readily, rather than just bouncing them around. This absorption creates the conditions needed for convection, and hence marks the beginning of the Sun's convection zone. The rising of hot plasma and sinking of cool plasma form a cycle that transports energy outward from the base of the convection zone to the photosphere (Figure 10.9a). There, we see bright blobs where hot gas is welling up from inside the Sun and darker borders around those blobs where the cooler gas is sinking (Figure 10.9b). Each bright blob lasts only a few minutes before being replaced by others.

In the photosphere, the density of the gas is low enough so that photons can escape to space. Thus, the energy produced hundreds of thousands of years earlier in the solar core finally emerges from the Sun as thermal radiation [Section 5.2] coming from the almost 6000 K gas of the photosphere. Once in space, the photons travel away at the speed of light, bathing the planets in sunlight.

• How do we know what is happening inside the Sun?

We cannot see inside the Sun, so you may wonder how we can claim to know so much about what goes on underneath its surface and in its core. In fact, we can study the Sun's interior in three different ways: through mathematical models of the Sun, observations of solar vibrations, and observations of solar neutrinos.

Mathematical Models The primary way we learn about the interior of the Sun (and other stars) is by creating *mathematical models* that use the laws of physics to predict internal conditions. A basic model starts with the Sun's observed composition and mass and then solves equations that describe gravitational equilibrium and the rate at which solar energy moves from the core to the photosphere. With the aid of a computer, we can use the model to calculate the Sun's temperature, pressure, and density at any depth. We can then predict the rate of nuclear fusion in the solar core by combining these calculations with knowledge about nuclear fusion gathered in laboratories here on Earth.

If a model is a good description of the Sun's interior, it should correctly predict the radius, surface temperature, luminosity, age, and other observable properties of the Sun. Current models do indeed predict these properties quite accurately, giving us confidence that we really do understand what is going on inside the Sun.

Solar Vibrations A second way to learn about the inside of the Sun is to observe vibrations of the Sun's surface that are somewhat similar to the vibrations that earthquakes cause on Earth. Gas moving around in the solar interior generates vibrations that travel through the Sun like sound waves moving through air. We can observe these vibrations on the Sun's surface by looking for Doppler shifts [Section 5.2]. Light from portions of the surface that are rising toward us is slightly blueshifted, while light

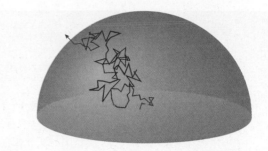

Figure 10.8

A photon in the solar interior bounces randomly among electrons, slowly working its way outward.

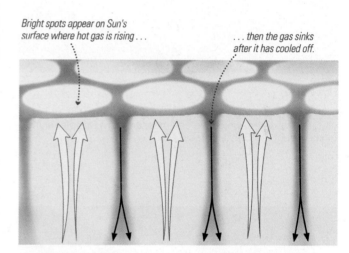

Bright spots appear on Sun's surface where hot gas is rising . . .

. . . then the gas sinks after it has cooled off.

a This diagram shows convection beneath the Sun's surface: hot gas (yellow arrows) rises while cooler gas (black arrows) descends around it.

Hot gas is rising here . . .   *. . . and cooler gas is sinking here.*

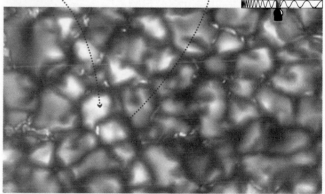

b This photograph shows the mottled appearance of the Sun's photosphere. The bright spots, each about 1000 kilometers across, correspond to the rising plumes of hot gas in the diagram in part (a).

Figure 10.9

The Sun's photosphere churns with rising hot gas and falling cool gas as a result of underlying convection.

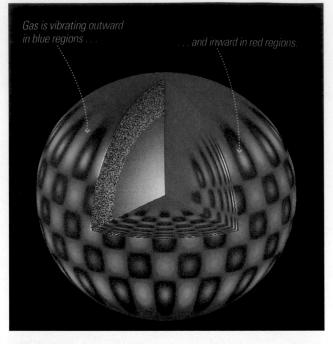

Gas is vibrating outward in blue regions . . .   *. . . and inward in red regions.*

Figure 10.10

This schematic diagram shows one of the Sun's many possible vibration patterns, which we can measure from Doppler shifts in spectra of the Sun's surface.

Figure 10.11

This tank of dry-cleaning fluid (visible underneath the catwalk), located deep within South Dakota's Homestake mine, was a solar neutrino detector. Some chlorine nuclei in the cleaning fluid turned into argon nuclei when they captured neutrinos from the Sun.

from portions that are falling away from us is slightly redshifted. The vibrations are relatively small but measurable (Figure 10.10).

The characteristics of solar vibrations support our mathematical models of the Sun's interior.

In principle, we can deduce a great deal about the solar interior by carefully analyzing these vibrations. (By analogy to seismology on Earth, this type of study of the Sun is called *helioseismology—helios* means "sun.") Results to date confirm that our mathematical models of the solar interior are on the right track. At the same time, they provide data that help improve the models further.

Solar Neutrinos   A third way to study the solar interior is to observe subatomic particles produced by fusion reactions in the core. If you look back at Figure 10.6, you'll see that some of the energy (about 2% of the total) released in nuclear fusion is carried off by subatomic particles called **neutrinos**. Neutrinos are a strange type of particle that rarely interacts with anything at all and can therefore pass through almost anything. For example, an inch of lead will stop an X ray, but stopping an average neutrino would require a slab of lead more than a light-year thick!

Neutrinos produced in the Sun's core therefore pass outward through the solar interior almost as though it were empty space. Traveling at nearly the speed of light, they reach us in minutes, in principle giving us a way to monitor what is happening in the core of the Sun. In practice, their elusiveness makes neutrinos dauntingly difficult to capture and count. About a thousand trillion solar neutrinos will zip through your body as you read this sentence, but don't panic—they will do no damage at all. In fact, nearly all of them will pass right through the entire Earth as well.

Nevertheless, neutrinos *do* occasionally interact with matter, and it is possible to capture a few solar neutrinos using a large enough detector. To distinguish neutrino captures from reactions caused by other particles, neutrino detectors are usually placed deep underground in mines. The overlying rock blocks most other particles, but the neutrinos have no difficulty passing through. The first major solar neutrino detector, built in the 1960s, was located 1500 m underground in the Homestake gold mine in South Dakota (Figure 10.11). The detector for this "Homestake experiment" consisted of a 400,000-liter vat of chlorine-based dry-cleaning fluid. On very rare occasions, a chlorine nucleus will capture a neutrino and change into a nucleus of radioactive argon. By looking for radioactive argon in the tank of cleaning fluid, experimenters could count the number of neutrinos captured in the detector.

From the many trillions of solar neutrinos that passed through the tank of cleaning fluid each second, experimenters expected to capture an average of just one neutrino per day. This predicted capture rate was based on measured properties of chlorine nuclei and models of nuclear fusion in the Sun. However, over a period of more than two decades, neutrinos were captured about once every three days, or only about one-third as often as predicted. This disagreement between model predictions and actual observations came to be called the *solar neutrino problem.*

For more than three decades, the solar neutrino problem was one of the great mysteries in astronomy. Other experiments found the same shortfall as the Homestake experiment. These results allowed only two possible conclusions: Either something was wrong with our understanding of fusion in the Sun, or some of the Sun's neutrinos were somehow escaping detection.

Scientists are now convinced that the neutrinos were going undetected. For decades, scientists have known that neutrinos come in three types, called *electron neutrinos, muon neutrinos,* and *tau neutrinos.* Fusion reactions in the Sun produce only electron neutrinos, and until recently most solar neutrino detectors could detect only electron neutrinos. We now think that some of the electron neutrinos produced by fusion change into neutrinos of the other two types on their way to Earth, explaining why past experiments counted fewer electron neutrinos than expected.

Subatomic particles called neutrinos provide a direct probe of nuclear fusion in the Sun, and recent results indicate that fusion occurs as our models predict.

Recent experiments support the idea that neutrinos change type on their way to Earth, probably during the portion of their trip from the Sun's core to its surface. Most significantly, a detector in Canada called the Sudbury Neutrino Observatory can detect all three neutrino types (Figure 10.12). Results from this observatory show that the total number of neutrinos of all types is indeed what we expect from our models of nuclear fusion in the Sun. The solar neutrino problem is now considered solved, meaning that neutrino counts verify that nuclear fusion occurs in the Sun in just the way our models predict.

10.3 The Sun–Earth Connection

Energy liberated by nuclear fusion in the Sun's core eventually reaches the solar surface. There, before it is ultimately released into space as sunlight, the energy helps create a wide variety of phenomena that we can observe from Earth. Sunspots are the most obvious of these phenomena. Because sunspots and other features of the Sun's surface change with time, they constitute what we call *solar weather* or **solar activity**. The "storms" associated with solar weather are not just of academic interest. Sometimes they are so violent that they affect our day-to-day life on Earth. In this section, we'll explore solar activity and its far-reaching effects.

• What causes solar activity?

Most of the Sun's surface churns constantly with rising and falling gas and thus looks like the close-up photo shown in Figure 10.9b. However, larger features sometimes appear, including sunspots, the huge explosions known as solar flares, and gigantic loops of hot gas extending high into the Sun's corona. All these features are created by magnetic fields, which form and change easily in the convecting plasma in the outer layers of the Sun.

Sunspots and Magnetic Fields Sunspots are the most striking features of the solar surface (Figure 10.13a). If you could look directly at a sunspot without damaging your eyes, you would find it blindingly bright. Sunspots appear dark in photographs only because they are less bright than the surrounding photosphere. They are less bright because they are cooler: The temperature of the plasma in sunspots is about 4000 K, significantly cooler than the 5800 K plasma that surrounds them.

If you think about this for a moment, you may wonder how sunspots can be so much cooler than their surroundings. Gas usually flows easily, so you might expect the hotter gas around a sunspot to mix with the cooler gas within it, quickly warming the sunspot. The fact that sunspots

Figure 10.12

This photograph shows the main tank of the Sudbury Neutrino Observatory in Canada, located at the bottom of a mine shaft more than 2 km underground. The large sphere, 12 m in diameter, contains 1000 tons of ultrapure "heavy water." (Heavy water is water in which the two hydrogen atoms are replaced by deuterium, making each molecule heavier than a molecule of ordinary water.) Neutrinos of all three types can cause changes in the heavy water molecules, and detectors surrounding the tank record these reactions when they occur.

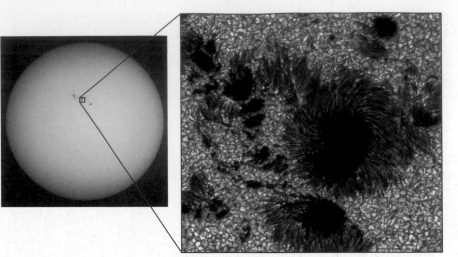

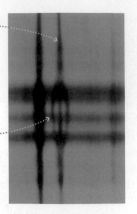

Outside a sunspot we see a single spectral line . . .

. . . but the strong magnetic field inside a sunspot splits that line into three lines.

a This close-up view of the Sun's surface shows two large sunspots and several smaller ones. Each of the big sunspots is roughly as large as Earth.

b Very strong magnetic fields split the absorption lines in spectra of sunspot regions. The dark vertical bands are absorption lines in a spectrum of the Sun. Notice that these lines split where they cross the dark horizontal bands corresponding to sunspots.

Figure 10.13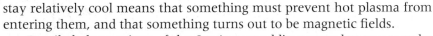

Sunspots are regions of strong magnetic fields.

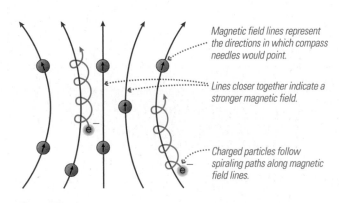

Magnetic field lines represent the directions in which compass needles would point.

Lines closer together indicate a stronger magnetic field.

Charged particles follow spiraling paths along magnetic field lines.

Figure 10.14

We draw magnetic field lines (red) to represent invisible magnetic fields.

stay relatively cool means that something must prevent hot plasma from entering them, and that something turns out to be magnetic fields.

Detailed observations of the Sun's spectral lines reveal sunspots to be regions with strong magnetic fields. These magnetic fields can alter the energy levels in atoms and ions and therefore can alter the spectral lines they produce. More specifically, magnetic fields cause some spectral lines to split into two or more closely spaced lines (Figure 10.13b). Wherever we see this effect (called the *Zeeman effect*), we know that magnetic fields must be present. Scientists can map magnetic fields on the Sun by looking for the splitting of spectral lines in light from different parts of the solar surface.

Sunspots are regions of strong magnetic fields, which keep the sunspots cooler than the surrounding photosphere.

To understand how sunspots stay cooler than their surroundings, we must investigate the nature of magnetic fields in a little more depth. Magnetic fields are invisible, but we can represent them by drawing **magnetic field lines** (Figure 10.14). These lines represent the directions in which compass needles would point if we placed them within the magnetic field. The lines are closer together where the field is stronger and farther apart where the field is weaker. Because these imaginary field lines are easier to visualize than the magnetic field itself, we usually discuss magnetic fields by talking about how the field lines would appear. Charged particles, such as the ions and electrons in the solar plasma, cannot easily move perpendicular to the field lines but instead follow spiraling paths along them.

Solar magnetic field lines act somewhat like elastic bands, being twisted into contortions and knots by turbulent motions in the solar atmosphere. Sunspots occur where tightly wound magnetic fields poke nearly straight out from the solar interior (Figure 10.15a). These tight magnetic field lines suppress convection within the sunspot and prevent surrounding plasma from entering the sunspot. With hot plasma unable to enter the region, the sunspot plasma becomes cooler than that of the rest of the photosphere. Individual sunspots typically last up to a few weeks, dissolving when their magnetic fields weaken and allow plasma to flow in.

Sunspots tend to occur in pairs, connected by a loop of magnetic field lines that can arc high above the Sun's surface (Figure 10.15b). Gas in the

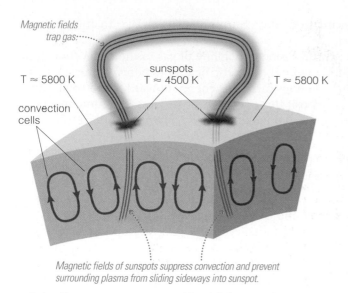

Magnetic fields trap gas

T ≈ 5800 K   sunspots   T ≈ 5800 K
                T ≈ 4500 K

convection
cells

Magnetic fields of sunspots suppress convection and prevent
surrounding plasma from sliding sideways into sunspot.

a Pairs of sunspots are connected by tightly wound magnetic
field lines.

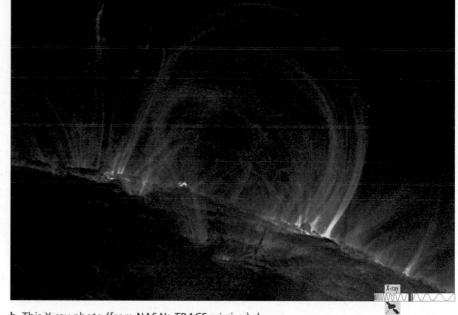

X-ray

b This X-ray photo (from NASA's *TRACE* mission) shows
hot gas trapped within looped magnetic field lines.

Figure 10.15

Strong magnetic fields keep sunspots cooler than the surrounding photosphere, while magnetic loops can arch from the
sunspots to great heights above the Sun's surface.

Sun's chromosphere and corona becomes trapped in these giant loops, called **solar prominences**. Some prominences, such as the one shown in the photo that opens this chapter (p. 284), rise to heights of more than 100,000 km above the Sun's surface. Individual prominences can last for days or even weeks.

Solar Storms The magnetic fields winding through sunspots and prominences sometimes undergo dramatic and sudden change, producing short-lived but intense storms on the Sun. The most dramatic of these storms are **solar flares**, which send bursts of X rays and fast-moving charged particles shooting into space (Figure 10.16).

> Energy released when magnetic field lines snap can lead to dramatic solar storms, which sometimes eject bursts of energetic particles into space.

Flares generally occur in the vicinity of sunspots, which is why we think they are created by changes in magnetic fields. The leading model for solar flares suggests that they occur when the magnetic field lines become so twisted and knotted that they can no longer bear the tension. They are thought to suddenly snap and reorganize themselves into a less twisted configuration. The energy released in the process heats the nearby plasma to 100 million K over the next few minutes to hours, generating X rays and accelerating some of the charged particles to nearly the speed of light.

Heating of the Chromosphere and Corona As we've seen, many of the most dramatic weather patterns and storms on the Sun involve the very hot gas of the Sun's chromosphere and corona. But why is that gas so hot in the first place? This question perplexed scientists for decades.

Remember that temperatures gradually decline as we move outward from the Sun's core to the top of its photosphere. We might expect the decline to continue in the Sun's atmosphere, but instead it reverses, making the chromosphere and corona much hotter than the Sun's surface. Some aspects of this atmospheric heating remain a mystery today, but we have at least a general explanation: The Sun's strong magnetic fields

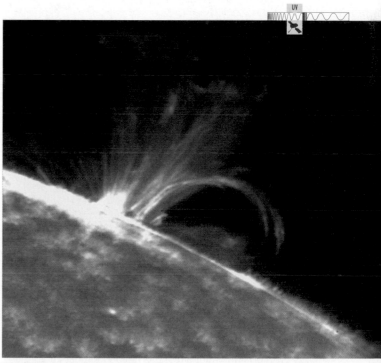

UV

Figure 10.16

This photo (from *TRACE*) of ultraviolet light emitted by hydrogen atoms shows a solar flare erupting from the Sun's surface.

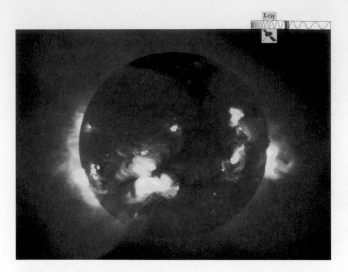

Figure 10.17 interactive photo

An X-ray image of the Sun reveals the million-degree gas of the corona. Brighter regions of this image (yellow) correspond to regions of stronger X-ray emission. The darker regions (such as near the north pole at the top of this photo) are the coronal holes from which the solar wind escapes. (From the Yohkoh Space Observatory.)

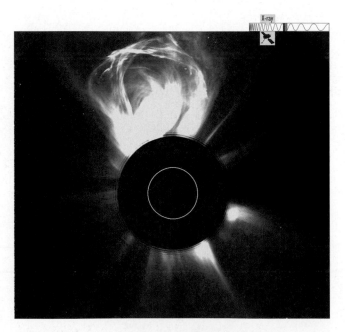

Figure 10.18

This X-ray image from the *SOHO* spacecraft shows a coronal mass ejection (the bright arc of gas headed almost straight upward as oriented here) during the solar storms of 2003. The central red disk blots out the Sun itself, and the white circle represents the size of the Sun in this picture.

carry energy upward from the churning solar surface to the chromosphere and corona.

More specifically, the rising and falling of gas in the convection zone probably shake tightly wound magnetic field lines beneath the solar surface. The magnetic field lines carry this energy upward to the solar atmosphere, where they deposit this energy as heat. Thus, the same magnetic fields that keep sunspots cool make the overlying plasma of the chromosphere and corona hot.

Magnetic fields deposit energy above the Sun's surface, heating the chromosphere and corona.

Observations confirm the connection between magnetic fields and the structure of the chromosphere and corona. The density of gas in the chromosphere and corona is so low that we cannot see this gas with our eyes except during a total eclipse, when we can see the faint visible light scattered by electrons in the corona (see Figure 2.24). However, we can observe the chromosphere and corona at any time with ultraviolet and X-ray telescopes in space: The roughly 10,000 K plasma of the chromosphere emits strongly in the ultraviolet, and the 1 million K plasma of the corona is the source of virtually all X rays coming from the Sun. Figure 10.17 shows an X-ray image of the Sun. The X-ray emission is brightest in regions where hot gas is being trapped and heated in magnetic field loops. The bright spots in the corona tend to be directly above sunspots in the photosphere, confirming that they are created by the same magnetic fields.

Notice that some regions of the corona barely show up in X-ray images; these regions, called **coronal holes**, are nearly devoid of hot, coronal gas. More detailed analyses show that the magnetic field lines in coronal holes project out into space like broken rubber bands, allowing particles spiraling along them to escape the Sun altogether. These particles streaming outward from the corona are the source of the solar wind.

• How does solar activity affect humans?

Flares and other solar storms sometimes eject large numbers of highly energetic charged particles from the Sun's corona. These particles travel outward from the Sun in huge bubbles that we call **coronal mass ejections** (Figure 10.18). These bubbles have strong magnetic fields and can reach Earth in a couple of days if they happen to be aimed in our direction. Once a coronal mass ejection reaches Earth, it can create a *geomagnetic storm* in Earth's magnetosphere. On the positive side, these storms can lead to unusually strong auroras (see Figure 7.6) that can be visible throughout much of the United States. On the negative side, they can also hamper radio communications, disrupt electrical power delivery, and damage the electronic components in orbiting satellites.

Particles ejected from the Sun during periods of high activity can hamper radio communications, disrupt power delivery, and damage orbiting satellites.

During a particularly powerful magnetic storm on the Sun in March 1989, the U.S. Air Force temporarily lost track of more than 2000 satellites, and powerful currents induced in the ground circuits of the Quebec hydroelectric system caused it to collapse for more than 8 hours. In January 1997, AT&T lost contact with a $200 million communications satellite, probably because of damage caused by particles coming from another powerful solar storm. More recently, in the fall of 2003, a series of extremely powerful solar flares once again threatened Earth's communications and electrical systems, but they escaped major damage, due in part to our improved preparedness for solar storms.

Satellites in low-Earth orbit are particularly vulnerable during periods of strong solar activity, when the increase in solar X rays and energetic particles heats Earth's upper atmosphere, causing it to expand. The density of the gas surrounding low-flying satellites therefore increases, exerting drag that saps their energy and angular momentum. If this drag proceeds unchecked, the satellites ultimately plummet back to Earth. Satellites in low orbits, including the Hubble Space Telescope and the Space Station, require occasional boosts to prevent them from falling out of the sky.

• How does solar activity vary with time?

Solar weather is just as unpredictable as weather on Earth. Individual sunspots can appear or disappear at almost any time, and we have no way of knowing that a solar storm is coming until we observe it through our telescopes. However, long-term observations have revealed overall patterns in solar activity that make sunspots and solar storms more common at some times than at others.

The Sunspot Cycle The most notable pattern in solar activity is the **sunspot cycle**—a cycle in which the average number of sunspots on the Sun gradually rises and falls (Figure 10.19). At the time of *solar maximum*, when sunspots are most numerous, we may see dozens of sunspots on the Sun at one time. In contrast, we may see few if any sunspots at the time of *solar minimum*. The frequency of prominences, flares, and coronal mass ejections also follows the sunspot cycle, with these events being most common at solar maximum and least common at solar minimum.

Notice that the sunspot cycle varies from one period to the next (Figure 10.19a). The average length of time between maximums is 11 years,

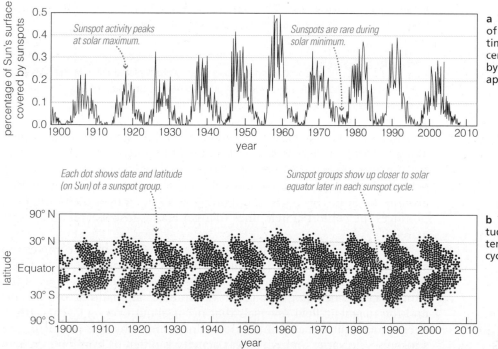

a This graph shows how the number of sunspots on the Sun changes with time. The vertical axis shows the percentage of the Sun's surface covered by sunspots. The cycle has a period of approximately 11 years.

b This graph shows how the latitudes at which sunspot groups appear tend to shift during a single sunspot cycle.

Figure 10.19

Sunspot cycle during the past century.

but we have observed it to be as short as 7 years and as long as 15 years. Observations going further back in time suggest that sunspot activity can

The average number of sunspots on the Sun rises and falls in an approximately 11-year cycle.

sometimes cease almost entirely. For example, astronomers observed virtually no sunspots between the years 1645 and 1715, a period sometimes called the *Maunder minimum* (after E. W. Maunder, who identified it in historical sunspot records).

The locations of sunspots on the Sun also vary with the sunspot cycle (Figure 10.19b). As a cycle begins at solar minimum, sunspots form primarily at mid-latitudes (30° to 40°) on the Sun. The sunspots tend to form at lower latitudes as the cycle progresses, appearing very close to the solar equator as the next solar minimum approaches. Then the sunspots of the next cycle begin to form near mid-latitudes again.

A less obvious feature of the sunspot cycle is that something peculiar happens to the Sun's magnetic field at each solar maximum: The Sun's entire magnetic field starts to flip, turning magnetic north into magnetic south and vice versa. We know this because the magnetic field lines connecting pairs of sunspots (see Figure 10.15) within each hemisphere all tend to point in the same direction throughout an 11-year cycle. For example, all compass needles might point from the easternmost sunspot to the westernmost sunspot in each sunspot pair north of the solar equator. However, by the time the cycle ends at solar minimum, the magnetic field has reversed: In the subsequent solar cycle, the field lines connecting pairs of sunspots point in the opposite direction. Thus, the Sun's *complete* magnetic cycle (sometimes called the *solar cycle*) really averages 22 years, since it takes two 11-year sunspot cycles for the magnetic field to return to the way it started.

The Cause of the Sunspot Cycle The precise reasons for the sunspot cycle are not fully understood, but the leading model ties it to a combination of convection and the Sun's rotation. Convection is thought to dredge up weak magnetic fields generated in the solar interior, amplifying them as they rise. The Sun's rotation—faster at its equator than near its poles—then stretches and shapes these fields.

think about it Suppose you observe two sunspots: one near the Sun's equator and one directly north of it at 20° latitude. Where would you expect to see the two spots if you observe the Sun again a few days later? Would one still be directly north of the other? Explain.

Imagine what happens to magnetic field lines that start out running along the Sun's surface from south to north (Figure 10.20). At the equator, the lines circle the Sun every 25 days, but at higher latitudes, they lag behind. As a result, the lines gradually get wound more and more tightly around the Sun. This process, operating at all times over the entire Sun, produces the contorted field lines that generate sunspots and other solar activity.

The detailed behavior of these magnetic fields is quite complex, so scientists attempt to study it with sophisticated computer models. Using these models, scientists have successfully replicated many features of the sunspot cycle, including changes in the number and latitude of sunspots and the magnetic field reversals that occur about every 11 years. However, much still remains mysterious, including why the period of the sunspot cycle varies and why solar activity is different from one cycle to the next.

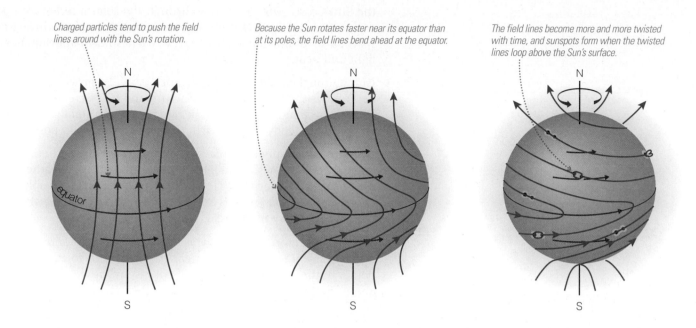

Charged particles tend to push the field lines around with the Sun's rotation.

Because the Sun rotates faster near its equator than at its poles, the field lines bend ahead at the equator.

The field lines become more and more twisted with time, and sunspots form when the twisted lines loop above the Sun's surface.

The Sunspot Cycle and Earth's Climate Despite the changes that occur during the sunspot cycle, the Sun's total output of energy barely changes at all—the largest measured changes have been less than 0.1% of the Sun's average luminosity. However, the ultraviolet and X-ray output of the Sun, which comes from the magnetically heated gas of the chromosphere and corona, can vary much more significantly. Could any of these changes affect the weather or climate on Earth?

Some data suggest connections between solar activity and Earth's climate. For example, the period from 1645 to 1715, when solar activity seems to have virtually ceased, was a time of exceptionally low temperatures in Europe and North America known as the *Little Ice Age*. However, no one knows whether the low solar activity caused these low temperatures, or whether it was just a coincidence. Similarly, some researchers have claimed that certain weather phenomena, such as drought cycles or frequencies of storms, are correlated with the 11- or 22-year cycle of solar activity. A few scientists have even claimed that changes in the Sun may be responsible for Earth's recent global warming [Section 7.5], though climate models indicate that the magnitude of the observed warming can be explained only by including human activity along with solar changes and other natural factors (see Figure 7.44). The bottom line is that we do not yet know whether or how much solar activity affects Earth's climate. Given the importance of the question, it is certain to remain a hot topic of scientific research.

Figure 10.20

The Sun rotates more quickly at its equator than it does near its poles. Because gas circles the Sun faster at the equator, it drags the Sun's north–south magnetic field lines into a more twisted configuration. The magnetic field lines linking pairs of sunspots, depicted here as green and black blobs, trace out the directions of these stretched and distorted field lines.

the big picture
Putting Chapter 10 into Context

In this chapter, we have examined our Sun, the nearest star. When you look back at this chapter, make sure you understand the following "big picture" ideas:

- The ancient riddle of why the Sun shines is now solved. The Sun shines with energy generated by fusion of hydrogen into helium in the Sun's core. After a journey through the solar interior lasting several hundred thousand years and an 8-minute journey through space, a small fraction of this energy reaches Earth and supplies sunlight and heat.

- Gravitational equilibrium, the balance between pressure and gravity, determines the Sun's interior structure and helps create a natural thermostat that keeps the fusion rate in the Sun steady. If the Sun were not so steady, life on Earth might not have been possible.

- The Sun's atmosphere displays its own version of weather and climate, governed by solar magnetic fields. Some solar weather, such as coronal mass ejections, clearly affects Earth's magnetosphere. Other claimed connections between solar activity and Earth's climate may or may not be real.

- The Sun is important not only as our source of light and heat, but also because it is the only star near enough for us to study in great detail. In the coming chapters, we will use what we've learned about the Sun to help us understand other stars.

summary of key concepts

10.1 A Closer Look at the Sun

• Why does the Sun shine?

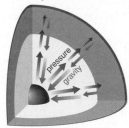
The Sun shines because **gravitational equilibrium** keeps its core hot and dense enough to release energy through nuclear fusion. The core originally became hot through the release of energy by **gravitational contraction**, as gravity made the Sun's birth cloud contract.

• What is the Sun's structure?

The Sun's interior layers, from the inside out, are the **core**, the **radiation zone**, and the **convection zone**. Atop the convection zone lies the **photosphere**, the layer from which photons can freely escape into space. Above the photosphere (essentially the surface of the Sun) are the warmer **chromosphere** and the very hot **corona**.

10.2 Nuclear Fusion in the Sun

• How does nuclear fusion occur in the Sun?

The core's extreme temperature and density are just right for fusion of hydrogen into helium, which occurs via the **proton–proton chain**. Because the fusion rate is so sensitive to temperature, gravitational equilibrium acts as a thermostat that keeps the rate of fusion steady.

• How does the energy from fusion get out of the Sun?
Energy moves through the deepest layers of the Sun—the core and the radiation zone—in the form of randomly bouncing photons. After energy emerges from the radiation zone, convection carries it the rest of the way to the photosphere, where it is radiated into space as sunlight. Energy produced in the core takes hundreds of thousands of years to reach the photosphere.

• How do we know what is happening inside the Sun?
We can construct theoretical models of the solar interior using known laws of physics and then check the models against observations of the Sun's size, surface temperature, and energy output. We also use studies of solar vibrations and solar **neutrinos**.

10.3 The Sun–Earth Connection

• What causes solar activity?

Convection combined with the rotation pattern of the Sun—faster at the equator than at the poles—causes **solar activity** because the gas motions stretch and twist the Sun's magnetic field. These contortions of the magnetic field are responsible for phenomena such as **sunspots**, **solar flares**, **solar prominences**, and **coronal mass ejections** and for heating the gas in the chromosphere and corona.

• How does solar activity affect humans?

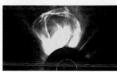

Bursts of charged particles ejected from the Sun during periods of high solar activity can hamper radio communications, disrupt electrical power generation, and damage orbiting satellites.

• How does solar activity vary with time?

The **sunspot cycle,** or the variation in the number of sunspots on the Sun's surface, has an average period of 11 years. The magnetic field flip-flops every 11 years or so, resulting in a 22-year magnetic cycle. Sunspots first appear at mid-latitudes at solar minimum, then become increasingly more common near the Sun's equator as the next minimum approaches, and sometimes seem to be absent altogether.

exercises and problems

For instructor-assigned homework go to www.masteringastronomy.com.

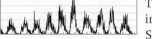

Review Questions

Short-Answer Questions Based on the Reading

1. Briefly describe how *gravitational contraction* generates energy. When was it important in the Sun's history? Explain.
2. What two forces are balanced in *gravitational equilibrium?* Describe how gravitational equilibrium makes the Sun hot and dense in its core.
3. State the Sun's luminosity, mass, radius, and average surface temperature, and put the numbers into a perspective that makes them meaningful.
4. Briefly describe the distinguishing features of each of the layers of the Sun shown in Figure 10.3.
5. What is the difference between nuclear *fission* and nuclear *fusion?* Which one is used in nuclear power plants? Which one is used by the Sun?
6. Why does nuclear fusion require high temperatures and pressures?
7. What is the overall nuclear fusion reaction in the Sun? Briefly describe the proton–proton chain.
8. Does the Sun's fusion rate remain steady or vary wildly? Describe the feedback process that regulates the fusion rate.
9. Why does the energy produced by fusion in the solar core take so long to reach the solar surface? Describe the processes by which energy generated by fusion makes its way to the Sun's surface.
10. Explain how mathematical models allow us to predict conditions inside the Sun. How can we be confident that the models are on the right track?
11. What are *neutrinos?* What was the *solar neutrino problem,* and why do we think it has now been solved?
12. What do we mean by *solar activity?* Describe some of the features of solar activity, including *sunspots, solar prominences, solar flares,* and *coronal mass ejections.*
13. Describe the appearance and temperature of the Sun's photosphere. Why does the surface look mottled? How are sunspots different from the surrounding photosphere?
14. How do magnetic fields keep sunspots cooler than the surrounding plasma? Explain.
15. Why is the chromosphere best viewed with ultraviolet telescopes? Why is the corona best viewed with X-ray telescopes? Briefly explain how we think the chromosphere and corona are heated.
16. What is the sunspot cycle? Describe how the Sun changes during the cycle, and how the changes are thought to be related to magnetic fields and the Sun's rotation. Does the sunspot cycle influence Earth's climate? Explain.

Test Your Understanding

Does It Make Sense?

Decide whether the statement makes sense (or is clearly true) or does not make sense (or is clearly false). Explain clearly; not all of these statements have definitive answers, so your explanation is more important than your chosen answer.

17. Before Einstein, gravitational contraction appeared to be a perfectly plausible mechanism for solar energy generation.
18. A sudden temperature rise in the Sun's core is nothing to worry about, because conditions in the core will soon return to normal.
19. If fusion in the solar core ceased today, worldwide panic would break out tomorrow as the Sun began to grow dimmer.
20. Astronomers have recently photographed magnetic fields churning deep beneath the solar photosphere.
21. Neutrinos probably can't harm me, but just to be safe I think I'll wear a lead vest.
22. There haven't been many sunspots this year, but there ought to be many more in about 5 years.
23. News of a major solar flare today caused concern among professionals in the fields of communications and electrical power generation.
24. By observing solar neutrinos, we can learn about nuclear fusion deep in the Sun's core.
25. If the Sun's magnetic field somehow disappeared, there would be no more sunspots on the Sun.
26. Scientists are currently building an infrared telescope designed to observe fusion reactions in the Sun's core.

Quick Quiz

Choose the best answer to each of the following. Explain your reasoning with one or more complete sentences.

27. Which of these groups of particles has the greatest mass?
    (a) a helium nucleus with two protons and two neutrons
    (b) four electrons (c) four individual protons

28. Which of these layers of the Sun is coolest? (a) photosphere (b) chromosphere (c) corona

29. Which of these layers of the Sun is coolest? (a) core (b) radiation zone (c) photosphere

30. Scientists estimate the central temperature of the Sun using (a) probes that measure changes in Earth's atmosphere. (b) mathematical models of the Sun. (c) laboratories that create miniature versions of the Sun.

31. Why do sunspots appear darker than their surroundings? (a) They are cooler than their surroundings. (b) They block some of the sunlight from the photosphere. (c) They do not emit any light.

32. At the center of the Sun, fusion converts hydrogen into (a) plasma. (b) radiation and elements like carbon and nitrogen. (c) helium, energy, and neutrinos.

33. Solar energy leaves the core of the Sun in the form of (a) photons. (b) rising hot gas. (c) sound waves.

34. How does the number of neutrinos passing through your body at night compare with the number passing through your body during the day? (a) about the same (b) much smaller (c) much larger

35. What is the most common kind of particle in the solar wind, in which the overall charge is neutral? (a) proton (b) electron (c) helium nucleus

36. Which of these things poses the greatest hazard to communications satellites? (a) photons from the Sun (b) solar magnetic fields (c) particles from the Sun

Process of Science

Examining How Science Works

37. *Inside the Sun.* Scientists claim to know what is going on inside the Sun, even though we cannot directly observe the Sun's interior. What is the basis for these claims, and how do they relate to the hallmarks of science outlined in Section 3.4?

38. *The Solar Neutrino Problem.* Early solar neutrino experiments detected only about a third of the number of neutrinos predicted by the theory of fusion in the Sun. Why didn't scientists simply abandon their models at this point? What features of the Sun did the model get right? What alternatives were there for explaining the mismatch between the predictions and the observations?

Investigate Further

In-Depth Questions to Increase Your Understanding

Short-Answer/Essay Questions

39. *The End of Fusion I.* Describe what would happen in the Sun if fusion reactions abruptly ceased.

40. *The End of Fusion II.* If fusion reactions in the Sun were to suddenly cease, would we be able to tell? If so, how?

41. *A Really Strong Force.* How would the interior temperature of the Sun be different if the strong force that binds nuclei together were 10 times stronger?

42. *Measuring the Sun's Rotation.* Suppose you observe a sunspot from Earth, taking a picture of the Sun each day. Using these photos, you measure the time it takes for the sunspot to return to the position it had in the first picture. Now suppose a friend of yours standing on Pluto does the same thing. Would you and your friend measure the same time for the sunspot to return to where it started? Explain your answer.

43. *Covered with Sunspots.* Describe what the Sun would look like from Earth if the entire photosphere were the same temperature as a sunspot.

44. *Inside the Sun.* Describe how scientists determine what the interior of the Sun is like. Why have we not yet sent a probe into the Sun to measure what is happening there?

45. *Solar Energy Output.* The Sun's output of visible light remains very steady, staying well within 1% of its current value for the past century. However, the Sun's maximum X-ray output can be as much as 10 times greater than its minimum output. Explain why the changes in X-ray output can be so much more pronounced than the changes in visible light output.

46. *An Angry Sun.* A *Time* magazine cover once suggested that an "angry Sun" is becoming more active as human activity is changing Earth's climate through global warming. It's certainly possible for the Sun to become more active at the same time that humans are affecting Earth, but is it possible that the Sun could be responding to human activity? Can humans affect the Sun in any significant way? Explain.

Quantitative Problems

Be sure to show all calculations clearly and state your final answers in complete sentences.

47. *The Color of the Sun.* The Sun's average surface temperature is about 5800 K. Use Wien's law (see Cosmic Calculations 5.1) to calculate the wavelength of peak thermal emission from the Sun. What color does this wavelength correspond to in the visible-light spectrum? Why do you think the Sun appears white or yellow to our eyes?

48. *The Color of a Sunspot.* The typical temperature of a sunspot is about 4000 K. Use Wien's law (see Cosmic Calculations 5.1) to calculate the wavelength of peak thermal emission from a sunspot. What color does this wavelength correspond to in the visible-light spectrum? How does this color compare with that of the Sun?

49. *Solar Mass Loss.* Estimate how much mass the Sun will lose through fusion reactions during its 10-billion-year life. You can simplify the problem by assuming the Sun's energy output remains constant. Compare the amount of mass lost with Earth's mass.

50. *Pressure of the Photosphere.* The gas pressure of the photosphere changes substantially from its upper levels to its lower levels. Near the top of the photosphere the temperature is about 4500 K and there are about 1.6×10^{16} gas particles per cubic centimeter. In the middle the temperature is about 5800 K and there are about 1.0×10^{17} gas particles per cubic centimeter. At the bottom of the photosphere the temperature is about 7000 K and there are about 1.5×10^{17} gas particles per cubic centimeter. Compare the pressures of each of these layers and explain the reason for the trend in pressure that you find. How do these gas pressures compare with Earth's atmospheric pressure at sea level? (*Hint:* See Cosmic Calculations 10.1.)

51. *The Lifetime of the Sun.* The total mass of the Sun is about 2×10^{30} kg, of which about 75% was hydrogen when the Sun formed. However, only about 13% of this hydrogen ever becomes available for fusion in the core. The rest remains in layers of the Sun where the temperature is too low for fusion.
    a. Based on the given information, calculate the total mass of hydrogen available for fusion over the lifetime of the Sun.

b. Combine your results from part (a) and the fact that the Sun fuses about 600 billion kg of hydrogen each second to calculate how long the Sun's initial supply of hydrogen can last. Give your answer in both seconds and years.

c. Given that our solar system is now about 4.6 billion years old, when will we need to start worrying about the Sun running out of hydrogen for fusion?

52. *Solar Power Collectors.* This problem leads you through the calculation and discussion of how much solar power can be collected by solar cells on Earth.

a. Imagine a giant sphere surrounding the Sun with a radius of 1 AU. What is the surface area of this sphere in square meters? (*Hint:* The formula for the surface area of a sphere is $4\pi r^2$.)

b. Because this imaginary giant sphere surrounds the Sun, the Sun's entire luminosity of 3.8×10^{26} watts must pass through it. Calculate the power passing through each square meter of this imaginary sphere in *watts per square meter.* Explain why this number represents the maximum power per square meter that a solar collector in Earth orbit can collect.

c. List several reasons why the average power per square meter collected by a solar collector on the ground will always be less than what you found in part (b).

d. Suppose you want to put a solar collector on your roof. If you want to optimize the amount of power you can collect, how should you orient the collector? (*Hint:* The optimum orientation depends on both your latitude and the time of year and day.)

Discussion Questions

53. *The Role of the Sun.* Briefly discuss how the Sun affects us here on Earth. Be sure to consider not only factors such as its light and warmth, but also how the study of the Sun has led us to new understandings in science and to technological developments. Overall, how important has solar research been to our lives?

54. *The Sun and Global Warming.* One of the most pressing environmental issues on Earth is the extent to which human emissions of greenhouse gases are warming our planet. Some people claim that part or all of the observed warming over the past century may be due to changes in the Sun rather than to anything humans have done. Discuss how a better understanding of the Sun might help us comprehend the threat posed by greenhouse gas emissions. Why is it so difficult to develop a clear understanding of how the Sun affects Earth's climate?

Web Projects

55. *Current Solar Weather.* Daily information about solar activity is available at numerous Web sites. Where are we in the sunspot cycle right now? When is the next solar maximum or minimum expected? Have there been any major solar storms in the past few months? If so, did they have any significant effects on Earth? Summarize your findings in a one- to two-page report.

56. *Sudbury Neutrino Observatory.* Visit the Web site for the Sudbury Neutrino Observatory (SNO) to learn how it has helped to solve the solar neutrino problem. Write a one- to two-page report describing recent results and expectations for future discoveries.

visual skills check

Use the following questions to check your understanding of some of the many types of visual information used in astronomy. Answers are provided in Appendix K. For additional practice, try the Chapter 10 Visual Quiz at **www.masteringastronomy.com.**

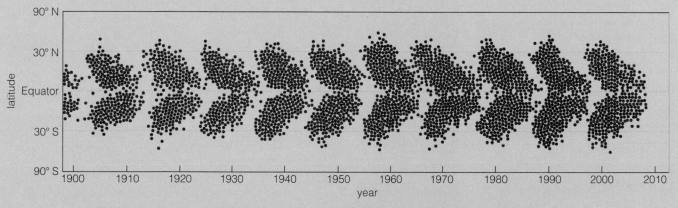

Figure 10.19b, repeated above, shows the latitudes at which sunspots appeared on the surface of the Sun during the 20th century. Answer the following questions, using the information provided in the figure.

1. Which of the following years had the least sunspot activity?
   a. 1930
   b. 1949
   c. 1961
   d. 1987

2. What is the approximate range in latitude over which sunspots appear?

3. According the figure, how do the positions of sunspots appear to change during one sunspot cycle? Do they get closer to or farther from the equator with time?

11

surveying the stars

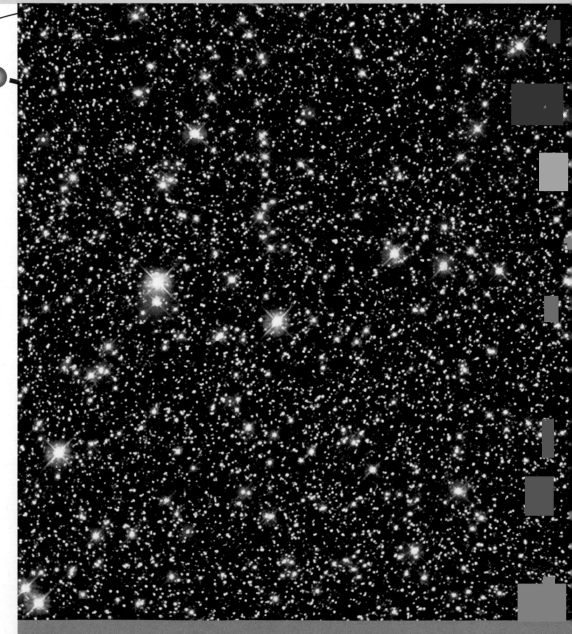

On a clear, dark night, a few thousand stars are visible to the naked eye. Many more become visible through binoculars, and a powerful telescope reveals so many stars that we could never hope to count them. Like each individual person, each individual star is unique. Like the human family, all stars have much in common.

Today, we know that stars are born from clouds of interstellar gas, shine brilliantly by nuclear fusion for millions to billions of years, and then die, sometimes in dramatic ways. In this chapter, we'll discuss how we study and categorize stars and how we have come to realize that stars, like people, change over their lifetimes.

essential preparation

1. How does Newton's law of gravity extend Kepler's laws? [Section 4.4]

2. How does light tell us what things are made of? [Section 5.2]

3. How does light tell us the temperatures of planets and stars? [Section 5.2]

4. How does light tell us the speed of a distant object? [Section 5.2]

11.1 Properties of Stars

Imagine that an alien spaceship flies by Earth on a simple but short mission: The visitors have just 1 minute to learn everything they can about the human race. In 60 seconds, they will see next to nothing of any individual person's life. Instead, they will obtain a collective "snapshot" of humanity showing people from all stages of life engaged in their daily activities. From this snapshot alone, they must piece together their entire understanding of human beings and their lives, from birth to death.

We face a similar problem when we look at the stars. Compared with stellar lifetimes of millions or billions of years, the few hundred years humans have spent studying stars with telescopes is rather like the aliens' 1-minute glimpse of humanity. We see only a brief moment in any star's life, and our collective snapshot of the heavens consists of such frozen moments for billions of stars. From this snapshot, we try to reconstruct the life cycles of stars.

Thanks to the efforts of hundreds of astronomers studying this snapshot of the heavens, we now know that all stars have a lot in common with the Sun. They all form in great clouds of gas and dust, and each one begins its life with roughly the same chemical composition as the Sun: About three-quarters of a star's mass at birth is hydrogen and about one-quarter is helium, with no more than about 2% consisting of elements heavier than helium. Nevertheless, stars are not all the same; they differ in size, age, brightness, and temperature. We'll devote most of this and the next chapter to understanding how and why stars differ. First, however, let's explore how we measure three of the most fundamental properties of stars: luminosity, surface temperature, and mass.

• How do we measure stellar luminosities?

If you go outside on any clear night, you'll immediately see that stars differ in brightness. Some stars are so bright that we can use them to identify constellations [Section 2.1]. Others are so dim that our naked eyes cannot see them at all. However, these differences in brightness do not by themselves tell us anything about how much light these stars are generating, because the brightness of a star depends on its distance as well as on how much light it actually emits. For example, the stars Procyon and Betelgeuse, which make up two of the three corners of the winter triangle (see Figure 2.2), appear about equally bright in our sky. However, Betelgeuse actually emits about 5000 times as much light as Procyon. It has about the same brightness in our sky because it is much farther away.

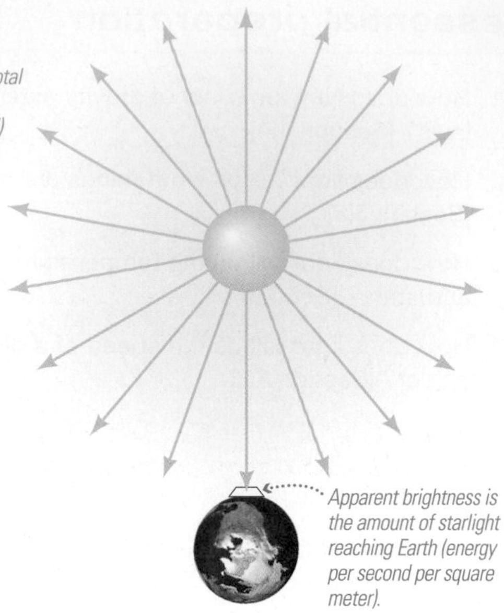

Luminosity is the total amount of power (energy per second) the star radiates into space.

Not to scale!

Apparent brightness is the amount of starlight reaching Earth (energy per second per square meter).

Figure 11.1

Luminosity is a measure of power, and apparent brightness is a measure of power per unit area.

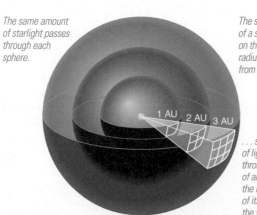

The same amount of starlight passes through each sphere.

The surface area of a sphere depends on the square of its radius (distance from the star) . . .

1 AU 2 AU 3 AU

. . . so the amount of light passing through each unit of area depends on the inverse square of its distance from the star.

Figure 11.2

The inverse square law for light: The apparent brightness of a star declines with the square of its distance.

Until the 20th century, people classified stars primarily by their brightness and location in our sky. On the next clear night, find your favorite constellation and visually rank its stars by brightness. Then look to see how that constellation is represented on the star charts in Appendix J. Why do the star charts use different size dots for different stars? Do the brightness rankings on the chart differ from what you see?

Because two similar-looking stars can be generating very different amounts of light, we need to distinguish clearly between a star's brightness in our sky and the actual amount of light that it emits into space (Figure 11.1):

- When we talk about how bright stars look in our sky, we are talking about **apparent brightness**. More specifically, we define the apparent brightness of any star in our sky as the amount of power (energy per second) reaching us *per unit area*.

- When we talk about how bright stars are in an absolute sense, regardless of their distance, we are talking about **luminosity**—the total amount of power that a star emits into space.

A star's *apparent brightness* in the sky depends on both its true light output, or *luminosity*, and its distance from us.

We can understand the difference between apparent brightness and luminosity by thinking about a 100-watt light bulb. The bulb always puts out the same amount of light, so its luminosity doesn't vary. However, its apparent brightness depends on your distance from the bulb: It will look quite bright if you stand very close to it, but quite dim if you are far away.

The Inverse Square Law for Light The apparent brightness of a star or any other light source obeys an *inverse square law* with distance, much like the inverse square law that describes the force of gravity [Section 4.4]. For example, if we viewed the Sun from twice Earth's distance, it would appear dimmer by a factor of $2^2 = 4$. If we viewed it from 10 times Earth's distance, it would appear $10^2 = 100$ times dimmer.

Figure 11.2 shows why apparent brightness follows an inverse square law. The same total amount of light must pass through each imaginary sphere surrounding the star. If we focus on the light passing through the small square on the sphere located at 1 AU, we see that the same amount of light must pass through *four* squares of the same size on the sphere located at 2 AU. Thus each square on the sphere at 2 AU receives only $\frac{1}{2^2} = \frac{1}{4}$ as much light as the square on the sphere at 1 AU. Similarly, the same amount of light passes through *nine* squares of the same size on the sphere located at 3 AU, so, each of these squares receives only $\frac{1}{3^2} = \frac{1}{9}$ as much light as the square on the sphere at 1 AU. Generalizing, we see that the amount of light received per unit area decreases with increasing distance by the square of the distance—an inverse square law.

Doubling the distance to a star would decrease its apparent brightness by a factor of 2^2, or 4.

This inverse square law leads to a very simple and important formula relating the apparent brightness, luminosity, and distance of any light source. We will call it the **inverse square law for light**:

$$\text{apparent brightness} = \frac{\text{luminosity}}{4\pi \times (\text{distance})^2}$$

Because the standard units of luminosity are watts [Section 10.1], the units of apparent brightness are *watts per square meter*. The 4π in the formula for the inverse square law for light comes from the fact that the surface area of a sphere is given by $4\pi \times (radius)^2$.

In principle, we can always measure a star's apparent brightness with a detector that records how much energy strikes its light-sensitive surface each second. We can therefore use the inverse square law to calculate a star's luminosity if we can first measure its distance, or to calculate a star's distance if we know its luminosity.

think about it Suppose Star A is four times as luminous as Star B. How will their apparent brightnesses compare if they are both the same distance from Earth? How will their apparent brightnesses compare if Star A is twice as far from Earth as Star B? Explain.

Measuring Cosmic Distances Tutorial, Lesson 2

Measuring Distance Through Stellar Parallax The most direct way to measure a star's distance is with *stellar parallax*, the small annual shifts in a star's apparent position caused by Earth's motion around the Sun [Section 2.4]. Astronomers measure stellar parallax by comparing observations of a nearby star made 6 months apart (Figure 11.3). The nearby star appears to shift against the background of more distant stars because we are observing it from two opposite points of Earth's orbit.

We can measure the distance to a nearby star by observing how its apparent location shifts as Earth orbits the Sun.

We can calculate a star's distance if we know the precise amount of the star's annual shift due to parallax. This means measuring the angle p in Figure 11.3, which we call the star's *parallax angle*. Notice that this angle would be smaller if the star were farther away, so we conclude that more distant stars have *smaller* parallax angles.

In fact, all stars are so far away that they have very small parallax angles, which explains why the ancient Greeks were never able to measure parallax. Even the nearest stars have parallax angles smaller than 1 arcsecond—well below the approximately 1 arcminute angular resolution of the naked eye [Section 5.3]. For increasingly distant stars, the parallax angles quickly become too small to measure even with our highest-resolution telescopes. Current technology allows us to measure parallax accurately only for stars within a few hundred light-years—not much farther than what we call our *local solar neighborhood* in the vast, 100,000-light-year-diameter Milky Way Galaxy.

By definition, the distance to an object with a parallax angle of 1 arcsecond is 1 parsec (pc), which is equivalent to 3.26 light-years. (The word *parsec* comes from combining the words *parallax* and *arcsecond*.) With a bit of geometry, we can find a simple formula for a star's distance: If we measure the parallax angle p in arcseconds, the star's distance d in parsecs is $d = \frac{1}{p}$; we just multiply by 3.26 to convert from parsecs to light-years. For example, a star with a parallax angle $p = \frac{1}{10}$ arcsecond is 10 parsecs away, or $10 \times 3.26 = 32.6$ light-years. You'll often hear astronomers state distances in parsecs, kiloparsecs (1000 parsecs), or megaparsecs (1 million parsecs), but in this book we'll stick to giving distances in light-years.

think about it Suppose Star A has a parallax angle of 0.2 arcsecond and Star B has a parallax angle of 0.4 arcsecond. How does the distance of Star A from Earth compare to that of Star B?

cosmic calculations 11.1

The Inverse Square Law for Light

If we use L for luminosity, d for distance, and b for apparent brightness, we can write the *inverse square law for light* as

$$b = \frac{L}{4\pi \times d^2}$$

As noted in the text, we can generally measure apparent brightness, which means we can use this formula to calculate luminosity if we know distance and to calculate distance if we know luminosity.

Example: The Sun's measured apparent brightness is 1.36×10^3 watts/m^2 at Earth's distance from the Sun, 1.5×10^{11} m. What is the Sun's luminosity?

Solution: In order to solve the inverse square law formula for the luminosity L, we first multiply both sides by $4\pi \times d^2$. Then we plug in the given values:

$$L = 4\pi \times d^2 \times b$$
$$= 4\pi \times (1.5 \times 10^{11} \text{ m})^2 \times \left(1.36 \times 10^3 \frac{\text{watts}}{\text{m}^2}\right)$$
$$= 3.8 \times 10^{26} \text{ watts}$$

Just by measuring the Sun's apparent brightness and distance, we can calculate the total power that it radiates into all directions in space.

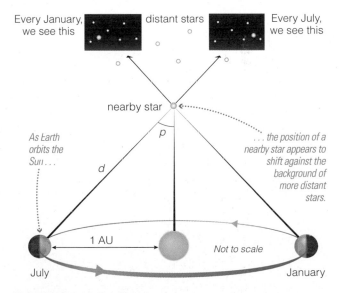

Figure 11.3 interactive figure

Parallax makes the apparent position of a nearby star shift back and forth with respect to distant stars over the course of each year. The angle p, called the *parallax angle*, represents half the total parallax shift each year. If we measure p in arcseconds, the distance d to the star in parsecs is $d = \frac{1}{p}$ (1 parsec = 3.26 light-years). The angle in this figure is greatly exaggerated: All stars have parallax angles of less than 1 arcsecond.

Photos of Stars

Photographs of stars, star clusters, and galaxies convey a great deal of information, but they also contain a few artifacts that are not real. For example, different stars seem to have different sizes in photographs such as that in Figure 11.4 , but stars are so far away that they should all appear as mere points of light. The sizes are an artifact of how our instruments record light. Bright stars tend to be overexposed in photographs, making them appear larger than dimmer stars. Overexposure also explains why the centers of globular clusters and galaxies usually look like big blobs in photographs: The central regions of these objects contain many more stars than the outskirts, and the combined light of so many stars tends to get overexposed, making a big blotch of light.

Spikes around bright stars in photographs, often making the pattern of a cross with a star at the center, are another such artifact. You can see these spikes around many of the brightest stars in Figure 11.4. These spikes are not real but rather are created by the interaction of starlight with the supports holding the secondary mirror in the telescope [Section 5.3]. The spikes generally occur only with point sources of light like stars, and not with larger objects like galaxies. When you look at a photograph showing many galaxies (for example, Figure 15.1), you can tell which objects are stars by looking for the spikes.

Parallax was the first reliable technique for measuring distances to stars, and it remains the only technique that tells us stellar distances without any assumptions about the nature of stars. If we know a star's distance from parallax, we can calculate its luminosity with the inverse square law for light. We now have parallax measurements for thousands of stars, which means that astronomers have been able to calculate the luminosities of all these stars. By discovering patterns among the stars with known luminosities, astronomers learned how to estimate luminosities for many more stars, even without first knowing their distances.

The Luminosity Range of Stars Now that we have discussed how we determine stellar luminosities, it's time to take a quick look at the results. We usually state stellar luminosities in comparison to the Sun's luminosity, which we write as L_{Sun} for short. For example, Proxima Centauri, the nearest of the three stars in the Alpha Centauri system and the nearest star besides our Sun, is only about 0.0006 times as luminous as the Sun, or $0.0006 L_{Sun}$. Betelgeuse, the bright left-shoulder star of Orion, has a luminosity of $38,000 L_{Sun}$, meaning that it is 38,000 times as luminous as the Sun. Overall, studies of the luminosities of many stars have taught us two particularly important lessons:

- Stars come in a wide range of luminosities, with our Sun somewhere in the middle. The dimmest stars have luminosities $\frac{1}{10,000}$ times that of the Sun ($10^{-4} L_{Sun}$), while the brightest stars are about 1 million times as luminous as the Sun ($10^6 L_{Sun}$).

- Dim stars are far more common than bright stars. For example, even though our Sun is roughly in the middle of the overall range of stellar luminosities, it is brighter than the vast majority of stars in our galaxy.

The Magnitude System If you study astronomy further, you'll find that many amateur and professional astronomers describe the apparent brightnesses and luminosities of stars in an alternative way: Instead of apparent brightness they quote **apparent magnitude**, and instead of luminosity they quote **absolute magnitude**. Although comparisons between stars are much easier when we think in terms of luminosity, the magnitude system is used often enough that it's worth understanding how it works.

The magnitude system was originally devised by the Greek astronomer Hipparchus (c. 190–120 B.C.). He gave the brightest stars the designation "first magnitude," the next brightest "second magnitude," and so on. The faintest visible stars were magnitude 6. Today we call these descriptions *apparent magnitudes* because they compare how bright different stars *appear* in the sky. Notice that the magnitude scale runs "backward": A larger apparent magnitude means a dimmer apparent brightness. For example, a star of magnitude 4 is dimmer than a star of magnitude 1.

In modern times, the magnitude system has been extended and more precisely defined: Each difference of five magnitudes is defined to represent a factor of exactly 100 in brightness. For example, a magnitude 1 star is 100 times as bright as a magnitude 6 star, and a magnitude 3 star is 100 times as bright as a magnitude 8 star. As a result of this precise definition, stars can have fractional apparent magnitudes and a few bright stars have apparent magnitudes *less than* 1—which means *brighter* than magnitude 1. For example, the brightest star in the night sky, Sirius, has an apparent magnitude of −1.46.

The modern magnitude system also defines a star's *absolute magnitude* as the apparent magnitude it would have *if* it were at a distance of 10 parsecs (32.6 light-years) from Earth. For example, the Sun's absolute magnitude is about 4.8, meaning that the Sun would have an apparent magnitude of 4.8 *if* it were 10 parsecs away from us—bright enough to be visible but not conspicuous on a dark night.

How do we measure stellar temperatures?

A second fundamental property of a star is its surface temperature. You might wonder why we emphasize *surface* temperature rather than interior temperature. The answer is that only surface temperature is directly measurable; interior temperatures are inferred from mathematical models of stellar interiors [Section 10.2]. Whenever you hear astronomers speak of the "temperature" of a star, you can be pretty sure they mean surface temperature unless they state otherwise.

Measuring a star's surface temperature is somewhat easier than measuring its luminosity, because the star's distance doesn't affect the measurement. Instead, we determine surface temperature directly from either the star's color or its spectrum.

Color and Temperature Take a careful look at Figure 11.4. Notice that stars come in almost every color of the rainbow. Simply looking at the colors tells us something about the surface temperatures of the stars. For example, a red star is cooler than a blue star.

Stars come in different colors because they emit thermal radiation [Section 5.2]. Recall that a thermal radiation spectrum depends only on the (surface) temperature of the object that emits it (see Figure 5.11). For example, the Sun's 5800 K surface temperature causes it to emit most strongly in the middle of the visible portion of the spectrum, which is why the Sun looks yellow or white in color. A cooler star, such as Betelgeuse (surface temperature 3400 K), looks red because it emits much more red light than blue light. A hotter star, such as Sirius (surface temperature 9400 K), emits a little more blue light than red light and therefore has a slightly blue color to it.

Astronomers can measure surface temperature fairly precisely by comparing a star's apparent brightness in two different colors of light. For example, by comparing the amount of blue light and red light coming from Sirius, astronomers can measure how much more blue light it emits than red light. Because thermal radiation spectra have a very distinctive shape (again, see Figure 5.11), this difference in blue and red light output allows astronomers to calculate a surface temperature.

Spectral Type and Temperature A star's spectral lines provide a second way to measure its surface temperature. Moreover, because interstellar dust can affect the apparent colors of stars, temperatures determined from spectral lines are generally more accurate than temperatures determined from colors alone. Stars displaying spectral lines of highly ionized elements must be fairly hot, because it takes a high temperature to ionize atoms. Stars displaying spectral lines of molecules must be relatively cool, because molecules break apart into individual atoms unless they are at relatively cool

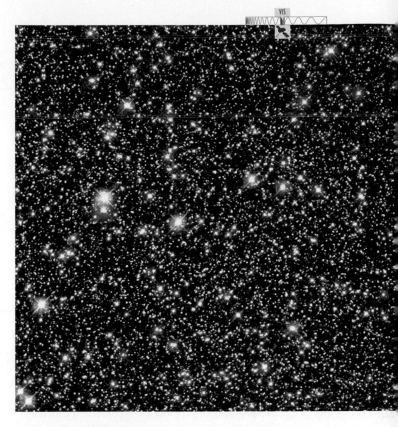

Figure 11.4 interactive photo

This Hubble Space Telescope photo shows a wide variety of stars that differ in color and brightness. Most of the stars in this photo are at roughly the same distance, about 2000 light-years, from the center of our galaxy. Clouds of gas and dust obscure our view of visible light from most of our galaxy's central regions, but a gap in the clouds allows us to see the stars in this photo.

TABLE 11.1   *The Spectral Sequence*

Spectral Type	Example(s)	Temperature Range	Key Absorption Line Features	Brightest Wavelength (color)	Typical Spectrum
O	Stars of Orion's Belt	>30,000 K	Lines of ionised helium, weak hydrogen lines	>97 nm (ultraviolet)*	
B	Rigel	30,000 K– 10,000 K	Lines of neutral helium, moderate hydrogen lines	97–290 nm (ultraviolet)*	
A	Sirius	10,000 K– 7500 K	Very strong hydrogen lines	290–390 nm (violet)*	
F	Polaris	7500 K– 6000 K	Moderate hydrogen lines, moderate lines of ionized calcium	390–480 nm (blue)*	
G	Sun, Alpha Centauri A	6000 K– 5000 K	Weak hydrogen lines, strong lines of ionized calcium	480–580 nm (yellow)	
K	Arcturus	5000 K– 3500 K	Lines of neutral and singly ionised metals, some molecules	580–830 nm (red)	
M	Betelgeuse, Proxima Centauri	<3500 K	Strong molecular lines	> 830 nm (infrared)	

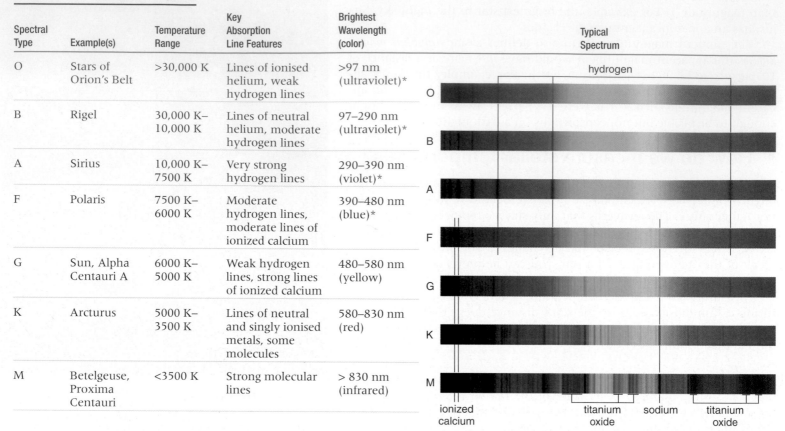

*All stars above 6000 K look more or less white to the human eye because they emit plenty of radiation at all visible wavelengths.

temperatures. The types of spectral lines present in a star's spectrum therefore provide a direct measure of the star's surface temperature.

Astronomers classify stars according to surface temperature by assigning a **spectral type** determined from the spectral lines present in a star's spectrum. The hottest stars, with the bluest colors, are called spectral type O, followed in order of declining surface temperature by spectral types B, A, F, G, K, and M. The time-honored mnemonic for remembering this sequence, OBAFGKM, is "Oh Be A Fine Girl/Guy, Kiss Me!" Table 11.1 summarizes the characteristics of each spectral type. (The sequence of spectral types has recently been extended beyond type M to include spectral types L and T, representing starlike objects—usually brown dwarfs [Section 12.1]—even cooler than stars of spectral type M.)

Spectra of stars show that their surface temperatures range from more than 40,000 K to less than 3000 K, corresponding to the sequence of spectral types OBAFGKM.

Each spectral type is subdivided into numbered subcategories (e.g., B0, B1, . . . , B9). The larger the number, the cooler the star. For example, the Sun is designated spectral type G2, which means it is slightly hotter than a G3 star but cooler than a G1 star.

The range of surface temperatures for stars is much narrower than the range of luminosities. The coolest stars, of spectral type M, have surface temperatures as low as 3000 K. The hottest stars, of spectral type O, have surface temperatures that can exceed 40,000 K. However, cool, red stars turn out to be much more common than hot, blue stars.

> **think about it** Invent your own mnemonic for the OBAFGKM sequence. To help get you thinking, here are two examples: (1) Only Bungling Astronomers Forget Generally Known Mnemonics and (2) Only Business Acts For Good, Karl Marx.

History of the Spectral Sequence You may wonder why the spectral types follow the peculiar order of OBAFGKM. The answer lies in the history of stellar spectroscopy.

Our current system of stellar classification began at Harvard College Observatory. The observatory director, Edward Pickering (1846–1919), had numerous assistants, whom he called "computers." Most of Pickering's computers were women who had studied physics or astronomy at women's colleges such as Wellesley and Radcliffe. Pickering's interest in studying and classifying stellar spectra provided plenty of work and opportunity for his computers, and many of the Harvard Observatory women ended up among the most prominent astronomers of the late 1800s and early 1900s.

One of the first computers was Williamina Fleming (1857–1911), who classified stellar spectra according to the strength of their hydrogen lines: type A for the strongest hydrogen lines, type B for slightly weaker hydrogen lines, and so on, to type O, for stars with the weakest hydrogen lines. Pickering published Fleming's classifications of more than 10,000 stars in 1890.

As more stellar spectra were obtained and the spectra were studied in greater detail, it became clear that a classification scheme based solely on hydrogen lines was inadequate. Ultimately, the task of finding a better classification scheme fell to Annie Jump Cannon (1863–1941), who joined Pickering's team in 1896 (Figure 11.5). Building on the work of Fleming and another of Pickering's computers, Antonia Maury (1866–1952), Cannon soon realized that the spectral classes fell into a natural order—but not the alphabetical order determined by hydrogen lines alone. Moreover, she found that some of the original classes overlapped others and could be eliminated. Cannon discovered that the natural sequence consisted of just a few of Pickering's original classes in the order OBAFGKM, and she also added the subdivisions by number.

The astronomical community adopted Cannon's system of stellar classification in 1910. However, no one at that time knew *why* spectra followed the OBAFGKM sequence. Many astronomers guessed, incorrectly, that the different sets of spectral lines indicated different compositions for the stars. The correct answer—that all stars are made primarily of hydrogen and helium and that a star's surface temperature determines the strength of its spectral lines—was discovered at Harvard Observatory in 1925 by Cecilia Payne-Gaposchkin (1900–1979). Relying on insights from what was then the newly developing science of quantum mechanics, Payne-Gaposchkin showed that the differences in spectral lines from star to star merely reflected changes in the ionization level of the emitting atoms. For example, O stars have weak hydrogen lines because, at their high surface temperatures, nearly all their hydrogen is ionized. Without an electron to "jump" between energy levels, ionized hydrogen can neither emit nor absorb its usual specific wavelengths of light. At the other end of the spectral sequence, M stars are cool enough for some

Figure 11.5

Edward Pickering and his "computers" pose at Harvard College Observatory in 1913. Annie Jump Cannon is fifth from the left in the back row.

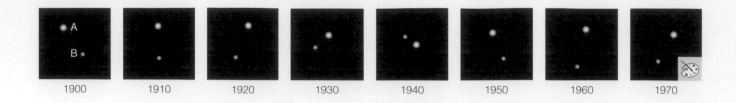

Figure 11.6

In this series of images, each frame represents the relative positions of Sirius A and Sirius B at 10-year intervals from 1900 to 1970. The back-and-forth "wobble" of Sirius A allowed astronomers to infer the existence of Sirius B even before the two stars could be resolved in telescopic photos. The average orbital separation of the binary system is about 20 AU.

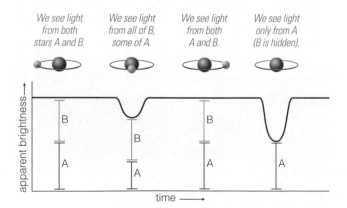

Figure 11.7 interactive figure

The apparent brightness of an eclipsing binary system drops when one star eclipses the other.

particularly stable molecules to form, explaining their strong molecular absorption lines.

• How do we measure stellar masses?

Mass is generally more difficult to measure than surface temperature or luminosity. The most dependable method for "weighing" a star relies on Newton's version of Kepler's third law [Section 4.4]. Recall that this law can be applied only when we can observe one object orbiting another, and it requires that we measure both the orbital period and the average orbital distance of the orbiting object. For stars, these requirements generally mean that we can apply the law to measure masses only in **binary star systems**—systems in which two stars continually orbit one another. Before we consider how we determine the orbital periods and distances needed to use Newton's version of Kepler's third law, let's look briefly at the different types of binary star systems that we can observe.

Types of Binary Star Systems   Surveys show that about half of all stars orbit a companion star of some kind and thus are members of binary star systems. These star systems fall into three classes:

- A *visual binary* is a pair of stars that we can see distinctly (with a telescope) as the stars orbit each other. Sometimes we observe a star slowly shifting position in the sky as if it were a member of a visual binary, but its companion is too dim to be seen. For example, slow shifts in the position of Sirius, the brightest star in the sky, revealed it to be a binary star long before its companion was discovered (Figure 11.6).

- An *eclipsing binary* is a pair of stars that orbit in the plane of our line of sight (Figure 11.7). When neither star is eclipsed, we see the combined light of both stars. When one star eclipses the other, the apparent brightness of the system drops because some of the light is blocked from our view. A *light curve*, or graph of apparent brightness against time, reveals the pattern of the eclipses. The most famous example of an eclipsing binary is Algol, the "demon star" in the constellation Perseus (*algol* is Arabic for "the ghoul"). Algol's brightness drops to only a third of its usual level for a few hours about every three days as the brighter of its two stars is eclipsed by its dimmer companion.

- If a binary system is neither visual nor eclipsing, we may be able to detect its binary nature by observing Doppler shifts in its spectral lines [Section 5.2]. Such a system is called a *spectroscopic binary*. If one star is orbiting another, it periodically moves toward us and away from us in its orbit. Its spectral lines show blueshifts and redshifts as a result of this motion (Figure 11.8). Sometimes we see two sets of lines shifting back and forth—one set from each of the two stars in the system (a *double-lined* spectroscopic binary). Other times we see a set of shifting lines from only one star because its companion is too dim to be detected (a *single-lined* spectroscopic binary).

Some star systems combine two or more of these binary types. For example, telescopic observations reveal Mizar (the second star in the handle of the Big Dipper) to be a visual binary. Spectroscopy then shows that each of the two stars in the visual binary is itself a spectroscopic binary (Figure 11.9).

Measuring Masses in Binary Systems Even for a binary system, we can apply Newton's version of Kepler's third law only if we can measure both the orbital period and the separation of the two stars. Measuring orbital period is fairly easy. In a visual binary, we simply observe how long each orbit takes. In an eclipsing binary, we measure the time between eclipses. In a spectroscopic binary, we measure the time it takes the spectral lines to shift back and forth.

> We can determine the masses of stars in binary systems if we can measure both their orbital period and the separation between them.

Determining the average separation of the stars in a binary system is usually much more difficult. In rare cases we can measure the separation directly; otherwise we can calculate the separation only if we know the actual orbital speeds of the stars from their Doppler shifts. Unfortunately, a Doppler shift tells us only the portion of a star's velocity that is directed toward us or away from us (see Figure 5.15). Because orbiting stars generally do not move directly along our line of sight, their actual velocities can be significantly greater than those we measure through the Doppler effect.

The exceptions are eclipsing binary stars. Because these stars orbit in the plane of our line of sight, their Doppler shifts can tell us their true orbital velocities. Eclipsing binaries are therefore particularly important to the study of stellar masses. As an added bonus, eclipsing binaries allow us to measure stellar radii directly. Because we know how fast the stars are moving across our line of sight as one eclipses the other, we can determine their radii by timing how long each eclipse lasts.

Through careful observations of eclipsing binaries and other binary star systems, astronomers have established the masses of many different kinds of stars. The overall range extends from as little as 0.08 times the mass of the Sun ($0.08M_{Sun}$) to about 150 times the mass of the Sun ($150M_{Sun}$). We'll discuss the reasons for this mass range in Chapter 12.

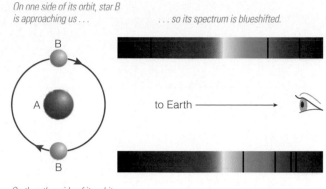

On one side of its orbit, star B is approaching us . . .   *. . . so its spectrum is blueshifted.*

to Earth →

On the other side of its orbit, star B is receding from us . . .   *. . . so its spectrum is redshifted.*

Figure 11.8 interactive figure ↖

The spectral lines of a star in a binary system are periodically blueshifted as it moves toward us in its orbit and redshifted as it moves away from us.

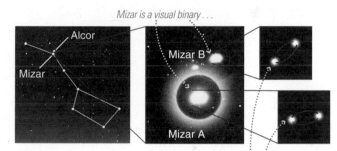

Mizar is a visual binary . . .

Alcor

Mizar B

Mizar

Mizar A

. . . and spectroscopy shows that each of the visual "stars" is itself binary.

Figure 11.9

Mizar looks like one star to the naked eye but is actually a system of four stars. Through a telescope, Mizar appears to be a visual binary made up of two stars, Mizar A and Mizar B, that gradually change positions, indicating that they orbit each other every few thousand years. Moreover, each of these two "stars" is itself a spectroscopic binary, for a total of four stars. (The ring around Mizar A is an artifact of the photographic process, not a real feature.)

(MA)™ **The Hertzsprung-Russell Diagram Tutorial, Lessons 1–3**

11.2 Patterns Among Stars

We have seen that stars come in a wide range of luminosities, surface temperatures, and masses. But are these characteristics randomly distributed among stars, or can we find patterns that might tell us something about stellar lives?

Before reading any further, take another look at Figure 11.4 and think about how you would classify these stars. Almost all of them are at nearly the same distance from Earth, so we can compare their true luminosities by looking at their apparent brightnesses in the photograph. If you look closely, you might notice a couple of important patterns:

- Most of the very brightest stars are reddish in color.

- If you ignore those relatively few bright red stars, there's a general trend to the luminosities and colors among all the rest of the stars: The brighter ones are white with a little bit of blue tint, the more modest ones are similar to our Sun in color with a yellowish white tint, and the dimmest ones are barely visible specks of red.

Keeping in mind that colors tell us about surface temperature—blue is hotter and red is cooler—we can see how these patterns tell us about relationships between surface temperature and luminosity.

Danish astronomer Ejnar Hertzsprung and American astronomer Henry Norris Russell discovered these relationships in the first decade of the 20th century. Building upon the work of Annie Jump Cannon and others, Hertzsprung and Russell independently decided to make graphs of stellar properties by plotting stellar luminosities on one axis and spectral types on the other. These graphs revealed previously unsuspected patterns among the properties of stars and ultimately unlocked the secrets of stellar life cycles.

• What is a Hertzsprung-Russell diagram?

Graphs of the type first made by Hertzsprung and Russell are now called **Hertzsprung-Russell (H-R) diagrams**. These diagrams quickly became one of the most important tools in astronomical research, and they remain central to the study of stars.

Basics of the H-R Diagram   Figure 11.10 displays an example of an H-R diagram. All you need to know to plot a star on an H-R diagram is its luminosity and its spectral type.

- The horizontal axis represents stellar surface temperature, which, as we've discussed, corresponds to spectral type. Temperature *decreases* from left to right because Hertzsprung and Russell based their diagrams on the spectral sequence OBAFGKM.

- The vertical axis represents stellar luminosity, in units of the Sun's luminosity (L_{Sun}). Stellar luminosities span a wide range, so we keep the graph compact by making each tick mark represent a luminosity 10 times as large as the prior tick mark.

An H-R diagram plots the surface temperatures of stars against their luminosities.

Each location on the H-R diagram represents a unique combination of spectral type and luminosity. For example, the dot representing the Sun in Figure 11.10 corresponds to the Sun's spectral type, G2, and its luminosity, $1L_{Sun}$. Because luminosity increases upward on the diagram and surface temperature increases leftward, stars near the upper left are hot and luminous. Similarly, stars near the upper right are cool and luminous, stars near the lower right are cool and dim, and stars near the lower left are hot and dim.

think about it   Explain how the colors of the stars in Figure 11.10 help indicate stellar surface temperature. Do these colors tell us anything about *interior* temperatures? Why or why not?

The H-R diagram also provides direct information about stellar radii, because a star's luminosity depends on both its surface temperature and its surface area or radius. If two stars have the same surface temperature, one can be more luminous than the other only if it is larger in size. Stellar radii therefore must increase as we go from the high-temperature, low-luminosity corner on the lower left of the H-R diagram to the low-temperature, high-luminosity corner on the upper right. Notice the diagonal lines that represent different stellar radii in Figure 11.10.

Patterns in the H-R Diagram
Stars do not fall randomly throughout an H-R diagram like Figure 11.10 but instead cluster into four major groups:

- Most stars fall somewhere along the **main sequence**, the prominent streak running from the upper left to the lower right on the H-R diagram. Notice that our Sun is one of these *main-sequence stars*.

- The stars in the upper right are called **supergiants** because they are very large in addition to being very bright.

- Just below the supergiants are the **giants**, which are somewhat smaller in radius and lower in luminosity (but still much larger and brighter than main-sequence stars of the same spectral type).

- The stars near the lower left are small in radius and appear white in color because of their high temperatures. We call these stars **white dwarfs**.

Luminosity Classes
In addition to the four major groups we've just listed, stars sometimes fall into "in-between" categories. For more precise work, astronomers therefore assign each star to a **luminosity class**, designated with a Roman numeral from I to V. The luminosity class describes the region of the H-R diagram in which the star falls; thus, despite the name, a star's luminosity class is more closely related to its size than to its luminosity. The basic luminosity classes are I for supergiants, III for giants, and V for main-sequence stars. Luminosity classes II and IV are intermediate to the others. For example, luminosity class IV represents stars with radii larger than those of main-sequence stars but not quite large enough to qualify them as giants. Table 11.2 summarizes the luminosity classes. White dwarfs fall outside this classification system and instead are often assigned the luminosity class "wd."

Complete Stellar Classification
We have now described two different ways of categorizing stars:

- A star's *spectral type*, designated by one of the letters OBAFGKM, tells us its surface temperature and color. O stars are the hottest and bluest, while M stars are the coolest and reddest.

- A star's *luminosity class*, designated by a Roman numeral, is based on its luminosity but also tells us about the star's radius. Luminosity class I stars have the largest radii, with radii decreasing to luminosity class V.

The full classification of a star includes both a spectral type (OBAFGKM) and a luminosity class.

We use both spectral type and luminosity class to fully classify a star. For example, the complete classification of our Sun is G2 V. The G2 spectral type means it is yellow-white in color, and the luminosity class V means it is a hydrogen-burning, main-sequence star. Betelgeuse is

Radius of a Star

Almost all stars are too distant for us to measure their radii directly. So how do we know that some are small and some are supergiants? We can calculate a star's radius from its luminosity and surface temperature. Recall from Cosmic Calculations 5.1 that the amount of thermal radiation emitted *per unit area* by a star of temperature T is σT^4, where the constant $\sigma = 5.7 \times 10^{-8}$ watt/(m^2 × Kelvin^4). The star's total luminosity L is equal to this power per unit area times the star's surface area, which is $4\pi r^2$ for a star of radius r:

$$L = 4\pi r^2 \times \sigma T^4$$

With a bit of algebra, we can solve this formula for the star's radius r:

$$r = \sqrt{\frac{L}{4\pi \sigma T^4}}$$

Example: The star Betelgeuse has a surface temperature of about 3400 K and a luminosity of 1.4×10^{31} watts, which is 38,000 times that of the Sun. What is its radius?

Solution: We use the formula given above with $L = 1.4 \times 10^{31}$ watts and $T = 3400$ K:

$$r = \sqrt{\frac{L}{4\pi \sigma T^4}}$$

$$= \sqrt{\frac{1.4 \times 10^{31} \text{ watts}}{4\pi \times \left(5.7 \times 10^{-8} \frac{\text{watt}}{\text{m}^2 \times \text{K}^4}\right) \times (3400 \text{ K})^4}}$$

$$= 3.8 \times 10^{11} \text{ m}$$

The radius of Betelgeuse is about 380 billion meters, which is the same as 380 million kilometers—more than twice the Earth–Sun distance of about 150 million kilometers. That is why we call it a supergiant.

TABLE 11.2 *Stellar Luminosity Classes*

Class	Description
I	Supergiants
II	Bright giants
III	Giants
IV	Subgiants
V	Main-sequence stars

Hertzsprung-Russell (H-R) diagrams are very important tools in astronomy because they reveal key relationships among the properties of stars. An H-R diagram is made by plotting stars according to their surface temperatures and luminosities. This figure shows a step-by-step approach to building an H-R diagram.

(1) **An H-R Diagram is a Graph:** A star's position along the horizontal axis indicates its surface temperature, which is closely related to its color and spectral type. Its position along the vertical axis indicates its luminosity.

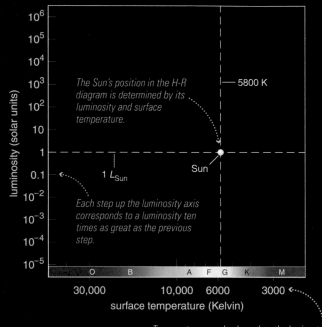

The Sun's position in the H-R diagram is determined by its luminosity and surface temperature.

5800 K

$1\ L_{Sun}$

Sun

Each step up the luminosity axis corresponds to a luminosity ten times as great as the previous step.

Temperature runs backward on the horizontal axis, with hot blue stars on the left and cool red stars on the right.

(2) **Main Sequence:** Our Sun falls along the main sequence, a line of stars extending from the upper left of the diagram to the lower right. Most stars are main-sequence stars, which shine by fusing hydrogen into helium in their cores.

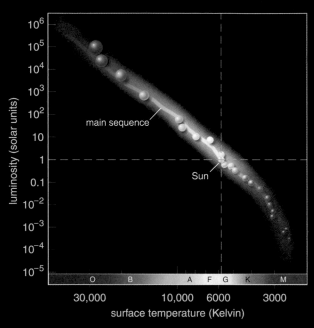

main sequence

Sun

(3) **Giants and Supergiants:** Stars in the upper right of an H–R diagram are more luminous than main-sequence stars of the same surface temperature. They must therefore be very large in radius, which is why they are known as *giants* and *supergiants*.

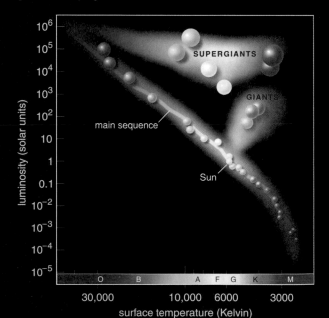

SUPERGIANTS

GIANTS

main sequence

Sun

(4) **White Dwarfs:** Stars in the lower left have high surface temperatures, dim luminosities, and small radii. These stars are known as *white dwarfs.*

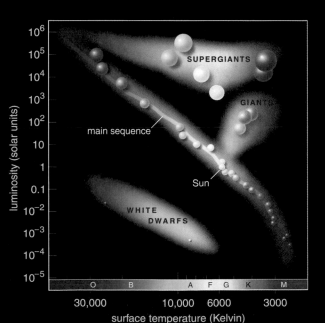

SUPERGIANTS

GIANTS

main sequence

Sun

WHITE DWARFS

(5) Multiple shell–burning supergiant: After the core runs out of helium, it shrinks and heats until fusion of heavier elements begins. Late in life, the star fuses many different elements in a series of shells while iron collects in the core.

(6) Supernova: Iron cannot provide fusion energy, so it accumulates in the core until degeneracy pressure can no longer support it. Then the core collapses, leading to the catastrophic explosion of the star.

(7) Neutron star or black hole: The core collapse forms a ball of neutrons, which may remain as a neutron star or collapse further to make a black hole.

Multiple shell–
burning supergiant

Supernova

Later stages: < 1 million years

TIME

Later stages: 1 billion years

Endpoint: White dwarf

Red giant star

Helium-burning star

Double shell–burning red giant

Planetary nebula

(4) Helium-burning star: Helium fusion begins when the core becomes hot enough to fuse helium into carbon. The core then expands, slowing the rate of hydrogen shell burning and allowing the star's outer layers to shrink.

(5) Double shell–burning red giant: Helium shell burning begins around the inert carbon core after the core helium is exhausted. The star then enters its second red giant phase, with fusion in both a hydrogen shell and a helium shell.

(6) Planetary nebula: The dying star expels its outer layers in a planetary nebula, leaving behind the exposed inert core.

(7) White dwarf: The remaining white dwarf is made primarily of carbon and oxygen because the core of the low-mass star never grows hot enough to produce heavier elements.

Algol shortly after its birth. The higher-mass star (left) evolved more quickly than its lower-mass companion (right).

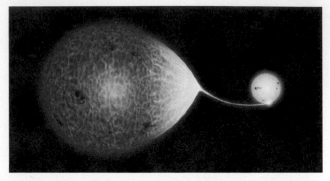

Algol at onset of mass transfer. When the more massive star expanded into a red giant, it began losing some of its mass to its normal, hydrogen-burning companion.

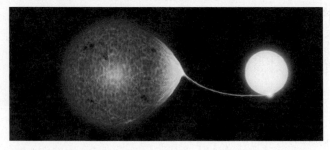

Algol today. As a result of the mass transfer, the red giant has shrunk to a subgiant, and the normal star on the right is now the more massive of the two stars.

Figure 12.23

Artist's conception of the development of the Algol close binary system.

In addition, the stars become *tidally locked* so that they always show the same face to each other, much as the Moon always shows the same face to Earth.

During the time that both stars are main-sequence stars, the tidal forces have little effect on their lives. However, when the more massive star (which exhausts its core hydrogen sooner) begins to expand into a red giant, gas from its outer layers can spill over onto its companion. This **mass exchange** occurs when the giant grows so large that its tidally distorted outer layers succumb to the gravitational attraction of the smaller companion star. The companion then begins to gain mass at the expense of the giant.

The solution to the Algol paradox should now be clear (Figure 12.23). The $0.8 M_{Sun}$ subgiant *used to be* much more massive. As the more massive star, it was the first to begin expanding into a red giant. As it expanded, however, so much of its matter spilled over onto its companion that it is now the less massive star.

> Stars in close binary systems can exchange mass with one another, altering their life histories.

The future may hold even more interesting events for Algol. The $3.7 M_{Sun}$ star is still gaining mass from its subgiant companion. Thus, its life cycle is actually accelerating as its increasing gravity raises its core hydrogen fusion rate. Millions of years from now, it will exhaust its hydrogen and begin to expand into a red giant itself. At that point, it can begin to transfer mass *back* to its companion. Even more amazing things can happen in other mass-exchange systems, particularly when one of the stars is a white dwarf or a neutron star. But that is a topic for the next chapter.

---

the big picture
Putting Chapter 12 into Context

In this chapter, we answered the question of the origin of elements that we first discussed in Chapter 1. As you look back over this chapter, keep in mind these "big picture" ideas:

- Virtually all elements in the universe besides hydrogen and helium were forged in stars. We and our planet are therefore made of stuff produced by stars that lived and died long ago.

- Low-mass stars like our Sun live long lives and die with the ejection of planetary nebulae, leaving behind white dwarfs.

- High-mass stars live fast and die young, exploding dramatically as supernovae and leaving behind neutron stars or black holes.

- Close binary stars can exchange mass, altering the usual course of stellar evolution.

summary of key concepts

12.1 Star Birth

• How do stars form?

Stars are born in cold, relatively dense **molecular clouds**. As a cloud fragment collapses under gravity, it becomes a rapidly rotating **protostar** surrounded by a spinning disk of gas in which planets may form. The protostar may also fire **jets** of matter outward along its poles.

• How massive are newborn stars?

Newborn stars come in a range of masses but cannot be more massive than about $150M_{Sun}$ or less massive than $0.08M_{Sun}$. Below this mass, **degeneracy pressure** prevents gravity from making the core hot enough for efficient hydrogen fusion, and the object becomes a "failed star" known as a **brown dwarf**.

12.2 Life as a Low-Mass Star

• What are the life stages of a low-mass star?

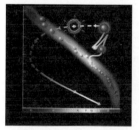

A low-mass star spends most of its life generating energy by fusing hydrogen in its core. When core hydrogen is exhausted, the core begins to shrink while the star as a whole expands to become a **red giant**, with **hydrogen shell burning** around an inert helium core. When the core becomes hot enough, a **helium flash** initiates **helium fusion** in the core, which fuses helium into carbon; the star shrinks somewhat in size and luminosity during this time. The core shrinks again when core helium burning ceases, while both helium and hydrogen fusion occur in shells around the inert carbon core and cause the outer layers to expand once more.

• How does a low-mass star die?

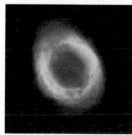

A low-mass star like the Sun never gets hot enough to fuse carbon in its core, because degeneracy pressure stops the gravitational collapse of the core. The star expels its outer layers into space as a **planetary nebula**, leaving behind its exposed core as a white dwarf, which is supported by degeneracy pressure.

12.3 Life as a High-Mass Star

• What are the life stages of a high-mass star?

A high-mass star lives a much shorter life than a low-mass star, fusing hydrogen into helium via the **CNO cycle**. After

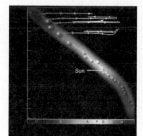

exhausting its core hydrogen, a high-mass star begins hydrogen shell burning and then goes through a series of stages burning successively heavier elements. The furious rate of this fusion makes the star swell in size to become a supergiant.

• How do high-mass stars make the elements necessary for life?

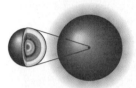

In its final stages of life, a high-mass star's core becomes hot enough to fuse carbon and other heavy elements. The variety of different fusion reactions produces a wide range of elements— including all the elements necessary for life—that are then released into space when the star dies.

• How does a high-mass star die?

A high-mass star dies in a cataclysmic explosion called a **supernova**, scattering newly produced elements into space and leaving behind a neutron star or black hole. The supernova occurs after fusion begins to pile up iron in the high-mass star's core. Because iron fusion cannot release energy, the core cannot hold off the crush of gravity for long. In the instant that gravity overcomes degeneracy pressure, the core collapses and the star explodes. The expelled gas may be visible for a few thousand years as a **supernova remnant**.

12.4 Summary of Stellar Lives

• How does a star's mass determine its life story?

A star's mass determines how it lives its life. Low-mass stars never get hot enough to fuse carbon or heavier elements in their cores and end their lives by expelling their outer layers and leaving white dwarfs behind. High-mass stars live short but brilliant lives, ultimately dying in supernova explosions.

• How are the lives of stars with close companions different?

When one star in a close binary system begins to swell in size at the end of its hydrogen-burning life, it can begin to transfer mass to its companion. This **mass exchange** can then change the remaining life histories of both stars.

exercises and problems

For instructor-assigned homework go to www.masteringastronomy.com.

Review Questions

Short-Answer Questions Based on the Reading

1. What is a *molecular cloud?* Briefly describe the process by which a protostar forms from gas in a molecular cloud.

2. Why do protostars rotate rapidly? How can a close binary star system form?

3. Why does a spinning disk of gas surround a protostar? Describe some of the phenomena seen among protostars, such as strong *winds* and *jets.*

4. What are the minimum and maximum masses for a star, and why do these limits occur? What is a *brown dwarf?*

5. What is *degeneracy pressure,* and how does it differ from *thermal pressure?* Explain why degeneracy pressure can support a stellar core against gravity even when the core becomes cold.

6. Briefly explain the changes that the Sun will go through after it exhausts its core hydrogen. Be sure to explain both the changes occurring in the Sun's core and the changes visible from outside the Sun. What do we mean by the stages we call *hydrogen shell burning, helium burning,* and *double shell burning?*

7. Why does helium fusion require much higher temperatures than hydrogen fusion? Briefly explain why helium fusion in the Sun will begin with a *helium flash.*

8. What is a *planetary nebula?* What happens to the core of a star after a planetary nebula occurs?

9. What will happen to Earth as the Sun changes in the future?

10. What do we mean by a star's *life track* on an H-R diagram? Summarize the stages of life that we see on the Sun's life track in Figure 12.12.

11. In broad terms, explain how the life of a high-mass star differs from that of a low-mass star.

12. Describe some of the nuclear reactions that can occur in high-mass stars after they exhaust their core helium. Why does this continued nuclear burning occur in high-mass stars but not in low-mass stars?

13. Why can't iron be fused to release energy?

14. Summarize some of the observational evidence supporting our ideas about how the elements form in massive stars.

15. What event initiates a *supernova?* Explain what happens during the explosion, and why a neutron star or black hole is left behind. What observational evidence supports our understanding of supernovae?

16. What is the *Algol paradox* and its resolution? How do the lives of close binary stars differ from those of single stars?

Test Your Understanding

Does It Make Sense?

Decide whether the statement makes sense (or is clearly true) or does not make sense (or is clearly false). Explain clearly; not all of these have definitive answers, so your explanation is more important than your chosen answer.

17. The iron in my blood came from a star that blew up more than 4 billion years ago.

18. I discovered stars being born within a patch of extremely low-density, hot interstellar gas.

19. Humanity will eventually have to find another planet to live on, because one day the Sun will blow up as a supernova.

20. I sure am glad hydrogen has a higher mass per nuclear particle than many other elements. If it had the lowest mass per nuclear particle, none of us would be here.

21. If the Sun had been born as a high-mass star some 4.6 billion years ago rather than as a low-mass star, the planet Jupiter would probably have Earth-like conditions today, while Earth would be hot like Venus.

22. If you could look inside the Sun today, you'd find that its core contains a much higher proportion of helium and a lower proportion of hydrogen than it did when the Sun was born.

23. I just discovered a $3.5M_{Sun}$ main-sequence star orbiting a $2.5M_{Sun}$ red giant. I'll bet that red giant was more massive than $3M_{Sun}$ when it was a main-sequence star.

24. Globular clusters generally contain lots of white dwarfs.

25. After hydrogen fusion stops in a low-mass star, its core cools off until the star becomes a red giant.

26. The gold in my new ring came from a supernova explosion.

Quick Quiz

Choose the best answer to each of the following. Explain your reasoning with one or more complete sentences.

27. Stars can form most easily in clouds that are (a) cold and dense. (b) warm and dense. (c) hot and low-density.

28. A brown dwarf is (a) an object not quite massive enough to be a star. (b) a white dwarf that has cooled off. (c) a star-like object that is less massive than Jupiter.

29. Which of these stars has the hottest *core?* (a) a blue main-sequence star (b) a red supergiant (c) a red main-sequence star

30. Which of these stars does *not* have fusion occurring in its core? (a) a red giant (b) a red main-sequence star (c) a blue main-sequence star

31. What happens to a low-mass star after helium flash? (a) Its luminosity goes up. (b) Its luminosity goes down. (c) Its luminosity stays the same.

32. What would stars be like if hydrogen had the smallest mass per nuclear particle? (a) Stars would be brighter. (b) All stars would be red giants. (c) Nuclear fusion would not occur in stars of any mass.

33. What would stars be like if carbon had the smallest mass per nuclear particle? (a) Supernovae would be more common. (b) Supernovae would never occur. (c) High-mass stars would be hotter.

34. What would you be most likely to find if you returned to the solar system in 10 billion years? (a) a neutron star (b) a white dwarf (c) a black hole

35. Which of these stars has the shortest life expectancy? (a) an isolated $1M_{Sun}$ star (b) a $1M_{Sun}$ star in a close binary system with a $0.8M_{Sun}$ star (c) a $1M_{Sun}$ star in a close binary system with a $2M_{Sun}$ star

36. What happens to the core of a high-mass star after it runs out of hydrogen? (a) It shrinks and heats up. (b) It shrinks and cools down. (c) Helium fusion begins right away.

Process of Science

37. *Predicting the Sun's Future.* Models of stellar evolution make detailed predictions about the fate of the Sun. Describe one piece of observational evidence that supports each of the following model predictions:
    a. The Sun cannot continue supplying Earth with light and heat forever.
    b. The Sun will become a red giant before the end of its life.
    c. The Sun will leave behind a white dwarf after it dies.
38. *Predicting the Properties of Brown Dwarfs.* Models of star formation predict that objects less massive than $0.08 M_{Sun}$ become brown dwarfs instead of true hydrogen-fusing stars. Once they have formed, these objects cool because no fusion is occurring to replace the thermal energy lost from their surfaces. How would you expect the properties of brown dwarfs in an older star cluster to compare with those of brown dwarfs in a younger one? Propose an observing program that could test your hypothesis.

Investigate Further

In-Depth Questions to Increase Your Understanding

Short-Answer/Essay Questions

39. *Brown Dwarfs.* How are brown dwarfs like jovian planets? In what ways are brown dwarfs like stars?
40. *Homes to Civilization?* We do not yet know how many stars have Earth-like planets, nor do we know the likelihood that such planets might harbor advanced civilizations like our own. However, some stars can probably be ruled out as candidates for advanced civilizations. For example, given that it took a few billion years for humans to evolve on Earth, it seems unlikely that advanced life would have had time to evolve around a star that is only a few million years old. For each of the following stars, decide whether you think it is possible that it could harbor an advanced civilization. Explain your reasoning in one or two paragraphs.
    a. $10 M_{Sun}$ main-sequence star
    b. $1.5 M_{Sun}$ main-sequence star
    c. $1.5 M_{Sun}$ red giant
    d. $1 M_{Sun}$ helium-burning star
    e. A red supergiant
41. *Rare Elements.* Lithium, beryllium, and boron are elements with atomic numbers 3, 4, and 5, respectively. Despite their being three of the five simplest elements, Figure 12.18 shows that they are rare compared to many heavier elements. Suggest a reason for their rarity. (*Hint:* Consider the process by which helium fuses into carbon.)
42. *Future Skies.* As a red giant, the Sun will have an angular size in Earth's sky of about 30°. What will sunset and sunrise be like? Do you think the color of the sky will be different from what it is today? Explain.
43. *Research: Historical Supernovae.* Historical accounts exist for supernovae in the years 1006, 1054, 1572, and 1604. Choose one of these supernovae and learn more about historical records of the event. Did the supernova influence human history in any way? Write a two- to three-page summary of your research findings.

Quantitative Problems

Be sure to show all calculations clearly and state your final answers in complete sentences.

44. *An Isolated Star-Forming Cloud.* Isolated molecular clouds can have a temperature as low as 10 K and a particle density as great as 100,000 particles per cubic centimeter. What is the minimum mass that a cloud with these properties needs in order to form a star?
45. *Density of a Red Giant.* Near the end of its life, the Sun's radius will extend nearly to the distance of Earth's orbit. Estimate the volume of the Sun at that time using the formula for the volume of a sphere ($V = 4\pi r^3/3$). Using that result, estimate the average matter density of the Sun at that time. How does that density compare with the density of water (1 g/cm^3)? How does it compare with the density of Earth's atmosphere at sea level (about 10^{-3} g/cm^3)?
46. *Supernova Betelgeuse.* The distance from Earth of the red supergiant Betelgeuse is approximately 427 light-years. If it were to explode as a supernova, it would be one of the brightest stars in the sky. Right now, the brightest star other than the Sun is Sirius, with a luminosity of $26 L_{Sun}$ and a distance of 8.6 light-years. How much brighter than Sirius would the Betelgeuse supernova be in our sky if it reached a maximum luminosity of $10^{10} L_{Sun}$?
47. *Construction of Elements.* Using the periodic table in Appendix D, determine which elements are made by the following nuclear fusion reactions. (You can assume the total number of protons in the reaction remains constant.)
    a. Fusion of a carbon nucleus with another carbon nucleus
    b. Fusion of a carbon nucleus with a neon nucleus
    c. Fusion of an iron nucleus with a helium nucleus
48. *Algol's Orbital Separation.* The Algol binary system consists of a $3.8 M_{Sun}$ star and a $0.8 M_{Sun}$ star with an orbital period of 2.87 days. Use Newton's version of Kepler's third law to calculate the orbital separation of the system. How does that separation compare with the typical size of a red giant star?

Discussion Questions

49. *Connections to the Stars.* In ancient times, many people believed that our lives were somehow influenced by the patterns of the stars in the sky. Modern science has not found any evidence to support this belief, but instead has found that we have a connection to the stars on a much deeper level: We are "star stuff." Discuss in some detail our real connections to the stars as established by modern astronomy. Do you think these connections have any philosophical implications in terms of how we view our lives and our civilization? Explain.
50. *Humanity in A.D. 5,000,000,000.* Do you think it is likely that humanity will survive until the Sun begins to expand into a red giant 5 billion years from now? Why or why not?

Web Projects

51. *Star Birth and the Spitzer Telescope.* The Spitzer Space Telescope is one of astronomers' best tools for learning about star birth. Go to the Spitzer Web site and look for the latest findings about star formation. Summarize your research in a one- to two-page report.

52. *Fireworks in Supernova 1987A.* The light show from Supernova 1987A is not yet over. Between now and about 2010, the expanding cloud of gas from the supernova is expected to continue ramming into the material surrounding the star, and the heat generated by the impact is expected to create a new light show. Learn more about how Supernova 1987A is changing and what we might expect to see from it in the future. Summarize your findings in a one- to two-page report.

53. *Picturing Star Birth and Death.* Photographs of stellar birthplaces (molecular clouds) and death places (e.g., planetary nebulae and supernova remnants) can be strikingly beautiful, but only a few such photographs are included in this chapter. Search the Web for at least 20 other examples, taken in different wavelengths of light. Put each photograph you find into a personal online journal, along with a one-paragraph description of what the photograph shows.

visual skills check

Use the following questions to check your understanding of some of the many types of visual information used in astronomy. Answers are provided in Appendix K. For additional practice, try the Chapter 12 Visual Quiz at **www.masteringastronomy.com.**

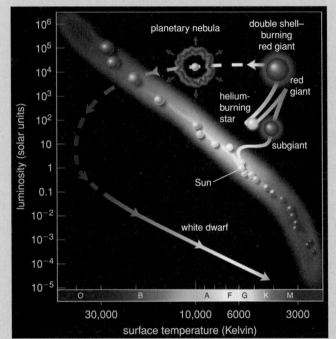

This figure, similar to the left side of Figure 12.12, shows the future life stages of the Sun on an H-R diagram. Answer the following questions, using the information provided in the figure.

1. What will the Sun's approximate luminosity be during the subgiant stage?
2. When the Sun is a red giant, what will its approximate surface temperature be?
3. Just before the Sun produces a planetary nebula, what will its approximate luminosity be?

4. When the Sun becomes a white dwarf with a surface temperature similar to its current surface temperature, what will its luminosity be?

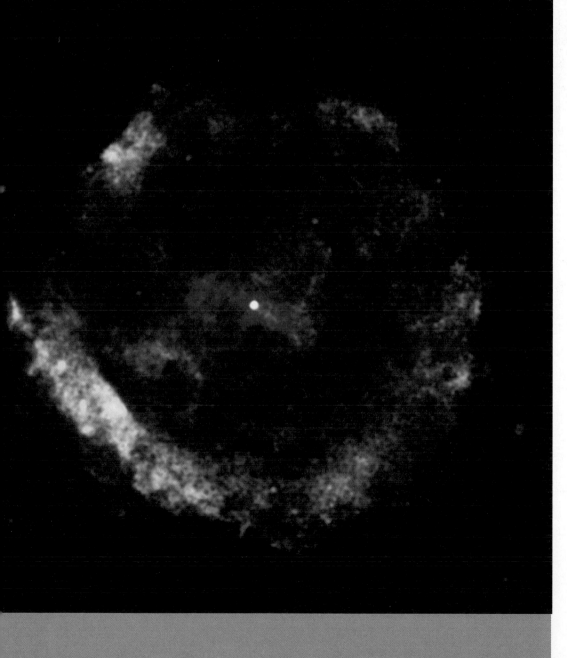

the bizarre stellar graveyard

learning goals

Welcome to the afterworld of stars, the fascinating domain of white dwarfs, neutron stars, and black holes. To scientists, these dead stars are ideal laboratories for testing the most extreme predictions of general relativity and quantum theory. To most other people, the eccentric behavior of stellar corpses demonstrates that the universe is stranger than they ever imagined.

Dead stars behave in unusual and unexpected ways that challenge our minds and stretch the boundaries of what we believe is possible. Stars that have finished nuclear burning have only one hope of staving off the crushing power of gravity: the quantum mechanical effect of degeneracy pressure. But even this strange pressure cannot save the most massive stellar cores. In this chapter, we will study the bizarre properties and occasional catastrophes of the stellar corpses known as white dwarfs, neutron stars, and black holes. Prepare to be amazed by the eerie inhabitants of the stellar graveyard!

(MA) **Stellar Evolution Tutorial, Lessons 1–2**

13.1 White Dwarfs

In the previous chapter, we saw that stars of different masses leave different types of stellar corpses. Low-mass stars like the Sun leave behind white dwarfs when they die. Higher-mass stars die in the titanic explosions known as supernovae, leaving behind neutron stars and black holes. Let's begin our study of stellar corpses with white dwarfs.

• What is a white dwarf?

As we discussed in Chapters 11 and 12, a white dwarf is essentially the exposed core of a star that has died and shed its outer layers in a planetary nebula [Section 12.2]. It is quite hot when it first forms, because it was the inside of a star, but it slowly cools with time. White dwarfs have masses like those of stars but sizes (radii) like that of Earth [Section 11.2], which is why they are generally quite dim compared to stars like the Sun. The hottest white dwarfs can shine quite brightly in high-energy light such as ultraviolet and X rays (Figure 13.1).

A white dwarf is the corpse of a low-mass star, supported against the crush of gravity by electron degeneracy pressure.

A white dwarf's combination of a starlike mass and a small size makes gravity very strong near its surface. If gravity were unopposed, it would crush the white dwarf to an even smaller size, so some sort of pressure must be pushing back equally hard to keep the white dwarf stable. Because there is no fusion to maintain heat and pressure inside a white dwarf, the pressure that opposes gravity must come from some other source. The source is *degeneracy pressure*—the same type of pressure that supports the "failed stars" that we call brown dwarfs and that arises when subatomic particles are packed as closely as the laws of quantum mechanics allow [Section 12.1]. More specifically, the degeneracy pressure in white dwarfs arises from closely

Sirius A is the brightest star in the night sky in visible and infrared light.

The hot white dwarf Sirius B is much less bright in visible and infrared light.

a Sirius as seen in infrared light by the Hubble Space Telescope.

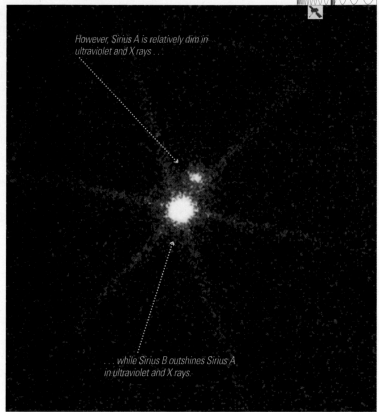

However, Sirius A is relatively dim in ultraviolet and X rays . . .

. . . while Sirius B outshines Sirius A in ultraviolet and X rays.

b Sirius as seen by the Chandra X-ray Telescope.

packed *electrons,* so we call it **electron degeneracy pressure**. A white dwarf exists in a state of balance because the outward push of electron degeneracy pressure matches the inward crush of gravity.

White Dwarf Composition, Density, and Size

Because a white dwarf is the core left over after a star has ceased nuclear fusion, its composition reflects the products of the star's final fusion stage. The white dwarf left behind by a $1M_{Sun}$ star like our Sun consists mostly of carbon, since stars like the Sun fuse helium into carbon in their final stage of life. The cores of very-low-mass stars never become hot enough to fuse helium and thus end up as helium white dwarfs.

A teaspoon of white dwarf matter would weigh several tons.

Despite its ordinary-sounding composition, a scoop of matter from a white dwarf would be unlike anything ever seen on Earth. A typical white dwarf has the mass of the Sun $(1M_{Sun})$ compressed into an object the size of Earth. If you recall that Earth is smaller than a typical sunspot, you'll realize that packing the entire mass of the Sun into the volume of Earth is no small feat. The density of a white dwarf is so high that a teaspoon of its material would weigh several tons—the weight of a small truck—if you could bring it to Earth.

More massive white dwarfs are actually smaller in size than less massive ones. For example, a $1.3M_{Sun}$ white dwarf is half the diameter of a $1.0M_{Sun}$ white dwarf (Figure 13.2). The more massive white dwarf is smaller because its greater gravity compresses matter to a much greater density. According to the laws of quantum mechanics, the electrons in a white dwarf respond to this compression by moving faster, which makes the degeneracy pressure strong enough to resist the greater force of gravity. The most massive white dwarfs are therefore the smallest.

Figure 13.1

The binary star system Sirius. Sirius B, the white dwarf in the binary, has a much hotter surface than its companion, Sirius A, and is therefore much brighter in ultraviolet and X-ray light. (The spikes emanating from the stars are not real. They are artifacts created by telescope optics.)

Earth   1.0M_{Sun} white dwarf   1.3M_{Sun} white dwarf

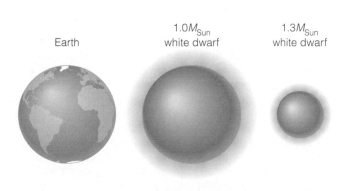

Figure 13.2

Contrary to what you might expect, more massive white dwarfs are actually *smaller* (and thus denser) than less massive white dwarfs. Earth is shown for scale.

Chapter 13 The Bizarre Stellar Graveyard **363**

The White Dwarf Limit The fact that electron speeds are higher in more massive white dwarfs leads to a fundamental limit on the maximum mass of a white dwarf. Theoretical calculations show that electron speeds would reach the speed of light in a white dwarf with a mass of about 1.4 times the mass of the Sun ($1.4M_{Sun}$). Because neither electrons nor anything else can travel faster than the speed of light (see Special Topic below), no white dwarf can have a mass greater than this $1.4M_{Sun}$ **white dwarf limit**. (The white dwarf limit is also called the *Chandrasekhar limit*, after its discoverer.)

special topic: --

Relativity and the Cosmic Speed Limit

THE IDEA THAT nothing can travel faster than the speed of light is one of several mind-boggling consequences of Einstein's *special theory of relativity*, published in 1905. Although relativity is often portrayed as being difficult, its basic ideas are easy to understand.

To understand what's "relative" about relativity, imagine a supersonic plane trip at a speed of 1670 km/hr from Nairobi, Kenya, to Quito, Ecuador. How fast is the plane going? At first, this question sounds trivial—we have just said that the plane is going 1670 km/hr. But wait. Nairobi and Quito are both nearly on Earth's equator, and the equatorial speed of Earth's rotation is the same speed, 1670 km/hr, at which the plane is flying. Moreover, the east-to-west flight from Nairobi to Quito is opposite the direction of Earth's rotation (see figure). If you lived on the Moon, the plane would appear to remain stationary *while Earth rotated beneath it*.

We have two alternative viewpoints about the plane's flight. People on Earth say that the plane travels westward across the surface of the Earth. Observers in space say that the plane is stationary while Earth rotates eastward beneath it. Einstein's theory tells us that both viewpoints are equally valid. That is, questions like "Who is really moving?" and "How fast are you going?" have no absolute answers, and everyone can agree only on the fact that the plane is traveling at 1670 km/hr *relative to* the surface of Earth. Indeed, the theory of relativity gets its name from the fact that measurements of motion (and of time and space) make sense only when we describe whom or what they are being measured relative to.

Note that while Einstein's theory says that motion is relative, it does *not* say that "everything" is relative. In fact, the theory states that two things in the universe are absolute:

1. The laws of nature are the same for everyone.
2. The speed of light is the same for everyone.

The first absolute, that the laws of nature are the same for everyone, is a more general version of the idea that all viewpoints on motion are equally valid. If they weren't, different observers would disagree about the laws of physics. The second absolute, that the speed of light is the same for everyone, is much more surprising. Ordinarily, we expect speeds to add and subtract. If you watch someone throw a ball forward from a moving car, you'd see the ball traveling at the speed it is thrown *plus* the speed of the car. But if a person shines a light beam from a moving car, you'd see it moving at precisely the speed of light (about 300,000 km/s), no matter how fast the car is going. This strange fact has been experimentally verified countless times.

The cosmic speed limit follows directly from this fact about the speed of light. To see why, imagine that you have just built the most incredible rocket possible, and you are taking it on a test ride. You

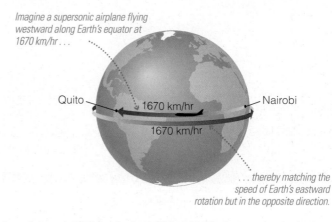

Imagine a supersonic airplane flying westward along Earth's equator at 1670 km/hr . . .

Quito   1670 km/hr   Nairobi

1670 km/hr

. . . thereby matching the speed of Earth's eastward rotation but in the opposite direction.

A plane flying at 1670 km/hr from Nairobi to Quito travels precisely opposite Earth's rotation.

push the acceleration button and keep going faster and faster and faster. With enough fuel, you might expect that you'd eventually be moving faster than the speed of light. But we must ask: What is your speed being measured relative to? Remember that you will always measure light to be traveling at the speed of light, 300,000 km/s. Thus, *you* will find that the light from your rocket's headlights is racing away from you at 300,000 km/s. This shouldn't be too surprising, as it's just another way of saying that you can't catch up with your own light. However, because *everyone* always measures the same speed of light, people back on Earth (or anyplace else) must also say that your headlight beams travel at a speed of 300,000 km/s. Because we already know that you can't keep up with your headlight beams, observers on Earth must find that you are traveling *slower* than the headlight beams, which means slower than the speed of light.

These facts about the relativity of motion and the absoluteness of the speed of light also lead to several other famous consequences of Einstein's theory. For example, they tell us that if you observe a person moving past you at a speed close to the speed of light, you'll see her time running slower than yours, you'll measure her size to be compressed (in the direction of motion) from what you'd measure if she were stationary (relative to you), and you'd conclude that her mass is greater than it would be if she were stationary. The theory also predicts that mass and energy should be equivalent, as stated by Einstein's famous formula $E = mc^2$. All these predictions of relativity have been experimentally tested and verified to high precision.

A white dwarf cannot have a mass greater than 1.4 times the mass of the Sun.

Strong observational evidence supports this theoretical limit on the mass of a white dwarf. Many known white dwarfs are members of binary systems, and hence we can measure their masses [Section 11.1]. In every observed case, the white dwarfs have masses below $1.4M_{Sun}$, just as our theory predicts.

• What can happen to a white dwarf in a close binary system?

Left to itself, a white dwarf will never again shine as brightly as the star it once was. With no source of fuel for fusion, it will simply cool with time into a cold, black dwarf. Its size will never change, because its electron degeneracy pressure will forever keep it stable against the crush of gravity. However, the situation can be quite different for a white dwarf in a close binary system.

A white dwarf in a close binary system can gradually gain mass if its companion is a main-sequence or giant star (Figure 13.3). When a clump of mass first spills over from the companion to the white dwarf, it has some small orbital velocity. The law of conservation of angular momentum dictates that the clump must orbit faster and faster as it falls toward

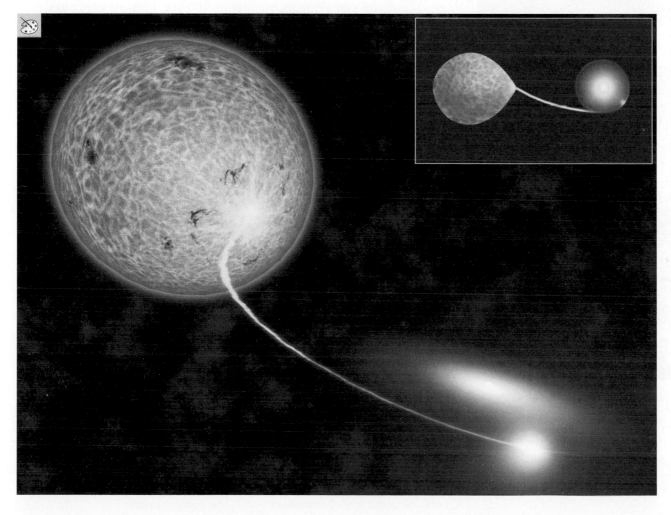

Figure 13.3

This artist's conception shows how mass spilling from a companion star (left) toward a white dwarf (right) forms an accretion disk around the white dwarf. The white dwarf itself is in the center of the accretion disk—too small to be seen on this scale. Matter streaming onto the disk creates a hot spot at the point of impact. The inset shows how the system looks from above rather than from the side.

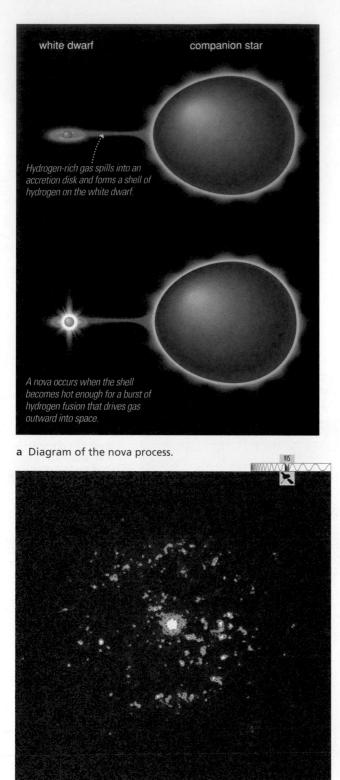

white dwarf   companion star

Hydrogen-rich gas spills into an accretion disk and forms a shell of hydrogen on the white dwarf.

A nova occurs when the shell becomes hot enough for a burst of hydrogen fusion that drives gas outward into space.

a Diagram of the nova process.

b Hubble Space Telescope image showing blobs of gas ejected from the nova T Pyxidis. The bright spot at the center of the blobs is the binary star system that generated the nova.

Figure 13.4

A nova occurs when hydrogen fusion ignites on the surface of a white dwarf in a binary star system.

the white dwarf's surface. The infalling matter therefore forms a whirlpool-like disk around the white dwarf. Because the process in which material falls onto another body is called *accretion,* this rapidly rotating disk is called an **accretion disk**.

> In a close binary system, gas from a companion star can spill toward a white dwarf, forming a swirling accretion disk around it.

Accretion can provide a "dead" white dwarf with a new energy source. The inward-spiraling gas in the accretion disk becomes quite hot as its gravitational potential energy is converted into thermal energy [Section 4.3], allowing it to shine with intense ultraviolet or X-ray radiation. More dramatic events can occur as fresh hydrogen gas from the companion star accumulates on the surface of a white dwarf.

Novae The hydrogen spilling toward the white dwarf from the companion gradually spirals inward through the accretion disk and eventually falls onto the surface of the white dwarf. The white dwarf's strong gravity compresses this hydrogen gas into a thin surface layer. Both the pressure and temperature rise as the layer builds up with more accreting gas. When the temperature at the bottom of the layer reaches about 10 million K, hydrogen fusion suddenly ignites.

The white dwarf blazes back to life as its hydrogen layer burns. This thermonuclear flash causes the binary system to shine for a few glorious weeks as a **nova** (Figure 13.4a). A nova is far less luminous than a supernova, but still can shine as brightly as 100,000 Suns. It generates heat and pressure, ejecting most of the material that has accreted onto the white dwarf. This material expands outward, creating a *nova remnant* that sometimes remains visible years after the nova explosion (Figure 13.4b).

> A nova is caused by hydrogen fusion on the surface of a white dwarf in a binary star system.

Accretion resumes after a nova explosion subsides, so the entire process can repeat itself. The time between successive novae in a particular system depends on the rate at which hydrogen accretes onto the white dwarf surface and on how highly compressed this hydrogen becomes. The compression of hydrogen is greatest for the most massive white dwarfs, which have the strongest surface gravities. In some cases, novae have been observed to repeat after just a few decades. More commonly, thousands of years can pass between nova outbursts.

White Dwarf Supernovae Each time a nova occurs, the white dwarf ejects some of its mass. Each time a nova subsides, the white dwarf begins to accrete matter again. Theoretical models cannot yet tell us whether the net result should be a gradual increase or decrease in the white dwarf's mass. Nevertheless, observations show that in at least some cases, accreting white dwarfs in binary systems continue to gain mass as time passes. If a white dwarf gains enough mass, it can one day approach the $1.4M_{Sun}$ white dwarf limit. This day is the white dwarf's last.

Remember that most white dwarfs are made largely of carbon. As the white dwarf's mass approaches $1.4M_{Sun}$, its interior temperature rises high enough for carbon fusion to ignite. When carbon fusion begins, it ignites almost instantly throughout the star, and the white dwarf explodes completely in what we will call a **white dwarf supernova**.

think about it According to our understanding of novae and white dwarf supernovae, can either of these events ever occur with a white dwarf that is *not* a member of a binary star system? Explain.

If a white dwarf gains enough matter to exceed the 1.4-solar-mass white dwarf limit, it will explode completely in a white dwarf supernova.

The "carbon bomb" detonation that creates a white dwarf supernova is quite different from the iron catastrophe that leads to a supernova ending the life of a high-mass star [Section 12.3], which we will call a **massive star supernova**.* Astronomers can distinguish between the two types of supernova by studying their light. Both types shine brilliantly, with peak luminosities about 10 billion times that of the Sun ($10^{10}L_{Sun}$), but the luminosities of white dwarf supernovac fade steadily, while the decline in brightness of a massive star supernova is often more complicated (Figure 13.5). In addition, spectra of white dwarf supernovae always lack hydrogen lines, because white dwarfs contain very little hydrogen. The spectra of massive star supernovae generally contain prominent hydrogen lines because massive stars usually have plenty of hydrogen in their outer layers when they explode.

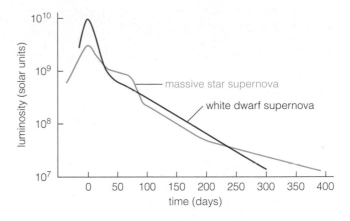

Figure 13.5

The curves on this graph show how the luminosities of the two types of supernovae fade with time. The white dwarf supernova fades quickly at first and then more gradually a few weeks after the peak, while the massive star supernova fades in a more complicated pattern. (Notice that the luminosity scale on the y-axis is expressed in powers of 10.)

 Stellar Evolution Tutorial, Lesson 3

13.2 Neutron Stars

White dwarfs with densities of several tons per teaspoon may seem incredible, but neutron stars are stranger still. The possibility that neutron stars might exist was first proposed in the 1930s, but most astronomers thought it preposterous that nature could truly make anything so bizarre. Nevertheless, a vast amount of evidence now confirms that neutron stars really exist. In this section, we'll examine their properties, their discovery, and the strange things that can happen to them in close binary star systems.

• What is a neutron star?

A *neutron star* is the ball of neutrons created by the collapse of the iron core in a massive star supernova (Figure 13.6). Typically just 10 kilometers in radius yet more massive than the Sun, neutron stars are essentially giant atomic nuclei made almost entirely of neutrons and held together by gravity. Like white dwarfs, neutron stars resist the crush of gravity with degeneracy pressure that arises when particles are packed as closely as nature allows. In the case of neutron stars, however, it is neutrons rather than electrons that are closely packed, so we say that **neutron degeneracy pressure** supports neutron stars.

A neutron star is a ball of neutrons just a few kilometers in radius but with a mass like that of the Sun.

The force of gravity at the surface of a neutron star is truly awe-inspiring. Escape velocity is about half the speed of light. If you foolishly chose to visit a neutron star's surface, your body would be squashed immediately into a microscopically thin pancake of subatomic particles.

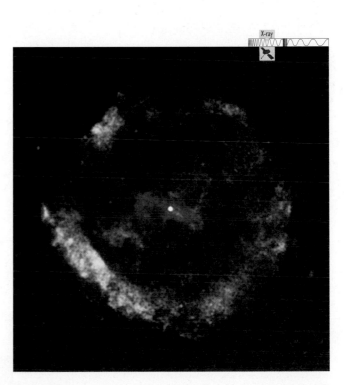

Figure 13.6

This X-ray image from the Chandra X-Ray Observatory shows the supernova remnant G11.2–03, the remains of a supernova observed by Chinese astronomers in A.D. 386. The white dot at the center is the neutron star left behind by the supernova. The different colors correspond to emission of X rays in different wavelength bands. The region pictured is about 23 light-years across.

---

*Observationally, astronomers classify supernovae as *Type II* if their spectra show hydrogen lines, and *Type I* otherwise. All Type II supernovae are assumed to be massive star supernovae. However, a Type I supernova can be either a white dwarf supernova or a massive star supernova in which the star blew away all its hydrogen before exploding. Type I supernovae fall into three classes, called *Type Ia*, *Type Ib*, and *Type Ic*. Only Type Ia supernovae are thought to be white dwarf supernovae.

Things would be only slightly less troubling if a bit of neutron star could somehow come to visit you. A paper clip with the density of neutron star material would outweigh Mount Everest. If such a paper clip magically appeared in your hand, you could not prevent it from falling. Down it would plunge, passing through the Earth like a rock falling through air. It would gain speed until it reached the center of the Earth, and its momentum would carry it onward until it slowed to a stop on the other side of our planet. Then it would fall back down again. If it came in from space, each plunge of the neutron star material would drill a different hole through the rotating Earth. In the words of astronomer Carl Sagan, the inside of the Earth would "look briefly like Swiss cheese" (until the melted rock flowed to fill in the holes) by the time friction finally brought the piece of neutron star to rest at the center of the Earth.

A neutron star could fit in your hometown, but its gravity would quickly destroy our planet.

In the unfortunate event that an *entire* neutron star came to visit you, it would not fall at all. Because it would be only about 10 kilometers in radius, the neutron star would probably fit in your hometown. Remember, however, that it would be 300,000 times as massive as Earth. As a result, the neutron star's immense surface gravity would quickly destroy your hometown and the rest of civilization. By the time the dust settled, the former Earth would be a shell no thicker than your thumb on the surface of the neutron star.

• How were neutron stars discovered?

The first observational evidence for neutron stars came in 1967, when a 24-year-old graduate student named Jocelyn Bell discovered a strange source of radio waves. The radio waves pulsed on and off at precise 1.337301-second intervals (Figure 13.7). At first, astronomers could not think of any natural process that could pulse in such a clockwork way, and they only half-jokingly called the new radio source "LGM," for Little Green Men. Today we refer to such rapidly pulsing radio sources as **pulsars**.

The mystery of pulsars was soon solved. By the end of 1968, astronomers had found two smoking guns: Pulsars sat at the centers of the Crab Nebula and the Vela Nebula, both gaseous remnants of supernovae (Figure 13.8). The pulsars are neutron stars left behind by the supernova explosions.

Neutron stars can spin rapidly and emit beams of radiation along their magnetic poles, which we detect as pulses of radiation if the beams sweep by Earth.

The pulsations arise because the neutron star is spinning rapidly as a result of the conservation of angular momentum [Section 4.3]: As an iron core collapses into a neutron star, its rotation rate must

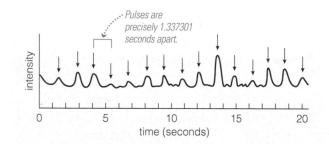

Figure 13.7

About 20 seconds of data from the first pulsar discovered by Jocelyn Bell in 1967.

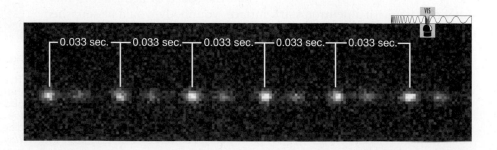

Figure 13.8

This time-lapse image of the pulsar at the center of the Crab Nebula, a supernova remnant, shows its main pulse recurring every 0.033 second. The fainter pulses are thought to come from the pulsar's other lighthouse-like beam. (Photo from the Very Large Telescope of the European Southern Observatory.)

increase as it shrinks in size. The collapse also bunches the magnetic field lines running through the core far more tightly, greatly amplifying the strength of the magnetic field. These intense magnetic fields direct beams of radiation out along the magnetic poles, although we do not yet know exactly how. If a neutron star's magnetic poles are not aligned with its rotation axis, the beams of radiation sweep round and round (Figure 13.9). Like lighthouses, the neutron stars actually emit a fairly steady beam of light, but we see a pulse of light each time the beam sweeps past Earth.

Pulsars are not quite perfect clocks. The continual twirling of a pulsar's magnetic field generates electromagnetic radiation that carries away energy and angular momentum, causing the neutron star's rotation rate to slow gradually. The pulsar in the Crab Nebula, for example, currently spins about 30 times per second. Two thousand years from now, it will spin less than half as fast. Eventually, a pulsar's spin slows so much and its magnetic field becomes so weak that we can no longer detect it. In addition, some spinning neutron stars may be oriented so that their beams do not sweep past our location. Thus, we have the following rule: *All pulsars are neutron stars, but not all neutron stars are pulsars.*

Figure 13.9

Radiation from a rotating neutron star can appear to pulse like the beam from a lighthouse.

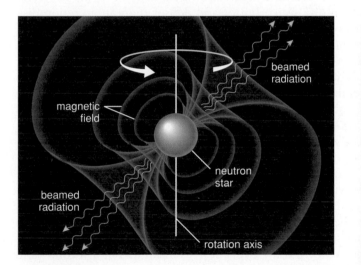

a A pulsar is a rotating neutron star that beams radiation along its magnetic axis.

b If the magnetic axis is not aligned with the rotation axis, the pulsar's beams sweep through space like lighthouse beams. Each time one of the pulsar's beams sweeps across Earth, we see a pulse of radiation.

think about it Suppose we do not see pulses from a particular neutron star and hence do not call it a pulsar. Is it possible that a civilization living in some other star system would see this neutron star as a pulsar? Explain.

We know that pulsars must be neutron stars because no other massive object could spin so fast. A white dwarf, for example, can spin no faster than about once per second. A spin rate much faster than this would tear the white dwarf apart because its surface would be rotating faster than the escape velocity from the surface of the star. Pulsars have been discovered that rotate as fast as 625 times per second. Only an object as small and dense as a neutron star could spin so fast without breaking apart.

• What can happen to a neutron star in a close binary system?

Like white dwarfs, neutron stars in close binary systems can brilliantly burst back to life. Just as in a white-dwarf binary system, gas overflowing from a companion star can create a hot, swirling accretion disk around the neutron star. However, the much stronger gravity of the neutron star makes its accretion disk much hotter and denser than an accretion disk around a white dwarf.

The high temperatures in the inner regions of the accretion disk around a neutron star make it radiate powerfully in X rays. Some close binaries with neutron stars emit 100,000 times more energy in X rays than our Sun emits in all wavelengths of light combined. Due to this intense X-ray emission, close binaries that contain accreting neutron stars are often called **X-ray binaries**. We have detected hundreds of X-ray binaries in the disk of the Milky Way Galaxy.

The emission from most X-ray binaries pulsates rapidly as the neutron star spins, much like the radio pulsations of the neutron stars we see as pulsars. However, the pulsation rates of X-ray binaries tend to accelerate with time, rather than slowing like radio pulsars. Apparently, matter accreting onto the neutron star adds angular momentum, speeding up the neutron star's rotation (Figure 13.10). Some of these neutron stars rotate so fast that they pulsate every few thousandths of a second (and hence are sometimes called *millisecond pulsars*).

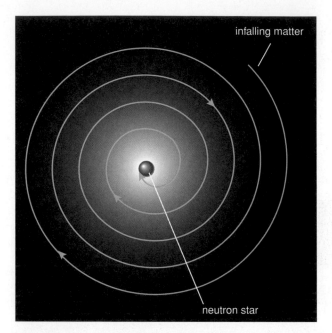

Figure 13.10

Matter accreting onto a neutron star adds angular momentum, increasing the neutron star's rate of spin. The spin increases because infalling matter just above the neutron star's surface is orbiting faster than the neutron star is spinning. When this matter hits the neutron star's surface, it boosts the star's spin rate.

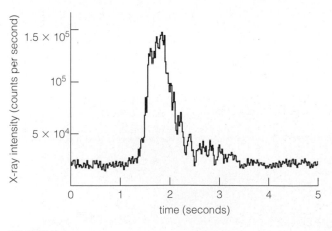

Figure 13.11

Light curve of an X-ray burst. In this particular burst, the X-ray luminosity of the neutron star spiked to over six times its usual value in a matter of seconds.

The hot accretion disks around neutron stars in close binary systems shine brightly with X rays.

Like accreting white dwarfs that occasionally erupt as novae, accreting neutron stars sporadically erupt with enormous luminosities. Hydrogen-rich material from the companion star builds up on the surface of the neutron star, forming a layer about a meter thick. Pressures at the bottom of this hydrogen layer are high enough to maintain steady fusion, which produces a layer of helium beneath the hydrogen. Helium fusion suddenly ignites when the temperature in this layer builds to 100 million K. The helium fuses rapidly to make carbon and heavier elements, generating a burst of energy that flows from the neutron star in the form of X rays. These **X-ray bursters** typically flare every few hours to every few days (Figure 13.11). Each burst lasts only a few seconds, but during those seconds the system radiates 100,000 times as much power as the Sun, all in X rays. Within a minute after a burst, the X-ray burster cools back down and resumes accretion.

13.3 Black Holes: Gravity's Ultimate Victory

The story of stellar corpses would be strange enough if it ended with white dwarfs and neutron stars, but it does not. Sometimes, the gravity in a stellar corpse becomes so strong that nothing can prevent the corpse from collapsing under its own weight. The stellar corpse collapses without end, crushing itself out of existence and forming perhaps the most bizarre type of object in the universe: a *black hole*.

special topic: -----

General Relativity and Curvature of Spacetime

EINSTEIN'S SPECIAL THEORY of relativity (see the Special Topic on p. 364) solved a lot of problems that had troubled physicists prior to its 1905 publication. However, Einstein knew the theory was incomplete; it is called *special* because it applies only to the special case of motion at constant velocity. It did not apply, for example, to objects accelerating under the influence of gravity. Einstein therefore sought to generalize the theory to include gravity and acceleration.

In 1907, Einstein hit upon what he later called "the happiest thought of my life." He realized that gravity and acceleration have identical effects on the laws of physics. This idea is called the *equivalence principle,* and it states: *The effects of gravity are exactly equivalent to the effects of acceleration.*

To clarify the meaning of this principle, imagine that you are sitting inside with no windows or doors when your room is magically removed from Earth and sent hurtling through space with the acceleration of gravity on Earth, or 9.8 m/s^2 [Section 4.1]. According to the equivalence principle, you would have no way of knowing that you'd left Earth. Any experiment you performed, such as dropping balls of different weights, would yield the same results you'd get on Earth. Likewise, experiments performed in a freely falling elevator would yield the same results as those performed in an elevator drifting at constant velocity through empty space.

Einstein's new point of view on motion, acceleration, and gravity brought about a radical revision of how we think about space and time. Instead of thinking about the three dimensions of space and the one dimension of time as separate, we learned to think of them as a seamless, four-dimensional entity known as *spacetime.* Einstein showed that a person would feel weightless, as though drifting through empty space, as long as the person's path through four-dimensional spacetime was as straight as possible.

So why do astronauts feel weightless even though they orbit Earth on a curved path? According to general relativity, they are still following the *straightest possible* path through four-dimensional spacetime. The path is curved only because spacetime itself is curved near the Earth. In other words, while Newton would have attributed the curved orbital path of the astronauts to the force of gravity, Einstein attributes it to curvature of spacetime. That is, gravity arises from curvature of spacetime.

The curvature of spacetime is caused by mass, and stronger gravity just means greater spacetime curvature. For example, the Sun's gravity curves spacetime more than Earth's gravity, and the strong gravity on the surface of a white dwarf curves spacetime more than the gravity on the surface of the Sun. Although we cannot visualize spacetime curvature, we can visualize two-dimensional analogies to it with rubber sheet diagrams like those shown in Figure 13.12. From this new perspective, planets orbit the Sun because of the way space is curved by the Sun: Each planet is moving as straight as it can, but space is curved in a way that keeps it going round and round like a marble in a salad bowl (see figure).

Given that we cannot actually perceive all four dimensions of spacetime at once, you may wonder why scientists think spacetime curvature is real. The answer is that Einstein's theory predicts measurable effects from this curvature. For example, if gravity really does arise from curvature of spacetime, then light paths ought to bend when they pass near large masses. Scientists have measured such bending of starlight as it passes near the Sun, and the amount of bending is precisely what general relativity predicts it should be. We have also observed this bending of light by distant galaxies, a phenomenon called *gravitational lensing* [Section 16.2]. Another key prediction of general relativity is that time should run slower in regions of stronger gravity (because curvature affects both space and time)—again, numerous observations and experimental tests verify Einstein's predictions. Strange as it may seem, we live in a four-dimensional universe in which space and time are intertwined and can never be disentangled.

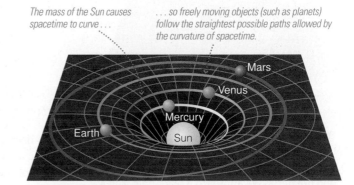

The mass of the Sun causes spacetime to curve . . .

. . . so freely moving objects (such as planets) follow the straightest possible paths allowed by the curvature of spacetime.

Mars
Venus
Mercury
Earth
Sun

According to general relativity, planets orbit the Sun for much the same reason that you can make a marble go around in a salad bowl: The planet is going as straight as it can, but the curvature of spacetime causes its path through space to curve.

• What is a black hole?

The "black" in the name *black hole* comes from the fact that nothing—not even light—can escape from a black hole. The escape velocity of any object depends on the strength of its gravity, which depends on its mass and size [Section 4.4]. Decreasing the size of an object of a particular mass makes its gravity stronger and increases its escape velocity. A black hole is so compact that it has an escape velocity greater than the speed of light. Because nothing can travel faster than the speed of light, neither light nor anything else can escape from within a black hole.

A black hole is a place where gravity is so strong that nothing—not even light—can escape from within it.

The "hole" part of the word *black hole* tells an even stranger story. Einstein discovered that space and time are not distinct, as we usually think of them, but instead are bound up together as four-dimensional **spacetime**. Moreover, in his general theory of relativity, Einstein showed that what we perceive as gravity arises from *curvature of spacetime* (see Special Topic: General Relativity and Curvature of Spacetime, p. 371). Curved spacetime is not easy to visualize, because we can see only three dimensions at once. However, we can understand the idea with a two-dimensional analogy.

Figure 13.12 represents all four dimensions of spacetime with a two-dimensional rubber sheet. In this analogy, the sheet is flat in a region far from any mass (Figure 13.12a). Near a massive object with strong gravity, the sheet becomes curved (Figure 13.12b)—the stronger the gravity, the more curved it gets. In this analogy, a black hole is like a bottomless pit in spacetime (Figure 13.12c): Gravity gets stronger and stronger as we get closer to the black hole, so the sheet curves more and more. Keep in mind that the illustration is only an analogy, and black holes actually are spherical, not funnel-shaped. Nevertheless, it captures the key idea that a black hole is like a hole in the observable universe in the following sense: If you enter a black hole, you leave our observable universe and can never return.

see it for yourself Curved space has different rules of geometry than "flat" space, a difference you can demonstrate with a plastic ball and a marker. Draw a straight line on the ball going one-quarter of the way around it, then make a right-angle turn and draw a line one-quarter of the way around the ball in this new direction, and finish the triangle with a straight line back to where you started. Measure all three angles. How does the sum of the three angles of the triangle on the ball compare to the 180° you would find for that sum in a triangle on a flat surface?

The Event Horizon The boundary between the inside of the black hole and the universe outside is called the **event horizon**. The event horizon essentially marks the point of no return for objects entering a black hole: It is the boundary around a black hole at which the escape velocity equals the speed of light. Nothing that passes within this boundary can ever escape. The event horizon gets its name because we have no hope of learning about any events that occur within it.

We usually think of the "size" of a black hole as the size of its event horizon. Our everyday understanding of "size" is hard to apply inside the event horizon because space and time are so distorted there. However, for someone outside the black hole, the event horizon is a sphere with a well-defined radius, called the **Schwarzschild radius** (after Karl Schwarzschild, who first computed it with Einstein's general theory of relativity).

The Schwarzschild radius of a black hole depends only on its mass. A black hole with the mass of the Sun has a Schwarzschild radius of about 3 km—only a little smaller than the radius of a neutron star of the same

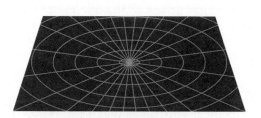

a A two-dimensional representation of "flat" spacetime. The distances between adjacent circles are the same.

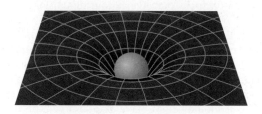

b Gravity arises from curvature of spacetime, represented here by a mass pushing down on the rubber sheet. Notice how the circles become more widely separated near the mass, showing that the curvature is greater as we approach the mass on the sheet.

event horizon

c The curvature of spacetime becomes greater and greater as we approach a black hole, and a black hole itself is a bottomless pit in spacetime.

Figure 13.12 interactive figure

We can use two-dimensional rubber sheets to show an analogy to curvature in four-dimensional spacetime.

mass. More massive black holes have larger Schwarzschild radii. For example, a black hole with 10 times the mass of the Sun has a Schwarzschild radius of about 30 km.

In essence, a collapsing stellar core becomes a black hole at the moment that it shrinks to a size smaller than its Schwarzschild radius. At that moment, the core disappears from view within its own event horizon. The black hole still contains all the mass and has the gravity associated with that mass, but its outward appearance tells us nothing about what fell in.

Singularity and the Limits to Knowledge We can't observe what happens to a stellar core once it collapses into a black hole, because no information can emerge from within the event horizon. Nevertheless, we can use our understanding of the laws of physics to predict what must occur inside a black hole. Because nothing can stop the crush of gravity in a black hole, we might expect that all the matter that forms a black hole must ultimately be crushed to an infinitely tiny and dense point in the black hole's center. We call this point a **singularity**.

Unfortunately, this idea of a singularity pushes up against the limits of scientific knowledge today. The problem is that that two very successful theories give different answers about the nature of a singularity. Einstein's general theory of relativity, which seems to explain successfully how gravity works throughout the universe, predicts that spacetime should grow infinitely curved as it enters the pointlike singularity. Quantum physics, which successfully explains the nature of atoms and the spectra of light, predicts that spacetime should fluctuate chaotically near the singularity. These are clearly different claims, and no theory that can reconcile them has yet been found.

• What would it be like to visit a black hole?

Imagine that you are a pioneer of the future, making the first visit to a black hole. You've selected a black hole with a mass of $10M_{Sun}$ and a Schwarzschild radius of 30 km. As your spaceship approaches the black hole, you fire its engines to put the ship on a circular orbit a few thousand kilometers above the event horizon. This orbit will be perfectly stable—there is no need to worry about getting "sucked in."

Your first task is to test Einstein's general theory of relativity. This theory predicts that time should run more slowly as the force of gravity grows stronger. It also predicts that light coming out of a strong gravitational field should show a redshift, called a *gravitational redshift*, that is due to gravity rather than to the Doppler effect. You test these predictions with the aid of two identical clocks whose numerals glow with blue light. You keep one clock aboard the ship and push the other one, with a small rocket attached, directly toward the black hole (Figure 13.13).

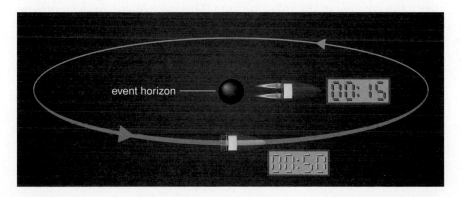

Figure 13.13 interactive figure

Time runs more slowly on the clock nearer to the black hole, and gravitational redshift makes its glowing blue numerals appear red from your orbiting spaceship.

The small rocket automatically fires its engines just enough so that the clock falls gradually toward the event horizon. Sure enough, the clock on the rocket ticks more slowly as it heads toward the black hole, and its light becomes increasingly redshifted. When the clock reaches a distance of about 10 km above the event horizon, you see it ticking only half as fast as the clock on your spaceship, and its numerals are red instead of blue.

The rocket has to expend fuel rapidly to keep the clock hovering in the strong gravitational field, and the fuel soon runs out. The clock plunges toward the black hole. From your safe vantage point inside your spaceship, you see the clock ticking more and more slowly as it falls. However, you soon need a radio telescope to "see" it, as the light from the clock face shifts from the red part of the visible spectrum to the infra-red, and on into the radio region. Finally, its light is so redshifted that no conceivable telescope could detect it. Just as the clock vanishes from view, you see that the time on its face has frozen to a stop.

Curiosity overwhelms the better judgment of one of your colleagues. He hurriedly climbs into a spacesuit, grabs the other clock, resets it, and jumps out of the airlock on a trajectory aimed straight for the black hole. Down he falls, clock in hand. He watches the clock, but because he and the clock are traveling together, its time seems to run normally and its numerals stay blue. From his point of view, time seems to neither speed up nor slow down. In fact, he'd say that *you* were the one with the strange time, as he would see your time running increasingly fast and your light becoming increasingly blueshifted. When his clock reads, say, 00:30, he and the clock pass through the event horizon. There is no bar-rier, no wall, no hard surface. The event horizon is a mathematical boundary, not a physical one. From his point of view, the clock keeps ticking. He is inside the event horizon, the first human being ever to leave our observable universe.

If you fell toward a black hole, you would rapidly accelerate and soon cross the event horizon. But to someone watching from afar, your fall would appear to take forever.

Back on the spaceship, you watch in horror as your overly cu-rious friend plunges to his death. Yet, from your point of view, he will *never* cross the event horizon. You'll see time come to a stop for him and his clock just as he vanishes from view due to the huge gravita-tional redshift of light. When you return home, you can play a video for the judges at your trial, proving that your friend is still outside the black hole. Strange as it may seem, all this is true according to Einstein's the-ory. From your point of view, your friend takes *forever* to cross the event horizon (even though he vanishes from view due to his ever-increasing redshift). From his point of view, it is but a moment's plunge before he passes into oblivion.

The truly sad part of this story is that your friend did not live to expe-rience the crossing of the event horizon. The force of gravity he felt as he approached the black hole grew so quickly that it actually pulled much harder on his feet than on his head, simultaneously stretching him lengthwise and squeezing him from side to side (Figure 13.14). In essence, your friend was stretched in the same way the oceans are stretched by the tides, except that the *tidal force* near the black hole is trillions of times stronger than the tidal force of the Moon on Earth [Section 4.4]. No human could survive it.

If he had thought ahead, your friend might have waited to make his jump until you visited a much larger black hole, like one of the *supermassive black holes* thought to reside in the centers of many galaxies [Section 15.4]. A black hole with a mass equal to $10^9 M_{Sun}$ has a Schwarzschild radius of

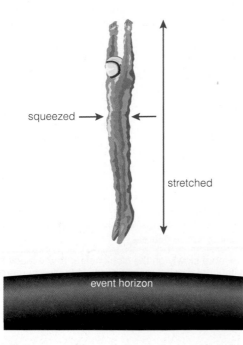

Figure 13.14

Tidal forces would be lethal near a black hole formed by the collapse of a star. The black hole's gravity would pull more strongly on the astronaut's feet than on his head, stretching him lengthwise and squeezing him from side to side.

3 billion km, the distance from our Sun to Uranus. Although the gravitational forces at the event horizon of all black holes are equally great, the larger size of a supermassive black hole makes its tidal forces much weaker and hence nonlethal. Your friend could safely plunge through the event horizon.

Again, from your point of view, the crossing would take forever, and you would see time come to a stop for him just as he vanished from sight because of the gravitational redshift. Again, he would experience time running normally and would see time in the outside universe running increasingly fast as he approached the event horizon. Unfortunately, anything he saw would do him little good as he plunged to oblivion inside the black hole.

• Do black holes really exist?

The idea of objects with escape velocities greater than the speed of light was first suggested in the late 1700s (by British philosopher John Mitchell and French physicist Pierre Laplace), though the bizarre implications of the idea were not understood until after Einstein published his general theory of relativity in 1915. As was the case for neutron stars, at first most astronomers who contemplated the idea of black holes thought them too strange to be true. Today, however, our understanding of physics gives us reason to think that black holes ought to be fairly common, and observational evidence strongly suggests that black holes really exist.

The Formation of a Black Hole The idea that black holes ought to exist comes from considering how they might form. Recall that white dwarfs cannot exceed $1.4M_{Sun}$, because gravity overcomes electron degeneracy pressure above that mass. Calculations show that the mass of a neutron star has a similar limit that lies somewhere between about 2 and 3 solar masses. Above this mass, neutron degeneracy pressure cannot hold off the crush of gravity in a collapsing stellar core.

A supernova occurs when the electron degeneracy pressure supporting the iron core of a massive star succumbs to gravity, causing the core to collapse catastrophically into a ball of neutrons [Section 12.3]. That is why most supernovae leave neutron stars behind. However, theoretical models show that the most massive stars might not succeed in blowing away all their upper layers. If enough matter falls back onto the neutron core, its mass may rise above the neutron star limit.

As soon as the core exceeds the neutron star limit, gravity overcomes the neutron degeneracy pressure and the core collapses once again. This time, no known force can keep the core from collapsing into oblivion as a black hole. Moreover, another effect of Einstein's theory of relativity makes it highly unlikely that any as-yet-unknown force could intervene either.

According to the known laws of physics, nothing can stop the collapse of a stellar corpse with a mass greater than about 3 solar masses.

Recall that Einstein's theory tells us that energy is equivalent to mass ($E = mc^2$) [Section 4.3], implying that energy, like mass, must exert some gravitational attraction. The gravity of pure energy usually is negligible, but not in a stellar core collapsing beyond the neutron star limit. Inside a star that collapses beyond the point where it could have been a neutron star, the energy associated with the enhanced internal temperature and pressure acts like additional mass, making the crushing power of gravity even stronger. The more the star collapses, the stronger gravity gets. To the best of our understanding, *nothing* can halt the crush of gravity at this point, so the star must inevitably collapse to form a black hole.

Observational Evidence for Black Holes

The fact that black holes emit no light might make it seem as if they should be impossible to detect. However, a black hole's gravity can influence its surroundings in a way that reveals its presence. Astronomers have discovered many objects that show the telltale signs of an unseen gravitational influence with a large enough mass to suggest that it is a black hole.

Compelling observational evidence for black holes formed by supernovae comes from studies of X-ray binaries. Recall that the accretion disks around neutron stars in close binary systems can emit powerful X-ray radiation, making an X-ray binary. The accretion disk forms because the neutron star's strong gravity pulls mass from its companion star. Because a black hole has even stronger gravity than a neutron star, a black hole in a close binary system should also be surrounded by a hot, X ray–emitting accretion disk. Some X-ray binaries may therefore contain black holes rather than neutron stars. The trick to learning which type of corpse resides in an X-ray binary depends on measuring the object's mass.

Some X-ray binaries probably contain accreting black holes rather than accreting neutron stars. One of the most promising black hole candidates is in an X-ray binary called Cygnus X-1 (Figure 13.15). This system contains an extremely luminous star with an estimated mass of $18M_{Sun}$. Based on Doppler shifts of its spectral lines, astronomers have concluded that this star orbits a compact, unseen companion with a mass of about $10M_{Sun}$. Although there is some uncertainty to these mass estimates, the mass of the invisible accreting object clearly exceeds the $3M_{Sun}$ neutron star limit. Because it is too massive to be a neutron star, by current knowledge, it cannot be anything other than a black hole.

think about it Recall that some X-ray binaries that contain neutron stars emit frequent X-ray bursts and are called X-ray bursters. Could an X-ray binary that contains a black hole exhibit the same type of X-ray bursts? Why or why not? (*Hint*: Where do the X-ray bursts occur in an X-ray binary with a neutron star?)

Figure 13.15 interactive figure

Artist's conception of the Cygnus X-1 system, named as such because it is the brightest X-ray source in the constellation Cygnus. The X rays come from the high-temperature gas in the accretion disk surrounding the black hole. The inset shows the location of Cygnus X-1 in the sky.

13.4 The Origin of Gamma-Ray Bursts

In the early 1960s, the United States began launching a series of top-secret satellites designed to look for gamma rays emitted by nuclear bomb tests. The satellites soon began detecting occasional bursts of gamma rays, typically lasting a few seconds (Figure 13.16). It took several years for military scientists to become convinced that these **gamma-ray bursts** were coming from space, not from some sinister human activity. They publicized the discovery in 1973. At first, astronomers assumed that these gamma-ray bursts were just a variation of the X-ray bursts that can arise from neutron stars in close binary systems. However, according to recent observations, at least some gamma-ray bursts seem to come from the formation of black holes at very great distances from Earth, although the process that produces the gamma rays still remains mysterious.

• Where do gamma-ray bursts come from?

When gamma-ray bursts were first discovered, astronomers had very little information about where they came from. The problem was that gamma-ray photons carry so much energy that they are very difficult to focus. Instead of reflecting off of a telescope's mirror, they tend to penetrate right through it.

Our ignorance about the origins of gamma-ray bursts began to lift with the 1991 launch of NASA's *Compton Gamma Ray Observatory*, or *Compton* for short, which operated in Earth orbit for almost a decade. Compton carried an array of eight detectors designed to study gamma-ray bursts. By comparing the data recorded by all eight detectors, scientists could determine the direction of a gamma-ray burst within about 1°. The results were stunning. Compton detected a gamma-ray burst about once a day and soon had found enough bursts to show their distribution on a map of the sky. To almost everyone's surprise, the bursts were *not* concentrated in the disk of the Milky Way Galaxy, where the X-ray binaries tend to be. This discovery ruled out the possibility that gamma-ray bursts come from the same types of systems as X-ray bursts. Instead, the gamma-ray bursts seemed to come randomly from all directions in space.

The even distribution of gamma-ray bursts across the sky suggested that they must come from far outside our own galaxy. If the bursts had been coming from objects distributed spherically about the Milky Way Galaxy, we would have seen a concentration of them in the direction of the galactic center, because we are located more than halfway out from the center of our galaxy.

Direct evidence that gamma-ray bursts do indeed come from far outside our galaxy arrived in 1997, thanks to a breakthrough in our observational capabilities: For the first time, it became possible to respond almost immediately to the detection of a gamma-ray burst with observations in other wavelength bands. When a space-based observatory detected a gamma-ray burst, the observers quickly trained X-ray telescopes on the region of sky from which the burst came and relayed its location to astronomers at visible-light telescopes on the ground. In this way, astronomers were able to observe X rays and visible light from the afterglow of the burst. The higher resolution possible in these other wavelengths allowed astronomers to pinpoint the location of the burst, confirming that it came from a distant galaxy. Since then, numerous other bursts have been traced

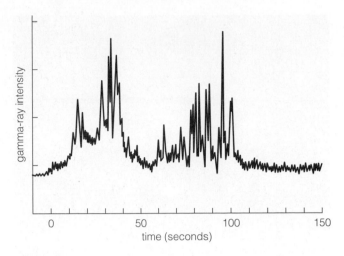

Figure 13.16

The intensity of a gamma-ray burst can fluctuate dramatically over time periods of just a few seconds.

Figure 13.17

The bright dot near the center of this photo is the visible-light afterglow of a gamma-ray burst, as seen by the Hubble Space Telescope. The elongated blob extending above the dot is the distant galaxy in which the burst occurred.

Figure 13.18

The bright object marked "source of gamma-ray burst" in the first panel of this sequence is the visible-light afterglow of a gamma-ray burst in the outskirts of a distant galaxy (marked "host galaxy"). The light detected from this object once the initial afterglow faded showed that it had the light curve of a massive star supernova.

to explosions in galaxies billions of light-years away (Figure 13.17). At least some gamma-ray bursts, but maybe not all of them, must therefore be extremely powerful explosions that occur at extremely large distances from Earth.

• What causes gamma-ray bursts?

Learning that gamma-ray bursts generally come from far outside our galaxy only deepened the mystery surrounding them. The afterglows of some gamma-ray bursts can be seen with binoculars, even though they are coming from galaxies billions of light-years away—making them by far the most powerful bursts of energy to occur in the universe. If these bursts shine their light equally in all directions, like a light bulb, then the total luminosity of a burst can briefly exceed the combined luminosity of a million galaxies like our Milky Way! Because such a high luminosity is very difficult to explain, some scientists speculate that gamma-ray bursts channel their energy into narrow searchlight beams, like pulsars, so that only some are visible from Earth. But even in this case, a burst's luminosity would still temporarily surpass that of thousands of galaxies combined.

What could cause such massive outbursts of energy? At least some gamma-ray bursts appear to come from extremely powerful supernova explosions. An ordinary supernova that forms a neutron star does not release enough energy to explain the luminosity of the brightest gamma-ray bursts. However, a supernova that forms a black hole crushes even more matter into an even smaller radius, releasing many times more gravitational potential energy than one that forms a neutron star. This kind of event (sometimes called a *hypernova*) might be powerful enough to explain the most extreme gamma-ray bursts.

The primary evidence linking gamma-ray bursts to exploding stars comes from observing them in other wavelengths of light, including visible light and X rays. Several recently launched gamma-ray telescopes—especially NASA's *Swift* satellite launched in 2004—are capable of rapidly pinpointing the location of a gamma-ray burst in the sky, thereby allowing other types of telescopes to be pointed at the gamma-ray source almost as soon as the burst is detected. Observations like these have shown that at least some gamma-ray bursts coincide in both time and location with powerful supernova explosions (Figure 13.18). In a number of other cases, the gamma-ray bursts are observed to come from distant galaxies that are actively forming new stars. Those observations also support the idea that

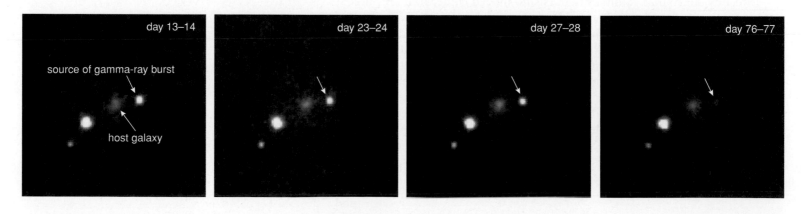

day 13–14 day 23–24 day 27–28 day 76–77

source of gamma-ray burst

host galaxy

gamma-ray bursts are produced in the explosions of extremely massive stars because such stars are very short-lived and therefore should be found only in places where stars are actively forming.

The association of a few gamma-ray bursts with powerful supernovae has not fully solved the mystery of their origin, because bursts come in at least two distinct types. Only one type has been observationally linked with explosions of massive stars. Bursts of the other type last only a few seconds, making them more difficult to catch in action with telescopes for other wavelengths. Nevertheless, in mid-2005 *Swift* and a satellite called *HETE* each captured a short burst in both gamma rays and X rays; the X rays from the burst were pinpointed well enough for follow-up observations with the Chandra X-Ray Observatory and the Hubble Space Telescope. Results show that these short bursts do *not* come from supernovae. What could they be coming from? The leading hypothesis is a collision in a binary system containing either two neutron stars or a neutron star and a black hole. Einstein's general theory of relativity predicts that such binary systems must lose substantial energy (to *gravitational waves*) over time, gradually causing the two objects to spiral in toward each other until they collide. Short gamma-ray bursts may be the by-product of such catastrophic collisions.

----

the big picture
Putting Chapter 13 into Context

We have now seen the mind-bending consequences of stellar death. As you think about the bizarre objects of the steller graveyard described in this chapter, try to keep in mind these "big picture" ideas:

- Despite the strange nature of stellar corpses, observational evidence exists for white dwarfs and neutron stars, and the case for black holes is very strong.

- White dwarfs, neutron stars, and black holes can all have close stellar companions from which they accrete matter. These binary systems produce some of the most spectacular events in the universe, including novae, white dwarf supernovae, and X-ray bursters.

- Black holes are holes in the observable universe that strongly warp space and time around them. The nature of black hole singularities remains beyond the frontier of current scientific understanding.

summary of key concepts

13.1 White Dwarfs

• What is a white dwarf?

A white dwarf is the core left over from a low-mass star, supported against the crush of gravity by **electron degeneracy pressure**. A white dwarf typically has the mass of the Sun compressed into a size no larger than Earth. No white dwarf can have a mass greater than $1.4M_{Sun}$.

• What can happen to a white dwarf in a close binary system?

A white dwarf in a close binary system can acquire hydrogen from its companion through an **accretion disk** that swirls toward the white dwarf's surface. As hydrogen builds up on the white dwarf's surface, it may begin nuclear fusion and cause a **nova** that, for a few weeks, may shine as brightly as 100,000 Suns. In extreme cases, accretion may continue until the white dwarf's mass exceeds the **white dwarf limit** of $1.4M_{Sun}$, at which point it will explode as a **white dwarf supernova**.

13.2 Neutron Stars

• What is a neutron star?

A neutron star is the ball of neutrons created by the collapse of the iron core in a massive star supernova. It resembles a giant atomic nucleus 10 kilometers in radius and with more mass than the Sun.

• How were neutron stars discovered?

Neutron stars spin rapidly when they are born, and their strong magnetic fields can direct beams of radiation that sweep through space as the neutron star spins. We see such neutron stars as **pulsars**, and these pulsars provided the first direct evidence for the existence of neutron stars.

• What can happen to a neutron star in a close binary system?

Neutron stars in close binary systems can accrete hydrogen from their companions, forming dense, hot accretion disks. The hot gas emits strongly in X rays, so we see these systems as **X-ray binaries**. In some of these systems, frequent bursts of helium fusion occur on the neutron star's surface, causing **X-ray bursts**.

13.3 Black Holes: Gravity's Ultimate Victory

• What is a black hole?

A black hole is a place where gravity has crushed matter into oblivion, creating a hole in the universe from which nothing can ever escape, not even light. The **event horizon** marks the boundary between our observable universe and the inside of the black hole; the black hole's **Schwarzschild radius** is the size of the event horizon.

• What would it be like to visit a black hole?

You could orbit a black hole just like you could any other object of the same mass. However, you'd see strange effects for an object falling toward the black hole: Time would seem to run slowly for the object, and its light would be increasingly redshifted as it approached the black hole. The object would never quite reach the event horizon, but it would soon disappear from view as its light became so redshifted that no instrument could detect it.

• Do black holes really exist?

No known force can stop the collapse of a stellar corpse with a mass above the neutron star limit of 2 to 3 solar masses, and theoretical studies of supernovae suggest that such objects should sometimes form. Observational evidence supports this idea: Some X-ray binaries include compact objects far too massive to be neutron stars, making it likely that they are black holes.

13.4 The Origin of Gamma-Ray Bursts

• Where do gamma-ray bursts come from?

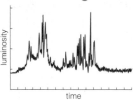

Gamma-ray bursts generally come from distant galaxies that can be billions of light-years from Earth. Because the sources of these bright bursts are both bright and very distant, they must be the most powerful bursts of energy we observe anywhere in the universe.

• What causes gamma-ray bursts?

No one knows exactly what causes gamma-ray bursts, but the energy required to make one suggests that gamma-ray bursts occur when certain kinds of black holes are formed. At least some gamma-ray bursts appear to come from unusually powerful supernova explosions that may create black holes.

exercises and problems

Review Questions

Short-Answer Questions Based on the Reading

1. What is degeneracy pressure, and how is it important to white dwarfs and neutron stars? What is the difference between electron degeneracy pressure and neutron degeneracy pressure?

2. Describe the mass, size, and density of a typical white dwarf. How does the size of a white dwarf depend on its mass?

3. What happens to the electron speeds in a more massive white dwarf, and how does this idea lead to a limit on the mass of a white dwarf? What is the *white dwarf limit*?

4. What are *accretion disks,* and why do we find them only in close binary systems? Explain how the accretion disk provides a white dwarf with a new source of energy that we can detect from Earth.

5. What is a *nova?* Describe the process that creates a nova and what a nova looks like.

6. What causes a *white dwarf supernova?* Observationally, how do we distinguish white dwarf and massive star supernovae?

7. Describe the mass, size, and density of a typical neutron star. What would happen if a neutron star came to your hometown?

8. How do we know that *pulsars* are neutron stars? Are all neutron stars also pulsars? Explain.

9. Explain how the presence of a neutron star can make a close binary star system appear to us as an *X-ray binary*. Why do some of these systems appear to us as *X-ray bursters?*

10. What do we mean when we say that a black hole is a hole in *spacetime?* What is the *event horizon* of a black hole? How is it related to the *Schwarzschild radius?*

11. What do we mean by the *singularity* of a black hole? How do we know that our current theories are inadequate to explain what happens at the singularity?

12. Suppose you are falling into a black hole. How will you perceive the passage of your own time? How will you perceive the passage of time in the universe around you? Briefly explain why your trip is likely to be lethal.

13. Why do we think that black holes should sometimes be formed by supernovae? What observational evidence supports the existence of black holes?

14. How do we know that gamma-ray bursts do *not* come from the same sources as X-ray bursts? Summarize current ideas about the cause of gamma-ray bursts.

Test Your Understanding

Does It Make Sense?

Decide whether the statement makes sense (or is clearly true) or does not make sense (or is clearly false). Explain clearly; not all these have definitive answers, so your explanation is more important than your chosen answer.

15. Most white dwarf stars have masses close to that of our Sun, but a few white dwarf stars are up to three times as massive as the Sun.

16. The radii of white dwarf stars in close binary systems gradually increase as they accrete matter.

17. If you want to find a pulsar, you might want to look near the remnant of a supernova described by ancient Chinese astronomers.

18. If a black hole 10 times as massive as our Sun were lurking just beyond Pluto's orbit, we'd have no way of knowing it was there.

19. If the Sun suddenly became a $1 M_{Sun}$ black hole, the orbits of the planets would not change at all.

20. We can detect black holes with X-ray telescopes because matter falling into a black hole emits X rays after it smashes into the event horizon.

21. If two black holes merge together, the resulting black hole is even smaller than the original ones because of its stronger gravity.

22. If gamma-ray bursts really channel their energy into narrow beams, then the total number of gamma-ray bursts that occur is probably far greater than the number we detect.

23. If the Sun suddenly became a $1 M_{Sun}$ black hole, Earth's tides would become much greater.

24. We know that gamma-ray bursts can't come from the disk of our galaxy because they are spread evenly over the entire sky.

Quick Quiz

Choose the best answer to each of the following. Explain your reasoning with one or more complete sentences.

25. Which of these objects has the smallest radius? (a) a $1.2 M_{Sun}$ white dwarf (b) a $0.6 M_{Sun}$ white dwarf (c) Jupiter

26. Which of these objects has the largest radius? (a) a $1.2 M_{Sun}$ white dwarf (b) a $1.5 M_{Sun}$ neutron star (c) a $3.0 M_{Sun}$ black hole

27. Which of these things has the smallest radius? (a) a $1.2 M_{Sun}$ white dwarf (b) the event horizon of a $3.0 M_{Sun}$ black hole (c) the event horizon of a $10 M_{Sun}$ black hole

28. What would happen if the Sun suddenly became a black hole without changing its mass? (a) The black hole would quickly suck in Earth. (b) Earth would gradually spiral into the black hole. (c) Earth's orbit would not change.

29. Which of these isolated neutron stars must have had a binary companion? (a) a pulsar inside a supernova remnant that pulses 30 times per second (b) an isolated pulsar that pulses 600 times per second (c) a neutron star that does not pulse at all

30. What would happen to a neutron star with an accretion disk orbiting in a direction opposite to the neutron star's spin? (a) Its spin would speed up. (b) Its spin would slow down. (c) Its spin would stay the same.

31. Which of these binary systems is most likely to contain a black hole? (a) an X-ray binary containing an O star and another object of equal mass (b) a binary with an X-ray burster (c) an X-ray binary containing a G star and another object of equal mass

32. How would a flashing red light appear as it fell into a black hole? (a) It would appear to flash more quickly. (b) Its flashes would appear bluer. (c) Its flashes would shift to the infrared part of the spectrum.

33. Which of these black holes exerts the weakest tidal forces on an object near its event horizon? (a) a $10 M_{Sun}$ black hole (b) a $100 M_{Sun}$ black hole (c) a $10^6 M_{Sun}$ black hole

34. Where do gamma-ray bursts tend to come from? (a) neutron stars in our galaxy (b) black holes in our galaxy (c) extremely distant galaxies

Process of Science

35. *Do Black Holes Really Exist?* It is difficult to prove beyond a doubt that black holes really exist because they emit no light. Thus, we cannot see them directly but instead must infer their existence from their gravitational effects on nearby objects. Review the evidence for the existence of black holes presented in this chapter, decide whether you find it convincing or unconvincing, and then defend your position in a one- to two-page essay.

36. *Unanswered Questions.* You have seen in this chapter that current theoretical models make numerous predictions about the nature of black holes but leave many questions unanswered. Briefly describe one important but unanswered question related to black holes. If you think it will be possible to answer this question in the future, describe how we could find an answer, being as specific as possible about the evidence needed. If you think the question will never be answered, explain why you think it is impossible to answer.

Investigate Further

In-Depth Questions to Increase Your Understanding

Short-Answer/Essay Questions

Life Stories of Stars. Write a one- to two-page life story for the scenarios in Problems 37 through 40. Each story should be detailed and scientifically correct but also creative. That is, it should be entertaining while at the same time prove that you understand stellar evolution. Be sure to state whether "you" are a member of a binary system.

37. You are a white dwarf of $0.8M_{Sun}$.
38. You are a neutron star of $1.5M_{Sun}$.
39. You are a black hole of $10M_{Sun}$.
40. You are a white dwarf in a close binary system and are accreting matter from your companion star.

41. *Census of Stellar Corpses.* Which kind of object do you think is most common in our galaxy: white dwarfs, neutron stars, or black holes? Explain your reasoning.

42. *Fate of an X-Ray Binary.* The X-ray bursts that happen on the surface of an accreting neutron star are not powerful enough to accelerate the exploding material to escape velocity. Predict what will happen in an X-ray binary system in which the companion star eventually feeds over 3 solar masses of matter into the neutron star's accretion disk.

43. *Why Black Holes Are Safe.* Explain why the principle of conservation of angular momentum makes it very difficult to fall into a black hole.

44. *Surviving the Plunge.* The tidal forces near a black hole with a mass similar to a star would tear a person apart before that person could fall through the event horizon. Black hole researchers have pointed out that a fanciful "black hole life preserver" could help counteract those tidal forces. The life preserver would need to have a mass similar to that of an asteroid and would need to be shaped like a flattened hoop placed around the person's waist. In what direction would the gravitational force from the hoop pull on the person's head? In what direction would it pull on the person's feet? Based on your answers, explain in general terms how the gravitational forces from the "life preserver" would help to counteract the black hole's tidal forces.

Quantitative Problems

Be sure to show all calculations clearly and state your final answers in complete sentences.

45. *Schwarzschild Radii.* Calculate the Schwarzschild radius (in kilometers) for each of the following.
    a. $10^8 M_{Sun}$ black hole in the center of a quasar
    b. $5M_{Sun}$ black hole that formed in the supernova of a massive star
    c. A mini–black hole with the mass of the Moon
    d. A mini–black hole formed when a superadvanced civilization decides to punish you (unfairly) by squeezing you until you become so small that you disappear inside your own event horizon

46. *The Crab Pulsar Winds Down.* Theoretical models of the slowing of pulsars predict that the age of a pulsar is approximately equal to $p/(2r)$, where p is the pulsar's current period and r is the rate at which the period is slowing with time. Observations of the pulsar in the Crab Nebula show that it pulses 30 times per second, so $p = 0.0333$ second, but the time interval between pulses is growing longer by 4.2×10^{-13} second with each passing second, so $r = 4.2 \times 10^{-13}$ second per second. Using that information, estimate the age of the Crab pulsar. How does your estimate compare with the true age of the pulsar, which was born in the supernova observed in A.D. 1054?

47. *A Water Black Hole.* A clump of matter does not need to be extraordinarily dense in order to have an escape velocity greater than the speed of light, as long as its mass is large enough. You can use the formula for the Schwarzschild radius R_s, to calculate the volume $\frac{4}{3} \pi R_s^3$ inside the event horizon of a black hole of mass M. What does the mass of a black hole need to be in order for its mass divided by its volume to be equal to the density of water (1 g/cm^3)?

48. *Energy of a Supernova.* In a massive star supernova explosion, a stellar core collapses down to form a neutron star roughly 10 kilometers in radius. The gravitational potential energy released in such a collapse is approximately equal to GM^2/r, where M is the mass of the neutron star, r is its radius, and the gravitational constant is $G = 6.67 \times 10^{-11} \text{ m}^3/\text{kg} \times \text{s}^2$. Using this formula, estimate the amount of gravitational potential energy released in a massive star supernova explosion. How does it compare with the amount of energy released by the Sun during its entire main-sequence lifetime?

Discussion Questions

49. *Too Strange to Be True?* Despite strong theoretical arguments for the existence of neutron stars and black holes, many scientists rejected the possibility that such objects could really exist until they were confronted with very strong observational evidence. Some people claim that this type of scientific skepticism demonstrates an unwillingness on the part of scientists to give up their deeply held scientific beliefs. Others claim that this type of skepticism is necessary for scientific advancement. What do you think? Defend your opinion.

50. *Black Holes in Popular Culture.* Phrases such as "it disappeared into a black hole" are now common in popular culture. Give a few examples in which the term *black hole* is used in popular culture but is not meant to be taken literally. In what ways are these

analogies to real black holes correct? In what ways are they incorrect? Why do you think such an esoteric scientific idea as that of a black hole has so strongly captured the public imagination?

Web Projects

51. *Gamma-Ray Bursts.* Go to the Web site for a mission studying gamma-ray bursts (such as *HETE, INTEGRAL,* or *Swift*) and find the latest information about these bursts. Write a one- to two-

page essay on recent discoveries and how they may shed light on the origin of gamma-ray bursts.

52. *Black Holes.* Andrew Hamilton, a professor at the University of Colorado, maintains a Web site with a great deal of information about black holes and what it would be like to visit one. Visit his site and investigate some aspect of black holes that you find particularly interesting. Write a short report on what you learn.

visual skills check

Use the following questions to check your understanding of some of the many types of visual information used in astronomy. Answers are provided in Appendix K. For additional practice, try the Chapter 13 Visual Quiz at **www.masteringastronomy.com.**

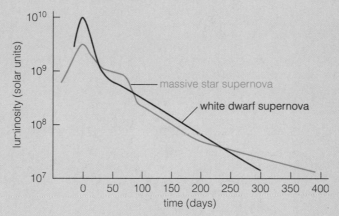

Figure 13.5, repeated above, shows how the luminosities of supernovae change with time. Answer the following questions, using the information provided in the figure.

1. Approximately how much more luminous is the peak brightness of the white dwarf supernova than the peak brightness of the massive star supernova?
   a. 1.5 times as luminous
   b. 3 times as luminous
   c. 10 times as luminous
   d. 100 times as luminous
2. How does the luminosity of a white dwarf supernova at its peak brightness compare with its luminosity 175 days later? Express your result as a percentage of the peak brightness.
   a. about 30% of the peak brightness
   b. about 10% of the peak brightness
   c. about 3% of the peak brightness
   d. about 1% of the peak brightness
3. Approximately how many days does it take for a white dwarf supernova to decline to 10% of its peak brightness?
   a. 3 days
   b. 30 days
   c. 170 days
   d. 300 days

4. Approximately how many days does it take for a massive star supernova to decline to 10% of its peak brightness?
   a. 10 days
   b. 30 days
   c. 100 days
   d. 300 days
5. Approximately how many days does it take for a massive star supernova to decline to 1% of its peak brightness?
   a. 3 days
   b. 30 days
   c. 170 days
   d. 300 days

We can understand the entire life cycle of a star in terms of the changing balance between pressure and gravity. This illustration shows how that balance changes over time and why those changes depend on a star's birth mass. *(Stars not to scale.)*

Key

pressure ⟶

gravity ⟵

(2) Thermal pressure comes into steady balance with gravity when the core becomes hot enough for hydrogen fusion to replace the thermal energy the star radiates from its surface [Section 10.1].

(3) The balance tips in favor of gravity after the core runs out of hydrogen. Fusion of hydrogen into helium temporarily stops supplying thermal energy in the core. The core again contracts and heats up. Hydrogen fusion begins in a shell around the core. The outer layers expand and cool, and the star becomes redder [Section 12.2].

Hydrogen shell-burning star

Pressure balances gravity at every point within a main-sequence star.

Main-sequence star

(1) Gravity overcomes pressure inside a protostar, causing the core to contract and heat up. A protostar cannot achieve steady balance between pressure and gravity because nuclear fusion is not replacing the thermal energy it radiates into space [Section 12.1].

> 0.08 M_{Sun}

The balance between pressure and gravity acts as a thermostat to regulate the core temperature:

A drop in core temperature decreases fusion rate, which lowers core pressure, causing the core to contract and heat up.

A rise in core temperature increases fusion rate, which raises core pressure, causing the core to expand and cool down.

Luminosity continually rises because core contraction causes the temperature and fusion rate in the hydrogen shell to rise.

Protostar

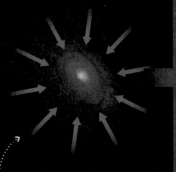

Contraction converts gravitational potential energy into thermal energy.

The balancing point between pressure and gravity depends on a star's mass:

Balance between pressure and gravity in high-mass stars results in a higher core temperature, a higher fusion rate, greater luminosity, and a shorter lifetime.

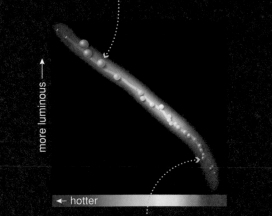

more luminous ⟶

← hotter

Balance between pressure and gravity in low-mass stars results in a cooler core temperature, a slower fusion rate, less luminosity, and a longer lifetime.

Degeneracy pressure balances gravity in objects of less than 0.08M_{Sun} before their cores become hot enough for steady fusion. These objects never become stars and end up as brown dwarfs.

< 0.08 M_{Sun}

④ Balance between thermal pressure and gravity is restored when the core temperature rises enough for helium fusion into carbon, which can once more replace the thermal energy radiated from the core [Section 12.2].

⑤ Gravity again gains the upper hand over pressure after the core helium is gone. Just as before, fusion stops replacing the thermal energy leaving the core. The core therefore resumes contracting and heating up, and helium fusion begins in a shell around the carbon core [Section 12.2].

⑥ In high mass stars, core contraction continues, leading to multiple shell burning that terminates with iron and a supernova explosion [Section 12.3].

⑦ At the end of a star's life, either degeneracy pressure has come into permanent balance with gravity or the star has become a black hole. The nature of the end state depends on the mass of the remaining core [Chapter 13].

Multiple shell–burning star

Black hole

Helium shell-burning star

Double shell–burning star

HIGH-MASS

Degeneracy pressure cannot balance gravity in a black hole.

Neutron star ($M < 3M_{Sun}$)

Luminosity remains steady because helium core fusion restores balance.

Neutron degeneracy pressure can balance gravity in a stellar corpse with less than about 2–3 M_{Sun}.

White dwarf ($M < 1.4M_{Sun}$)

LOW-MASS

Electron degeneracy pressure balances gravity in the core of a low-mass star before it gets hot enough to fuse heavier elements. The star ejects its outer layers and ends up as a white dwarf.

Electron degeneracy pressure can balance gravity in a stellar corpse of mass < 1.4 M_{Sun}.

Brown dwarf ($M < 0.08M_{Sun}$)

Degeneracy pressure keeps a brown dwarf stable in size even as it cools steadily with time.

14

our galaxy

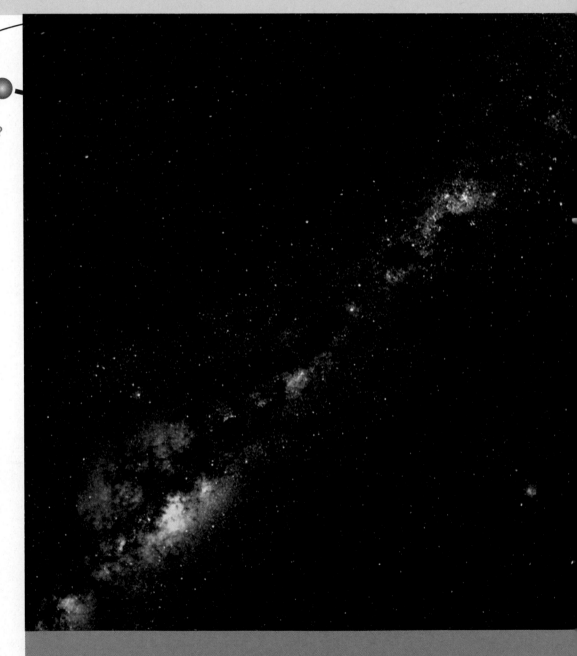

In previous chapters, we saw how stars forge new elements and expel them into space. We also studied how interstellar gas clouds enriched with these stellar by-products form new stars and planetary systems. These processes do not occur in isolation. Instead, they are part of a dynamic system that acts throughout our Milky Way Galaxy.

You are probably familiar with the idea that all living species on Earth interact with one another and with the environment to form a large, interconnected ecosystem. In a similar way, but on a much larger scale, our galaxy is a nearly self-contained system that cycles matter from stars into interstellar space and back into stars again. The birth of our solar system and the evolution of life on Earth would not have been possible without this "galactic ecosystem."

In this chapter, we will study our Milky Way Galaxy. We will examine the structure and motion of the galaxy, investigate the galactic processes that maintain an ongoing cycle of stellar life and death, probe the history of the galaxy, and explore the mysteries of the galactic center. Through it all, we will see that we are not only "star stuff" but also "galaxy stuff"—the product of eons of complex recycling and reprocessing of matter and energy in the Milky Way Galaxy.

essential preparation

1. How does Newton's law of gravity extend Kepler's laws? [Section 4.4]

2. What is light? [Section 5.1]

3. Where did the solar system come from? [Section 6.3]

4. How do stars form? [Section 12.1]

5. How does a high-mass star die? [Section 12.3]

14.1 The Milky Way Revealed

On a dark night, you can see a faint band of light slicing across the sky through several constellations, including Sagittarius, Cygnus, Perseus, and Orion. This band of light looked like a flowing ribbon of milk to the ancient Greeks, and we now call it the *Milky Way* (see Figure 2.1). In the early seventeenth century, Galileo used his telescope to prove that the light of the Milky Way comes from myriad individual stars. Together these stars make up the kind of stellar system we call a *galaxy,* echoing the Greek word for "milk," *galactos.*

The true size and shape of our Milky Way Galaxy are hard to guess from the way it looks in our night sky. Because we live inside the galaxy, trying to determine its structure is somewhat like trying to draw a picture of your house without ever leaving your bedroom. The fact that much of our galaxy's visible light is hidden from our view makes this task even more difficult. Only recently have we had the technology to observe the galaxy in other wavelengths of light. Nevertheless, by carefully observing our galaxy and comparing it to others that we see from the outside, we now have a good understanding of the processes that shape our galaxy. In this section, we'll begin our exploration of the galaxy by investigating its basic structure and orbital motion.

• What does our galaxy look like?

Our Milky Way Galaxy holds over 100 billion stars and is just one among tens of billions of galaxies in the observable universe [Section 1.1]. It is a vast **spiral galaxy**, so named because of the spectacular **spiral arms** illustrated in Figure 14.1a. If we viewed it from the side, as shown in Figure 14.1b, we'd see that the spiral arms are part of a fairly flat **disk** of stars surrounding a bright central **bulge**. The entire disk is surrounded by a dimmer, rounder **halo**. Most of the galaxy's bright stars reside in its disk.

a Artist's conception of the Milky Way viewed from the outside.

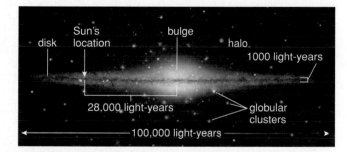

b Edge-on schematic view of the Milky Way.

Figure 14.1 interactive figure

The Milky Way Galaxy.

The most prominent stars in the halo are found in about 200 *globular clusters* of stars [Section 11.3].

Our galaxy consists of a flat disk with spiral arms surrounding a central bulge, with a roughly spherical halo surrounding everything.

The entire galaxy is about 100,000 light-years in diameter, but the disk is only about 1000 light-years thick. Our Sun is located in the disk about 28,000 light-years from the galactic center—a little more than halfway out from the center to the edge of the disk. Remember that this distance is incredibly huge [Section 1.2]: The few thousand stars visible to the naked eye together fill only a tiny dot in a picture like that in Figure 14.1.

It took us a long time to learn these facts about the Milky Way's true size and shape. Clouds of interstellar gas and dust known collectively as the **interstellar medium** fill the galactic disk, obscuring our view when we try to peer directly through it. The dusty, smoglike nature of the interstellar medium hides most of our galaxy from us when we try to observe it in visible light, and as a result it long fooled astronomers into believing that we lived near our galaxy's center (see Special Topic below).

Astronomer Harlow Shapley finally proved otherwise in the 1920s, when he demonstrated that the Milky Way's globular clusters are centered on a point tens of thousands of light-years from our Sun. He concluded that this point, not our Sun, must be the center of the galaxy, establishing that our Sun holds no special place in the galaxy or the universe.

The Milky Way is a relatively large galaxy. Within our Local Group of galaxies (see Figure 1.1), only the Andromeda Galaxy is comparable in size. The Milky Way's strong gravity influences smaller galaxies in its vicinity. For instance, two small galaxies known as the *Large Magellanic Cloud* and the *Small Magellanic Cloud* orbit the Milky Way at distances of some 150,000

special topic: --

How Did We Learn the Structure of the Milky Way?

FOR MOST OF HUMAN HISTORY, we knew the Milky Way only as an indistinct river of light in the sky. In 1610, Galileo used his telescope to discover that the Milky Way is made up of innumerable stars, but we still did not know the size or shape of our galaxy.

In the late eighteenth century, British astronomers William and Caroline Herschel (brother and sister) tried to determine the shape of the Milky Way more accurately by counting how many stars lie in each direction. Their approach suggested that the Milky Way's width was five times its thickness. In the early twentieth century, Dutch astronomer Jacobus Kapteyn and his colleagues used a more sophisticated star-counting method to gauge the size and shape of the Milky Way. Their results seemed to confirm the general picture found by the Herschels and suggested that the Sun lies near the center of the galaxy.

Kapteyn's results made astronomers with a sense of history nervous. Only four centuries earlier, before Copernicus challenged the Ptolemaic system, astronomers had believed Earth was the center of the universe. Kapteyn's placement of the Sun near the Milky Way's center seemed to give Earth a central place again. Kapteyn knew that obscuring material could be hiding much of the galaxy like some kind of interstellar fog, but he found no evidence for such a fog.

While Kapteyn was counting stars, American astronomer Harlow Shapley was studying globular clusters. He found that these clusters appeared to be centered on a point tens of thousands of light-years from the Sun. (IIis original estimate was that the center of the galaxy was 45,000 light-years from the Sun, which was later refined to the current 28,000 light-years.) Shapley concluded that this point marked the true center of our galaxy and that Kapteyn must be wrong.

Today we know that Shapley was right. The Milky Way's interstellar medium is the "fog" that misled Kapteyn. Robert Trumpler, working at California's Lick Observatory in the 1920s, established the existence of this dusty gas by studying open clusters of stars. By assuming that all open clusters had about the same diameter, he estimated their distances from their apparent sizes in the sky, much as you might estimate the distances of cars at night from the apparent separation of their headlights. He found that stars in distant clusters appeared dimmer than expected based on their estimated distance, just as a car's headlights might appear in foggy weather. Trumpler concluded that light-absorbing material fills the spaces between the stars, partially obscuring the distant clusters and making them appear fainter than they would otherwise. Thus, we learned that interstellar material had misled earlier astronomers, and that the stars visible in the night sky occupy a minuscule portion of the observable universe.

Subsequent studies established the spiral nature of our Milky Way Galaxy. Once astronomers understood the effects of interstellar dust, they were able to learn the size and shape of our galaxy. And from the 1950s onward, careful studies of the motions of stars and gas clouds have gradually uncovered the detailed structure of the galactic disk.

and 200,000 light-years, respectively. Both Magellanic Clouds are visible to the naked eye from the Southern Hemisphere.

The Milky Way is a large galaxy, and several smaller galaxies orbit it. Two other small galaxies lie even closer to the Milky Way than the Magellanic Clouds, but these small galaxies were discovered more recently because their visible light is obscured from view by the gas and dust in the Milky Way's disk. These two galaxies—the Sagittarius Dwarf and Canis Major Dwarf—are in the process of colliding with the Milky Way's disk. Our galaxy's tidal forces will ultimately rip both of these small galaxies apart.

Keep in mind that while these four nearby galaxies are quite small as galaxies go, they are still vast objects, each containing up to a few billion stars. This makes them a thousand or more times the size of typical globular clusters. These galaxies are small only when we compare them to the enormous size of our own Milky Way Galaxy, which is hundreds of times as large as its smaller companions.

• How do stars orbit in our galaxy?

Now that we have discussed the basic structure of our galaxy, let's turn our attention to the orbital motions of the stars within it. A spiral galaxy like ours may look like it should rotate like a giant pinwheel, but the galaxy is not a solid structure. Instead, each individual star follows its own orbital path around the center of the galaxy.

Nearly all stars in the Milky Way follow one of two basic orbital patterns. Stars in the disk orbit in roughly circular paths that all go in the same direction in nearly the same plane. In contrast, stars in the bulge and halo soar high above and below the disk on randomly oriented orbits.

Orbits of Disk Stars If you could stand outside the Milky Way and watch it for a few billion years, the disk would resemble a huge merry-go-round. Like horses on a merry-go-round, individual stars bob up and down through the disk as they orbit. The general orbit of a star around the galaxy arises from its gravitational attraction toward the galactic center, while the bobbing arises from the localized pull of gravity within the disk itself (Figure 14.2). A star that is "too far" above the disk is pulled back into the disk by gravity. Because the density of interstellar gas is too low to slow the star, it flies through the disk until it is "too far" *below* the disk on the other side. Gravity then pulls it back in the other direction. This ongoing process produces the bobbing of the stars.

Disk stars orbit the galaxy's center in orderly circles that all go in the same direction, except for a slight up-and-down bobbing as they orbit. The up-and-down motions of the disk stars give the disk its thickness of about 1000 light-years—a great distance by human standards, but only about 1% of the disk's 100,000-light-year diameter. In the vicinity of our Sun, each star's orbit takes more than 200 million years, and each up-and-down "bob" takes a few tens of millions of years.

The galaxy's rotation is unlike that of a merry-go-round in one important respect: On a merry-go-round, horses near the edge move much faster than those near the center. But in our galaxy's disk, the orbital velocities of stars near the edge and those near the center are about the same. Stars closer to the center therefore complete each orbit in less time than stars farther out.

Halo stars travel high above and far below the disk on orbits with random orientations.

Bulge stars also have orbits with random orientations.

Disk stars orbit in circles with the same orientation, except for a little up-and-down motion.

Figure 14.2

Characteristic orbits of disk stars (yellow), bulge stars (red), and halo stars (green) around the galactic center. (The yellow path exaggerates the up-and-down motion of the disk star orbits.)

Orbits of Halo and Bulge Stars The orbits of stars in the halo and bulge are much less organized than the orbits of stars in the disk (see Figure 14.2). Individual bulge and halo stars travel around the galactic center on more or less elliptical paths, but the orientations of these paths are relatively random. Neighboring halo stars can circle the galactic center in opposite directions. They swoop from high above the disk to far below it and back up again, plunging through the disk at velocities so high that the disk's gravity hardly alters their trajectories. Several fast-moving halo stars are currently passing through the disk not too far from our own solar system.

Halo stars swoop high above and below the disk on randomly oriented orbits.

These swooping orbits explain why the bulge and halo are much puffier than the disk. Halo and bulge stars soar to heights above the disk far greater than the up-and-down bobbing of the disk stars. The differences between orbits in the disk and orbits in the bulge and halo indicate that disk stars and halo stars have different origins. As we'll discuss in Section 14.3, these orbital differences provide an important clue to how our galaxy formed.

think about it Is there much danger that a halo star swooping through the disk of the galaxy will someday hit the Sun or Earth? Why or why not? (*Hint*: Think about the typical distances between stars, as illustrated by the use of the 1-to-10-billion scale in Chapter 1.)

(MA) Detecting Dark Matter in a Spiral Galaxy Tutorial, Lessons 1–2

Using Stellar Orbits to Measure the Mass of the Galaxy The Sun's orbital path is fairly typical of disk stars located near our 28,000-light-year distance from the galactic center. By measuring the speeds of globular clusters relative to the Sun, we've determined that the Sun and its neighbors orbit the center of the Milky Way at a speed of about 220 kilometers per second (800,000 km/hr [Section 1.3]). Even at this speed, it takes the Sun about 230 million years to complete one orbit around the galactic center. Early dinosaurs were just emerging on Earth when our Sun last visited this side of the galaxy.

special topic: --

How Do We Determine Stellar Orbits?

ASTRONOMERS LEARNED HOW STARS ORBIT in the Milky Way by measuring the motions of many different stars relative to the Sun. Although these measurements are easy in principle, they can be difficult in practice.

Determining a star's precise motion relative to the Sun requires knowing the star's true velocity through space. However, our primary means of measuring speeds in the universe—the Doppler effect—can tell us only a star's *radial velocity*, the component of its velocity directed toward or away from us (see Figure 5.15). If we want to know the star's true velocity, we must also measure its *tangential velocity*, the component of its velocity directed across our line of sight.

Tangential velocity is difficult to measure because of the vast distances to stars. Over tens of thousands of years, the tangential velocities of stars cause their apparent positions in our sky to change, which changes the shapes of the constellations. These changes are far too small for human eyes to notice. However, we can measure tangential velocities for many stars by comparing telescopic photographs taken years or decades apart. For example, if photographs taken 10 years apart show that a star has moved across our sky by an angle of 1 arcsecond, we know the star is moving at an angular rate of 0.1 arcsecond per year. We can convert this angular rate of motion (often called the star's *proper motion*) to a tangential velocity if we also know the star's distance. For a given angular rate, the tangential velocity is greater for more distant stars.

Because the earliest telescopic photographs date only to the late nineteenth century, we can measure tangential velocities only for objects that have moved noticeably since that time. That is why proper motion measurements are currently possible only for stars within a few hundred light-years of Earth. We can therefore determine precise stellar orbits only for relatively nearby stars. For more distant stars (and galaxies), we can usually measure only radial velocities.

The orbital motion of the Sun and other stars gives us a way to determine the mass of the galaxy. Recall that Newton's law of gravity determines how quickly objects orbit one another. This fact, embodied in Newton's version of Kepler's third law [Section 4.4], allows us to determine the mass of a relatively large object when we know the period and average distance of a much smaller object in orbit around it.

We can calculate the mass of the galaxy *within* the Sun's orbit using the Sun's orbital properties and Newton's version of Kepler's third law.

For example, we can use the Sun's orbital velocity and its distance from the galactic center to determine the mass of our galaxy lying *within* the Sun's orbit. To understand why the Sun's orbital motion allows us to calculate only the mass within the Sun's orbit, rather than the total mass of the galaxy, we need to consider the difference between the gravitational effects of mass within the Sun's orbit and of mass beyond its orbit. Every part of the galaxy exerts gravitational forces on the Sun as it orbits, but the net force from matter outside the Sun's orbit is relatively small because the pulls from opposite sides of the galaxy virtually cancel one another. In contrast, the net gravitational forces from mass within the Sun's orbit all pull the Sun in the same direction—toward the galactic center. The Sun's orbital velocity therefore responds almost exclusively to the gravitational pull of matter inside its orbit. By using the Sun's 28,000-light-year distance from the galactic center and its 220-km/s orbital velocity in Newton's version of Kepler's third law, we find that the total amount of mass within the Sun's orbit is about 2×10^{41} kg, or about 100 billion times the mass of the Sun.

The orbits of the Milky Way's stars reveal that most of the galaxy's mass consists of invisible dark matter in the halo.

Similar calculations based on the orbits of more distant stars in the Milky Way have revealed one of the greatest mysteries in astronomy—one that we first encountered in Chapter 1. Photographs of spiral galaxies make it appear that most of their mass is concentrated near their centers. However, orbital motions tell us that just the opposite is true. If most of the mass were concentrated near the galaxy's center, the orbital speeds of more distant stars would be slower, just as the orbital speeds of the planets decline with distance from the Sun. Instead, we find that orbital speeds remain about the same out to great distances from the galactic center, telling us that most of the galaxy's mass resides far from the center and is distributed throughout the halo. Because we see few stars and virtually no gas or dust in the halo, we conclude that most of the galaxy's mass must not give off any light that we can detect, and hence we refer to it as *dark matter*. We will discuss the evidence for dark matter in more detail in Chapter 16, along with ideas about its nature and its implications for the history and fate of the universe.

The Orbital Velocity Law

Newton's laws of motion and gravity can be used to derive a law—the orbital velocity law—that allows us to calculate the amount of mass contained *within* any object's orbit. For an object on a circular orbit, the *orbital velocity law* is

$$M_r = \frac{r \times v^2}{G}$$

where v is the object's orbital velocity, r is its orbital radius, M_r is the amount of mass contained within its orbit, and $G = 6.67 \times 10^{-11} \frac{m^3}{kg \times s^2}$ is the gravitational constant. Notice that for a given radius, a larger orbital velocity indicates a larger amount of mass, reflecting the fact that a stronger gravitational pull is necessary to hold a faster-moving object in orbit.

Example: Calculate the mass of the Milky Way Galaxy within the Sun's orbit using the orbital velocity law.

Solution: The Sun orbits the center of the Milky Way Galaxy with velocity $v = 220$ km/s $= 2.2 \times 10^5$ m/s. The Sun's orbital radius (distance) is 28,000 light-years, which is equivalent to $(28,000 \text{ ly}) \times (9.46 \times 10^{15} \text{ m/ly}) = 2.6 \times 10^{20}$ m. Substituting these values for v and r into the orbital velocity law, we obtain

$$M_r = \frac{r \times v^2}{G}$$

$$= \frac{(2.6 \times 10^{20} \text{ m}) \times \left(2.2 \times 10^5 \frac{m}{s}\right)^2}{6.67 \times 10^{-11} \frac{m^3}{kg \times s^2}}$$

$$= 1.9 \times 10^{41} \text{ kg}$$

The mass of the Milky Way Galaxy within the Sun's orbit is about 2×10^{41} kg. Dividing this mass by the Sun's mass of 2×10^{30} kg, we find that the mass of the galaxy within the Sun's orbit is about 10^{11} (100 billion) solar masses.

14.2 Galactic Recycling

The Milky Way Galaxy is home to our solar system, but its importance to our existence runs much deeper. The birth of our solar system could not have occurred without the galactic recycling that takes place within the disk of the galaxy and its interstellar medium.

In addition to making new generations of stars possible, galactic recycling gradually changes the chemical composition of the interstellar medium, although different regions of the galaxy change in composition at different rates. Recall that the universe was born with only the elements hydrogen and helium. All the elements heavier than helium, often

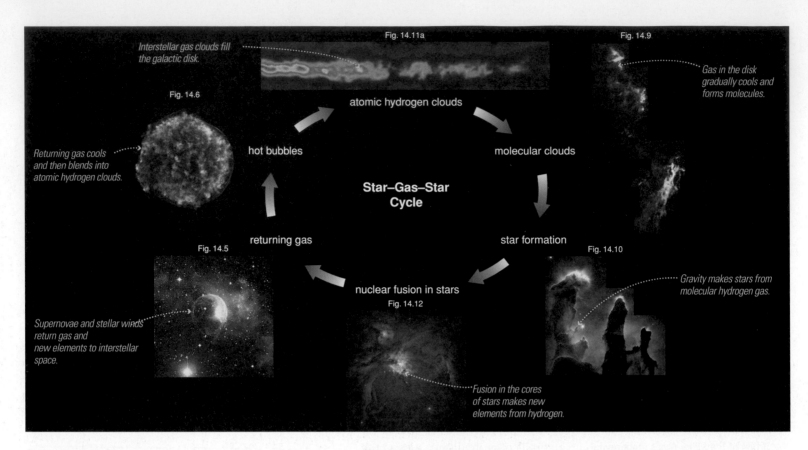

Fig. 14.11a

Fig. 14.9

Interstellar gas clouds fill the galactic disk.

Gas in the disk gradually cools and forms molecules.

atomic hydrogen clouds

Fig. 14.6

hot bubbles

molecular clouds

Star–Gas–Star Cycle

Returning gas cools and then blends into atomic hydrogen clouds.

returning gas

star formation

Fig. 14.10

Fig. 14.5

nuclear fusion in stars

Fig. 14.12

Gravity makes stars from molecular hydrogen gas.

Supernovae and stellar winds return gas and new elements to interstellar space.

Fusion in the cores of stars makes new elements from hydrogen.

Figure 14.3 interactive figure

A pictorial representation of the star–gas–star cycle. These photos appear individually later in the chapter. Their figure numbers are indicated.

called *heavy elements* by astronomers, have been made by fusion in stars. These elements are scattered into space when the stars die, and the galaxy then recycles them into new generations of stars (Figure 14.3). Today, thanks to more than 10 billion years of galactic recycling, elements heavier than helium constitute about 2% of the galaxy's gaseous content by mass. The remaining 98% still consists of hydrogen (about 70%) and helium (about 28%).

think about it Recall from Chapter 11 that stars in our galaxy's globular clusters are all extremely old, while stars in open clusters are relatively young. Based on this fact, which stars would you expect to contain a higher proportion of heavy elements: stars in globular clusters or stars in open clusters? Explain.

The galactic recycling process may sound simple enough, but it leaves some important questions unanswered. In particular, the supernova explosions that scatter most heavy elements into space send debris flying out at speeds of several thousand kilometers per second—much faster than the escape velocity from the galaxy. So how have the chemical riches produced by stars managed to remain in our galaxy? The answer turns out to depend on interactions between the matter expelled by supernovae and the interstellar medium that fills the galactic disk. In this section, we'll examine these interactions in more detail and learn why our existence owes as much to the cycling of gas through our galaxy as it does to the manufacturing of heavy elements by stars.

• How is gas recycled in our galaxy?

The galactic recycling process proceeds in several stages, making up what we will call the **star–gas–star cycle**, summarized in Figure 14.3.

We have already discussed the general idea of galactic recycling in Chapters 6 and 12. Stars are born when gravity causes the collapse of molecular clouds. Stars shine for millions or billions of years with energy produced by nuclear fusion, dying only when they've exhausted their fuel for fusion. Stars ultimately return much of their material back to the interstellar medium through supernovae and stellar winds. These steps are shown in the lower three frames of Figure 14.3. Let's now complete the loop and look at the remainder of the cycle, starting with the gas ejected by dying stars and finishing back at star birth.

Gas from Dying Stars All stars return much of their original mass to interstellar space in two basic ways: through stellar winds that blow throughout their lives, and through "death events" of planetary nebulae (for low-mass stars) or supernovae (for high-mass stars). Low-mass stars generally have weak stellar winds while they are on the main sequence. Their winds grow stronger and carry more material into space when they become red giants. By the time a low-mass star like the Sun ends its life with the ejection of a planetary nebula [Section 12.2], it has returned almost half its original mass to the interstellar medium (Figure 14.4).

High-mass stars lose mass much more dynamically and explosively. The powerful winds from supergiants and massive O and B stars recycle large amounts of matter into the galaxy (Figure 14.5). At the ends of their lives, these stars explode as supernovae. The high-speed gas ejected into space by supernovae or powerful winds sweeps up surrounding interstellar material, creating a **bubble** of hot, ionized gas (gas in which atoms are missing some of their electrons [Section 5.1]) around the exploding star. These hot, tenuous bubbles are quite common in the disk of the galaxy, but they are not always easy to detect. While some emit strongly in visible light and others are hot enough to emit profuse amounts of X rays, many bubbles are evident only through their emission of radio waves.

Supernovae and high-speed stellar winds can produce hot bubbles of gas in the interstellar medium.

The bubbles created by supernovae can have even more dramatic effects on the interstellar medium than those created by fast-moving stellar winds. Supernovae generate *shock waves*—waves of pressure that move faster than the speed of sound. A shock wave sweeps up surrounding gas as it travels, creating a "wall" of fast-moving gas on its leading edge. When we observe a *supernova remnant* [Section 12.3], we are seeing the aftermath of such a shock wave, which compresses, heats, and ionizes all the interstellar gas it encounters.

Figure 14.6 shows a young supernova remnant whose shocked gas is hot enough to emit X rays. In contrast, the older supernova remnant shown in Figure 14.7 is cooler because its shock wave has swept up more material, thereby distributing its energy more widely. Eventually, the shocked gas will radiate away most of its original energy, and the expanding wall of gas will slow to subsonic speeds and merge with the surrounding interstellar medium.

Giant bubbles of hot gas can erupt out of the disk, spreading their contents over a large region of the galaxy.

In the regions of the galactic disk where many supernovae have recently exploded, we find giant, elongated bubbles extending from young clusters of stars to distances of 3000 light-years or more above or below the disk. These probably are places where the bubbles from many individual supernovae have merged to make a giant bubble so large that it cannot be contained within the Milky Way's disk. Once the top of a giant bubble breaks out

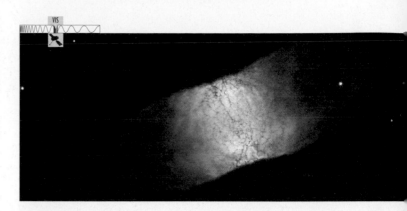

Figure 14.4

A dying low-mass star, like this one photographed by the Hubble Space Telescope, returns gas to the interstellar medium in a planetary nebula. This particular planetary nebula is known as the Retina Nebula. It is about 1 light-year across in the longer direction.

The wind from a hot star blows a bubble in the interstellar medium.

Figure 14.5

This photo shows a bubble of hot, ionized gas blown by a wind from the hot star near its center. Although it looks much like a soap bubble, it is actually an expanding shell of hot gas about 10 light-years in diameter. It glows where gas piles up as the bubble sweeps outward through the interstellar medium.

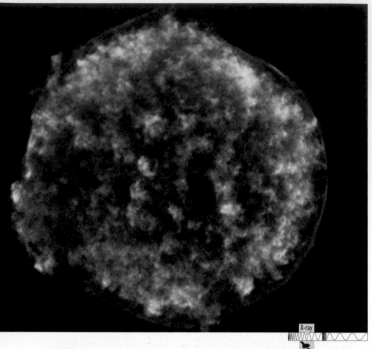

Figure 14.6

This image shows X-ray emission from hot gas in a young supernova remnant. The most energetic X rays (blue) come from 20-million-degree gas just behind the expanding shock wave. Less energetic X rays (green and red) come from the 10-million-degree debris ejected by the exploded star. The remnant is about 20 light-years across.

of the disk, where nearly all of the Milky Way's gas resides, nothing remains to slow its expansion except gravity. The result is a *blowout* that is similar to a volcanic eruption, but on a galactic scale: Hot gas erupts from the disk, spreading out as it shoots upward into the galactic halo (Figure 14.8). The gravity of the galactic disk slows the rise of the gas, eventually pulling it back down. Near the top of its trajectory, the ejected gas starts to cool and form clouds. These clouds can then rain back down into the disk, where their contents mix with the gas in a large region of the galaxy.

In addition to their effects on the interstellar medium, supernovae may also affect life on Earth by generating **cosmic rays** that can cause genetic mutations in living organisms. Cosmic rays are made of electrons, protons, and atomic nuclei that zip through interstellar space at close to the speed of light. Some cosmic rays penetrate Earth's atmosphere and reach Earth's surface. On average, about one cosmic-ray particle strikes your body each second. The cosmic-ray bombardment rate is 100 times greater at the high altitudes at which jet planes fly, and even more cosmic rays funnel along magnetic-field lines to Earth's magnetic poles.

Cooling and Cloud Formation   The hot, ionized gas in bubbles heated by supernovae is dynamic and widespread, but it is a relatively small fraction of the gas in the Milky Way. Most of this gas is much cooler—cool enough so that hydrogen atoms remain neutral rather than being ionized. We therefore refer to this gas as **atomic hydrogen gas**, although the hydrogen is mixed with neutral atoms of helium and heavy elements in the usual proportions for the galaxy (70% hydrogen, 28% helium, and 2% heavy elements by mass). After the gas that makes bubbles or superbubbles cools, it becomes part of this widespread atomic hydrogen gas in the galaxy.

We can map the distribution of atomic hydrogen gas in the Milky Way with radio observations. Atomic hydrogen emits a spectral line with

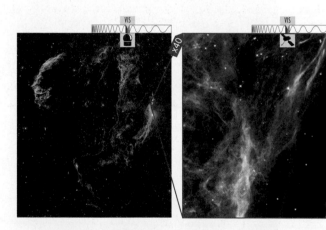

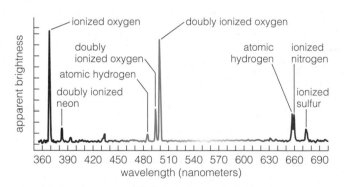

a This photograph shows the entire supernova remnant glowing in visible light. The angular size of this remnant in our sky is six times that of the Moon, and it is about 130 light-years across.

b A close-up view from the Hubble Space Telescope shows the fine filamentary structure in a small piece of the remnant. The spectacular colors correspond to the emission lines of the atoms and ions indicated in the spectrum in part (c).

c A visible-light spectrum from the Cygnus Loop shows the strong emission lines that account for the distinct colors in the photo in part (b).

Figure 14.7

Emission of visible light from an older supernova remnant, the Cygnus Loop.

a wavelength of 21 centimeters, which lies in the radio portion of the electromagnetic spectrum (see Figure 5.2). We detect the radio emission of this **21-centimeter line** coming from all directions, telling us that atomic hydrogen gas is distributed throughout the galactic disk. Based on the overall strength of the 21-centimeter emission, the total amount of atomic hydrogen gas in our galaxy is about 5 billion solar masses, which is a few percent of the galaxy's total mass.

Matter remains in the atomic hydrogen stage of the star–gas–star cycle for millions of years. Gravity slowly draws blobs of this gas together into tighter clumps, which radiate energy more efficiently as they grow denser. The blobs therefore cool and contract, forming clouds of cooler and denser gas. This process takes a much longer time than the other steps in the cycle from star death to star birth, which is why so much of the Milky Way's gas is in the atomic hydrogen stage of the star–gas–star cycle.

Clouds of atomic hydrogen also contain a small amount of interstellar dust. Interstellar **dust grains** are tiny, solid flecks of carbon and silicon minerals that resemble particles of smoke and form in the winds of red giant stars [Section 12.2]. Once formed, dust grains remain in the interstellar medium until they are heated and destroyed by a passing shock wave or incorporated into a protostar. Dust grains make up only about 1% of the mass of atomic hydrogen clouds, but they are responsible for the absorption of visible light that prevents us from seeing through the disk of the galaxy.

Gas heated by supernovae first cools into atomic hydrogen clouds, then cools further into molecular clouds.

As the temperature drops further in the center of a cool cloud of atomic hydrogen, hydrogen atoms combine into molecules, making a *molecular cloud* [Section 12.1]. Molecular clouds are the coldest, densest collections of gas in the interstellar medium. They often congregate into *giant molecular clouds* that hold up to a million solar masses of gas. The total mass of molecular clouds in the Milky Way is somewhat uncertain, but it is probably about the same as the total mass of atomic hydrogen gas—about 5 billion solar masses. Throughout much of this molecular gas, temperatures hover only a few degrees above absolute zero.

Molecular hydrogen (H_2) is by far the most abundant molecule in molecular clouds, but it is difficult to detect because temperatures are usually too cold for the gas to produce H_2 emission lines. Most of what we know about molecular clouds comes from observing spectral lines of molecules that make up only a tiny fraction of a cloud's mass. Carbon monoxide (CO) is the most abundant of these molecules. It produces strong emission lines in the radio portion of the spectrum at the 10–30 K temperatures of molecular clouds (Figure 14.9). More than 125 other molecules have been identified in molecular clouds by their radio emission lines, including water (H_2O), ammonia (NH_3), and ethyl alcohol (C_2H_5OH).

Because molecular clouds are heavy and dense compared to the rest of the interstellar gas, they tend to settle toward the central layers of the Milky Way's disk. This tendency creates a phenomenon you can see with your own eyes: the dark lanes running through the luminous band of light in our sky that we call the Milky Way (see Figure 2.1).

Completing the Cycle

A large molecular cloud gives birth to a cluster of stars. Once a few stars form in a newborn cluster, their radiation begins to erode the surrounding gas in the molecular cloud. Ultraviolet photons from high-mass stars heat and ionize the gas, and winds and radiation pressure push the ionized gas away. This kind of feedback prevents much of the gas in a molecular cloud from turning into stars.

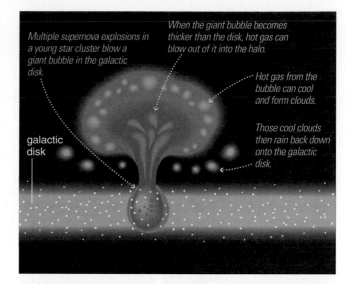

Multiple supernova explosions in a young star cluster blow a giant bubble in the galactic disk.

When the giant bubble becomes thicker than the disk, hot gas can blow out of it into the halo.

Hot gas from the bubble can cool and form clouds.

Those cool clouds then rain back down onto the galactic disk.

galactic disk

Figure 14.8

Hot gas erupting from a giant bubble out into the galactic halo eventually cools into gas clouds that rain back down on the disk. This process may be an important part of the galaxy-wide recycling system that incorporates the products of supernova explosions into new generations of stars and planets.

common misconceptions

The Sound of Space

In many science fiction movies, a thunderous sound accompanies the demolition of a spaceship. If the moviemakers wanted to be more realistic, they would silence the explosion. On Earth, we perceive sound when sound waves—which are waves of alternately rising and falling pressure—cause trillions of gas atoms to push our eardrums back and forth. Although sound waves can and do travel through interstellar gas, the extremely low density of this gas means that only a handful of atoms per second would collide with something the size of a human eardrum. It would therefore be impossible for a human ear (or a similar-size microphone) to register any sound, which is why the real sound of space is silence.

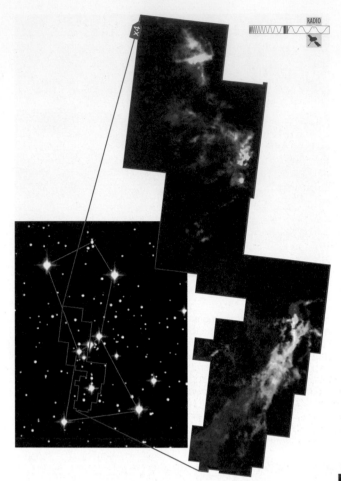

Ultraviolet radiation from newly forming stars can erode the molecular clouds
that gave them birth.

The process of molecular cloud erosion is vividly illustrated in the Eagle Nebula, a complex of clouds where new stars are forming (Figure 14.10). The dark, lumpy columns are molecular clouds. To the upper right (outside the picture), newly formed massive stars glow with ultraviolet radiation. This radiation sears the surface of the molecular clouds, destroying molecules and stripping electrons from atoms. As a result, matter "evaporates" from the molecular clouds and joins the hotter, ionized gas encircling them.

We have arrived back where we started in the star–gas–star cycle. The most massive stars now forming in the Eagle Nebula will explode within a few million years, filling the region with bubbles of hot gas and newly formed heavy elements. The expanding bubbles will slow and cool as their gas merges with the widespread atomic hydrogen gas in the galaxy. Eventually, this gas will cool further and coalesce into molecular clouds, forming new stars, new planets, and maybe even new civilizations.

Despite the recycling of matter from one generation of stars to the next, the star–gas–star cycle cannot go on forever. With each new generation of stars, some of the galaxy's gas becomes permanently locked away in brown dwarfs that never return material to space and in stellar corpses left behind when stars die. The interstellar medium therefore is slowly running out of gas, and the rate of star formation will gradually taper off over the next 50 billion years or so. Eventually, star formation will cease.

Putting It All Together: The Distribution of Gas in the Milky Way

Different regions of the galaxy are in different stages of the star–gas–star cycle. Because the cycle proceeds over such a long period of time

Figure 14.9

This image shows the complex structure of a molecular cloud in the constellation Orion. The picture was made by measuring Doppler shifts of emission lines from carbon monoxide molecules, and the colors indicate gas motions: Bluer parts are moving toward us and redder parts are moving away from us (relative to the cloud as a whole). This enormous cloud is about 1600 light-years distant and several hundred light-years across.

Radiation from nearby stars is eroding the surfaces of these clouds and causing them to glow . . .

. . . but the densest knots of gas resist that erosion and continue to form stars.

Figure 14.10

A portion of the Eagle Nebula, as seen by the Hubble Space Telescope. The dark columns of gas are molecular clouds, and stars are currently forming in the densest parts of these clouds. The region pictured here is about 5 light-years across.

compared to a human lifetime, each stage appears to us as a snapshot. We therefore see the interstellar medium in a wide variety of manifestations, ranging from the tenuous million-degree gas of bubbles to the cold, dense gas of molecular clouds. Table 14.1 summarizes the different states in which we observe interstellar gas in the galactic disk.

To get a complete picture of the star–gas–star cycle of the Milky Way, we need to observe its gas in many different kinds of light. We can see how these different states of gas are arranged in our galaxy by observing it in different wavelengths of light. Figure 14.11 shows seven views of the disk of the Milky Way Galaxy. Each view represents a panorama in a particular wavelength band, made by photographing the Milky Way's disk in every direction from Earth.

- Figure 14.11a shows variations in the intensity of radio emission from the 21-centimeter line of atomic hydrogen. It therefore maps the distribution of atomic hydrogen gas, which fills much of the galactic disk.

- Figure 14.11b shows variations in the intensity of radio emission lines from carbon monoxide (CO) and therefore maps the distribution of molecular clouds. These cold, dense clouds are concentrated in a narrow layer near the midplane of the galactic disk.

- Figure 14.11c shows variations in the intensity of long-wavelength infrared emission from interstellar dust grains. Notice that the regions of strongest emission from dust correspond to the locations of the molecular clouds in Figure 14.11b.

TABLE 14.1 *Typical States of Gas in the Interstellar Medium*

	State of Gas		
	Hot Bubbles	Atomic Hydrogen Clouds	Molecular Clouds
Primary Constituent	Ionized hydrogen	Atomic hydrogen	Molecular hydrogen
Approximate Temperature	1,000,000 K	100–10,000 K	30 K
Approximate Density (atoms per cm³)	0.01	1–100	300
Description	Pockets of gas heated by stellar winds or supernovae	The most common form of gas, filling much of the galactic disk	Regions of star formation

Figure 14.11 interactive photo

Panoramic views of the Milky Way in different bands of the spectrum. The center of the galaxy is in the center of each strip. The rest of each strip shows all other directions in the Milky Way disk as seen from Earth. (Imagine attaching the left and right ends of each strip to form a circular band that corresponds to the 360-degree band of the Milky Way in our sky.)

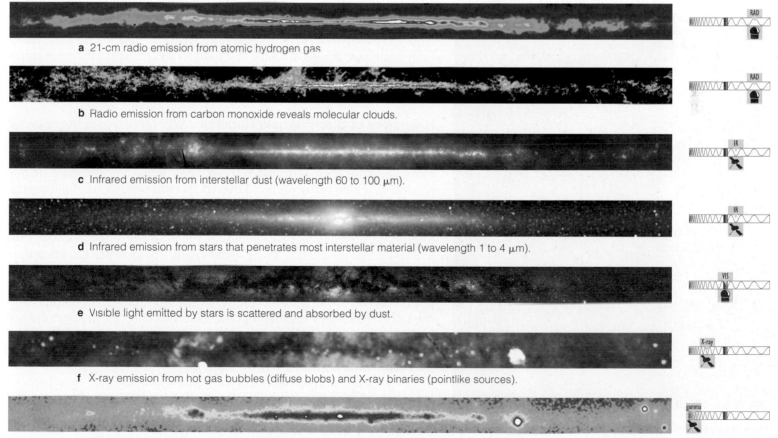

a 21-cm radio emission from atomic hydrogen gas

b Radio emission from carbon monoxide reveals molecular clouds.

c Infrared emission from interstellar dust (wavelength 60 to 100 μm).

d Infrared emission from stars that penetrates most interstellar material (wavelength 1 to 4 μm).

e Visible light emitted by stars is scattered and absorbed by dust.

f X-ray emission from hot gas bubbles (diffuse blobs) and X-ray binaries (pointlike sources).

g Gamma-ray emission from collisions of cosmic rays with atomic nuclei in interstellar clouds.

- Figure 14.11d shows shorter-wavelength infrared light from stars at wavelengths that penetrate clouds of gas and dust. This image therefore shows how our galaxy would look if there were no dust blocking our view. The galactic bulge is clearly evident at the center.

- Figure 14.11e shows the galactic disk in visible light. Because visible light cannot penetrate interstellar dust, the dark blotches correspond closely to the bright patches of molecular radio emission and infrared dust emission in Figures 14.11b and c.

- Figure 14.11f shows X-ray light from the galactic disk. The point-like blotches in this view are mostly X-ray binaries [Section 13.2]. The rest of the X-ray emission comes primarily from hot gas bubbles. Because hot gas tends to rise into the halo, it is less concentrated in the midplane than the atomic and molecular gases are.

- Figure 14.11g shows gamma-ray emission from the Milky Way. Most of this emission is produced by collisions between cosmic-ray particles and atomic nuclei in interstellar clouds. Such collisions happen most often where gas densities are highest, so the gamma-ray emission corresponds to the locations of molecular and atomic gases.

think about it Carefully compare and contrast the views of the Milky Way's disk in Figure 14.11. Why do regions that appear dark in some views appear bright in others? What general patterns do you notice?

• Where do stars tend to form in our galaxy?

The star–gas–star cycle has operated continuously since the Milky Way's birth, yet new stars are not spread evenly across the galaxy. Some regions seem much more fertile than others. Regions rich in molecular clouds tend to spawn new stars easily, while gas-poor regions do not. However, molecular clouds are dark and hard to see. Certain other signatures of star formation are much more obvious. A quick tour of some star-forming galactic environments will help you spot where the action is.

Hallmarks of Star-Forming Regions Wherever we see hot, massive stars, we know that we have spotted a region of active star formation. Because these stars live fast and die young, they never get a chance to move far from their birthmates. Thus, they signal the presence of star clusters in which many of their lower-mass companions are still forming.

Hot, massive stars and ionization nebulae are found only near clouds that are actively forming stars.

Regions of active star formation can be extraordinarily picturesque. Near hot stars we often find colorful, wispy blobs of glowing gas known as **ionization nebulae** (also called *emission nebulae* or *H II regions*). These nebulae glow because electrons in their atoms are raised to high energy levels [Section 5.2] or ionized when they absorb ultraviolet photons from the hot stars, so they emit light as the electrons fall back to lower energy levels. The Orion Nebula, about 1500 light-years away in the "sword" of the constellation Orion, is among the most famous. Few astronomical objects can match its beauty (Figure 14.12).

Most of the striking colors in an ionization nebula come from particular spectral lines produced by particular atomic transitions. For example, the transition in which an electron falls from energy level 3 to energy level 2 in a hydrogen atom generates a red photon with a wavelength of 656 nanometers (see Figure 5.10). Ionization nebulae appear predominantly red in photographs because of all the red photons released by this

Figure 14.12 interactive photo

A Hubble Space Telescope photo of the Orion Nebula, an ionization nebula energized by ultraviolet photons from hot stars.

particular transition. Transitions in other elements produce other spectral lines of different colors (Figure 14.13).

The blue and black tints in some star-forming regions have a different origin. Starlight reflected from dust grains produces the blue colors, because interstellar dust grains scatter blue light more easily than red light (Figure 14.14). These *reflection nebulae* are always bluer in color than the stars supplying the light. (The effect is similar to the scattering of sunlight in our atmosphere that makes the sky blue [Section 7.1].) The black regions of nebulae are dark, dusty gas clouds that block our view of the stars beyond them. Figure 14.15 shows a multicolored nebula characteristic of a hot-star neighborhood.

think about it In Figure 14.15, identify the red ionized regions, the blue reflecting regions, and the dark obscuring regions. Briefly explain the origin of the colors in each region.

Spiral Arms Taking a broader view of our galaxy, we can see that the spiral arms must be full of newly forming stars because they bear all the hallmarks of star formation. They are home to both molecular clouds and numerous clusters of young, bright, blue stars surrounded by ionization nebulae. Detailed images of other spiral galaxies show these characteristics more clearly (Figure 14.16). Hot, blue stars, sometimes surrounded by large, red ionization nebulae, trace out the arms while the stars between

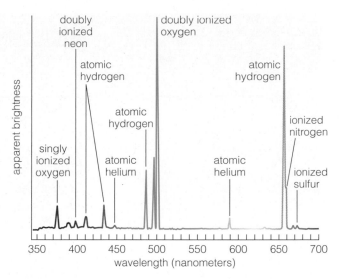

Figure 14.13

A spectrum of the Orion Nebula. The prominent emission lines reveal the atoms and ions that emit most of the light. Through careful study of these lines, we can determine the nebula's chemical composition.

Figure 14.14

The blue tints in this nebula in the constellation Scorpius are produced by reflected light.

Figure 14.15

A photo of the Horsehead Nebula and its surroundings. (The region pictured is about 150 light-years across.)

Chapter 14 Our Galaxy **399**

Dark patches on inner edge of spiral arm show where gas clouds are packing together . . .

. . . and compression of these clouds triggers star formation in the arm.

Blue specks are young stars that formed in the spiral arm.

Red patches are ionization nebulae around the hottest, youngest stars.

Flow of gas and stars through spiral arm

Figure 14.16 interactive photo

Spiral arms in the Galaxy M51. This photo from the Hubble Space Telescope shows M51's two magnificent spiral arms along with a smaller galaxy that is currently interacting with one of those arms. Notice that the spiral arms are much bluer in color than the central bulge. Because massive blue stars live only for a few million years, the relative blueness of the spiral arms tells us that stars must be forming more actively within them than elsewhere in the galaxy. (The large image shows a region roughly 90,000 light-years across.)

common misconceptions

What Is a Nebula?

The term *nebula* means "cloud," but in astronomy it can refer to many different kinds of objects, which sometimes leads to misconceptions. Many astronomical objects look "cloudy" through small telescopes, and in past centuries astronomers called any such object a nebula as long as they were sure it wasn't a comet. For example, galaxies were called nebulae because they looked like fuzzy round or spiral blobs.

Using the term *nebula* to refer to a galaxy now sounds somewhat dated, given the enormous differences between these distant star systems and the much smaller clouds of gas that populate the interstellar medium. Nevertheless, some people still refer to spiral galaxies as "spiral nebulae." Today, we generally use the term *nebula* to refer only to true interstellar clouds, but be aware that the term is still sometimes used in other ways.

the arms are generally redder and older. We also see enhanced amounts of molecular and atomic gas in the spiral arms, and streaks of interstellar dust often obscure the inner sides of the arms themselves. These features show that spiral arms contain both young stars and the material necessary to make new stars.

At first glance, spiral arms look as if they ought to move with the stars, like the fins of a giant pinwheel in space. However, we know that spiral arms cannot be fixed patterns of stars that rotate along with the galaxy, because stars near the center of the galaxy complete an orbit in much less time than stars far from the center. If the spiral arms simply moved along with the stars, the central parts of the arms would complete several orbits around the galaxy as the outer parts orbited just once. This difference in orbital periods would eventually wind up the spiral arms into a tight coil. We generally don't see such tightly wound spiral arms in galaxies, so we conclude that spiral arms are more like swirling ripples in a whirlpool than like the fins of a giant pinwheel.

In fact, we now believe that spiral arms are enormous waves of star formation that propagate through the gaseous disk of a spiral galaxy like the Milky Way. Theoretical models suggest that disturbances called **spiral density waves** produce the spiral arms. According to these models, spiral arms are places in a galaxy's disk where stars and gas clouds get more densely packed. Pushing the stars closer together has little effect on the stars themselves—they are still much too widely separated to collide with each other. However, the large gas clouds do collide, and packing the clouds closer together enhances the force of gravity within them, triggering the formation of many new stars (Figure 14.17).

To visualize how spiral density waves propagate through a galaxy's disk, consider how traffic backs up behind a slow-moving tractor on a rural highway. Cars approaching the tractor slow down and bunch together. After cars pass the tractor, they speed up and spread out again. Thus, there are always some cars bunched up behind the tractor, even though the cars themselves are gradually flowing past it.

Spiral arms are waves of star formation that spread through our galaxy's disk.

In a spiral density wave, gravity plays the role of the tractor, while stars and gas clouds play the role of

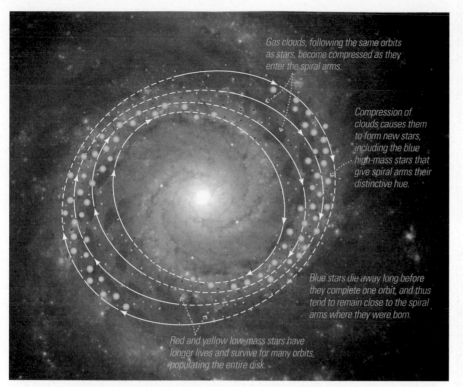

Gas clouds, following the same orbits as stars, become compressed as they enter the spiral arms.

Compression of clouds causes them to form new stars, including the blue high-mass stars that give spiral arms their distinctive hue.

Blue stars die away long before they complete one orbit, and thus tend to remain close to the spiral arms where they were born.

Red and yellow low-mass stars have longer lives and survive for many orbits, populating the entire disk.

Figure 14.17 interactive figure

Schematic illustration of star formation produced by spiral density waves. Stars and gas clouds pass through spiral arms as they orbit the galaxy. Each arm is a self-sustaining pattern because gravity slows down stars and gas clouds as they pass through an arm, causing them to be more densely packed there.

the cars. The stars and gas clouds in a galaxy's disk are constantly flowing through its spiral arms, but the extra density of matter in the spiral arm alters that flow. The extra matter exerts a gravitational force that pulls stars and gas clouds into the arm and tries to halt their escape as they move out the other side. This gravitational pull is not strong enough to trap the stars and gas clouds. However, like the tractor, it temporarily slows them down, producing a long-lasting pattern that gets stretched into a spiral shape by the rotation of the disk. We call this kind of propagating disturbance a wave because, like a wave in water, it moves through matter without carrying matter along with it. And like a water wave, some sort of disturbance, like a gravitational tug from another galaxy, is needed to generate it.

14.3 The History of the Milky Way

Now that we have discussed the basic properties of the Milky Way, we are ready to turn our attention to the history of our galaxy. All the galaxy's properties provide clues to its history. However, some of the most important clues come from a detailed comparison of disk stars with halo stars. We'll begin with this comparison and then discuss a basic model of galaxy formation that explains many of the differences between these two groups of stars.

• What do halo stars tell us about our galaxy's history?

We have already seen how the disorderly orbits of halo stars differ from the generally circular orbits of disk stars. Two other differences between halo stars and disk stars give us clues to their origins. First, we don't find any young stars in the halo, while in the disk we see stars of many different ages. Second, the spectra of halo stars show that they contain fewer

A protogalactic cloud contains only hydrogen and helium gas.

Halo stars begin to form as the protogalactic cloud collapses.

Conservation of angular momentum ensures that the remaining gas flattens into a spinning disk.

Billions of years later, the star–gas–star cycle supports ongoing star formation within the disk. The lack of gas in the halo precludes star formation outside the disk.

Figure 14.18 interactive figure

This four-picture sequence illustrates a simple schematic model of galaxy formation, showing how a spiral galaxy might develop from a protogalactic cloud of hydrogen and helium gas.

heavy elements than do disk stars. Because of these differences, astronomers divide the Milky Way's stars into two distinct populations.

1. The **disk population** (sometimes called *Population I*) contains both young stars and old stars, all of which have heavy-element proportions of about 2%, like our Sun.
2. The **spheroidal population** (or *Population II*) consists of stars in the halo and the bulge, both of which are roughly spherical in shape. Stars in this population are always old and therefore low in mass, and those in the halo can have heavy-element proportions as low as 0.02%—meaning that heavy elements are about 100 times rarer in these stars than in the disk population.

We can understand why bulge and halo stars differ from disk stars by looking at how the Milky Way's gas is distributed. The halo does not contain the cold, dense molecular clouds required for star formation. In fact, the halo contains almost no gas at all, and that small amount of gas is generally quite hot. Because star-forming molecular clouds are found only in the disk, new stars can be born only in the disk and not in the halo.

Halo stars are all old with a very low proportion of heavy elements, while disk stars come in all ages and contain a higher proportion of heavy elements.

The relative lack of heavy elements in halo stars indicates that they must have formed early in the galaxy's history—before many supernovae had exploded and added heavy elements to star-forming clouds. We therefore conclude that the halo has lacked the gas needed for star formation for a very long time. Apparently, all the Milky Way's cool gas settled into the disk long ago. The only stars that still survive in the halo are long-lived, low-mass stars.

think about it How does the halo of our galaxy resemble the distant future fate of the galactic disk? Explain.

• How did our galaxy form?

Any model for our galaxy's formation must account for the differences between disk stars and halo stars. The most basic model proposes that our galaxy began as a giant **protogalactic cloud** containing all the hydrogen and helium gas that eventually turned into stars. Gravity would have caused such a cloud to contract and fragment, just as in present-day star-forming clouds [Section 12.1].

According to this model, the stars of the spheroidal population (bulge and halo) formed first. Early on, the gravity associated with our protogalactic cloud drew in matter from all directions, creating a cloud that was blobby in shape and had little or no measurable rotation. The orbits of stars forming within such a cloud could have had any orientation, accounting for the randomly oriented orbits of stars in the spheroidal population.

Later, the remaining gas contracted under the force of gravity and settled into a flattened, spinning disk because of conservation of angular momentum (Figure 14.18). This process was much like the process that leads to the formation of spinning disks of gas around young stars [Sections 6.3, 12.1], but on a much larger scale. Collisions among gas particles tended to average out their random motions, leading them to acquire orbits in the same direction and in the same plane. Stars that form within this spinning disk orbit with the same speed and direction as their neighbors and thus become members of the disk population of stars.

Halo stars formed when our galaxy's protogalactic cloud was still large and blobby, and disk stars formed after the gas had settled into a spinning disk.

We can test this basic model by studying the stars of the Milky Way. The clues found to date support the basic picture but suggest that the full story of galaxy formation may be somewhat more complex.

All available evidence confirms that the stars in the Milky Way's halo are very old. The main-sequence turnoff points in H-R diagrams of globular clusters show that their stars were born at least 12 billion years ago [Section 11.3]. Individual halo stars and some of the bulge stars appear similarly old. Furthermore, the proportions of heavy elements in halo stars are quite low, indicating that they formed before many generations of supernovae had a chance to enrich the Milky Way's interstellar medium.

However, careful study of heavy-element proportions suggests that our galaxy formed from a few different gas clouds. If the Milky Way had formed from a single protogalactic cloud, it would have steadily accumulated heavy elements during its inward collapse as stars formed and exploded within it. In that case, the outermost stars in the halo would be the oldest and the most deficient in heavy elements. Stars belonging to different globular clusters in the Milky Way's halo do differ in age and heavy-element content, but these variations do not seem to depend on the stars' distance from the galactic center.

Our galaxy's halo stars may have formed in several small protogalactic clouds that later merged to form a single, larger protogalactic cloud.

The easiest way to account for the variations is to suppose that the Milky Way's earliest stars formed in relatively small protogalactic clouds, each with a few globular clusters, and that these clouds later collided and combined to create the full protogalactic cloud that became the Milky Way (Figure 14.19). Similar processes may still be happening. As we discussed earlier, the Sagittarius and Canis Major dwarf galaxies are currently crashing through the Milky Way's disk and being torn apart in the process. A billion years from now, their stars will be indistinguishable from halo stars because they will all be circling the Milky Way on orbits that carry them high above the disk. In fact, recent observations of halo-star motions have shown that some halo stars move in organized streams that seem to be the remnants of dwarf galaxies torn apart long ago by the Milky Way's gravity.

think about it   If the preceding scenario is true, then the Milky Way suffered several collisions early in its history. Explain why we think that galaxy collisions (or collisions between protogalactic clouds) were common in the distant past. (*Hint:* How did the average separations of galaxies in the past compare to their average separations today?)

Once the Milky Way's full protogalactic cloud was in place, its collapse and heavy-element enrichment should have continued in a more orderly fashion than previously. Support for this scenario comes from a layer of stars intermediate between the disk and the halo. The heavy-element content of stars and globular clusters in this intermediate layer does indeed depend on their distances from the galactic center. These stars are nearly as old as halo stars but formed just before the spinning protogalactic cloud finished flattening into a disk.

The distribution of heavy elements in our galaxy suggests that the Milky Way's disk began with only about 10% as many heavy elements as it has today. However, that proportion of heavy elements differed from

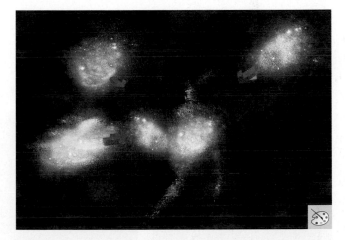

Figure 14.19

This painting shows a model of how the Milky Way's halo may have formed. The characteristics of stars in the Milky Way's halo suggest that several smaller gas clouds, already bearing some stars and globular clusters, may have merged to form the Milky Way's protogalactic cloud. These stars and star clusters remained in the halo while the gas settled into the Milky Way's disk.

place to place in the galaxy. Observations of old stars in the bulge of our galaxy show that some of them have a chemical composition similar to that of our Sun, even though their ages exceed 10 billion years. Many of the heavy elements produced during the initial collapse of our galaxy must have ended up near the center and became incorporated into the bulge stars that subsequently formed.

After the formation of the disk, the star–gas–star cycle increased the abundance of heavy elements in the disk more gradually than in the bulge. Because the star–gas–star cycle has been operating continuously in the Milky Way's disk ever since the disk formed, the ages of disk stars range from newly born to 10 billion or more years old. New stars will continue to be born in the disk as long as enough gas remains within it.

(MA) Black Holes Tutorial, Lessons 1–2

14.4 The Mysterious Galactic Center

The center of the Milky Way Galaxy lies in the direction of the constellation Sagittarius. This region of the sky does not look particularly special to our unaided eyes. However, if we could remove the interstellar dust that obscures our view, the galaxy's central bulge would be one of the night sky's most spectacular sights. And, deep within the bulge, a gigantic black hole with a mass as great as 4 million times the mass of our Sun may lurk at the center of our galaxy.

• What lies in the center of our galaxy?

Although the Milky Way's clouds of gas and dust prevent us from seeing visible light from the center of the galaxy, we can peer into the heart of our galaxy with radio, infrared, and X-ray telescopes. Figure 14.20 shows a series of infrared and radio views, looking ever deeper into the galaxy's center. Within about 1000 light-years of the center, we find swirling clouds of gas and a cluster of several million stars. Bright radio emissions trace out the magnetic fields that thread this turbulent region. In the exact center we find a source of radio emission named Sagittarius A* (pronounced "Sagittarius A-star"), or Sgr A* for short, that is quite unlike any other radio source in our galaxy.

Figure 14.20 interactive photo ◄

Zooming into the galactic center at infrared and radio wavelengths.

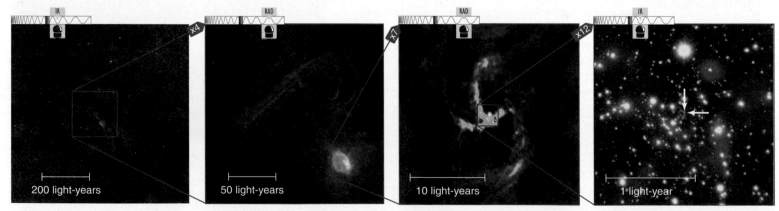

| 200 light-years | 50 light-years | 10 light-years | 1 light-year |

a This infrared image shows stars and gas clouds within 1000 light-years of the center of the Milky Way.

b This radio image shows vast threads of emissions tracing magnetic field lines near the galactic center.

c This radio image zooms in on gas swirling around the radio source Sgr A* (marked by the white dot), suspected to contain a very massive black hole.

d This infrared image shows stars within about 1 light-year of Sgr A*. The two arrows point to the precise location of Sgr A*.

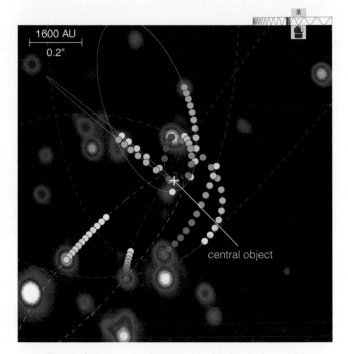
see it for yourself Use the star charts in Appendix J to find the constellation Sagittarius, which is easily visible on summer and fall evenings and looks like a teapot with a handle on the left and a spout on the right. The center of the Milky Way Galaxy is near the tip of the spout. Can you see the Milky Way's faint band of light passing through the constellations Sagittarius, Cygnus, and Casseopeia?

Several hundred stars crowd the region within a light-year of Sgr A*, and the motions of these stars and nearby gas indicate that there are a few million solar masses within this tiny region of space. Observations show that there are not nearly enough stars in this region to account for so much mass, even though stars there are much more crowded together than in our region of the galaxy. As a result, astronomers suspect that Sgr A* contains an enormously massive black hole. These suspicions were confirmed in 2002 when astronomers monitoring infrared light from the galactic center observed stars swooping within a few light-hours of this massive object in their orbits (Figure 14.21). By applying Newton's version of Kepler's third law to the orbits of these stars, they concluded that the central object must have a mass of 3 to 4 million solar masses, all packed into a region of space just a little larger than our solar system. An object that massive within such a small space is almost certainly a black hole.

Figure 14.21

This diagram shows stellar positions observed with the Keck telescope and calculated orbits for several stars near the center of the galaxy. By applying Newton's version of Kepler's third law, we infer that the central object has a mass 3 to 4 million times that of our Sun, packed into a space so small that it is almost certainly a black hole.

Stars quite close to our galaxy's center orbit a nearly invisible and tiny object more than three million times as massive as our Sun—probably a huge black hole.

However, the behavior of this suspected black hole is puzzling. Most suspected black holes are thought to accumulate matter through accretion disks that radiate brightly in X rays. These include black holes in binary star systems like Cygnus X-1 [Section 13.3] and some giant black holes at the centers of other galaxies [Section 15.4]. If the black hole at the center of our galaxy had an accretion disk, its X-ray light would easily penetrate the dusty gas of our galaxy and appear fairly bright to our X-ray telescopes. Yet the X-ray emission from Sgr A* has usually been relatively faint.

Observations of Sgr A* made with the Chandra X-Ray Observatory are helping us understand this surprising behavior. In October 2000, an enormous X-ray flare lasting 3 hours was observed coming from the location of the suspected black hole (Figure 14.22). This sudden change in X-ray brightness probably came from energy released by a comet-size lump of matter that was torn apart by tidal forces just before it disappeared into the black hole. If we continue to observe similar X-ray flares from Sgr A*, then the explanation for its generally low X-ray brightness may be that matter falls into it in big chunks instead of from a smooth, swirling accretion disk. Until we better understand Sgr A*, it is sure to remain a favorite target for observation.

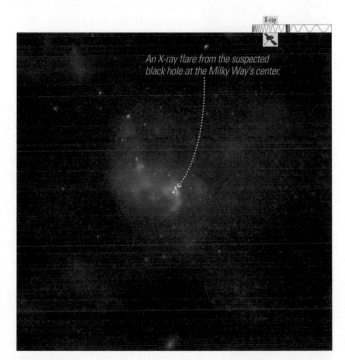

X-ray

An X-ray flare from the suspected black hole at the Milky Way's center.

Figure 14.22

This X-ray image shows the central 60 light-years of our galaxy. The circled white dot is an X-ray flare from the suspected black hole at the Milky Way's center. The flare probably came from energy released by a comet-size lump of matter that was torn apart by tidal forces just before it disappeared into the black hole.

the big picture
Putting Chapter 14 into Context

In this chapter, we have explored the structure, motion, and history of our galaxy, along with the recycling of gas that has made our existence possible. As you review, keep in mind these "big picture" ideas:

- The inability of visible light to pass through interstellar gas and dust concealed the true nature of our galaxy until recent times. Modern astronomical instruments reveal the Milky Way Galaxy to be a dynamic system of stars and gas that continually gives birth to new stars and planetary systems.

- Stellar winds and explosions make interstellar space a violent place. Hot gas tears through the atomic hydrogen that fills much of the galactic disk, leaving expanding bubbles and fast-moving clouds in its wake. All this violence might seem dangerous, but it performs the great service of mixing new heavy elements into the gas of the Milky Way.

- Although the elements from which we are made were forged in stars, we could not exist if stars were not organized into galaxies. The Milky Way Galaxy acts as a giant recycling plant, converting gas expelled from each generation of stars into the next generation and allowing some of the heavy elements to solidify into planets like our own.

summary of key concepts

14.1   The Milky Way Revealed

• What does our galaxy look like?

The Milky Way Galaxy consists of a thin **disk** about 100,000 light-years in diameter with a central bulge and a spherical region called the **halo** that surrounds the entire disk. The disk contains most of the gas and dust of the **interstellar medium**, while the halo contains only a small amount of hot gas and virtually no cold gas.

• How do stars orbit in our galaxy?

Stars in the disk all orbit the galactic center in about the same plane and in the same direction. Halo and bulge stars also orbit the center of the galaxy, but their orbits are randomly inclined to the disk of the galaxy. Orbital motions of stars allow us to determine the distribution of mass in our galaxy.

14.2   Galactic Recycling

• How is gas recycled in our galaxy?

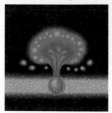

Stars are born from the gravitational collapse of gas clumps in **molecular clouds**. Massive stars explode as supernovae when they die, creating hot **bubbles** in the interstellar medium that contain the new elements made by these stars. Eventually, this gas cools and mixes with the interstellar medium, turning into atomic hydrogen gas and then cooling further, producing star-forming molecular clouds. We call this process the **star–gas–star cycle**.

• Where do stars tend to form in our galaxy?

Active star-forming regions, marked by the presence of hot, massive stars and ionization nebulae, are found mostly in **spiral arms**. In each arm a **spiral density wave** has caused gas clouds to crash into each other and to form clusters of new stars.

14.3   The History of the Milky Way

• What do halo stars tell us about our galaxy's history?
The halo generally contains only old, low-mass stars that have a much smaller proportion of heavy elements than stars in the disk. Halo stars must therefore have formed early in the galaxy's history, before the gas settled into a disk.

• How did our galaxy form?

The galaxy probably began as a huge blob of gas called a **protogalactic cloud**. Gravity caused the cloud to shrink in size, and conservation of angular momentum caused the gas to form the spinning disk of our galaxy. Stars in the halo formed before the gas finished collapsing into the disk.

14.4   The Mysterious Galactic Center

• What lies in the center of our galaxy?

Motions of stars near the center of our galaxy suggest that it contains a black hole about three to four million times as massive as the Sun. The black hole appears to be powering a bright source of radio emission known as Sgr A*.

exercises and problems

Review Questions

Short-Answer Questions Based on the Reading

1. Draw simple sketches of our galaxy as it would appear face-on and edge-on. Identify the *disk, bulge, halo*, and *spiral arms*, and indicate the galaxy's approximate dimensions.
2. What are the Large and Small Magellanic Clouds, and the Sagittarius and Canis Major Dwarfs?
3. Describe the basic characteristics of stellar orbits in the bulge, disk, and halo of our galaxy.
4. How can we use orbital properties to learn about the mass of the galaxy? What have we learned?
5. Summarize the stages of the *star–gas–star cycle* in Figure 14.3.
6. What creates a *bubble* of hot, ionized gas? What happens to the gas in the bubble over time?
7. What are *cosmic rays*, and where are they thought to come from?
8. What do we mean by *atomic hydrogen gas*? How common is it, and how do we map its distribution in the galaxy?
9. Briefly summarize the different types of gas present in the disk of the galaxy, and describe how they appear when we view the galaxy in different wavelengths of light.
10. What are *ionization nebulae*, and why are they found near hot, massive stars?
11. How do we know that spiral arms do not rotate like giant pinwheels? What makes spiral arms bright?
12. What triggers star formation within a spiral arm? How do we think spiral arms are maintained?
13. Briefly describe the characteristics that distinguish the galaxy's *disk population* of stars from its *spheroidal population* of stars.
14. How do the different ages of disk stars and halo stars support the idea that our galaxy formed from the gravitational collapse of a *protogalactic cloud?*
15. Why do we think that the Milky Way's full protogalactic cloud may have been formed from the merger of several smaller protogalactic clouds?
16. What is Sgr A*? What evidence suggests that it contains a massive black hole?

Test Your Understanding

Does It Make Sense?

Decide whether the statement makes sense (or is clearly true) or does not make sense (or is clearly false). Explain clearly; not all of these have definitive answers, so your explanation is more important than your chosen answer.

17. We did not understand the true size and shape of our galaxy until NASA launched satellites into the galactic halo, enabling us to see what the Milky Way looks like from the outside.
18. Planets like Earth probably didn't form around the first stars because there were so few heavy elements back then.
19. If I could see infrared light, the galactic center would look much more impressive.
20. Many spectacular ionization nebulae are seen throughout the Milky Way's halo.
21. The carbon in my diamond ring was once part of an interstellar dust grain.

22. The Sun's velocity around the Milky Way tells us that most of our galaxy's dark matter lies near the center of the galactic disk.
23. We know that a black hole lies at our galaxy's center because numerous stars near it have vanished over the past several years, telling us that they've been sucked in.
24. If we could watch a time-lapse movie of a spiral galaxy over millions of years, we'd see many stars being born and dying within the spiral arms.
25. The star–gas–star cycle will keep the Milky Way looking just as bright in 100 billion years as it looks now.
26. Halo stars orbit the center of our galaxy much faster than the disk stars.

Quick Quiz

Choose the best answer to each of the following. Explain your reasoning with one or more complete sentences.

27. Where are most of the Milky Way's globular clusters found? (a) in the disk (b) in the bulge (c) in the halo
28. Why do disk stars bob up and down as they orbit the galaxy? (a) because the gravity of other disk stars always pulls them toward the disk (b) because of friction with the interstellar medium (c) because the halo stars keep knocking them back into the disk
29. How do we determine the Milky Way's mass outside the Sun's orbit? (a) from the Sun's orbital velocity and its distance from the center of our galaxy (b) from the orbits of halo stars near the Sun (c) from the orbits of stars and gas clouds orbiting the galactic center at greater distances than the Sun
30. Which part of the galaxy has gas with the hottest average temperature? (a) the disk (b) the halo (c) the bulge
31. What is the typical hydrogen content of stars that are forming right now in the vicinity of the Sun? (a) 100% hydrogen (b) 75% hydrogen (c) 70% hydrogen
32. Which of these forms of radiation passes most easily through the disk of the Milky Way? (a) red light (b) blue light (c) infrared light
33. Where would you least expect to find an ionization nebula? (a) in the halo (b) in the disk (c) in a spiral arm
34. Where would be the most likely place to find an ionization nebula? (a) in the halo (b) in the bulge (c) in a spiral arm
35. Which kind of star is most likely to be part of the spheroidal population? (a) an O star (b) an A star (c) an M star
36. We measure the mass of the black hole at the galactic center from (a) the orbits of stars in the galactic center. (b) the orbits of gas clouds in the galactic center. (c) the amount of radiation coming from the galactic center.

Process of Science

Examining How Science Works

37. *Discovering the Structure of the Milky Way.* The story of how we came to learn the structure of the Milky Way (see Special Topic, p. 388) is an excellent demonstration of how science progresses. What features of the Milky Way's appearance in our sky led scientists to conclude that its width is much larger than its thickness? Why did they originally believe that the Sun was near the Milky Way's

center? What key observations forced scientists to change their views about the location of the Sun within the Milky Way?

38. *Formation of the Milky Way.* Figure 14.18 outlines a simple model for the formation of the Milky Way Galaxy. Explain how observational evidence supports each step of the model. Why do we think the Milky Way's protogalactic cloud contained only hydrogen and helium? What evidence suggests that the halo stars formed first? Why do we think that the disk stars formed after the protogalactic cloud collapsed into a spinning disk? What is the evidence that stars have been forming steadily in the disk for billions of years?

Investigate Further

In-Depth Questions to Increase Your Understanding

Short-Answer/Essay Questions

39. *Unenriched Stars.* Suppose you discover a star made purely of hydrogen and helium. How old do you think it would be? Explain.

40. *Enrichment of Star Clusters.* The gravitational pull of an isolated globular cluster is rather weak—a single supernova explosion can blow all the interstellar gas out of a globular cluster. How might this fact relate to observations indicating that stars ceased to form in globular clusters long ago? How might it relate to the fact that globular clusters are deficient in elements heavier than hydrogen and helium? Summarize your answers in one or two paragraphs.

41. *High-Velocity Star.* The average speed of stars relative to the Sun in the solar neighborhood (i.e., the speed at which we see stars moving toward or away from the Sun—*not* their orbital speed around the galaxy) is about 20 km/s. Suppose you discover a star in the solar neighborhood that is moving relative to the Sun at a much higher speed, say, 200 km/s. What kind of orbit does this star probably have around the Milky Way? In what part of the galaxy does it spend most of its time? Explain.

42. *Research: Discovering the Milky Way.* Humans have been looking at the Milky Way since long before recorded history, but only in the past century did we verify the true shape of the galaxy and our location within it. Learn more about how conceptions of the Milky Way developed through history. What names did different cultures give the band of light they saw? What stories did they tell about it? How have ideas about the galaxy changed in the past few centuries? Try to locate diagrams that illustrate these changes. Write a two- to three-page summary of your findings.

43. *Future of the Milky Way.* Describe how the Milky Way would look from the outside if you could watch it for the next 100 billion years. How would its appearance change?

44. *Orbits at the Galactic Center.* Using the information in Figure 14.21, identify which two stars reach the highest orbital speeds. Explain how the orbits of those two stars illustrate Kepler's first two laws.

45. *A Nonspinning Galaxy.* How would the development of the Milky Way Galaxy have been different if its original protogalactic cloud had no angular momentum? Describe how you think our galaxy would look today and explain your reasoning.

46. *Gas Distribution in the Milky Way.* Make a sketch of the gas distribution in the plane of the Milky Way based on the photographs in Figure 14.11. In your sketch, map out where you would find molecular clouds, atomic hydrogen clouds, and bubbles of hot gas. Explain why each of those components of the interstellar medium is found in the location where you have drawn it.

Quantitative Problems

Be sure to show all calculations clearly and state your final answers in complete sentences.

47. *Mass of the Milky Way's Halo.* The Large Magellanic Cloud is a small galaxy that orbits the Milky Way. It is currently orbiting the Milky Way at a distance of roughly 160,000 light-years from the galactic center at a velocity of about 300 km/s. Use these values in the orbital velocity law to get an estimate of the Milky Way's mass within 160,000 light-years from the center.

48. *Mass of the Central Black Hole.* Suppose you observe a star orbiting the galactic center at a speed of 1000 km/s in a circular orbit with a radius of 20 light-days. What would your estimate be for the mass of the object that the star is orbiting?

49. *Mass of a Globular Cluster.* Stars in a typical globular cluster orbit the center at speeds of about 10 km/s. Stars at the outskirts of a globular cluster are about 50 light-years from the center. Use these facts to estimate the mass of a typical globular cluster.

50. *Mass of Saturn.* The innermost rings of Saturn orbit in a circle with a radius of 67,000 kilometers at a speed of 23.8 km/s. Use the orbital velocity law to compute the mass contained within the orbit of those rings. Compare your answer with the mass of Saturn listed in Appendix E.

Discussion Questions

51. *Galactic Ecosystem.* We have likened the star–gas–star cycle in our Milky Way to the ecosystem that sustains life on Earth. Here on our planet, water molecules cycle from the sea to the sky to the ground and back to the sea. Our bodies convert atmospheric oxygen molecules into carbon dioxide, and plants convert carbon dioxide back into oxygen molecules. How are the cycles of matter on Earth similar to the cycles of matter in the galaxy? How do they differ? Do you think the term *ecosystem* is appropriate in discussions of the galaxy?

52. *Galaxy Stuff.* In the chapters on stars, we learned why we are "star stuff." Explain why we are also "galaxy stuff." Does the fact that the entire galaxy was involved in bringing forth life on Earth change your perspective on Earth or on life? If so, how? If not, why not?

Web Projects

53. *Images of the Star–Gas–Star Cycle.* Find pictures on the Web of ionization nebulae and other forms of interstellar gas in different stages of the star–gas–star cycle. Assemble the pictures into a sequence that tells the story of interstellar recycling, with a one-paragraph explanation of each image.

54. *The Galactic Center.* Search the Web for recent images of the galactic center, along with information on the question of whether the center hides a massive black hole. Write a two- to three-page report, with pictures, on the latest findings about the center of the Milky Way Galaxy.

Use the following questions to check your understanding of some of the many types of visual information used in astronomy. Answers are provided in Appendix K. For additional practice, try the Chapter 14 Visual Quiz at **www.masteringastronomy.com.**

Visible light from stars

Infrared light from stars

Radio emission from molecules

X-ray emission from hot gas

The images above, taken from Figure 14.1, show what the central region of our galaxy looks like from our viewpoint in different wavelengths of light. Each image shows the same region of the sky. Use the following questions to check your understanding of what these images show and what we can learn by comparing them.

1. The image of radio emission uses different colors of light to represent different levels of brightness. Which color represents the brightest radio emission? Which color represents the lowest levels of brightness?
2. The image of X-ray emission uses different colors of light to represent different levels of brightness. The dark blue color represents the lowest levels of X-ray brightness. Which color represents the greatest levels of X-ray brightness?
3. How do regions showing strong radio emission from molecules look in the visible light image? Are they bright or are they dark?
4. How do regions showing strong radio emission from molecules look in the infrared light image? Are they bright or are they dark?

5. Compare the radio, infrared, and visible light images. Which of the following conclusions is best supported by your comparison?
   a. Gas clouds containing molecules absorb roughly equal amounts of infrared starlight and visible starlight.
   b. Gas clouds containing molecules absorb substantial amounts of visible starlight but don't absorb infrared starlight at all.
   c. Gas clouds containing molecules absorb infrared starlight less effectively than they absorb visible starlight.
6. Compare the radio and X-ray images. Can you conclude from this comparison that gas clouds containing molecules absorb X-ray light?

galaxies and the foundation of modern cosmology

learning goals

15.1 Islands of Stars

- What are the three major types of galaxies?
- How are galaxies grouped together?

15.2 Distances of Galaxies

- How do we measure the distances to galaxies?
- What is Hubble's law?
- How do distance measurements tell us the age of the universe?

15.3 Galaxy Evolution

- How do we observe the life histories of galaxies?
- How did galaxies form?
- Why do galaxies differ?

15.4 Quasars and Other Active Galactic Nuclei

- What are quasars?
- What is the power source for quasars and other active galactic nuclei?
- Do supermassive black holes really exist?

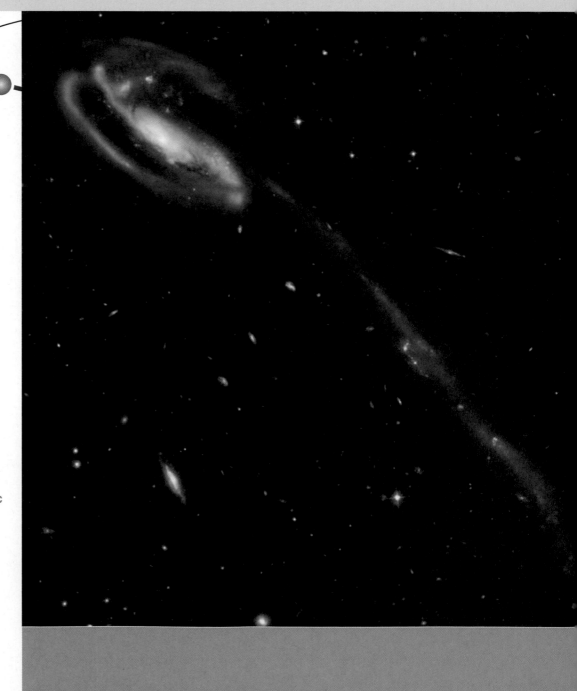

Far beyond the Milky Way, we see billions of other galaxies scattered throughout space. Some look similar to our own galaxy, while others look quite different. The sight of all these galaxies inspires us not only to wonder how they came to be, but also to ask fundamental questions about our universe: How old is the universe? How big is it? How is the universe changing with time? Such questions might have seemed ridiculously speculative a century ago. Today, we believe we know the answers to these questions with respectable accuracy.

Edwin Hubble, the man for whom the Hubble Space Telescope is named, provided the key discovery when he proved conclusively that galaxies exist beyond the Milky Way. The distances he measured to those galaxies revealed an astonishing fact: The more distant a galaxy is, the faster it moves away from us. Hubble's discovery dealt a mortal blow to the traditional belief in a static, eternal, and unchanging universe. The motions of the galaxies away from one another imply that the entire universe is expanding and that its age is finite.

In this chapter, we will get acquainted with the islands of stars we call galaxies. We will discuss how we measure their distances and how learning their distances has helped us learn about the age and size of our universe. Finally, we will study galaxies themselves, looking for clues to how they have evolved through time and what they can tell us about the history of our universe as a whole.

15.1 Islands of Stars

Figure 15.1 shows an amazing image of a tiny patch of the sky taken by the Hubble Space Telescope. The telescope pointed in a single direction in the sky and collected all the light it could for 10 days. If you held a grain of sand at arm's length, the angular size of the grain would match the angular size of everything in this picture. Almost every blob of light in this image is a galaxy—an island of stars bound together by gravity. Like our own Milky Way, each galaxy is a dynamic system that has cycled hydrogen gas through stars for billions of years, producing new elements for future generations of stars. We can use this photo to estimate the total number of galaxies in the observable universe: We simply count the number of galaxies in this photo and multiply by the number of such photos it would take to make a montage of the entire sky. Careful counts from this photo and the even more detailed photo shown on page 1 tell us that the observable universe contains well over 100 billion galaxies.

Even without knowing the distances to the individual galaxies in Figure 15.1, we can see that galaxies come in many sizes, colors, and shapes. Some look large, some small. Some are reddish, some whitish. Some appear round, and some appear flat. We would like to understand why galaxies differ in these ways, but it is not easy to learn their histories. Just as with stars, our observations capture only the briefest instant

Figure 15.1

The Hubble Deep Field (far right) is an image composed of 10 days of exposures taken with the Hubble Space Telescope. Some of the galaxies pictured are three-quarters of the way across the observable universe. The zoom-in sequence shows the location of the field within the Big Dipper, recognizable in the lower left frame.

in any galaxy's life, leaving us to piece together the life story of a typical galaxy from pictures of different galaxies at various life stages.

The task is made even more difficult because we see young galaxies only at great distances—distances at which we are looking far back into the universe's past (see Figure 1.4). This fact means that the study of galaxies is intimately connected with what we call **cosmology**—the study of the overall structure and evolution of the universe. We'll therefore need to study not only galaxies in this chapter, but also our modern understanding of how universal expansion affects the lives of galaxies. We begin with an attempt to categorize the galaxies we see nearby, whose details are easier to observe.

• What are the three major types of galaxies?

Astronomers classify galaxies into three major categories:

- **Spiral galaxies,** such as our own Milky Way, look like flat white disks with yellowish bulges at their centers. The disks are filled with cool gas and dust, interspersed with hotter ionized gas, and usually display beautiful spiral arms.

- **Elliptical galaxies** are redder, rounder, and often longer in one direction than in the other, like a football. Compared with spiral galaxies, elliptical galaxies contain very little cool gas and dust, though they often contain very hot ionized gas.

- **Irregular galaxies** appear neither disklike nor rounded.

Galaxies come in three major types: spiral, elliptical, and irregular.

The differing colors of galaxies arise from the different kinds of stars that populate them: Spiral and irregular galaxies look white because they contain stars of all different colors and ages, while elliptical galaxies look redder because old, reddish stars produce most of their light. Galaxies also come in a wide range of sizes, from *dwarf galaxies* containing as few as 100 million (10^8) stars to *giant galaxies* with more than 1 trillion (10^{12}) stars.

think about it Take a moment and try to classify the larger galaxies in Figure 15.1. How many appear spiral? Elliptical? Irregular? Do the colors of galaxies seem related to their shapes?

Spiral Galaxies Like the Milky Way, other spiral galaxies also have a thin *disk* extending outward from a central *bulge* (Figure 15.2). The bulge merges smoothly into a *halo* that can extend to a radius of more than 100,000 light-years. However, the halo is difficult to see in photographs because its stars are generally dim and spread over a large volume.

Spiral galaxies have a disk, bulge, and halo like the Milky Way.

Recall that the Milky Way is made up of two distinct populations of stars [Section 14.3]. The *disk population* (sometimes called Population I) includes stars of all ages and masses that orbit in the disk of the galaxy. The *spheroidal population* (sometimes called Population II) consists of halo and bulge stars, with the halo stars generally being old and low in mass. We find the same two populations in other spiral galaxies, and we use them to define two primary components of galaxies:

- The **disk component** is the flat disk in which stars follow orderly, nearly circular orbits around the galactic center. The disk component always contains an *interstellar medium* of gas and dust, but the amounts and proportions of molecular, atomic, and ionized gases in this medium differ from one spiral galaxy to the next.

- The bulge and halo together make up the **spheroidal component**, named for its rounded shape. Stars in the spheroidal component have orbits with many different inclinations, and the spheroidal component generally contains little cool gas and dust. Figure 15.3 shows a spiral galaxy with an unusually large bulge that illustrates the general shape of the spheroidal component.

All spiral galaxies have both a disk and a spheroidal component, but there are some variations on the general theme. Some spiral galaxies appear to have a straight bar of stars cutting across the center, with spiral arms curling away from the ends of the bar. Such galaxies are known as *barred spiral galaxies* (Figure 15.4). Astronomers suspect that the Milky Way itself is a barred spiral galaxy, because our galaxy's bulge appears to be somewhat elongated.

Other galaxies have disk and spheroidal components like spiral galaxies but appear to lack spiral arms. These *lenticular galaxies* (*lenticular* means "lens-shaped") are sometimes considered an intermediate class between spirals and ellipticals, because they tend to have less cool gas than normal spirals but more than ellipticals. Among large galaxies in the universe, most (75–85%) are spiral or lenticular. Galaxies with obvious disks are much rarer among small galaxies.

VIS

Figure 15.2 interactive photo

The giant spiral galaxy M101. It is about 170,000 light-years in diameter.

VIS

Figure 15.3

NGC 4594 (the Sombrero Galaxy) is a spiral galaxy with a large bulge and a dusty disk that we see almost edge-on. A much larger but nearly invisible halo surrounds the entire galaxy. The bulge and halo together make up the spheroidal component of the galaxy. The region shown in this image is about 82,000 light-years across.

Figure 15.4

NGC 1300, a barred spiral galaxy about 110,000 light-years in diameter.

Elliptical Galaxies Elliptical galaxies differ from spiral galaxies primarily in that they have *only* a spheroidal component and lack a significant disk component. That is, elliptical galaxies look much like the bulge and halo of a spiral galaxy that is missing its disk. (For this reason, elliptical galaxies are sometimes called *spheroidal galaxies*.) Although the most massive galaxies are the relatively rare *giant elliptical galaxies* (Figure 15.5), the vast majority of elliptical galaxies are small. These small elliptical galaxies are the most common type of galaxy in the universe. Particularly small ellipticals with less than about a billion stars, known as *dwarf elliptical galaxies,* are often found near larger spiral galaxies. For example, over a dozen dwarf elliptical galaxies belong to our Local Group.

> Elliptical galaxies differ from spiral galaxies in that they do not have significant disks.

Elliptical galaxies usually contain very little dust or cool gas, although some have relatively small and cold gaseous disks rotating at their centers. However, some large elliptical galaxies contain substantial amounts of very hot gas. This low-density, X-ray–emitting gas is much like the gas in the hot bubbles created by supernovae and powerful stellar winds in the Milky Way [Section 14.2].

The lack of cool gas in elliptical galaxies means that, like the halo of the Milky Way, they generally have little or no ongoing star formation. Elliptical galaxies therefore look red or yellow in color because they do not have any of the hot, young, blue stars found in the disks of spiral galaxies.

Irregular Galaxies Some of the galaxies we see nearby fall into neither of the two major categories. This class of *irregular galaxies* is a miscellaneous class, encompassing small galaxies such as the Magellanic Clouds (Figure 15.6) and larger "peculiar" galaxies that appear to be in disarray.

> Irregular galaxies appear to be in disarray.

These blobby star systems are usually white and dusty, like the disks of spirals. Their colors tell us that they contain young, massive stars. Among nearby galaxies, only a small percentage of galaxies as large as the Milky Way are irregular.

Figure 15.5

M87, a giant elliptical galaxy in the Virgo Cluster, is one of the most massive galaxies in the universe. The region shown is more than 120,000 light-years across.

Telescopic observations probing deeper into the universe show that distant galaxies are more likely to be irregular in shape than nearby galaxies. Because the light of more distant galaxies has taken longer to reach us, these observations tell us that irregular galaxies were more common when the universe was younger.

Hubble's Galaxy Classes   Edwin Hubble invented a system for classifying galaxies that organizes the galaxy types into a diagram shaped like a tuning fork (Figure 15.7). Elliptical galaxies appear on the "handle" at the left, designated by the letter *E* and a number. The larger the number, the flatter the elliptical galaxy: An E0 galaxy is a sphere, and the numbers increase to the highly elongated type E7. The two forks show spiral galaxies, designated by the letter *S* for ordinary spirals and *SB* for barred spirals, followed by a lowercase *a*, *b*, or *c:* The bulge size decreases from *a* to *c*, while the amount of dusty gas increases. Lenticular galaxies are designated S0, and irregular galaxies are designated Irr.

Astronomers had once hoped that the classification of galaxies might yield deep insights, just as the classification of stars did in the early twentieth century. The Hubble classification scheme itself was suspected for a time to be an evolutionary sequence in which galaxies flattened and spread out as they aged. Unfortunately for astronomers, galaxies turn out to be far more complex than stars, and classification schemes like this one have not led to any major insights into their nature.

Figure 15.6

The Large Magellanic Cloud, an irregular galaxy that is a small companion to the Milky Way. It is about 30,000 light-years across.

• How are galaxies grouped together?

Although some galaxies travel solo through the universe, many are gravitationally bound together with neighboring galaxies. Spiral galaxies are often found in loose collections of up to a few dozen galaxies, called

Figure 15.7

This "tuning fork" diagram illustrates Hubble's galaxy classes.

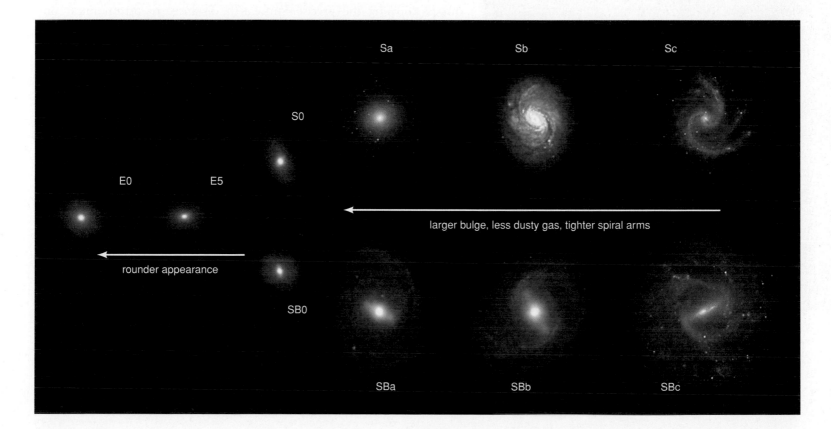

VIS

Figure 15.8

Hickson Compact Group 87, a small group of galaxies consisting of a large edge-on spiral galaxy (bottom), two smaller spiral galaxies (center and upper left), and an elliptical galaxy (right). The whole group is about 170,000 light-years in diameter. (The other objects in this photograph are foreground stars in our own galaxy.)

Figure 15.9

Central part of the galaxy cluster Abell 1689. Almost every object in this photograph is a galaxy belonging to the cluster. Yellowish elliptical galaxies outnumber the whiter spiral galaxies. The region pictured is about 2 million light-years across. (A few stars from our own galaxy appear in the foreground and look like white dots with four spikes in the form of a cross.)

groups. Our Local Group is one example (see Figure 1.1). Figure 15.8 shows another galaxy group.

Spiral galaxies tend to congregate in small groups, while elliptical galaxies are primarily found in large clusters. Elliptical galaxies are particularly common in clusters of galaxies, which can contain hundreds to thousands of galaxies extending over more than 10 million light-years (Figure 15.9). Elliptical galaxies make up about half the large galaxies in the central regions of clusters, while they represent only a small minority (about 15%) of the large galaxies found outside clusters.

(MA) Measuring Cosmic Distances Tutorial, Lessons 1–4

15.2 Distances of Galaxies

To learn more about galaxies than just their shape, color, and type, we need to know how far away they are. Measuring the distances to galaxies is one of the most challenging tasks we face when trying to understand galaxies and the universe as a whole, but the payoff is enormous. Besides telling us where galaxies are located, such measurements also reveal the size and age of the observable universe.

• How do we measure the distances to galaxies?

Our determinations of astronomical distances depend on a chain of methods in which each step allows us to measure greater distances in the universe. We have already discussed the measurement of distances to nearby stars by parallax [Section 11.1]. Because parallax is the apparent shift in a star's position as Earth orbits the Sun, measuring distances by parallax requires knowing the precise Sun–Earth distance, or astronomical unit (AU). Astronomers measure the AU with a technique called **radar ranging**, in which radio waves are transmitted from Earth and bounced off Venus. Radio waves travel at the speed of light, so the round-trip travel time for the radar signals tells us Venus's distance from Earth. We can then use Kepler's laws and a little geometry to calculate the length of an AU. Radar ranging measurements of the AU represent the first link in the distance chain, and parallax measurements of distances to nearby stars represent the second link. We will now follow the rest of this chain, link by link, to the outermost reaches of the observable universe.

Standard Candles Once we have measured distances to nearby stars through parallax, we can begin to measure distances to other stars in the same way that we might estimate the distance to a street lamp at night. If the street lamp does not look very bright, then it's probably far away. If it looks very bright, then it's probably nearby.

We can determine distance by measuring the apparent brightness of an object whose luminosity we already know and applying the inverse square law for light. We can determine the lamp's distance more accurately if we can measure its apparent brightness. For example, suppose we see a distant street lamp and know that every street lamp of its type emits 1000 watts of light. If we then measure its apparent brightness, we can calculate its distance using the inverse square law for light [Section 11.1].

An object such as a street lamp, for which we are likely to know the true luminosity, represents what astronomers call a **standard candle**—a light source of a known, standard luminosity. Unlike light bulbs, however, astronomical objects do not come marked with wattage. An astronomical object can serve as a standard candle only if we have some way of knowing its true luminosity without first measuring its apparent brightness and distance. Fortunately, many astronomical objects meet this requirement. For example, any star that is a twin of our Sun—that is, a main-sequence star with spectral type G2—should have about the same luminosity as the Sun. Thus, if we measure the apparent brightness of a Sun-like star, we can assume it has the same luminosity as the Sun and use the inverse square law for light to estimate its distance.

Beyond the few hundred light-years for which we can measure distances by parallax, we use standard candles for most cosmic distance measurements. These distance measurements always have some uncertainty, because no astronomical object is a perfect standard candle. The challenge of measuring astronomical distances comes down to the challenge of finding the objects that make the best standard candles. The more confidently we know an object's true luminosity, the more certain we are of its distance.

Main-Sequence Fitting Although Sun-like stars are reasonably good standard candles, they are of limited use because Sun-like stars are relatively dim, making them difficult to detect beyond relatively short distances. To measure distances beyond 1000 light-years or so, we need brighter standard candles.

Brighter main-sequence stars represent an obvious first choice, because all main-sequence stars of a particular spectral type have about the same luminosity [Section 11.2]. However, before we can use any main-sequence star as a standard candle, we must first have some way of knowing its true luminosity. We therefore need to follow two steps to use bright main-sequence stars as standard candles:

1. We identify a star cluster that is close enough for us to determine its distance by parallax and plot its H-R diagram. Because we know the distances to the cluster stars, we can use the inverse square law for light to establish their true luminosities from their apparent brightnesses.
2. We can look at stars in other clusters that are too far away for parallax measurements and measure their apparent brightnesses. If we assume that main-sequence stars in other clusters have the same luminosities as their counterparts in the nearby cluster, we can calculate their distances from the inverse square law for light.

Twentieth-century astronomers laid the groundwork for this technique by calibrating the luminosities on an H-R diagram. This calibration relied largely on a single, nearby star cluster—the Hyades Cluster in the constellation Taurus, whose distance is now known from its parallax. We can find the distances to other star clusters by comparing the apparent brightnesses of their main-sequence stars with those in the Hyades Cluster and assuming that all main-sequence stars of the same color have the same luminosity (Figure 15.10). This technique of determining distances by comparing main sequences in different star clusters is called **main-sequence fitting**.

Cepheid Variables Main-sequence fitting works well for measuring distances to star clusters throughout the Milky Way, but not for measuring distances to other galaxies. Most main-sequence stars are too faint to be seen in other galaxies, even with our largest telescopes. Instead, we need

cosmic calculations 15.1

Standard Candles

The inverse square law for light tells us how an object's apparent brightness depends on its luminosity and distance (see Cosmic Calculations 11.1). With a little algebra, we can rewrite this law in a form that enables us to calculate distance if we know luminosity and apparent brightness, as we do for any object that qualifies as a *standard candle*.

$$\text{distance} = \sqrt{\frac{\text{luminosity}}{4\pi \times (\text{apparent brightness})}}$$

For a standard candle of a particular luminosity, the formula tells us that a smaller apparent brightness means a greater distance.

Example: You observe a main-sequence star of spectral type G2, measuring its apparent brightness to be 1.0×10^{-12} watt/m^2. How far away is it?

Solution: The star has the same spectral type as our Sun, so we assume that its luminosity is approximately equal to the Sun's luminosity of 3.8×10^{26} watts. We use this luminosity and the star's apparent brightness in the formula:

$$\text{distance} = \sqrt{\frac{3.8 \times 10^{26} \text{ watts}}{4\pi \times \left(1.0 \times 10^{-12} \frac{\text{watt}}{\text{m}^2}\right)}} \approx 5.5 \times 10^{18} \text{ m}$$

The distance to the star is about 5.5×10^{18} m. When we divide this distance by 9.5×10^{16} m/ly, we find that it is equivalent to about 580 light-years. Of course, if the star's actual luminosity is different from the Sun's, the distance we found will also be in error.

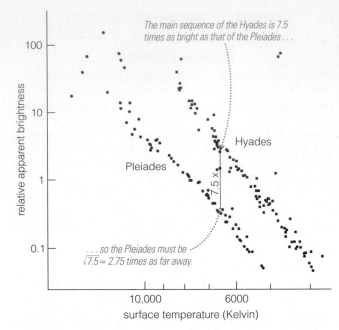

Hyades

Pleiades

7.5 ×

. . . so the Pleiades must be
$\sqrt{7.5} \approx 2.75$ times as far away.

relative apparent brightness

surface temperature (Kelvin)

Figure 15.10

To use the technique of main-sequence fitting, we compare the apparent brightness of the main sequence of a cluster of unknown distance (the Pleiades, in this case) to that of a cluster whose distance we already know (such as the Hyades, whose distance is known from parallax).

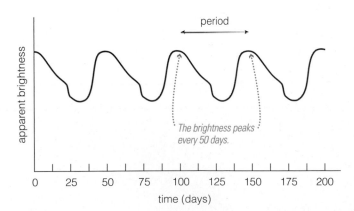

period

The brightness peaks every 50 days.

apparent brightness

time (days)

Figure 15.11

This graph shows how the brightness of a Cepheid varies with time. The period is the time from one peak of brightness to the next.

very bright stars to serve as standard candles for distance measurements beyond the Milky Way.

The most useful bright stars for measuring the distances to galaxies are called **Cepheid variable stars**, or **Cepheids** for short. These stars vary in brightness in our sky, alternately becoming dimmer and brighter with periods ranging from a few days to a few months. Each Cepheid has its own particular time period between peaks in luminosity. Figure 15.11 shows the brightness variations of a Cepheid with a period of about 50 days.

> Cepheid variable stars are useful for measuring distances because we can determine a Cepheid's luminosity from the period between its peaks of brightness.

In 1912, Henrietta Leavitt discovered that the periods of Cepheids are very closely related to their luminosities: The longer the period, the more luminous the star (Figure 15.12). We say that Cepheids obey a **period–luminosity relation** that allows us to determine (within about 10%) a Cepheid's luminosity simply by measuring the time period over which its brightness varies. Figure 15.12 shows that a Cepheid variable whose brightness peaks every 30 days is effectively screaming out, "Hey, everybody, my luminosity is 10,000 times that of the Sun!" Once we measure a Cepheid's period, we know its luminosity and we can use the inverse square law for light to determine its distance.

Leavitt discovered the period–luminosity relation with careful observations of Cepheids in the Magellanic Clouds but she did not know why Cepheids vary in this special way. We now know that Cepheids vary because they have a peculiar problem in matching the amount of energy their surfaces radiate with the amount welling up from the core. In a futile quest for a steady equilibrium, the upper layers of a Cepheid variable star alternately expand and contract, causing the star's luminosity to rise and fall. The period–luminosity relation holds because larger (more luminous) Cepheids take longer to pulsate in and out in size.

Cepheids have been used for almost a century to measure distances to nearby galaxies. As we'll discuss shortly, they played a critical role in Edwin Hubble's discoveries. More recently, one of the main missions of the Hubble Space Telescope was to measure accurate distances to galaxies up to 100 million light-years away by studying Cepheids within them.

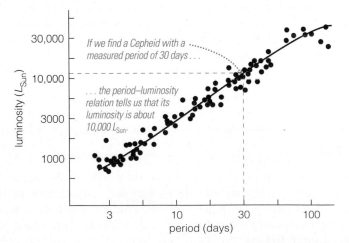

If we find a Cepheid with a measured period of 30 days . . .

. . . the period–luminosity relation tells us that its luminosity is about 10,000 L_{Sun}.

luminosity (L_{Sun})

period (days)

Figure 15.12

Cepheid period–luminosity relation. The data show that all Cepheids of a particular period have very nearly the same luminosity. Measuring a Cepheid's period therefore allows us to determine its luminosity and calculate its distance. (Cepheids actually come in two types with two different period–luminosity relations. The relation here is for Cepheids with heavy-element content similar to that of our Sun, or Type I Cepheids.)

This distance may sound very large, but it is still quite small compared with the distances of the galaxies in Figure 15.1. To go further, we use the distances determined with these Cepheids to learn the luminosities of even brighter standard candles.

Distant Standard Candles Astronomers have discovered several techniques for estimating distances beyond those for which we can observe Cepheids. The most valuable of these techniques has proved to be the use of white dwarf supernovae as standard candles. Recall that white dwarf supernovae are thought to be exploding white dwarf stars that have reached the $1.4M_{Sun}$ limit [Section 13.1]. These supernovae should all have nearly the same luminosity, because they all come from stars of the same mass that explode in the same way. Although white dwarf supernovae are rare in any individual galaxy, several have been detected during the past century in galaxies within about 50 million light-years of the Milky Way. Astronomers kept careful records of those events, so that today we can determine the true luminosities of these supernovae by using Cepheids to measure the distances to the galaxies in which they occurred. These measurements confirm that the luminosities of all white dwarf supernovae are about the same.

> White dwarf supernovae are useful for measuring large distances because they are bright and all have about the same peak luminosity.

Because white dwarf supernovae are so bright—about 10 billion solar luminosities at their peak—we can detect them even when they occur in galaxies billions of light-years away (Figure 15.13). We can therefore use them to measure the distances of galaxies in the far reaches of the observable universe, completing the distance chain. Although the number of galaxies whose distances we can measure with this technique is relatively small, because white dwarf supernovae occur only once every few hundred years in a typical galaxy, these galaxies have allowed us to calibrate another technique—one that relies on the expansion of the universe.

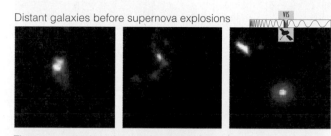
Distant galaxies before supernova explosions

The same galaxies after supernova explosions

Figure 15.13

White dwarf supernovae in galaxies approximately 10 billion light-years away. White arrows in the lower images indicate the supernovae, and the upper images show what these galaxies looked like without supernovae.

(MA) **Hubble's Law Tutorial, Lessons 1–3**

• What is Hubble's law?

The ability to measure distances to galaxies is the key to much of our modern understanding of the size and age of the universe. We can trace the beginning of this understanding directly back to discoveries made by Edwin Hubble. Let's explore how Hubble discovered the famous law that bears his name and how this law has helped us answer fundamental questions about the universe in which we live.

Edwin Hubble and the Andromeda Galaxy Before Edwin Hubble's groundbreaking work in the 1920s, no one knew for certain whether the spiral-shaped objects they saw in the sky were merely clouds of gas within the Milky Way—and therefore that the Milky Way represented the entire universe—or distant and distinct galaxies. The opinions of astronomers were split on this issue, which became a subject of great debate. The problem was that neither side could prove its case, because the techniques available at the time could not distinguish objects within the Milky Way from those beyond it.

Hubble put the debate to rest in 1924. Using the new, 100-inch telescope atop southern California's Mount Wilson (Figure 15.14)—the largest telescope in the world at the time—he identified Cepheid variables in the Andromeda Galaxy by comparing photographs of the galaxy taken days apart. He then used those Cepheid observations and Henrietta Leavitt's period–luminosity relation to estimate the galaxy's distance, proving that the Andromeda Galaxy resides far beyond the outer reaches of the Milky Way.

Edwin Hubble used Cepheids to prove that the Andromeda Galaxy lies beyond the Milky Way.

This single scientific discovery dramatically changed our view of the universe. Rather than thinking the Milky Way was the entire universe, we suddenly knew that it is just one among many galaxies in an enormous universe. The stage was set for an even greater discovery.

Distance and Redshift Astronomers had known since the 1910s that the spectra of most spiral galaxies tended to be *redshifted* (Figure 15.15). Recall that redshifts occur when the object emitting the radiation is moving away from us [Section 5.2]. Because Hubble had not yet proved that the spiral galaxies were separate from the Milky Way, no one understood the true significance of their motions.

Following his discovery of Cepheids in Andromeda, Hubble and his coworkers spent the next few years measuring the redshifts of galaxies and estimating their distances. Because even Cepheids were too dim to be seen in most of these galaxies, Hubble needed brighter standard candles for his distance estimates. One of his favorite techniques was to use the brightest object he could see in each galaxy as a standard candle, because he assumed these objects to be very bright stars that would always have about the same luminosity.

A galaxy's redshift tells us how fast it is moving away from us, and the relationship between redshift and distance shows that the universe is expanding.

In 1929, Hubble announced his conclusion: The more distant a galaxy, the greater its redshift and hence the faster it moves away from us. As we discussed in Chapter 1 (see Figure 1.15), this discovery implies that the entire universe is expanding. Thus, Hubble discovered the expansion of the universe.

Hubble's original assertion was based on an amazingly small sample of galaxies. Even more incredibly, he had grossly underestimated the luminosities of his standard candles. The "brightest stars" he had been using as standard candles were really entire *clusters* of bright stars. Fortunately, Hubble was both bold and lucky. Subsequent studies of much larger samples of galaxies showed that they are indeed receding from us, but they are even farther away than Hubble thought.

Hubble's Law We express the idea that more distant galaxies move away from us faster with a very simple formula, now known as **Hubble's law**:

$$v = H_0 \times d$$

where v stands for a galaxy's velocity away from us (sometimes called the *recession velocity*), d stands for its distance, and H_0 (pronounced "H-naught") is a number called **Hubble's constant**. We usually write Hubble's law in this form to express the idea that the speed of a galaxy depends on its distance from us. However, astronomers more often use the law in

Figure 15.14

Edwin Hubble at the Mount Wilson Observatory.

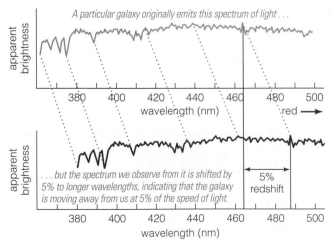

Figure 15.15

Redshifted galaxy spectrum.

reverse—measuring a galaxy's speed from its redshift and then using Hubble's law to estimate its distance.

Hubble's law expresses a relationship between galaxy speeds and distances, and hence allows us to determine a galaxy's distance from its speed.

Because Hubble's law in principle applies to all distant galaxies for which we can measure a redshift, it is the most useful technique for determining distances to galaxies that are very far away. Nevertheless, we encounter two practical difficulties when we try to use Hubble's law to measure galactic distances:

1. Galaxies do not obey Hubble's law perfectly. Hubble's law gives an exact distance only for a galaxy whose speed is determined solely by the expansion of the universe. In reality, nearly all galaxies experience gravitational tugs from other galaxies, and these tugs alter their speeds from the values predicted by Hubble's law.

2. Even when galaxies obey Hubble's law well, the distances we find with it are only as accurate as our best measurement of Hubble's constant.

The first problem is most serious for nearby galaxies. Within the Local Group, for example, Hubble's law does not work at all: The galaxies in the Local Group are gravitationally bound together with the Milky Way and therefore are *not* moving away from us in accord with Hubble's law. However, Hubble's law works well for more distant galaxies, which have recession velocities so great that any motions caused by the gravitational tugs of neighboring galaxies are tiny in comparison.

The second problem means that, even for distant galaxies, we can know only *relative* distances until we pin down the true value of H_0. For example, Hubble's law tells us that a galaxy moving away from us at 20,000 km/s is twice as far away as one moving at 10,000 km/s, but we can determine the actual distances of the two galaxies only if we know H_0.

The quest to measure Hubble's constant was one of the main missions of the Hubble Space Telescope.

The Hubble Space Telescope has helped us obtain an accurate value of H_0. Astronomers used the telescope to find Cepheid variables in galaxies out to about 60 million light-years and then used those distances to determine the luminosities of distant standard candles such as white dwarf supernovae. Plotting galactic distances measured with those distant standard candles against the velocities indicated by their redshifts has pinned down the value of H_0 to somewhere between 20 and 24 *kilometers per second per million light-years* (Figure 15.16). That is, a galaxy's speed away from us is between 20 and 24 km/s for every million light-years it is from us. For example, with this range of values for Hubble's constant, Hubble's law predicts that a galaxy located 100 million light-years away would be moving away from us at a speed between 2000 and 2400 km/s.

Distance Chain Summary Figure 15.17 summarizes the chain of measurements that allows us to determine ever greater distances. However, each link in the distance chain adds more uncertainty to the estimated distance. As a result, although we know the Earth–Sun distance to the estimated the beginning of the chain extremely accurately, distances to the farthest reaches of the observable universe remain uncertain by about 10%.

Hubble's Law

Page 420 shows Hubble's law written $v = H_0 \times d$. However, astronomers usually measure a galaxy's recession velocity v, then use the law to calculate the galaxy's distance d. To do this, it's more useful to divide both sides of the law by Hubble's constant H_0 to put it in the form

$$d = \frac{v}{H_0}$$

In this form, we can calculate a galaxy's distance in light-years if we have its measured velocity in kilometers per second and Hubble's constant in units of *kilometers per second per million light-years* (km/s/Mly).

Example: Estimate the distance to a galaxy whose red-shift indicates that it is moving away from us at a speed of 22,000 km/s. Assume that Hubble's constant is $H_0 = 22$ km/s/Mly.

Solution: Putting the given values into our formula, we find

$$d = \frac{v}{H_0} = \frac{22,000 \text{ km/s}}{22 \frac{\text{km/s}}{\text{Mly}}} = 1000 \text{ Mly}$$

The galaxy's distance is 1000 million light-years, which is the same as 1 billion light-years.

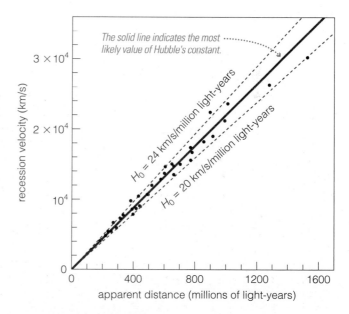

Figure 15.16 interactive figure

White dwarf supernovae can be used as standard candles to establish Hubble's law out to very large distances. The points on the graph show the apparent distances of white dwarf supernovae and the recession velocities of the galaxies in which they exploded. The fact that these points all fall close to a straight line demonstrates that these supernovae are good standard candles.

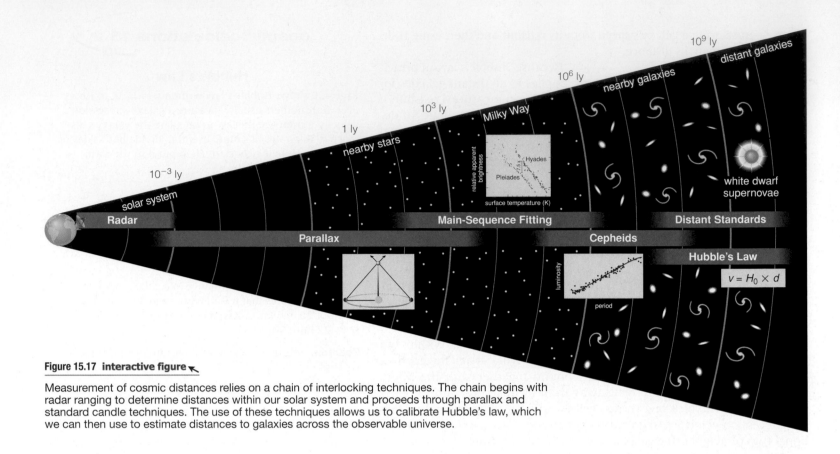

Figure 15.17 interactive figure

Measurement of cosmic distances relies on a chain of interlocking techniques. The chain begins with radar ranging to determine distances within our solar system and proceeds through parallax and standard candle techniques. The use of these techniques allows us to calibrate Hubble's law, which we can then use to estimate distances to galaxies across the observable universe.

- ***Radar ranging:*** We measure the Earth–Sun distance by bouncing radio waves off Venus and using some geometry.

- ***Parallax:*** We measure the distances to nearby stars by observing how their positions appear to change as Earth orbits the Sun. These distances rely on our knowledge of the Earth–Sun distance, determined with radar ranging.

- ***Main-sequence fitting:*** We know the distance to the Hyades star cluster in our Milky Way Galaxy through parallax. Comparing the apparent brightnesses of its main-sequence stars to those of stars in other clusters gives us the distances to these other star clusters in our galaxy.

- ***Cepheid variables:*** By studying Cepheids in star clusters with distances measured by main-sequence fitting, we learn the precise period–luminosity relation for Cepheids. When we find a Cepheid in a more distant star cluster or galaxy, we can determine its true luminosity by measuring the period between its peaks in brightness and then use this true luminosity to determine the distance.

- ***Distant standards:*** By measuring distances to relatively nearby galaxies with Cepheids, we learn the true luminosities of white dwarf supernovae and other distant standard candles, enabling us to measure great distances throughout the universe.

- ***Hubble's law:*** Distances measured to galaxies with white dwarf supernovae and other distant standards allow us to measure Hubble's constant, H_0. Once we know H_0, we can use Hubble's law to determine a galaxy's distance from its redshift.

• How do distance measurements tell us the age of the universe?

Hubble's law is a remarkably powerful tool for understanding the universe. Not only does it tell us that the universe is expanding and give us a way to measure galactic distances, but it also helps us determine the age and size of the observable universe. To see how, we must first consider the expansion of the universe in a little more detail.

Universal Expansion Galaxies all across the universe are moving away from one another, and this fact implies that galaxies must have been closer together in the past. Tracing this convergence back in time, we reason that all the matter in the observable universe started very close together and that the entire universe came into being at a single moment—the Big Bang [Section 1.1].

The expansion of the universe implies that the universe came into being at a single moment in time.

It may be tempting to think of the expanding universe as a ball of galaxies expanding into a void, but this impression is mistaken. To the best of our knowledge, the universe is not expanding into anything. As far as we can tell, there is no edge to the distribution of galaxies in the universe. On very large scales, the distribution of galaxies appears to be relatively smooth, meaning that the overall appearance of the universe around you would look more or less the same no matter where you are located. The idea that the matter in the universe is evenly distributed, without a center or an edge, is often called the **Cosmological Principle**. Although we cannot prove it to be true, it is consistent with all our observations of the universe [Section 16.3].

So how can the universe be expanding if it's not expanding into anything? Back in Chapter 1, we likened the expanding universe to a raisin cake baking [Section 1.3], but a cake has a center and edges that grow into empty space as it bakes. A better analogy is something that can expand but that has no center and no edges. The surface of a balloon can fit the bill, as can an infinite surface such as a sheet of rubber that extends to infinity in all directions.

Because it's hard to visualize infinity, let's use the surface of a balloon in our analogy to the expanding universe (Figure 15.18). Note that this analogy uses the balloon's two-dimensional surface to represent all three dimensions of space. The surface of the balloon represents the entire universe, and the spaces inside and outside the balloon have no meaning in this analogy. Aside from the reduced number of dimensions, the analogy works well because the balloon's spherical *surface* has no center and no edges. We can represent galaxies with plastic polka dots attached to the balloon, and we can make our model universe expand by inflating the balloon.

see it for yourself Make a one-dimensional model of the expanding universe with a rubber band and some paper clips. Cut the rubber band so you can stretch it out along a line, then attach four paper clips along it. Pin down the ends of the rubber band and measure the distances from one paper clip to each of the others. Then unpin one of the ends, stretch the rubber band more, pin it down again, and remeasure the distances. How much have they changed? How do your measurements illustrate Hubble's law?

The Age of the Universe We can now see how Hubble's law leads us to an age for the universe. Imagine that some miniature scientists are

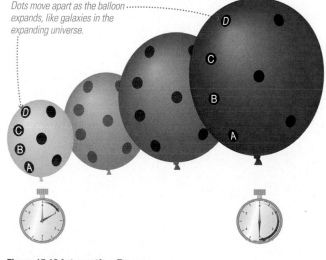

Dots move apart as the balloon expands, like galaxies in the expanding universe.

Figure 15.18 interactive figure

As the balloon expands, dots move apart in the same way that galaxies move apart in our expanding universe.

living on Dot B in Figure 15.18. Suppose that, 3 seconds after the balloon begins to expand, they measure the following:

Dot A is 3 cm away and moving at 1 cm/s.
Dot C is 3 cm away and moving at 1 cm/s.
Dot D is 6 cm away and moving at 2 cm/s.

They could summarize these observations as follows: *Every dot is moving away from our home with a speed that is 1 cm/s for each 3 cm of distance.* Because the expansion of the balloon is uniform, scientists living on any other dot would come to the same conclusion. Each scientist living on the balloon would determine that the following formula relates the distances and velocities of other dots on the balloon:

$$v = \left(\frac{1}{3\,\text{s}} \right) \times d$$

where v and d are the velocity and distance of any dot, respectively.

think about it Confirm that this formula gives the correct values for the speeds of Dots C and D, as seen from Dot B, 3 seconds after the balloon began expanding. How fast would a dot located 9 cm from Dot B move, according to the scientists on Dot B?

If the miniature scientists think of their balloon as a bubble, they might call the number relating distance to velocity—the term $\frac{1}{3\,\text{s}}$ in the preceding formula—the "bubble constant." An especially insightful miniature scientist might flip over the "bubble constant" and find that it is exactly equal to the time elapsed since the balloon started expanding. That is, the "bubble constant" $\frac{1}{3\,\text{s}}$ tells them that the balloon has been expanding for 3 seconds. Perhaps you see where we are heading.

Just as the inverse of the "bubble constant" tells the miniature scientists that their balloon has been expanding for 3 seconds, the inverse of the Hubble constant, or $\frac{1}{H_0}$, tells us something about how long our universe has been expanding. The "bubble constant" for the balloon depends on when it is measured, but it is always equal to 1 divided by the time elapsed since the balloon started expanding. Similarly, the Hubble constant actually changes with time but stays roughly equal to 1 divided by the age of the universe. We call it a constant because it is the same at all locations in the universe at a particular moment in time, and because its value does not change noticeably on the time scale of human civilization.

The rate at which the universe expands tells us how old it is—about 14 billion years.

Current estimates based on the value of Hubble's constant put the age of the universe between about 12 and 15 billion years. To derive a more precise value for the universe's age, we need to know whether the expansion has been speeding up or slowing down over time, a question we will examine more closely in Chapter 16. If the gravitational pull of each galaxy on every other galaxy has significantly slowed the expansion rate, then the universe's age is somewhat less than $\frac{1}{H_0}$. If some mysterious force has accelerated the expansion rate, then the universe's age is somewhat more than $\frac{1}{H_0}$. The best available evidence suggests that the universe is about 14 billion years old.

Lookback Time The expansion of the universe leads to a complication in discussing galaxy distances that we have ignored up to this point. Imagine observing a supernova in a distant galaxy. Suppose the supernova occurred 400 million years ago but we are only just now seeing it. The supernova's light must have traveled 400 million light-years to reach Earth, but this simple statement about how far the light has traveled does not translate easily into a distance for the galaxy in which the supernova occurred. The problem is that the universe is expanding, making the distance between Earth and the supernova greater today than it was at the time of the supernova event. If we simply say that the galaxy is "400 million light-years away," it's not clear whether we mean its distance now, its distance at the time of the supernova, or something in between.

An object's lookback time is the time it took for the object's light to reach us. Because distances between galaxies are always changing, it is easier to speak about faraway objects in terms of how much time their light takes to reach us—400 million years in the case of the supernova. We call this the **lookback time** to the supernova. In other words, a distant object's lookback time is the difference between the current age of the universe and the age of the universe when the light left the object. Unlike a statement about distance, the meaning of a lookback time is unambiguous: If the lookback time is 400 million years, it means the light traveled through space for a period of 400 million years to reach us.

An object's lookback time is directly related to its redshift. Recall that the redshifts of galaxies tell us how quickly they are moving away from us. In the context of an expanding universe, redshifts have an additional, more fundamental interpretation. Let's return again to the balloon universe. Suppose you draw wavy lines on the balloon's surface to represent light waves. As the balloon inflates, these wavy lines stretch out, and their wavelengths increase (Figure 15.19). This stretching closely resembles what happens to photons in an expanding universe. The expansion of the universe stretches out all the photons within it, shifting them to longer, redder wavelengths. We call this effect a **cosmological redshift.**

In a sense, we have a choice when we interpret the redshift of a distant galaxy: We can think of the redshift either as being caused by the Doppler effect as the galaxy moves away from us or as being a cosmological redshift due to the photon-stretching expansion of the universe. However, as we look to very distant galaxies, the ambiguity in the meaning of *distance* also makes it difficult to specify precisely what we mean by a galaxy's *speed*. Thus, it becomes preferable to interpret the redshift as being due to photon stretching in an expanding universe.

From this perspective, it is better to think of space itself as expanding, carrying the galaxies along for the ride, than to think of the galaxies as projectiles flying through a static universe. The cosmological redshift of a galaxy thus tells us how much space has expanded during the time since light from the galaxy left on its journey to us.

The Horizon of the Universe When we began our discussion of the expanding universe, we stressed that the universe as a whole does not seem to have an edge. However, the universe does have a *horizon* beyond which we cannot see. The **cosmological horizon** that marks the limits of the observable universe is a boundary in time, not in space.

The size of the observable universe is determined by the age of the universe. Our quest to study galaxies and measure their distances has brought us to the very limits of the observable universe. As we discussed in Chapter 1, the age of the universe limits the size of our observable universe (see Figure 1.4): In a universe that is 14 billion years old, we cannot see any object with a lookback time greater than 14 billion years.

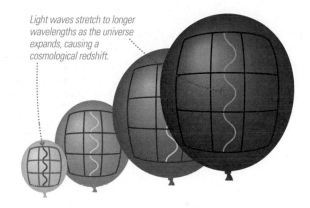

Light waves stretch to longer wavelengths as the universe expands, causing a cosmological redshift.

Figure 15.19 interactive figure

As the universe expands, photon wavelengths stretch like the wavy lines on this expanding balloon.

common misconceptions

Beyond the Horizon

Perhaps you're thinking there must be something beyond the cosmological horizon. After all, we can see farther and farther with each passing second. The new matter we see had to come from somewhere, didn't it? The problem with this reasoning is that the cosmological horizon, unlike a horizon on Earth, is a boundary in time, not in space.

At any one moment, in whatever direction we look, the cosmological horizon lies at the beginning of time and encompasses a certain volume of the universe. At the next moment, the cosmological horizon still lies at the beginning of time, but it encompasses a slightly larger volume. When we peer into the distant universe, we are looking back in both space and time. We cannot look "past the horizon" because we cannot look back to a time before the universe began.

15.3 Galaxy Evolution

Now that we understand how galactic distances are measured, we are ready to turn our attention back to understanding galaxies themselves. We understand less about the formation and development of galaxies, or **galaxy evolution**, than we do about the lives of stars. Nevertheless, we've made great progress in understanding galaxy evolution in recent decades, thanks largely to the same powerful telescopes that have allowed us to measure distances more accurately.

• How do we observe the life histories of galaxies?

Powerful telescopes can be used like time machines to observe the life histories of galaxies—the farther we look into the universe, the further back we look in time [Section 1.1]. The most distant galaxies already had some stars in place by about 13 billion years ago, which means these stars match the ages of the oldest stars in our own galaxy. It therefore seems safe to assume that most galaxies began to form at about this time, in which case most galaxies would be roughly the same age today. A galaxy's lookback time is therefore linked directly to its age. Most of the galaxies we see nearby are around 13 billion years old. Observations of galaxies with a lookback time of 10 billion years show us what they were like when they were no more than 3 billion years old.

Images of the deep universe allow us to study galaxies at many different distances and therefore many different ages. The linkage between a galaxy's distance and its age gives us a remarkable ability: Simply by photographing galaxies at different distances, we can assemble "family albums" showing them in different stages of development. Pictures of the farthest galaxies show them in their childhood, and pictures of the nearest show mature galaxies as they are today. Figure 15.20 shows partial family albums for elliptical, spiral, and irregular galaxies. Each individual photograph shows a single galaxy at a single stage in its life. Grouping these photographs by galaxy type allows us to see how galaxies of a particular type have changed through time.

think about it In the preceding paragraph, we use the term *today* in a very broad sense. For example, if we look at a relatively nearby galaxy—one located, say, 20 million light-years away—we see it as it was 20 million years ago. In what sense is this "today"? (*Hint:* How does 20 million years compare to the age of the universe? To the lifetime of a massive star?)

The photographs in Figure 15.20, taken by the Hubble Space Telescope, show galaxies extending back to a time when the universe was just 1 or 2 billion years old. However, the first stars and galaxies appear to have formed even earlier than this. Observing these first stars and galaxies is a challenge not even Hubble can meet. Detecting such faraway galaxies will require larger telescopes, and the extreme redshifts expected for such galaxies mean the telescopes will need to be particularly sensitive to infrared light. Within a decade, NASA hopes to launch a much larger, infrared-sensitive successor to the Hubble Space Telescope (called the James Webb Space Telescope), but for now we have little direct information about galaxy birth.

• How did galaxies form?

Because our telescopes cannot yet see back to the birth of galaxies, we must use theoretical models to learn about galaxy formation. The most successful models assume the following:

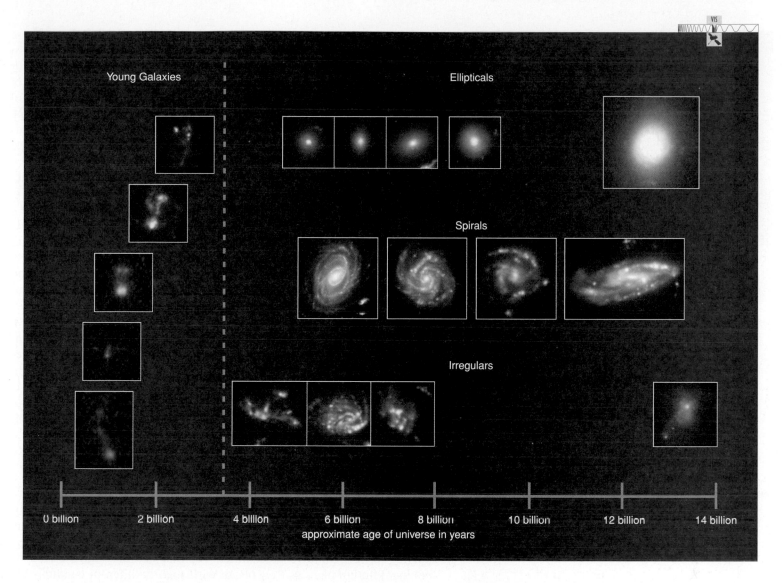

Young Galaxies

Ellipticals

Spirals

Irregulars

0 billion   2 billion   4 billion   6 billion   8 billion   10 billion   12 billion   14 billion

approximate age of universe in years

- Hydrogen and helium gas filled all of space more or less uniformly when the universe was very young—say, in the first million years after its birth.

- The distribution of matter in the universe was not perfectly uniform—certain regions of the universe started out slightly more dense than others.

Beginning with these assumptions, which a mounting body of evidence supports [Section 17.2], we can model galaxy formation using well-established laws of physics to trace how the denser regions in the early universe grew into galaxies (Figure 15.21). The models show that the regions of enhanced density originally expanded along with the rest of the universe. However, the slightly greater pull of gravity in these regions gradually slowed their expansion. Within about a billion years, the expansion of these denser regions halted and reversed, and the material within them began to contract into *protogalactic clouds* like the cloud of matter that eventually formed our Milky Way [Section 14.3].

Our most successful models of galaxy formation suggest that protogalactic clouds formed in regions of slightly enhanced density in the early universe.

According to the models, the protogalactic clouds that eventually formed spiral galaxies initially cooled as they contracted, radiating away their thermal energy, and the

Figure 15.20

Family albums for elliptical, spiral, and irregular galaxies of different ages, plus some very young galaxies shown on the far left. These photos are all zoomed-in images of galaxies shown in Figure 15.1. We see more distant galaxies as they were when they were younger, indicated by the approximate age of the universe along the horizontal axis. The younger galaxies appear smaller because they are more distant.

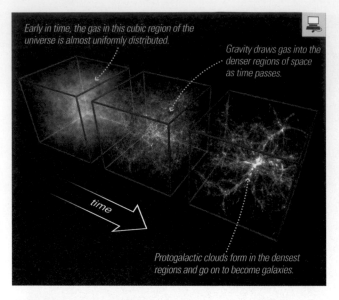

Early in time, the gas in this cubic region of the universe is almost uniformly distributed.

Gravity draws gas into the denser regions of space as time passes.

time

Protogalactic clouds form in the densest regions and go on to become galaxies.

Figure 15.21 interactive figure

A computer simulation of the formation of protogalactic clouds. The simulated region of space is about 500 million light-years wide and will go on to form numerous galaxies.

first generation of stars grew from the densest, coldest clumps of gas. These first-generation stars were probably quite massive, living and dying within just a few million years—a short time compared to the time required for the collapse of a protogalactic cloud into a spiral disk. The supernovae of these massive stars seeded each collapsing cloud with its first sprinkling of heavy elements and generated shock waves that heated the surrounding interstellar gas. This heating slowed the collapse of the protogalactic clouds and the rate at which stars formed within them, allowing time for the rest of the gas in each newly forming galaxy to settle into a rotating disk.

This picture explains many of the basic features of galaxies. In particular, it explains why other spiral galaxies have the same basic structure as the Milky Way: a *disk population* of stars that orbit the galactic center in a fairly flat plane, and a *spheroidal population* of stars with more randomly oriented orbits (see Figure 14.2). The spheroidal population consists of stars that were born before the galaxy's rotation became organized, which is why they have orbits oriented in many different ways around the galactic center. The disk population consists of stars born after the gas in the protogalactic cloud settled into a rotating disk, which is why they all have similar orbits around the center of the galaxy (see Figure 14.18).

However, this basic picture leaves at least two major questions unanswered. First, our models assume that galaxies formed in regions of slightly enhanced density, but they do not tell us where these density enhancements came from. The origin of density enhancements in the early universe is one of the major puzzles in astronomy; we'll revisit it in Chapters 16 and 17. Second, our basic picture explains the origin of spiral galaxies quite well, but it does not tell us why some galaxies are elliptical and others irregular. That is the question to which we turn our attention next.

• Why do galaxies differ?

All protogalactic clouds are thought to have begun in the same basic way, with gravity collecting matter around a pre-existing density enhancement. How, then, did galaxies end up in the different types we see today? Our models suggest two possibilities: (1) Galaxies may have ended up looking different because they began with slightly different conditions in their protogalactic clouds, or (2) galaxies may have begun their lives similarly but later changed due to interactions with other galaxies. To help differentiate the role of conditions in the protogalactic cloud from that of subsequent interactions, we will explore a single key question: Why do spiral galaxies have gas-rich disks, while elliptical galaxies do not?

Conditions in the Protogalactic Cloud Two plausible explanations for the differences between spiral galaxies and elliptical galaxies trace a galaxy's type back to the protogalactic cloud from which it formed:

- *Protogalactic spin.* A galaxy's type might be determined by the spin of the protogalactic cloud from which it formed. If the original cloud had a significant amount of angular momentum, it would have rotated quickly as it collapsed. The galaxy it produced would therefore have tended to form a disk, and the resulting galaxy would be a spiral. If the protogalactic cloud had little or no angular momentum, its gas might not have formed a disk at all, and the resulting galaxy would be elliptical.

- *Protogalactic density.* A galaxy's type might be determined by the density of the protogalactic cloud from which it formed. A protogalactic cloud with relatively high gas density would have radiated energy more effectively and cooled more quickly, thereby allowing more rapid star formation. If the star formation proceeded fast enough, all the gas could have been turned into stars before any of it had time to settle into a disk, making it an elliptical galaxy. In contrast, a lower-density cloud would have formed stars more slowly, leaving plenty of gas to form the disk of a spiral galaxy.

Elliptical galaxies may have formed from protogalactic clouds that were spinning more slowly or were denser than those that formed spiral galaxies.

Evidence for the latter scenario comes from a few giant elliptical galaxies at very great distances. These galaxies look very red even after we have accounted for their large redshifts. They apparently have no blue or white stars at all, indicating that new stars no longer form within these galaxies—even though we are seeing them as they were when the universe was only a few billion years old. This finding suggests that all the stars in these elliptical galaxies were born at almost the same time, which is consistent with the idea that all the stars formed before a disk could develop.

Galactic Collisions The two previous scenarios, in which the formation of a gas-rich disk depends on the angular momentum or density of the protogalactic cloud, probably describe important parts of the overall story. However, they ignore one key fact: Galaxies rarely evolve in isolation.

Think back to our scale model solar system in Chapter 1. Using a scale on which the Sun was the size of a grapefruit, the nearest star was like another grapefruit a few thousand kilometers away. Because the average distances between stars are so huge compared to the sizes of stars, collisions between stars are extremely rare. However, if we rescale the universe so that our *galaxy* is the size of a grapefruit, the Andromeda Galaxy is like another grapefruit only about 3 meters away, and a few smaller galaxies lie considerably closer. The average distances between galaxies are not much larger than the sizes of galaxies, meaning that collisions between galaxies are inevitable.

Galaxy collisions are spectacular events that unfold over hundreds of millions of years (Figure 15.22). During our short lifetimes, we can see only a snapshot of a collision in progress, distorting the shapes of the colliding galaxies. Collisions must have been even more frequent in the distant past, when the universe was smaller and galaxies were closer together. Observations confirm that distorted-looking galaxies—probably galaxy collisions in progress—were more common in the early universe than they are today.

Computer models show that a collision between two spiral galaxies can form an elliptical galaxy.

We can learn much more about galactic collisions with the aid of computer simulations that allow us to "watch" collisions that in nature take hundreds of millions of years to unfold. These models show that a collision between two spiral galaxies can create an elliptical galaxy (Figure 15.23). Tremendous tidal forces between the colliding galaxies tear apart the two disks, randomizing the orbits of their stars. Meanwhile, a large fraction of their gas sinks to the center of the collision and rapidly forms new stars. Supernovae and stellar winds eventually blow away the rest of the gas. When the cataclysm finally settles down, the merger of the two

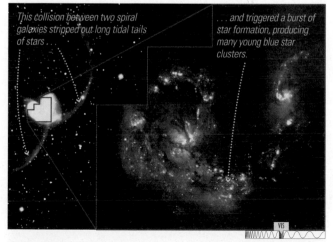

This collision between two spiral galaxies stripped out long tidal tails of stars . . .

. . . and triggered a burst of star formation, producing many young blue star clusters.

Figure 15.22

A pair of colliding spiral galaxies known as the Antennae (or NGC 4038/4039). The image taken from the ground (left) reveals their vast tidal tails, while the Hubble Space Telescope image (right) shows the burst of star formation at the center of the collision.

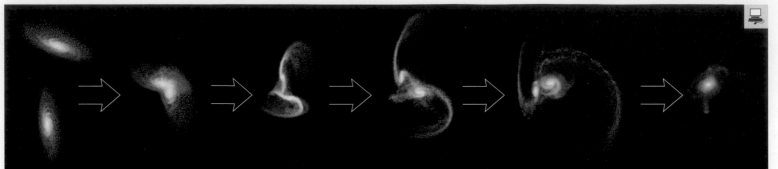

Two simulated spiral galaxies approach each other on a collision course.

The first encounter begins to disrupt the two galaxies and sends them into orbit around each other.

As the collision continues, much of the gas in the disk of each galaxy collapses toward the center.

Gravitational forces between the two galaxies tear out long streamers of stars called tidal tails.

The centers of the two galaxies approach each other and begin to merge.

The single galaxy resulting from the collision and merger is an elliptical galaxy surrounded by debris.

Figure 15.23 interactive figure

Several stages in a supercomputer simulation of a collision between two spiral galaxies that results in an elliptical galaxy. At least some of the elliptical galaxies in the present-day universe formed in this way. The whole sequence spans about 1.5 billion years.

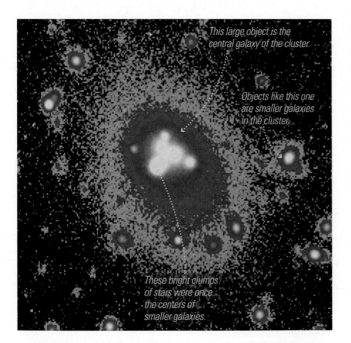

This large object is the central galaxy of the cluster.

Objects like this one are smaller galaxies in the cluster.

These bright clumps of stars were once the centers of smaller galaxies.

Figure 15.24

The central dominant galaxy of the cluster Abell 3827 contains multiple clumps of stars that probably once were the centers of individual galaxies that it consumed. The colors in this image are not real but instead represent levels of brightness, with yellow representing the brightest clumps.

spirals has produced a single elliptical galaxy. Little gas is left for a disk, and the orbits of the stars have random orientations.

Observations support the idea that at least some elliptical galaxies result from collisions and subsequent mergers. Elliptical galaxies dominate the galaxy populations at the cores of dense clusters of galaxies, where collisions should be most frequent. This fact may mean that any spirals once present became ellipticals through collisions.

Stronger evidence comes from structural details of elliptical galaxies, which often attest to a violent past. Some elliptical galaxies have stars and gas clouds with orbits suggesting that they are leftover pieces of galaxies that merged in a past collision. Others are surrounded by shells of stars that may have formed in gas stripped out of a galaxy during a past collision.

Observations show that at least some elliptical galaxies have experienced past collisions.

The most decisive evidence that collisions affect the evolution of elliptical galaxies comes from observations of the **central dominant galaxies** found at the centers of many dense clusters. Central dominant galaxies are giant elliptical galaxies that apparently grew to a huge size by consuming other galaxies through collisions. They frequently contain several tightly bound clumps of stars that probably were the centers of individual galaxies before being swallowed (Figure 15.24). This process of *galactic cannibalism* can create central dominant galaxies more than 10 times as massive as the Milky Way, making them the most massive galaxies in the universe.

Starbursts   We have discussed two general ways in which elliptical galaxies may have formed: (1) as a result of birth in an unusually slow-rotating or high-density protogalactic cloud, and (2) as a result of later collisions and mergers of spiral galaxies. The first scenario makes it easy to account for the lack of hot, young stars and ongoing star formation in elliptical galaxies, since all the cool gas needed for star formation is used up before a spiral disk can form. But how can we account for the lack of young stars and cool gas in ellipticals that formed from more recent collisions?

Computer models suggest that galaxy collisions should ignite huge bursts of rapid star formation—so rapid that all the cool gas would be quickly transformed into stars, leaving little left over to form a disk.

Observational support for this idea comes from a relatively small number of galaxies that appear to be in the midst of such bursts of star formation. We call these objects **starburst galaxies**.

Most galaxies with ongoing star formation appear comfortably settled, steadily forming new stars at modest rates. The Milky Way Galaxy produces an average of about one new star per year. At this rate, the Milky Way won't exhaust the interstellar gas in its disk until long after the Sun has died. Starburst galaxies are not so thrifty. Some form new stars at rates exceeding 100 stars per year, more than 100 times the star formation rate in the Milky Way. These galaxies cannot possibly sustain such a rapid pace of star formation for long. At their current rates of star formation, starburst galaxies would consume *all* their interstellar gas in just a few hundred million years. Because this is a relatively short time compared to the billions of years since galaxies first formed, a starburst must be only a temporary phase in a galaxy's life.

> Starburst galaxies are forming stars so quickly that they will run out of star-forming clouds in just a few hundred million years.

Such a high rate of star formation means that there will also be a high incidence of supernovae. The bubbles of hot gas blown out by these many individual supernovae merge into a gigantic bubble so large that it rapidly bursts through any gaseous disk. The hot gas then erupts outward into intergalactic space, creating a **galactic wind** (Figure 15.25).

Many of the most luminous starbursts do indeed appear to be occurring in colliding galaxies. For example, the colliding galaxy pair shown in Figure 15.22 is undergoing a starburst. Starbursts may therefore be the answer to the mystery of why elliptical galaxies lack young stars and cool gas. The starburst uses up most of the cool gas, the galactic wind blows away what remains, and all the hot, massive stars die out within just a few hundred million years after the starburst ends. By the time the collision ends and the merger into an elliptical galaxy is complete, there is simply no cool gas left to support ongoing star formation.

The causes of smaller-scale starbursts are not yet clear. At least some irregular galaxies look irregular because they are currently undergoing collisions and starbursts, but not all irregular galaxies are colliding. For example, the Large Magellanic Cloud is an irregular galaxy that is also undergoing a period of rapid star formation. The starburst leading to this galaxy's irregular appearance might have been triggered by a close encounter with the Milky Way. No matter what their exact causes turn out to be, starbursts clearly represent an important piece in the overall puzzle of galaxy evolution.

 Black Holes Tutorial, Lessons 1, 2

15.4 Quasars and Other Active Galactic Nuclei

Starbursts may be spectacular, but some galaxies display even more incredible phenomena: extreme amounts of radiation, and sometimes powerful jets of material, emanating from deep in their centers (Figure 15.26). These unusually bright galactic centers are usually called **active galactic nuclei**. The most luminous active galactic nuclei are known as **quasars**, which in some cases produce more light than 1000 galaxies the size of the Milky Way.

a This visible light photograph from the Subaru telescope shows violently disturbed gas (red) blowing out both above and below the disk.

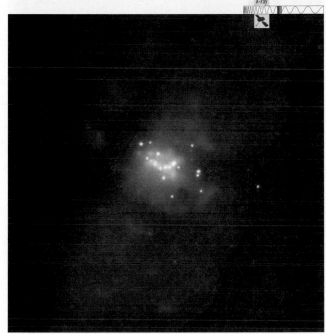

b This X-ray image from the Chandra X-Ray Observatory shows the same region as the visible-light photograph in (a). The reddish region represents X-ray emission from very hot gas blowing out of the disk. The bright dots in the galactic disk probably represent X-ray emission from accretion disks around black holes or neutron stars produced by recent supernovae.

Figure 15.25 interactive photo

Visible-light and X-ray views of a starburst galaxy called M82 and its galactic wind. Both images show the same region, which is about 16,000 light-years across.

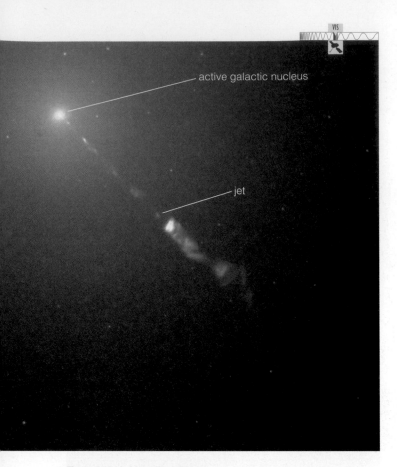

active galactic nucleus

jet

Figure 15.26 interactive photo

The active galactic nucleus in the elliptical galaxy M87. The bright yellow spot is the active nucleus, and the blue streak is a jet of particles shooting outward from the nucleus at nearly the speed of light.

The glory days of quasars are long past. We find quasars primarily at great distances, telling us that these blazingly luminous objects were most common billions of years ago, when galaxies were in their youth. Because we find no nearby quasars and relatively few nearby galaxies with any type of active galactic nucleus, we conclude that the objects that shine as quasars in young galaxies must become dormant as the galaxies age. Many nearby galaxies that now look quite normal must have centers that once shone brilliantly as quasars.

• What are quasars?

What could possibly be the source of the incredible luminosities of quasars, and why did quasars fade away? A growing body of evidence points to a single answer: The energy output of quasars comes from gigantic accretion disks surrounding **supermassive black holes** with masses millions to billions of times that of our Sun. Before we study how these incredible powerhouses work, let's investigate the evidence that points to their existence.

The Discovery of Quasars   In the early 1960s, a young professor at the California Institute of Technology named Maarten Schmidt was busy identifying cosmic sources of radio-wave emission. Radio astronomers would tell him the coordinates of newly discovered radio sources, and he would try to match them with objects seen through visible-light telescopes. Usually the radio sources turned out to be normal-looking galaxies, but one day he discovered a major mystery: A radio source called 3C273 looked like a blue star through a telescope but had strong emission lines at wavelengths that did not appear to correspond to any known chemical element.

After months of puzzlement, Schmidt suddenly realized that the emission lines were not coming from an unfamiliar element at all. Instead, they were emission lines of hydrogen that were hugely redshifted from their normal wavelengths. Schmidt calculated that the expansion of the universe was carrying 3C273 away from us at 17% of the speed of light. He then computed 3C273's distance using Hubble's law, and from its distance and apparent brightness he estimated its luminosity [Section 11.1]. What he found was astonishing: 3C273 has a luminosity of about 10^{39} watts, or well over a trillion (10^{12}) times that of our Sun—making it hundreds of times more luminous than the entire Milky Way Galaxy. Discoveries of similar but even more distant objects soon followed. Because the first few of these objects were strong sources of radio emission that looked like stars through visible-light telescopes, they were named "quasi-stellar radio sources," or *quasars* for short. Later, astronomers learned that most quasars are not such powerful radio emitters, but the name has stuck.

Quasars can be more than a trillion times as luminous as the Sun.

For many years, a debate raged among astronomers over whether Hubble's law could really be used to determine quasar distances. Some argued that quasars might have high redshifts for reasons other than distance and therefore might be much nearer to us than Hubble's law suggested. The vast majority of astronomers now consider this debate settled. Improved images show that quasars are indeed the centers of extremely distant galaxies and often are members of very distant galaxy clusters (Figure 15.27).

Most quasars lie more than halfway to the cosmological horizon. The lines in typical quasar spectra are shifted to more than three times their rest wavelengths, which tells us that the light from these quasars emerged when the universe was less than a third of its present age. The light we see from the farthest known quasars began its journey when the universe was less than 1 billion years old.

Evidence from Nearby Active Galactic Nuclei

Quasars are difficult to study in detail because they are so far away. Luckily, some quasarlike objects are much closer to home. About 1% of nearby galaxies have active galactic nuclei that look very much like quasars, except that they are less powerful. (These galaxies are often called *Seyfert galaxies* after astronomer Carl Seyfert, who in 1943 grouped galaxies with active galactic nuclei into a special class.)

The light-emitting regions of active galactic nuclei are so small that even the sharpest images do not resolve them. Our best visible-light images show only that active galactic nuclei must be smaller than 100 light-years across. Radio-wave images made with the aid of *interferometry* [Section 5.3] show that these nuclei are even smaller: less than 3 light-years across. Rapid changes in the luminosities of some active galactic nuclei point to an even smaller size.

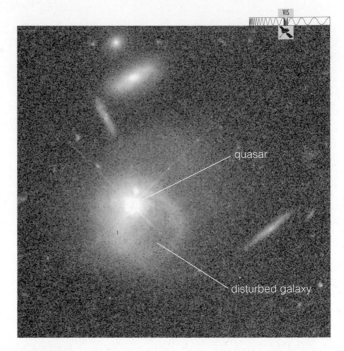

Figure 15.27

Photographs like this one help confirm that quasars are extremely far away. The quasar not only appears to be located in the middle of a distant galaxy but also has the same redshift as the rest of the galaxy and the other galaxies in the photo. The fact that the quasar has the same redshift implies that it really is the extremely luminous center of a very distant galaxy.

The immense luminosities of quasars and other active galactic nuclei come from regions no larger than our solar system.

To understand how variations in luminosity tell us about an object's size, imagine that you are a master of the universe and you want to signal one of your fellow masters a billion light-years away. An active galactic nucleus would make an excellent signal beacon, because it is so bright. However, suppose the smallest nucleus you can find is 1 light-year across. Each time you flash it on, the photons from the front end of the source reach your fellow master a full year before the photons from the back end. If you flash it on and off more than once a year, your signal will be smeared out. Similarly, if you find a source that is 1 light-day across, you can transmit signals that flash on and off no more than once a day. If you want to send signals just a few hours apart, you need a source no more than a few light-hours across.

Occasionally, the luminosity of an active galactic nucleus doubles in a matter of hours. The fact that we see a clear signal indicates that the source must be less than a few light-hours across. In other words, *the incredible luminosities of active galactic nuclei and quasars are apparently being generated in a volume of space not much bigger than our solar system.*

Radio Galaxies and Jets

Another clue to the nature of quasars came in the early 1950s, a decade before their discovery. Radio astronomers noticed that certain galaxies, now called **radio galaxies**, emit unusually strong radio waves. Upon closer inspection, they learned that much of the radio emission comes not from the galaxies themselves but rather from pairs of huge *radio lobes*, one on either side of the galaxy. Today, radio telescopes resolve the structure of radio galaxies in vivid detail (Figure 15.28).

Active galactic nuclei in radio galaxies can propel huge jets of particles moving at nearly the speed of light.

At the center of a radio galaxy we see emission from an active galactic nucleus less than a few light-years across. We often see two gigantic **jets** of plasma shooting out of the active galactic nucleus in opposite directions. Using time-lapse radio images taken several years apart, we can track the motions of various plasma blobs in the jets. Some of these

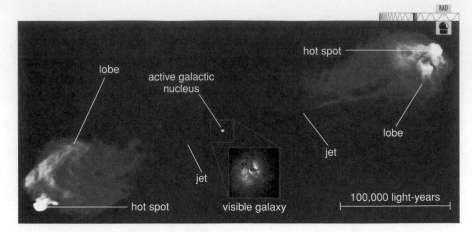

Figure 15.28

This image, made with the Very Large Array in New Mexico, shows radio-wave emission from the radio galaxy Cygnus A. Brighter regions represent stronger radio emission. Notice that the strongest emission comes from two radio lobes that lie far beyond the bounds of the galaxy that we see with visible light (inset photo), but the lobes are clearly connected to the central active galactic nucleus by two long jets of particles.

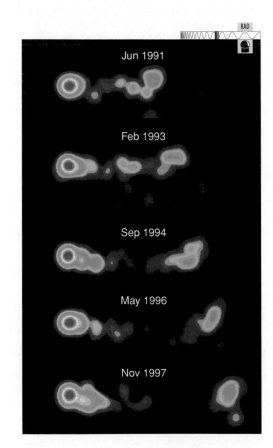

Figure 15.29

These radio images, taken over a period of several years, show a blob of plasma moving at almost the speed of light in a jet extending from the quasar 3C279; the quasar is at the far left in each image.

blobs move at close to the speed of light. The lobes lie at the ends of the jets, which are sometimes as far as a million light-years from the galactic center.

We now suspect that quasars and radio galaxies are the same types of objects viewed in slightly different ways. We have discovered that many quasars have jets and radio lobes like those seen in radio galaxies (Figure 15.29). Moreover, the active galactic nuclei of many radio galaxies seem to be concealed beneath donut-shaped rings of dark molecular clouds (Figure 15.30). Such structures may look like quasars when they are oriented so that we can see the active galactic nucleus at the center and look like the nuclei of radio galaxies when the ring of dusty gas dims our view of the central object. (A subset of active galactic nuclei called *BL Lac objects* probably represents the centers of radio galaxies whose jets happen to point directly at us.)

• What is the power source for quasars and other active galactic nuclei?

Astronomers have worked hard to envision physical processes that might explain how radio galaxies, quasars, and other active galactic nuclei release so much energy within such small central volumes. Only one explanation seems to fit: The energy comes from matter falling into a supermassive black hole. The gravitational potential energy of matter falling toward the black hole is converted into kinetic energy, and collisions between infalling particles convert the kinetic energy into thermal energy. The resulting heat causes this matter to emit the intense radiation we observe. As in X-ray binaries, we expect that the infalling matter swirls through an accretion disk before it disappears within the event horizon of the black hole [Section 13.3].

Quasars are probably powered by matter falling into supermassive black holes. If the supermassive black hole model is correct, it should be able to explain the major observed features of quasars and other active galactic nuclei. In particular, it should explain their extreme luminosities, the fact that they emit radiation over a broad range of wavelengths from radio waves to X rays, and the presence of their powerful jets.

Explaining the extreme luminosities of quasars was the main motivation for the supermassive black hole model. Matter falling into a black hole can generate awesome amounts of energy. During its fall to the event horizon of a black hole, as much as 10–40% of the mass-energy ($E = mc^2$) of a chunk of matter can be converted into thermal energy and ultimately to radiation. Accretion by black holes can therefore produce light far more efficiently than nuclear fusion, which converts less than 1% of mass-energy into photons. Remember that the light is coming not from the black hole itself but rather from the hot gas in the accretion disk.

The environment surrounding a supermassive black hole explains why active galactic nuclei emit light across such a broad wavelength range. Hot gas in and above the accretion disk produces enormous amounts of ultraviolet and X-ray photons. This radiation ionizes surrounding interstellar gas, creating ionization nebulae that emit visible light. (The emission lines produced by these nebulae are the same ones Maarten Schmidt used to measure the first quasar redshifts.) Dust grains in the molecular clouds that encircle the active galactic nucleus (see Figure 15.30) absorb high-energy light and reemit it as infrared light. Finally, the fast-moving electrons that we sometimes see jetting from these nuclei at nearly the speed of light can produce the radio emission from active galactic nuclei.

The powerful jets emerging from active galactic nuclei are more difficult to explain, but certainly there is plenty of energy available for flinging material outward at nearly the speed of light. One plausible model for jet production relies on the twisted magnetic fields thought to accompany accretion disks (Figure 15.31). As an accretion disk spins, it pulls the magnetic field lines that thread it around in circles. Charged particles fly outward along the field lines like beads on a twirling string, forming a jet that shoots out into space.

On the whole, the supermassive black hole model seems to explain the major observed features of quasars and other active galactic nuclei. However, several important mysteries remain unsolved. For example, we do not yet know why quasars eventually run out of gas to accrete and stop shining so brightly. We also do not know how those black holes formed in the first place. Nevertheless, the hypothesis that gigantic black holes are responsible for quasars, nearby active galactic nuclei, and radio galaxies is so far withstanding the tests of time and thousands of observations.

• Do supermassive black holes really exist?

The idea that such monster black holes exist has been hard for some astronomers to swallow. Proving that we have found a black hole is tricky. Black holes themselves do not emit any light, so we need to infer their existence from the ways in which they alter their surroundings. In the vicinity of a black hole, matter should be orbiting at high speed around something invisible. We have already examined the evidence for a black hole at the center of our Milky Way [Section 14.4], but what about other galaxies?

Hunting for Supermassive Black Holes Detailed observations of matter orbiting at the centers of nearby galaxies suggest that supermassive black holes are quite common. In fact, it is possible that *all* galaxies might contain supermassive black holes at their centers. One prominent example is the relatively nearby galaxy M87, which features a bright, active galactic nucleus and a jet that emits both radio and visible light

Figure 15.30

Artist's conception of the central few hundred light-years of a radio galaxy. The active galactic nucleus, obscured by a ring of dusty molecular clouds, lies at the point from which the jets emerge. If viewed along a direction closer to the jet axis, the active nucleus would not be obscured and would look more like a quasar.

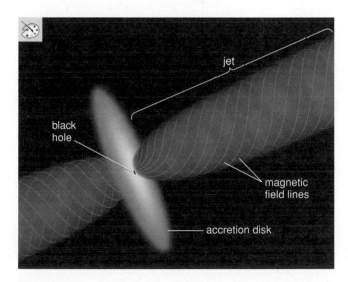

Figure 15.31

This schematic drawing illustrates one model that might explain how supermassive black holes create jets. The model relies on the magnetic field lines thought to thread the accretion disk surrounding a black hole. As the accretion disk spins, it twists the magnetic field lines. Charged particles at the accretion disk's surface can then fly outward along the twisted magnetic field lines.

Figure 15.32

This Hubble Space Telescope photo shows gas near the center of the galaxy M87, and the graph shows Doppler shifts of spectra from gas 60 light-years from the center on opposite sides (the circled regions in the photo). The Doppler shifts tell us that gas is orbiting around the galactic center, and precise measurements tell us that orbital speed is about 800 km/s. From the orbital speed and Newton's laws, we find that the central object must have a mass 2–3 billion times that of the Sun.

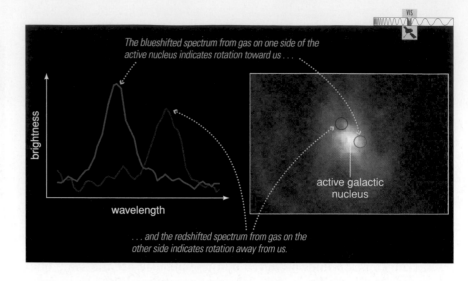

The blueshifted spectrum from gas on one side of the active nucleus indicates rotation toward us . . .

brightness

wavelength

active galactic nucleus

. . . and the redshifted spectrum from gas on the other side indicates rotation away from us.

(see Figure 15.26). It was already a prime black hole suspect when astronomers pointed the Hubble Space Telescope at its core (Figure 15.32). The spectra they gathered showed blueshifted emission lines on one side of the nucleus and redshifted emission lines on the other. This pattern of Doppler shifts is the characteristic signature of orbiting gas: On one side of the orbit the gas is coming toward us and hence is blueshifted, while on the other side it is moving away from us and is redshifted. The magnitude of these Doppler shifts shows that the gas, located up to 60 light-years from the center, is orbiting an unseen object at a speed of hundreds of kilometers per second. This high-speed orbital motion indicates that the central object has a mass some 2–3 billion times that of our Sun.

> Observations of rapidly orbiting material at the centers of galaxies indicate that at least some galaxies, and perhaps all of them, harbor supermassive black holes.

Observations of NGC 4258, another galaxy with a visible jet, have delivered even more persuasive evidence. A ring of molecular clouds orbits the nucleus of this galaxy in a circle less than 1 light-year in radius. We can pinpoint these clouds because they amplify the microwave emission lines of water molecules, generating beams of microwaves similar to laser beams. The Doppler shifts of these emission lines allow us to determine the orbits of the clouds very precisely. Their orbital motion tells us that the clouds are circling a single, invisible object with a mass of 36 million solar masses.

As the search for black holes in active galactic nuclei progresses, we are continuing to find examples like these. In each case, a supermassive black hole seems to be the only explanation for the enormous orbital speeds. We may never be 100% certain that these objects are indeed giant black holes. The best we can do is rule out all other possibilities. However, a supermassive black hole is the only thing we know of that could be so massive while remaining unseen.

Black Holes and Galaxy Formation Evidence for supermassive black holes is also found in galaxies whose centers are not currently active. The masses of those black holes follow a very interesting pattern:

The mass of the black hole at the center of a galaxy appears to be closely related to the properties of the galaxy's spheroidal component. Detailed studies of the orbital speeds of stars and gas clouds in the centers of nearby galaxies show that the mass of the central black hole is typically about $\frac{1}{500}$ of the mass of the galaxy's bulge (Figure 15.33). Because this relationship holds for galaxies with a wide range of properties, from small spiral galaxies with a bulge mass of less than $10^8 M_{Sun}$ to giant elliptical galaxies whose spheroidal component exceeds $10^{11} M_{Sun}$, we conclude that the growth of a central black hole must be closely linked with the process of galaxy formation.

Astronomers have long suspected that galaxy evolution goes hand in hand with the formation of supermassive black holes because quasars were so much more common early in time, when galaxies were growing rapidly. Unfortunately, we do not yet know how that process works. Some scientists have suggested that the black holes formed first out of gas at the center of a protogalactic cloud and that their energy output regulated the growth of the galaxy around them. Other scientists have suggested that clusters of neutron stars resulting from extremely dense starbursts at the centers of young galaxies might have somehow coalesced to form an enormous black hole. For now, the origins of supermassive black holes and their connection to galaxy evolution remain a mystery.

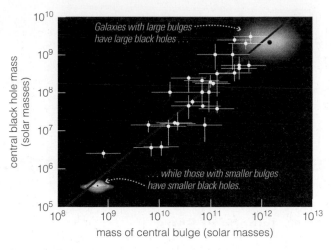

Figure 15.33

The relationship between bulge mass and the mass of a supermassive black hole.

---

the big picture
Putting Chapter 15 into Context

The picture could hardly get any bigger than it has in this chapter. Looking back through both space and time, we have seen a wide variety of galaxies extending nearly to the limits of the observable universe. As you look back, keep sight of these "big picture" ideas:

- The universe is filled with galaxies that come in a variety of shapes and sizes. In order to learn the histories of these galaxies, we must also consider how the universe itself has evolved through time.

- Much of our current understanding of the structure and evolution of the universe is based on measurements of distances to faraway galaxies. These measurements rely on a chain of techniques in which each link in the chain builds upon the links that come before it.

- It has been less than a century since Hubble first proved that the Milky Way is only one of billions of galaxies in the universe. This discovery, and his subsequent discovery of the universe's expansion, provided the foundation upon which modern cosmology has been built. Measurements of the rate of expansion tell us that the universe began about 14 billion years ago.

- Although we do not yet know the complete story of galaxy evolution, we are rapidly learning more. Galaxies probably all began as protogalactic clouds, but they do not always evolve peacefully. Some galaxies collide with neighbors, often with dramatic results.

- The tremendous energy outputs of quasars and other active galactic nuclei, including those of radio galaxies, are probably powered by gas accreting onto supermassive black holes. The centers of many present-day galaxies must still contain the supermassive black holes that once enabled them to shine as quasars.

summary of key concepts

15.1 Islands of Stars

• What are the three major types of galaxies?

(1) **Spiral galaxies** have prominent disks and spiral arms.

(2) **Elliptical galaxies** are rounder and redder than spiral galaxies and contain less cool gas and dust.

(3) **Irregular galaxies** are neither disk-like nor rounded in appearance.

• How are galaxies grouped together?

Spiral galaxies tend to collect in groups that contain up to several dozen galaxies. Elliptical galaxies are more common in clusters of galaxies, which contain hundreds to thousands of galaxies, all bound together by gravity.

15.2 Distances of Galaxies

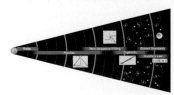

• How do we measure the distances to galaxies?

Our measurements of galaxy distances depend on a chain of methods. The chain begins with **radar ranging** in our own solar system and parallax measurements of distances to nearby stars, then relies on **standard candles** to measure greater distances.

• What is Hubble's law?

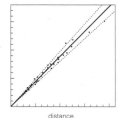

Hubble's law tells us that more-distant galaxies are moving away faster: $v = H_0 \times d$, where H_0 is **Hubble's constant.** It allows us to determine a galaxy's distance from the speed at which it is moving away from us, which we can measure from its Doppler shift.

• How do distance measurements tell us the age of the universe?

Combining distance measurements with velocity measurements tells us Hubble's constant, and the inverse of Hubble's constant tells us how long it would have taken the universe to reach its present size *if* the expansion rate had never changed. Based on Hubble's constant and estimates of how it has changed with time, we now estimate the age of the universe at about 14 billion years, which restricts our view of the universe to objects with **lookback times** smaller than that age.

15.3 Galaxy Evolution

• How do we observe the life histories of galaxies?

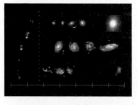

Today's telescopes enable us to observe galaxies of many different ages because they are powerful enough to detect light from objects with lookback times almost as large as the age of the universe. We can therefore assemble "family albums" of galaxies at different distances and lookback times.

• How did galaxies form?

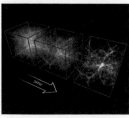

The most successful models of galaxy formation assume that galaxies formed as gravity pulled together regions of the universe that were ever so slightly denser than their surroundings. Gas collected in protogalactic clouds, and stars began to form as the gas cooled.

• Why do galaxies differ?

Differences between present-day galaxies probably arise both from conditions in their protogalactic clouds and from collisions with other galaxies. Slowly rotating or high-density protogalactic clouds may form elliptical rather than spiral galaxies. Ellipticals may also form through the merger of spiral galaxies.

15.4 Quasars and Other Active Galactic Nuclei

• What are quasars?

Some galaxies have unusually bright centers known as **active galactic nuclei**. A **quasar** is a particularly bright active galactic nucleus. Quasars are generally found at great distances from us, telling us that they were more common early in the history of the universe.

• What is the power source for quasars and other active galactic nuclei?

Supermassive black holes are thought to be the power sources for active galactic nuclei. As matter falls into a supermassive black hole through an accretion disk, its gravitational potential energy is transformed with enormous efficiency into thermal energy and then into light.

• Do supermassive black holes really exist?

Observations of orbiting stars and gas clouds in the nuclei of galaxies suggest that *all* galaxies may harbor supermassive black holes at their centers. The masses we measure for those black holes are closely related to the properties of the galaxy around them, suggesting that the growth of these black holes is closely tied to the process of galaxy evolution.

exercises and problems

For instructor-assigned homework go to **www.masteringastronomy.com**.

Review Questions

Short-Answer Questions Based on the Reading

1. What are the three major types of galaxies, and how do they look different from one another?
2. Describe the differences between normal spiral galaxies, barred spiral galaxies, and lenticular galaxies.
3. Distinguish between the *disk component* and the *spheroidal component* of a spiral galaxy. Which component includes cool gas and active star formation?
4. What is the major difference between spiral and elliptical galaxies? Answer in terms of the presence or absence of disk and spheroidal components. How does this difference explain the lack of hot, young stars in elliptical galaxies?
5. How are galaxy types in clusters of galaxies different from those among smaller groups and isolated galaxies?
6. What do we mean by a *standard candle?* Explain how we can use standard candles to measure distances.
7. Summarize each of the major links in the distance chain. Why are *Cepheid variable stars* so important? Why are white dwarf supernovae so useful, even though they are quite rare?
8. Explain how Hubble used Cepheid variable stars to prove that the Andromeda Galaxy lies beyond the bounds of the Milky Way.
9. What is *Hubble's law?* What is *Hubble's constant?* Explain what we mean when we say that Hubble's constant is between 20 and 24 kilometers per second per million light-years.
10. What is the *Cosmological Principle*, and how is it important to our understanding of the universe?
11. How is the expansion of the surface of an inflating balloon similar to the expansion of the universe? Use the balloon analogy to explain why Hubble's constant is related to the age of the universe.
12. What do we mean by the *lookback time* to a distant galaxy? Briefly explain why lookback times are less ambiguous than distances when discussing objects very far away.
13. What do we mean by a *cosmological redshift?* How does our interpretation of a distant galaxy's redshift differ if we think of it as a cosmological redshift rather than as a Doppler shift?
14. What is the *cosmological horizon,* and what determines how far away it lies?
15. What do we mean by *galaxy evolution?* How do telescopic observations allow us to study galaxy evolution?
16. Briefly describe the starting assumptions for models of galaxy formation and the basic scenario that leads to the formation of a spiral galaxy.
17. Describe two ways in which conditions in a protogalactic cloud might lead to the birth of an elliptical rather than a spiral galaxy. Is there any other general way in which some elliptical galaxies might have formed? Explain.
18. Briefly explain why we expect that collisions between galaxies should be relatively common, while collisions between stars are extremely rare. Why should galaxy collisions have been more common in the past than they are today? What evidence supports the idea that galaxy collisions sometimes occur?
19. What is a *starburst galaxy?* How do observations of starbursts help us understand why the stars in elliptical galaxies are so old?
20. Briefly describe the discovery of *quasars.* What evidence convinced astronomers that the high redshifts of quasars really do imply great distances? Why can we learn more about quasars by studying nearby *active galactic nuclei* and *radio galaxies?*
21. Briefly explain how we can use variations in luminosity to set limits on the size of an object's emitting region. For example, if an object doubles its luminosity in 1 hour, how big can it be?

22. Summarize the supermassive black hole model for the energy output of quasars and other active galactic nuclei. What evidence suggests that such black holes really exist?

Test Your Understanding

Does It Make Sense?

Decide whether the statement makes sense (or is clearly true) or does not make sense (or is clearly false). Explain clearly; not all of these have definitive answers, so your explanation is more important than your chosen answer.

23. If you want to find a lot of elliptical galaxies, you'll have better luck looking in clusters of galaxies than elsewhere in the universe.
24. Cepheids make good standard candles because they all have exactly the same luminosity.
25. After measuring the galaxy's redshift, I used Hubble's law to estimate its distance.
26. The center of the universe is more crowded with galaxies than any other place in the universe.
27. I'd love to live in one of the galaxies near our cosmological horizon, because then I could see the black void into which the universe is expanding.
28. If someone in a galaxy with a lookback time of $4\frac{1}{2}$ billion years had a superpowerful telescope, they could see our solar system in the process of its formation.
29. Galaxies that are more than 10 billion years old are too far away to see even with our most powerful telescopes.
30. If the Andromeda Galaxy someday collides and merges with the Milky Way, the resulting galaxy may be elliptical.
31. NGC 9645 is a starburst galaxy that has been forming stars at the same furious pace for some 10 billion years.
32. Astronomers proved that quasar 3C473 contains a supermassive black hole when they discovered that its center is dark.

Quick Quiz

Choose the best answer to each of the following. Explain your reasoning with one or more complete sentences.

33. Which of these galaxies is most likely to be oldest? (a) a galaxy in the Local Group (b) a galaxy observed at a distance of 5 billion light-years (c) a galaxy observed at a distance of 10 billion light-years
34. Which of these galaxies would you most likely find at the *center* of a large cluster of galaxies? (a) a large spiral galaxy (b) a giant elliptical galaxy (c) a small irregular galaxy
35. If all the main-sequence stars of a star cluster are typically only one-hundredth as bright as their main-sequence counterparts in the Hyades Cluster, then that cluster's distance from us is (a) 100 times as far as the Hyades' distance. (b) 30 times as far as the Hyades' distance. (c) 10 times as far as the Hyades' distance.
36. Which kind of object is the best standard candle for measuring distances to extremely distant galaxies? (a) a white dwarf (b) a Cepheid variable star (c) a white dwarf supernova
37. Why do virtually all the galaxies in the universe appear to be moving away from our own? (a) because we are located near where the Big Bang happened (b) because we are located near the center of the universe (c) because expansion causes all galaxies to move away from nearly all others.
38. When we observe a distant galaxy whose photons have traveled for 10 billion years before reaching Earth, we are seeing that galaxy as it was when the universe was about (a) 10 billion years old. (b) 7 billion years old. (c) 4 billion years old.

39. Which of these items is a key assumption in our most successful models for galaxy formation? (a) The distribution of matter was perfectly uniform early in time. (b) Some regions of the universe were slightly denser than others. (c) Galaxies formed around supermassive black holes.
40. The luminosity of a quasar is generated in a region the size of (a) the Milky Way. (b) a star cluster. (c) the solar system.
41. The rate at which supernovae explode in a starburst galaxy that is forming stars 10 times faster than the Milky Way is (a) about the same as in the Milky Way. (b) about 10 times higher than in the Milky Way. (c) about 100 times higher than in the Milky Way.
42. The primary source of a quasar's energy is (a) chemical energy. (b) nuclear energy. (c) gravitational potential energy.

Process of Science

Examining How Science Works

43. *Deviations from Hubble's Law.* Suppose you are measuring distances and velocities of galaxies in order to test Hubble's law. You find that 90% of the galaxies have velocities that are within 200 kilometers per second of the predictions of Hubble's law but 10% have velocities that deviate from the predictions by up to 1000 kilometers per second. Propose a hypothesis that would explain these deviations from Hubble's law, and outline a set of observations that could test your hypothesis.
44. *Why Do Galaxies Differ?* Which explanation for why galaxies differ seems most convincing to you? Explain in two paragraphs why you think that explanation is most convincing, using evidence presented in the chapter to support your position.
45. *The Quasar Controversy.* For many years, some astronomers argued that quasars are not really as distant as Hubble's law indicates. Research the history of the discovery of quasars and the debates that followed. What evidence led some astronomers to think quasars might be nearer than Hubble's law suggests? Why did most astronomers eventually conclude that quasars really are far away? Write a one- to two-page report summarizing your findings.

Investigate Further

In-Depth Questions to Increase Your Understanding

Short-Answer/Essay Questions

46. *The Hubble Deep Fields.* Both the Hubble Deep Field in Figure 15.1 and the Hubble Ultra Deep Field shown at the beginning of Chapter 1 were chosen for observation in large part because they are completely ordinary parts of the sky. Why do you think astronomers would want to devote so much precious telescope time to observing totally ordinary regions of the sky in such great detail? Explain your reasoning.
47. *Supernovae in Other Galaxies.* In which type of galaxy would you be most likely to observe a massive star supernova, in a giant elliptical galaxy or in a large spiral galaxy? Explain your reasoning.
48. *Hubble's Galaxy Types.* How would you classify the following galaxies using the system illustrated in Figure 15.7? Justify your answers.
   a. Galaxy M101 (Figure 15.2)
   b. Galaxy NGC 4594 (Figure 15.3)
   c. Galaxy NGC 1300 (Figure 15.4)
   d. Galaxy M87 (Figure 15.5).

49. *Cepheids as Standard Candles.* Suppose you are observing Cepheids in a nearby galaxy. You observe one Cepheid with a period of 8 days between peaks in brightness and another with a period of 35 days. Estimate the luminosity of each star. Explain how you arrived at your estimate. (*Hint:* See Figure 15.12.)

50. *Galaxies at Great Distances.* The most distant galaxies that astronomers have observed are much easier to see in infrared light than in visible light. Explain why that is the case.

51. *Universe on a Balloon.* In what ways is the surface of a balloon a good analogy for the universe? In what ways is this analogy limited? Explain why a miniature scientist living in a polka dot on the balloon would observe all other dots to be moving away, with more distant dots moving away faster.

52. *Life Story of a Spiral.* Imagine that you are a spiral galaxy. Describe your life history from birth to the present day. Your story should be detailed and scientifically consistent, but also creative. That is, it should be entertaining while at the same time incorporating current scientific ideas about the formation of spiral galaxies.

53. *Life Story of an Elliptical.* Imagine that you are an elliptical galaxy. Describe your life history from birth to the present. There are several possible scenarios for the formation of elliptical galaxies, so choose one and stick to it. Be creative while also incorporating scientific ideas that demonstrate your understanding.

54. *The Color of an Elliptical Galaxy.* Explain how the color of an old elliptical galaxy has changed during the last 10 billion years. How might you be able to use the galaxy's color to determine when it formed?

55. *Quasar Redshifts.* Some astronomers did not immediately accept that the large redshifts of quasars means that they must be at extremely large distances. What properties would an object need in order to have a large redshift but still be no farther than a billion light-years from the Milky Way? How might you be able to test the hypothesis that redshifts of quasars are not cosmological in origin?

Quantitative Problems

Be sure to show all calculations clearly and state your final answers in complete sentences.

56. *Counting Galaxies.* Estimate how many galaxies are pictured in Figure 15.1. Explain the method you used to arrive at this estimate. This picture shows about $\frac{1}{30,000,000}$ of the sky, so multiply your estimate by 30,000,000 to obtain an estimate of how many galaxies like these fill the entire sky.

57. *Distances to Star Clusters.* The distance of the Hyades Cluster is known from parallax to be 151 light-years. Estimate the distance to the Pleiades Cluster using the information in Figure 15.10 and the formula from Cosmic Calculations 15.1.

58. *Cepheids in M100.* Scientists using the Hubble Space Telescope have observed Cepheids in the galaxy M100. Here are the actual data for three Cepheids in M100:
   - Cepheid 1: luminosity = 3.9×10^{30} watts, brightness 9.3×10^{-19} watt/m^2
   - Cepheid 2: luminosity = 1.2×10^{30} watts, brightness 3.8×10^{-19} watt/m^2
   - Cepheid 3: luminosity = 2.5×10^{30} watts, brightness 8.7×10^{-19} watt/m^2

Compute the distance to M100 with data from each of the three Cepheids. Do all three distance computations agree? Based on your results, estimate the uncertainty in the distance you have found.

59. *Distances from Hubble's Law.* Imagine that you have obtained spectra for several galaxies and have measured the redshift of each galaxy to determine its speed away from us. Here are your results:
   - Galaxy 1: Speed away from us is 15,000 km/s.
   - Galaxy 2: Speed away from us is 20,000 km/s.
   - Galaxy 3: Speed away from us is 25,000 km/s.

Estimate the distance to each galaxy from Hubble's law. Assume that $H_0 = 22$ km/s/Mly.

60. *Your Last Hurrah.* Suppose you fell into an accretion disk that swept you into a supermassive black hole. On your way down, the disk radiates 10% of your mass-energy, $E = mc^2$.
   a. What is your mass in kilograms? (Recall that 1 kg = 2.2 pounds.) Calculate how much radiative energy will be produced by the accretion disk as a result of your fall into the black hole.
   b. Calculate approximately how long a 100-watt light bulb would have to burn to radiate this same amount of energy.

Discussion Questions

61. *Cosmology and Philosophy.* One hundred years ago, many scientists believed that the universe was infinite and eternal, with no beginning and no end. When Einstein first developed his general theory of relativity, he found that it predicted that the universe should be either expanding or contracting. He believed so strongly in an eternal and unchanging universe that he modified the theory, a modification he would later call his "greatest blunder." Why do you think Einstein and others assumed that the universe had no beginning? Do you think that a universe with a definite beginning in time, some 14 billion or so years ago, has any important philosophical implications? Explain.

62. *The Case for Supermassive Black Holes.* The evidence for supermassive black holes at the center of galaxies is strong. However, it is very difficult to prove absolutely that they exist because the black holes themselves emit no light. We can only infer their existence from their powerful gravitational influences on surrounding matter. How compelling is the evidence? Do you think astronomers have proved the case for black holes beyond a reasonable doubt? Defend your opinion.

Web Projects

63. *Galaxy Gallery.* Many fine images of galaxies are available on the Web. Collect several images of each major type and build a galaxy gallery of your own. Supply a descriptive paragraph about each galaxy.

64. *Greatest Lookback Time.* Look for recent discoveries of objects with the largest lookback times (or redshifts). What is the most distant known object to date? What kind of object is it?

65. *Future Observatories.* The subject of galaxy evolution is a very active area of research. Look for information on current and future observatories involved in investigating galaxy evolution (such as the James Webb Space Telescope). How big are the planned telescopes? What wavelengths will they look at? When will they be built? Write a short summary of a proposed mission.

Use the following questions to check your understanding of some of the many types of visual information used in astronomy. Answers are provided in Appendix K. For additional practice, try the Chapter 15 Visual Quiz at **www.masteringastronomy.com**.

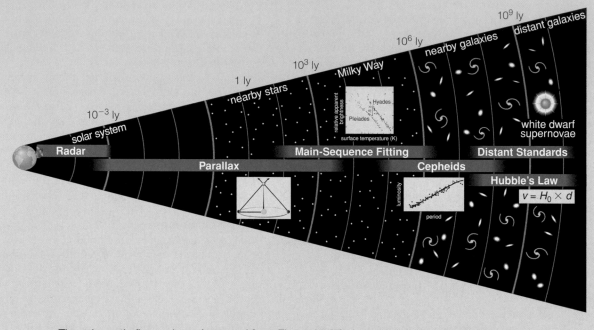

The schematic figure above (repeated from Figure 15.17) shows the distance measurement techniques that are best suited to measuring various astronomical distances. Answer the following questions based on the information in the figure.

1. Which distance measurement technique is most suited to measuring objects at a distance of 10 million light-years?
2. Which distance measurement technique is most suited to measuring at a distance of 10 light-years?

3. What range of astronomical distances can we currently measure using both parallax and main-sequence fitting?

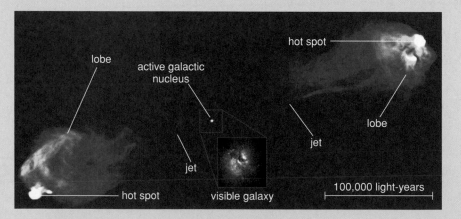

In the figure above (repeated from Figure 15.28), different colors are used to represent different levels of brightness in radio waves, and the scale bar shows a distance of 100,000 light-years. Use the information in the figure to answer the following questions.

4. What color is used to represent the brightest regions of radio emission?
5. What color is used to represent an absence of detectable radio waves?

6. What is the distance between the two radio hot spots?
7. What is the size of the visible galaxy shown in the inset image?

dark matter, dark energy, and the fate of the universe

learning goals

Essential Preparation

1. What determines the strength of gravity? [Section 4.4]

2. How does Newton's law of gravity extend Kepler's laws? [Section 4.4]

3. How do stars orbit in our galaxy? [Section 14.1]

4. What is Hubble's law? [Section 15.2]

In Chapter 15, we explored current understanding of the evolution of galaxies in an expanding universe. However, we left at least two crucial questions unaddressed: First, what is the source of the gravity that causes galaxies to form and holds them together? Second, what will happen to the expansion of the universe in the future? Both questions are interesting in their own right, but they've taken on even greater importance because our attempts to answer them have led us to two of the greatest mysteries in science.

If we are interpreting the data correctly, then the dominant source of gravity in the universe is an unidentified type of mass, known as *dark matter,* that is completely invisible to our eyes and telescopes. The gravity of dark matter must therefore be the "glue" that holds galaxies like our own together. The fate of the universe itself seems to hinge on the total amount of dark matter and on the existence of an even more mysterious force or energy—often called *dark energy*—that may counteract the effects of gravity on large scales.

In this chapter, we will explore the evidence for dark matter and dark energy and discuss why their nature remains so mysterious. We'll also investigate how dark matter determines the current structure of the universe and the implications of dark energy for the fate of the universe, if dark matter and dark energy indeed prove to be real.

16.1 Unseen Influences in the Cosmos

What is the universe made of? Ask an astronomer this seemingly simple question, and you might see a professional scientist blush with embarrassment. Based on all the available evidence today, the answer to this simple question is "We do not know."

It might seem incredible that we still do not know the composition of most of the universe, but you might also wonder why this should be so. After all, we can measure the chemical composition of distant stars and galaxies from their spectra, so we know that stars and gas clouds are made mostly of hydrogen and helium, with small amounts of heavier elements mixed in. But notice the key words "chemical composition." When we say these words, we are talking about the composition of material built from atoms of elements such as hydrogen, helium, carbon, and iron. While it is true that all familiar objects—including people, planets, and stars—are built from atoms, the same may not be true of the universe as a whole.

In fact, we now have good reason to think that the vast majority of the universe is *not* composed of atoms. Instead, the universe may consist largely of a mysterious form of mass known as *dark matter* and a mysterious form of energy known as *dark energy*. As we'll discuss in Chapter 17, some recent observations apparently tell us the precise percentages of the universe that consist of dark energy, dark matter, and ordinary atoms. Nevertheless, the actual nature of dark matter and dark energy remains completely unknown.

• What do we mean by dark matter and dark energy?

It's easy to talk about dark matter and dark energy, but what do these terms really mean? They are nothing more than names given to unseen influences in the cosmos. In both cases we have been led to think that there is something out there, even though we cannot identify it, because we have observed phenomena that otherwise do not make sense.

Dark matter is the name given to mass that emits no detectable radiation; we infer its existence from its gravitational effects.

You might at first guess that the major source of gravity that holds together galaxies is the same gas and dust that make up their stars. However, decades of observations suggest otherwise: By carefully observing gravitational effects on matter that we can see, such as stars or glowing clouds of gas, we've learned that there must be far more matter than meets the eye. Because this matter apparently gives off little or no light, we call it **dark matter**. Therefore, *dark matter* is simply a name we give to whatever unseen influence is causing the observed gravitational effects. We've already discussed dark matter briefly in Chapters 1 and 14, noting that studies of the Milky Way's rotation suggest that most of our galaxy's mass is distributed throughout its halo while most of the galaxy's light comes from stars and gas clouds in the thin galactic disk (see Figure 1.14).

We infer the existence of dark energy from careful studies of the expansion of the universe. From the time that Edwin Hubble first discovered the expansion, it was generally assumed that gravity must slow the expansion with time. In just the past few years, however, mounting evidence has suggested that the expansion of the universe is actually accelerating [Section 16.4]. If so, some mysterious influence must be able to counteract the effects of gravity on very large scales.

Dark energy is the name given to the unseen influence that may be causing the expansion of the universe to accelerate with time.

Dark energy is the most common name given to whatever it is that may be causing the expansion to accelerate, but it is not the only name; you may occasionally hear the same unseen influence attributed to *quintessence* or to a *cosmological constant*. The term *dark energy* has become popular because it echoes the term *dark matter,* but there's nothing unusually "dark" about it—after all, we don't expect to see light from the mere presence of a force or energy field. Despite the similarity in their names, dark matter and dark energy are thought to exist for completely different reasons.

Astronomers are generally more comfortable with the notion that dark matter exists than with the notion that dark energy exists. Scientific confidence in dark matter has been building for decades and is now at the point where dark matter seems almost indispensable to explaining the current structure of the universe. That is why we will devote most of this chapter to a discussion of dark matter and its presumed role as the dominant source of gravity in our universe—there is simply a lot more we can say about it. Our discussion of dark energy will wait until the end of the chapter, where we will present the evidence that it also exists and that its nature will determine the fate of the universe.

Before we continue, it's important to think about dark matter and dark energy in the context of science. Strange as these concepts may seem, they have emerged from careful scientific study conducted in accordance with the hallmarks of science discussed in Chapter 3 (see Figure 3.21).

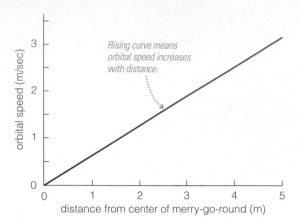

a A rotation curve for a merry-go-round.

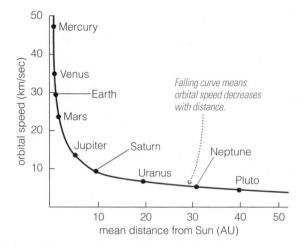

b The rotation curve for the planets in our solar system.

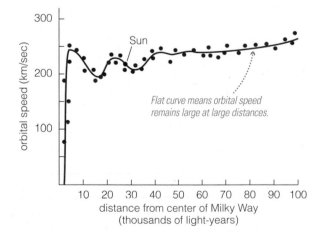

c The rotation curve for the Milky Way Galaxy. Dots represent stars or gas clouds whose orbital speeds have been measured.

Figure 16.1 interactive figure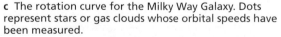

Rotation curves show how the orbital speed of a system depends on distance from its center. The solar system's rotation curve declines with distance because its mass is concentrated at the center. The Milky Way's rotation curve is flat, indicating that the galaxy's mass extends well beyond the Sun's orbit.

Dark matter and dark energy were each proposed to exist because they seemed the simplest way to explain observed motions in the universe. They've each gained credibility because models of the universe that assume their existence make testable predictions, and, at least so far, observations have borne out some of those predictions. Even if we some-day conclude that we were wrong to infer the existence of dark matter or dark energy, we will still need alternative explanations for the observations made to date. One way or the other, what we learn as we explore the mysteries of these unseen influences will forever change our view of the universe.

(MA) Detecting Dark Matter in a Spiral Galaxy Tutorial, Lessons 1–3

16.2 Evidence for Dark Matter

We are now ready to begin investigating dark matter in greater detail. In this section, we'll examine the evidence for the existence of dark matter and what the evidence indicates about its nature.

• What is the evidence for dark matter in galaxies?

Let's begin our discussion of the evidence for dark matter by examining the case for dark matter in our own Milky Way Galaxy. We'll then proceed to other galaxies and clusters of galaxies.

Distribution of Mass in the Milky Way In Chapter 14 we saw how the Sun's motion around the galaxy reveals the total amount of mass within its orbit. Similarly, we can use the orbital motion of any other star to measure the mass of the Milky Way within that star's orbit. In principle, we could determine the complete distribution of mass in the Milky Way by doing the same thing with the orbits of stars at every different distance from the galactic center.

In practice, interstellar dust obscures our view of disk stars more than a few thousand light-years away from us, making it very difficult to measure stellar velocities. However, radio waves penetrate this dust, and clouds of atomic hydrogen gas emit a spectral line at the radio wavelength of 21 centimeters [Section 14.2]. Measuring the Doppler shift of this 21-centimeter line tells us a cloud's velocity toward or away from us. With the help of a little geometry, we can then determine the cloud's orbital velocity.

A rotation curve plots the orbital speeds of objects in a galaxy against their distances from the center of the galaxy.

We can summarize the results of orbital velocity measurements with a diagram called a **rotation curve,** which plots the *orbital speed* of objects against their orbital distance. As a simple example, consider the rotation curve for a merry-go-round. Every object on a merry-go-round goes around the center with the same rotational period, but objects farther from the center move in larger circles. Objects farther from the center therefore move at faster speeds, and the rotation curve for a merry-go-round is a straight line that rises steadily outward (Figure 16.1a).

In contrast, the rotation curve for our solar system drops off with distance from the Sun, because inner planets orbit at faster speeds than outer planets (Figure 16.1b). This drop-off in speed with distance occurs because virtually all the mass of the solar system is concentrated in the Sun. The gravitational force holding a planet in its orbit decreases with distance from the Sun, and a smaller force means a lower orbital speed. The rotation curve of any astronomical system whose mass is concentrated toward the center must drop similarly.

Figure 16.1c shows the rotation curve for the Milky Way Galaxy. Each individual dot represents the distance from the galactic center and the orbital speed of a particular star or gas cloud. The curve running through the dots represents a best fit to the data. Notice that the orbital speeds remain approximately constant beyond the inner few thousand light-years, making the rotation curve look flat. This behavior contrasts sharply with the steeply declining rotation curve of the solar system. Therefore, most of the mass of the Milky Way must *not* be concentrated at its center. Instead, the orbits of progressively more distant gas clouds must encircle more and more mass. The Sun's orbit encompasses about 100 billion solar masses, but a circle twice as large surrounds twice as much mass, and a larger circle surrounds even more mass.

The flatness of the Milky Way's rotation curve indicates that a large amount of dark matter lies beyond our galaxy's visible regions.

The flatness of the Milky Way Galaxy's rotation curve therefore implies that most of its mass lies well beyond our Sun, tens of thousands of light-years from the galactic center. A more detailed analysis suggests that most of this mass is located in the spherical halo that surrounds the disk of our galaxy and that the total amount of this mass might be *10 times* the total mass of all the stars in the disk. Because we have detected very little radiation coming from this enormous amount of mass, it qualifies as dark matter. If we are interpreting the evidence correctly, the luminous part of the Milky Way's disk must be rather like the tip of an iceberg, marking only the center of a much larger clump of mass (Figure 16.2).

Dark Matter in Other Spiral Galaxies

Other galaxies also seem to contain vast quantities of dark matter. We can determine the amount of dark matter in a galaxy by comparing the galaxy's mass to its luminosity. More formally, astronomers calculate the galaxy's *mass-to-light ratio* (see Cosmic Calculations 16.1). First, we use the galaxy's luminosity to estimate the amount of mass that the galaxy contains in the form of stars. Next, we determine the galaxy's total mass by applying the law of gravity to observations of the orbital velocities of stars and gas clouds. If this total mass is larger than the mass that we can attribute to stars, then we infer that the excess mass must be dark matter.

We can measure a galaxy's luminosity as long as we can determine its distance with one of the techniques discussed in Chapter 15 (see Figure 15.17). We simply point a telescope at the galaxy in question, measure its apparent brightness, and calculate its luminosity from its distance using the inverse square law for light [Section 11.1]. Measuring the galaxy's total mass requires measuring orbital speeds of stars or gas clouds as far from the galaxy's center as possible. Atomic hydrogen gas clouds can be found in spiral galaxies at greater distances from the center than stars, so most of our data come from radio observations of the 21-centimeter line from these clouds. We use Doppler shifts of the 21-centimeter line to determine how fast a cloud is moving toward us or away from us (Figure 16.3). We then combine observations of clouds at varying orbital distances to construct the galaxy's rotation curve.

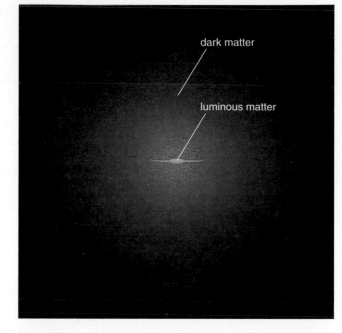

Figure 16.2

The dark matter associated with a spiral galaxy like the Milky Way occupies a much larger volume than the galaxy's luminous matter. The radius of this dark-matter halo may be 10 times as large as the galaxy's halo of stars.

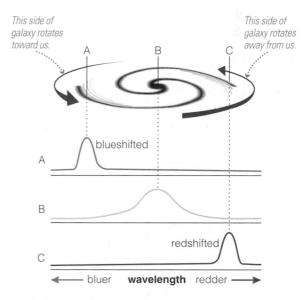

Figure 16.3

Measuring the rotation of a spiral galaxy with the 21-centimeter line of atomic hydrogen. Blueshifted lines on the left side of the disk show how fast that side is rotating toward us. Redshifted lines on the right side show how fast that side is rotating away from us.

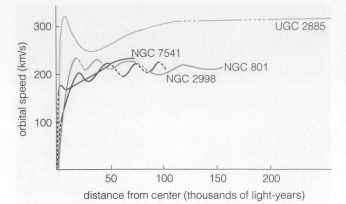

Figure 16.4

Actual rotation curves of four spiral galaxies. They are all nearly flat over a wide range of distances from the center, indicating that dark matter is common in spiral galaxies.

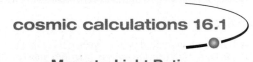

cosmic calculations 16.1

Mass-to-Light Ratio

We define an object's mass-to-light ratio as its total mass in units of *solar masses* divided by its total *visible* luminosity in units of *solar luminosities*. Thus, by definition, the mass-to-light ratio of the Sun is

$$\frac{1 M_{Sun}}{1 L_{Sun}} = 1 \frac{M_{Sun}}{L_{Sun}}$$

We read this answer with its units as "1 solar mass per solar luminosity."

Example 1: What is the mass-to-light ratio of a $1 M_{Sun}$ red giant with luminosity $100 L_{Sun}$?

Solution: The red giant's mass-to-light ratio is

$$\frac{1 M_{Sun}}{100 L_{Sun}} = 0.01 \frac{M_{Sun}}{L_{Sun}}$$

The ratio is *less* than 1 because a red giant puts out *more* light per unit mass than the Sun.

Example 2: The part of the galaxy within the Sun's orbit contains about 100 billion (10^{11}) solar masses of material and has a total luminosity of about 15 billion (1.5×10^{10}) solar luminosities. What is the mass-to-light ratio of the matter in our galaxy within the Sun's orbit?

Solution: We divide the region's mass by its luminosity, both in solar units:

$$\frac{10^{11} M_{Sun}}{1.5 \times 10^{10} L_{Sun}} = 6.7 \frac{M_{Sun}}{L_{Sun}}$$

The mass-to-light ratio of the matter within the Sun's orbit is about 7 solar masses per solar luminosity. Because this mass-to-light ratio is *greater* than the Sun's ratio of 1 solar mass per solar luminosity, it tells us that most matter in this region is *dimmer* per unit mass than our Sun.

The flat rotation curves of spiral galaxies other than the Milky Way indicate that they, too, harbor lots of dark matter.

The rotation curves of most spiral galaxies turn out to be remarkably flat (Figure 16.4). Just as in the Milky Way, these flat rotation curves imply that a great deal of matter lies far out in the haloes of these other spiral galaxies. A detailed analysis tells us that these other spiral galaxies also have at least 10 times as much mass in dark matter as they do in stars. In other words, the composition of typical spiral galaxies is 90% or more dark matter and 10% or less matter in stars.

Dark Matter in Elliptical Galaxies We must use a different technique to measure the masses of elliptical galaxies, because most of them contain very little atomic hydrogen gas and hence do not produce detectable 21-centimeter radiation. We generally measure mass in the inner parts of elliptical galaxies by observing the motions of the stars themselves.

The motions of stars in an elliptical galaxy are disorganized, so we cannot assemble their velocities into a sensible rotation curve. Nevertheless, the velocity of each individual star still depends on the mass inside the star's orbit. At any particular distance from an elliptical galaxy's center, some stars are moving toward us and some are moving away from us. As a result, every star has a slightly different Doppler shift. When we look at spectral lines from the galaxy as a whole, we see the combined effect of all these Doppler shifts. Together, they change the spectral line from a nice sharp line at a particular wavelength to a *broadened* line spanning a range of wavelengths (Figure 16.5). The greater the broadening of the spectral line, the faster the stars must be moving.

The widths of the absorption lines in an elliptical galaxy's spectrum tell us how much matter it contains, and how much of that matter is dark.

When we compare spectral lines from different regions of an elliptical galaxy, we find that the speeds of the stars remain fairly constant as we look farther from the galaxy's center. Just as in spirals, most of the matter in elliptical galaxies

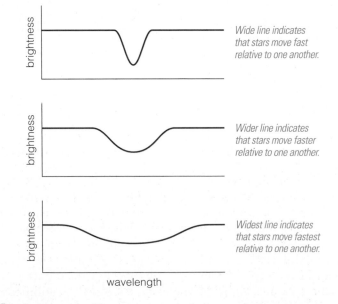

Wide line indicates that stars move fast relative to one another.

Wider line indicates that stars move faster relative to one another.

Widest line indicates that stars move fastest relative to one another.

Figure 16.5

The broadening of absorption lines in an elliptical galaxy's spectrum tells us how fast its stars move relative to one another.

must lie beyond the stars we can see and hence must be dark matter. The evidence for dark matter is even more convincing for cases in which we can measure the speeds of globular star clusters orbiting at large distances from the center of an elliptical galaxy. These measurements suggest that elliptical galaxies, like spirals, contain far more matter than we can see in the form of stars.

• What is the evidence for dark matter in clusters of galaxies?

The evidence we have discussed so far suggests that stars make up only about 10% of a galaxy's mass—the remaining mass consists of dark matter. Observations of galaxy clusters suggest that the total proportion of dark matter in clusters is even greater.

The evidence for dark matter in clusters comes from three different ways of measuring cluster masses: measuring the speeds of galaxies orbiting the center of the cluster, studying the X-ray emission from hot gas between the cluster galaxies, and observing how the clusters bend light as *gravitational lenses*. Let's investigate each of these techniques more closely.

Orbits of Galaxies in Clusters The problem of dark matter in astronomy is not particularly new. In the 1930s, astronomer Fritz Zwicky was already arguing that clusters of galaxies held enormous amounts of this mysterious stuff (Figure 16.6). Few of his colleagues paid attention, but later observations supported Zwicky's claims.

Zwicky was one of the first astronomers to think of galaxy clusters as huge swarms of galaxies bound together by gravity. It seemed natural to him that galaxies clumped closely in space should all be orbiting one another, just like the stars in a star cluster. He therefore assumed that he could measure cluster masses by observing galaxy motions and applying Newton's laws of motion and gravitation.

Figure 16.6

Fritz Zwicky, discoverer of dark matter in clusters of galaxies. Zwicky had an eccentric personality, but some of his ideas that seemed strange in the 1930s proved correct many decades later.

---

special topic: -

Pioneers of Science

SCIENTISTS ALWAYS TAKE a risk when they publish what they think are ground-breaking results. If their results turn out to be in error, their reputations may suffer. In the case of dark matter, the pioneers in its discovery risked their entire careers. A case in point is Fritz Zwicky, with his proclamations in the 1930s about dark matter in clusters of galaxies. Most of his colleagues considered him an eccentric who leapt to premature conclusions.

Another pioneer in the discovery of dark matter was Vera Rubin, an astronomer at the Carnegie Institution. Working in the 1960s, she became the first woman to observe under her own name at California's Palomar Observatory, then the largest telescope in the world. (Another woman, Margaret Burbidge, was permitted to observe at Palomar earlier but was required to apply for time under the name of her husband, also an astronomer.) Rubin first saw the gravitational signature of dark matter in spectra of stars in the Andromeda Galaxy. She noticed that stars in the outskirts of Andromeda moved at surprisingly high speeds, suggesting a stronger gravitational attraction than could be explained by the mass of the galaxy's stars alone. In other words, she found that the rotation curve

for Andromeda is relatively flat out to great distances from the center, as we now know is also the case for the Milky Way.

Working with her colleague Kent Ford, Rubin constructed rotation curves for the hydrogen gas in many other spiral galaxies (by studying Doppler shifts in the spectra of hydrogen gas) and discovered that flat rotation curves are common. Although Rubin and Ford did not immediately recognize the significance of the results, they were soon arguing that the universe must contain substantial quantities of dark matter.

For a while, many astronomers had trouble believing the results. Some astronomers suspected that the bright galaxies studied by Rubin and Ford were unusual for some reason. So Rubin and Ford went back to work, obtaining rotation curves for fainter galaxies. They found flat rotation curves—a signature of dark matter—in these galaxies as well. By the 1980s, the evidence compiled by Rubin, Ford, and others was so overwhelming that even the critics came around. Either the theory of gravity was wrong, or the astronomers measuring rotation curves had discovered dark matter in spiral galaxies.

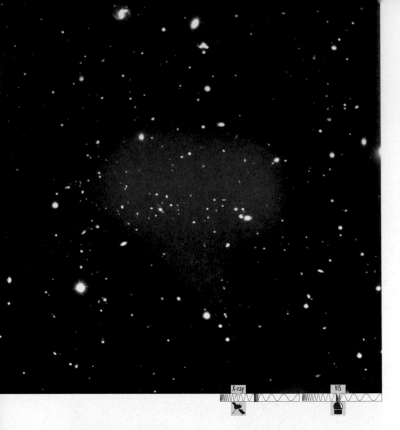

Figure 16.7

A distant cluster of galaxies in both visible light and X-ray light. The visible light photo shows the individual galaxies as yellowish blobs. The blue overlay represents X-ray emission from extremely hot gas (around 100 million K) in the cluster. Evidence for dark matter comes both from the observed motions of the visible galaxies and from the temperature of the hot gas. (The photo shows a region about 8 million light-years across.)

Using a spectrograph, Zwicky measured the redshifts of the galaxies in a cluster and used these redshifts to calculate the velocities at which the individual galaxies are moving away from us. He determined the velocity of the cluster as a whole by averaging the velocities of its individual galaxies. He then estimated the orbital speed of a galaxy around the cluster by subtracting this average velocity from the individual galaxy's velocity. Finally, he used these orbital speeds to estimate the cluster's mass and compared this mass to the cluster's luminosity.

The orbits of galaxies in clusters tell us that galaxy clusters contain huge amounts of dark matter.

To his surprise, Zwicky found that clusters of galaxies have much greater masses than their luminosities would suggest. When he estimated the total mass of stars necessary to account for the overall luminosity of a cluster, he found that it was far less than the mass he measured by studying galaxy speeds. He concluded that most of the matter within these clusters must not be in the form of stars and instead must be almost entirely dark. Many astronomers disregarded Zwicky's result, believing that he must have done something wrong to arrive at such a strange answer. Today, sophisticated measurements of galaxy orbits in clusters confirm Zwicky's original finding.

Hot Gas in Clusters A second method for measuring the mass of a galaxy cluster relies on observations of hot gas that fills the space between the galaxies in the cluster (Figure 16.7). This gas (sometimes called the *intracluster medium*) is so hot—typically tens of millions of Kelvin and sometimes above 100 million K—that it emits primarily X rays. It was therefore unknown until the 1960s, when X-ray telescopes were first launched above Earth's atmosphere. Some large clusters have up to seven times as much mass in the form of X-ray emitting gas as they do in the form of stars.

The hot gas can tell us about dark matter because its temperature depends on the total mass of the cluster. The gas in most clusters is nearly in a state of *gravitational equilibrium*—that is, its outward pressure balances gravity's inward pull [Section 10.1]. In this state of balance, the average kinetic energies of the gas particles are determined primarily by the strength of gravity and hence by the amount of mass within the cluster. Because the temperature of a gas reflects the average kinetic energies of its particles, the gas temperatures we measure with X-ray telescopes tell us the average speeds of the X-ray-emitting particles. We can then use these particle speeds to determine the cluster's total mass.

Temperature measurements of hot gas can also tell us the amount of dark matter in clusters and give results that agree with those we infer from galaxy velocities.

The results obtained with this method agree with the results found by studying the orbital motions of a cluster's galaxies. Even after we account for the mass of the hot gas, we find that the amount of dark matter in a typical galaxy cluster is up to 50 times that of the combined mass of the stars in the cluster's galaxies. In other words, the gravity of dark matter seems to be binding the galaxies of a cluster together in much the same way that it binds matter within an individual galaxy.

Gravitational Lensing Until recently, astronomers relied exclusively on methods based on Newton's laws to measure galaxy and cluster masses. These laws keep telling us that the universe holds far more matter than we can see. Can we trust these laws? One way to check is to measure masses in a different way. Today, astronomers have an additional tool for measuring masses: *gravitational lensing.*

Gravitational lensing occurs because masses distort the space around them [Section 13.3]. Massive objects can therefore act as **gravitational lenses** that bend light beams passing nearby. This prediction of Einstein's general theory of relativity was first verified in 1919 during an eclipse of the Sun. Because the light-bending angle of a gravitational lens depends on the mass of the object doing the bending, we can measure the masses of objects by observing how strongly they distort light paths.

Gravity's light-bending effects distort the images of galaxies lying behind a cluster, enabling us to measure the cluster's mass without relying on Newton's laws.

Figure 16.8 shows a striking example of how a cluster of galaxies can act as a gravitational lens. Many of the yellow elliptical galaxies concentrated toward the center of the picture belong to the cluster, but at least one of the galaxies pictured does not. The multiple blue ovals are actually all images of the same galaxy. This galaxy lies almost directly behind the center of the cluster, at a much greater distance. We see multiple images of this single galaxy because photons do not follow straight paths as they travel from the galaxy to Earth. Instead, the cluster's gravity bends the photon paths, causing light from the galaxy to arrive at Earth from a few slightly different directions (Figure 16.9). Each alternative path produces a separate, distorted image of the blue galaxy.

Multiple images of a gravitationally lensed galaxy are rare. They occur only when a distant galaxy lies directly behind the lensing cluster. However, single, distorted images of gravitationally lensed galaxies are quite common. Figure 16.10 shows a typical example. This picture shows numerous normal-looking galaxies and several arc-shaped galaxies. The oddly curved galaxies are not members of the cluster, nor are they really curved. They are normal galaxies lying far beyond the cluster whose images have been distorted by the cluster's gravity.

Cluster masses measured through gravitational lensing agree with those measured from galaxy velocities and gas temperatures.

Careful analyses of the distorted images created by clusters enable us to measure cluster masses without resorting to Newton's laws. Instead, Einstein's theory of general relativity tells us how massive these clusters must be to generate the observed distortions. It is reassuring that cluster masses derived in this way generally agree with those derived from galaxy velocities and X-ray temperatures. The three different methods all indicate that clusters of galaxies hold very substantial amounts of dark matter.

• Does dark matter really exist?

Astronomers have made a strong case for the existence of dark matter, but could they be completely off base? Is it possible that dark matter is a figment of human imagination and that there's a completely different explanation for the observations we've discussed?

All the evidence for dark matter rests on our understanding of gravity. For individual galaxies, the case for dark matter rests primarily on applying Newton's laws of motion and gravity to observations of the orbital

Figure 16.8

This Hubble Space Telescope photo shows a galaxy cluster acting as a gravitational lens. The yellow elliptical galaxies are cluster members. The many small blue ovals (such as those indicated by the arrows) are multiple images of a single galaxy that lies almost directly behind the cluster's center. (The picture shows a region about 1.4 million light-years across.)

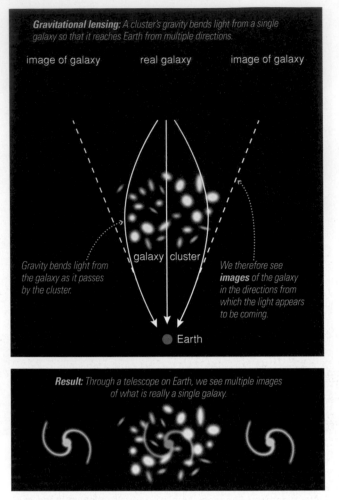

Gravitational lensing: *A cluster's gravity bends light from a single galaxy so that it reaches Earth from multiple directions.*

image of galaxy   real galaxy   image of galaxy

Gravity bends light from the galaxy as it passes by the cluster.

galaxy cluster

*We therefore see **images** of the galaxy in the directions from which the light appears to be coming.*

● Earth

Result: *Through a telescope on Earth, we see multiple images of what is really a single galaxy.*

Figure 16.9 interactive figure

A cluster's powerful gravity bends light paths from background galaxies to Earth. If light can arrive from several different directions, we see multiple images of the same galaxy.

Figure 16.10

Hubble Space Telescope photo of the cluster Abell 2218. The thin, elongated galaxies around the main clump of galaxies on the left side of the picture are the images of background galaxies distorted by the cluster's gravity. By measuring these distortions, astronomers can determine the total amount of mass in the cluster. (The region pictured is about 1.4 million light-years across.)

speeds of stars and gas clouds. We've used the same laws to make the case for dark matter in clusters, along with additional evidence based on gravitational lensing predicted by Einstein's general theory of relativity. It therefore seems that one of the following must be true:

1. Dark matter really exists, and we are observing the effects of its gravitational attraction.
2. There is something wrong with our understanding of gravity, which is causing us to mistakenly infer the existence of dark matter.

We cannot yet rule out the second possibility, but most astronomers consider it very unlikely. Newton's laws of motion and gravity are among the most trustworthy tools in science. We have used them time and again to measure masses of celestial objects from their orbital properties. We found the masses of Earth and the Sun by applying Newton's version of Kepler's third law to objects that orbit them [Section 4.4]. We used this same law to calculate the masses of stars in binary star systems, revealing the general relationships between the masses of stars and their outward appearances. Newton's laws have also told us the masses of things we can't see directly, such as the masses of orbiting neutron stars in X-ray binaries and of black holes in active galactic nuclei. Einstein's general theory of relativity likewise stands on solid ground, having been repeatedly tested and verified in many observations and experiments. We therefore have good reason to trust our current understanding of gravity.

Either dark matter exists or our current understanding of gravity is incorrect.

Many scientists have already made valiant efforts to come up with alternate theories of gravity that could account for the observations without assuming the existence of dark matter. (After all, there's a Nobel Prize waiting for anyone who can substantiate a new theory of gravity.) So far, no one has succeeded in doing so in a way that can also explain the many other observations accounted for by our current theories of gravity.

In essence, our high level of confidence in our current understanding of gravity gives us equally high confidence that dark matter really exists. While we should always keep an open mind about the possibility of future changes in our understanding, we will proceed for now under the assumption that dark matter is real. With that assumption in mind, let's turn our attention to the nature of what seems to be the most common form of matter in the universe.

Distorted images of background galaxies

VIS

think about it Should the fact that we have three different ways of measuring cluster masses give us greater confidence that we really do understand gravity and that dark matter really does exist? Why or why not?

What might dark matter be made of?

What is all this dark stuff in galaxies and clusters of galaxies? We don't yet know. Nevertheless, we can make educated guesses.

At least some of the dark matter is likely to be *ordinary*, made of protons, neutrons, and electrons. The only unusual thing about this dark matter is that it doesn't emit much detectable radiation. However, as we'll discuss shortly, it's also likely that some of the dark matter is *extraordinary*, made of particles that we have yet to discover.

A bit of terminology will be useful. Because the protons and neutrons that make up most of the mass of ordinary matter belong to a category of particles called **baryons**, ordinary matter is sometimes called **baryonic matter**. By extension, extraordinary matter is called **nonbaryonic matter**.

Ordinary Dark Matter Matter need not be extraordinary to be dark. In astronomy, "dark" merely means not as bright as a normal star and therefore not visible across vast distances of space. Your body is dark matter, because you would be far too dim for our telescopes to detect if you were somehow flung into the halo of our galaxy. Everything you own is dark matter. Earth and the rest of the planets are dark matter as well. In fact, the "failed stars" known as *brown dwarfs* [Section 12.1] and even faint red main-sequence stars of spectral type M [Section 11.2] are too dim for current telescopes to see in the halo and therefore qualify as dark matter. It's conceivable that trillions of faint red stars, brown dwarfs, and Jupiter-size objects left over from the Milky Way's formation still roam our galaxy's halo, providing much of its mass. These objects are sometimes called **MACHOs**, for *massive compact halo objects*, although they are better thought of as dim, starlike (or planetlike) objects.

MACHOs such as dim, red stars and brown dwarfs are too faint for us to see directly, but there are other ways to search for them. One innovative technique takes advantage of gravitational lensing on a much smaller scale than the examples we studied for clusters of galaxies. If trillions of these dim stars and similar objects roam the halo of the galaxy, we should occasionally see one pass in front of a more distant star. When the object lies almost directly between us and the farther star, its gravity will focus more of the star's light directly toward the Earth. The distant star will appear much brighter than usual for several days or weeks as the lensing object passes in front of it (Figure 16.11). We cannot see the object itself, but the duration of the lensing event reveals its mass.

> Gravitational lensing by starlike objects in our galaxy's halo does not turn up enough objects to account for all the dark matter.

Gravitational lensing events such as these are rare, happening to about one star in a million each year. To observe such lensing events, we therefore must monitor huge numbers of stars. Current large-scale monitoring projects now record numerous lensing events annually. These events demonstrate that dim starlike objects (MACHOs) do indeed populate our galaxy's halo, but not in large enough numbers to account for all the Milky Way's dark matter. Similar measurements rule out the possibility that

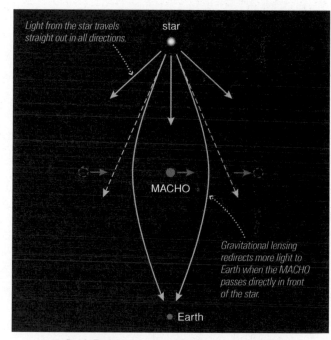

Light from the star travels straight out in all directions.

star

MACHO

Gravitational lensing redirects more light to Earth when the MACHO passes directly in front of the star.

Earth

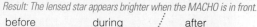

Result: The lensed star appears brighter when the MACHO is in front.

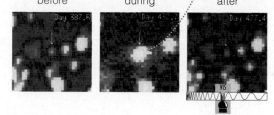

before   during   after

Figure 16.11

When a small, starlike object (MACHO) passes in front of a more distant star, gravitational lensing temporarily makes the star appear brighter. Searches for such events show that our galaxy's halo does indeed contain dim, starlike objects, but that these objects do not constitute the majority of the galaxy's dark matter.

the dark matter consists of large numbers of black holes formed by the deaths of massive stars. Something else must lurk unseen in the outer reaches of our galaxy.

Extraordinary Dark Matter A more exotic possibility is that most of the dark matter in galaxies and clusters of galaxies is not made of ordinary, baryonic matter at all. Let's begin to explore this possibility by taking another look at those nonbaryonic particles we discussed in Section 10.2: neutrinos. These unusual particles are dark by their very nature, because they have no electrical charge and hence cannot emit electromagnetic radiation of any kind. Moreover, they are never bound together with charged particles in the way that neutrons are bound in atomic nuclei, so their presence cannot be revealed by associated light-emitting particles.

Particles like neutrinos interact with other forms of matter only through the force of gravity and the *weak force,* which governs some nuclear reactions. For this reason, they are said to be *weakly interacting particles.* If you recall that trillions of neutrinos from the Sun are passing through your body at this very moment without doing any damage, you'll see why the name *weakly interacting* fits well.

The dark matter in galaxies cannot be made of neutrinos, because these very low mass particles travel through the universe at enormous speeds and can easily escape a galaxy's gravitational pull. (However, neutrinos make up a small amount of the dark matter outside galaxies.) What if other weakly interacting particles exist that are similar to neutrinos but considerably heavier? They too would evade direct detection, but they would move more slowly so that their mutual gravity could hold together a large collection of them. Such hypothetical particles are called **WIMPs**, for *weakly interacting massive particles.* Note that WIMPs are subatomic particles, so the "massive" in their name is relative—they are massive only in comparison to lightweight particles like neutrinos. (They are also often called *cold dark matter* to set them apart from the faster-moving neutrinos.) WIMPs could make up most of the mass of a galaxy or cluster of galaxies, but they would be completely invisible in all wavelengths of light. Most astronomers now consider it likely that WIMPs make up the majority of dark matter and hence the majority of all matter in the universe.

Scientists suspect that dark matter is mostly made up of weakly interacting particles that are like neutrinos but more massive.

It might surprise you that scientists suspect the universe to be filled with particles they haven't yet discovered. However, this hypothesis would also explain why dark matter seems to be distributed throughout spiral galaxy halos rather than concentrated in flattened disks like the visible matter. Recall that galaxies are thought to have formed as gravity pulled together matter in regions of slightly enhanced density in the early universe [Section 15.3]. This matter would have consisted mostly of dark matter mixed with some ordinary hydrogen and helium gas. The ordinary gas could collapse to form a rotating disk because individual gas particles could lose orbital energy: Collisions among many gas particles can convert some of their orbital energy into radiative energy that escapes from the galaxy in the form of photons. In contrast, WIMPs cannot produce photons, and they rarely interact and exchange energy with other particles. As the gas collapsed to form a disk, WIMPs would therefore have remained stuck in orbits far out in the galactic halo—just where most dark matter seems to be located.

By itself, the agreement between the measured distribution of dark matter in galaxies and what we'd expect from dark matter made of WIMPs doesn't prove that extraordinary dark matter exists. However, as we'll discuss in the next chapter, there are additional reasons why many astronomers believe that baryons represent only a minority of the universe's mass and hence that WIMPs are the most common form of matter in the universe.

think about it　What do you think of the idea that much of the universe is made of as-yet-undiscovered particles? Can you think of other instances in the history of science in which the existence of something was predicted before it was discovered?

16.3 Structure Formation

Dark matter remains enigmatic, but every year we are learning more about its role in the universe. Because galaxies and clusters of galaxies seem to contain much more dark matter than luminous matter, dark matter's gravitational pull must be the primary force holding these structures together. Therefore, we suspect that the gravitational attraction of dark matter is what pulled galaxies and clusters together in the first place.

• What is the role of dark matter in galaxy formation?

Stars, galaxies, and clusters of galaxies are all *gravitationally bound systems*— their gravity is strong enough to hold them together. In most of the gravitationally bound systems we have discussed so far, gravity has completely overwhelmed the expansion of the universe. That is, while the universe as a whole is expanding, space is *not* expanding within star systems, galaxies, or clusters of galaxies.

Our best guess at how galaxies formed, outlined in Section 15.3, envisions them growing from slight density enhancements that were present in the very early universe. During the first few million years after the Big Bang, the universe expanded everywhere. Gradually, the stronger gravity in regions of enhanced density pulled in matter until these regions stopped expanding and became protogalactic clouds, even as the universe as a whole continued to expand.

The gravity of dark matter was probably the main force that caused protogalactic clouds to become galaxies and galaxies to group into clusters.

If dark matter is indeed the most common form of mass in galaxies, it must have provided most of the gravitational attraction responsible for creating protogalactic clouds. The hydrogen and helium gas in the protogalactic clouds collapsed inward and gave birth to stars, while weakly interacting dark matter remained in the outskirts because of its inability to radiate away its orbital energy. According to this model, the luminous matter in each galaxy must still be nestled inside the larger cocoon of dark matter that initiated the galaxy's formation (see Figure 16.2), just as observational evidence seems to suggest.

The formation of galaxy clusters probably echoes the formation of galaxies. Early on, all the galaxies that will eventually constitute a cluster fly apart with the expansion of the universe, but the gravity of the dark matter associated with the cluster eventually reverses the trajectories of

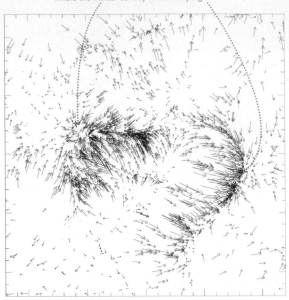

Gravity pulls galaxies into regions of the universe where the matter density is relatively high.

Figure 16.12

This diagram represents the motions of galaxies due to effects of gravity. Each arrow represents the amount by which a galaxy's actual velocity (inferred from a combination of observations and modeling) differs from the velocity we'd expect it to have from Hubble's law alone. The Milky Way is at the center of the picture, which shows an area about 600 million light-years across. (Only a representative sample of galaxies is shown.) Notice how the galaxies tend to flow into regions where the density of galaxies is already high. These vast, high-density regions are probably superclusters in the process of formation.

these galaxies. The galaxies ultimately fall back inward and start orbiting each other randomly, like the stars in the halo of our galaxy.

Some clusters of galaxies apparently have not yet finished forming, because their immense gravity is still drawing in new members. For example, the Virgo Cluster of galaxies (about 60 million light-years away) appears to be drawing in the Milky Way and other galaxies of the Local Group. The evidence comes from careful study of galaxy speeds. Plugging the Virgo Cluster's distance into Hubble's law tells us the speed at which the Milky Way and the Virgo Cluster should be drifting apart due to the universal expansion [Section 15.2]. However, the measured speed is about 400 km/s slower than the speed we predict from Hubble's law alone. We conclude that this 400 km/s discrepancy (sometimes called a *peculiar velocity*) arises because the cluster's gravity is pulling us back against the flow of universal expansion. In other words, while the Milky Way and other galaxies of our Local Group are still moving away from the Virgo Cluster with the expansion of the universe, the rate at which we are separating from the cluster is slowing with time. Eventually, the cluster's gravity may stop the separation altogether, at which point the cluster will begin pulling in the galaxies of our Local Group, ultimately making them members of the cluster.

Similar processes are taking place on the outskirts of other large clusters of galaxies, where we see many galaxies whose speeds indicate that the cluster's gravity is pulling on them. Eventually, some or perhaps all of these galaxies will fall into the cluster. That means many clusters are still attracting galaxies, adding to the hundreds they already contain. On even larger scales, clusters themselves seem to be tugging on one another, hinting that they are parts of even bigger gravitationally bound systems, called **superclusters**, that are still in the early stages of formation (Figure 16.12). But there are structures even larger than superclusters.

think about it State whether each of the following is a gravitationally bound system, and explain why: (a) Earth; (b) a hurricane (on Earth); (c) the Orion Nebula; (d) a supernova.

• What are the largest structures in the universe?

Beyond about 300 million light-years from Earth, deviations from Hubble's law owing to gravitational tugs are insignificant compared with the universal expansion, so Hubble's law becomes our primary method for measuring galaxy distances [Section 15.2]. Using this law, astronomers can make maps of the distribution of galaxies in space. Such maps reveal **large-scale structures** much vaster than clusters of galaxies.

Mapping Large-Scale Structures Making maps of galaxy locations requires an enormous amount of data. A long-exposure photo showing galaxy positions is not enough, because it does not tell us the galaxy distances. Instead, we must measure the redshift of each individual galaxy so that we can estimate its distance by applying Hubble's law. Current technology allows redshift measurements for hundreds of galaxies during a single night of telescopic observation. As a result, we now have redshift measurements—and hence estimated distances—for hundreds of thousands of galaxies.

Figure 16.13 shows the distribution of galaxies in three slices of the universe, each extending farther out in distance. Our Milky Way Galaxy

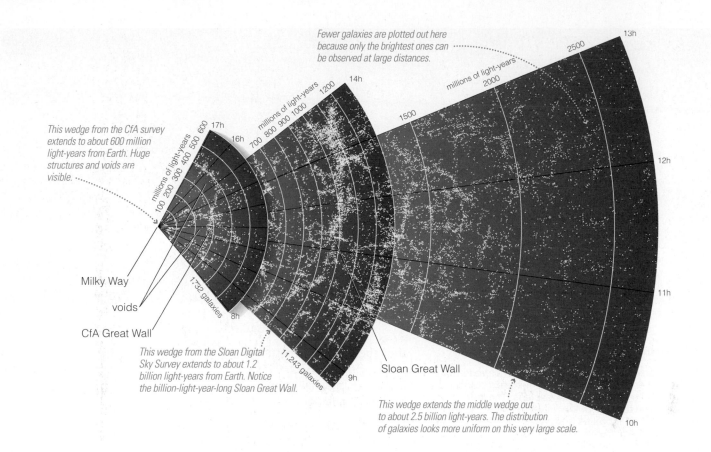

Fewer galaxies are plotted out here because only the brightest ones can be observed at large distances.

This wedge from the CfA survey extends to about 600 million light-years from Earth. Huge structures and voids are visible.

Milky Way

voids

CfA Great Wall

This wedge from the Sloan Digital Sky Survey extends to about 1.2 billion light-years from Earth. Notice the billion-light-year-long Sloan Great Wall.

1,732 galaxies

11,243 galaxies

Sloan Great Wall

This wedge extends the middle wedge out to about 2.5 billion light-years. The distribution of galaxies looks more uniform on this very large scale.

is located at the vertex at the far left, and each dot represents another galaxy.

Galaxies appear to be arranged in immense structures hundreds of millions of light-years across.

The slice at the left comes from one of the first surveys of large-scale structures performed in the 1980s. This map, which required years of effort by many astronomers, revealed the complex structure of our corner of the universe. It showed that galaxies are not scattered randomly through space but instead are arranged in huge chains and sheets that span many millions of light-years. Clusters of galaxies are located at the intersections of these chains. Between these chains and sheets of galaxies lie giant empty regions called **voids**. The other two slices show data from the more recent Sloan Digital Sky Survey, which measured redshifts for nearly a million galaxies spread across one-fourth of the sky.

Some of the structures in these pictures are amazingly large. For example, the Sloan Great Wall, clearly visible in the center slice, extends more than 1 billion light-years from end to end. Immense structures such as this one apparently have not yet collapsed into randomly orbiting, gravitationally bound systems.

In these surveys, there also appears to be a limit to the size of the largest structures. If you look closely at the right-most slice in Figure 16.13, you'll notice that the overall distribution of galaxies appears nearly uniform on scales larger than about a billion light-years. On such scales, the universe looks much the same everywhere, in agreement with what we expect from the *Cosmological Principle* [Section 15.2].

The Origin of Large Structures Why is gravity collecting matter on such enormous scales? Just as we suspect that galaxies formed from regions of slightly enhanced density in the early universe, we suspect

Figure 16.13

Each of these three wedges shows a "slice" of the universe a few angular degrees in thickness that extends outward from our own Milky Way Galaxy. The dots represent galaxies, shown at their measured distances from Earth. We see that galaxies trace out long chains and sheets surrounded by huge voids containing very few galaxies.

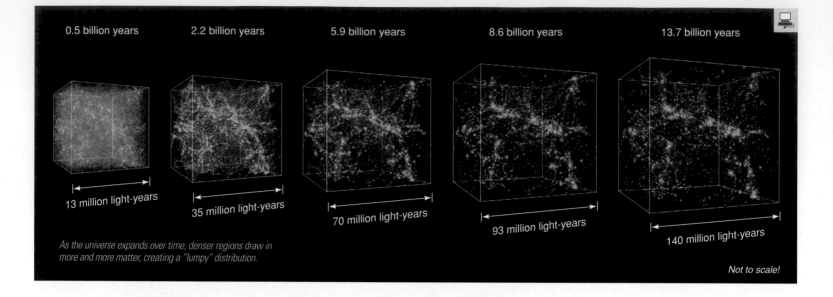

0.5 billion years   2.2 billion years   5.9 billion years   8.6 billion years   13.7 billion years

13 million light-years

35 million light-years

70 million light-years

93 million light-years

140 million light-years

As the universe expands over time, denser regions draw in more and more matter, creating a "lumpy" distribution.

Not to scale!

Figure 16.14 interactive figure

Frames from a supercomputer simulation of structure formation. These five boxes depict the development of a cubical region that is now 140 million light-years across. The labels above the boxes give the age of the universe and the labels below give the size of the box as it expands with time. Notice that the distribution of matter is only slightly lumpy when the universe is young (left frame). Structures grow more pronounced with time as the densest lumps draw in more and more matter.

that these larger structures were also regions of enhanced density. Galaxies, clusters, superclusters, and the Sloan Great Wall probably all started as higher-density regions of different sizes. The voids in the distribution of galaxies probably started as lower-density regions.

The structure we see in today's universe probably mirrors the distribution of dark matter when the universe was very young.

If this model of structure formation is correct, then the structures we see in today's universe should mirror the original distribution of dark matter very early in time. Supercomputer models of structure formation can now simulate the growth of galaxies, clusters, and larger structures from tiny density enhancements as the universe evolves (Figure 16.14). The results of these models look remarkably similar to the slices of the universe in Figure 16.13, bolstering our confidence in this scenario. However, the models do not tell us *why* the universe started with these slight density enhancements—that is a topic for the next chapter. Nevertheless, it seems increasingly clear that these "lumps" in the early universe were the seeds of all the marvelous structures we see in the universe today.

(MA) Fate of the Universe Tutorial, Lessons 1–3

16.4 The Universe's Fate

We now arrive at one of the ultimate questions in astronomy: How will the universe end? Edwin Hubble's work established that galaxies are rapidly flying away from one another [Section 15.2], but the gravitational pull of each galaxy on every other galaxy acts to slow the expansion. The possible outcomes seem to fall into two general categories. If gravity is strong enough, the expansion will someday halt and the universe will begin collapsing, eventually ending in a cataclysmic crunch. Alternatively, if the expansion can overcome the pull of gravity, the universe will continue to expand forever, growing ever colder as its galaxies grow ever farther apart. The fate of the universe thus seems to boil down to a simple question: Is the universe expanding fast enough to escape its own gravitational pull and keep on expanding forever?

Some say the world will end in fire,
Some say in ice.
From what I've tasted of desire
I hold with those who favor fire.
But if it had to perish twice,
I think I know enough of hate
To say that for destruction ice
Is also great
And would suffice.

***Fire and Ice,* by Robert Frost**

Will the universe continue expanding forever?

Let's begin by considering the fate of the universe as it seemed just over a decade ago, before the discoveries that suggested the presence of dark energy in the universe. In the absence of dark energy, we would expect gravity to slow the expansion of the universe with time. In that case, the fate of the universe hinges on the overall strength of the universe's gravitational pull. The strength of this pull depends on the density of matter in the universe: The greater the density, the greater the overall strength of gravity and the higher the likelihood that gravity will someday halt the expansion.

Precise calculations show that gravity can win out over expansion if the current density of the universe exceeds a seemingly minuscule 10^{-29} gram per cubic centimeter, which is roughly equivalent to a few hydrogen atoms in a volume the size of a closet. The precise density marking the dividing line between eternal expansion and eventual collapse is called the **critical density**. (Remember that, for the moment, we are considering a universe without dark energy.)

Observations of the luminous matter in galaxies show that the mass contained in stars falls far short of the critical density. The visible parts of galaxies contribute about 0.5% of the matter density needed to halt the universe's expansion. The fate of the universe would therefore seem to rest with the dark matter. Is there enough dark matter to halt the expansion of the universe?

Because stars do not provide enough gravity, the expansion could halt only if the total mass of dark matter were at least 200 times that of the mass in stars. Our studies of individual galaxies suggest that they contain at least 10 times as much dark matter as matter in stars, and studies of clusters of galaxies raise that number further to about 50 times as much dark matter as matter in stars. However, this is still only about a quarter of the amount of dark matter needed to halt the expansion. If the proportion of dark matter in the universe at large is indeed similar to that in clusters, the universe seems destined to expand forever. For gravity to reverse the expansion and pull the universe back together, even more dark matter would have to lie beyond the boundaries of clusters.

The universe does not appear to contain enough dark matter to prevent it from expanding forever.

If large-scale structures really did contain a higher proportion of dark matter than do clusters, the influence of that extra dark matter would show up in the velocities of galaxies near those large-scale structures: Larger amounts of dark matter would cause greater deviations from Hubble's law. However, studies of galaxy velocities are holding the line near the value we infer from clusters, which is about 25% of the critical density required to reverse the expansion. If that is the case, the universe seems destined to expand forever, even without dark energy.

Is the expansion of the universe accelerating?

In the past decade, observations of distant white dwarf supernovae have enabled us to study the expansion of the universe in an entirely new way. Because white dwarf supernovae are such good standard candles [Section 15.2], we can use them to determine whether gravity has been slowing the universe's expansion, as it must if the universe is destined to end in a cataclysmic crunch.

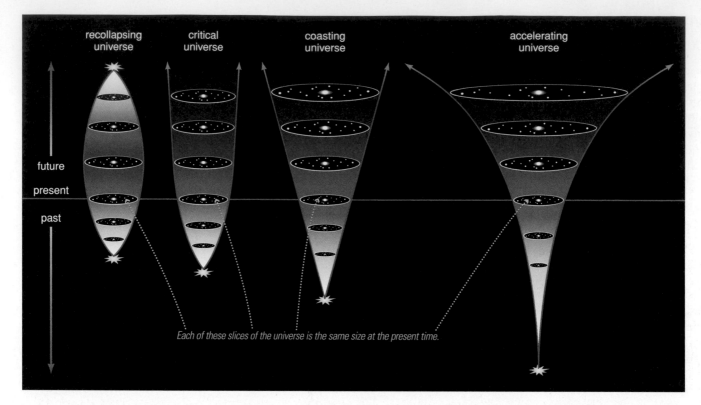

recollapsing universe   critical universe   coasting universe   accelerating universe

future
present
past

Each of these slices of the universe is the same size at the present time.

Figure 16.15

Four models for the fate of the universe. Each diagram shows how the size of a circular slice of the universe changes with time in a particular model. The slices are the same size at the present time, marked by the red line, but the models make different predictions about the sizes of the slices in the past and future. The first three cases assume that there is no dark energy, so that the fate of the universe depends only on how its actual density compares to the critical density. The last case assumes that a repulsive force—perhaps from dark energy—is accelerating the expansion with time. (The diagram assumes continuous acceleration, but observations suggest that the expansion of our universe actually slowed down for a few billion years before acceleration began.)

However, the astronomers who set out to measure gravity's influence over the universe by observing supernovae discovered something quite unexpected. Instead of slowing because of gravity, the expansion of the universe appears to be speeding up, suggesting that some mysterious repulsive force—possibly produced by *dark energy*—is pushing all the universe's galaxies apart. This discovery, if it holds up to further scrutiny, has far-reaching implications for both the fate of the universe and our understanding of the forces that govern its behavior on large scales. To understand the evidence for an accelerating expansion, we must become more familiar with the possible futures of the universe.

Four Expansion Patterns Astronomers subdivide the two general possibilities for the fate of the universe, expanding forever or someday collapsing, into four broad categories. Each represents a particular pattern of change in the future expansion rate (Figure 16.15). We will call these four possible expansion patterns *recollapsing, critical, coasting,* and *accelerating.* The first three possibilities assume that gravity is the only force that affects the expansion rate of the universe, while the fourth adds a repulsive force (from dark energy) that opposes the gravity of matter.

> The expansion rate of the universe depends on the balance between gravity, which acts to slow the expansion, and dark energy, which acts to accelerate it.

- **A recollapsing universe:** If there is no dark energy and the matter density of the universe is *larger* than the critical density, the collective gravity of all its matter will eventually halt the universe's expansion and reverse it. All matter will come crashing back together, and the entire universe will end in a fiery "Big Crunch." We call this a recollapsing universe because the final state, with all the matter collapsed together, would look much like the state in which the universe began. (A recollapsing universe is sometimes called a *closed universe*, because mathematical calculations show that it must have an overall geometry that closes in upon itself like the surface of a sphere, but in more dimensions [Section 17.3].)

- **A critical universe:** If there is no dark energy and the matter density of the universe *equals* the critical density, the collective gravity of all its matter is exactly the amount needed to balance the expansion. The universe will never collapse but will expand more and more slowly as time progresses. We call this a critical universe because its density is the critical density. (Mathematically speaking, a critical universe's overall geometry is "flat"—like the surface of a table but in more dimensions. Thus, a critical universe is one example of what astronomers call a *flat universe*.)

- **A coasting universe:** If there is no dark energy and the matter density of the universe is *smaller* than the critical density, the collective gravity of all its matter cannot halt the expansion. The universe will keep expanding (coasting) forever, with little change in its rate of expansion. (A coasting universe is sometimes called an *open universe*, because its overall geometry is more like the open surface of a saddle than like the closed surface of a sphere.)

- **An accelerating universe:** If dark energy exerts a repulsive force that causes the expansion of the universe to *accelerate* with time, then the expansion rate will grow with time. Galaxies will recede from one another with increasing speed, and the universe will become cold and dark more quickly than it would in a coasting universe. (Depending on the strength of gravity relative to the repulsive force, the overall geometry of an accelerating universe could be flat, open, or closed. As we'll discuss in Chapter 17, current evidence suggests a flat geometry.)

think about it Do you think that one of the potential fates of the universe is preferable to the others? If so, why? If not, why not?

special topic: ---

What Did Einstein Consider His Greatest Blunder?

SHORTLY AFTER EINSTEIN completed his general theory of relativity in 1915, he found that it predicted that the universe could not be standing still: The mutual gravitational attraction of all the matter would make the universe collapse. Because Einstein thought at the time that the universe should be eternal and static, he decided to alter his equations. In essence, he inserted a "fudge factor" called the *cosmological constant* that acted as a repulsive force to counteract the attractive force of gravity.

Had he not been so convinced that the universe should be standing still, Einstein might instead have come up with the correct explanation for why the universe is not collapsing: because it is still expanding from the event of its birth. After Hubble discovered the universal expansion, Einstein called his invention of the cosmological constant "the greatest blunder" of his career.

However, astronomers have begun to take the idea of a cosmological constant more seriously. In the mid-1990s, a few observations suggested that the oldest stars were slightly older than the age of the universe derived from Hubble's constant under the assumption that gravity is the only force affecting the universe's expansion. Clearly, stars cannot be older than the universe. If these observations were being interpreted correctly, the universe had to be older than the age implied by Hubble's constant. If the expansion rate has accelerated, so that the universe is expanding faster today than it was in the past, then the age of the universe would be greater than that ordinarily found from Hubble's constant (see Figure 16.16). What could cause the expansion of the universe to accelerate over time? The repulsive force represented by a cosmological constant, of course.

Further study of these confusing observations eventually showed that the oldest stars probably are *not* older than the age of the universe derived from Hubble's constant. However, measurements of distances to high-redshift galaxies using white dwarf supernovae as standard candles now show that the expansion *is* accelerating. The measured acceleration is just what one would expect from a cosmological constant arising from dark energy. Einstein's greatest blunder, it seems, might not have been such a blunder after all.

Figure 16.16

Data from white dwarf supernovae are shown along with four possible models for the expansion of the universe. Each curve shows how the average distance between galaxies changes with time for a particular model. A rising curve means that the universe is expanding, and a falling curve means that the universe is contracting. Notice that the supernova data fit the accelerating universe better than the other models.

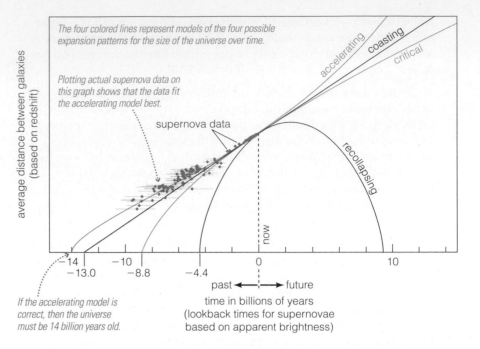

The four colored lines represent models of the four possible expansion patterns for the size of the universe over time.

Plotting actual supernova data on this graph shows that the data fit the accelerating model best.

supernova data

accelerating
coasting
critical
recollapsing

average distance between galaxies (based on redshift)

−14
−13.0
−10
−8.8
−4.4
0
now
10

If the accelerating model is correct, then the universe must be 14 billion years old.

past ← | → future

time in billions of years (lookback times for supernovae based on apparent brightness)

Evidence for Acceleration

Figure 16.16 illustrates how the average distance between galaxies should change with time for each possibility. The lines for the accelerating, coasting, and critical universes always continue upward as time increases, because in these cases the universe is always expanding. The steeper the slope, the faster the expansion. In the recollapsing case, the line begins on an upward slope but eventually turns around and declines as the universe contracts. All the lines pass through the same point and have the same slope at the moment labeled "now," because the current separation between galaxies and the current expansion rate in each case must agree with observations of the present-day universe.

The age that we infer for the universe from its expansion rate differs in each case. A recollapsing universe requires the least amount of time to arrive at the current separation between galaxies—the example in Figure 16.16 goes from zero separation to the current separation in less than 5 billion years. The cases for which gravity is less important require more time to achieve the current separation between galaxies. The ages we would infer from the examples in Figure 16.16 are 8.8 billion years for a critical universe, 13 billion years for a coasting universe, and around 14 billion years for an accelerating universe.

see it for yourself   Toss a ball in the air, and observe how it rises and falls. Then make a graph to illustrate your observations, with time on the horizontal axis and height on the vertical axis. Which universe model does your graph most resemble? What is the reason for that resemblance? How would your graph look different if Earth's gravity were not as strong?

This relationship between the age of the universe and its expansion pattern enables us to determine the expansion pattern from observations of white dwarf supernovae. Because these supernovae are so bright and make such excellent standard candles, we can identify them and measure their distances and redshifts even when they are more than halfway across the observable universe. The distance we measure essentially tells us the lookback time to the supernova, while its redshift tells us how much the universe has expanded since the time of the supernova explosion. Combining these two pieces of information for supernova explosions at different times in the past tells us how the expansion rate has changed with time.

Distances measured to faraway white dwarf supernovae indicate that the expansion of the universe is now speeding up.

Observations of such distant supernovae are still very difficult, but we have some data that are plotted as dots in Figure 16.16. These data fit the curve for an accelerating universe better than any of the other models. In other words, observational evidence seems to support an accelerating universe.

Exactly why the expansion of the universe might be accelerating remains a deep mystery. No known force would push the universe's galaxies apart, and an enormous amount of energy would be required to do so. Of course, our lack of understanding does not stop us from giving a name to whatever is causing the acceleration, and we have already discussed why it is often called *dark energy*. Keep in mind, however, that we do not yet have any idea of what the dark energy might actually be. Nevertheless, if dark energy really exists, it is the most prevalent form of energy in the universe, outstripping the total mass-energy of all the matter in the universe—including the dark matter. Only continued observations will tell us whether the dark energy is real or an artifact of our still-limited data.

A Never-Ending Expansion? Whether or not dark energy exists, and whatever dark energy might turn out to be, it now seems likely that the universe is indeed doomed to expand forever, its galaxies receding ever more quickly into an icy, empty future. After all, our examination of the strength of gravity showed it to be too weak to stop the expansion even without dark energy, and the acceleration due to dark energy would only seem to seal this fate. Some scientists even hypothesize that the dark energy could eventually cause galaxies, stars, and planets to break apart and disperse.

Based on current data, the universe seems destined to expand forever.

However, before we convince ourselves that we now know the fate of the universe, we should bear in mind that forever is a very long time. The universe may hold other surprises that we haven't yet discovered, surprises that might force us to rethink what might happen between now and the end of time.

This is the way the world ends
This is the way the world ends
This is the way the world ends
Not with a bang but a whimper.

From *The Hollow Men*, by T. S. Eliot

---

the big picture
Putting Chapter 16 into Context

We have found that there may be much more to the universe than meets the eye. Dark matter too dim for us to see seems to far outweigh the stars, and a mysterious dark energy may be even more prevalent (see Figure 16.17 for a review). Here are some key "big picture" ideas to remember from this chapter:

- Dark matter and dark energy sound very similar, but they are each hypothesized to explain different observations. *Dark matter* is thought to exist because we detect its gravitational influence.

Scientists suspect that most of the matter in the universe is *dark matter* we cannot see, and that the expansion of the universe is accelerating because of a *dark energy* we cannot directly detect. Both dark matter and dark energy have been proposed to exist because they bring our models of the universe into better agreement with observations, in accordance with the process of science. This figure presents some of the evidence supporting the existence of dark matter and dark energy.

(1) **Dark Matter in Galaxies:** Applying Newton's laws of gravity and motion to the orbital speeds of stars and gas clouds suggests that galaxies contain much more matter than we observe in the form of stars and glowing gas.

(2) **Dark Matter in Clusters:** Further evidence for dark matter comes from studying galaxy clusters. Observations of galaxy motions, hot gas, and gravitational lensing all suggest that galaxy clusters contain far more matter than we can directly observe in the form of stars and gas.

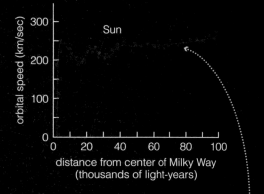

Orbital speeds of stars and gas clouds remain high even quite far from our galaxy's center . . .

This cluster of galaxies acts as a gravitational lens to bend light from a single galaxy behind it into the multiple blue shapes in this photo. The amount of bending allows astronomers to calculate the total amount of matter in the cluster.

dark matter

luminous matter

. . . indicating that the visible portion of our galaxy lies at the center of a much larger volume of dark matter.

HALLMARK OF SCIENCE   **A scientific model must seek explanations for observed phenomena that rely solely on natural causes.** Orbital motions within galaxies demand a natural explanation, which is why scientists proposed the existence of dark matter.

HALLMARK OF SCIENCE   **Science progresses through creation and testing of models of nature that explain the observations as simply as possible.** Dark matter accounts for our observations of galaxy clusters more simply than alternative hypotheses.

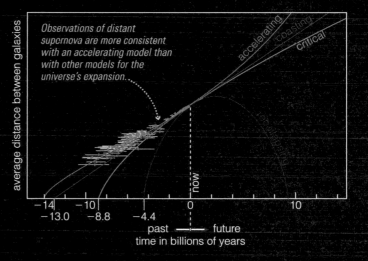

(3) **Structure Formation:** If dark matter really is the dominant source of gravity in the universe, then its gravitational force must have been what assembled galaxies and galaxy clusters in the first place. We can test this prediction using supercomputers to model the formation of large-scale structures both with and without dark matter. Models with dark matter provide a better match to what we observe in the real universe.

(4) **Universal Expansion and Dark Energy:** The expansion of a universe consisting primarily of dark matter would slow down over time (due to gravity), but observations suggest that the expansion is actually speeding up. Scientists hypothesize that a mysterious *dark energy* is causing the expansion to accelerate. Models that include both dark matter and dark energy agree more closely with observations than models containing dark matter alone.

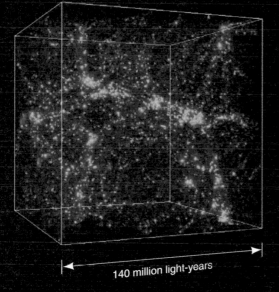

140 million light-years

Supercomputer models in which dark matter is the dominant source of gravity show galaxies organized into strings and sheets similar in size and shape to those we observe in the real universe.

HALLMARK OF SCIENCE A scientific model makes testable predictions about natural phenomena. If predictions do not agree with observations, the model must be revised or abandoned. Observations of the universe's expansion have forced us to further modify our models of the universe to include dark energy along with dark matter.

Dark energy is a term given to the source of the force that may be accelerating the expansion of the universe.

- Either dark matter exists, or we do not understand how gravity operates on large scales. There are many reasons to be confident in our understanding of gravity, so most astronomers conclude that dark matter is real.

- Dark matter seems by far to be the most abundant form of mass in the universe. We still do not know what it is, but we suspect it is largely made up of some type of as-yet-undiscovered subatomic particles.

- If dark matter is indeed the dominant source of gravity in the universe, then it is the glue that binds together galaxies, clusters, superclusters, and other large-scale structures in the universe. All this structure has probably grown from regions where the density of dark matter in the early universe was slightly enhanced.

- The fate of the universe depends on whether gravity can halt the expansion. The total strength of gravity seems too weak to do so even when we account for dark matter, and the evidence that the expansion is accelerating because of dark energy only reinforces the suggestion that the expansion will never cease.

summary of key concepts

16.1 Unseen Influences in the Cosmos

• **What do we mean by dark matter and dark energy?**
Dark matter and **dark energy** have never been directly observed, but each has been proposed to exist because it seems the simplest way to explain a set of observed motions in the universe. *Dark matter* is the name given to the unseen mass whose gravity governs the observed motions of stars and gas clouds. *Dark energy* is the name given to whatever may be causing the expansion of the universe to accelerate.

16.2 Evidence for Dark Matter

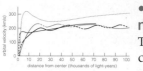

• **What is the evidence for dark matter in galaxies?**
The orbital velocities of stars and gas clouds in galaxies do not change much with distance from the center of the galaxy. Applying Newton's laws of gravitation and motion to these orbits leads to the conclusion that the total mass of a galaxy is far larger than the mass of its stars. Because no detectable visible light is coming from this matter, we call it *dark matter*.

• **What is the evidence for dark matter in clusters of galaxies?**

We have three different ways of measuring the amount of dark matter in clusters of galaxies: from galaxy orbits, from the temperature of the hot gas in clusters, and from the **gravitational lensing** predicted by Einstein. All of these methods agree, indicating that the total mass of a cluster is about 50 times the mass of its stars, implying huge amounts of dark matter.

• **Does dark matter really exist?**
We infer that dark matter exists from its gravitational influence on the matter we can see, leaving two possibilities: Either dark matter exists, or there is something wrong with our understanding of gravity. We cannot rule out the latter possibility, but we have good reason to be confident that our current understanding of gravity is correct and dark matter is real.

• **What might dark matter be made of?**
Some of the dark matter could be ordinary matter, or **baryonic matter**, in the form of dim stars or planetlike objects, but there does not appear to be enough ordinary matter to account

for all the dark matter. Most of it is probably extraordinary matter, or **nonbaryonic matter**, consisting of yet undiscovered particles that we call **WIMPs**.

16.3   Structure Formation

• **What is the role of dark matter in galaxy formation?**

Because most of a galaxy's mass is in the form of dark matter, the gravity of that dark matter is probably what formed protogalactic clouds and then galaxies from slight density enhancements in the early universe.

• **What are the largest structures in the universe?**

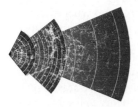
Galaxies appear to be distributed in gigantic chains and sheets that surround great voids. These giant structures trace their origin directly back to regions of slightly enhanced density early in time.

16.4   The Universe's Fate

• **Will the universe continue expanding forever?**

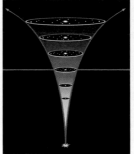
Even before we consider the possibility of a mysterious dark energy, the evidence points to eternal expansion. The **critical density** is the average matter density the universe would need for the strength of gravity to eventually halt the expansion. The overall matter density of the universe appears to be only about 25% of the critical density.

• **Is the expansion of the universe accelerating?**

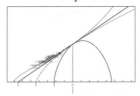
Observations of distant supernovae indicate that the expansion of the universe is speeding up. No one knows the nature of the mysterious force (due to dark energy) that could be causing this acceleration.

exercises and problems

For instructor-assigned homework go to www.masteringastronomy.com.

Review Questions

Short-Answer Questions Based on the Reading

1. Define *dark matter* and *dark energy,* and clearly distinguish between them. What types of observations have led scientists to propose the existence of each of these unseen influences?
2. What is a *rotation curve?* Describe the rotation curve of the Milky Way, and explain how it indicates the presence of large amounts of dark matter.
3. How do we construct rotation curves for other spiral galaxies? What do they tell us about the galaxy masses and dark matter?
4. How do we measure the masses of elliptical galaxies? What do these masses lead us to conclude about dark matter in elliptical galaxies?
5. Briefly describe the three different ways of measuring the mass of a cluster of galaxies. Do the results from the different methods agree? What do they tell us about dark matter in galaxy clusters?
6. What is *gravitational lensing?* Why does it occur? How can we use it to estimate the masses of lensing objects?
7. Briefly explain why the conclusion that dark matter exists rests on assuming that we understand gravity correctly. Is it possible that our understanding of gravity is not correct? Explain.
8. In what sense is dark matter "dark"? Briefly explain why objects like you, planets, and even dim stars could qualify as dark matter.
9. What are *MACHOs?* How can we search for them? Briefly describe why these searches suggest that starlike (or planetlike) objects and black holes *cannot* account for all the dark matter in the halo of our galaxy.
10. Explain what we mean when we say that a neutrino is a *weakly interacting particle.* Why can't the dark matter in galaxies be made of neutrinos?

11. What are *WIMPs?* Why does it seem possible that dark matter consists of these particles, even though we do not yet know what they are?
12. Briefly explain why dark matter is thought to have played a major role in the formation of galaxies and larger structures in the universe. What evidence suggests that large structures are still forming?
13. What does the large-scale structure of the universe look like? Explain why we think this structure reflects the density patterns of the early universe.
14. What do we mean by the *critical density* of the universe? According to current evidence, how does the actual density of matter in the universe compare to the critical density?
15. Describe and compare four possible patterns to the expansion of the universe: recollapsing, critical, coasting, and accelerating. Observationally, how can we decide which of the four possible expansion models is the right one? Based on the current evidence, which model is favored?
16. Assuming the acceleration is real, what does it imply for the fate of the universe? What does current evidence suggest for the fate if the acceleration is not real? Explain.

Test Your Understanding

Does It Make Sense?

Decide whether the statement makes sense (or is clearly true) or does not make sense (or is clearly false). Explain clearly; not all of these have definitive answers, so your explanation is more important than your chosen answer.

17. Strange as it may sound, the majority of mass and energy in the universe may take forms that we are unable to detect directly.

18. A cluster of galaxies is held together by the mutual gravitational attraction of all the stars in the cluster's galaxies.

19. We can estimate the total mass of a cluster of galaxies by studying the distorted images of galaxies whose light passes through the cluster.

20. Clusters of galaxies are the largest structures that we have so far detected in the universe.

21. The primary evidence for an accelerating universe comes from observations of young stars in the Milky Way.

22. There is no doubt remaining among astronomers that the fate of the universe is to expand forever.

23. Dark matter is called "dark" because it blocks light from traveling between the stars.

24. If the universe has more dark matter than we think, then it is also younger than we think.

25. The distance to a white dwarf supernova with a particular redshift is larger in an accelerating universe than in a universe with no acceleration.

26. If dark matter consists of WIMPs, then we should be able to observe photons produced by collisions between these particles.

Quick Quiz

Choose the best answer to each of the following. Explain your reasoning with one or more complete sentences.

27. Dark matter is inferred to exist because (a) we see lots of dark patches in the sky. (b) it explains how the expansion of the universe can be accelerating. (c) we can observe its gravitational influence on visible matter.

28. Dark energy has been hypothesized to exist in order to explain (a) observations suggesting that the expansion of the universe is accelerating. (b) the high orbital speeds of stars far from the center of our galaxy. (c) the giant voids between large-scale structures in the universe.

29. The flat part of the Milky Way's rotation curve tells us that stars in the outskirts of the galaxy (a) orbit the galactic center just as fast as stars closer to the center. (b) rotate rapidly on their axes. (c) travel in straight lines rather than elliptical orbits.

30. Strong evidence for the existence of dark matter comes from observations of (a) our solar system. (b) the center of the Milky Way. (c) clusters of galaxies.

31. A photograph of a cluster of galaxies shows distorted images of galaxies that lie behind it at greater distances. This is an example of what astronomers call (a) dark energy. (b) spiral density waves. (c) a gravitational lens.

32. Based on the observational evidence, is it possible that dark matter doesn't really exist? (a) No, there is too much evidence to think that it could be in error. (b) Yes, but only if there is something wrong with our current understanding of how gravity should work on large scales. (c) Yes, but only if all the observations themselves are in error.

33. Based on current evidence, which of the following is considered a likely candidate for the majority of the dark matter in galaxies? (a) subatomic particles that we have not yet detected (b) swarms of dim, red stars (c) supermassive black holes

34. Which region of the early universe was most likely to become a galaxy? (a) a region whose matter density was lower than average (b) a region whose matter density was higher than average (c) a region with a high concentration of dark energy

35. The major evidence for the idea that the expansion of the universe is accelerating comes from observations of (a) white dwarf supernovae. (b) the orbital speeds of stars within galaxies. (c) the evolution of quasars.

36. Which of these possible types of universe would *not* expand forever? (a) a critical universe (b) an accelerating universe (c) a recollapsing universe

Process of Science

Examining How Science Works

37. *Dark Matter.* Overall, how convincing do you find the case for the existence of dark matter? Write a short essay in which you first describe what we mean by dark matter and what evidence there is for its existence and then give your opinion about the strength of the evidence.

38. *Dark Energy.* Overall, how convincing do you find the case for the existence of dark energy? Write a short essay in which you first describe what we mean by dark energy and what evidence there is for its existence and then give your opinion about the strength of the evidence.

39. *Alternative Gravity.* Suppose someone proposes a new theory of gravity that claims to explain observed motion in galaxies and clusters of galaxies without dark matter. Briefly describe at least one other test that you would expect the new theory to be able to pass if it is, in fact, a better theory of gravity than general relativity.

Investigate Further

In-Depth Questions to Increase Your Understanding

Short-Answer/Essay Questions

40. *The Future Universe.* Based on current evidence concerning the growth of structure in the universe, briefly describe what you would expect the universe to look like on large scales about 10 billion years from now.

41. *Dark Matter and Life.* State and explain at least two reasons why dark matter is (or was) essential for life to exist on Earth.

42. *Rotation Curves.* Draw and label a rotation curve for each of the following hypothetical situations. Make sure the radius axis has approximate distances labeled.
    a. All the mass of the galaxy is concentrated in the center.
    b. The galaxy has a constant mass density inside 20,000 light-years and zero density outside of this region.
    c. The galaxy has a constant mass density inside 20,000 light-years, and its enclosed mass increases proportionally to the distance outside of this region.

43. *Dark Energy and Supernova Brightness.* When astronomers began measuring the brightnesses and redshifts of distant white dwarf supernovae, they expected to find that expansion of the universe was slowing down. Instead they found that it was speeding up! Were the distant supernovae brighter or fainter than expected? Explain why. (*Hint:* In Figure 16.16, the position of a supernova point on the vertical axis is proportional to its redshift. Its position on the horizontal axis is proportional to its brightness—supernovae seen farther back in time are not as bright as those seen closer in time.)

44. *What Is Dark Matter?* Describe at least three possible constituents of dark matter. Explain how we would expect each to interact with light, and how we might go about detecting its existence.

45. *Alternative Gravity.* How would gravity have to be different in order to explain the rotation curves of galaxies without the need for dark matter? Would gravity need to be stronger or weaker than expected at very large distances? Explain.

Quantitative Problems

Be sure to show all calculations clearly and state your final answers in complete sentences.

46. *White Dwarf M/L.* What is the mass-to-light ratio of a $1M_{Sun}$ white dwarf with luminosity $0.001L_{Sun}$?

47. *Supergiant M/L.* What is the mass-to-light ratio of a $30M_{Sun}$ supergiant star with luminosity $300,000L_{Sun}$?

48. *Solar System M/L.* What is the mass-to-light ratio of the solar system?

49. *Mass from Rotation Curve.* Study the rotation curve for the spiral galaxy NGC 7541, which is shown in Figure 16.4.
    a. Use the orbital velocity law from Cosmic Calculations 14.1 to determine the mass (in solar masses) of NGC 7541 enclosed within a radius of 30,000 light-years from its center. (*Hint:* 1 light-year = 9.461×10^{15} m.)
    b. Use the orbital velocity law to determine the mass of NGC 7541 enclosed within a radius of 60,000 light-years from its center.
    c. Based on your answers to parts (a) and (b), what can you conclude about the distribution of mass in this galaxy?

50. *Weighing a Cluster.* A cluster of galaxies has a radius of about 6.7 million light-years (6.2×10^{22} m) and its hot gas has a temperature of 8×10^7 K, corresponding to an orbital speed for the galaxies of 1080 km/s. Estimate the mass of the cluster using the orbital velocity law from Cosmic Calculations 14.1. Give your answer in both kilograms and solar masses. Suppose the combined luminosity of all the stars in the cluster is $8 \times 10^{12}L_{Sun}$. What is the cluster's mass-to-light ratio?

Discussion Questions

51. *Dark Matter or Revised Gravity.* One possible explanation for the evidence we find for dark matter is that we are currently using the wrong law of gravity to measure the masses of very large objects. If we really do misunderstand gravity, then many fundamental theories of physics, including Einstein's theory of general relativity, will need to be revised. Which explanation for our observations do you find more appealing, dark matter or revised gravity? Explain why. Why do you suppose most astronomers find dark matter more appealing?

52. *Our Fate.* Scientists, philosophers, and poets alike have speculated on the fate of the universe. How would you prefer the universe as we know it to end, in a "Big Crunch" or through eternal expansion? Explain the reasons behind your preference.

Web Projects

53. *Gravitational Lenses.* Gravitational lensing occurs in numerous astronomical situations. Compile a catalog of 5–10 examples from the Web, including pictures of lensed stars, quasars, and galaxies. Give a one-paragraph explanation of what's happening in each picture.

54. *Accelerating Universe.* Search for the most recent information about the possible acceleration of the expansion of the universe. Write a one- to three-page report on your findings.

55. *The Nature of Dark Matter.* Use the Web to find recent reports on the possible nature of dark matter. Write a one- to three-page report that summarizes the latest ideas about what dark matter is made of.

visual skills check

Use the following questions to check your understanding of some of the many types of visual information used in astronomy. Answers are provided in Appendix K. For additional practice, try the Chapter 16 Visual Quiz at **www.masteringastronomy.com.**

The schematic figure to the right shows a more complicated expansion history than the four idealized models shown in Figure 16.15. Answer the questions, using the information given in this figure.

1. At Time A, is the expansion of the universe accelerating, coasting, or decelerating?
2. At Time B, is the expansion of the universe accelerating, coasting, or decelerating?
3. At Time C, is the expansion of the universe accelerating, coasting, or decelerating?
4. At Time D, is the expansion of the universe accelerating, coasting, or decelerating?

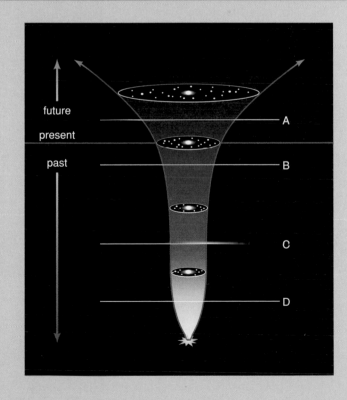

17

the beginning of time

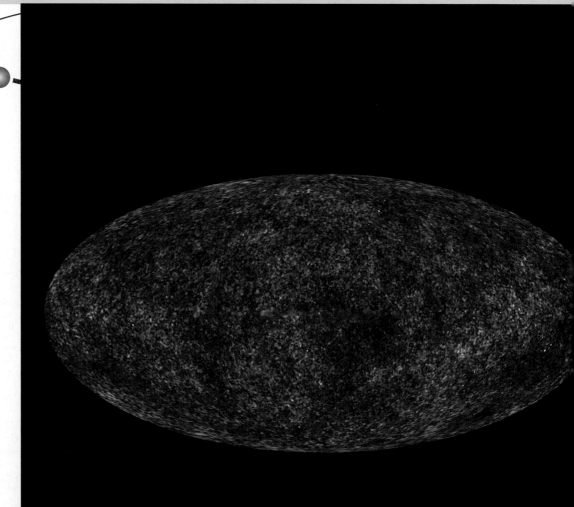

The universe has been expanding for about 14 billion years. During that time, matter collected into galaxies. Stars formed in those galaxies, producing heavy elements that were recycled into later generations of stars. One of these late-coming stars formed about $4\frac{1}{2}$ billion years ago, in the outskirts of a galaxy called the Milky Way. This star was born with a host of planets that formed in a flattened disk surrounding it. One of these planets became covered with life that evolved into ever more complex forms. Today, the most advanced lifeform on this planet, human beings, can look back on this series of events and marvel at how the universe created conditions suitable for life.

Up to this point in the book, we have discussed how the matter produced in the early universe gradually assembled into planets, stars, and galaxies. However, we have not yet answered one big question: Where did *matter itself* come from? To answer this question, we must go far beyond the most distant galaxies. We must go back not only to the origins of matter and energy but to the beginning of time itself.

MA Hubble's Law Tutorial, Lessons 1–3

17.1 The Big Bang

Is it really possible to study the origin of the entire universe? Not long ago, questions about the origin of everything we see were considered unfit for scientific study. That attitude began to change with Hubble's discovery that the universe is expanding. This discovery led to the insight that all things very likely sprang into being at a single moment in time, in an event that we have come to call the *Big Bang*. Today, powerful telescopes allow us to view how galaxies have changed over the past 14 billion years, and at great distances we see young galaxies still in the process of forming [Section 15.3]. These observations confirm that the universe is gradually aging, just as we should expect if the entire universe really was born some 14 billion years ago.

Unfortunately, we cannot see back to the very beginning of time. Light from the most distant galaxies shows us what the universe looked like when it was 1 or 2 billion years old. Beyond these galaxies, we have not yet found any objects shining brightly enough for us to see them. Ultimately, we face an even more fundamental problem. The universe is filled with a faint glow of radiation that appears to be leftover heat from the Big Bang. This faint glow is light that has traveled freely through space since the universe was about 380,000 years old, which is when the universe first became transparent to light. Before that time, light could not pass freely through the universe, so there is no possibility of seeing light from earlier times. Just as we must rely on mathematical modeling to determine what the Sun is like on the inside, we must also use modeling to investigate what the universe was like during its earliest moments.

essential preparation

1. How can we know what the universe was like in the past? [Section 1.1]

2. Where do objects get their energy? [Section 4.3]

3. What is matter? [Section 5.1]

4. How does light tell us the temperatures of planets and stars? [Section 5.2]

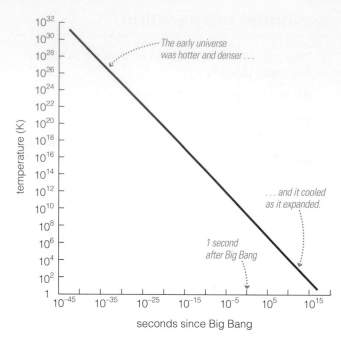

Figure 17.1

The universe cools as it expands. We can calculate past temperatures by using the laws of physics and the current temperature of the universe (about 3 K). This graph shows the results. Notice that both axis scales use powers of 10. (The graph extends to the present: 14 billion years $\approx 4 \times 10^{17}$ seconds.)

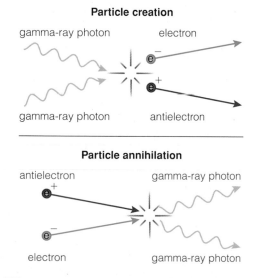

Figure 17.2

Electron–antielectron creation and annihilation. Reactions like these constantly converted photons to particles and vice versa in the early universe.

• What were conditions like in the early universe?

Scientific models of the conditions that prevailed in the early universe are based on fundamental principles of physics. The universe is cooling and becoming less dense as it expands, so it must have been hotter and denser in the past. Calculating exactly how hot and dense the universe must have been when it was more compressed is similar to calculating the temperature and density of gas in a balloon when you squeeze it, except that the conditions become much more extreme. Figure 17.1 shows the results of such calculations.

Particle Creation and Annihilation   The universe was so hot during its first few seconds that photons could transform themselves into matter, and vice versa, in accordance with Einstein's formula $E = mc^2$ [Section 4.3]. Reactions that create and destroy matter are now relatively rare in the universe, but physicists can reproduce such reactions in laboratories.

One such reaction is the creation or destruction of an *electron–antielectron pair* (Figure 17.2). When two photons collide with a total energy greater than twice the mass-energy of an electron (the electron's mass times c^2), they can create two brand-new particles: a negatively charged electron and its positively charged twin, the *antielectron* (also known as a *positron*). The electron is a particle of **matter**, and the antielectron is a particle of **antimatter**. The reaction that creates an electron–antielectron pair also runs in reverse. When an electron and an antielectron meet, they *annihilate* each other, transforming all their mass-energy back into photon energy.

The very early universe was so hot that energy could be transformed into matter and vice versa.

Similar reactions can produce or destroy any particle–antiparticle pair, such as a proton and antiproton or a neutron and antineutron. The early universe therefore was filled with an extremely hot and dense blend of photons, matter, and antimatter, converting furiously back and forth. Despite all these vigorous reactions, describing conditions in the early universe is straightforward, at least in principle. We simply need to use the laws of physics to calculate the proportions of the various forms of radiation and matter at each moment in the universe's early history. The only difficulty is our incomplete understanding of the laws of physics.

To date, physicists have investigated the behavior of matter and energy at temperatures as high as those that existed in the universe just *one ten-billionth* (10^{-10}) of a second after the Big Bang, giving us confidence that we actually understand what was happening at that early time. Our understanding of physics is less certain under the more extreme conditions that prevailed even earlier, but we have some ideas about what the universe was like when it was a mere 10^{-38} second old, and perhaps a glimmer of what it was like at the age of just 10^{-43} second. These tiny fractions of a second are so small that, for all practical purposes, we are studying the very moment of creation—the Big Bang itself.

Fundamental Forces   To understand the changes that occurred in the early universe, it helps to think in terms of *forces*. Everything that happens in the universe today is governed by four distinct forces: *gravity, electromagnetism*, the *strong force*, and the *weak force*. We have already encountered each of these forces individually.

Gravity is the most familiar of the four forces, acting as the "glue" that holds planets, stars, and galaxies together. The electromagnetic force, which depends on the electrical charge of a particle instead of its mass, is far stronger than gravity. It is therefore the dominant force between particles in atoms and molecules, responsible for all chemical and biological reactions. However, the existence of both positive and negative electrical charge causes the electromagnetic force to lose out to gravity on large scales, even though both forces decline with distance following an inverse square law. Most astronomical objects are electrically neutral overall, making the electromagnetic force unimportant on that scale. Gravity therefore becomes the dominant force for such objects, because more mass always means more gravity.

The strong and weak forces operate only over extremely short distances, making them important within atomic nuclei but not on larger scales. The strong force binds atomic nuclei together [Section 10.2]. The weak force plays a crucial role in nuclear reactions such as fission and fusion, and it is the only force besides gravity that affects weakly interacting particles such as neutrinos and WIMPs (weakly interacting massive particles [Section 16.2]).

Although the four forces behave quite differently from one another, scientists now think that they are actually just different aspects of a smaller number of more fundamental forces, probably only one or two (Figure 17.3). At the high temperatures that prevailed in the early universe, the four forces were not as distinct as they are today.

As an analogy, think about ice, liquid water, and water vapor. These three substances are quite different from one another in appearance and behavior, yet they are just different phases of the single substance H_2O. In a similar way, experiments have shown that the electromagnetic and weak forces lose their separate identities under conditions of very high temperature or energy and merge together into a single **electroweak force**. At even higher temperatures and energies, the electroweak force may merge with the strong force and ultimately with gravity. Theories that predict the merger of the electroweak and strong forces are called **grand unified theories**, or **GUTs** for short. The merger of the strong, weak, and electromagnetic forces is therefore often called the *GUT force*. Many physicists suspect that at even higher energies, the GUT force and gravity merge into a single "super force" that governs the behavior of everything. (Among the names you may hear for theories linking all four forces are *supersymmetry, superstrings,* and *supergravity*.)

The four forces that operate in the universe today may have been unified early in time and later became distinct as the universe expanded and cooled.

If these ideas are correct, then the universe was governed solely by the "super force" in the first instant after the Big Bang. As the universe expanded and cooled, the super force split into gravity and the GUT force, which then split further into the strong and electroweak forces. Ultimately, all four forces became distinct. As we'll see shortly, these changes in the fundamental forces probably occurred before the universe was one ten-billionth of a second old.

• What is the history of the universe according to the Big Bang theory?

The **Big Bang theory**—the scientific theory of the universe's earliest moments—is based on applying known and tested laws of physics to the idea that everything we see today, from Earth to the cosmic horizon,

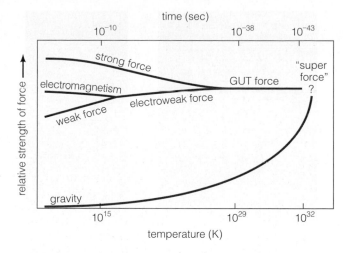

Figure 17.3

The four forces are distinct at low temperatures but may merge at very high temperatures, such as those that prevailed during the first fraction of a second after the Big Bang.

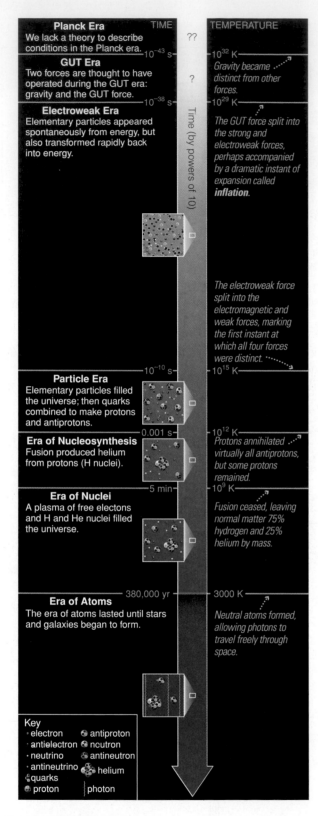

Planck Era
We lack a theory to describe conditions in the Planck era.

GUT Era
Two forces are thought to have operated during the GUT era: gravity and the GUT force.

Electroweak Era
Elementary particles appeared spontaneously from energy, but also transformed rapidly back into energy.

Particle Era
Elementary particles filled the universe; then quarks combined to make protons and antiprotons.

Era of Nucleosynthesis
Fusion produced helium from protons (H nuclei).

Era of Nuclei
A plasma of free electrons and H and He nuclei filled the universe.

Era of Atoms
The era of atoms lasted until stars and galaxies began to form.

TIME
??
10^{-43} s
?
10^{-38} s

Time (by powers of 10)

10^{-10} s
0.001 s
5 min
380,000 yr

TEMPERATURE

10^{32} K
Gravity became distinct from other forces.

10^{29} K
The GUT force split into the strong and electroweak forces, perhaps accompanied by a dramatic instant of expansion called **inflation**.

The electroweak force split into the electromagnetic and weak forces, marking the first instant at which all four forces were distinct.

10^{15} K

10^{12} K
Protons annihilated virtually all antiprotons, but some protons remained.

10^9 K
Fusion ceased, leaving normal matter 75% hydrogen and 25% helium by mass.

3000 K
Neutral atoms formed, allowing photons to travel freely through space.

Key
- electron
- antielectron
- neutrino
- antineutrino
- quarks
- proton
- antiproton
- neutron
- antineutron
- helium
- photon

Figure 17.4 interactive figure

A time line for the eras of the early universe. The only era not shown is the era of galaxies, which began with the birth of stars and galaxies when the universe was a few hundred million years old.

began as an incredibly tiny, hot, and dense collection of matter and radiation. The Big Bang theory describes how expansion and cooling of this unimaginably intense mixture of particles and photons could have led to the present universe of stars and galaxies, and it explains several aspects of today's universe with impressive accuracy. We will discuss the evidence supporting the Big Bang theory later in this chapter. First, in order to help you understand the significance of the evidence, we'll examine the history of the universe according to this theory.

To help make sense of the universe's early history, we will divide it into a series of *eras*, or time periods, each distinguished from the next by some major change in physical conditions as the universe cools. As you can see in Figure 17.1, the temperature dropped extremely rapidly during the first second after the Big Bang: While the temperature was above 10^{32} K during the first instant of time, it had already dropped below 10^{10} K by the time the universe was just 1 second old. Because the behavior of matter and energy depends on temperature, this enormous drop in temperature led to dramatic changes in the universe.

Many of the key events that shaped today's universe occurred within the first few minutes after the Big Bang.

The rest of this section describes the conditions and transitions that marked the eras of the early universe. You'll find it useful to refer to the time line shown in Figure 17.4 as you read along. Notice that most of the key events in the history of the universe occurred in a very short period of time. It will take you longer to read this chapter than it took the universe to progress through the first five eras we will discuss.

The Planck Era As we work our way back through time, we ultimately reach the limit of our current scientific ability to understand the physical conditions when the universe was an incomprehensibly young 10^{-43} second old and the temperature was above 10^{32} K. This instant in time is called the *Planck time* after physicist Max Planck, one of the founders of the science of quantum mechanics. We refer to all times prior to the Planck time (the first 10^{-43} second) as the **Planck era**.

We do not yet understand the physics of the universe well enough to describe what it was like during the Planck era.

Current theories of physics cannot adequately describe the extreme conditions that must have existed during the Planck era. According to the laws of quantum mechanics, there must have been substantial energy fluctuations from point to point in the very early universe. Because energy and mass are equivalent, Einstein's general theory of relativity tells us that these energy fluctuations must have generated a rapidly changing gravitational field that randomly warped space and time. During the Planck era, these random energy fluctuations were so large that our current theories are inadequate to describe what might have been happening. The problem is that we do not yet have a theory that links quantum mechanics (our successful theory of the very small) and general relativity (our successful theory of the very big). Perhaps someday we will be able to merge these theories of the very small and the very big into a single "theory of everything." Until that happens, science cannot describe the universe of the Planck era.

Nevertheless, while we can't say much about the Planck era itself, we have some idea of how it ended and what followed. If you look back at Figure 17.3, you'll see that all four forces are thought to merge into the single, unified "super force" at temperatures above 10^{32} K. In that case, the Planck era would have been a time of ultimate simplicity, when just a single force operated in nature. The Planck era came to an end at the instant when the temperature dropped below 10^{32} K, allowing gravity to become distinct from the other three forces, which were still merged as the GUT force. By analogy to ice crystals forming as a liquid cools, we say that gravity "froze out" at the end of the Planck era.

The GUT Era   The universe subsequently entered the **GUT era**, when two forces operated in the universe: gravity and the GUT force. Recall that the GUT force is a unified force representing the merger of the strong, weak, and electromagnetic forces. According to grand unified theories, these three forces merge together only at temperatures above 10^{29}K (see Figure 17.3). The GUT era therefore lasted only until the temperature fell to 10^{29} K, at which point the GUT force split into the strong and electroweak forces. The universe reached this temperature at an age of a mere 10^{-38} second, which means the entire GUT era lasted less than a trillion-trillion-trillionth of a second.

Energy released near the end of the GUT era may have caused a dramatic expansion of the universe known as inflation.

Our current understanding of physics allows us to say only slightly more about the GUT era than the Planck era, and none of our ideas about the GUT era have been sufficiently well-tested to give us great confidence about what occurred during that time. However, if the grand unified theories are correct, the freezing out of the strong and electroweak forces at the end of the GUT era may have released an enormous amount of energy, causing a sudden and dramatic expansion of the universe that we call **inflation**. In a mere 10^{-36} second, pieces of the universe the size of an atomic nucleus may have grown to the size of our solar system. Inflation sounds bizarre, but as we will discuss later, it explains several important features of today's universe.

The Electroweak Era   Once the GUT force split at the end of the GUT era, the universe entered an era during which three distinct forces operated: gravity, the strong force, and the electroweak force. We call this time the **electroweak era**, because the electromagnetic and weak forces were still unified in the electroweak force. Intense radiation filled all of space, as it had since the Planck era, spontaneously producing matter and antimatter particles that almost immediately annihilated each other and turned back into photons.

The universe continued to expand and cool throughout the electroweak era, dropping to a temperature of 10^{15} K when it reached an age of 10^{-10} second. This temperature is still 100 million times hotter than the temperature in the core of the Sun, but it was low enough for the electromagnetic and weak forces to freeze out from the electroweak force. After this instant (10^{-10} second), all four forces were forever distinct in the universe.

The end of the electroweak era marks an important transition not only in the physical universe, but also in human understanding of the universe. The theory that unified the weak and electromagnetic forces, developed in the 1970s, predicted the emergence of new types of particles (called the W and Z bosons, or *weak bosons*) at temperatures above the 10^{15} K that pervaded the universe when it was 10^{-10} second old. In 1983, experiments performed in a huge particle accelerator near the French/Swiss border reached energies equivalent to such high temperatures for the first time. The new particles showed up just as predicted, produced from the extremely high energy in accord with $E = mc^2$.

In other words, we have direct experimental evidence concerning the conditions in the universe at the end of the electroweak era. We do *not* have any direct experimental evidence of conditions prior to that time. Our theories concerning the earlier parts of the electroweak era and the GUT era consequently are much more speculative than our theories describing the universe from the end of the electroweak era to the present.

The Particle Era As long as the universe was hot enough for the spontaneous creation and annihilation of particles, the total number of particles was roughly in balance with the total number of photons. Once it became too cool for this spontaneous exchange of matter and energy to continue, photons became the dominant form of energy in the universe. We refer to the time between the end of the electroweak era and the moment when spontaneous particle production ceased as the **particle era**, to emphasize the role of subatomic particles during this period.

During the first part of the particle era (and earlier eras), photons turned into all sorts of exotic particles that we no longer find freely existing in the universe today, including *quarks*—the building blocks of protons and neutrons. By the end of the particle era, all quarks had combined into protons and neutrons, which shared the universe with other particles such as electrons, neutrinos, and perhaps WIMPs. The particle era came to an end when the universe reached an age of 1 millisecond (0.001 second) and the temperature had fallen to 10^{12} K. At this point, it was no longer hot enough to produce protons and antiprotons (or neutrons and antineutrons) spontaneously from pure energy.

If the universe had contained equal numbers of protons and antiprotons (or neutrons and antineutrons) at the end of the particle era, all of the pairs would have annihilated each other, creating photons and leaving essentially no matter in the universe. From the obvious fact that the universe contains matter, we conclude that protons must have slightly outnumbered antiprotons at the end of the particle era.

We can estimate the size of the imbalance between matter and antimatter by comparing the present numbers of protons and photons in the universe. The two numbers should have been similar in the very early universe, but today photons outnumber protons by about a billion to one. We conclude that for every billion antiprotons in the early universe, there were about a billion and one protons. As a result, for each 1 billion protons and antiprotons that annihilated each other at the end of the particle era, a single proton was left over. This slight excess of matter over antimatter makes up all the ordinary matter in the present-day universe. Some of the protons (and neutrons) left over from when the universe was 0.001 second old are the very ones that make up our bodies.

The Era of Nucleosynthesis So far, everything we have discussed occurred within the first 0.001 second of the universe's existence—a time span shorter than the time it takes you to blink an eye. At this point, the protons and neutrons left over after the annihilation of antimatter began to fuse into heavier nuclei. However, the heat of the universe remained so high that most nuclei broke apart as fast as they formed. This dance of fusion and breakup marked the **era of nucleosynthesis**.

Most of the helium in the universe was made during the first 5 minutes.

The era of nucleosynthesis ended when the universe was about 5 minutes old. After this time, the density in the expanding universe had dropped so much that fusion no longer occurred, even though the temperature was still about a billion Kelvin (10^9 K)—much hotter than the temperature at the center of the Sun today. When fusion ceased, about 75% of the mass of the ordinary (baryonic) matter in the universe remained as individual protons, or hydrogen nuclei. The other 25% of this mass had fused into helium nuclei, with trace amounts of deuterium (hydrogen with a neutron) and lithium (the next heaviest element after hydrogen and helium). Except for the relatively small amount of matter that stars later forged into heavier elements, the chemical composition of the universe remains the same today.

The Era of Nuclei At the end of the era of nucleosynthesis, the universe consisted of a very hot plasma of hydrogen nuclei, helium nuclei, and electrons. This basic picture held for the next 380,000 years as the universe continued to expand and cool. The fully ionized nuclei moved independently of electrons during this period (rather than being bound with electrons in neutral atoms), which we call the **era of nuclei**. Throughout this era, photons bounced rapidly from one electron to the next, just as they do deep inside the Sun today [Section 10.2], never managing to travel far between collisions. Any time a nucleus captured an electron to form a complete atom, one of the photons quickly ionized it.

Photons began to travel freely through the universe about 380,000 years after the Big Bang, when electrons first combined with nuclei to make atoms.

The era of nuclei ended when the universe was about 380,000 years old and the temperature had fallen to about 3000 K—roughly half the temperature of the Sun's surface today. Hydrogen and helium nuclei captured electrons for good, forming stable, neutral atoms for the first time. With electrons bound into atoms, the universe became transparent, as if a thick fog had lifted. Photons, formerly trapped among the electrons, began to stream freely across the universe. We still see these photons today as the *cosmic microwave background*, which we will discuss shortly.

The Era of Atoms and the Era of Galaxies We've already discussed the rest of the story in earlier chapters. The end of the era of nuclei marked the beginning of the **era of atoms**, when the universe consisted of a mixture of neutral atoms and plasma (ions and electrons), along with a large number of photons. Thanks to the slight density enhancements present in the universe at this time and the gravitational attraction of dark matter, the atoms and plasma slowly assembled into protogalactic clouds. Stars formed in these clouds, transforming the gas clouds into galaxies. The first full-fledged galaxies had formed by the time the universe was about 1 billion years old, beginning what we call the **era of galaxies**.

The first galaxies formed by the time the universe was a billion years old.

The era of galaxies continues to this day. Generation after generation of star formation in galaxies steadily produces elements heavier than helium and incorporates them into new star systems. Some of these star systems develop planets, and on at least one of these planets, life burst into being a few billion years ago. Now here we are, thinking about it all.

Early Universe Summary Figure 17.5 (on pages 478–479) summarizes the major ideas from our brief overview of the history of the universe as

The Big Bang theory is a scientific model that explains how the present-day universe developed from an extremely hot and dense beginning. This schematic diagram shows how conditions in the early universe changed as the universe expanded and cooled with time.

① Our expanding universe must have started out much hotter and denser than it is today because the expansion caused matter and energy to cool down and spread out with time.

② As the universe cooled down, it may have undergone a brief period of very rapid expansion known as inflation that could account for several key properties of today's universe.

Big Bang

Planck Era

10^{-43} second

10^{-38} second

GUT Era

hotter

10^{32} K

10^{29} K

Electroweak Era

Eras of the Early Universe

Time steps on this strip are in powers of 10. For example, the electroweak era looks wide because it spans 28 powers of 10 in time, even though the entire era lasted only one ten billionth of a second.

This illustration depicts how a small portion of the entire universe changes as it expands with time, but the actual expansion is much greater than shown.

This bright spot represents the instant of the Big Bang, when the universe came into existence.

This dramatic widening represents inflation—the rapid expansion that may have happened at the end of the GUT era.

The early universe was filled with bright light everywhere. The gradually changing color represents the gradually cooling temperature over time.

This blotchy surface at 380,000 years marks the moment when photons first streamed freely through the universe. We can still see those photons today as the cosmic microwave background.

After the release of the cosmic microwave background, the universe was dark until the birth of stars and galaxies.

The era of galaxies was underway by the time the universe was about a billion years old, and it continues to this day.

TIME

space

space

14 billion years

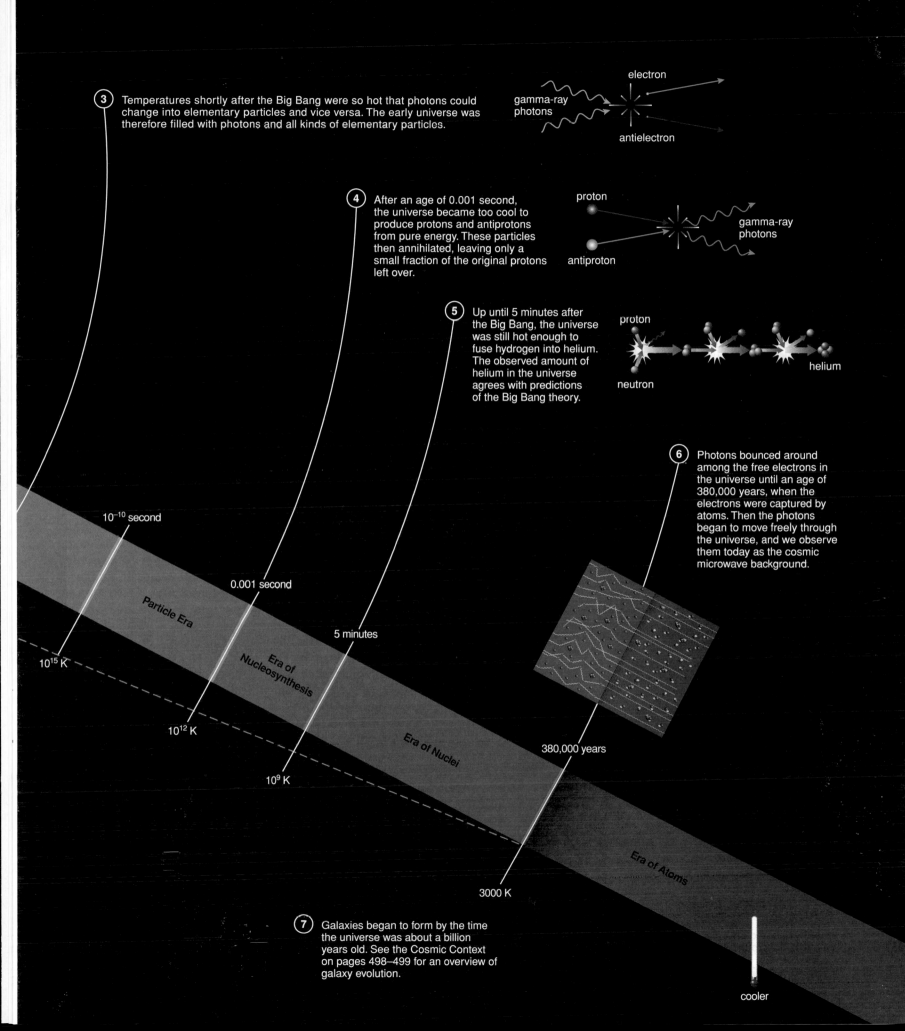

3 Temperatures shortly after the Big Bang were so hot that photons could change into elementary particles and vice versa. The early universe was therefore filled with photons and all kinds of elementary particles.

gamma-ray photons

electron

antielectron

4 After an age of 0.001 second, the universe became too cool to produce protons and antiprotons from pure energy. These particles then annihilated, leaving only a small fraction of the original protons left over.

proton

antiproton

gamma-ray photons

5 Up until 5 minutes after the Big Bang, the universe was still hot enough to fuse hydrogen into helium. The observed amount of helium in the universe agrees with predictions of the Big Bang theory.

proton

neutron

helium

6 Photons bounced around among the free electrons in the universe until an age of 380,000 years, when the electrons were captured by atoms. Then the photons began to move freely through the universe, and we observe them today as the cosmic microwave background.

10^{-10} second

0.001 second

Particle Era

5 minutes

Era of Nucleosynthesis

10^{15} K

10^{12} K

10^{9} K

Era of Nuclei

380,000 years

Era of Atoms

3000 K

7 Galaxies began to form by the time the universe was about a billion years old. See the Cosmic Context on pages 498–499 for an overview of galaxy evolution.

cooler

35. What is the earliest time in the universe that we can directly observe? (a) a few hundred million years after the Big Bang (b) a few hundred thousand years after the Big Bang (c) a few minutes after the Big Bang

36. Which of these options is the best explanation for why the night sky is dark? (a) The universe is not infinite in space. (b) The universe has not always looked the way it looks today. (c) The distribution of matter in the universe is not uniform on very large scales.

Process of Science

Examining How Science Works

37. *Unanswered Questions.* We have seen in this chapter that our understanding of the early universe remains incomplete. Briefly describe one important but unanswered question about the events that happened shortly after the Big Bang. If you think it will be possible to answer that question in the future, describe how we could find an answer, being as specific as possible about the evidence necessary to do so. If you think the question will never be answered, explain why you think it is impossible to answer.

38. *Darkness at Night.* Suppose you are Kepler, pondering the darkness of the night sky without any knowledge of the Big Bang or the expanding universe. Develop a hypothesis to explain the darkness of the night sky that does not depend on the Big Bang theory. Propose an experiment that scientists could perform today to test that hypothesis.

Investigate Further

In-Depth Questions to Increase Your Understanding

Short-Answer/Essay Questions

39. *Life Story of a Proton.* Tell the life story of a proton from its formation shortly after the Big Bang to its presence in the nucleus of an oxygen atom you have just inhaled. Your story should be creative and imaginative, but it should also demonstrate your scientific understanding of as many stages in the proton's life as possible. You can draw on material from the entire book, and your story should be three to five pages long.

40. *Creative History of the Universe.* The story of creation as envisioned by the Big Bang theory is quite dramatic, but it is usually told in a fairly straightforward, scientific way. Write a more dramatic telling of the story, in the form of a short story, play, or poem. Be as creative as you wish, but be sure to remain accurate according to the science as it is understood today.

41. *Re-Creating the Big Bang.* Particle accelerators on Earth can push particles to extremely large speeds. When these particles collide, the amount of energy associated with the colliding particles is much greater than the mass-energy these particles have when at rest. As a result, these collisions can produce many other particles out of pure energy. Explain in your own words how the conditions that occur in these accelerators are similar to the conditions that prevailed shortly after the Big Bang. Also, point out some of the differences between what happens in particle accelerators and what happened in the early universe.

42. *Betting on the Big Bang Theory.* If you had $100, how much money would you wager on the proposition that we have a reasonable scientific understanding of what the universe was like when it was 1 minute old? Explain the reasoning behind your choice in terms of the scientific evidence presented in this chapter.

43. *"Observing" the Early Universe.* The only way we have of studying the era of nucleosynthesis is through the abundances of the nuclei that it left behind. Explain why we will never be able to observe that era through direct detection of the radiation emitted at that time.

44. *Element Production in the Big Bang.* Nucleosynthesis in the early universe was unable to produce more than trace amounts of elements heavier than helium. Using the information in Figure 12.17, which shows the mass per nuclear particle for many different elements, explain why producing elements like lithium (3 protons), boron (4 protons), and beryllium (5 protons) was so difficult.

45. *Evidence for the Big Bang.* Alternatives to the Big Bang theory need to propose alternative explanations for many of the observations scientists have made of the universe. Make a list of at least seven observed features of the universe that are satisfactorily explained by the Big Bang theory when it is combined with the idea of inflation.

Quantitative Problems

Be sure to show all calculations clearly and state your final answers in complete sentences.

46. *Energy from Antimatter.* The total annual U.S. power consumption is about 2×10^{20} joules. Suppose you could supply that energy by combining pure matter with pure antimatter. Estimate the total mass of matter–antimatter fuel you would need to supply the United States with energy for 1 year. How does that mass compare with the amount of matter in your car's gas tank? (A gallon of gas has a mass of about 4 kg.)

47. *Uniformity of the Cosmic Microwave Background.* The temperature of the cosmic microwave background differs by only a few parts in 100,000 across the sky. Compare that level of uniformity to the surface of a table in the following way. Consider a table that is 1 meter in size. How big would the largest bumps on that table be if its surface were smooth to the level of one part in 100,000? Could you see bumps of that size on the table's surface?

48. *Daytime at "Night."* According to Olbers' paradox, the entire sky would be as bright as the surface of a typical star if the universe were infinite in space, unchanging in time, and the same everywhere. However, conditions would not need to be quite that extreme for the "nighttime" sky to be as bright as the daytime sky.

    a. Using the inverse square law for light from Cosmic Calculations 11.1, determine the apparent brightness of the Sun in our sky.

    b. Using the inverse square law for light, determine the apparent brightness our Sun would have if it were at a distance of 10 billion light-years.

    c. From your answers to parts (a) and (b), estimate how many stars like the Sun would need to exist at a distance of 10 billion light-years for their total apparent brightness to equal that of our Sun.

    d. Compare your answer to part (c) with the estimate of 10^{22} stars in our observable universe given in Section 1.2. Use your answer to explain why the night sky is much darker than the daytime sky. How much larger would the total number of stars need to be for "night" to be as bright as day?

Discussion Questions

49. *The Moment of Creation.* You've probably noticed that, when discussing the Big Bang theory, we never talk about the very first instant. Even our most speculative theories at present take us back only to within 10^{-43} second of creation. Do you think it will *ever* be possible for science to consider the moment of creation itself? Will we ever be able to answer questions such as *why* the Big Bang happened? Defend your opinions.

50. *The Big Bang.* How convincing do you find the evidence for the Big Bang model of the universe's origin? What are the strengths of the theory? What does it fail to explain? Overall, do *you* think the Big Bang really happened? Defend your opinion.

Web Projects

51. *New Tests of the Big Bang Theory.* The satellites *COBE* and *WMAP* have provided striking confirmation of several predictions of the Big Bang theory. An upcoming mission called *Planck* will test the Big Bang theory further. Use the Web to gather pictures and information about *COBE*, *WMAP*, and *Planck*. Write a one- to two-page report about the strength of the evidence compiled by *COBE* and *WMAP* and what we might learn from upcoming missions.

52. *New Ideas in Inflation.* The idea of inflation solves many of the puzzles associated with the standard Big Bang theory, but we are still a long way from finding evidence confirming that inflation really occurred. Read a few recent articles that discuss some of the latest ideas about inflation and how we might test these ideas. Write a two- to three-page summary of your findings.

visual skills check

Use the following questions to check your understanding of some of the many types of visual information used in astronomy. Answers are provided in Appendix K. For additional practice, try the Chapter 17 Visual Quiz at **www.masteringastronomy.com.**

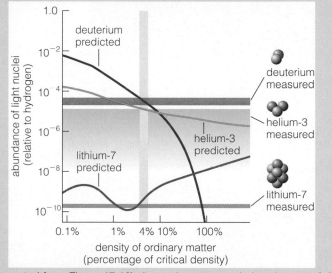

This graph (repeated from Figure 17.12) shows the measured abundances of deuterium, helium-3, and lithium-7, along with the amounts predicted by the Big Bang model. These predictions depend on the density of ordinary matter in the universe. Answer the following questions based on the information provided in the graph.

1. How does the measured abundance of deuterium compare with the measured abundance of hydrogen?
   a. Deuterium is 40,000 times as abundant as hydrogen.
   b. Deuterium and hydrogen have roughly the same abundance.
   c. Hydrogen is between 4 and 5 times as abundant as deuterium.
   d. Hydrogen is 40,000 times as abundant as deuterium.

2. How does the measured abundance of lithium compare with the measured abundance of deuterium?
   a. Deuterium is between 3 and 4 times as abundant as lithium.
   b. Deuterium is between 30 and 40 times as abundant as lithium.
   c. Deuterium is between 300 and 400 times as abundant as lithium.
   d. Deuterium is between 1000 and 100,000 times as abundant as lithium.

3. What abundance of deuterium is predicted by a model in which the density of ordinary matter is 4% of the critical density?
   a. Deuterium is predicted to be 40,000 times as abundant as hydrogen.

   b. Deuterium is predicted to have the same abundance as hydrogen.
   c. The deuterium abundance is predicted to be about 4% of the hydrogen abundance.
   d The deuterium abundance is predicted to be about 2.5×10^{-5} times that of hydrogen.

4. If you measured the abundance of deuterium in the universe to be 1% of the abundance of hydrogen, what would you conclude about the density of ordinary matter?
   a. The density of ordinary matter is slightly less than 0.1% of the critical density.
   b. The density of ordinary matter is slightly greater than 1% of the critical density.
   c. The density of ordinary matter is about 4% of the critical density.
   d. The density of ordinary matter is around 100% of the critical density.

All galaxies, including our Milky Way, developed as gravity pulled together matter in regions of the universe that started out slightly denser than surrounding regions. The central illustration depicts how galaxies formed over time, starting from the Big Bang in the upper left and proceeding to the present day in the lower right, as space gradually expanded according to Hubble's law.

380,000 years

1 billion years

Big Bang

TIME

1 Dramatic inflation early in time is thought to have produced large-scale ripples in the density of the universe. All the structure we see today formed as gravity drew additional matter into the peaks of these ripples. [Section 17.3]

Inflation may have stretched tiny quantum fluctuations into large-scale ripples.

2 Observations of the cosmic microwave background show us what the regions of enhanced density were like about 380,000 years after the Big Bang. [Section 17.2]

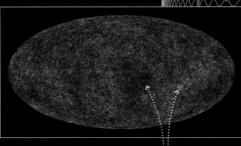

Photo by *WMAP*

Variations in the cosmic microwave background show that regions of the universe differed in density by only a few parts in 100,000.

3 Large-scale surveys of the universe show that gravity has gradually shaped early regions of enhanced density into a web-like structure, with galaxies arranged in huge chains and sheets. [Section 16.3]

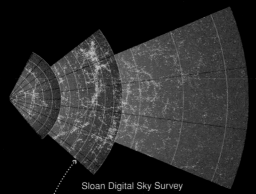

Sloan Digital Sky Survey

The web-like patterns of structure observed in large-scale galaxy surveys agree with those seen in large-scale computer simulations of structure formation.

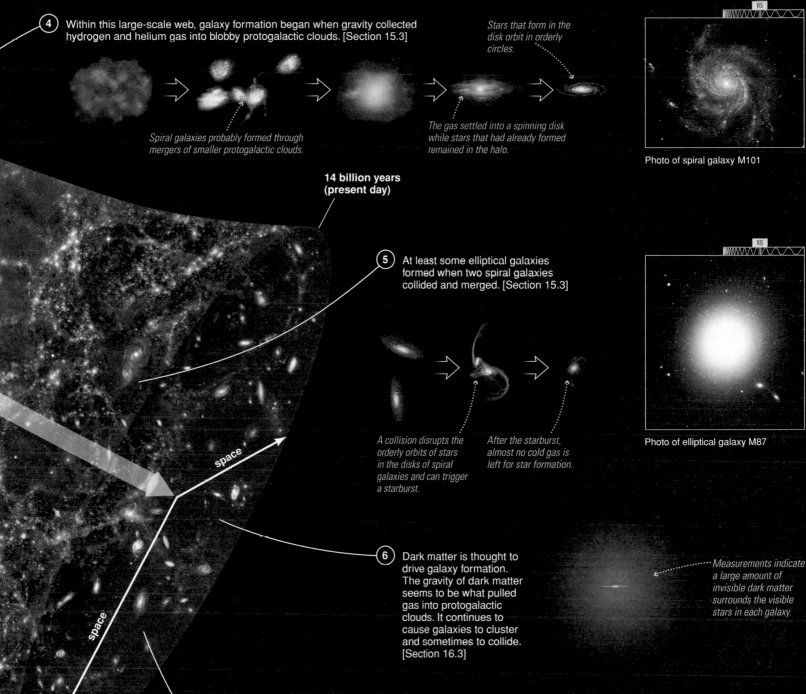

4 Within this large-scale web, galaxy formation began when gravity collected hydrogen and helium gas into blobby protogalactic clouds. [Section 15.3]

Stars that form in the disk orbit in orderly circles.

Spiral galaxies probably formed through mergers of smaller protogalactic clouds.

The gas settled into a spinning disk while stars that had already formed remained in the halo.

Photo of spiral galaxy M101

14 billion years (present day)

5 At least some elliptical galaxies formed when two spiral galaxies collided and merged. [Section 15.3]

A collision disrupts the orderly orbits of stars in the disks of spiral galaxies and can trigger a starburst.

After the starburst, almost no cold gas is left for star formation.

Photo of elliptical galaxy M87

space

space

6 Dark matter is thought to drive galaxy formation. The gravity of dark matter seems to be what pulled gas into protogalactic clouds. It continues to cause galaxies to cluster and sometimes to collide. [Section 16.3]

Measurements indicate a large amount of invisible dark matter surrounds the visible stars in each galaxy.

7 Today, in the disks of spiral galaxies like the Milky Way, the star-gas-star cycle continues to produce new stars and planets from matter that was once in protogalactic clouds. [Section 14.2]

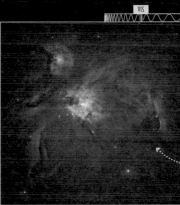

Photo of Orion Nebula

New stars and planetary systems—some perhaps much like our own—are currently forming in the Orion Nebula, 1500 light-years from Earth.

life in the universe

learning goals

We have covered a lot of ground in this book, discussing fundamental questions about the nature and origin of our planet, our star, our galaxy, and our universe. But we have not yet discussed one of the most profound questions of all: Are we alone? The universe seems to be filled with worlds beyond imagination—more than 100 billion star systems in our galaxy alone, and some 100 billion galaxies in the observable universe—yet we do not know whether any world other than our own has ever been home to life.

In this chapter, we will discuss the possibility of life beyond Earth. We'll begin by considering the history of life on Earth, which will help us understand the prospects of finding life elsewhere. We'll then consider the possibility of finding microbial life elsewhere in our solar system or beyond, and examine efforts to search for extraterrestrial intelligence (SETI). Finally, we'll discuss the astonishing implications that the search for life may hold for the future of our own civilization.

essential preparation

1. **How do we detect planets around other stars?** [Section 6.5]

2. **What unique features of Earth are important for life?** [Section 7.5]

3. **Why are Jupiter's Galilean moons so geologically active?** [Section 8.2]

4. **What geological activity do we see on Titan and other moons?** [Section 8.2]

5. **Why is a star's mass its most important property?** [Section 11.2]

18.1 Life on Earth

It may seem that aliens are everywhere. Popular shows and movies, such as *Star Trek* and *Star Wars*, show aliens from many different worlds traveling easily among the stars. Closer to home, supermarket tabloids routinely carry headlines about the latest alien atrocities or about alien corpses hidden by the government at "Area 51."

Despite their media popularity, any alien visitors have been sadly negligent in leaving scientific evidence of their trespass. Decades of scientific observation and study have not turned up a single piece of undeniable evidence that aliens have been here (see Special Topic, p. 523). Why, then, do so many scientists believe it is worth the effort to search for life beyond Earth?

For a long time, the answer was simply that it seemed natural for other worlds to be inhabited. Belief in life on other worlds was common even among ancient Greek philosophers, and many famous scientists of the past few centuries took it as a given that intelligent beings exist on other planets. For example, Kepler suggested that the Moon was inhabited, and William Herschel spoke of life on virtually all the planets in our solar system. Most famously, in the late 19th century, Percival Lowell claimed to see networks of canals on Mars, which he argued were the mark of an advanced civilization. Lowell's ideas became the basis for H. G. Wells's novel *The War of the Worlds.*

Today, the question of whether there is life beyond Earth has become a topic of serious scientific research. This research has been spurred in part by the discovery of planets around other stars [Section 6.5], which confirms that there are plenty of places to look for life beyond our own solar system. But new discoveries about life on Earth have played an even more important role. In particular, three developments in the study of life on Earth have made it seem much more likely that life might exist elsewhere:

- We have learned that life arose quite early in Earth's history, suggesting that life might also form quickly on other worlds with the right conditions.

- Laboratory experiments have shown that the chemical constituents thought to have been common on the young Earth combine readily into complex organic molecules. These experiments suggest that life might have arisen through naturally occurring chemistry—in which case the same chemistry could have given rise to life on many other worlds.

- We have discovered microscopic living organisms that probably could survive in conditions similar to those on at least some other worlds in our solar system, suggesting that the necessities of life may be common in the universe.

Because these three ideas are so important to understanding the modern science of life in the universe, often called *astrobiology*, let's examine each of them in greater detail. Along the way, we'll also discuss the nature and history of life on Earth. We will then be prepared to consider how we might actually search for life on other worlds.

• When did life arise on Earth?

Our planet was born about $4\frac{1}{2}$ billion years ago, but life might have had a difficult time during the first several hundred million years of Earth's history. This was the time of the *heavy bombardment* [Section 6.4], when Earth should have been repeatedly struck by large asteroids or comets. A few of these impacts may have had sufficient energy to vaporize the early oceans completely, likely killing off any life that might already have been present on Earth. The ancestors of living organisms today probably could not have arisen until the end of the heavy bombardment.

Life probably could not have taken permanent hold on Earth until the end of the heavy bombardment, between about 4.2 and 3.9 billion years ago.

Studies of craters on the Moon suggest that the last major impacts of the heavy bombardment occurred between about 4.2 and 3.9 billion years ago, with some evidence suggesting a *late heavy bombardment* toward the end of that time period. Remarkably, we now have evidence suggesting that life was thriving prior to 3.85 billion years ago. If we are interpreting the evidence correctly, life arose in a geological blink of the eye from the moment when conditions first allowed it. To understand this evidence and its importance, we need to discuss how scientists study the history of life on Earth.

The Geological Time Scale We learn about the history of life on Earth through the study of **fossils**, relics of organisms that lived and died long ago. Most fossils form when dead organisms fall to the bottom of a

Figure 18.1

Formation of sedimentary rock. Each layer represents a particular time and place in Earth's history and is characterized by fossils of organisms that lived in that time and place.

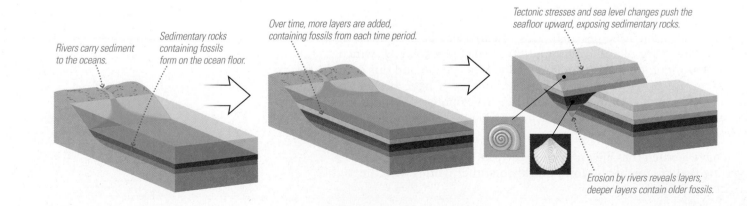

Rivers carry sediment to the oceans.

Sedimentary rocks containing fossils form on the ocean floor.

Over time, more layers are added, containing fossils from each time period.

Tectonic stresses and sea level changes push the seafloor upward, exposing sedimentary rocks.

Erosion by rivers reveals layers; deeper layers contain older fossils.

sea (or other body of water) and are gradually buried by layers of sediment. The sediments are produced by erosion on land and carried by rivers to the sea. Over millions of years, sediments pile up on the seafloor, and the weight of the upper layers compresses underlying layers into rock. Erosion or tectonic activity can later expose the fossils (Figure 18.1). In some places, such as the Grand Canyon, the sedimentary layers record hundreds of millions of years of Earth's history (Figure 18.2).

The key to reconstructing the history of life is to determine the dates at which fossil organisms lived. The *relative* ages of fossils found in different layers are easy to determine: Deeper layers formed earlier and contain more ancient fossils. Radiometric dating [Section 6.4] confirms these relative ages and gives us fairly precise absolute ages for fossils. Based on the layering of rocks and fossils, geologists divide Earth's history into a set of distinct intervals that make up what we call the **geological time scale**. Figure 18.3 shows the names of the various intervals on a timeline, along with numerous important events in Earth's history.

think about it How does the length of time during which animals and plants have lived on land compare to the length of time during which life has existed? How does the length of time during which humans have existed compare to the length of time since mammals and dinosaurs first arose?

Fossil Evidence for the Early Origin of Life You might wonder why the geological time scale shows so much more detail in the last few hundred million years than it does for earlier times. The answer is that fossils become increasingly difficult to find as we look deeper into Earth's history, for three major reasons: First, older rocks are much rarer than younger rocks, because most of Earth's surface is geologically young. Second, even when we find very old rocks, they often turn out to have been

Figure 18.2

The rock layers of the Grand Canyon record more than 500 million years of Earth's history.

Figure 18.3

The geological time scale. Notice that the lower timeline is an expanded view of the last portion of the upper timeline. The eons, eras, and periods are defined by changes observed in the fossil record. The absolute ages come from radiometric dating. (The *K–T event* is the geological term for the impact linked to the mass extinction of the dinosaurs [Section 9.4].)

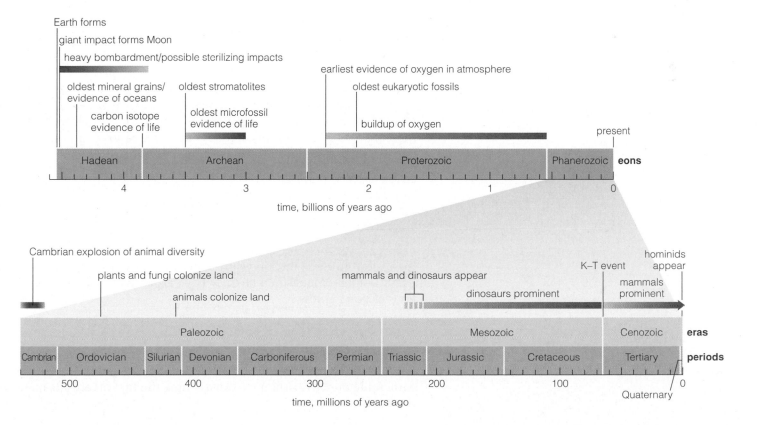

a These large mats at Shark Bay, Western Australia, are colonies of microbes known as "living stromatolites"; they stand about knee-high. Microbes near the top generate energy through photosynthesis.

b The bands visible in this section of a modern-day mat are formed by layers of sediment adhering to different types of microbes.

c This section of a 3.5-billion-year-old stromatolite shows a structure nearly identical to that of a living mat.

Figure 18.4

Rocks called stromatolites offer evidence of microbial life as early as 3.5 billion years ago.

subject to transformations (caused by heat and pressure) that would have destroyed any fossil evidence they may have contained. Third, all life prior to a few hundred million years ago was microscopic, and microscopic fossils are much more difficult to identify than dinosaur bones.

Fossil evidence suggests that life on Earth was already thriving by 3.5 billion years ago.

Despite these difficulties, geologists have found a few very old rocks that suggest life was already thriving on Earth by 3.5 billion years ago, and possibly for several hundred million years before that. One strong line of evidence comes from rocks called *stromatolites*, which look strikingly similar to large microbial mats today (Figure 18.4). Careful analysis suggests that stromatolites were made by living organisms, and radiometric dating shows that some of them are 3.5 billion years old.

If organisms were already advanced enough to build stromatolites by 3.5 billion years ago, more primitive organisms must have lived even earlier. Some evidence supports this idea. More ancient rocks that have undergone too much change to leave fossils intact may still hold carbon that was once part of living organisms.

Carbon has two stable isotopes: carbon-12, with six protons and six neutrons in its nucleus, and carbon-13, which has one extra neutron (see Figure 5.6). Living organisms incorporate carbon-12 slightly more easily than carbon-13. As a result, the fraction of carbon-13 is always a bit lower in fossils than in rock samples that lack fossils. All life and all fossils tested to date show the same characteristic ratio of the two carbon isotopes. Dating very ancient rocks can be difficult, but some rocks that are more than 3.85 billion years old show the same ratio of carbon-12 to carbon-13, suggesting that these rocks contain remnants of life.

Carbon isotope evidence suggests that life arose more than 3.85 billion years ago—quite soon after the end of the heavy bombardment.

To summarize, fossil evidence points to life already thriving by 3.5 billion years ago, and carbon isotope evidence suggests that life was present more than 3.85 billion years ago. We therefore conclude that life on Earth arose within no more than a few hundred million years. If life couldn't start until the end of the heavy bombardment, then life may have arisen within a few tens of millions of years or less.

• How did life arise on Earth?

We have discussed *when* life arose on Earth, but *how* did it arise? To answer this question, we must first consider how life today differs from the earliest life on Earth. Then we can ask how the first organisms may have originated.

The Theory of Evolution The fossil record clearly shows that life has gone through great changes over time (see Figure 18.3). If we are going to understand how life arose, we must understand what causes these changes, so that we can trace life back to its origin. The unifying theory through which scientists understand the history of life on Earth is the **theory of evolution**, proposed by Charles Darwin (1809–1882).

> The fossil record shows us how life has changed through time, and the theory of evolution explains how these changes have occurred.

Evolution simply means "change with time." Many scientists had recognized the strong evidence for evolution in the fossil record, but no one before Darwin had successfully explained how species might undergo change. In essence, the fossil record provides strong evidence that evolution *has* occurred, while Darwin's theory of evolution explains *how* it occurs.

Darwin's theory tells us that evolution proceeds through a process called **natural selection**. The many individuals of any species always differ in certain small ways. If an individual possesses a trait that gives it an advantage in survival and reproduction, then the trait is likely to be passed on to future generations. We say that nature "selects" the advantageous trait, which is why the process is called natural selection. Over time, natural selection can help individuals of a species become better able to compete for scarce resources. If enough small individual variations accumulate, natural selection can even give rise to an entirely new species.

Darwin compiled a tremendous body of evidence supporting the idea that natural selection is the primary mechanism by which evolution proceeds. Some 150 years of ongoing research has only given further support to Darwin's idea, which is why we call it the *theory* of evolution [Section 3.4]. Perhaps the strongest support for the theory has come from the discovery of DNA, the genetic material of all life on Earth. DNA (short for *deoxyribonucleic acid*) has allowed scientists to understand how evolution occurs on a molecular level. Living organisms reproduce by copying DNA and passing these copies on to their descendants. A molecule of DNA consists of two long strands—somewhat like the interlocking strands of a zipper—wound together in the spiral shape known as a double helix (Figure 18.5). The instructions for assembling a living organism are written in the precise order of four chemical bases that make up the interlocking portions of the DNA "zipper." (The four chemical bases are represented in Figure 18.5 by the letters A, T, G, and C, which stand for the first letters of their chemical names.) These bases pair up in a way that ensures that both strands of a DNA molecule contain the same genetic information. By unwinding and allowing new strands to form alongside the original ones (with the new strand made from chemicals floating around inside a cell), a single DNA molecule can give rise to two identical copies of itself. This is how genetic material is copied and passed on to future generations.

Evolution occurs because the passing of genetic information from one generation to the next is not always perfect. An organism's DNA may be occasionally altered by copying errors or by external influences, such as ultraviolet light from the Sun or exposure to toxic or radioactive chemicals. Any change in an organism's DNA is called a **mutation**. Many mutations are lethal, killing the cell in which the mutation occurs. Some, however, may improve a cell's ability to survive and reproduce. The cell then passes on this improvement to its offspring.

Our understanding of the molecular mechanism of natural selection has put the theory of evolution on a stronger foundation than ever.

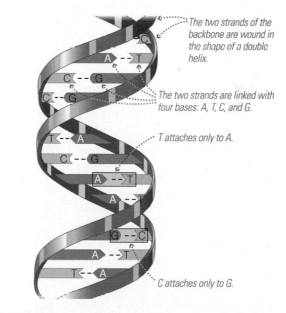

The two strands of the backbone are wound in the shape of a double helix.

The two strands are linked with four bases: A, T, C, and G.

T attaches only to A.

C attaches only to G.

Figure 18.5

This diagram represents a small piece of a DNA molecule, which looks much like a zipper twisted into a spiral. Hereditary information is contained in the "teeth" linking the strands. These "teeth" are the DNA bases. Only four DNA bases are used, and they can link up between the two strands only in specific ways: T attaches only to A, and C attaches only to G. (The color coding is arbitrary and is used only to represent different types of chemical groups; in the backbone, blue and yellow represent sugar and phosphate groups, respectively.)

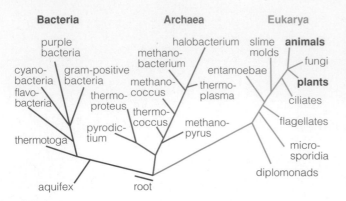

Figure 18.6

The tree of life, showing evolutionary relationships determined by comparison of DNA sequences in different organisms. Just two small branches represent *all* plant and animal species.

Figure 18.7

This photograph shows a black smoker—a volcanic vent on the ocean floor that spews out hot, mineral-rich water. DNA studies indicate that the microbes living near these vents are evolutionarily older than nearly all other living organisms, suggesting that early life may have arisen in similar environments.

While no theory can ever be proved true beyond all doubt, the theory of evolution is as solid as any theory in science, including the theory of gravity and the theory of atoms. Biologists routinely witness evolution occurring before their eyes among laboratory microorganisms, or over periods of just a few decades among plants and animals subjected to environmental stress. Moreover, the theory of evolution has become the underpinning of virtually all modern biology, medicine, and agriculture. For example, agricultural scientists apply the idea of natural selection to develop pest control strategies that reduce the populations of harmful insects without harming the populations of beneficial ones; medical researchers test new drugs on animals that are genetically similar to humans, because the theory of evolution tells us that genetically similar species should have at least somewhat similar physiological responses; and biologists study the relationships between organisms by comparing their DNA.

The First Living Organisms The basic chemical nature of DNA is virtually identical among all living organisms. This fact, along with other biochemical similarities shared by all living organisms, tells us that all life on Earth today can trace its origins to a common ancestor that lived sometime before about 3.85 billion years ago.

All living organisms today evolved from a common ancestor that lived long ago.

What did the first living organisms look like? We are unlikely to find fossils of the earliest organisms, but we can learn about early life through careful studies of the DNA of living organisms. Biologists can determine the evolutionary relationships among living species by comparing the sequences of bases in their DNA. For example, two organisms whose DNA sequences differ in five places for a particular gene are probably more distantly related than two organisms whose gene sequences differ in only one place. Many such DNA comparisons suggest that all living organisms are related in a way depicted schematically by the "tree of life" in Figure 18.6. Although details in the structure of this tree remain uncertain, it confirms that living organisms share a common ancestry. The tree of life as it is known today also shows that life on Earth is divided into three major groupings, or *domains*, called Bacteria, Archaea, and Eukarya. Notice that plants and animals represent only two tiny branches of the great diversity of life on Earth.

Organisms on branches located closer to the root of the tree of life must contain DNA that is evolutionarily older, suggesting that they more closely resemble the organisms that lived early in Earth's history. The modern-day organisms that appear to be evolutionarily oldest are microbes that live in very hot water around seafloor volcanic vents called *black smokers* (because of the dark, mineral-rich water that flows out of them) in the deep oceans (Figure 18.7). Unlike most life at Earth's surface, which depends on sunlight, these organisms get energy from chemical reactions in water heated by undersea volcanoes. We conclude that the most likely home of early life on Earth was around such deep-sea vents.

DNA studies suggest that the earliest organisms probably lived in hot water near sources of volcanic heat.

The idea that early organisms might have lived in such "extreme" conditions may seem surprising, but it makes sense when we think about it. The deep ocean environment would have been protected from the harmful ultraviolet radiation that bathed Earth's surface before our atmosphere had oxygen or an ozone layer. Moreover, the chemical pathways used to extract energy from mineral-rich hot water are simpler than any other chemical pathways (such as photosynthesis) used by living organisms to obtain energy, making it reasonable to assume that early life obtained energy in this way.

The Transition from Chemistry to Biology The theory of evolution explains how the earliest organisms evolved into the great diversity of life on Earth today. But where did the first organisms come from? We may never know for sure, because it is unlikely that any fossil evidence could have recorded the transition from nonlife to life. However, over the past several decades, scientists have conducted many laboratory experiments designed to mimic the conditions that existed on the young Earth. These experiments suggest that life could have arisen through natural chemical reactions.

The first such experiments were performed in the 1950s (Figure 18.8), and they have been refined and improved since that time. In essence, the experiments mix chemical ingredients thought to have been present on the early Earth and then "spark" the chemicals with electricity to simulate lightning or other energy sources. The chemical reactions that follow have produced all the major molecules of life, including all the amino acids and DNA bases. Many of these same molecules are found today in meteorites, suggesting that some organic molecules may have arrived from space.

Figure 18.8

Stanley Miller poses with a reproduction of the experimental setup he first used in the 1950s to study pathways to the origin of life. (He worked with Harold Urey, so the experiment is called the Miller-Urey experiment.)

special topic: --

What Is Life?

YOU MAY HAVE NOTICED that while we've been talking about life, we haven't actually defined the term. In fact, it's surprisingly difficult to draw a clear boundary between life and nonlife. Life can be so difficult to define that we may be tempted to fall back on the famous words of Supreme Court Justice Potter Stewart, who, in avoiding the difficulty of defining pornography, wrote: "I shall not today attempt further to define [it] ... But I know it when I see it." If living organisms on other worlds turn out to be much like those found on Earth, it may prove true that we'll know them when we see them. But if the organisms are fairly different from those on Earth, we'll need clearer guidelines to decide whether they are truly "living."

One approach seeks to identify distinguishing features common to all known life. For example, nearly all living organisms on Earth appear to share the following six key properties:

1. *Order:* Living organisms are not random collections of molecules but rather have molecules arranged in orderly patterns that form cell structures.
2. *Reproduction:* Living organisms are capable of reproducing.
3. *Growth and development:* Living organisms grow and develop in patterns determined at least in part by heredity.
4. *Energy utilization:* Living organisms use energy to fuel their many activities.
5. *Response to the environment:* Living organisms actively respond to changes in their surroundings. For example, organisms may alter their chemistry or movements in the presence of a food source.

6. *Evolutionary adaptation:* Life evolves through natural selection, as organisms pass on traits that make them better adapted to survival in their local environments.

These six properties are all important, but biologists today regard evolution as the most fundamental and unifying of them. Evolutionary adaptation is the only property that can explain the great diversity of life on Earth. Moreover, understanding how evolution works allows us to understand how all the other properties came to be. As a result, the simplest definition of life might be "something that can reproduce and evolve through natural selection."

For most practical purposes, this definition of life would probably suffice. However, some cases may still challenge this definition. For example, computer scientists can now write programs (that is, lines of computer code) that can reproduce themselves (that is, create additional sets of identical lines of code). By adding to the programs instructions that allow random changes, they can even make "artificial life" that evolves on a computer. Should this "artificial life," which consists of nothing but electronic signals processed by computer chips, be considered alive?

The fact that we have such difficulty distinguishing the living from the nonliving on Earth suggests that we should be very cautious about constraining our search for life elsewhere. No matter what definition of life we choose, there's always the possibility that we'll someday encounter something that challenges it. Nevertheless, the properties of reproduction and evolution seem likely to be shared by most if not all life in the universe and therefore provide a useful starting point as we consider how to explore the possibility of extraterrestrial life.

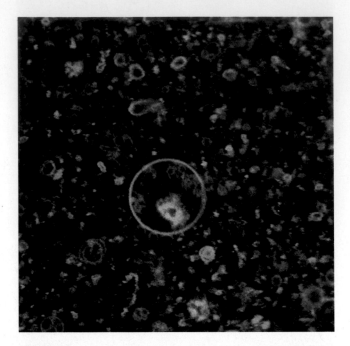

Figure 18.9

This microscopic photo (made with the aid of fluorescent dyes) shows short strands of RNA (red) contained within an enclosed membrane (green circle), both of which formed spontaneously with the aid of clay minerals beneath them.

Figure 18.10

A summary of the steps by which chemistry on the early Earth may have led to the origin of life.

Laboratory experiments show that the molecules of life form easily under conditions that existed on the early Earth.

Laboratory experiments also show that mixing a warm, dilute solution of organic molecules with naturally occurring clay allows the molecules to assemble themselves into much more complex molecules. Strands of RNA, a molecule that looks much like a single strand of DNA, have been produced in the laboratory with lengths of up to nearly 100 bases. Because some RNA molecules are capable of self-replication, many biologists now presume that RNA was the original genetic material of life on Earth, with DNA coming later.

Perhaps even more significantly, experiments have shown that microscopic enclosed membranes also form on the surface of the same clay minerals that help assemble RNA molecules, sometimes with RNA inside them (Figure 18.9). Self-replicating RNA molecules enclosed in such membranes would have in essence made "pre-cells" in which the RNA might have undergone rapid changes. Pre-cells in which RNA replicated faster and more accurately were more likely to spread, leading to a type of positive feedback that would have encouraged even faster and more accurate replication. Given millions of years of chemical reactions occurring all over Earth, RNA might eventually have evolved into DNA to make true living organisms.

Figure 18.10 summarizes the steps by which chemistry might have become biology on Earth. We may never know whether life really arose in this way, but it certainly seems possible.

Could Life Have Migrated to Earth? Our scenario suggests that life could have arisen naturally here on Earth. However, an alternative possibility is that life arose somewhere else first—perhaps on Venus or Mars—and then migrated to Earth on meteorites. Remember that we have collected meteorites that were blasted by impacts from the surfaces of the Moon and Mars [Section 9.1]. Calculations suggest that Venus, Earth, and Mars all should have exchanged many tons of rock, especially in the early days of the solar system when impacts were more common.

The idea that life could travel through space to land on Earth once seemed outlandish. After all, it's hard to imagine a more forbidding environment than that of space, with no air, no water, and constant bombardment by dangerous radiation from the Sun and stars. However, the presence of organic molecules in meteorites and comets tells us that the building blocks of life can survive in space, and tests have shown that some microbes can survive in space for years.

In a sense, Earth, Venus, and Mars have been "sneezing" on each other for billions of years. Life could conceivably have originated on any of these three planets and been transported to the others. It's an intrigu-

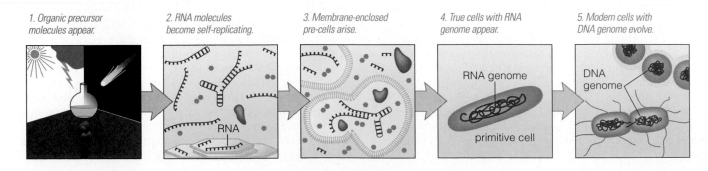

1. Organic precursor molecules appear.

2. RNA molecules become self-replicating.

RNA

3. Membrane-enclosed pre-cells arise.

4. True cells with RNA genome appear.

RNA genome

primitive cell

5. Modern cells with DNA genome evolve.

DNA genome

ing thought, but it does not change our basic scenario for the origin of life—it simply moves it from one planet to another.

A Brief History of Life on Earth   Earth formed about $4\frac{1}{2}$ billion years ago, and the giant impact thought to have formed the Moon [Section 6.4] probably happened quite early. Mineral evidence suggests that Earth had oceans by 4.3 to 4.4 billion years ago, and these early oceans would have been natural laboratories for chemistry that could have led to life. Once the major impacts of the heavy bombardment subsided, life took hold on Earth. Once life started, evolution rapidly diversified it.

Despite the rapid pace of evolution, the most complex life-forms remained single-celled for at least a billion years after life first arose. Some 2 billion years ago, the land was still inhospitable because of the lack of a protective ozone layer [Section 7.1]. Continents much like those of today were surrounded by oceans teeming with life, but, despite pleasant temperatures and plentiful rainfall, the land itself was probably as barren as Mars is now. Things began to change only when oxygen started building up in Earth's atmosphere.

Nearly all the oxygen in our atmosphere was originally released through photosynthesis by single-celled organisms known as *cyanobacteria* (Figure 18.11). These microbes may have been producing oxygen through photosynthesis as early as 3.5 billion years ago. However, the oxygen did not immediately begin to accumulate in the atmosphere. For more than a billion years, chemical reactions with surface rocks pulled oxygen back out of the atmosphere as fast as the cyanobacteria could produce it. But these tiny organisms were abundant and persistent, and eventually the surface rock was so saturated with oxygen that the rate of oxygen removal slowed down. At that point, some 2 billion years ago, oxygen began to accumulate in the atmosphere, though it may not have reached a level that we could have breathed until just a few hundred million years ago.

Today, we often think of oxygen as a necessity for life. However, oxygen was probably poisonous to most organisms living before about 2 billion years ago (and remains a poison to many microbes still living today). The rise of atmospheric oxygen therefore caused tremendous evolutionary pressure and may have been a major factor in the evolution of complex plants and animals.

There were undoubtedly many crucial changes as primitive microbes gradually evolved into multicellular organisms and early plants and animals, but the fossil record does not allow us to pinpoint the times at which all these changes occurred. However, we see a dramatic change in the fossil record dating to the *Cambrian period,* from about 540 million to 500 million years ago (see Figure 18.3). During these 40 million years, animal life evolved from tiny and primitive organisms into all the basic body types (phyla) that we find on Earth today. This remarkable diversification occurred in such a short time relative to the history of Earth that it is often called the *Cambrian explosion.*

Living organisms remained single-celled for most of Earth's history, with larger plants and animals arising only in the past few hundred million years.

Early dinosaurs and mammals arose some 225 to 250 million years ago, but dinosaurs at first proved more successful and dominated for well over 100 million years. Their sudden demise 65 million years ago [Section 9.4] paved the way for the evolution of large mammals—including humans. The earliest humans appeared on the scene only a few million years ago, or after 99.9% of Earth's history to date had already gone by. Our few centuries of industry and technology have come after 99.99999% of Earth's history.

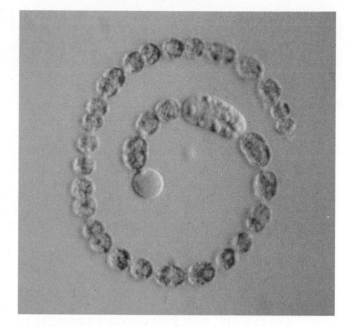

Figure 18.11

This microscopic photo shows a chain of modern cyanobacteria. The ancestors of these living organisms produced essentially all the oxygen in Earth's atmosphere.

• What are the necessities of life?

If we are going to look for other worlds that might harbor life, the first step is to understand the necessities of life. While it is possible that life on other worlds could be quite different from life on Earth, it's easiest to begin the search for life by looking for conditions in which organisms from Earth could live. So what exactly does Earth life need to survive?

If we think about ourselves, the requirements for life seem fairly stringent: We need abundant oxygen in an atmosphere that is otherwise not poisonous, we need temperatures in a fairly narrow range of conditions, and we need abundant and varied food sources. However, the discovery of life in "extreme" environments—such as in the hot water near black smokers—shows that many microbes (often called *extremophiles*) can survive in a much wider range of conditions.

The organisms living in hot water prove that at least some microbes can survive in much higher temperatures than we would have guessed. Other organisms live in other extremes. In the freezing cold but very dry valleys of Antarctica, scientists have found microbes that live *inside* rocks, surviving on tiny droplets of liquid water and energy from sunlight. Microscopic life has also been found deep underground in water that fills pores within subterranean rock. We have found life thriving in environments so acidic, alkaline, or salty that humans would be poisoned almost instantly. We have even found microbes that can survive high doses of radiation, making it possible for them to survive for many years in the radiation-filled environment of space.

If we compare all the different forms of life on Earth, we find that life as a whole has only three basic requirements:

- A source of nutrients (atoms and molecules) from which to build living cells

- Energy to fuel the activities of life, whether from sunlight, from chemical reactions, or from the heat of Earth itself

- Liquid water

These requirements give us a basic road map for the search for life elsewhere. If we want to find life on another world, it makes sense to start by searching for worlds that offer these basic necessities.

Life on Earth requires nutrients, energy, and liquid water. Of the three requirements, liquid water is the only one that is not common on other worlds. Interestingly, only the third requirement (liquid water) seems to pose much of a constraint. Organic molecules are present almost everywhere—even on meteorites and comets. Many worlds are large enough to retain internal heat that could provide energy for life, and virtually all worlds have sunlight (or starlight) bathing their surfaces, although the inverse square law for light [Section 11.1] means that light provides less energy to worlds farther from their star. Nutrients and energy should therefore be available to some degree on almost every planet and moon. In contrast, liquid water is relatively rare, and the search for liquid water therefore drives the search for life in our universe.

18.2 Life in the Solar System

If we assume that life elsewhere would be at least a little bit like life on Earth, then the search for life begins with a search for habitable worlds—worlds that contain the basic necessities for life as we know it, including liquid water. The liquid water requirement rules out most of the worlds in our solar system. For example, Mercury and the Moon are barren and dry, Venus is too hot for liquid water, and most of the small bodies of the outer solar system are too cold. The jovian planets may have droplets of liquid water in some of their clouds, but the strong vertical winds on these planets make their clouds unlikely homes for life. That leaves two major possibilities besides Earth for habitability in our solar system: (1) Mars and (2) a few of the large moons orbiting jovian planets, most notably Europa [Section 8.2]. Let's discuss the current evidence about the potential habitability of these worlds.

think about it   Is it possible for a world to be habitable but not actually have-life? Is it possible for a world to have life but not be habitable? Explain.

• Could there be life on Mars?

We now have sufficiently detailed images of Mars to be quite confident that no civilizations have ever existed there. The canals seen by Percival Lowell were some type of mirage, presumably formed by a combination of real Martian features and his own imagination. Nevertheless, we have good reason to believe that liquid water flowed on Mars in the past [Section 7.3], making it seem possible that life could once have found a home there. Moreover, since Mars contains subsurface ice today, it's conceivable that life could still survive near sources of volcanic heat, where pockets of liquid water might persist underground.

Missions to Mars   Mars is not only the best candidate in our solar system for life beyond Earth, but also the only place where we've begun an actual search for life. Our first attempt came with the *Viking* missions to Mars in the 1970s. Two *Viking* landers arrived on the surface of Mars in 1976. Each was equipped with a robotic arm for scooping up soil samples, which were fed into several on-board, robotically controlled experiments.

Three of the *Viking* experiments were designed expressly to look for signs of life. None of these experiments could actually "see" life but rather looked for chemical changes that could be attributed to living organisms. Although all three experiments gave results that initially seemed consistent with life, further study suggested that chemical reactions could have produced the same results. Moreover, a fourth experiment, which analyzed the content of Martian soil, found no measurable level of organic molecules—the opposite of what we would expect if life were present. As a result, most scientists have concluded that no life was present at the locations where *Viking* sampled the soil.

More recent missions have taken a more measured approach to the search for life on Mars (Figure 18.12). Rather than conducting direct experiments like *Viking*, these missions have been designed to help us better understand Martian conditions, so that we'll have a better idea of how and where to search for life with future missions.

Mars may have been habitable in the past and could have underground liquid water today, making it a prime target for space missions searching for life.

Within a decade or so, NASA hopes to launch a mission to Mars that will bring back surface samples for study in laboratories on Earth. Later, we may send more advanced robots or humans to Mars, where they could search for fossils or living organisms in deep canyons like Valles Marineris, in ancient

Figure 18.12

The *Spirit* rover studies a rock on Mars (the photograph was taken by a camera aboard the rover). Such studies help scientists learn if and when Mars may have been habitable.

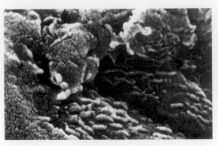

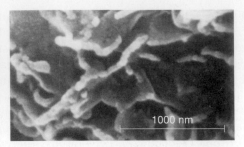

1000 nm

a The Martian meteorite ALH84001, before it was cut open for detailed study. The block on the right provides a sense of scale; it measures 1 cubic centimeter, about the size of a typical sugar cube.

b This photo shows rod-shaped structures found in a highly magnified slice of ALH84001. They measure about 100 nanometers in length and are as small as 10–20 nanometers in width.

c This photo shows terrestrial nanobacteria in a sample of volcanic rock from Sicily. They are close in size to the structures seen in ALH84001.

Figure 18.13

Controversial meteorite evidence for life on Mars.

valley bottoms and dried-up lake beds, or in underground pockets of water near not-quite-dead volcanoes. The search will not be easy, but we should eventually learn whether life has ever existed on Mars.

The Debate over Martian Meteorites An entirely different approach to the search for life on Mars relies on studies of the couple of dozen known meteorites whose chemical composition suggests they came from Mars. One of these meteorites, designated ALH84001, generated particular excitement when a group of scientists claimed that it contained evidence of past life on Mars (Figure 18.13).

ALH84001 was found on the Antarctic ice in 1984. Careful study of the meteorite shows that it landed in Antarctica about 13,000 years ago, following a 16-million-year journey through space after being blasted from Mars by an impact. The rock itself dates to 4.5 billion years ago, which means that it solidified shortly after Mars formed and remained on Mars during times when the climate may have been warmer and wetter. Analysis of the meteorite reveals several lines of evidence that could indicate the past presence of life on Mars. For example, the rock contains layered carbonate minerals and complex organic molecules (polycyclic aromatic hydrocarbons, or PAHs) that are associated with life when found in Earth rock, as well as microscopic chains of magnetite crystals quite similar to chains made in Earth rocks by living bacteria. Most intriguingly, highly magnified images reveal rod-shaped structures that look much like ordinary bacteria, except that they are much smaller. Could these structures be fossil life from Mars?

A meteorite from Mars shows tantalizing evidence of past life, but the evidence can also be explained in a nonbiological way.

It's possible, but each of the tantalizing hints of Martian life can also be explained in a nonbiological way. Subsequent studies have shown that chemical and geological processes can produce structures very similar to those found in the Martian meteorite. In addition, terrestrial bacteria have been found living inside the meteorite, indicating that it was contaminated by Earth life during the 13,000 years it resided in Antarctica. This contamination may explain the presence of the complex molecules found in the rock.

On balance, most scientists now doubt that the Martian meteorite shows true evidence of Martian life. Nevertheless, studies of ALH84001 and other Martian meteorites are continuing, and they may yet turn up surprises.

• Could there be life on Europa or other jovian moons?

After Mars, the next most likely candidates for life in our solar system are some of the moons of the jovian planets—especially Jupiter's moons Europa, Ganymede, and Callisto, and Saturn's moons Titan and Enceladus [Section 8.2]. Europa is the strongest candidate, because we have good reason to think it contains a deep ocean beneath its icy crust (see Figure 8.17). The ice and rock from which Europa formed undoubtedly included the necessary chemical ingredients for life, and Europa's internal heating (primarily due to tidal heating) is strong enough to power volcanic vents on the sea bottom. It's fairly easy to imagine places on Europa's ocean floor that look much like black smokers on Earth. If life on Earth really first arose near such undersea volcanic vents, Europa would seem to have everything needed for life to originate.

If Europa's ocean really exists, it may have seafloor vents much like the black smokers where life on Earth may have first arisen. The possibility of life on Europa is especially interesting because, unlike any potential life on Mars, it would not necessarily have to be microscopic. After all, the several kilometers of surface ice that hide Europa's ocean (if it exists) could also hide large creatures swimming within it. However, the potential energy sources for life on Europa are far more limited than the energy sources for life on Earth (mainly because sunlight could not fuel photosynthesis in the subsurface ocean). As a result, most scientists suspect that any life that might exist on Europa would probably be quite small and primitive.

As we discussed in Chapter 8, some evidence suggests that Jupiter's moons Ganymede and Callisto may also have subsurface oceans. However, these moons would have even less energy for life than Europa. If they have life at all, it is almost certainly small and primitive. Nevertheless, Europa, Ganymede, and Callisto offer the astonishing possibility that Jupiter alone could be orbited by more worlds with life than we find in all the rest of the solar system.

Titan offers another enticing place to look for life. Its surface is far too cold for liquid water, but it appears to have lakes and rivers of liquid methane or ethane [Section 8.2]. Although many biologists think it unlikely, it is possible that this liquid could support life as water does on Earth. Titan might also have liquid water or a colder ammonia–water mixture deep underground, so even water-based life is possible, if unlikely.

Other possible homes to life in our solar system include Jupiter's moons Ganymede and Callisto and Saturn's moons Titan and Enceladus. The discovery of ice fountains on Saturn's moon Enceladus suggests that it, too, could be habitable if these fountains are powered by subsurface liquids. If this proves to be the case, it would open up the possibility that similar liquids—and potentially life—could exist on other solar system bodies, including Neptune's moon Triton and some of the moons of Uranus.

 **Detecting Extrasolar Planets Tutorial, Lessons 1–3**

18.3 Life Around Other Stars

We already know of planets orbiting hundreds of stars besides our Sun, and it's likely that billions of planetary systems inhabit our galaxy. These numbers make prospects for life elsewhere seem quite good, but numbers alone don't tell the whole story. In this section, we'll consider the prospects for life on worlds orbiting other stars.

Before we begin, we must distinguish between *surface* life like that on Earth and *subsurface* life like that we envision as a possibility on Mars or Europa. While large telescopes could in principle allow us to discover surface life on extrasolar planets, no foreseeable technology will allow us to find life that is hidden deep underground in other star systems (unless the subsurface life has a noticeable effect on the planet's atmosphere). We therefore will focus on the search for life on planets with habitable surfaces—surfaces with temperatures and pressures that could allow liquid water to exist.

• Are habitable planets likely?

We have not yet discovered any extrasolar planets that seem likely to be habitable. Presumably because of the limitations of current detection techniques [Section 6.5], nearly all the extrasolar planets found to date are much more massive than Earth, suggesting that they are more like the jovian planets of our solar system than the terrestrial planets. Such planets are unlikely candidates for life, though some of them have orbits that could conceivably allow them to have moons with life.

However, the existence of these jovian planets makes it reasonable to suppose that terrestrial planets are also common around other stars. Although our technology is not yet up to the task of finding them, we can already begin to address the questions of where we might expect them to exist and whether they are likely to be habitable.

Constraints on Star Systems Before we consider planets themselves, it's useful to ask how many stars could potentially have planets with life. In other words, which stars would make good "Suns," providing heat and light to the surfaces of terrestrial planets that happen to orbit them?

The first requirement for a star to have life-bearing worlds is that it be old enough that life could have arisen. Stars of greater mass live shorter lives, and the most massive stars live no more than a few million years [Section 12.3]. Given that life on Earth did not arise for hundreds of millions of years after our solar system was born, we can rule out any star with more than a few times the mass of our Sun. However, because low-mass stars are far more common than high-mass stars, the lifetime constraint rules out only about 1% of all stars.

A second requirement is that the star allow planets to have stable orbits. About half of all stars are in binary or multiple star systems, around which stable planetary orbits are less likely than around single stars. If life is not possible in such systems, then we can rule out about half the stars in our galaxy as potential homes to life. Of course, the other half—still some 100 billion stars or more—remain possible homes for life. Also, under some circumstances stable planetary orbits are possible in multiple star systems, so we shouldn't entirely rule out life in such systems.

A third constraint on the likelihood of finding habitable planets is the size of a star's **habitable zone**—the region in which a terrestrial planet of the right size could have a surface temperature that might allow for liquid water and life. Figure 18.14 shows the approximate sizes, to scale, of the habitable zones around our Sun, around a star with about $\frac{1}{2}$ the mass of the Sun (spectral type K), and around a star with about $\frac{1}{10}$ the mass

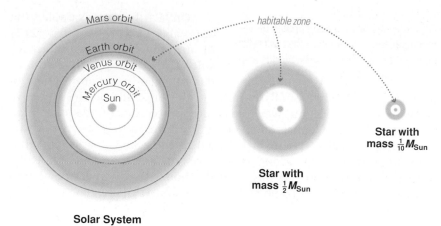

Solar System

Star with
mass $\frac{1}{2}M_{Sun}$

Star with
mass $\frac{1}{10}M_{Sun}$

Figure 18.14 interactive figure ↖

The approximate habitable zones around our Sun, a star with $\frac{1}{2}$ the mass of the Sun (spectral type K), and a star with $\frac{1}{10}$ the mass of the Sun (spectral type M), shown to scale. The habitable zone becomes increasingly smaller and closer in for stars of lower mass and luminosity.

of our Sun (spectral type M). Although habitable planets seem possible in all three cases, the smaller size of the habitable zones around the less massive stars makes it less likely that suitable planets would have formed in these regions.

Although not all star systems are potential homes for life, many billions of stars in our galaxy could be orbited by habitable planets.

All in all, it seems that the vast majority of stars are at least potentially capable of having life-bearing planets. Moreover, even very conservative assumptions suggest enormous numbers of possibilities. For example, limiting the search for habitable planets to stars very similar to our Sun (that is, spectral type G) would still mean billions of potential other suns in the Milky Way Galaxy.

Finding Habitable Planets Finding Earth-size planets is a daunting technological challenge [Section 6.5]. Recall that looking for an Earth-like planet around a nearby star is like standing on the East Coast of the United States and looking for a pinhead on the West Coast. Nevertheless, advancing technology should soon put Earth-like planets within our telescopic reach. The *Kepler* mission, scheduled for launch in 2009, will look for transits of Earth-size planets in front of their stars (see Figure 6.29). Scientists hope that *Kepler* will detect hundreds of Earth-size planets, and orbital properties measured by *Kepler* will tell us whether these planets lie within their stars' habitable zones.

We will need images or spectra to determine whether the planets really are habitable or have life. Scientists are actively working on technologies that may provide such data. If all goes well, within about a decade NASA hopes to launch an orbiting telescope that will be capable of obtaining low-resolution spectra and crude images (a few pixels) of Earth-like planets around nearby stars.

Signatures of Life Images from future telescopes may tell us whether extrasolar planets have continents and oceans like Earth and perhaps even allow us to monitor seasonal changes. Spectra should prove even more important to the search for life. Moderate-resolution infrared spectra can reveal the presence and abundance of many atmospheric gases, including carbon dioxide, ozone, methane, and water vapor (Figure 18.15). Careful analysis of atmospheric makeup might tell us whether a planet has life.

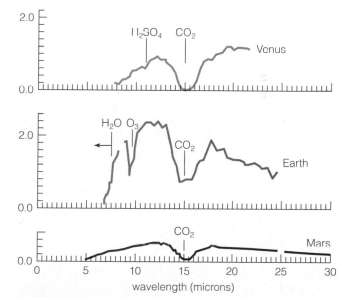

Figure 18.15

The infrared spectra of Venus, Earth, and Mars, as they might be seen from afar, showing absorption features that point to the presence of carbon dioxide (CO_2), ozone (O_3), and sulfuric acid (H_2SO_4) in their atmospheres. While carbon dioxide is present in all three spectra, only our own planet has appreciable oxygen (and hence ozone)—a product of photosynthesis. If we could make similar spectral analyses of distant planets, we might detect atmospheric gases that would indicate life.

Future telescopes may be capable of obtaining images and spectra that will tell us whether distant terrestrial worlds are habitable or have life.

On Earth, for example, the large abundance of oxygen (21% of our atmosphere) is a direct result of photosynthetic life. Abundant oxygen in the atmosphere of a distant world might similarly indicate the presence of life, since we know of no nonbiological way to produce an oxygen abundance as high as Earth's. Other evidence might come from the ratio of oxygen to the other detected gases. Scientists engaged in the search for life in other planetary systems are working to improve our understanding of how life influences atmospheric chemistry in the hopes that we will be able to recognize particular gas combinations as signatures of life.

• Are Earth-like planets rare or common?

We will not know for certain whether habitable planets exist or how common such planets may be until we survey many star systems with telescopes capable of detecting such small planets. Nevertheless, the existence of more than one Earth-size planet in our own solar system and our understanding of planetary formation make it seem reasonable to think that many Earth-size planets exist within the habitable zones of other stars. But should we expect these planets to have Earth-like conditions in which life could arise and evolve?

Most scientists think so, but a few have raised some interesting questions. In essence, these scientists suggest that Earth's hospitality is the result of several rare kinds of planetary luck. According to this idea, sometimes called the "rare Earth hypothesis," the specific circumstances that have allowed life on Earth to survive and evolve into complex forms (such as oak trees and people) might be so rare that ours could be the only planet in the galaxy that harbors anything but the simplest life. Let's briefly examine some of the key issues in the rare Earth hypothesis.

Galactic Constraints Proponents of the rare Earth hypothesis suggest that Earth-like planets can form in only a relatively small region of the Milky Way Galaxy, making the number of potential homes for life far smaller than we might otherwise expect it to be. In essence, they argue that there is a fairly narrow ring at about our solar system's distance from the center of the Milky Way Galaxy that makes up a *galactic habitable zone* analogous to the habitable zone around an individual star (Figure 18.16).

According to the arguments for a galactic habitable zone, outer regions of our galaxy are unlikely to have terrestrial planets because of a low abundance of elements other than hydrogen and helium. Recall that the fraction of heavy elements varies among stars, from less than 0.1% among the old stars in globular clusters to more than 2% among young stars in the galactic disk [Section 14.3]. Even within the galactic disk, the abundance tends to decline with distance from the center of the galaxy. Because terrestrial planets are made almost entirely of heavy elements, a lower abundance of these elements might lessen the chance that terrestrial planets could form. The inner regions of the galaxy are ruled out primarily through an argument concerning supernova rates. Supernovae are more common in the more crowded, inner regions of the galactic disk, making it more likely that a terrestrial planet would be exposed to the intense radiation from a nearby supernova. By assuming that this radiation would be detrimental to life, the proponents of a galactic habitable zone argue against finding habitable planets in the inner regions of

Figure 18.16

The highlighted green ring in this painting of the Milky Way Galaxy represents what some scientists suspect to be a galactic habitable zone—the only region of the galaxy in which Earth-like planets are likely to be found. However, other scientists think that Earth-like planets could be far more widespread.

the galaxy. Together, the constraints on finding Earth-like planets in the inner and outer regions of the galaxy leave the galactic habitable zone as a relatively narrow ring encompassing no more than about 10% of the stars in the galactic disk.

However, other scientists offer counterarguments to both sets of galactic constraints. For the heavy element abundance, they note that Earth's mass is less than $\frac{1}{100,000}$ of the mass of the Sun. Even a very small heavy element abundance could be enough to make one or more Earth-like planets. Unless there is something about the planetary accretion process that prevents terrestrial planets from forming in systems with low heavy-element abundances, then we might find terrestrial planets around most any star. Regarding the radiation danger from supernovae, we do not really know whether such radiation would be detrimental to life. A planet's atmosphere might protect life against the effects of this radiation. It is even possible that the radiation could be beneficial to life by increasing the rate of mutations and thereby accelerating the pace of evolution. If these counterarguments are correct, then Earth-like planets might be found throughout much or all of the galaxy.

Impact Rates and Jupiter   Another issue raised by rare Earth proponents is the impact rate on planets in other star systems. We have seen that Earth was probably subjected to numerous large impacts—some large enough to vaporize the oceans and sterilize the planet—during the heavy bombardment that went on during the first half-billion years or so after our planet's birth. In our solar system, the impact rate lessened dramatically after that. But could the impact rate remain high much longer in other planetary systems?

The most numerous small objects in our solar system are the trillion or so comets of the distant Oort cloud [Section 9.2]. Fortunately for us, nearly all these myriad objects are essentially out of reach, posing no threat to our planet. However, the reason they are out of reach can be traced directly to Jupiter: Recall that the Oort cloud's comets are thought to have formed among the jovian planets, later being "kicked out" to their current orbits by close encounters with these planets, especially Jupiter (see Figure 9.25). If Jupiter did not exist, many of these comets might have remained in regions of the solar system where they could pose a danger to Earth. In that case, the heavy bombardment might never have ended, and huge impacts would continue to this day. From this viewpoint, our existence on Earth has been possible only because of the "luck" of having Jupiter as a planetary neighbor.

The primary question in this case is just how "lucky" this situation might be. Our discoveries of extrasolar planets so far suggest that Jupiter-size planets are quite common. However, we've also found that many large planets migrate inward [Section 6.5], perhaps disrupting terrestrial planet orbits along the way. Whether most star systems have a Jupiter in the right place for ejecting comets remains an open question.

Climate Stability   Another issue affecting the rarity of Earth-like planets is climate stability. Earth's climate has been stable enough for liquid water to exist throughout the past 4 billion years. This climate stability has almost certainly played a major role in allowing complex life to evolve. If our planet had frozen over like Mars or overheated like Venus, we would not be here today. Advocates of the rare Earth hypothesis point to at least two pieces of "luck" related to Earth's stable climate.

The first piece of "luck" is the existence of plate tectonics. Recall that plate tectonics plays a major role in regulating Earth's climate through the carbon dioxide cycle [Section 7.5]. Plate tectonics probably was not necessary to the origin of life, but it seems to have been important in keeping the climate stable enough for the subsequent evolution of plants and animals. But are we really "lucky" to have plate tectonics, or should this geological process be common on similar-size planets elsewhere? We do not yet know. The lack of plate tectonics on Venus, which is quite similar in size to Earth, might seem to argue for plate tectonics being rare. On the other hand, some scientists suspect that the lack of plate tectonics on Venus can be traced to its runaway greenhouse effect, which may have baked out subsurface water and caused Venus's lithosphere to thicken too much to allow plate movements [Section 7.4]. Venus suffered its runaway greenhouse effect because it is not quite far enough from the Sun to be within the Sun's habitable zone. In that case, it's possible that any Earth-size planet within a star's habitable zone would have plate tectonics.

The second piece of "luck" in climate stability is the existence of Earth's relatively large Moon. Models of Earth's rotation and orbit show that if the Moon did not exist, gravitational tugs from other planets would cause large swings in Earth's axis tilt over periods of tens to hundreds of thousands of years. (Such swings in axis tilt are thought to occur on Mars.) Changes in axis tilt would affect the severity of the seasons, which in turn could cause deeper ice ages and more intense periods of warmth. Given that the Moon formed as a result of a random, giant impact [Section 6.4], we might seem to be very lucky to have the Moon and the climate stability it brings.

Again, there are other ways to look at the issue. Changes in axis tilt might warm or cool different parts of the planet dramatically, but the changes would probably occur slowly enough for life to adapt or migrate as the climate changed. In addition, our Moon's presumed formation in a random giant impact does not necessarily mean that large moons are rare. At least a few giant impacts should be expected in any planetary system. Indeed, Earth may not be the only planet in our own solar system that ended up with a large moon through a giant impact. Pluto's moon Charon may have formed in the same way [Section 9.3]. Luck was likely involved in Earth's having a large moon, but it might not have been a very rare kind of luck.

The Bottom Line The bottom line is that while the rare Earth hypothesis offers some intriguing arguments, it is too early to say whether any of them will hold up over time. For each potential argument that Earth has been lucky, we've seen counterarguments suggesting otherwise. There's no doubt that our solar system and our world have "personality"—they exhibit properties that might be found only occasionally in other star systems—but we have no clear reason to think that these special properties are essential to the existence of complex or even intelligent life.

The rare Earth hypothesis suggests that Earth has been the beneficiary of several instances of rare planetary luck, but it remains controversial.

Indeed, it may be that we have missed out on some helpful phenomena that could have sped evolution on Earth. We might be less lucky than we recognize, and creatures on other worlds might regard the nature of our planet with disappointment. Until we learn much more about other planets in the universe, we cannot know whether Earth-like planets and complex life are common or rare.

18.4 The Search for Extraterrestrial Intelligence

So far, we have focused on search strategies for microbial or other nonintelligent life. However, if intelligent beings and civilizations exist elsewhere, we might be able to find them with a completely different type of strategy. Instead of searching for hard-to-find spectroscopic signs of life, we might simply listen for signals that intelligent beings are sending into interstellar space, either in deliberate attempts to contact other civilizations or as a means of communicating among themselves. The search for signals from other civilizations is generally known as the **search for extraterrestrial intelligence**, or **SETI** for short.

• How many civilizations are out there?

SETI efforts have a chance to succeed only if other advanced civilizations are broadcasting signals that we could receive. To judge the chances of SETI success, we'd need to know how many civilizations are broadcasting such signals right now.

Given that we do not even know whether microbial life exists anywhere beyond Earth, we certainly don't know whether other civilizations exist, let alone how many there might be. Nevertheless, for the purposes of planning a search for extraterrestrial intelligence, it is useful to have an organized way of thinking about the number of civilizations that might be out there. To keep our discussion simple, let's consider only the number of potential civilizations in our own galaxy. We can always extend our estimate to the rest of the universe by simply multiplying the result we find for our galaxy by 100 billion, the approximate number of galaxies in our universe.

The Drake Equation In 1961, astronomer Frank Drake wrote a simple equation designed to summarize the factors that would determine the number of civilizations we might contact (Figure 18.17).

> The Drake equation summarizes the factors that determine the number of civilizations in our galaxy with whom we could potentially communicate.

This equation is now known as the **Drake equation**, and in principle it gives us a simple way to calculate the number of civilizations capable of interstellar communication that are currently sharing the Milky Way Galaxy with us. In a form slightly modified from the original, the Drake equation looks like this:

$$\text{Number of civilizations} = N_{\text{HP}} \times f_{\text{life}} \times f_{\text{civ}} \times f_{\text{now}}$$

This equation will make sense once you understand the meaning of each factor:

- N_{HP} is the number of habitable planets in the galaxy; that is, the number of planets that are *potentially* capable of having life.

- f_{life} is the fraction of habitable planets that actually *have* life. For example, if $f_{\text{life}} = 1$ it would mean that all habitable planets have life, and if $f_{\text{life}} = \frac{1}{1,000,000}$ it would mean that only 1 in a million habitable planets have life. Thus, the product $N_{\text{HP}} \times f_{\text{life}}$ tells us the number of life-bearing planets in the galaxy.

- f_{civ} is the fraction of the life-bearing planets upon which a civilization capable of interstellar communication *has at some time* arisen. For example, if $f_{\text{civ}} = \frac{1}{1000}$ it would mean that such a civilization has existed on 1 out of 1000 planets with life, while the other 999 out of

Figure 18.17

Astronomer Frank Drake, with the equation he first wrote in 1961. (With Dr. Drake's approval, we use a slightly modified form of his equation in this book.)

1000 have not had a species that learns to build radio transmitters, high-powered lasers, or other devices for interstellar conversation. When we multiply this factor by the first two factors to form the product $N_{HP} \times f_{life} \times f_{civ}$, we get the total number of planets upon which intelligent beings have evolved and developed a communicating civilization at some time in the galaxy's history.

- f_{now} is the fraction of these civilization-bearing planets that happen to have a civilization *now*, as opposed to, say, millions or billions of years in the past. This factor is important because we can hope to contact only civilizations that are broadcasting signals we could receive at present (assuming we take into account the light-travel time for signals from other stars).

Because the product of the first three factors tells us the total number of civilizations that have *ever* arisen in the galaxy, multiplying by f_{now} tells us how many civilizations we could potentially make contact with today. In other words, the result of the Drake equation is the number of civilizations that we might hope to contact.

We do not know the values of any of the factors in the Drake equation, but it is still a useful tool for organizing our thinking.

Unfortunately, we don't yet know the value of any of the factors in the Drake equation, so we cannot actually calculate the number of civilizations in our galaxy. Nevertheless, the equation is a useful way of organizing our thinking, as we can see by thinking about the potential values for each of its factors.

think about it Try the following sample numbers in the Drake equation. Suppose that there are 1000 habitable planets in our galaxy, that 1 in 10 habitable planets has life, that 1 in 4 planets with life has at some point had an intelligent civilization, and that 1 in 5 civilizations that have ever existed is in existence now. How many civilizations would exist at present? Explain.

The Number of Life-Bearing Planets Let's begin with the first two factors in the Drake equation, whose product ($N_{HP} \times f_{life}$) tells us the number of life-bearing planets in our galaxy. We can make a reasonably educated guess only about the first factor, the number of habitable planets (N_{HP}). Current understanding of solar system formation (see Figure 6.24) suggests that terrestrial planets ought to form fairly easily and, as we discussed earlier, there ought to be many billions of stars in our galaxy with habitable zones large enough to have Earth-like planets. Unless some of the "rare Earth" ideas prove to be correct, it seems entirely reasonable to suppose our galaxy has billions of habitable planets.

Our galaxy is likely to have billions of habitable planets, but we have no reliable way to estimate the fraction of them that actually have life.

The factor f_{life} presents more difficulty. At present, we have no reliable way to estimate the fraction of habitable planets upon which life actually arose. The problem is that we cannot generalize when we have only one example to study—our own Earth. Still, the fact that life arose rapidly on Earth suggests that the origin of life was fairly "easy." In that case, we might expect most or all habitable planets to also have life, making the fraction f_{life} close to 1. Of course, until we have solid evidence that life arose anywhere else, such as on Mars, it is also possible that Earth has been very lucky and that f_{life} might be so close to zero that life has never arisen on any other planet in our galaxy.

The Question of Intelligence Even if life-bearing planets are very common, civilizations capable of interstellar communication might not be. The fraction of life-bearing planets that at some time have such civilizations, f_{civ}, depends on at least two things: First, a planet would have to have a species evolve with sufficient intelligence to develop interstellar communication. In other words, the planet needs a species at least as smart as we are. Second, that species would have to develop a civilization with technology at least as advanced as ours.

Although we cannot really be sure, most scientists suspect that only the first requirement is difficult to meet. A fundamental assumption in nearly all of science today is that we are not "special" in any particular way. We live on a fairly typical planet orbiting an ordinary star in a normal galaxy, and we assume that living creatures elsewhere—whether they prove to be rare or common—would be subjected to evolutionary pressures similar to those that have operated on Earth. Thus, if species with intelligence similar to ours have arisen elsewhere, we assume that they would have similar sociological drives that would eventually lead them to develop the technology necessary for interstellar communication.

If this assumption is correct, then the fraction f_{civ} depends primarily on the question of whether sufficient intelligence is rare or common among life-bearing planets. As with the question of life of any kind, the short answer is that we just don't know, but we can get at least some insight by considering what happened on Earth.

> Life on Earth arose quickly, but it took nearly 4 billion years to get a species intelligent enough to learn about it.

Look again at Figure 18.3. While life arose quite quickly on Earth, nearly all life remained microbial until just a few hundred million years ago, and it took nearly 4 billion years for humans to arrive on the scene. This slow progress toward intelligence might suggest that producing a civilization is very difficult even when life is present. On the other hand, roughly half the stars in the Milky Way are older than our Sun, so if Earth's case is typical, then plenty of planets have existed long enough for intelligence to arise.

Another way to address the question is by considering our level of intelligence in comparison to that of other animals on Earth. We can get a rough measure of intelligence by comparing brain mass to total body mass (a measure sometimes called the *encephalization quotient*, or EQ). Figure 18.18 shows the brain weights for a sampling of birds and mammals (including primates) plotted against their body weight. There is a clear and expected trend in that heavier animals have heavier brains. By drawing a straight line that fits these data, we can define an average value of brain mass for each body mass. Animals whose brain mass falls above the line are smarter than average, while animals whose brain mass falls below the line are less mentally agile. Keep in mind that it is the *vertical* distance above the line that tells us how much smarter a species is than the average, and that the scale goes in powers of 10 on both axes. If you look closely, you'll see that the data point for humans lies significantly farther above the line than the data point for any other species. By this measure of intelligence, we are far smarter than any other species that has ever existed on Earth.

Some people use this fact to argue that even on a planet with complex life, a species as intelligent as we are would be very rare. They say that even if there is an evolutionary drive toward intelligence in general, it takes extreme luck to reach our level of intelligence. After all, while it's evolutionarily useful to have enough intelligence to capture

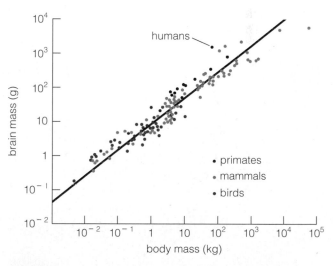

Figure 18.18

This graph shows how brain mass compares to body mass for some mammals (including primates) and birds. The straight line represents an average of the ratio of brain mass to body mass, so that animals that fall above the line are smarter than average and animals that fall below the line are less smart. Note that the scale uses powers of 10 on both axes. (Adapted from Sagan, 1977.)

prey and evade other predators, it's not clear why natural selection would lead to brains big enough to build spacecraft. However, the same data can be used to reach an opposite conclusion. The scatter in the levels of intelligence among different animals tells us that some variation should be expected, and statistical analysis shows that we are not unreasonably far above the average. It might therefore be inevitable that some species would develop our level of intelligence on any planet with complex life.

Technological Lifetimes For the sake of argument, let's assume that life and intelligence are reasonably likely, so that thousands or millions of planets in our galaxy have at some time given birth to a civilization. In that case, the final factor in the Drake equation, f_{now}, determines the likelihood of there being someone whom we could contact now. The value of this factor depends on how long civilizations survive.

Consider our own example. In the roughly 12 billion years during which our galaxy has existed, we have been capable of interstellar communication via radio for only about 60 years. If we were to destroy ourselves tomorrow (saving you the unpleasantness of a final exam), then other civilizations could have received signals from us during only 60 years out of the galaxy's 12-billion-year existence, equivalent to 1 part in 200 million of the galaxy's history. If such a short technological lifetime is typical of civilizations, then f_{now} would be only $\frac{1}{200,000,000}$, and some 200 million civilization-bearing planets would need to have existed at one time or another in the Milky Way in order for us to have a decent chance of finding another civilization out there now.

Even if civilizations arise on many planets, they are unlikely to exist today unless they avoid early self-destruction.

However, we'd expect f_{now} to be so small only if we are on the brink of self-destruction—after all, the fraction will grow larger for as long as our civilization survives. Thus, if civilizations are at all common, survivability is the key factor in whether any are out there now. If most civilizations self-destruct shortly after achieving the technology for interstellar communication, then we are almost certainly alone in the galaxy at present. But if most survive and thrive for thousands or millions of years, the Milky Way may be brimming with civilizations—most of them far more advanced than our own.

think about it Describe a few reasons why a civilization capable of interstellar communication would also be capable of self-destruction. Overall, do you believe our civilization can survive for thousands or millions of years? Defend your opinion.

• How does SETI work?

If there are indeed other civilizations out there, then in principle we ought to be able to make contact with them. Based on our current understanding of physics, it seems likely that even very advanced civilizations would communicate much as we do—by encoding signals in radio waves or other forms of light. Most SETI researchers use large radio telescopes to search for alien radio signals (Figure 18.19). A few researchers are beginning to check other parts of the electromagnetic spectrum as well. For example, some scientists use visible light telescopes to search for communications encoded as laser pulses. Of course, advanced civilizations may well have invented communication technologies that we cannot even imagine, let alone detect.

Figure 18.19

The Allen Telescope Array, now under construction in Hat Creek, California, is being used to search for radio signals from extraterrestrial civilizations.

In principle, aliens within about 50 light-years could watch past television broadcasts. Our own current SETI efforts could pick up only much stronger signals.

A good way to think about our chances of picking up an alien signal is to imagine what aliens would need to do to pick up signals from us. We have been sending relatively high-power transmissions into space since about the 1950s in the form of television broadcasts. In principle, anyone within about 50 light-years of Earth could watch our old television shows (perhaps a frightening thought). However, in order to detect our broadcasts, they would need far larger and more sensitive radio telescopes than we have today. If their technology were at the same level as ours, they could receive a signal from us only if we deliberately broadcast an unusually high-powered transmission.

special topic:

Are Aliens Already Here?

IN THIS CHAPTER, we have discussed contact with intelligent aliens as a possibility, not a reality. However, public opinion polls suggest that up to half the American public believes that aliens are already visiting us. What can science say about this remarkable notion?

The bulk of the claimed evidence for alien visitation consists of sightings of UFOs—unidentified flying objects. Many thousands of UFOs are reported each year, and no one doubts that unidentified objects are being seen. The question is whether they are alien spacecraft.

Aliens have long been a staple of science fiction, but modern interest in UFOs began with a widely reported sighting in 1947. While flying a private plane near Mount Rainier in Washington State, businessman Kenneth Arnold saw nine mysterious objects streaking across the sky. He told a reporter that the objects "flew erratic, like a saucer if you skip it across the water." (In fact, he may have seen meteors skipping across the atmosphere, though no one knows for sure.) He did not say that the objects were saucer-shaped, but the reporter nevertheless wrote up Arnold's experience as a sighting of "flying saucers." The story was front-page news throughout America, and within a decade, "flying saucers" had invaded popular culture, if not our planet.

The flying saucer reports also interested the U.S. Air Force, largely out of concern that the UFOs might represent new types of aircraft developed by the Soviet Union. For two decades, the Air Force hired teams of academics to study UFO reports. In most cases, these experts were able to specify a plausible identification of the UFO. The explanations included bright stars and planets, aircraft and gliders, rocket launches, balloons, birds, meteors, atmospheric phenomena, and the occasional hoax. For a minority of the sightings, the investigators could not deduce what was seen, but their overall conclusion was that there was no reason to believe the UFOs were either highly advanced Soviet craft or visitors from other worlds. The Air Force eventually dropped its investigations of UFOs.

Believers discounted the Air Force denials, claiming to have other evidence of alien visitation. So far, none of this evidence has withstood scientific scrutiny. Photographs and film clips are nearly always too fuzzy to clearly show alien spacecraft, except in cases that are obviously faked. UFO witnesses are frequently credible (they include seasoned pilots), but generally there are several possible explanations for what they've seen besides alien spacecraft. Crop circles are easily made by

pranksters. Stories of alien abductions are dramatic but cannot be verified. Pieces of metal that "UFO experts" say could not have been made by humans have turned out to be pieces of cars or refrigerators. Champions of alien visitation generally explain away the lack of clear evidence in one of two ways: government cover-ups or a failure of the mainstream scientific community to take the relevant phenomena seriously. Neither explanation is compelling.

It's certainly conceivable that a secretive government might *try* to put the lid on evidence of alien visits, though the motivation for doing so is unclear. The usual explanations are that the public couldn't handle the news or that the government is taking secret advantage of the alien materials to design new military hardware (via "reverse-engineering"). Both explanations are silly. Half the population already believes in alien visitors and would hardly be shocked if newspapers announced that aliens were stacked up in government warehouses. As for reverse-engineering extraterrestrial spacecraft, we should keep in mind how difficult it is to travel from star to star. Any society that could do so routinely would be technologically far beyond our own. Reverse-engineering their spaceships is as unlikely as expecting Neanderthals to construct personal computers just because a laptop somehow landed in their cave. In addition, while a government might successfully hide evidence for a short time, does it really seem possible that evidence could remain secret for decades? After all, talk show fame and riches would await anyone who uncovered the conspiracy. Moreover, unless the aliens landed only in the United States, can we seriously believe that *every* government on Earth has cooperated in hiding the evidence?

Alleged disinterest on the part of the scientific community is an equally unimpressive claim. Scientists are constantly competing with one another to be the first to make a great discovery, and clear evidence of alien visitors would be one of the most important discoveries of all time. Countless researchers would work evenings and weekends, without pay, if they thought they could make such a discovery. The fact that few scientists are engaged in such study reflects not a lack of interest, but a lack of evidence worthy of study.

Of course, absence of evidence is not evidence of absence. Most scientists are open to the possibility that we might someday find evidence of alien visits, and many would welcome aliens with open arms. So far, however, we have no hard evidence to support the belief that aliens are already here.

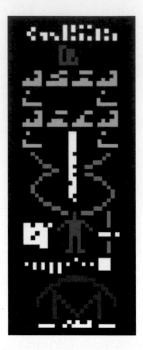

Figure 18.20

In 1974, a short message was broadcast to the globular cluster M13 using the Arecibo radio telescope. The picture shown here was encoded by using two different radio frequencies, one for "on" and one for "off" (the colors shown here are arbitrary). To decode the message, the aliens would need to realize that the bits are meant to be arranged in a rectangular grid as shown, but that should not be difficult: The grid has 73 rows and 23 columns, and aliens would presumably know that these are both prime numbers. The picture represents the Arecibo radio dish, our solar system, a human stick figure, and a schematic of DNA and the eight simple molecules used in its construction.

To date, humans have made only a few attempts to broadcast our existence in this way. The most powerful of these transmissions was made in 1974 and lasted only 3 minutes. The powerful planetary radar transmitter on the Arecibo radio telescope (see Figure 5.20) was used to send a simple pictorial message to the globular cluster M13 (Figure 18.20). This target was chosen in part because it contains a few hundred thousand stars, seemingly offering a good chance that at least one has a civilization around it. However, M13 is about 21,000 light-years from Earth, so it will take some 21,000 years for our signal to get there and another 21,000 years for any response to make its way back to Earth.

Several SETI projects under way or in development would be capable of detecting signals like the one we broadcast from Arecibo if they came from civilizations within a few hundred light-years. These SETI efforts scan millions of radio frequency bands simultaneously. If anyone nearby is deliberately broadcasting on an ongoing basis, we have a good chance of detecting the signals.

think about it SETI efforts are often controversial because of their cost and uncertain chance of success, but supporters say the cost is justified because contact with an extraterrestrial intelligence would be such an important discovery. Do you agree? Defend your opinion.

18.5 Interstellar Travel and Its Implications to Civilization

So far, we have discussed ways of detecting distant civilizations without ever leaving the comfort of our own planet. Could we ever actually visit other worlds in other star systems? A careful analysis of this question turns out to have profound implications for the future of our civilization. To see why, we first need to consider the prospects for achieving interstellar travel.

• How difficult is interstellar travel?

In many science fiction movies, our descendants travel among the stars as routinely as we jet about the Earth in airplanes. They race around the galaxy in starships of all sizes and shapes, circumventing nature's prohibition on faster-than-light travel by entering hyperspace, wormholes, or warp drive. They witness firsthand incredible cosmic phenomena, such as stars and planets in all stages of development, accretion disks around white dwarfs and neutron stars, and the distortion of spacetime near black holes. Along the way they encounter numerous alien species, most of which happen to look and act a lot like us.

Unfortunately, Einstein's theory of relativity tells us that real interstellar travel must be at speeds slower than the speed of light, and even to approach that speed we will need to overcome huge technological hurdles. Nevertheless, we have already sent out our first emissaries to the stars, and there's no reason to believe that we won't develop better technologies in the future.

The Challenge of Interstellar Travel To date, we have launched five spacecraft that will leave our solar system and eventually travel among the stars: the planetary probes *Pioneer 10*, *Pioneer 11*, *Voyager 1*, *Voyager 2*, and the *New Horizons* spacecraft currently en route to Pluto. These spacecraft are traveling about as fast as anything ever built by humans, but their speeds are still less than $\frac{1}{10,000}$ the speed of light. It would take each of them some 100,000 years just to reach the next nearest star system (Alpha Centauri), but their trajectories won't take them anywhere near it. Instead, they will simply continue their journey without passing close to any nearby stars, wandering the Milky Way for millions or even billions of years to come. The *Pioneer* and *Voyager* spacecraft carry greetings from Earth, just in case someone comes across one of them someday (Figure 18.21).

High-speed interstellar travel would require thousands of times as much energy as the entire world currently uses each year.

If we want to make interstellar journeys within human lifetimes, we will need starships that can travel at speeds close to the speed of light. We will need entirely new types of engines to reach such high speeds. The energy requirements of interstellar spacecraft may pose an even more daunting challenge. For example, the energy needed to accelerate a single ship the size of *Star Trek*'s *Enterprise* to just half the speed of light would be more than 2000 times the total annual energy use of the world today. Clearly, interstellar travel will require vast new sources of energy. In addition, fast-moving starships will require new types of shielding to protect crew members from instant death. As a starship travels through interstellar gas at near-light speed, ordinary atoms and ions will hit it like a deadly flood of high-energy cosmic rays.

If we succeed in building starships capable of traveling at speeds close to the speed of light, the crews will face significant social challenges. According to well-tested principles of Einstein's theory of relativity (see Special Topic: Relativity and the Cosmic Speed Limit, p. 364), time will run much slower on a spaceship that travels at high speed to the stars than it does here on Earth. For example, in a ship traveling at an average speed of 99.9% of the speed of light, the 50-light-year round trip to the star Vega would take the travelers aboard only about 2 years—but more than 50 years would pass on Earth while they were gone. The crew would therefore need only 2 years' worth of provisions and would age only 2 years during the voyage, but they would return to a world quite different from the one they left. Family and friends would be older or deceased, new technologies might have made their knowledge and skills obsolete, and many political and social changes may have occurred in their absence. The crew would face a difficult adjustment when they came home to Earth.

Starship Design Despite all the difficulties, some scientists and engineers have already proposed designs that could in principle take us to nearby stars. In the 1960s, a group of scientists proposed *Project Orion*, which envisioned accelerating a spaceship with repeated detonations of relatively small hydrogen bombs. Each explosion would take place a few tens of meters behind the spaceship and would propel the ship forward as the vaporized debris impacted a "pusher plate" on the back of the spacecraft (Figure 18.22). Calculations showed that a spaceship accelerated by the rapid-fire detonation of a million H-bombs could reach Alpha Centauri in just over a century. In principle, we could build an *Orion* spacecraft with existing technology, though it would be very expensive and would require an exception to the international treaty banning nuclear detonations in space.

A century to the nearest star isn't bad, but it still wouldn't make interstellar travel easy. Unfortunately, no available technology could go much faster. The problem is mass: Making a rocket faster requires more fuel, but adding fuel adds mass and makes it more difficult for the rocket

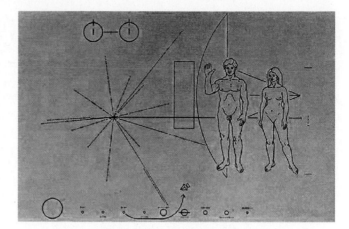

a The *Pioneer* plaque, about the size of an automobile license plate. The human figures are shown in front of a drawing of the spacecraft to give a sense of scale. The "prickly" graph to their left shows the Sun's position relative to nearby pulsars, and Earth's location around the Sun is shown below. Binary code indicates the pulsar periods; because pulsars slow with time, the periods allow someone reading the plaque to determine when the spacecraft was launched.

b *Voyagers 1* and *2* carry a phonograph record—a 12-inch gold-plated copper disk containing music, greetings, and images from Earth.

Figure 18.21

Messages aboard the *Pioneer* and *Voyager* spacecraft, which are bound for the stars.

Figure 18.22

Artist's conception of the *Project Orion* starship, showing one of the small hydrogen bomb detonations that would propel it. Debris from the detonation strikes the flat disk, called the pusher plate, at the back of the spaceship. The central sections (enclosed in a lattice) hold the bombs, and the front sections house the crew.

to accelerate. Calculations show that even in the best case, rockets carrying nuclear fuel could achieve speeds no more than a few percent of the speed of light. Nevertheless, we can envision some possible future technologies that might get around this problem.

One idea suggests powering starships with engines that generate energy through matter–antimatter annihilation. Whereas nuclear fusion converts less than 1% of the mass of atomic nuclei into energy, matter–antimatter annihilation [Section 17.1] converts *all* the annihilated mass into energy. Starships with matter–antimatter engines could probably reach speeds of 90% or more of the speed of light. At these speeds, the slowing of time predicted by relativity becomes noticeable, putting many nearby stars within a few years' journey for the crew members. However, because no natural reservoirs of antimatter exist, we would have to be able to manufacture many tons of antimatter and then store it safely for the trip—capabilities that are far beyond our present means.

High-speed interstellar travel remains well beyond our current capabilities, but we can envision future technologies that could make it possible.

An even more speculative and futuristic design, known as an *interstellar ramjet*, would collect interstellar hydrogen with a gigantic scoop, using the collected gas as fuel for its nuclear engines (Figure 18.23). By collecting fuel along the way, the ship would not need to carry the weight of fuel on board. However, because the density of interstellar gas is so low, the scoop would need to be enormous. As astronomer Carl Sagan said, we are talking about "spaceships the size of worlds."

The bottom line is that while we face enormous obstacles to achieving interstellar travel, there's no reason to think it's impossible. If we can avoid self-destruction and if we continue to explore space, our descendants might well journey to the stars.

• Where are the aliens?

Imagine that we survive long enough to become interstellar travelers and that we begin colonizing habitable planets around nearby stars. As the colonies grow at each new location, some of the people may decide to set

Figure 18.23

Artist's conception of a spaceship powered by an interstellar ramjet. The giant scoop in the front (left) collects interstellar hydrogen for use as fusion fuel.

out for other star systems. Even if our starships traveled at relatively low speeds—say, a few percent of the speed of light—we could have dozens of outposts around nearby stars within a few centuries. In 10,000 years, our descendants would be spread among stars within a few hundred light-years of Earth. In a few million years, we could have outposts throughout the Milky Way Galaxy. We will have become a true galactic civilization.

Now, if we take the idea that *we* could develop a galactic civilization within a few million years and combine it with the reasonable (though unproved) idea that civilizations ought to be common, we are led to an astonishing conclusion: Someone else should already have created a galactic civilization. In fact, it should have been done a long time ago.

To see why, let's take some sample numbers. Suppose the overall odds of a civilization arising around a star are about the same as your odds of winning the lottery, or 1 in a million. Taking a low estimate of 100 billion stars in the Milky Way Galaxy, this would mean there are some 100,000 civilizations in our galaxy alone. Further, suppose we are a fairly typical civilization, so that civilizations generally arise when their stars are approaching 5 billion years old. Given that the galaxy is some 12 billion years old, the first of these 100,000 civilizations would have arisen at least 7 billion years ago.

> If we can develop interstellar travel in the future, then it would seem that other civilizations should have developed the capability long ago.

Others would have arisen, on average, about every 70,000 years. Under these assumptions, we would expect the youngest civilization besides ourselves to be some 70,000 years ahead of us technologically, and most would be millions or billions of years ahead of us.

We thereby encounter a strange paradox: Plausible arguments suggest that a galactic civilization should already exist, yet we have so far found no evidence of such a civilization. This paradox is often called *Fermi's paradox*, after the Nobel Prize–winning physicist Enrico Fermi. During a 1950 conversation with other scientists about the possibility of extraterrestrial intelligence, Fermi responded to speculations by asking, "So where is everybody?"

This paradox has many possible solutions, but broadly speaking we can group them into three categories:

1. *We are alone.* There is no galactic civilization because civilizations are extremely rare—so rare that we are the first to have arisen on the galactic scene, perhaps even the first in the universe.
2. *Civilizations are common, but no one has colonized the galaxy.* There are at least three possible reasons why this might be the case. Perhaps interstellar travel is much harder or more expensive than we have guessed, and civilizations are unable to venture far from their home worlds. Perhaps the desire to explore is unusual, and other societies either never leave their home star systems or stop exploring before they've colonized much of the galaxy. Most ominously, perhaps many civilizations have arisen, but they have all destroyed themselves before achieving the ability to colonize the stars.
3. *There IS a galactic civilization,* but it has not yet revealed its existence to us.

We do not know which, if any, of these explanations is the correct solution to the question "Where are the aliens?" However, each category of solution has astonishing implications for our own species.

Consider the first solution—that we are alone. If this is true, then our civilization is a remarkable achievement. It implies that through all of cosmic evolution, among countless star systems, we are the first piece of our galaxy or the universe ever to know that the rest of the universe exists. Through us, the universe has attained self-awareness. Some philosophers and many religions argue that the ultimate purpose of life is to become truly self-aware. If so, and if we are alone, then the destruction of our civilization and the loss of our scientific knowledge would represent an inglorious end to something that took the universe some 14 billion years to achieve. From this point of view, humanity becomes all the more precious, and the collapse of our civilization would be all the more tragic.

The second category of solutions has much more terrifying implications. If thousands of civilizations before us have all failed to achieve interstellar travel on a large scale, what hope do we have? Unless we somehow think differently than all other civilizations, this solution says that we will never go far in space. Because we have always explored when the opportunity arose, this solution almost inevitably leads to the conclusion that failure will come about because we destroy ourselves. We can only hope that this answer is wrong.

The third solution is perhaps the most intriguing. It says that we are newcomers on the scene of a galactic civilization that has existed for millions or billions of years before us. Perhaps this civilization is deliberately leaving us alone for the time being and will someday decide the time is right to invite us to join it.

No matter what the answer turns out to be, learning it will surely mark a turning point in the brief history of our species. Moreover, this turning point is likely to be reached within the next few decades or centuries. We already have the ability to destroy our civilization. If we do so, then our fate is sealed. But if we survive long enough to develop technology that can take us to the stars, the possibilities seem almost limitless.

the big picture
Putting Chapter 18 into Context

Throughout our study of astronomy, we have taken the "big picture" view of trying to understand how we fit into the universe. Here, at last, we have returned to Earth and examined the role of our own generation in the big picture of human history. Tens of thousands of past human generations have walked this Earth. Ours is the first generation with the technology to study the far reaches of our universe, to search for life elsewhere, and to travel beyond our home planet. It is up to us to decide whether we will use this technology to advance our species or to destroy it.

Imagine for a moment the grand view, a gaze across the centuries and millennia from this moment forward. Picture our descendants living among the stars, having created or joined a great galactic civilization. They will have the privilege of experiencing ideas, worlds, and discoveries far beyond our wildest imagination. Perhaps, in their history lessons, they will learn of our generation—the generation that history placed at the turning point and that managed to steer its way past the dangers of self-destruction and onto the path to the stars.

summary of key concepts

18.1 Life on Earth

• When did life arise on Earth?

Fossil evidence puts the origin of life at least 3.5 billion years ago, and carbon isotope evidence pushes this date to more than 3.85 billion years ago. Life therefore arose within a few hundred million years after the last major impact of the heavy bombardment, and possibly in a much shorter time.

• How did life arise on Earth?

Genetic evidence suggests that all life on Earth evolved from a common ancestor, which was probably similar to microbes that live today in hot water near undersea volcanic vents or hot springs. We do not know how this first organism arose, but laboratory experiments suggest that it may have been the result of natural chemical processes on the early Earth. Once life arose, it rapidly diversified and evolved through **natural selection**.

• What are the necessities of life?

Life on Earth thrives in a wide range of environments and in general seems to require only three things: a source of nutrients, a source of energy, and liquid water.

18.2 Life in the Solar System

• Could there be life on Mars?

Mars once had conditions that may have been conducive to an origin of life. If life arose, it might still survive in pockets of liquid water underground.

• Could there be life on Europa or other jovian moons?

Europa probably has a subsurface ocean of liquid water and may have undersea volcanoes on its ocean floor. If so, it has conditions much like those in which life on Earth probably arose, making it a good candidate for life. Ganymede and Callisto might have oceans as well. Titan may have other liquids on its surface, though it is too cold for liquid water. Perhaps life can survive in these other liquids, or perhaps Titan has liquid water deep underground. Enceladus also shows evidence of subsurface liquids, offering yet another possibility for life.

18.3 Life Around Other Stars

• Are habitable planets likely?

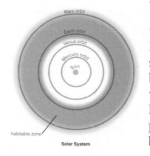
Billions of stars have at least moderate-size **habitable zones** in which life-bearing planets might exist. We do not yet have the technology to search for habitable planets directly, but several planned missions should be able to begin the search soon.

• **Are Earth-like planets rare or common?**
We don't know. Arguments can be made on both sides of the question, and at present we lack the data to determine which answer is correct.

18.4 The Search for Extraterrestrial Intelligence

• **How many civilizations are out there?**
We don't know, but the **Drake equation** gives us a way to organize our thinking about the question. The equation (in a modified form) says that the number of civilizations in the Milky Way Galaxy with whom we could potentially communicate is $N_{HP} \times f_{life} \times f_{civ} \times f_{now}$, where N_{HP} is the number of habitable planets in the galaxy, f_{life} is the fraction of habitable planets that actually have life on them, f_{civ} is the fraction of life-bearing planets upon which a civilization capable of interstellar communication has at some time arisen, and f_{now} is the fraction of all these civilizations that exist now.

• **How does SETI work?**

SETI, the search for extraterrestrial intelligence, generally refers to efforts to detect signals—such as radio or laser communications—coming from civilizations on other worlds.

18.5 Interstellar Travel and Its Implications to Civilization

• **How difficult is interstellar travel?**

Convenient interstellar travel remains well beyond our technological capabilities, because of the technological requirements for engines, the enormous energy needed to accelerate spacecraft to speeds near the speed of light, and the difficulties of shielding the crew from radiation. Nevertheless, people have proposed ways around all these difficulties, and it seems reasonable to think that we will someday achieve interstellar travel if we survive long enough.

• **Where are the aliens?**
A civilization capable of interstellar travel ought to be able to colonize the galaxy in a few million years or less, and the galaxy was around for at least 7 billion years before Earth was born. It therefore seems that someone should have colonized the galaxy long ago—yet we have no evidence of other civilizations. Every possible explanation for this surprising fact has astonishing implications for our species and our place in the universe.

exercises and problems

For instructor-assigned homework go to **www.masteringastronomy.com**.

Review Questions

Short-Answer Questions Based on the Reading

1. Describe three recent developments in the study of life on Earth that make it seem much more likely that we could find life elsewhere.
2. How do we study the history of life on Earth? Describe the *geological time scale* and a few of the major events along it.
3. Summarize the evidence pointing to an early origin of life on Earth. How far back in Earth's history did life exist?
4. Why is the *theory of evolution* so critical to our understanding of the history of life on Earth? Explain how evolution proceeds by *natural selection*, and what happens to DNA that allows species to evolve.
5. Give a brief overview of the history of life on Earth. What evidence points to a common ancestor for all life? How and when did oxygen accumulate in Earth's atmosphere? When did larger animals diversify on Earth?
6. Where do we think life on Earth first arose, and why?
7. How are laboratory experiments helping us study the origin of life on Earth? Is it possible that life migrated to Earth from elsewhere? Explain.
8. Describe the range of environments in which life thrives on Earth. What three basic requirements apply to life in all these environments?
9. What is a *habitable world?* Which worlds in our solar system seem potentially habitable, and why?

10. Briefly summarize the debate over possible fossil evidence of life in a meteorite from Mars.
11. What do we mean by a star's *habitable zone?* Do we expect many stars to be capable of having habitable planets? Explain.
12. What is the "rare Earth hypothesis"? Summarize the arguments on both sides regarding the validity of this hypothesis.
13. What is the *Drake equation?* Define each of its factors, and describe the current state of understanding about the potential values of each factor.
14. What is *SETI?* Describe the capabilities of current SETI efforts.
15. Summarize the factors that make interstellar travel difficult, and describe a few technologies that might someday make it possible.
16. What is *Fermi's paradox?* Describe several potential solutions to the paradox, and the implications of each to our civilization.

Test Your Understanding

Fantasy or Science Fiction?

Each of the following describes some futuristic scenario that may or may not be plausible. In each case, decide whether the scenario is plausible or implausible according to our present understanding of science. Explain clearly; not all these have definitive answers, so your explanation is more important than your chosen answer.

17. The first human explorers on Mars discover the ruins of an ancient civilization, including remnants of tall buildings and temples.

18. The first human explorers on Mars drill a hole into a Martian volcano to collect a sample of soil from several meters underground. Upon analysis of the soil, they discover that it holds living microbes resembling terrestrial bacteria but with a different biochemistry.

19. In 2020, a spacecraft lands on Europa and melts its way through the ice into the Europan ocean. It finds numerous strange, living microbes, along with a few larger organisms that feed on the microbes.

20. It's the year 2075. A giant telescope on the Moon, consisting of hundreds of small telescopes linked together across a distance of 500 km, has just captured a series of images of a planet around a distant star that clearly show seasonal changes in vegetation.

21. A century from now, after completing a careful study of planets around stars within 100 light-years of Earth, astronomers discover that the most diverse life exists on a planet orbiting a young star that formed just 100 million years ago.

22. In 2030, a brilliant teenager working in her garage builds a coal-powered rocket that can travel at half the speed of light.

23. In the year 2750, we receive a signal from a civilization around a nearby star telling us that the *Voyager 2* spacecraft recently crash-landed on its planet.

24. Crew members of the matter–antimatter spacecraft *Star Apollo*, which left Earth in the year 2165, return to Earth in the year 2450, looking only a few years older than when they left.

25. Aliens from a distant star system invade Earth with the intent to destroy us and occupy our planet, but we successfully fight them off when their technology proves no match for ours.

26. A single great galactic civilization exists. It originated on a single planet long ago but is now made up of beings from many different planets, assimilated into the galactic culture.

Quick Quiz

Choose the best answer to each of the following. Explain your reasoning with one or more complete sentences.

27. Fossil evidence suggests that life on Earth arose (a) almost immediately after Earth formed. (b) very soon after the end of the heavy bombardment. (c) about a billion years before the rise of the dinosaurs.

28. The theory of evolution is (a) a scientific theory, supported by extensive evidence. (b) one of several competing scientific models that all seem equally successful in explaining the nature of life on Earth. (c) essentially just a guess about how life changes through time.

29. Plants and animals are (a) the two major forms of life on Earth. (b) the only organisms that have DNA. (c) just two small branches of the diverse "tree of life" on Earth.

30. Which of the following is a reason why early living organisms on Earth could not have survived on the surface? (a) the lack of an ozone layer (b) the lack of oxygen in the atmosphere (c) the fact that these organisms were single-celled

31. According to current understanding, a key requirement for life is (a) photosynthesis. (b) liquid water. (c) an ozone layer.

32. Which of the following worlds is not considered a candidate for harboring life? (a) Europa (b) Mars (c) the Moon

33. How does the habitable zone around a star of spectral type G compare to that around a star of spectral type M? (a) It is larger. (b) It is hotter. (c) It is closer to its star.

34. In the Drake equation, suppose that the term $f_{life} = \frac{1}{2}$. What would this mean? (a) Half the stars in the Milky Way Galaxy have a planet with life. (b) Half of all life forms in the universe

are intelligent. (c) Half of the habitable worlds in the galaxy actually have life, while the other half don't.

35. The amount of energy that would be needed to accelerate a large spaceship to half the speed of light is (a) about 100 times as much energy as is needed to launch the Space Shuttle. (b) more than 2000 times the current annual global energy consumption. (c) more than the amount of energy released by a supernova.

36. According to current scientific understanding, the idea that the Milky Way Galaxy might be home to a civilization millions of years more advanced than ours is (a) a virtual certainty. (b) extremely unlikely. (c) one reasonable answer to Fermi's paradox.

Process of Science

Examining How Science Works

37. *Extraordinary Claims.* As discussed in the chapter, both the *Viking* results and the study of a Martian meteorite led some scientists to think we had found evidence of life on Mars, even while most scientists disagreed. Those who disagree often point to Carl Sagan's dictum that "extraordinary claims require extraordinary evidence." Briefly discuss one of these cases, and decide whether you think it should require extraordinary evidence before being accepted. What follow-up evidence might lead scientists to re-evaluate their positions?

38. *Unanswered Questions.* In a sense, this entire chapter was about one big, unanswered question: Are we alone in the universe? But as we attempt to answer this "big" question, there are many smaller questions that we might wish to answer along the way. Describe one currently unanswered question about life in the universe that we might be able to answer with new missions or experiments over the next couple of decades. What kinds of evidence will we need to answer the question? How will we know when it is answered?

Investigate Further

In-Depth Questions to Increase Your Understanding

Short-Answer/Essay Questions

39. *Most Likely to Have Life.* Suppose you were asked to vote in a contest to name the world in our solar system (besides Earth) "most likely to have life." Which world would you cast your vote for? Explain and defend your choice in a one-page essay.

40. *Likely Suns.* Study the stellar data for nearby stars given in Appendix F, Table F.1. Which star on the list would you expect to have the largest habitable zone? Which would have the second-largest habitable zone? If we rule out multiple-star systems, which star would you expect to have the highest probability of having a habitable planet? Explain your answers.

41. *Are Earth-like Planets Common?* Based on what you have learned in this book, do you think Earth-like planets will ultimately prove to be rare, common, or something in between? Write a one- to two-page essay explaining and defending your opinion.

42. *Solution to the Fermi Paradox.* Among the various possible solutions to the question "Where are the aliens?," which do you think is most likely? Write a one- to two-page essay in which you explain why you favor this solution.

43. *What's Wrong with This Picture?* Many science fiction stories have imagined the galaxy divided into a series of empires, each having arisen from a different civilization on a different world, that hold

each other at bay because they are all at about the same level of military technology. Is this a realistic scenario? Explain.

44. *Aliens in the Movies.* Choose a science fiction movie (or television show) that involves an alien species. Do you think aliens like this could really exist? Write a one- to two-page critical review of the movie, focusing primarily on the question of whether the movie portrays the aliens in a scientifically reasonable way.

Quantitative Problems

Be sure to show all calculations clearly and state your final answers in complete sentences.

45. *SETI Search.* Suppose there are 10,000 civilizations broadcasting radio signals in the Milky Way Galaxy right now. On average, how many stars would we have to search before we would expect to hear a signal? Assume there are 500 billion stars in the galaxy. How does your answer change if there are only 100 civilizations instead of 10,000?

46. *SETI Signal.* Consider a civilization broadcasting a signal with a power of 10,000 watts. The Arecibo radio telescope, which is about 300 m in diameter, could detect this signal if it were coming from as far away as 100 light-years. Suppose instead that the signal is being broadcast from the other side of the Milky Way Galaxy, about 70,000 light-years away. How large a radio telescope would we need to detect this signal? (*Hint:* Use the inverse square law for light.)

47. *Cruise Ship Energy.* Suppose we have a spaceship about the size of a typical ocean cruise ship today, which means it has a mass of about 100 million kg, and we want to accelerate the ship to a speed of 10% of the speed of light.
   a. How much energy would be required? (*Hint:* You can find the answer simply by calculating the kinetic energy of the ship when it reaches its cruising speed; because 10% of the speed of light is still small compared to the speed of light, you can use the formula kinetic energy $= \frac{1}{2} \times m \times v^2$.)
   b. How does your answer compare to total worldwide energy use at present, which is about 5×10^{22} joules per year?
   c. The typical cost of energy today is roughly 5¢ per 1 million joules. Using this price, how much would it cost to generate the energy needed by this spaceship?

48. *Matter–Antimatter Engine.* Consider the spaceship from Problem 47. Suppose you want to generate the energy to get it to cruising speed using matter–antimatter annihilation. How much antimatter would you need to produce and take on the ship? (*Hint:* Remember that when matter and antimatter meet, they turn all their mass into energy equivalent to mc^2.)

Discussion Questions

49. *Funding the Search for Life.* Imagine that you are a member of Congress who decides how much government funding goes to research in different areas of science. How much would you allot to the search for life in the universe compared to the amount allotted to research in other areas of astronomy and planetary science? Why?

50. *Distant Dream or Near-Reality?* Considering all the issues surrounding interstellar flight, when (if ever) do you think we are likely to begin traveling among the stars? Why?

51. *The Turning Point.* Discuss the idea that our generation has acquired a greater responsibility for the future than any previous generation. Do you agree with this assessment? If so, how should we deal with this responsibility? Defend your opinion.

Web Projects

52. *Astrobiology News.* Go to NASA's astrobiology home page and read some of the articles about the search for life in the universe. Choose one recent article, and write a one- to two-page summary of the research and how it relates to the question of life in the universe.

53. *Martian Meteorites.* Research the latest discoveries concerning Martian meteorites. Choose one discovery and write a short summary of how you think it alters the debate about the habitability of Mars.

54. *The Search for Extraterrestrial Intelligence.* Go to the home page for the SETI Institute. Learn more about how SETI is funded and carried out, and summarize your findings in one page or less.

55. *Starship Design.* Read about a proposal for starship propulsion or design. How would the proposed starship work? What new technologies would be needed, and what existing technologies could be applied? Write a one- to two-page report on your research.

56. *Advanced Spacecraft Technologies.* NASA supports many efforts to incorporate new technologies into spaceships. Although few of them may be suitable for interstellar colonization, most are innovative and fascinating. Learn about one such NASA project, and write a short overview of your findings.

Use the following questions to check your understanding of some of the many types of visual information used in astronomy. Answers are provided in Appendix K. For additional practice, try the Chapter 18 Visual Quiz at **www.masteringastronomy.com.**

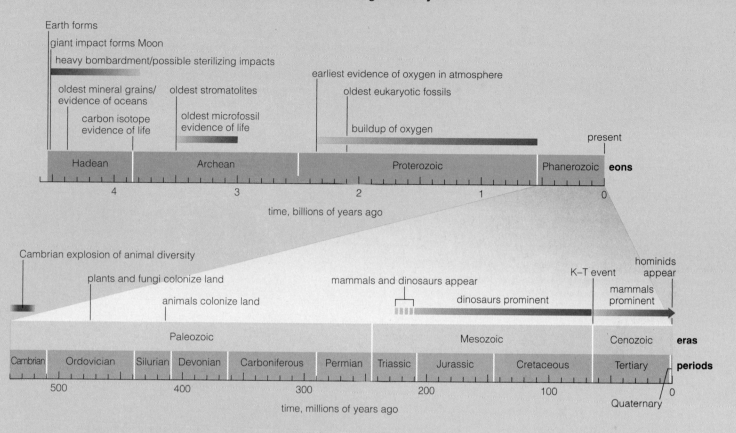

The figure above, which repeats Figure 18.3, shows the geological time scale. Use this figure to answer the following questions.

1. List the following events in the order in which they occurred, from first to last.
   a. earliest humans
   b. earliest animals
   c. impact causes extinction of dinosaurs
   d. earliest mammals
   e. earliest plants living on land
   f. first time there is significant oxygen in Earth's atmosphere
   g. first life on Earth
2. List the following time frames in order by how long they lasted, from the longest one to the shortest one.
   a. Hadean eon
   b. Proterozoic eon
   c. Paleozoic era
   d. Cretaceous period
3. Which of the following are time frames in which we are living today? More than one may apply.
   a. Quaternary period
   b. Tertiary period
   c. Cenozoic era
   d. Phanerozoic eon
   e. Paleozoic era

4. How long did the Cambrian explosion last?
   a. less than 1 year
   b. about a decade
   c. about 10,000 years
   d. about 40 million years
   e. about 500 million years
5. When did the heavy bombardment end?
   a. about 4.5 billion years ago
   b. between about 4.3 and 4.5 billion years ago
   c. between about 3.8 and 4.0 billion years ago
   d. exactly 3.85 billion years ago
6. How long have mammals been present on Earth?
   a. about 1 million years
   b. about 65 million years
   c. about 225 million years
   d. about 510 million years

cosmic context   Part VI at a Glance. A Universe of Life?

Throughout this book, we have seen that the history of the universe has proceeded in a way that has made our existence on Earth possible. This figure summarizes some of the key ideas, and leads us to ask: If life arose here, shouldn't it also have arisen on many other worlds? We do not yet know the answer, but scientists are actively seeking to learn whether life is rare or common in the universe.

(4) Our planet and all the life on it is made primarily of elements formed by nuclear fusion in high-mass stars and dispersed into space by supernovae [Section 12.3].

(3) The attractive force of gravity pulls together the matter that makes galaxies, stars, and planets [Section 4.4].

(2) Ripples in the density of the early universe were necessary for life to form later on. Without those ripples, matter would never have collected into galaxies, stars, and planets [Section 17.3].

(1) The protons, neutrons, and electrons in the atoms that make up Earth and life were created out of pure energy during the first few moments after the Big Bang, leaving the universe filled with hydrogen and helium gas [Section 17.1].

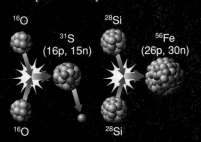

^{16}O ^{28}Si

^{31}S
(16p, 15n)

^{56}Fe
(26p, 30n)

^{16}O ^{28}Si

$$F_g = G\,\frac{M_1 M_2}{d^2}$$

M_1 M_2

d

Every piece of matter in the universe pulls on every other piece.

High-mass stars have cores hot enough to make elements heavier than carbon.

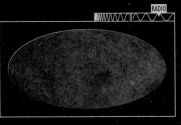

RADIO

We observe the seeds of structure formation in the cosmic microwave background.

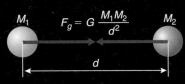

electron

gamma-ray photons

antielectron

Matter can be created from energy: $E = mc^2$.

5 Our galaxy is large enough to retain the elements ejected by supernovae, and it recycles them into new stars and planetary systems [Section 14.2].

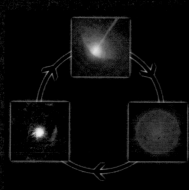

New elements mix with the interstellar medium, which then forms new stars and planets.

6 Planets can form in gaseous disks of material around newly formed stars. Earth was built from heavy elements that condensed from the gas as particles of metal and rock, which then gradually accreted to become our planet [Section 6.4].

Terrestrial planets formed in warm, inner regions of the solar nebula; jovian planets formed in cooler, outer regions.

7 Life as we know it requires liquid water, so we define the habitable zone around a star to be the zone in which a suitably large planet can have liquid water on its surface [Section 18.3].

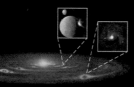

Earth orbit

Sun

Habitable zone

The Sun's habitable zone (green) occupies a region from beyond the orbit of Venus to near the orbit of Mars.

8 Early life has had the time needed to evolve into complex forms—including us—because the solar thermostat has kept the Sun shining steadily for billions of years [Section 10.2].

Solar Thermostat: Gravitational Equilibrium

The solar thermostat keeps the Sun's fusion rate stable.

appendixes

A useful numbers

Astronomical Distances

1 AU $\approx 1.496 \times 10^8$ km $= 1.496 \times 10^{11}$ m

1 light-year $\approx 9.46 \times 10^{12}$ km $= 9.46 \times 10^{15}$ m

1 parsec (pc) $\approx 3.09 \times 10^{13}$ km ≈ 3.26 light-years

1 kiloparsec (kpc) $= 1000$ pc $\approx 3.26 \times 10^3$ light-years

1 megaparsec (Mpc) $= 10^6$ pc $\approx 3.26 \times 10^6$ light-years

Astronomical Times

1 solar day (average) $= 24^h$

1 sidereal day $\approx 23^h56^m4.09^s$

1 synodic month (average) ≈ 29.53 solar days

1 sidereal month (average) ≈ 27.32 solar days

1 tropical year ≈ 365.242 solar days

1 sidereal year ≈ 365.256 solar days

Universal Constants

Speed of light: $c = 3.00 \times 10^5$ km/s $= 3 \times 10^8$ m/s

Gravitational constant: $G = 6.67 \times 10^{-11} \dfrac{m^3}{kg \times s^2}$

Planck's constant: $h = 6.63 \times 10^{-34}$ joule $\times$ s

Stefan–Boltzmann constant: $\sigma = 5.67 \times 10^{-8} \dfrac{watt}{m^2 \times Kelvin^4}$

Mass of a proton: $m_p = 1.67 \times 10^{-27}$ kg

Mass of an electron: $m_e = 9.11 \times 10^{-31}$ kg

Useful Sun and Earth Reference Values

Mass of the Sun: $1 M_{Sun} \approx 2 \times 10^{30}$ kg

Radius of the Sun: $1 R_{Sun} \approx 696{,}000$ km

Luminosity of the Sun: $1 L_{Sun} \approx 3.8 \times 10^{26}$ watts

Mass of Earth: $1 M_{Earth} \approx 5.97 \times 10^{24}$ kg

Radius (equatorial) of Earth: $1 R_{Earth} \approx 6378$ km

Acceleration of gravity on Earth: $g = 9.8$ m/s^2

Escape velocity from surface of Earth: $v_{escape} = 11$ km/s $= 11{,}000$ m/s

Energy and Power Units

Basic unit of energy: 1 joule $= 1\dfrac{kg \times m^2}{s^2}$

Basic unit of power: 1 watt $= 1$ joule/s

Electron-volt: 1 eV $= 1.60 \times 10^{-19}$ joule

B useful formulas

- Universal law of gravitation for the force between objects of mass M_1 and M_2, with distance d between their centers:

$$F = G\frac{M_1 M_2}{d^2}$$

- Newton's version of Kepler's third law; p and a are period and semimajor axis, respectively, of either orbiting mass:

$$p^2 = \frac{4\pi^2}{G(M_1 + M_2)} a^3$$

- Escape velocity at distance R from center of object of mass M:

$$v_{escape} = \sqrt{\frac{2GM}{R}}$$

- Relationship between a photon's wavelength (λ), frequency (f), and the speed of light (c):

$$\lambda \times f = c$$

- Energy of a photon of wavelength λ or frequency f:

$$E = hf = \frac{hc}{\lambda}$$

- Stefan–Boltzmann law for thermal radiation at temperature T (in Kelvin):

$$\text{emitted power per unit area} = \sigma T^4$$

- Wien's law for the peak wavelength (λ_{max}) thermal radiation at temperature T (in Kelvin):

$$\lambda_{max} = \frac{2{,}900{,}000}{T}\ \text{nm}$$

- Doppler shift (radial velocity is positive if the object is moving away from us and negative if it is moving toward us):

$$\frac{\text{radial velocity}}{\text{speed of light}} = \frac{\text{shifted wavelength} - \text{rest wavelength}}{\text{rest wavelength}}$$

- Angular separation (α) of two points with an actual separation s, viewed from a distance d (assuming d is much larger than s):

$$\alpha = \frac{s}{2\pi d} \times 360°$$

- Inverse square law for light (d is the distance to the object):

$$\text{apparent brightness} = \frac{\text{luminosity}}{4\pi d^2}$$

- Parallax formula (distance d to a star with parallax angle p in arcseconds):

$$d\ (\text{in parsecs}) = \frac{1}{p\ (\text{in arcseconds})}$$

$$\text{or } d\ (\text{in light-years}) = 3.26 \times \frac{1}{p\ (\text{in arcseconds})}$$

- The orbital velocity law, to find the mass M_r contained within the circular orbit of radius r for an object moving at speed v:

$$M_r = \frac{r \times v^2}{G}$$

C a few mathematical skills

THIS APPENDIX REVIEWS the following mathematical skills: powers of 10, scientific notation, working with units, the metric system, and finding a ratio. You should refer to this appendix as needed while studying the textbook.

C.1 Powers of 10

Powers of 10 simply indicate how many times to multiply 10 by itself. For example:

$$10^2 = 10 \times 10 = 100$$

$$10^6 = 10 \times 10 \times 10 \times 10 \times 10 \times 10 = 1,000,000$$

Negative powers are the reciprocals of the corresponding positive powers. For example:

$$10^{-2} = \frac{1}{10^2} = \frac{1}{100} = 0.01$$

$$10^{-6} = \frac{1}{10^6} = \frac{1}{1,000,000} = 0.000001$$

Table C.1 lists powers of 10 from 10^{-12} to 10^{12}. Note that powers of 10 follow two basic rules:

1. A positive exponent tells how many zeros follow the 1. For example, 10^0 is a 1 followed by no zeros, and 10^8 is a 1 followed by eight zeros.

2. A negative exponent tells how many places are to the right of the decimal point, including the 1. For example, $10^{-1} = 0.1$ has one place to the right of the decimal point; $10^{-6} = 0.000001$ has six places to the right of the decimal point.

Multiplying and Dividing Powers of 10

Multiplying powers of 10 simply requires adding exponents, as the following examples show:

$$10^4 \times 10^7 = \underbrace{10,000}_{10^4} \times \underbrace{10,000,000}_{10^7} = \underbrace{100,000,000,000}_{10^{4+7} = 10^{11}} = 10^{11}$$

$$10^5 \times 10^{-3} = \underbrace{100,000}_{10^5} \times \underbrace{0.001}_{10^{-3}} = \underbrace{100}_{10^{5+(-3)} = 10^2} = 10^2$$

$$10^{-8} \times 10^{-5} = \underbrace{0.00000001}_{10^{-8}} \times \underbrace{0.00001}_{10^{-5}} = \underbrace{0.0000000000001}_{10^{-8+(-5)} = 10^{-13}} = 10^{-13}$$

TABLE C.1 *Powers of 10*

Zero and Positive Powers			Negative Powers		
Power	Value	Name	Power	Value	Name
10^0	1	One			
10^1	10	Ten	10^{-1}	0.1	Tenth
10^2	100	Hundred	10^{-2}	0.01	Hundredth
10^3	1000	Thousand	10^{-3}	0.001	Thousandth
10^4	10,000	Ten thousand	10^{-4}	0.0001	Ten-thousandth
10^5	100,000	Hundred thousand	10^{-5}	0.00001	Hundred-thousandth
10^6	1,000,000	Million	10^{-6}	0.000001	Millionth
10^7	10,000,000	Ten million	10^{-7}	0.0000001	Ten-millionth
10^8	100,000,000	Hundred million	10^{-8}	0.00000001	Hundred-millionth
10^9	1,000,000,000	Billion	10^{-9}	0.000000001	Billionth
10^{10}	10,000,000,000	Ten billion	10^{-10}	0.0000000001	Ten-billionth
10^{11}	100,000,000,000	Hundred billion	10^{-11}	0.00000000001	Hundred-billionth
10^{12}	1,000,000,000,000	Trillion	10^{-12}	0.000000000001	Trillionth

Dividing powers of 10 requires subtracting exponents, as in the following examples:

$$\frac{10^5}{10^3} = \underbrace{100,000}_{10^5} \div \underbrace{1000}_{10^3} = \underbrace{100}_{10^{5-3} = 10^2} = 10^2$$

$$\frac{10^3}{10^7} = \underbrace{1000}_{10^3} \div \underbrace{10,000,000}_{10^7} = \underbrace{0.0001}_{10^{3-7} = 10^{-4}} = 10^{-4}$$

$$\frac{10^{-4}}{10^{-6}} = \underbrace{0.0001}_{10^{-4}} \div \underbrace{0.000001}_{10^{-6}} = \underbrace{100}_{10^{-4-(-6)} = 10^2} = 10^2$$

Powers of Powers of 10

We can use the multiplication and division rules to raise powers of 10 to other powers or to take roots. For example:

$$(10^4)^3 = 10^4 \times 10^4 \times 10^4 = 10^{4+4+4} = 10^{12}$$

Note that we can get the same end result by simply multiplying the two powers:

$$(10^4)^3 = 10^{4 \times 3} = 10^{12}$$

Because taking a root is the same as raising to a fractional power (e.g., the square root is the same as the $\frac{1}{2}$ power, the cube root is the same as the $\frac{1}{3}$ power, etc.), we can use the same procedure for roots, as in the following example:

$$\sqrt{10^4} = (10^4)^{1/2} = 10^{4 \times (1/2)} = 10^2$$

Adding and Subtracting Powers of 10

Unlike for multiplying and dividing powers of 10, there is no shortcut for adding or subtracting powers of 10. The values must be written in longhand notation. For example:

$$10^6 + 10^2 = 1{,}000{,}000 + 100 = 1{,}000{,}100$$

$$10^8 + 10^{-3} = 100{,}000{,}000 + 0.001 = 100{,}000{,}000.001$$

$$10^7 - 10^3 = 10{,}000{,}000 - 1000 = 9{,}999{,}000$$

Summary

We can summarize our findings using n and m to represent any numbers:

- To *multiply* powers of 10, *add* exponents: $10^n \times 10^m = 10^{n+m}$

- To *divide* powers of 10, *subtract* exponents: $\dfrac{10^n}{10^m} = 10^{n-m}$

- To *raise* powers of 10 to other powers, multiply exponents: $(10^n)^m = 10^{n \times m}$

C.2 Scientific Notation

When we are dealing with large or small numbers, it's generally easier to write them with powers of 10. For example, it's much easier to write the number 6,000,000,000,000 as 6×10^{12}. This format, in which a number *between* 1 and 10 is multiplied by a power of 10, is called **scientific notation.**

Converting a Number to Scientific Notation

We can convert numbers written in ordinary notation to scientific notation with a simple two-step process:

1. Move the decimal point to come after the *first* nonzero digit.

2. The number of places the decimal point moves tells you the power of 10; the power is *positive* if the decimal point moves to the left and *negative* if it moves to the right.

   Examples:

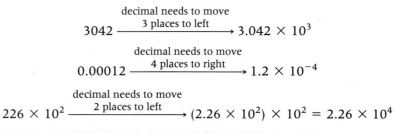

Converting a Number from Scientific Notation

We can convert numbers written in scientific notation to ordinary notation by the reverse process:

1. The power of 10 indicates how many places to move the decimal point; move it to the *right* if the power of 10 is positive and to the *left* if it is negative.

2. If moving the decimal point creates any open places, fill them with zeros.

Examples:

$$4.01 \times 10^2 \xrightarrow[\text{2 places to right}]{\text{move decimal}} 401$$

$$3.6 \times 10^6 \xrightarrow[\text{6 places to right}]{\text{move decimal}} 3{,}600{,}000$$

$$5.7 \times 10^{-3} \xrightarrow[\text{3 places to left}]{\text{move decimal}} 0.0057$$

Multiplying or Dividing Numbers in Scientific Notation

Multiplying or dividing numbers in scientific notation simply requires operating on the powers of 10 and the other parts of the number separately.

Examples:

$$(6 \times 10^2) \times (4 \times 10^5) = (6 \times 4) \times (10^2 \times 10^5) = 24 \times 10^7 = (2.4 \times 10^1) \times 10^7 = 2.4 \times 10^8$$

$$\frac{4.2 \times 10^{-2}}{8.4 \times 10^{-5}} = \frac{4.2}{8.4} \times \frac{10^{-2}}{10^{-5}} = 0.5 \times 10^{-2-(-5)} = 0.5 \times 10^3 = (5 \times 10^{-1}) \times 10^3 = 5 \times 10^2$$

Note that, in both these examples, we first found an answer in which the number multiplied by a power of 10 was *not* between 1 and 10. We therefore followed the procedure for converting the final answer to scientific notation.

Addition and Subtraction with Scientific Notation

In general, we must write numbers in ordinary notation before adding or subtracting.

Examples:

$$(3 \times 10^6) + (5 \times 10^2) = 3{,}000{,}000 + 500 = 3{,}000{,}500 = 3.0005 \times 10^6$$

$$(4.6 \times 10^9) - (5 \times 10^8) = 4{,}600{,}000{,}000 - 500{,}000{,}000 = 4{,}100{,}000{,}000 = 4.1 \times 10^9$$

When both numbers have the *same* power of 10, we can factor out the power of 10 first.

Examples:

$$(7 \times 10^{10}) + (4 \times 10^{10}) = (7 + 4) \times 10^{10} = 11 \times 10^{10} = 1.1 \times 10^{11}$$

$$(2.3 \times 10^{-22}) - (1.6 \times 10^{-22}) = (2.3 - 1.6) \times 10^{-22} = 0.7 \times 10^{-22} = 7.0 \times 10^{-23}$$

C.3 Working with Units

Showing the units of a problem as you solve it usually makes the work much easier and also provides a useful way of checking your work. If an answer does not come out with the units you expect, you probably did something wrong. In general, working with units is very similar to working with numbers, as the following guidelines and examples show.

Five Guidelines for Working with Units

Before you begin any problem, think ahead and identify the units you expect for the final answer. Then operate on the units along with the numbers as you solve the problem. The following five guidelines may be helpful when you are working with units:

1. Mathematically, it doesn't matter whether a unit is singular (e.g., meter) or plural (e.g., meters); we can use the same abbreviation (e.g., m) for both.

2. You cannot add or subtract numbers unless they have the *same* units. For example, 5 apples + 3 apples = 8 apples, but the expression 5 apples + 3 oranges cannot be simplified further.

3. You *can* multiply units, divide units, or raise units to powers. Look for key words that tell you what to do.

   - *Per* suggests division. For example, we write a speed of 100 kilometers per hour as

   $$100\,\frac{\text{km}}{\text{hr}} \quad \text{or} \quad \frac{100\ \text{km}}{1\ \text{hr}}$$

   - *Of* suggests multiplication. For example, if you launch a 50-kg space probe at a launch cost *of* $10,000 per kilogram, the total cost is

   $$50\ \text{kg} \times \frac{\$10,000}{\text{kg}} = \$500,000$$

   - *Square* suggests raising to the second power. For example, we write an area of 75 square meters as 75 m^2.

   - *Cube* suggests raising to the third power. For example, we write a volume of 12 cubic centimeters as 12 cm^3.

4. Often the number you are given is not in the units you wish to work with. For example, you may be given that the speed of light is 300,000 km/s but need it in units of m/s for a particular problem. To convert the units, simply multiply the given number by a *conversion factor:* a fraction in which the numerator (top of the fraction) and denominator (bottom of the fraction) are equal, so that the value of the fraction is 1; the number in the denominator must have the units that you wish to change. In the case of changing the speed of light from units of km/s to m/s, you need a conversion factor for kilometers to meters. Thus, the conversion factor is

$$\frac{1000\ \text{m}}{1\ \text{km}}$$

Note that this conversion factor is equal to 1, since 1000 meters and 1 kilometer are equal, and that the units to be changed (km) appear in the denominator. We can now convert the speed of light from units of km/s to m/s simply by multiplying by this conversion factor:

$$\underbrace{300,000\,\frac{\text{km}}{\text{s}}}_{\substack{\text{speed of light} \\ \text{in km/s}}} \times \underbrace{\frac{1000\ \text{m}}{1\ \text{km}}}_{\substack{\text{conversion from} \\ \text{km to m}}} = \underbrace{3 \times 10^8\,\frac{\text{m}}{\text{s}}}_{\substack{\text{speed of light} \\ \text{in m/s}}}$$

Note that the units of km cancel, leaving the answer in units of m/s.

5. It's easier to work with units if you replace division with multiplication by the reciprocal. For example, suppose you want to know how many minutes are represented by 300 seconds. We can find the answer by dividing 300 seconds by 60 seconds per minute:

$$300\ \text{s} \div 60\,\frac{\text{s}}{\text{min}}$$

However, it is easier to see the unit cancellations if we rewrite this expression by replacing the division with multiplication by the reciprocal (this process is easy to remember as "invert and multiply"):

$$300 \text{ s} \div 60\frac{\text{s}}{\text{min}} = 300 \cancel{\text{s}} \times \underbrace{\frac{1 \text{ min}}{60 \cancel{\text{s}}}}_{\substack{\text{invert} \\ \text{and multiply}}} = 5 \text{ min}$$

We now see that the units of seconds (s) cancel in the numerator of the first term and the denominator of the second term, leaving the answer in units of minutes.

More Examples of Working with Units

Example 1. How many seconds are there in 1 day?

Solution: We can answer the question by setting up a *chain* of unit conversions in which we start with 1 *day* and end up with *seconds*. We use the facts that there are 24 hours per day (24 hr/day), 60 minutes per hour (60 min/hr), and 60 seconds per minute (60 s/min):

$$\underbrace{1 \cancel{\text{day}}}_{\substack{\text{starting} \\ \text{value}}} \times \underbrace{\frac{24 \cancel{\text{hr}}}{\cancel{\text{day}}}}_{\substack{\text{conversion} \\ \text{from} \\ \text{day to hr}}} \times \underbrace{\frac{60 \cancel{\text{min}}}{\cancel{\text{hr}}}}_{\substack{\text{conversion} \\ \text{from} \\ \text{hr to min}}} \times \underbrace{\frac{60 \text{ s}}{\cancel{\text{min}}}}_{\substack{\text{conversion} \\ \text{from} \\ \text{min to s}}} = 86,400 \text{ s}$$

Note that all the units cancel except *seconds*, which is what we want for the answer. There are 86,400 seconds in 1 day.

Example 2. Convert a distance of 10^8 cm to km.

Solution: The easiest way to make this conversion is in two steps, since we know that there are 100 centimeters per meter (100 cm/m) and 1000 meters per kilometer (1000 m/km):

$$\underbrace{10^8 \text{ cm}}_{\substack{\text{starting} \\ \text{value}}} \times \underbrace{\frac{1 \text{ m}}{100 \text{ cm}}}_{\substack{\text{conversion} \\ \text{from} \\ \text{cm to m}}} \times \underbrace{\frac{1 \text{ km}}{1000 \text{ m}}}_{\substack{\text{conversion} \\ \text{from} \\ \text{m to km}}} = 10^8 \cancel{\text{cm}} \times \frac{1 \cancel{\text{m}}}{10^2 \cancel{\text{cm}}} \times \frac{1 \text{ km}}{10^3 \cancel{\text{m}}} = 10^3 \text{ km}$$

Alternatively, if we recognize that the number of kilometers should be smaller than the number of centimeters (because kilometers are larger), we might decide to do this conversion by dividing as follows:

$$10^8 \text{ cm} \div \frac{100 \text{ cm}}{\text{m}} \div \frac{1000 \text{ m}}{\text{km}}$$

In this case, before carrying out the calculation, we replace each division with multiplication by the reciprocal:

$$10^8 \text{ cm} \div \frac{100 \text{ cm}}{\text{m}} \div \frac{1000 \text{ m}}{\text{km}} = 10^8 \text{ cm} \times \frac{1 \text{ m}}{100 \text{ cm}} \times \frac{1 \text{ km}}{1000 \text{ m}}$$
$$= 10^8 \cancel{\text{cm}} \times \frac{1 \cancel{\text{m}}}{10^2 \cancel{\text{cm}}} \times \frac{1 \text{ km}}{10^3 \cancel{\text{m}}}$$
$$= 10^3 \text{ km}$$

Note that we again get the answer that 10^8 cm is the same as 10^3 km, or 1000 km.

Example 3. Suppose you accelerate at 9.8 m/s² for 4 seconds, starting from rest. How fast will you be going?

Solution: The question asked "how fast?" so we expect to end up with a speed. Therefore, we multiply the acceleration by the amount of time you accelerated:

$$9.8 \, \frac{m}{s^2} \times 4 \, s = (9.8 \times 4) \, \frac{m \times \cancel{s}}{s^{\cancel{2}}} = 39.2 \, \frac{m}{s}$$

Note that the units end up as a speed, showing that you will be traveling 39.2 m/s after 4 seconds of acceleration at 9.8 m/s².

Example 4. A reservoir is 2 km long and 3 km wide. Calculate its area, in both square kilometers and square meters.

Solution: We find its area by multiplying its length and width:

$$2 \, km \times 3 \, km = 6 \, km^2$$

Next we need to convert this area of 6 km² to square meters, using the fact that there are 1000 meters per kilometer (1000 m/km). Note that we must square the term 1000 m/km when converting from km² to m²:

$$6 \, km^2 \times \left(1000 \, \frac{m}{km}\right)^2 = 6 \, km^2 \times 1000^2 \, \frac{m^2}{km^2} = 6 \, \cancel{km^2} \times 1,000,000 \, \frac{m^2}{\cancel{km^2}}$$

$$= 6,000,000 \, m^2$$

The reservoir area is 6 km², which is the same as 6 million m².

C.4 The Metric System (SI)

The modern version of the metric system, known as *Système Internationale d'Unites* (French for "International System of Units") or **SI**, was formally established in 1960. Today, it is the primary measurement system in nearly every country in the world with the exception of the United States. Even in the United States, it is the system of choice for science and international commerce.

The basic units of length, mass, and time in the SI are

- The **meter** for length, abbreviated m

- The **kilogram** for mass, abbreviated kg

- The **second** for time, abbreviated s

Multiples of metric units are formed by powers of 10, using a prefix to indicate the power. For example, *kilo* means 10^3 (1000), so a kilometer is 1000 meters; a microgram is 0.000001 gram, because *micro* means 10^{-6}, or one millionth. Some of the more common prefixes are listed in Table C.2.

TABLE C.2 *SI (Metric) Prefixes*

Small Values			Large Values		
Prefix	Abbreviation	Value	Prefix	Abbreviation	Value
Deci	d	10^{-1}	Deca	da	10^1
Centi	c	10^{-2}	Hecto	h	10^2
Milli	m	10^{-3}	Kilo	k	10^3
Micro	μ	10^{-6}	Mega	M	10^6
Nano	n	10^{-9}	Giga	G	10^9
Pico	p	10^{-12}	Tera	T	10^{12}

Metric Conversions

Table C.3 lists conversions between metric units and units used commonly in the United States. Note that the conversions between kilograms and pounds are valid only on Earth, because they depend on the strength of gravity.

Example 1. International athletic competitions generally use metric distances. Compare the length of a 100-meter race to that of a 100-yard race.

Solution: Table C.3 shows that 1 m = 1.094 yd, so 100 m is 109.4 yd. Note that 100 meters is almost 110 yards; a good "rule of thumb" to remember is that distances in meters are about 10% longer than the corresponding number of yards.

Example 2. How many square kilometers are in 1 square mile?

Solution: We use the square of the miles-to-kilometers conversion factor:

$$(1 \text{ mi}^2) \times \left(\frac{1.6093 \text{ km}}{1 \text{ mi}} \right)^2 = (1 \text{ mi}^2) \times \left(1.6093^2 \, \frac{\text{km}^2}{\text{mi}^2} \right) = 2.5898 \text{ km}^2$$

Therefore, 1 square mile is 2.5898 square kilometers.

C.5 Finding a Ratio

Suppose you want to compare two quantities, such as the average density of Earth and the average density of Jupiter. The way we do such a comparison is by dividing, which tells us the *ratio* of the two quantities. In this case, Earth's average density is 5.52 g/cm^3 and Jupiter's average density is 1.33 g/cm^3 (see Figure 8.1), so the ratio is

$$\frac{\text{average density of Earth}}{\text{average density of Jupiter}} = \frac{5.52 \, \text{g/cm}^3}{1.33 \, \text{g/cm}^3} = 4.15$$

Notice how the units cancel on both the top and the bottom of the fraction. We can state our result in two equivalent ways:

- The ratio of Earth's average density to Jupiter's average density is 4.15.

- Earth's average density is 4.15 times Jupiter's average density.

    Sometimes, the quantities that you want to compare may each involve an equation. In such cases, you could, of course, find the ratio by first calculating each of the two quantities individually and then dividing. However, it is much easier if you first express the ratio as a fraction, putting the equation for one quantity on top and the other on the bottom. Some of the terms in the equation may then cancel out, making any calculations much easier.

Example 1. Compare the kinetic energy of a car traveling at 100 km/hr to that of a car traveling at 50 km/hr.

TABLE C.3 *Metric Conversions*

To Metric	From Metric
1 inch = 2.540 cm	1 cm = 0.3937 inch
1 foot = 0.3048 m	1 m = 3.28 feet
1 yard = 0.9144 m	1 m = 1.094 yards
1 mile = 1.6093 km	1 km = 0.6214 mile
1 pound = 0.4536 kg	1 kg = 2.205 pounds

Solution: We do the comparison by finding the ratio of the two kinetic energies, recalling that the formula for kinetic energy is $\frac{1}{2}mv^2$. Since we are not told the mass of the car, you might at first think that we don't have enough information to find the ratio. However, notice what happens when we put the equations for each kinetic energy into the ratio, calling the two speeds v_1 and v_2:

$$\frac{\text{K.E. car at } v_1}{\text{K.E. car at } v_2} = \frac{\frac{1}{2}m_{\text{car}}v_1^2}{\frac{1}{2}m_{\text{car}}v_2^2} = \frac{v_1^2}{v_2^2} = \left(\frac{v_1}{v_2}\right)^2$$

All the terms cancel except those with the two speeds, leaving us with a very simple formula for the ratio. Now we put in 100 km/hr for v_1 and 50 km/hr for v_2:

$$\frac{\text{K.E. car at 100 km/hr}}{\text{K.E. car at 50 km/hr}} = \left(\frac{100 \text{ km/hr}}{50 \text{ km/hr}}\right)^2 = 2^4 = 4$$

The ratio of the car's kinetic energies at 100 km/hr and 50 km/hr is 4. That is, the car has four times as much kinetic energy at 100 km/hr as it has at 50 km/hr.

Example 2. Compare the strength of gravity between Earth and the Sun to the strength of gravity between Earth and the Moon.

Solution: We do the comparison by taking the ratio of the Earth–Sun gravity to the Earth–Moon gravity. In this case, each quantity is found from the equation of Newton's law of gravity. (See Section 4.4.) Thus, the ratio is

$$\frac{\text{Earth–Sun gravity}}{\text{Earth–Moon gravity}} = \frac{G\dfrac{M_{\text{Earth}}M_{\text{Sun}}}{(d_{\text{Earth–Sun}})^2}}{G\dfrac{M_{\text{Earth}}M_{\text{Moon}}}{(d_{\text{Earth–Moon}})^2}} = \frac{M_{\text{Sun}}}{(d_{\text{Earth–Sun}})^2} \times \frac{(d_{\text{Earth–Moon}})^2}{M_{\text{Moon}}}$$

Note how all but four of the terms cancel; the last step comes from replacing the division with multiplication by the reciprocal (the "invert and multiply" rule for division). We can simplify the work further by rearranging the terms so that we have the masses and distances together:

$$\frac{\text{Earth–Sun gravity}}{\text{Earth–Moon gravity}} = \frac{M_{\text{Sun}}}{M_{\text{Moon}}} \times \frac{(d_{\text{Earth–Moon}})^2}{(d_{\text{Earth–Sun}})^2}$$

Now it is just a matter of looking up the numbers (see Appendix E) and calculating:

$$\frac{\text{Earth–Sun gravity}}{\text{Earth–Moon gravity}} = \frac{1.99 \times 10^{30} \text{ kg}}{7.35 \times 10^{22} \text{ kg}} \times \frac{(384.4 \times 10^3 \text{ km})^2}{(149.6 \times 10^6 \text{ km})^2} = 179$$

In other words, the Earth–Sun gravity is 179 times as strong as the Earth–Moon gravity.

D the periodic table of the elements

Key

12
Mg
Magnesium
24.305

— Atomic number
— Element's symbol
— Element's name
— Atomic mass*

*Atomic masses are fractions because they represent a weighted average of atomic masses of different isotopes— in proportion to the abundance of each isotope on Earth.

1 **H** Hydrogen 1.00794																	2 **He** Helium 4.003
3 **Li** Lithium 6.941	4 **Be** Beryllium 9.01218											5 **B** Boron 10.81	6 **C** Carbon 12.011	7 **N** Nitrogen 14.007	8 **O** Oxygen 15.999	9 **F** Fluorine 18.988	10 **Ne** Neon 20.179
11 **Na** Sodium 22.990	12 **Mg** Magnesium 24.305											13 **Al** Aluminum 26.98	14 **Si** Silicon 28.086	15 **P** Phosphorus 30.974	16 **S** Sulfur 32.06	17 **Cl** Chlorine 35.453	18 **Ar** Argon 39.948
19 **K** Potassium 39.098	20 **Ca** Calcium 40.08	21 **Sc** Scandium 44.956	22 **Ti** Titanium 47.88	23 **V** Vanadium 50.94	24 **Cr** Chromium 51.996	25 **Mn** Manganese 54.938	26 **Fe** Iron 55.847	27 **Co** Cobalt 58.9332	28 **Ni** Nickel 58.69	29 **Cu** Copper 63.546	30 **Zn** Zinc 65.39	31 **Ga** Gallium 69.72	32 **Ge** Germanium 72.59	33 **As** Arsenic 74.922	34 **Se** Selenium 78.96	35 **Br** Bromine 79.904	36 **Kr** Krypton 83.80
37 **Rb** Rubidium 85.468	38 **Sr** Strontium 87.62	39 **Y** Yttrium 88.9059	40 **Zr** Zirconium 91.224	41 **Nb** Niobium 92.91	42 **Mo** Molybdenum 95.94	43 **Tc** Technetium (98)	44 **Ru** Ruthenium 101.07	45 **Rh** Rhodium 102.906	46 **Pd** Palladium 106.42	47 **Ag** Silver 107.868	48 **Cd** Cadmium 112.41	49 **In** Indium 114.82	50 **Sn** Tin 118.71	51 **Sb** Antimony 121.75	52 **Te** Tellurium 127.60	53 **I** Iodine 126.905	54 **Xe** Xenon 131.29
55 **Cs** Cesium 132.91	56 **Ba** Barium 137.34		72 **Hf** Hafnium 178.49	73 **Ta** Tantalum 180.95	74 **W** Tungsten 183.85	75 **Re** Rhenium 186.207	76 **Os** Osmium 190.2	77 **Ir** Iridium 192.22	78 **Pt** Platinum 195.08	79 **Au** Gold 196.967	80 **Hg** Mercury 200.59	81 **Ti** Thallium 204.383	82 **Pb** Lead 207.2	83 **Bi** Bismuth 208.98	84 **Po** Polonium (209)	85 **At** Astatine (210)	86 **Rn** Hadon (222)
87 **Fr** Francium (223)	88 **Ra** Radium 226.0254		104 **Rf** Rutherfordium (263)	105 **Db** Dubnium (262)	106 **Sg** Seaborgium (266)	107 **Bh** Bohrium (267)	108 **Hs** Hassium (277)	109 **Mt** Meitnerium (268)	110 **Ds** Darmstadtium (281)	111 **Rg** Roentgenium (272)	112 **Uub** Ununbium (285)	113 (284)	114 (289)	115 (288)	116 (292)		

Lanthanide Series

57 **La** Lanthanum 138.906	58 **Ce** Cerium 140.12	59 **Pr** Praseodymium 140.908	60 **Nd** Neodymium 144.24	61 **Pm** Promethium (145)	62 **Sm** Samarium 150.36	63 **Eu** Europium 151.96	64 **Gd** Gadolinium 157.25	65 **Tb** Terbium 158.925	66 **Dy** Dysprosium 162.50	67 **Ho** Holmium 164.93	68 **Er** Erbium 167.26	69 **Tm** Thulium 168.934	70 **Yb** Ytterbium 173.04	71 **Lu** Lutetium 174.967

Actinide Series

89 **Ac** Actinium 227.028	90 **Th** Thorium 232.038	91 **Pa** Protactinium 231.036	92 **U** Uranium 238.029	93 **Np** Neptunium 237.048	94 **Pu** Plutonium (244)	95 **Am** Americium (243)	96 **Cm** Curium (247)	97 **Bk** Berkelium (247)	98 **Cf** Californium (251)	99 **Es** Einsteinium (252)	100 **Fm** Fermium (257)	101 **Md** Mendelevium (258)	102 **No** Nobelium (259)	103 **Lr** Lawrencium (260)

E planetary data

TABLE E.1 *Physical Properties of the Sun and Planets*

Name	Radius (Eq[a]) (km)	Radius (Eq) (Earth units)	Mass (kg)	Mass (Earth units)	Average Density (g/cm^3)	Surface Gravity (Earth = 1)	Escape Velocity (km/s)
Sun	695,000	109	1.99×10^{30}	333,000	1.41	27.5	—
Mercury	2440	0.382	3.30×10^{23}	0.055	5.43	0.38	4.43
Venus	6051	0.949	4.87×10^{24}	0.815	5.25	0.91	10.4
Earth	6378	1.00	5.97×10^{24}	1.00	5.52	1.00	11.2
Mars	3397	0.533	6.42×10^{23}	0.107	3.93	0.38	5.03
Jupiter	71,492	11.19	1.90×10^{27}	317.9	1.33	2.36	59.5
Saturn	60,268	9.46	5.69×10^{26}	95.18	0.70	0.92	35.5
Uranus	25,559	3.98	8.66×10^{25}	14.54	1.32	0.91	21.3
Neptune	24,764	3.81	1.03×10^{26}	17.13	1.64	1.14	23.6
Pluto[b]	1160	0.181	1.31×10^{22}	0.0022	2.05	0.07	1.25
Eris[b]	1430	0.22	1.66×10^{22}	0.0028	2.30	0.08	1.4

[a] Eq = equatorial.

[b] Under the IAU definitions of August 2006, Pluto and Eris are officially designated "dwarf planets."

TABLE E.2 *Orbital Properties of the Sun and Planets*

Name	Distance from Sun[a] (AU)	Distance from Sun[a] (10⁶ km)	Orbital Period (years)	Orbital Inclination[b] (degrees)	Orbital Eccentricity	Sidereal Rotation Period (Earth days)[c]	Axis Tilt (degrees)
Sun	—	—	—	—	—	25.4	7.25
Mercury	0.387	57.9	0.2409	7.00	0.206	58.6	0.0
Venus	0.723	108.2	0.6152	3.39	0.007	−243.0	177.3
Earth	1.00	149.6	1.0	0.00	0.017	0.9973	23.45
Mars	1.524	227.9	1.881	1.85	0.093	1.026	25.2
Jupiter	5.203	778.3	11.86	1.31	0.048	0.41	3.08
Saturn	9.539	1427	29.42	2.48	0.056	0.44	26.73
Uranus	19.19	2870	84.01	0.77	0.046	−0.72	97.92
Neptune	30.06	4497	164.8	1.77	0.010	0.67	29.6
Pluto	39.48	5906	248.0	17.14	0.248	−6.39	112.5
Eris	67.67	10,120	557	44.19	0.442	15.8	78

[a] Semimajor axis of the orbit.

[b] With respect to the ecliptic.

[c] A negative sign indicates rotation is backward relative to other planets.

Table E.3 Satellites of the Solar System (as of 2007)[a]

Planet / Satellite	Radius or Dimensions[b] (km)	Distance from Planet (10³ km)	Orbital Period[c] (Earth days)	Mass[d] (kg)	Density[d] (g/cm³)	Notes About the Satellite
Earth						**Earth**
Moon	1738	384.4	27.322	7.349×10^{22}	3.34	Moon: Probably formed in giant impact.
Mars						**Mars**
Phobos	$13 \times 11 \times 9$	9.38	0.319	1.3×10^{16}	1.9	Phobos, Deimos: Probable captured asteroids.
Deimos	$8 \times 6 \times 5$	23.5	1.263	1.8×10^{15}	2.2	
Jupiter						**Jupiter**
Small inner moons (4 moons)	8 to 83	128–222	0.295–0.674	—	—	Metis, Adrastea, Amalthea, Thebe: Small moonlets within and near Jupiter's ring system.
Io	1821	421.6	1.769	8.933×10^{22}	3.57	Io: Most volcanically active object in the solar system.
Europa	1565	670.9	3.551	4.797×10^{22}	2.97	Europa: Possible oceans under icy crust.
Ganymede	2634	1070.0	7.155	1.482×10^{23}	1.94	Ganymede: Largest satellite in solar system; unusual ice geology.
Callisto	2403	1883.0	16.689	1.076×10^{23}	1.86	Callisto: Cratered iceball.
Irregular group 1 (7 moons)	4–85	7200–17,000	130–457	—	—	Themisto, Leda, Himalia, Lysithea, Elara, and others: Probable captured moons with inclined orbits.
Irregular group 2 (48 moons)	1–30	15,900–29,500	−490 to −983	—	—	Ananke, Carme, Pasiphae, Sinope, and others: Probable captured moons in inclined backward orbits.
Saturn						**Saturn**
Small inner moons (11)	3–89	134–212	0.574–1.1	—	—	Pan, Atlas, Prometheus, Pandora, Epimetheus, Janus, and others: Small moonlets within and near Saturn's ring system.
Mimas	199	185.52	0.942	3.70×10^{19}	1.17	Mimas, Enceladus, Tethys: Small and medium-size iceballs, many with interesting geology.
Enceladus	249	238.02	1.370	1.2×10^{20}	1.24	
Tethys	503	294.66	1.888	6.17×10^{20}	1.26	
Calypso and Telesto	8–12	294.66	1.888	—	—	Calypso and Telesto: Small moonlets sharing Tethys's orbit.
Dione	559	377.4	2.737	1.08×10^{21}	1.44	Dione: Medium-size iceball, with interesting geology.
Helene and Polydeuces	2–16	377.4	2.737	1.6×10^{16}	—	Helene and Polydeuces: Small moonlets sharing Dione's orbit.
Rhea	764	527.04	4.518	2.31×10^{21}	1.33	Rhea: Medium-size iceball, with interesting geology.
Titan	2575	1221.85	15.945	1.35×10^{23}	1.88	Titan: Dense atmosphere shrouds surface; ongoing geological activity.
Hyperion	$180 \times 140 \times 112$	1481.1	21.277	2.8×10^{19}	—	Hyperion: Only satellite known not to rotate synchronously.
Iapetus	718	3561.3	79.331	1.59×10^{21}	1.21	Iapetus: Bright and dark hemispheres show greatest contrast in the solar system.

Satellite	Radius (km)	Distance (10³ km)	Period (days)	Mass (kg)	Density	Notes
Phoebe	110	12,952	−550.4	1×10^{19}	—	*Phoebe:* Very dark; material ejected from Phoebe may coat one side of Iapetus.
Irregular groups (25 moons)	2–16	11,400–23,400	450–930 / −550 to −1320	—	—	Probable captured moons with highly inclined and/or backward orbits.
Uranus						**Uranus**
Small inner moons (13 moons)	5–81	49–98	0.4–0.9	—	—	*Cordelia, Ophelia, Bianca, Cressida, Desdemona, Juliet, Portia, Rosalind, Cupid, Belinda, Perdita, Puck, Mab, 1986 U10, 2003 U1, 2003 U3:* Small moonlets within and near Uranus's ring system.
Miranda	236	129.8	1.413	6.6×10^{19}	1.26	*Miranda, Ariel, Umbriel, Titania, Oberon:* Small and medium-size iceballs, with some interesting geology.
Ariel	579	191.2	2.520	1.35×10^{21}	1.65	
Umbriel	584.7	266.0	4.144	1.17×10^{21}	1.44	
Titania	788.9	435.8	8.706	3.52×10^{21}	1.59	
Oberon	761.4	582.6	13.463	3.01×10^{21}	1.50	
Irregular group (9 moons)	5–95	4280–21,000	580–2820	—	—	*Francisco, Caliban, Stephano, Trinculo, Sycorax, Margaret, Prospero, Setebos, Ferdinand, 2001 U2, 2001 U3, 2003 U3:* Probable captured moons; several in backward orbits.
Neptune						**Neptune**
Small inner moons (5 moons)	29–86	48–74	0.30–0.55	—	—	*Naiad, Thalassa, Despina, Galatea, Larissa:* Small moonlets within and near Neptune's ring system.
Proteus	218 × 208 × 201	117.6	1.121	6×10^{19}	—	
Triton	1352.6	354.59	−5.875	2.14×10^{22}	2.0	*Triton:* Probable captured Kuiper belt object—largest captured object in solar system.
Nereid	170	5588.6	360.125	3.1×10^{19}	—	*Nereid:* Small, icy moon; very little known.
Irregulars (5 moons)	15–27	16,600–48,600	1870–9412	—	—	*2002 N1, N2, N3, N4, 2003 N1:* Possible captured moons in inclined or backward orbit.
Pluto						**Pluto**
Charon	593	19.6	6.38718	1.56×10^{21}	1.6	*Charon:* Unusually large compared to Pluto; may have formed in giant impact.
Nix	50	48,680	24.9	—	—	*Nix and Hydra:* Newly discovered moons outside Charon's orbit.
Hydra	75	64,780	38.2	—	—	
Eris						**Eris**
Dysnomia	50	37,000	15.8	—	—	*Dysnomia:* Approximate properties determined in June 2007.

[a] Note: Authorities differ substantially on many of the values in this table.

[b] a × b × c values for the dimensions are the approximate lengths of the axes (center to edge) for irregular moons.

[c] Negative sign indicates backward orbit.

[d] Masses and densities are most accurate for those satellites visited by a spacecraft on a flyby. Masses for the smallest moons have not been measured but can be estimated from the radius and an assumed density.

Table E.4 Fifty Extrasolar Planets of Note (listed in order of distance from their star)

Name	Detection Methods	Minimum Mass (Jupiter masses)	Semimajor Axis (AU)	Period (days)	Radius (Jupiter radii)	Stellar Mass (Solar masses)	Notes
Gliese 876 d	radial velocity	0.018	0.02081	1.93776	—	0.32	"Hot Jupiter"; sub-Uranus mass; least massive planet confirmed as of 2007
OGLE-TR-56 b	transit, radial velocity	1.29	0.0225	1.21191	1.30	1.17	"Hot Jupiter"; first planet discovered by transit; planet with the shortest confirmed period as of 2007
GJ 436 b	radial velocity	0.0713	0.0285	2.64385	—	0.44	"Hot Jupiter"; sub-Saturn mass
SWEEPS-11	transit, radial velocity	9.7	0.03	1.796	1.13	1.10	"Hot Jupiter"
OGLE-TR-132 b	transit, radial velocity	1.14	0.0306	1.68986	1.18	1.26	"Hot Jupiter"
WASP-2 b	transit, radial velocity	0.88	0.0307	2.15223	1.04	0.79	"Hot Jupiter"
TrES-2	transit, radial velocity	1.98	0.0367	2.47063	1.22	0.98	"Hot Jupiter"
55 Cnc e	radial velocity	0.045	0.038	2.81	—	1.03	"Hot Jupiter"; sub-Neptune mass
WASP-1 b	transit, radial velocity	0.89	0.0382	2.51997	1.44	1.15	"Hot Jupiter"; "puffed-up planet"
TrES-1	transit, radial velocity, eclipse	0.61	0.0393	3.03007	1.081	0.87	"Hot Jupiter"; sub-Saturn mass
HD 46375 b	radial velocity	0.249	0.041	3.024	—	0.91	"Hot Jupiter"; sub-Saturn mass
Gliese 581 b	radial velocity	0.0492	0.041	5.3683	—	0.31	—
OGLE-TR-10 b	transit, radial velocity	0.63	0.04162	3.10129	1.26	1.18	"Hot Jupiter"
HD 149026 b	radial velocity, transit	0.36	0.042	2.8766	0.725	1.3	"Hot Jupiter"
HD 209458 b	radial velocity, transit, eclipse	0.69	0.045	3.52475	1.32	1.01	"Hot Jupiter"; "puffed-up planet"; first planet and first atmosphere successfully detected by transit
HD 88133 b	radial velocity	0.22	0.047	3.41	—	1.20	"Hot Jupiter"; sub-Saturn mass
OGLE-TR-111 b	transit, radial velocity	0.53	0.047	4.01445	1.067	0.82	"Hot Jupiter"
XO-1 b	transit, radial velocity	0.9	0.0488	3.94153	1.184	1.00	"Hot Jupiter"
51 Peg b	radial velocity	0.468	0.052	4.23077	—	1.06	"Hot Jupiter"; first exoplanet discovered around Sun-like star
SWEEPS-04	transit, radial velocity	3.8	0.055	4.2	0.81	1.24	"Hot Jupiter"
HAT—P-1 b	transit, radial velocity	0.53	0.0551	4.46529	1.36	1.12	"Hot Jupiter"; "puffed-up planet"
Ups And b	radial velocity	0.69	0.059	4.61708	—	1.27	"Hot Jupiter"; in first multiplanet system discovered around Sun-like star
Gliese 581 c	radial velocity	0.0158	0.073	12.932	—	0.31	—
HD 160691 d	radial velocity	0.044	0.09	9.55	—	1.08	"Hot Jupiter"; sub-Uranus mass
55 Cnc b	radial velocity	0.784	0.115	14.67	—	1.03	—
Gliese 876 c	radial velocity	0.56	0.13	30.1	—	0.32	"Eccentric"

HD 102117 b	radial velocity	0.172	0.1532	20.67	—	0.95	sub-Saturn mass
Gliese 876 b	radial velocity	1.935	0.20783	60.94	—	0.32	First exoplanet discovered orbiting a red dwarf
55 Cnc c	radial velocity	0.217	0.24	43.93	—	1.03	"Eccentric"; sub-Saturn mass
Gliese 581 d	radial velocity	0.0243	0.25	83.60	—	0.31	—
HD 16141 b	radial velocity	0.23	0.35	75.56	—	1.00	sub-Saturn mass
HD 80606 b	radial velocity	3.41	0.439	111.78	—	0.9	"Eccentric"; highest known planetary eccentricity (0.927)
HD 82943 c	radial velocity	2.01	0.746	219.0	—	1.18	"Eccentric"
Ups And c	radial velocity	1.98	0.83	241.52	—	1.27	"Eccentric"; in first multiplanet system discovered around Sun-like star
HR 810 b	radial velocity	1.94	0.91	311.288	—	1.11	—
HD 210277 b	radial velocity	1.23	1.10	442.1	—	0.92	"Eccentric"; planet mass partially inferred from surrounding disk
HD 27442 b	radial velocity	1.28	1.18	423.841	—	1.2	—
HD 41004 A b	radial velocity	2.3	1.31	655.0	—	0.7	"Eccentric"; planet in a system with two stars and a brown dwarf
HD 4208 b	radial velocity	0.80	1.67	812.197	—	0.93	—
HD 45350 b	radial velocity	1.79	1.92	890.76	—	1.02	—
Gamma Cephei b	radial velocity	1.60	2.044	902.9	—	1.4	First extrasolar planet discovered in close stellar binary system
HD 187085 b	radial velocity	0.75	2.05	986.0	—	1.22	"Eccentric"
47 Uma b	radial velocity	2.60	2.11	1083.2	—	1.03	—
HD 10697 b	radial velocity	6.12	2.13	1077.906	—	1.15	—
Ups And d	radial velocity	3.95	2.51	1274.6	—	1.27	In first multiplanet system discovered around Sun-like star
HD 202206 c	radial velocity	2.44	2.55	1383.4	—	1.13	"Eccentric"
HD 37124 c	radial velocity	0.683	3.19	2295.0	—	0.91	—
Epsilon Eridani b	radial velocity, astrometry	1.55	3.39	2502.0	—	0.83	"Eccentric"; star surrounded by dust disk; closest in distance exoplanet to Earth
HD 38529 c	radial velocity	12.7	3.68	2174.3	—	1.39	"Eccentric"
HD 72659 b	radial velocity	2.96	4.16	3177.4	—	0.95	"Eccentric"
55 Cnc d	radial velocity	3.92	5.257	4517.4	—	1.03	"Eccentric"; largest confirmed semimajor axis as of 2007
2M1207 b	direct imaging	~5	~41–51	—	1.5	0.025	Only confirmed image detection of exoplanet; orbit very uncertain but mass well-constrained

Notes:
1. The list includes all planets detected by two methods, most planets in multiple systems, most "Hot Jupiters," and a representative sample of other extrasolar planets. More than 200 known extrasolar planets are not listed.

2. Where two detection methods are listed, the discovery method is given first.

3. "Eccentric" means eccentricity > 0.25.

TABLE F.1 *Stars Within 12 Light-Years*

Star	Distance (ly)	Spectral Type		RA h	RA m	Dec °	Dec ′	Luminosity (L/L_Sun)
Sun	0.000016	G2	V	—	—	—	—	1.0
Proxima Centauri	4.2	M5.5	V	14	30	−62	41	0.0006
α Centauri A	4.4	G2	V	14	40	−60	50	1.6
α Centauri B	4.4	K0	V	14	40	−60	50	0.53
Barnard's Star	6.0	M4	V	17	58	+04	42	0.005
Wolf 359	7.8	M6	V	10	56	+07	01	0.0008
Lalande 21185	8.3	M2	V	11	03	+35	58	0.03
Sirius A	8.6	A1	V	06	45	−16	42	26.0
Sirius B	8.6	DA2	White dwarf	06	45	−16	42	0.002
UV Ceti	8.7	M5.5	V	01	39	−17	57	0.0009
BL Ceti	8.7	M6	V	01	39	−17	57	0.0006
Ross 154	9.7	M3.5	V	18	50	−23	50	0.004
Ross 248	10.3	M5.5	V	23	42	+44	11	0.001
ε Eridani	10.5	K2	V	03	33	−09	28	0.37
Lacaille 9352	10.7	M1.5	V	23	06	−35	51	0.05
Ross 128	10.9	M4	V	11	48	+00	49	0.003
EZ Aquarii A	11.3	M5	V	22	39	−15	18	0.0006
EZ Aquarii B	11.3	—	—	22	39	−15	18	0.0004
EZ Aquarii C	11.3	—	—	22	39	−15	18	0.0003
Procyon A	11.4	F5	IV–V	07	39	+05	14	8.6
Procyon B	11.4	DA	White dwarf	07	39	+05	14	0.0005
61 Cygni A	11.4	K5	V	21	07	+38	42	0.17
61 Cygni B	11.4	K7	V	21	07	+38	42	0.10
Gliese 725 A	11.5	M3	V	18	43	+59	38	0.02
Gliese 725 B	11.5	M3.5	V	18	43	+59	38	0.01
GX Andromedae	11.6	M1.5	V	00	18	+44	01	0.03
GQ Andromedae	11.6	M3.5	V	00	18	+44	01	0.003
ε Indi A	11.8	K5	V	22	03	−56	45	0.30
ε Indi B	11.8	T1.0	Brown dwarf	22	04	−56	46	—
ε Indi C	11.8	T1.0	Brown dwarf	22	04	−56	46	—
DX Cancri	11.8	M6.5	V	08	30	+26	47	0.0003
τ Ceti	11.9	G8	V	01	44	−15	57	0.67
GJ 1061	12.0	M5.5	V	03	36	−44	31	0.001

Note: These data were provided by the RECONS project, courtesy of Dr. Todd Henry (June, 2007). The luminosities are all total (bolometric) luminosities. The DA stellar types are white dwarfs. The coordinates are for the year 2000. The bolometric luminosity of the brown dwarfs is primarily in the infrared and has not been measured accurately yet.

TABLE F.2 *Twenty Brightest Stars*

Star	Constellation	RA h	RA m	Dec °	Dec '	Distance (ly)	Spectral Type		Apparent Magnitude	Luminosity (L/L_{Sun})
Sirius	Canis Major	6	45	−16	42	8.6	A1	V	−1.46	26
Canopus	Carina	6	24	−52	41	313	F0	Ib-II	−0.72	13,000
α Centauri	Centaurus	14	40	−60	50	4.4	G2	V	−0.01	1.6
							K0	V	1.3	0.53
Arcturus	Boötes	14	16	+19	11	37	K2	III	−0.06	170
Vega	Lyra	18	37	+38	47	25	A0	V	0.04	60
Capella	Auriga	5	17	+46	00	42	G0	III	0.75	70
							G8	III	0.85	77
Rigel	Orion	5	15	−08	12	772	B8	Ia	0.14	70,000
Procyon	Canis Minor	7	39	+05	14	11.4	F5	IV–V	0.37	7.4
Betelgeuse	Orion	5	55	+07	24	427	M2	Iab	0.41	38,000
Achernar	Eridanus	1	38	−57	15	144	B5	V	0.51	3600
Hadar	Centaurus	14	04	−60	22	525	B1	III	0.63	100,000
Altair	Aquila	19	51	+08	52	17	A7	IV–V	0.77	10.5
Acrux	Crux	12	27	−63	06	321	B1	IV	1.39	22,000
							B3	V	1.9	7500
Aldebaran	Taurus	4	36	+16	30	65	K5	III	0.86	350
Spica	Virgo	13	25	−11	09	260	B1	V	0.91	23,000
Antares	Scorpio	16	29	−26	26	604	M1	Ib	0.92	38,000
Pollux	Gemini	7	45	+28	01	34	K0	III	1.16	45
Fomalhaut	Piscis Austrinus	22	58	−29	37	25	A3	V	1.19	18
Dencb	Cygnus	20	41	+45	16	2500	A2	Ia	1.26	170,000
β Crucis	Crux	12	48	−59	40	352	B0.5	IV	1.28	37,000

Note: Three of the stars on this list, Capella, α Centauri, and Acrux, are binary systems with members of comparable brightness. They are counted as single stars because that is how they appear to the naked eye. All the luminosities given are total (bolometric) luminosities. The coordinates are for the year 2000.

G galaxy data

TABLE G.1 *Galaxies of the Local Group*

Galaxy Name	Distance (millions of ly)	Type[a]	RA h	RA m	Dec °	Dec ′	Luminosity (millions of L$_{Sun}$)
Milky Way	—	Sbc	—	—	—	—	15,000
WLM	3.0	Irr	00	02	−15	30	50
IC 10	2.7	dIrr	00	20	+59	18	160
Cetus	2.5	dE	00	26	−11	02	0.72
NGC 147	2.4	dE	00	33	+48	30	131
And III	2.5	dE	00	35	+36	30	1.1
NGC 185	2.0	dE	00	39	+48	20	120
NGC 205	2.7	E	00	40	+41	41	370
And VIII	2.7	dE	00	42	+40	37	240
M32	2.6	E	00	43	+40	52	380
M31	2.5	Sb	00	43	+41	16	21,000
And I	2.6	dE	00	46	+38	00	4.7
SMC	0.19	Irr	00	53	−72	50	230
And IX	2.9	dE	00	52	+43	12	—
Sculptor	0.26	dE	01	00	−33	42	2.2
LGS 3	2.6	dIrr	01	04	+21	53	1.3
IC 1613	2.3	Irr	01	05	+02	08	64
And V	2.9	dE	01	10	+47	38	—
And II	1.7	dE	01	16	+33	26	2.4
M33	2.7	Sc	01	34	+30	40	2800
Phoenix	1.5	dIrr	01	51	−44	27	0.9
Fornax	0.45	dE	02	40	−34	27	15.5
EGB0427 + 63	4.3	dIrr	04	32	+63	36	9.1
LMC	0.16	Irr	05	24	−69	45	1300
Carina	0.33	dE	06	42	−50	58	0.4
Canis Major	0.025	dIrr	07	15	−28		—
Leo A	2.2	dIrr	09	59	+30	45	3.0
Sextans B	4.4	dIrr	10	00	+05	20	41
NGC 3109	4.1	Irr	10	03	−26	09	160
Antlia	4.0	dIrr	10	04	−27	19	1.7
Leo I	0.82	dE	10	08	+12	18	4.8
Sextans A	4.7	dIrr	10	11	−04	42	56
Sextans	0.28	dE	10	13	−01	37	0.5
Leo II	0.67	dE	11	13	+22	09	0.6
GR 8	5.2	dIrr	12	59	+14	13	3.4
Ursa Minor	0.22	dE	15	09	+67	13	0.3
Draco	2.7	dE	17	20	+57	55	0.3
Sagittarius	0.08	dE	18	55	−30	29	18
SagDIG	3.5	dIrr	19	30	−17	41	6.8
NGC 6822	1.6	Irr	19	45	−14	48	94
DDO 210	2.6	dIrr	20	47	−12	51	0.8
IC 5152	5.2	dIrr	22	03	−51	18	70
Tucana	2.9	dE	22	42	−64	25	0.5
UKS2323-326	4.3	dE	23	26	−32	23	5.2
And VII	2.6	dE	23	38	+50	35	—
Pegasus	3.1	dIrr	23	29	+14	45	12
And VI	2.8	dE	23	52	+24	36	—

[a]Types beginning with S are spiral galaxies classified according to Hubble's system (see Chapter 15). Type E galaxies are elliptical or spheroidal. Type Irr galaxies are irregular. The prefix d denotes a dwarf galaxy. This list is based on a list originally published by M. Mateo in 1998 and augmented by discoveries of Local Group galaxies made between 1998 and 2005.

TABLE G.2   *Nearby Galaxies in the Messier Catalog*[a,b]

Galaxy Name (M / NGC)[c]	RA h	RA m	Dec °	Dec '	RV$_{hel}$[d]	RV$_{gal}$[e]	Type[f]	Nickname
M31 / NGC 224	00	43	+41	16	−300 ± 4	−122	Spiral	Andromeda
M32 / NGC 221	00	43	+40	52	−145 ± 2	32	Elliptical	
M33 / NGC 598	01	34	+30	40	−179 ± 3	−44	Spiral	Triangulum
M49 / NGC 4472	12	30	+08	00	997 ± 7	929	Elliptical/ Lenticular/Seyfert	
M51 / NGC 5194	13	30	+47	12	463 ± 3	550	Spiral/Interacting	Whirlpool
M58 / NGC 4579	12	38	+11	49	1519 ± 6	1468	Spiral/Seyfert	
M59 / NGC 4621	12	42	+11	39	410 ± 6	361	Elliptical	
M60 / NGC 4649	12	44	+11	33	1117 ± 6	1068	Elliptical	
M61 / NGC 4303	12	22	+04	28	1566 ± 2	1483	Spiral/Seyfert	
M63 / NGC 5055	13	16	+42	02	504 ± 4	570	Spiral	Sunflower
M64 / NGC 4826	12	57	+21	41	408 ± 4	400	Spiral/Seyfert	Black Eye
M65 / NGC 3623	11	19	+13	06	807 ± 3	723	Spiral	
M66 / NGC 3627	11	20	+12	59	727 ± 3	643	Spiral/Seyfert	
M74 / NGC 628	01	37	+15	47	657 ± 1	754	Spiral	
M77 / NGC 1068	02	43	−00	01	1137 ± 3	1146	Spiral/Seyfert	
M81 / NGC 3031	09	56	+69	04	−34 ± 4	73	Spiral/Seyfert	
M82 / NGC 3034	09	56	+69	41	203 ± 4	312	Irregular/Starburst	
M83 / NGC 5236	13	37	−29	52	516 ± 4	385	Spiral/Starburst	
M84 / NGC 4374	12	25	+12	53	1060 ± 6	1005	Elliptical	
M85 / NGC 4382	12	25	+18	11	729 ± 2	692	Spiral	
M86 / NGC 4406	12	26	+12	57	−244 ± 5	−298	Elliptical/Lenticular	
M87 / NGC 4486	12	30	+12	23	1307 ± 7	1254	Elliptical/Central Dominant/Seyfert	Virgo A
M88 / NGC 4501	12	32	+14	25	2281 ± 3	2235	Spiral/Seyfert	
M89 / NGC 4552	12	36	+12	33	340 ± 4	290	Elliptical	
M90 / NGC 4569	12	37	+13	10	−235 + 4	−282	Spiral/Seyfert	
M91 / NGC 4548	12	35	+14	30	486 ± 4	442	Spiral/Seyfert	
M94 / NGC 4736	12	51	+41	07	308 ± 1	360	Spiral	
M95 / NGC 3351	10	44	+11	42	778 ± 4	677	Spiral/Starburst	
M96 / NGC 3368	10	47	+11	49	897 ± 4	797	Spiral/Seyfert	
M98 / NGC 4192	12	14	+14	54	−142 ± 4	−195	Spiral/Seyfert	
M99 / NGC 4254	12	19	+14	25	2407 ± 3	2354	Spiral	
M100 / NGC 4321	12	23	+15	49	1571 ± 1	1525	Spiral	
M101 / NGC 5457	14	03	+54	21	241 ± 2	360	Spiral	
M104 / NGC 4594	12	40	−11	37	1024 ± 5	904	Spiral/Seyfert	Sombrero
M105 / NGC 3379	10	48	+12	35	911 ± 2	814	Elliptical	
M106 / NGC 4258	12	19	+47	18	448 ± 3	507	Spiral/Seyfert	
M108 / NGC 3556	11	09	+55	57	695 ± 3	765	Spiral	
M109 / NGC 3992	11	55	+53	39	1048 ± 4	1121	Spiral	
M110 / NGC 205	00	38	+41	25	−241 ± 3	−61	Elliptical	

[a]Galaxies identified in the catalog published by Charles Messier in 1781; these galaxies are relatively easy to observe with small telescopes.

[b]Data obtained from NED: NASA/IPAC Extragalactic Database (http://ned.ipac.caltech.edu). The original Messier list of galaxies was obtained from SED, and the list data were updated to 2001 and M102 was dropped.

[c]The galaxies are identified by the Messier number (M followed by a number) and NGC number, which comes from the *New General Catalog* published in 1888.

[d]Radial velocity in kilometers per second, with respect to the Sun (heliocentric). Positive values mean motion away from the Sun; negative values are toward the Sun.

[e]Radial velocity in kilometers per second, with respect to the Milky Way Galaxy, calculated from the RV$_{hel}$ values with a correction for the Sun's motion around the galactic center.

[f]Galaxies are first listed by their primary type (spiral, elliptical, or irregular) and then by any other special categories that apply (see Chapter 15).

TABLE G.3 *Nearby X-Ray Bright Clusters of Galaxies*

Cluster Name	Redshift	Distance[a] (billions of ly)	Temperature of Intracluster Medium (millions of K)	Average Orbital Velocity of Galaxies[b] (km/s)	Cluster Mass[c] ($10^{15}M_{Sun}$)
Abell 2142	0.0907	1.26	101. ± 2	1132 ± 110	1.6
Abell 2029	0.0766	1.07	100. ± 3	1164 ± 98	1.5
Abell 401	0.0737	1.03	95.2 ± 5	1152 ± 86	1.4
Coma	0.0233	0.32	95.1 ± 1	821 ± 49	1.4
Abell 754	0.0539	0.75	93.3 ± 3	662 ± 77	1.4
Abell 2256	0.0589	0.82	87.0 ± 2	1348 ± 86	1.4
Abell 399	0.0718	1.00	81.7 ± 7	1116 ± 89	1.1
Abell 3571	0.0395	0.55	81.1 ± 3	1045 ± 109	1.1
Abell 478	0.0882	1.23	78.9 ± 2	904 ± 281	1.1
Abell 3667	0.0566	0.79	78.5 ± 6	971 ± 62	1.1
Abell 3266	0.0599	0.84	78.2 ± 5	1107 ± 82	1.1
Abell 1651a	0.0846	1.18	73.1 ± 6	685 ± 129	0.96
Abell 85	0.0560	0.78	70.9 ± 2	969 ± 95	0.92
Abell 119	0.0438	0.61	65.6 ± 5	679 ± 106	0.81
Abell 3558	0.0480	0.67	65.3 ± 2	977 ± 39	0.81
Abell 1795	0.0632	0.88	62.9 ± 2	834 ± 85	0.77
Abell 2199	0.0314	0.44	52.7 ± 1	801 ± 92	0.59
Abell 2147	0.0353	0.49	51.1 ± 4	821 ± 68	0.56
Abell 3562	0.0478	0.67	45.7 ± 8	736 ± 49	0.48
Abell 496	0.0325	0.45	45.3 ± 1	687 ± 89	0.47
Centaurus	0.0103	0.14	42.2 ± 1	863 ± 34	0.42
Abell 1367	0.0213	0.30	41.3 ± 2	822 ± 69	0.41
Hydra	0.0126	0.18	38.0 ± 1	610 ± 52	0.36
C0336	0.0349	0.49	37.4 ± 1	650 ± 170	0.35
Virgo	0.0038	0.05	25.7 ± 0.5	632 ± 41	0.20

Note: This table lists the 25 brightest clusters of galaxies in the X-ray sky from a catalog by J. P. Henry (2000).

[a]Cluster distances were computed using a value for Hubble's constant of 21.5 km/s/million light-years.

[b]The average orbital velocities given in this column are the velocity component along our line of sight. This velocity should be multiplied by the square root of 2 to get the average orbital velocity.

[c]This column gives each cluster's mass within the largest radius at which the intracluster medium can be in gravitational equilibrium. Because our estimates of that radius depend on Hubble's constant, these masses are inversely proportional to Hubble's constant, which we have assumed to be 21.5 km/s/million light-years.

H selected astronomical Web sites

The Web contains a vast amount of astronomical information. For all your astronomical Web surfing, the best starting point is the Web site for this textbook:

MasteringAstronomy™
www.masteringastronomy.com

The following are some other sites that may be of particular use. In case any of the links change, you can always find live links to these sites, and many more, on the MasteringAstronomy Web site.

Key Mission Sites

The following table lists the Web pages for major current astronomy missions.

Site	Description	Web Address
NASA's Office of Space Science Missions Page	**Direct links to all past, present, and planned NASA space science missions**	**http://science.hq.nasa.gov/ missions/phase.html**
Cassini-Huygens	Mission arrived at Saturn in 2004	http://saturn.jpl.nasa.gov/index.cfm
Chandra X-Ray Observatory	Latest discoveries, educational activities, and other information from the Chandra X-Ray Observatory	http://chandra.harvard.edu
Dawn	Study of large asteroids	http://dawn.jpl.nasa.gov
Hubble Space Telescope	Latest discoveries, educational activities, and other information from the Hubble Space Telescope	http://hubblesite.org
Mars Express	European mission orbiting Mars	http://www.esa.int/SPECIALS/ Mars_Express
MESSENGER	Study of Mercury	http://messenger.jhuapl.edu/
NASA Mars Exploration Program	Information on current and planned Mars missions	http://mars.jpl.nasa.gov
New Horizons	Will reach Pluto in 2015	http://pluto.jhuapl.edu/
Swift	Study of gamma-ray bursts	http://swift.gsfc.nasa.gov
Venus Express	European mission orbiting Venus	http://www.csa.int/SPECIALS/ Venus_Express
Wilkinson Microwave Anisotropy Probe (WMAP)	Mission to study the cosmic microwave background	http://map.gsfc.nasa.gov
Spitzer Space Telescope	Infrared observatory launched in 2003	http://spitzer.caltech.edu
Stratospheric Observatory for Infrared Astronomy (SOFIA)	Airborne observatory scheduled to begin science observations in 2009	http://sofia.arc.nasa.gov

Key Observatory Sites

The following table lists the Web pages leading to major ground-based observatories.

Site	Description	Web Address
World's Largest Optical Telescopes	**Direct links to most of the world's major optical observatories**	**http://astro.nineplanets.org/ bigeyes.html**
Arecibo Observatory (Puerto Rico)	World's largest single-dish radio telescope	http://www.naic.edu
Cerro Tololo Inter-American Observatory	Links to major observatories on site in Cerro Tololo, Chile	http://www.ctio.noao.edu
Allen Telescope Array	Used in search for extraterrestrial intelligence	http://www.seti.org/ata/
European Southern Observatory	Links to European telescope projects in Chile, including the Very Large Telescope	http://www.eso.org
Mauna Kea Observatories	Links to major observatories in Hawaii, including Keck, Gemini, Subaru, CFHT, and others	http://www.ifa.hawaii.edu/mko
National Optical Astronomy Observatory	Home page for U.S. national observatories in Arizona, Hawaii, and Chile	http://www.noao.edu
National Radio Astronomy Observatory	Home page for U.S. national radio observatories, including the Very Large Array (VLA)	http://www.nrao.edu

More Astronomical Web Sites

The following Web sites are some of the authors' favorites among many other noncommercial resources for astronomy.

Site	Description	Web Address
MasteringAstronomy	**Don't forget to start here for all your astronomical Web surfing**	**http://www.masteringastronomy.com**
American Association of Variable Star Observers (AAVSO)	One of the largest organizations of amateur astronomers in the world. Check this site if you are interested in serious amateur astronomy	http://www.aavso.org
Astronomical Society of the Pacific	An organization for both professional astronomers and the general public, devoted largely to astronomy education	http://www.astrosociety.org
Astronomy Picture of the Day	An archive of beautiful pictures, updated daily	http://antwrp.gsfc.nasa.gov/apod
AstroWeb	Listing of major resources for astronomy on the Web	http://www.stsci.edu/resources/
Canadian Space Agency	Home page for Canada's space program	http://www.space.gc.ca
European Space Agency (ESA)	Home page for this international agency	http://www.esa.int
The Extrasolar Planets Encyclopaedia	Information about the search for and discoveries of extrasolar planets	http://exoplanet.eu
NASA Home Page	Learn almost anything you want about NASA	http://www.nasa.gov
NASA Science News	Read the latest news from NASA; has option to subscribe to e-mail notices of news releases	http://science.nasa.gov
The ~~Nine~~ 8 Planets (University of Arizona)	A multimedia tour of the solar system	http://www.nineplanets.org
The Planetary Society	Has more than 100,000 members who are interested in planetary exploration and the search for life in the universe	http://planetary.org
The SETI Institute	Devoted to the search for other civilizations	http://www.seti.org
Voyage Scale Model Solar System	Take a virtual tour of the Voyage Scale Model Solar System	http://www.jeffreybennett.com/virtual_tour/index.html

Constellation Names (English Equivalent in Parentheses)

Andromeda (The Chained Princess)
Antlia (The Air Pump)
Apus (The Bird of Paradise)
Aquarius (The Water Bearer)
Aquila (The Eagle)
Ara (The Altar)
Aries (The Ram)
Auriga (The Charioteer)
Boötes (The Herdsman)
Caelum (The Chisel)

Camelopardalis (The Giraffe)
Cancer (The Crab)
Canes Venatici (The Hunting Dogs)
Canis Major (The Great Dog)
Canis Minor (The Little Dog)
Capricornus (The Sea Goat)
Carina (The Keel)
Cassiopeia (The Queen)
Centaurus (The Centaur)
Cepheus (The King)

Cetus (The Whale)
Chamaeleon (The Chameleon)
Circinus (The Drawing Compass)
Columba (The Dove)
Coma Berenices (Berenice's Hair)
Corona Australis (The Southern Crown)
Corona Borealis (The Northern Crown)
Corvus (The Crow)
Crater (The Cup)
Crux (The Southern Cross)

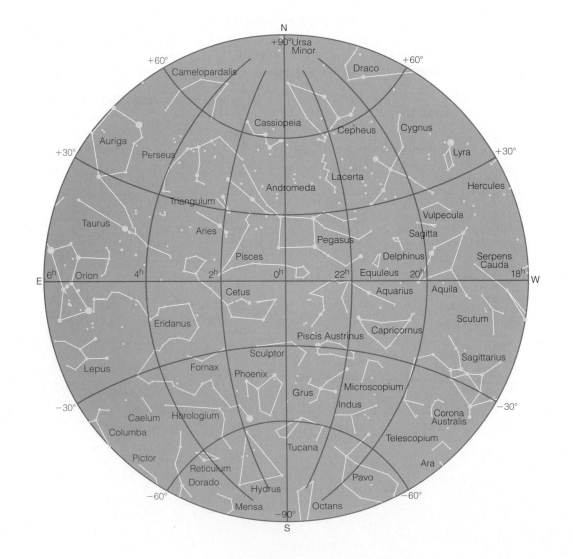

Cygnus (The Swan)
Delphinus (The Dolphin)
Dorado (The Goldfish)
Draco (The Dragon)
Equuleus (The Little Horse)
Eridanus (The River)
Fornax (The Furnace)
Gemini (The Twins)
Grus (The Crane)
Hercules
Horologium (The Clock)
Hydra (The Sea Serpent)
Hydrus (The Water Snake)
Indus (The Indian)
Lacerta (The Lizard)
Leo (The Lion)
Leo Minor (The Little Lion)
Lepus (The Hare)
Libra (The Scales)
Lupus (The Wolf)
Lynx (The Lynx)
Lyra (The Lyre)
Mensa (The Table)

Microscopium (The Microscope)
Monoceros (The Unicorn)
Musca (The Fly)
Norma (The Level)
Octans (The Octant)
Ophiuchus (The Serpent Bearer)
Orion (The Hunter)
Pavo (The Peacock)
Pegasus (The Winged Horse)
Perseus (The Hero)
Phoenix (The Phoenix)
Pictor (The Painter's Easel)
Pisces (The Fish)
Piscis Austrinus (The Southern Fish)
Puppis (The Stern)
Pyxis (The Compass)
Reticulum (The Reticle)
Sagitta (The Arrow)
Sagittarius (The Archer)
Scorpius (The Scorpion)
Sculptor (The Sculptor)
Scutum (The Shield)
Serpens (The Serpent)

Sextans (The Sextant)
Taurus (The Bull)
Telescopium (The Telescope)
Triangulum (The Triangle)
Triangulum Australe (The Southern Triangle)
Tucana (The Toucan)
Ursa Major (The Great Bear)
Ursa Minor (The Little Bear)
Vela (The Sail)
Virgo (The Virgin)
Volans (The Flying Fish)
Vulpecula (The Fox)

Constellation Locations

Each of the charts on these pages shows half of the celestial sphere in projection, so you can use them to learn the approximate locations of the constellations. The grid lines are marked by right ascension and declination.

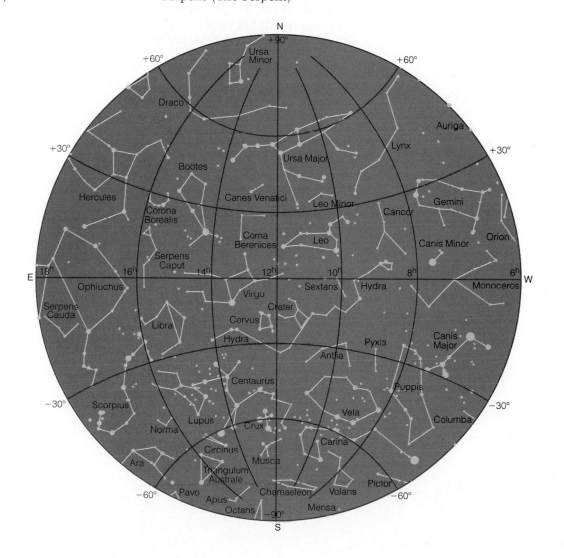

star charts

How to use the star charts:

Check the times and dates under each chart to find the best one for you. Take it outdoors within an hour or so of the time listed for your date. Bring a dim flashlight to help you read it.

On each chart, the round outside edge represents the horizon all around you. Compass directions around the horizon are marked in yellow. Turn the chart around so that the edge marked with the direction you're facing (for example, north, southeast) is down. The stars above this horizon now match the stars you are facing. Ignore the rest until you turn to look in a different direction.

The center of the chart represents the sky overhead, so a star plotted on the chart halfway from the edge to the center can be found in the sky halfway from the horizon to straight up.

The charts are drawn for 40°N latitude (for example, Denver, New York, Madrid). If you live far south of there, stars in the southern part of your sky will appear higher than on the chart and stars in the north will be lower. If you live far north of there, the reverse is true.

© 1999 *Sky & Telescope*

Use this chart January, February, and March.

Early January — 1 A.M.	Early February — 11 P.M.	Early March — 9 P.M.
Late January — Midnight	Late February — 10 P.M.	Late March — Dusk

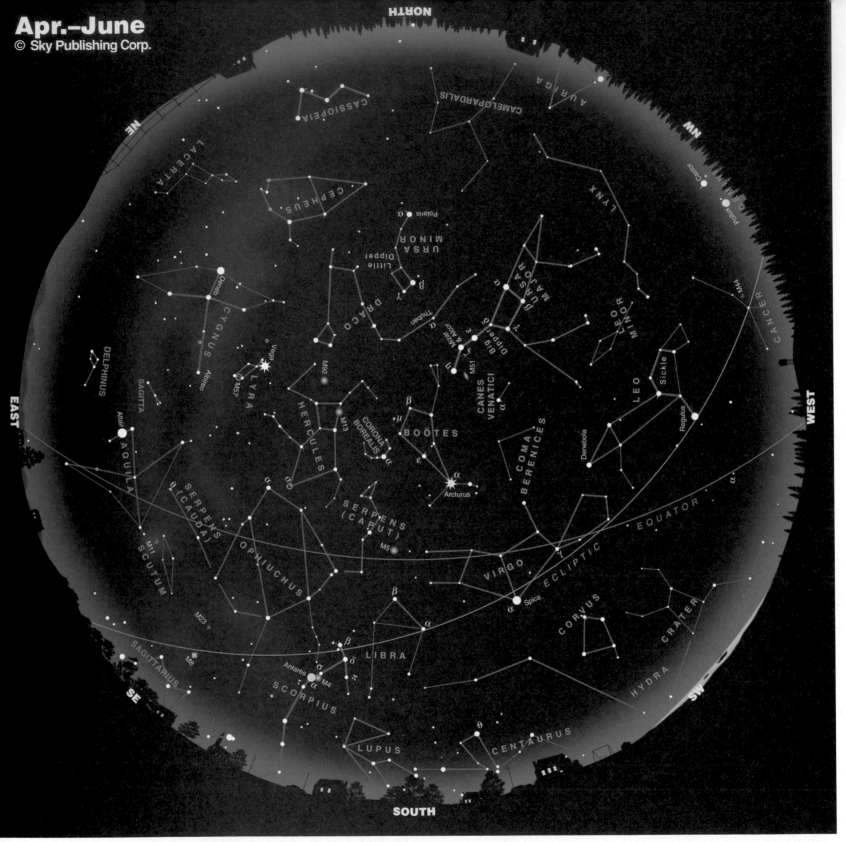

© 1999 *Sky & Telescope*

Use this chart April, May, and June.

Early April — 3 A.M.*   Early May — 1 A.M.*   Early June — 11 P.M.*
Late April — 2 A.M.*   Late May — Midnight*   Late June — Dusk

*Daylight Saving Time

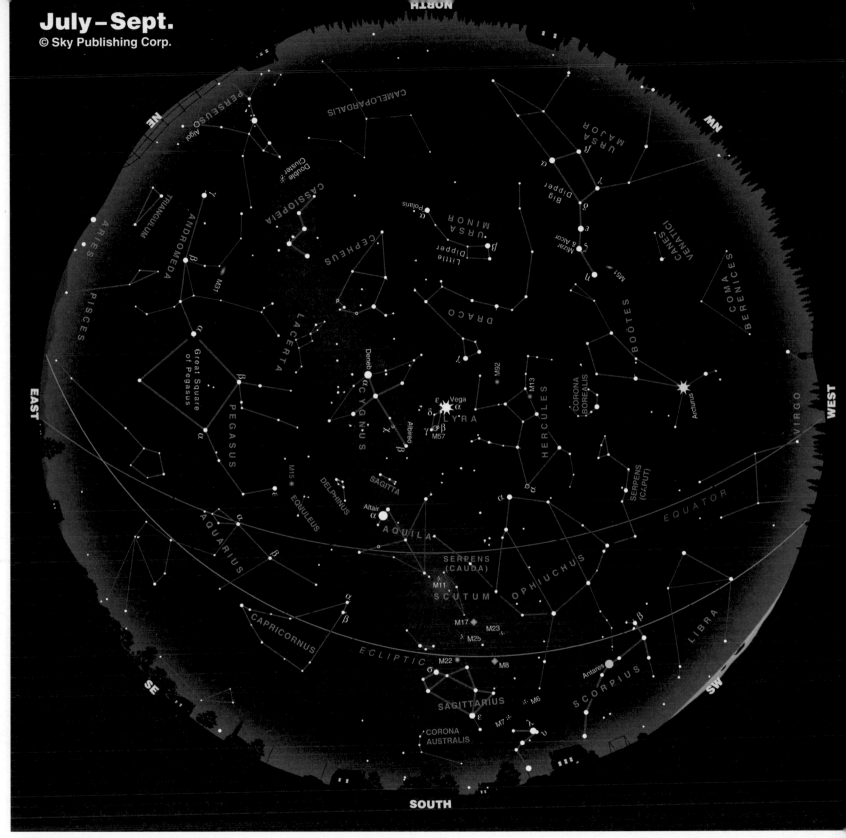

Use this chart July, August, and September.

Early July — 1 A.M.*   Early August — 11 P.M.*   Early September — 9 P.M.*
Late July — Midnight   Late August — 10 P.M.*   Late September — Dusk

*Daylight Saving Time

© 1999 *Sky & Telescope*

Use this chart October, November, and December.

Early October — 1 A.M.*   Early November — 10 P.M.   Early December — 8 P.M.
Late October — Midnight*   Late November — 9 P.M.   Late December — 7 P.M.

*Daylight Saving Time

K solutions to visual skills checks

Chapter 1

1. b   2. c   3. c
4. The nearest stars would not fit on Earth on this scale.

Chapter 2

1. B   2. D   3. A   4. C
5. d   6. d   7. c   8. c

Chapter 3

1. d   2. b   3. d   4. a
5. a   6. a   7. a

Chapter 4

1. b   2. d   3. a   4. d   5. c

Chapter 5

1. 5   2. 1   3. b   4. b   5. c

Chapter 6

1. About 4 days; about 50 meters per second
2. 1: d, 2: a, 3: c, 4: b
3. 1: c, 2: b, 3: d, 4: a
4. b

Chapter 7

1. a   2. c   3. b   4. b, c, a

Chapter 8

1. c   2. e   3. b   4. b

Chapter 9

1. About 5 km, though its unusual shape could lead to answers between 4 and 8 km
2. b   3. a   4. c   5. b

Chapter 10

1. d
2. Sunspots appear over a range of 40–50° North latitude to 40–50° South latitude.
3. Sunspots get closer to the equator during a sunspot cycle.

Chapter 11

1. b   2. d   3. c
4. Luminosity: about $10,000L_{Sun}$; lifetime: slightly longer than 10 million years
5. Luminosity: about $100L_{Sun}$; lifetime: slightly shorter than 1 billion years
6. Luminosity: about $30L_{Sun}$; lifetime: approximately 1 billion years

Chapter 12

1. Approximately $10L_{Sun}$
2. Approximately 3500 K
3. Approximately $10^4 L_{Sun}$
4. Approximately $10^{-4} L_{Sun}$

Chapter 13

1. b   2. d   3. b   4. c   5. d

Chapter 14

1. Brightest: white; lowest brightness: black/dark blue
2. White
3. Regions with strong radio emission are dark in the visible-light image.
4. Regions with strong radio emission are brighter in the infrared image than they are in the visible-light image.
5. c
6. Yes

Chapter 15

1. Cepheids
2. Parallax
3. Approximately 100–10,000 light-years
4. White
5. Purple
6. Approximately 400,000 light-years
7. Approximately 20,000 light-years

Chapter 16

1. accelerating
2. accelerating
3. coasting
4. decelerating

Chapter 17

1. d   2. d   3. d   4. a

Chapter 18

1. g, f, b, e, d, c, a
2. b, a, c, d
3. a, c, d
4. d   5. c   6. c

absolute magnitude A measure of an object's luminosity; defined to be the apparent magnitude the object would have if it were located exactly 10 parsecs away.

absolute zero The coldest possible temperature, which is 0 K.

absorption (of light) The process by which matter absorbs radiative energy.

absorption line spectrum A spectrum that contains absorption lines.

accelerating universe The possible fate of our universe in which a repulsive force (*see* cosmological constant) causes the expansion of the universe to accelerate with time. Its galaxies will recede from one another increasingly faster, and it will become cold and dark more quickly than a coasting universe.

acceleration The rate at which an object's velocity changes. Its standard units are m/s^2.

acceleration of gravity The acceleration of a falling object. On Earth, the acceleration of gravity, designated by g, is 9.8 m/s^2.

accretion The process by which small objects gather together to make larger objects.

accretion disk A rapidly rotating disk of material that gradually falls inward as it orbits a starlike object (e.g., white dwarf, neutron star, or black hole).

active galactic nuclei The unusually luminous centers of some galaxies, thought to be powered by accretion onto supermassive black holes. Quasars are the brightest type of active galactic nuclei; radio galaxies also contain active galactic nuclei.

active galaxy A term sometimes used to describe a galaxy that contains an *active galactic nucleus*.

adaptive optics A technique in which telescope mirrors flex rapidly to compensate for the bending of starlight caused by atmospheric turbulence.

albedo Describes the fraction of sunlight reflected by a surface; albedo = 0 means no reflection at all (a perfectly black surface); albedo = 1 means all light is reflected (a perfectly white surface).

Algol Paradox A paradox concerning the binary star Algol, which contains a sub-giant star that is less massive than its main-sequence companion.

altitude (above horizon) The angular distance between the horizon and an object in the sky.

amino acids The building blocks of proteins.

analemma The figure-8 path traced by the Sun over the course of a year when viewed at the same place and the same time each day;

represents the discrepancies between apparent and mean solar time.

Andromeda Galaxy (M31; the Great Galaxy in Andromeda) The nearest large spiral galaxy to the Milky Way.

angular momentum Momentum attributable to rotation or revolution. The angular momentum of an object moving in a circle of radius r is the product $m \times v \times r$.

angular resolution (of a telescope) The smallest angular separation that two pointlike objects can have and still be seen as distinct points of light (rather than as a single point of light).

angular size (or **angular distance**) A measure of the angle formed by extending imaginary lines outward from our eyes to span an object (or between two objects).

annihilation *See* matter–antimatter annihilation

annular solar eclipse A solar eclipse during which the Moon is directly in front of the Sun but its angular size is not large enough to fully block the Sun; thus, a ring (or *annulus*) of sunlight is still visible around the Moon's disk.

Antarctic Circle The circle on Earth with latitude 66.5°S.

antielectron *See* positron

antimatter Refers to any particle with the same mass as a particle of ordinary matter but whose other basic properties, such as electrical charge, are precisely opposite.

aphelion The point at which an object orbiting the Sun is farthest from the Sun.

apogee The point at which an object orbiting Earth is farthest from Earth.

apparent brightness The amount of light reaching us *per unit area* from a luminous object; often measured in units of watts/m^2.

apparent magnitude A measure of the apparent brightness of an object in the sky, based on the ancient system developed by Hipparchus.

apparent retrograde motion Refers to the apparent motion of a planet, as viewed from Earth, during the period of a few weeks or months when it moves westward relative to the stars in our sky.

apparent solar time Time measured by the actual position of the Sun in your local sky; defined so that noon is when the Sun is *on* the meridian.

arcminutes (or **minutes of arc**) One arcminute is 1/60 of 1°.

arcseconds (or **seconds of arc**) One arcsecond is 1/60 of an arcminute, or 1/3600 of 1°.

Arctic Circle The circle on Earth with latitude 66.5°N.

asteroid A relatively small and rocky object that orbits a star; asteroids are officially considered part of a category known as "small solar system bodies."

asteroid belt The region of our solar system between the orbits of Mars and Jupiter in which asteroids are heavily concentrated.

astrobiology The study of life on Earth and beyond; emphasizes research into questions of the origin of life, the conditions under which life can survive, and the search for life beyond Earth.

astrometric technique The detection of extrasolar planets through the side-to-side motion of a star caused by gravitational tugs from the planet.

astronomical unit (AU) The average distance (semimajor axis) of Earth from the Sun, which is about 150 million km.

atmosphere A layer of gas that surrounds a planet or moon, usually very thin compared to the size of the object.

atmospheric pressure The surface pressure resulting from the overlying weight of an atmosphere.

atmospheric structure The layering of a planetary atmosphere due to variations in temperature with altitude. For example, Earth's atmospheric structure from the ground up consists of the troposphere, stratosphere, thermosphere, and exosphere.

atomic mass number The combined number of protons and neutrons in an atom.

atomic number The number of protons in an atom.

atoms Consist of a nucleus made from protons and neutrons surrounded by a cloud of electrons.

aurora Dancing lights in the sky caused by charged particles entering our atmosphere; called the *aurora borealis* in the Northern Hemisphere and the *aurora australis* in the Southern Hemisphere.

autumnal equinox *See* fall equinox

axis tilt (of a planet in our solar system) The amount by which a planet's axis is tilted with respect to a line perpendicular to the ecliptic plane.

azimuth (usually called *direction* in this book) Direction around the horizon from due north, measured clockwise in degrees. For example, the azimuth of due north is 0°, due east is 90°, due south is 180°, and due west is 270°.

bar The standard unit of pressure, approximately equal to Earth's atmospheric pressure at sea level.

baryonic matter Refers to ordinary matter made from atoms (because the nuclei of atoms contain protons and neutrons, which are both baryons).

baryons Particles, including protons and neutrons, that are made from three quarks.

basalt A type of dark, high-density volcanic rock that is rich in iron and magnesium-based silicate minerals; it forms a runny (easy flowing) lava when molten.

belts (on a jovian planet) Dark bands of sinking air that encircle a jovian planet at a particular set of latitudes.

Big Bang The name given to the event thought to mark the birth of the universe.

Big Bang theory The scientific theory of the universe's earliest moments, stating that all the matter in our observable universe came into being at a single moment in time as an extremely hot, dense mixture of subatomic particles and radiation.

Big Crunch If gravity ever reverses the universal expansion, the universe will someday begin to collapse and presumably end in a Big Crunch.

binary star system A star system that contains two stars.

biosphere Refers to the "layer" of life on Earth.

BL Lac objects The name given to a class of active galactic nuclei that probably represent the centers of radio galaxies whose jets happen to be pointed directly at us.

blackbody radiation *See* thermal radiation

black hole A bottomless pit in spacetime. Nothing can escape from within a black hole, and we can never again detect or observe an object that falls into a black hole.

black smokers Structures around seafloor volcanic vents that support a wide variety of life.

blowout Ejection of the hot, gaseous contents of a superbubble when it grows so large that it bursts out of the cooler layer of gas filling the galaxy's disk.

blueshift A Doppler shift in which spectral features are shifted to shorter wavelengths, caused when an object is moving toward the observer.

bosons Particles, such as photons, to which the exclusion principle does not apply.

bound orbits Orbits on which an object travels repeatedly around another object; bound orbits are elliptical in shape.

brown dwarf An object too small to become an ordinary star because electron degeneracy pressure halts its gravitational collapse before fusion becomes self-sustaining; brown dwarfs have mass less than $0.08 M_{Sun}$.

bubble (interstellar) The surface of a bubble is an expanding shell of hot, ionized gas driven by stellar winds or supernovae; inside the bubble, the gas is very hot and has very low density.

bulge (of a spiral galaxy) The central portion of a spiral galaxy that is roughly spherical (or football shaped) and bulges above and below the plane of the galactic disk.

Cambrian explosion The dramatic diversification of life on Earth that occurred between about 540 and 500 million years ago.

carbonate rock A carbon-rich rock, such as limestone, that forms underwater from chemical reactions between sediments and carbon dioxide. On Earth, most of the outgassed carbon dioxide currently resides in carbonate rocks.

carbon dioxide cycle The process that cycles carbon dioxide between Earth's atmosphere and surface rocks.

carbon stars Stars whose atmospheres are especially carbon-rich, thought to be near the ends of their lives; carbon stars are the primary sources of carbon in the universe.

Cassini division A large, dark gap in Saturn's rings, visible through small telescopes on Earth.

CCD (charge coupled device) A type of electronic light detector that has largely replaced photographic film in astronomical research.

celestial coordinates The coordinates of right ascension and declination that fix an object's position on the celestial sphere.

celestial equator (CE) The extension of Earth's equator onto the celestial sphere.

celestial navigation Navigation on the surface of the Earth accomplished by observations of the Sun and stars.

celestial sphere The imaginary sphere on which objects in the sky appear to reside when observed from Earth.

Celsius (temperature scale) The temperature scale commonly used in daily activity internationally. Defined so that, on Earth's surface, water freezes at 0°C and boils at 100°C.

center of mass (of orbiting objects) The point at which two or more orbiting objects would balance if they were somehow connected; it is the point around which the orbiting objects actually orbit.

central dominant galaxy A giant elliptical galaxy found at the center of a dense cluster of galaxies, apparently formed by the merger of several individual galaxies.

Cepheid *See* Cepheid variable

Cepheid variable A particularly luminous type of pulsating variable star that follows a period–luminosity relation and hence is very useful for measuring cosmic distances.

Chandrasekhar limit *See* white dwarf limit

charged particle belts Zones in which ions and electrons accumulate and encircle a planet.

chemical enrichment The process by which the abundance of heavy elements (heavier than helium) in the interstellar medium gradually increases over time as these elements are produced by stars and released into space.

chondrites Another name for primitive meteorites, named for the round chodrules within them. Achondrites, meaning "without chondrules," is another name for processed meteorites.

chromosphere The layer of the Sun's atmosphere below the corona; most of the Sun's ultraviolet light is emitted from this region, in which the temperature is about 10,000 K.

circulation cells (also called *Hadley cells*) Large-scale cells (similar to convection cells) in a planet's atmosphere that transport heat between the equator and the poles.

circumpolar star A star that always remains above the horizon for a particular latitude.

climate Describes the long-term average of weather.

close binary A binary star system in which the two stars are very close together.

closed universe The universe is closed if its average density is greater than the critical density, in which case spacetime must curve back on itself to the point where its overall shape is analogous to that of the surface of a sphere. In the absence of a repulsive force (*see* cosmological constant), a closed universe would someday stop expanding and begin to contract.

cluster of galaxies A collection of a few dozen or more galaxies bound together by gravity; smaller collections of galaxies are simply called *groups*.

cluster of stars A group of anywhere from several hundred to a million or so stars; star clusters come in two types—open clusters and globular clusters.

CNO cycle The cycle of reactions by which intermediate- and high-mass stars fuse hydrogen into helium.

coasting universe The possible fate of our universe in which the mass density of the universe is *smaller* than the critical density, so that the collective gravity of all matter cannot halt the expansion. In the absence of a repulsive force (*see* cosmological constant), such a universe would keep expanding forever with little change in its rate of expansion.

coma (of a comet) The dusty atmosphere of a comet created by sublimation of ices in the nucleus when the comet is near the Sun.

comet A relatively small, icy object that orbits a star. Like asteroids, comets are officially considered part of a category known as "small solar system bodies."

comparative planetology The study of the solar system by examining and understanding the similarities and differences among worlds.

compound (chemical) A substance made from molecules consisting of two or more atoms with different atomic numbers.

condensates Solid or liquid particles that condense from a cloud of gas.

condensation The formation of solid or liquid particles from a cloud of gas.

conduction (of energy) The process by which thermal energy is transferred by direct contact from warm material to cooler material.

conjunction (of a planet with the Sun) When a planet and the Sun line up in the sky.

conservation of angular momentum (law of) The principle that, in the absence of net torque (twisting force), the total angular momentum of a system remains constant.

conservation of energy (law of) The principle that energy (including mass-energy) can be neither created nor destroyed, but can only change from one form to another.

conservation of momentum (law of) The principle that, in the absence of net force, the total momentum of a system remains constant.

constellation A region of the sky; 88 official constellations cover the celestial sphere.

continental crust The thicker lower-density crust that makes up Earth's continents. It is made when remelting of seafloor crust allows lower-density rock to separate and erupt to the surface. Continental crust ranges in age from very young to as old as about 4 billion years (or more).

convection The energy transport process in which warm material expands and rises, while cooler material contracts and falls.

convection cell An individual small region of convecting material.

convection zone (of a star) A region in which energy is transported outward by convection.

Copernican revolution The dramatic change, initiated by Copernicus, that occurred when we learned that Earth is a planet orbiting the Sun rather than the center of the universe.

core (of a planet) The dense central region of a planet that has undergone differentiation.

core (of a star) The central region of a star, in which nuclear fusion can occur.

Coriolis effect Causes air or objects moving on a rotating planet to deviate from straight-line trajectories.

corona (solar) The tenuous uppermost layer of the Sun's atmosphere; most of the Sun's X rays are emitted from this region, in which the temperature is about 1 million K.

coronal holes Regions of the corona that barely show up in X-ray images because they are nearly devoid of hot coronal gas.

coronal mass ejections Bursts of charged particles from the Sun's corona that travel outward into space.

cosmic microwave background The remnant radiation from the Big Bang, which we detect using radio telescopes sensitive to microwaves (which are short-wavelength radio waves).

cosmic rays Particles such as electrons, protons, and atomic nuclei that zip through interstellar space at close to the speed of light.

cosmological constant The name given to a term in Einstein's equations of general relativity. If it is not zero, then it represents a repulsive force or a type of energy (sometimes called *dark energy* or *quintessence*) that might cause the expansion of the universe to accelerate with time.

cosmological horizon The boundary of our observable universe, which is where the lookback time is equal to the age of the universe. Beyond this boundary in spacetime, we cannot see anything at all.

Cosmological Principle The idea that matter is distributed uniformly throughout the universe on very large scales, meaning that the universe has neither a center nor an edge.

cosmological redshift Refers to the redshifts we see from distant galaxies, caused by the fact that expansion of the universe stretches all the photons within it to longer, redder wavelengths.

cosmology The study of the overall structure and evolution of the universe.

cosmos An alternate name for the universe.

critical density The precise average density for the entire universe that marks the dividing line between a recollapsing universe and one that will expand forever.

critical universe The possible fate of our universe in which the mass density of the universe *equals* the critical density. The universe will never collapse, but in the absence of a repulsive force it will expand more and more slowly as time progresses. *See* cosmological constant

crust (of a planet) The low-density surface layer of a planet that has undergone differentiation.

curvature of spacetime A change in the geometry of space produced in the vicinity of a massive object and that is responsible for the force we call gravity. The overall geometry of the universe may also be curved, depending on its overall mass-energy content.

cycles per second Units of frequency for a wave; describes the number of peaks (or troughs) of a wave that pass by a given point each second. Equivalent to *hertz*.

dark energy Name sometimes given to energy that could be causing the expansion of the universe to accelerate. *See* cosmological constant.

dark matter Matter that we infer to exist from its gravitational effects but from which we have not detected any light; dark matter apparently dominates the total mass of the universe.

daylight saving time Standard time plus 1 hour, so that the Sun appears on the meridian around 1 P.M. rather than around noon.

declination (dec) Analogous to latitude, but on the celestial sphere; it is the angular north-south distance between the celestial equator and a location on the celestial sphere.

degeneracy pressure A type of pressure unrelated to an object's temperature, which arises when electrons (electron degeneracy pressure) or neutrons (neutron degeneracy pressure) are packed so tightly that the exclusion and uncertainty principles come into play.

degenerate object An object in which degeneracy pressure is the primary pressure pushing back against gravity, such as a brown dwarf, white dwarf, or neutron star.

deuterium A form of hydrogen in which the nucleus contains a proton and a neutron, rather than only a proton (as is the case for most hydrogen nuclei).

differential rotation Describes the rotation of an object in which the equator rotates at a different rate than the poles.

differentiation The process in which gravity separates materials according to density, with high-density materials sinking and low-density materials rising.

diffraction grating A finely etched surface that can split light into a spectrum.

diffraction limit The angular resolution that a telescope could achieve if it were limited only by the interference of light waves; it is smaller (i.e., better angular resolution) for larger telescopes.

dimension (mathematical) Describes the number of independent directions in which movement is possible; e.g., the surface of the Earth is two-dimensional because only two independent directions of motion are possible (north-south and east-west).

direction (in local sky) One of the two coordinates (the other is altitude) needed to pinpoint an object in the local sky. It is the direction, such as north, south, east, or west, in which you must face to see the object. *See also* azimuth

disk component (of a galaxy) The portion of a spiral galaxy that looks like a disk and contains an interstellar medium with cool gas and dust; stars of many ages are found in the disk component.

disk population Refers to stars that orbit within the disk of a spiral galaxy. Sometimes called Population I.

DNA (deoxyribonucleic acid) The molecule that represents the genetic material of life on Earth.

Doppler effect (shift) The effect that shifts the wavelengths of spectral features in objects that are moving toward or away from the observer.

Doppler technique The detection of extrasolar planets through the motion of a star toward and away from the observer caused by gravitational tugs from the planet.

double-shell burning star A star that is fusing helium into carbon in a shell around an inert carbon core and is fusing hydrogen into helium in a shell at the top of the helium layer.

down quark One of the two quark types (the other is the up quark) found in ordinary protons and neutrons. Has a charge of $-\frac{1}{3}$.

Drake equation An equation that lays out the factors that play a role in determining the number of communicating civilizations in our galaxy.

dust (or dust grains) Tiny solid flecks of material; in astronomy, we often discuss interplanetary dust (found within a star system) or interstellar dust (found between the stars in a galaxy). *See also* interstellar dust grains

dust tail (of a comet) One of two tails seen when a comet passes near the Sun (the other is the plasma tail); composed of small solid particles pushed away from the Sun by the radiation pressure of sunlight.

dwarf elliptical galaxy A small elliptical galaxy with less than about a billion stars.

dwarf galaxies Relatively small galaxies, consisting of less than about ten billion stars.

dwarf planet An object that orbits the Sun and is massive enough for its gravity to have made it nearly round in shape, but that does not qualify as an official planet because it has not cleared its orbital neighborhood. The dwarf planets of our solar system include the asteroid Ceres and the Kuiper belt objects Pluto, Eris, Haumea, and Makemake.

Earth-orbiters (spacecraft) Spacecraft designed to study the Earth or the universe from Earth orbit.

eccentricity A measure of how much an ellipse deviates from a perfect circle; defined as the center-to-focus distance divided by the length of the semimajor axis.

eclipse Occurs when one astronomical object casts a shadow on another or crosses our line of sight to the other object.

eclipse seasons Periods during which lunar and solar eclipses can occur because the nodes of the Moon's orbit are aligned with Earth and Sun.

eclipsing binary A binary star system in which the two stars happen to be orbiting in the plane of our line of sight, so that each star will periodically eclipse the other.

ecliptic The Sun's apparent annual path among the constellations.

ecliptic plane The plane of Earth's orbit around the Sun.

ejecta (from an impact) Debris ejected by the blast of an impact.

electrical charge A fundamental property of matter that is described by its amount and as either positive or negative; more technically, it is a measure of how a particle responds to the electromagnetic force.

electromagnetic field An abstract concept used to describe how a charged particle would affect other charged particles at a distance.

electromagnetic force One of the four fundamental forces; it is the force that dominates atomic and molecular interactions.

electromagnetic radiation Another name for light of all types, from radio waves through gamma rays.

electromagnetic spectrum The complete spectrum of light, including radio waves, infrared, visible light, ultraviolet light, X rays, and gamma rays.

electromagnetic wave A synonym for light, which consists of waves of electric and magnetic fields.

electron degeneracy pressure Degeneracy pressure exerted by electrons, as in brown dwarfs and white dwarfs.

electrons Fundamental particles with negative electric charge; the distribution of electrons in an atom gives the atom its size.

electron-volt (eV) A unit of energy equivalent to 1.60×10^{-19} joule.

electroweak era The era of the universe during which only three forces operated (gravity, strong force, and electroweak force), lasting from 10^{-38} second to 10^{-10} second after the Big Bang.

electroweak force The force that exists at high energies when the electromagnetic force and the weak force exist as a single force.

element (chemical) A substance made from individual atoms of a particular atomic number.

ellipse A type of oval that happens to be the shape of bound orbits. An ellipse can be drawn by moving a pencil along a string whose ends are tied to two tacks; the locations of the tacks are the foci (singular, focus) of the ellipse.

elliptical galaxies Galaxies that appear rounded in shape, often longer in one direction, like a football. They have no disks and contain very little cool gas and dust compared to spiral galaxies, though they often contain very hot, ionized gas.

elongation (greatest) For Mercury or Venus, the point at which it appears farthest from the Sun in our sky.

emission (of light) The process by which matter emits energy in the form of light.

emission line spectrum A spectrum that contains emission lines.

emission nebula Another name for an ionization nebula. *See also* ionization nebula

energy Broadly speaking, energy is what can make matter move. The three basic types of energy are kinetic, potential, and radiative.

equation of time Describes the discrepancies between apparent and mean solar time.

equinox *See* fall equinox *and* spring equinox

equivalence principle The fundamental starting point for general relativity, which states that the effects of gravity are exactly equivalent to the effects of acceleration.

era of atoms The era of the universe lasting from about 500,000 years to about 1 billion years after the Big Bang, during which it was cool enough for neutral atoms to form.

era of galaxies The present era of the universe, which began with the formation of galaxies when the universe was about 1 billion years old.

era of nuclei The era of the universe lasting from about 3 minutes to about 380,000 years after the Big Bang, during which matter in the universe was fully ionized and opaque to light. The cosmic background radiation was released at the end of this era.

era of nucleosynthesis The era of the universe lasting from about 0.001 second to about 3 minutes after the Big Bang, by the end of which virtually all of the neutrons and about one-seventh of the protons in the universe had fused into helium.

erosion The wearing down or building up of geological features by wind, water, ice, and other phenomena of planetary weather.

eruption The process of releasing hot lava on the planet's surface.

escape velocity The speed necessary for an object to completely escape the gravity of a large body such as a moon, planet, or star.

evaporation The process by which atoms or molecules escape into the gas phase from a liquid.

event Any particular point along a worldline represents a particular event; all observers will agree on the reality of an event but may disagree about its time and location.

event horizon The boundary that marks the "point of no return" between a black hole and the outside universe; events that occur within the event horizon can have no influence on our observable universe.

evolution (biological) The gradual change in populations of living organisms responsible for transforming life on Earth from its primitive origins to the great diversity of life today.

exchange particle According to the standard model of physics, each of the four fundamental forces is transmitted by the transfer of particular types of exchange particles.

excited state (of an atom) Any arrangement of electrons in an atom that has more energy than the ground state.

exclusion principle The law of quantum mechanics that states that two fermions cannot occupy the same quantum state at the same time.

exosphere The hot, outer layer of an atmosphere, where the atmosphere "fades away" to space.

expansion (of universe) The idea that the space between galaxies or clusters of galaxies is growing with time.

exposure time The amount of time for which light is collected to make a single image.

extrasolar planet A planet orbiting a star other than our Sun.

Fahrenheit (temperature scale) The temperature scale commonly used in daily activity in the United States. Defined so that, on Earth's surface, water freezes at 32°F and boils at 212°F.

fall equinox (autumnal equinox) Refers both to the point in Virgo on the celestial sphere where the ecliptic crosses the celestial equator and to the moment in time when the Sun appears at that point each year (around September 21).

false-color image An image displayed in colors that are *not* the true, visible-light colors of an object.

fault (geological) A place where rocks slip sideways relative to one another.

feedback processes Processes in which a small change in some property (such as temperature) leads to changes in other properties that either amplify or diminish the original small change.

fermions Particles, such as electrons, neutrons, and protons, that obey the exclusion principle.

Fermi's paradox The question posed by Enrico Fermi about extraterrestrial intelligence—"So where is everybody?"—which asks why we have not observed other civilizations even though simple arguments would suggest that some ought to have spread throughout the galaxy by now.

field An abstract concept used to describe how a particle would interact with a force. For example, the idea of a *gravitational field* describes how a particle would react to the local strength of gravity, and the idea of an *electromagnetic field* describes how a charged particle would respond to forces from other charged particles.

filter (for light) A material that transmits only particular wavelengths of light.

fireball A particularly bright meteor.

fission The process by which one atomic nucleus breaks into two smaller nuclei. It releases energy if the two smaller nuclei together are less massive than the original nucleus.

flare star A small, spectral type M star that displays particularly strong flares on its surface.

flat (or Euclidean) geometry Refers to any case in which the rules of geometry for a flat plane hold, such as that the shortest distance between two points is a straight line.

flat universe A universe in which the overall geometry of spacetime is flat (Euclidean), as would be the case if the density of the universe is equal to the critical density.

flybys (spacecraft) Spacecraft that fly past a target object (such as a planet), usually just once, as opposed to entering a bound orbit of the object.

focal plane The place where an image created by a lens or mirror is in focus.

foci Plural of *focus.*

focus (of a lens or mirror) The point at which rays of light that were initially parallel (such as light from a distant star) converge.

focus (of an ellipse) One of two special points within an ellipse that lie along the major axis; these are the points around which we could stretch a pencil and string to draw an ellipse. When one object orbits a second object, the second object lies at one focus of the orbit.

force Anything that can cause a change in momentum.

formation properties (of planets) In this book, for the purpose of understanding geological processes, planets are defined to be born with four formation properties: size (mass and radius), distance from the Sun, composition, and rotation rate.

fossil Any relic of an organism that lived and died long ago.

frame of reference (in relativity) Two (or more) objects share the same frame of reference if they are *not* moving relative to each other.

free-fall Refers to conditions in which an object is falling without resistance; objects are weightless when in free-fall.

free-float frame A frame of reference in which all objects are weightless and hence float freely.

frequency Describes the rate at which peaks of a wave pass by a point; measured in units of 1/s, often called *cycles per second* or *hertz.*

frost line The boundary in the solar nebula beyond which ices could condense; only metals and rocks could condense within the frost line.

fundamental forces There are four known fundamental forces in nature: gravity, the electromagnetic force, the strong force, and the weak force.

fundamental particles Subatomic particles that cannot be divided into anything smaller.

fusion The process by which two atomic nuclei fuse together to make a single, more massive nucleus. It releases energy if the final nucleus is less massive than the two nuclei that went into the reaction.

galactic cannibalism The term sometimes used to describe the process by which large galaxies merge with other galaxies in collisions. *Central dominant galaxies* are products of galactic cannibalism.

galactic disk (of a spiral galaxy) *See* disk component

galactic fountain Refers to a model for the cycling of gas in the Milky Way Galaxy in which fountains of hot, ionized gas rise from the disk into the halo and then cool and form clouds as they sink back into the disk.

galactic wind A wind of low-density but extremely hot gas flowing out from a starburst galaxy, created by the combined energy of many supernovae.

galaxy A huge collection of anywhere from a few hundred million to more than a trillion stars, all bound together by gravity.

galaxy cluster *See* cluster of galaxies.

galaxy evolution The formation and development of galaxies.

Galilean moons The four moons of Jupiter that were discovered by Galileo: Io, Europa, Ganymede, and Callisto.

gamma-ray burst A sudden burst of gamma rays from deep space; such bursts apparently come from distant galaxies, but their precise mechanism is unknown.

gamma rays Light with very short wavelengths (and hence high frequencies)—shorter than those of X rays.

gap moons Tiny moons located within a gap in a planet's ring system. The gravity of a gap moon helps clear the gap.

gas phase The phase of matter in which atoms or molecules can move essentially independently of one another.

gas pressure Describes the force (per unit area) pushing on any object due to surrounding gas. *See also* pressure

general theory of relativity Einstein's generalization of his special theory of relativity so that the general theory also applies when we consider effects of gravity or acceleration.

genetic code The "language" that living cells use to read the instructions chemically encoded in DNA.

geocentric model Any of the ancient Greek models that were used to predict planetary positions under the assumption that Earth lay in the center of the universe.

geocentric universe (ancient belief in) The idea that the Earth is the center of the entire universe.

geological activity Processes that change a planet's surface long after formation, such as volcanism, tectonics, and erosion.

geological processes The four basic geological processes are impact cratering, volcanism, tectonics, and erosion.

geological time scale The time scale used by scientists to describe major eras in Earth's past.

geology The study of surface features (on a moon, planet, or asteroid) and the processes that create them.

geostationary satellite A satellite that appears to stay stationary in the sky as viewed from Earth's surface, because it orbits in the same time it takes Earth to rotate and orbits in Earth's equatorial plane.

geosynchronous satellite A satellite that orbits Earth in the same time it takes Earth to rotate, or one sidereal day.

giant impact A collision between a forming planet and a very large planetesimal, such as is thought to have formed our Moon.

giant molecular cloud A very large cloud of cold, dense interstellar gas, typically containing up to a million solar masses worth of material. *See also* molecular clouds

giants (luminosity class III) Stars that appear just below the supergiants on the H-R diagram because they are somewhat smaller in radius and lower in luminosity.

global positioning system (GPS) A system of navigation by satellites orbiting Earth.

global warming An expected increase in Earth's global average temperature caused by human input of carbon dioxide and other greenhouse gases into the atmosphere.

global wind patterns (or **global circulation**) Wind patterns that remain fixed on a global scale, determined by the combination of surface heating and the planet's rotation.

globular cluster A spherically shaped cluster of up to a million or more stars; globular clusters are found primarily in the halos of galaxies and contain only very old stars.

gluons The exchange particles for the strong force.

grand unified theory (GUT) A theory that unifies three of the four fundamental forces—the strong force, the weak force, and the electromagnetic force (but not gravity)—in a single model.

granulation (on the Sun) The bubbling pattern visible in the photosphere, produced by the underlying convection.

gravitation (law of) *See* universal law of gravitation

gravitational constant The experimentally measured constant G that appears in the law of universal gravitation:

$$G = 6.67 \times 10^{-11} \frac{m^3}{kg \times s^2}$$

gravitational contraction The process in which gravity causes an object to contract, thereby converting gravitational potential energy into thermal energy.

gravitational encounter Occurs when two (or more) objects pass near enough so that each can feel the effects of the other's gravity and can therefore exchange energy.

gravitational equilibrium Describes a state of balance in which the force of gravity pulling inward is precisely counteracted by pressure pushing outward.

gravitational lensing The magnification or distortion (into arcs, rings, or multiple images) of an image caused by light bending through a gravitational field, as predicted by Einstein's general theory of relativity.

gravitationally bound system Any system of objects, such as a star system or a galaxy, that is held together by gravity.

gravitational potential energy Energy that an object has by virtue of its position in a gravitational field; an object has more gravitational potential energy when it has a greater distance that it can potentially fall.

gravitational redshift A redshift caused by the fact that time runs slow in gravitational fields.

gravitational time dilation The slowing of time that occurs in a gravitational field, as predicted by Einstein's general theory of relativity.

gravitational waves Predicted by Einstein's general theory of relativity, these waves travel at the speed of light and transmit distortions of space through the universe. Although not yet observed directly, we have strong indirect evidence that they exist.

gravitons The exchange particles for the force of gravity.

gravity One of the four fundamental forces; it is the force that dominates on large scales.

grazing incidence (in telescopes) Reflections in which light grazes a mirror surface and is deflected at a small angle; commonly used to focus high-energy ultraviolet light and X rays.

great circle A circle on the surface of a sphere whose center is at the center of the sphere.

greatest elongation *See* elongation (greatest).

Great Red Spot A large, high-pressure storm on Jupiter.

greenhouse effect The process by which greenhouse gases in an atmosphere make a planet's surface temperature warmer than it would be in the absence of an atmosphere.

greenhouse gases Gases, such as carbon dioxide, water vapor, and methane, that are particularly good absorbers of infrared light but are transparent to visible light.

Gregorian calendar Our modern calendar, introduced by Pope Gregory in 1582.

ground state (of an atom) The lowest possible energy state of the electrons in an atom.

group (of galaxies) A few to a few dozen galaxies bound together by gravity. *See also* cluster of galaxies

GUT era The era of the universe during which only two forces operated (gravity and the grand-unified-theory or GUT force), lasting from 10^{-43} second to 10^{-38} second after the Big Bang.

GUT force The proposed force that exists at very high energies when the strong force, the weak force, and the electromagnetic force (but not gravity) all act as one.

H II region Another name for an ionization nebula. *See* ionization nebula

habitable world A world with environmental conditions under which life could *potentially* arise or survive.

habitable zone The region around a star in which planets could potentially have surface temperatures at which liquid water could exist.

Hadley cells *See* circulation cells.

half-life The time it takes for half of the nuclei in a given quantity of a radioactive substance to decay.

halo (of a galaxy) The spherical region surrounding the disk of a spiral galaxy.

Hawking radiation Radiation predicted to arise from the evaporation of black holes.

heavy bombardment The period in the first few hundred million years after the solar system formed during which the tail end of planetary accretion created most of the craters found on ancient planetary surfaces.

heavy elements In astronomy, *heavy elements* generally refers to all elements *except* hydrogen and helium.

helium-burning star A star that is currently fusing helium into carbon in its core.

helium-capture reactions Fusion reactions that fuse a helium nucleus into some other nucleus; such reactions can fuse carbon into oxygen, oxygen into neon, neon into magnesium, and so on.

helium flash The event that marks the sudden onset of helium fusion in the previously inert helium core of a low-mass star.

helium fusion The fusion of three helium nuclei into one carbon nucleus; also called the *triple-alpha reaction*.

hertz (Hz) The standard unit of frequency for light waves; equivalent to units of 1/s.

Hertzsprung-Russell (H-R) diagram A graph plotting individual stars as points, with stellar luminosity on the vertical axis and spectral type (or surface temperature) on the horizontal axis.

high-mass stars Stars born with masses above about $8M_{Sun}$; these stars will end their lives by exploding as supernovae.

horizon A boundary that divides what we can see from what we cannot see.

horizontal branch The horizontal line of stars that represents helium-burning stars on an H-R diagram for a cluster of stars.

horoscope A predictive chart made by an astrologer; in scientific studies, horoscopes have never been found to have any validity as predictive tools.

hot spot (geological) A place within a plate of the lithosphere where a localized plume of hot mantle material rises.

hour angle (HA) The angle or time (measured in hours) since an object was last on the meridian in the local sky. Defined to be 0 hours for objects that *are* on the meridian.

Hubble's constant A number that expresses the current rate of expansion of the universe; designated H_0, it is usually stated in units of km/s/Mpc. The reciprocal of Hubble's constant is the age the universe would have *if* the expansion rate had never changed.

Hubble's law Mathematically expresses the idea that more distant galaxies move away from us faster; its formula is $v = H_0 \times d$, where v is a galaxy's speed away from us, d is its distance, and H_0 is Hubble's constant.

hydrogen compounds Compounds that contain hydrogen and were common in the solar nebula, such as water (H_2O), ammonia (NH_3), and methane (CH_4).

hydrogen shell burning Hydrogen fusion that occurs in a shell surrounding a stellar core.

hydrosphere Refers to the "layer" of water on the Earth consisting of oceans, lakes, rivers, ice caps, and other liquid water and ice.

hydrostatic equilibrium *See* gravitational equilibrium

hyperbola The precise mathematical shape of one type of unbound orbit (the other is a parabola) allowed under the force of gravity; at great distances from the attracting object, a hyperbolic path looks like a straight line.

hypernova A term sometimes used to describe a supernova (explosion) of a star so massive that it leaves a black hole behind.

hyperspace Any space with more than three dimensions.

hypothesis A tentative model proposed to explain some set of observed facts, but which has not yet been rigorously tested and confirmed.

ice ages Periods of global cooling during which the polar caps, glaciers, and snow cover extend closer to the equator.

ices (in solar system theory) Materials that are solid only at low temperatures, such as the hydrogen compounds water, ammonia, and methane.

ideal gas law The law relating the pressure, temperature, and number density of particles in an ideal gas.

image A picture of an object made by focusing light.

imaging (in astronomical research) The process of obtaining pictures of astronomical objects.

impact The collision of a small body (such as an asteroid or comet) with a larger object (such as a planet or moon).

impact basin A very large impact crater often filled by a lava flow.

impact crater A bowl-shaped depression left by the impact of an object that strikes a planetary surface (as opposed to burning up in the atmosphere).

impact cratering The excavation of bowl-shaped depressions (*impact craters*) by asteroids or comets striking a planet's surface.

impactor The object responsible for an impact.

inflation (of the universe) A sudden and dramatic expansion of the universe thought to have occurred at the end of the GUT era.

infrared light Light with wavelengths that fall in the portion of the electromagnetic spectrum between radio waves and visible light.

inner solar system Generally considered to encompass the region of our solar system out to about the orbit of Mars.

intensity (of light) A measure of the amount of energy coming from light of specific wavelength in the spectrum of an object.

interferometry A telescopic technique in which two or more telescopes are used in tandem to produce much better angular resolution than the telescopes could achieve individually.

intermediate-mass stars Stars born with masses between about 2 and 8 M_{Sun}; these stars end their lives by ejecting a planetary nebula and becoming a white dwarf.

interstellar cloud A cloud of gas and dust between the stars.

interstellar dust grains Tiny solid flecks of carbon and silicon minerals found in cool interstellar clouds; they resemble particles of smoke and form in the winds of red giant stars.

interstellar medium Refers to gas and dust that fills the space between stars in a galaxy.

interstellar ramjet A hypothesized type of spaceship that uses a giant scoop to sweep up interstellar gas for use in a nuclear fusion engine.

interstellar reddening The change in the color of starlight as it passes through dusty gas. The light appears redder because dust grains absorb and scatter blue light more effectively than red light.

intracluster medium Hot, X-ray-emitting gas found between the galaxies within a cluster of galaxies.

inverse square law Any quantity that decreases with the square of the distance between two objects is said to follow an inverse square law.

inverse square law for light The law stating that an object's apparent brightness depends on its actual luminosity and the inverse square of its distance from the observer:

$$\text{apparent brightness} = \frac{\text{luminosity}}{4\pi \times (\text{distance})^2}$$

inversion (atmospheric) A local weather condition in which air is colder near the surface than higher up in the troposphere—the opposite of the usual condition, in which the troposphere is warmer at the bottom.

ionization The process of stripping an electron from an atom.

ionization nebula A colorful, wispy cloud of gas that glows because neighboring hot stars irradiate it with ultraviolet photons that can ionize hydrogen atoms.

ionosphere A portion of the thermosphere in which ions are particularly common (due to ionization by X rays from the Sun).

ions Atoms with a positive or negative electrical charge.

Io torus A donut-shaped charged-particle belt around Jupiter that approximately traces Io's orbit.

irregular galaxies Galaxies that look neither spiral nor elliptical.

isotopes Each different isotope of an element has the *same* number of protons but a *different* number of neutrons.

jets High-speed streams of gas ejected from an object into space.

joule The international unit of energy, equivalent to about 1/4000 of a Calorie.

jovian nebulae The clouds of gas that swirled around the jovian planets, from which the moons formed.

jovian planets Giant gaseous planets similar in overall composition to Jupiter.

Julian calendar The calendar introduced in 46 B.C. by Julius Caesar and used until it was replaced by the Gregorian calendar.

Kelvin (temperature scale) The most commonly used temperature scale in science, defined such that absolute zero is 0 K and water freezes at 273.15 K.

Kepler's first law States that the orbit of each planet about the Sun is an ellipse with the Sun at one focus.

Kepler's laws of planetary motion Three laws discovered by Kepler that describe the motion of the planets around the Sun.

Kepler's second law States that, as a planet moves around its orbit, it sweeps out equal areas in equal times. This tells us that a planet moves faster when it is closer to the Sun (near perihelion) than when it is farther from the Sun (near aphelion) in its orbit.

Kepler's third law States that the square of a planet's orbital period is proportional to the cube of its average distance from the Sun (semimajor axis), which tells us that more distant planets move more slowly in their orbits. In its original form, written $p^2 = a^3$. *See also* Newton's version of Kepler's third law

kinetic energy Energy of motion, given by the formula $\frac{1}{2}mv^2$.

Kirchhoff's laws A set of rules that summarizes the conditions under which objects produce thermal, absorption line, or emission line spectra. In brief: (1) An opaque object produces thermal radiation. (2) An absorption line spectrum occurs when thermal radiation passes through a thin gas that is cooler than the object emitting the thermal radiation. (3) An emission line spectrum occurs when we view a cloud of gas that is warmer than any background source of light.

Kirkwood gaps On a plot of asteroid semimajor axes, regions with few asteroids as a result of orbital resonances with Jupiter.

K–T event (impact) The collision of an asteroid or comet 65 million years ago that caused the mass extinction best known for wiping out the dinosaurs. K–T stands for the initials of the geological layers above and below the event.

Kuiper belt The comet-rich region of our solar system that spans distances of about 30–100 AU from the Sun; Kuiper belt comets have orbits that lie fairly close to the plane of planetary orbits and travel around the Sun in the same direction as the planets.

Kuiper belt object Any object orbiting the Sun within the region of the Kuiper belt, although the term is most often used for relatively large objects. For example, Pluto and Eris are considered large Kuiper belt objects.

Large Magellanic Cloud One of two small, irregular galaxies (the other is the Small Magellanic Cloud) located about 150,000 light-years away; it probably orbits the Milky Way Galaxy.

large-scale structure (of the universe) Generally refers to structure of the universe on size scales larger than that of clusters of galaxies.

latitude The angular north-south distance between the Earth's equator and a location on the Earth's surface.

leap year A calendar year with 366 rather than 365 days; our current calendar (the Gregorian calendar) has a leap year every 4 years (by adding February 29) except in century years that are not divisible by 400.

length contraction Refers to the effect in which you observe lengths to be shortened in reference frames moving relative to you.

lenticular galaxies Galaxies that look lens-shaped when seen edge-on, resembling spiral galaxies without arms. They tend to have less cool gas than normal spiral galaxies but more gas than elliptical galaxies.

leptons Fermions *not* made from quarks, such as electrons and neutrinos.

life track A track drawn on an H-R diagram to represent the changes in a star's surface temperature and luminosity during its life; also called an *evolutionary track*.

light-collecting area (of a telescope) The area of the primary mirror or lens that collects light in a telescope.

light curve A graph of an object's intensity against time.

light gases (in solar system theory) Refers to hydrogen and helium, which never condense under solar nebula conditions.

light pollution Human-made light that hinders astronomical observations.

light-year The distance that light can travel in 1 year, which is 9.46 trillion km.

liquid phase The phase of matter in which atoms or molecules are held together but move relatively freely.

lithosphere The relatively rigid outer layer of a planet; generally encompasses the crust and the uppermost portion of the mantle.

Local Bubble (interstellar) The bubble of hot gas in which our Sun and other nearby stars apparently reside. *See also* bubble (interstellar)

Local Group The group of about 40 galaxies to which the Milky Way Galaxy belongs.

local sidereal time (LST) Sidereal time for a particular location, defined according to the position of the spring equinox in the local sky. More formally, the local sidereal time at any moment is defined to be the hour angle of the spring equinox.

local sky The sky as viewed from a particular location on Earth (or another solid object). Objects in the local sky are pinpointed by the coordinates of *altitude* and *direction* (or azimuth).

local solar neighborhood The portion of the Milky Way Galaxy that is located relatively close (within a few hundred to a couple thousand light-years) to our Sun.

Local Supercluster The supercluster of galaxies to which the Local Group belongs.

longitude The angular east-west distance between the prime meridian (which passes through Greenwich) and a location on the Earth's surface.

lookback time Refers to the amount of time since the light we see from a distant object was emitted. That is, if an object has a lookback time of 400 million years, we are seeing it as it looked 400 million years ago.

low-mass stars Stars born with masses less than about $2M_{Sun}$; these stars end their lives by ejecting a planetary nebula and becoming a white dwarf.

luminosity The total power output of an object, usually measured in watts or in units of solar luminosities ($L_{Sun} = 3.8 \times 10^{26}$ watts).

luminosity class Describes the region of the H-R diagram in which a star falls. Luminosity class I represents supergiants, III represents giants, and V represents main-sequence stars; luminosity classes II and IV are intermediate to the others.

lunar eclipse Occurs when the Moon passes through Earth's shadow, which can occur only at full moon; may be total, partial, or penumbral.

lunar maria The regions of the Moon that look smooth from Earth and actually are impact basins.

lunar month *See* synodic month

lunar phase Describes the appearance of the Moon as seen from Earth.

MACHOs Stands for *massive compact halo objects* and represents one possible form of dark matter in which the dark objects are relatively large, like planets or brown dwarfs.

magma Underground molten rock.

magnetic braking The process by which a star's rotation slows as its magnetic field transfers its angular momentum to the surrounding nebula.

magnetic field Describes the region surrounding a magnet in which it can affect other magnets or charged particles in its vicinity.

magnetic field lines Lines that represent how the needles on a series of compasses would point if they were laid out in a magnetic field.

magnetosphere The region surrounding a planet in which charged particles are trapped by the planet's magnetic field.

magnitude system A system of describing stellar brightness by using numbers, called *magnitudes,* based on an ancient Greek way of describing the brightnesses of stars in the sky. This system uses *apparent magnitude* to describe a star's apparent brightness and *absolute magnitude* to describe a star's luminosity.

main sequence (luminosity class V) The prominent line of points running from the upper left to the lower right on an H-R diagram; main-sequence stars shine by fusing hydrogen in their cores.

main-sequence fitting A method for measuring the distance to a cluster of stars by comparing the apparent brightness of the cluster's main sequence with the standard main sequence.

main-sequence lifetime The length of time for which a star of a particular mass can shine by fusing hydrogen into helium in its core.

main-sequence stars Stars whose temperature and luminosity place them on the main sequence of the H-R diagram. Main-sequence stars are all releasing energy by fusing hydrogen into helium in their cores.

main-sequence turnoff A method for measuring the age of a cluster of stars from the point on its H-R diagram where its stars turn off from the main sequence; the age of the cluster is equal to the main-sequence lifetime of stars at the main-sequence turnoff point.

mantle (of a planet) The rocky layer that lies between a planet's core and crust.

Martian meteorite This term is used to describe meteorites found on Earth that are thought to have originated on Mars.

mass A measure of the amount of matter in an object.

mass-energy The potential energy of mass, which has an amount $E = mc^2$.

mass exchange (in close binary star systems) The process in which tidal forces cause matter to spill from one star to a companion star in a close binary system.

mass extinction An event in which a large fraction of the species living on Earth go extinct, such as the event in which the dinosaurs died out about 65 million years ago.

mass increase (in relativity) Refers to the effect in which an object moving past you seems to have a mass greater than its rest mass.

massive star supernova A supernova that occurs when a massive star dies, initiated by the catastrophic collapse of its iron core; often called a Type II supernova.

mass-to-light ratio The mass of an object divided by its luminosity, usually stated in units of solar masses per solar luminosity. Objects with high mass-to-light ratios must contain substantial quantities of dark matter.

matter–antimatter annihilation Occurs when a particle of matter and a particle of antimatter meet and convert all of their mass-energy to photons.

mean solar time Time measured by the average position of the Sun in your local sky over the course of the year.

meridian A half-circle extending from your horizon (altitude 0°) due south, through your zenith, to your horizon due north.

metallic hydrogen Hydrogen that is so compressed that the hydrogen atoms all share electrons and thereby take on properties of metals, such as conducting electricity. Occurs only under very high-pressure conditions, such as those found deep within Jupiter.

metals (in solar system theory) Elements, such as nickel, iron, and aluminum, that condense at fairly high temperatures.

meteor A flash of light caused when a particle from space burns up in our atmosphere.

meteorite A rock from space that lands on Earth.

meteor shower A period during which many more meteors than usual can be seen.

Metonic cycle The 19-year period, discovered by the Babylonian astronomer Meton, over which the lunar phases occur on the same dates.

microwaves Light with wavelengths in the range of micrometers to millimeters. Microwaves are generally considered to be a subset of the radio wave portion of the electromagnetic spectrum.

mid-ocean ridges Long ridges of undersea volcanoes on Earth, along which mantle material erupts onto the ocean floor and pushes apart the existing seafloor on either side. These ridges are essentially the source of new seafloor crust, which then makes its way along the ocean bottom for millions of years before returning to the mantle at a subduction zone.

Milankovitch cycles The cyclical changes in Earth's axis tilt and orbit that can change the climate and cause ice ages.

Milky Way Used both as the name of our galaxy and to refer to the band of light we see in the sky when we look into the plane of the Milky Way Galaxy.

millisecond pulsars Pulsars with rotation periods of a few thousandths of a second.

minor planets An alternate name for *asteroids*.

model (scientific) A representation of some aspect of nature that can be used to explain and predict real phenomena without invoking myth, magic, or the supernatural.

molecular bands The tightly bunched lines in an object's spectrum that are produced by molecules.

molecular cloud fragments (or *molecular cloud cores*) The densest regions of molecular clouds, which usually go on to form stars.

molecular clouds Cool, dense interstellar clouds in which the low temperatures allow hydrogen atoms to pair up into hydrogen molecules (H_2).

molecular dissociation The process by which a molecule splits into its component atoms.

molecule Technically the smallest unit of a chemical element or compound; in this text, the term refers only to combinations of two or more atoms held together by chemical bonds.

momentum The product of an object's mass and velocity.

moon An object that orbits a planet.

mutations Errors in the copying process when a living cell replicates itself.

natural selection The process by which mutations that make an organism better able to survive get passed on to future generations.

neap tides The lower-than-average tides on Earth that occur at first- and third-quarter moon, when the tidal forces from the Sun and Moon oppose one another.

nebula A cloud of gas in space, usually one that is glowing.

nebular capture The process by which icy planetesimals capture hydrogen and helium gas to form jovian planets.

nebular theory The detailed theory that describes how our solar system formed from a cloud of interstellar gas and dust.

net force The overall force to which an object responds; the net force is equal to the rate of change in the object's momentum, or equivalently to the object's mass × acceleration.

neutrino A type of fundamental particle that has extremely low mass and responds only to the weak force; neutrinos are leptons and come in three types—electron neutrinos, mu neutrinos, and tau neutrinos.

neutron degeneracy pressure Degeneracy pressure exerted by neutrons, as in neutron stars.

neutron star The compact corpse of a high-mass star left over after a supernova; typically contains a mass comparable to the mass of the Sun in a volume just a few kilometers in radius.

neutrons Particles with no electrical charge found in atomic nuclei, built from three quarks.

newton The standard unit of force in the metric system:

$$1 \text{ newton} = 1\frac{\text{kg} \times \text{m}}{\text{s}^2}$$

Newton's first law of motion States that, in the absence of a net force, an object moves with constant velocity.

Newton's laws of motion Three basic laws that describe how objects respond to forces.

Newton's second law of motion States how a net force affects an object's motion. Specifically: force = rate of change in momentum, or force = mass × acceleration.

Newton's third law of motion States that, for any force, there is always an equal and opposite reaction force.

Newton's universal law of gravitation *See* universal law of gravitation

Newton's version of Kepler's third law This generalization of Kepler's third law can be used to calculate the masses of orbiting objects from measurements of orbital period and distance. Usually written as:

$$p^2 = \frac{4\pi^2}{G(M_1 + M_2)}a^3$$

nodes (of Moon's orbit) The two points in the Moon's orbit where it crosses the ecliptic plane.

nonbaryonic matter Refers to exotic matter that is not part of the normal composition of atoms, such as neutrinos or the hypothetical WIMPs.

nonscience As defined in this book, nonscience is any way of searching for knowledge that makes no claim to follow the scientific method, such as seeking knowledge through intuition, tradition, or faith.

north celestial pole (NCP) The point on the celestial sphere directly above Earth's North Pole.

nova The dramatic brightening of a star that lasts for a few weeks and then subsides; occurs when a burst of hydrogen fusion ignites in a shell on the surface of an accreting white dwarf in a binary star system.

nuclear fission The process in which a larger nucleus splits into two (or more) smaller particles.

nuclear fusion The process in which two (or more) smaller nuclei slam together and make one larger nucleus.

nucleus (of a comet) The solid portion of a comet, and the only portion that exists when the comet is far from the Sun.

nucleus (of an atom) The compact center of an atom made from protons and neutrons.

observable universe The portion of the entire universe that, at least in principle, can be seen from Earth.

Occam's razor A principle often used in science, holding that scientists should prefer the simpler of two models that agree equally well with observations. Named after the medieval scholar William of Occam (1285–1349).

Olbers' paradox Asks the question of how the night sky can be dark if the universe is infinite and full of stars.

Oort cloud A huge, spherical region centered on the Sun, extending perhaps halfway to the nearest stars, in which trillions of comets orbit the Sun with random inclinations, orbital directions, and eccentricities.

opacity A measure of how much light a material absorbs compared to how much it transmits; materials with higher opacity absorb more light.

opaque (material) Describes a material that absorbs light.

open cluster A cluster of up to several thousand stars; open clusters are found only in the disks of galaxies and often contain young stars.

open universe The universe is open if its average density is less than the critical density, in which case spacetime has an overall shape analogous to the surface of a saddle.

opposition The point at which a planet appears opposite the Sun in our sky.

optical quality Describes the ability of a lens, mirror, or telescope to obtain clear and properly focused images.

orbital energy The sum of an orbiting object's kinetic and gravitational potential energies.

orbital resonance Describes any situation in which one object's orbital period is a simple ratio of another object's period, such as 1/2, 1/4, or 5/3. In such cases, the two objects periodically line up with each other, and the extra gravitational attractions at these times can affect the objects' orbits.

orbital velocity law This law (a variation on Newton's version of Kepler's third law) allows us to use a star's orbital speed and distance from the galactic center to determine the total mass of the galaxy contained *within* the star's orbit. Mathematically, it is written:

$$M_r = \frac{r \times v^2}{G}$$

where M_r is the mass contained within the star's orbit, r is the star's distance from the galactic center, v is the star's orbital velocity, and G is the gravitational constant.

orbiters (of other worlds) Spacecraft that go into orbit of another world for long-term study.

outer solar system Generally considered to encompass the region of our solar system beginning at about the orbit of Jupiter.

outgassing The process of releasing gases from a planetary interior, usually through volcanic eruptions.

oxidation Refers to chemical reactions, often with the surface of a planet, that remove oxygen from the atmosphere.

ozone The molecule O_3, which is a particularly good absorber of ultraviolet light.

ozone depletion Refers to the declining levels of atmospheric ozone found worldwide on Earth, especially in Antarctica, in recent years.

ozone hole A place where the concentration of ozone in the stratosphere is dramatically lower than is the norm.

pair production The process in which a concentration of energy spontaneously turns into a particle and its antiparticle.

parabola The precise mathematical shape of a special type of unbound orbit allowed under the force of gravity; if an object in a parabolic orbit loses only a tiny amount of energy, it will become bound.

paradigm (in science) Refers to general patterns of thought that tend to shape scientific beliefs during a particular time period.

paradox A situation that, at least at first, seems to violate common sense or contradict itself. Resolving paradoxes often leads to deeper understanding.

parallax The apparent shifting of an object against the background, due to viewing it from different positions. *See also* stellar parallax

parallax angle Half of a star's annual back-and-forth shift due to stellar parallax; related to the star's distance according to the formula

$$\text{distance in parsecs} = \frac{1}{p}$$

where p is the parallax angle in arcseconds.

parsec (pc) Approximately equal to 3.26 light-years; it is the distance to an object with a parallax angle of 1 arcsecond.

partial lunar eclipse A lunar eclipse in which the Moon becomes only partially covered by the Earth's umbral shadow.

partial solar eclipse A solar eclipse during which the Sun becomes only partially blocked by the disk of the Moon.

particle accelerator A machine designed to accelerate subatomic particles to high speeds in order to create new particles or to test fundamental theories of physics.

particle era The era of the universe lasting from 10^{-10} second to 0.001 second after the Big Bang, during which subatomic particles were continually created and destroyed and ending when matter annihilated antimatter.

peculiar velocity (of a galaxy) The component of a galaxy's velocity relative to the Milky Way that deviates from the velocity expected by Hubble's law.

penumbra The lighter, outlying regions of a shadow.

penumbral (lunar) eclipse A lunar eclipse in which the Moon passes only within the Earth's penumbral shadow and does not fall within the umbra.

perigee The point at which an object orbiting Earth is nearest to Earth.

perihelion The point at which an object orbiting the Sun is closest to the Sun.

period–luminosity relation The relation that describes how the luminosity of a Cepheid variable star is related to the period between peaks in its brightness; the longer the period, the more luminous the star.

phase (of matter) Describes the way in which atoms or molecules are held together; the common phases are solid, liquid, and gas.

photon An individual particle of light, characterized by a wavelength and a frequency.

photosphere The visible surface of the Sun, where the temperature averages just under 6000 K.

pixel An individual "picture element" on a CCD.

Planck era The era of the universe prior to the Planck time.

Planck's constant A universal constant, abbreviated h, with value $h = 6.626 \times 10^{-34}$ joule $\times$ s.

Planck time The time when the universe was 10^{-43} second old, before which random energy fluctuations were so large that our current theories are powerless to describe what might have been happening.

planet A moderately large object that orbits a star and shines primarily by reflecting light from its star. More precisely, according to a definition approved in 2006, a planet is an object that (1) orbits a star (but is itself neither a star nor a moon); (2) is massive enough for its own gravity to give it a nearly round shape; and (3) has cleared the neighborhood around its orbit. Objects that meet the first two criteria but not the third, including Ceres, Pluto, and Eris, are designated *dwarf planets.*

planetary geology The extension of the study of Earth's surface and interior to apply to other solid bodies in the solar system such as terrestrial planets and jovian planet moons.

planetary nebula The glowing cloud of gas ejected from a low-mass star at the end of its life.

planetesimals The building blocks of planets, formed by accretion in the solar nebula.

plasma A gas consisting of ions and electrons.

plasma tail (of a comet) One of two tails seen when a comet passes near the Sun (the other is the dust tail); composed of ionized gas blown away from the Sun by the solar wind.

plates (on a planet) Pieces of a lithosphere that apparently float upon the denser mantle below.

plate tectonics The geological process in which plates are moved around by stresses in a planet's mantle.

polarization (of light) Light is polarized when the electric and magnetic fields of light waves are all aligned in some particular way.

Population I *See* disk population

Population II *See* spheroidal population

positron The antimatter equivalent of an electron. It is identical to an electron in virtually all respects, except it has a positive rather than a negative electrical charge.

potential energy Energy stored for later conversion into kinetic energy; includes gravitational potential energy, electrical potential energy, and chemical potential energy.

power The rate of energy usage, usually measured in watts (1 watt = 1 joule/s).

precession The gradual wobble of the axis of a rotating object around a vertical line.

precipitation Condensed atmospheric gases that fall to the surface in the form of rain, snow, or hail.

pressure Describes the force (per unit area) pushing on an object. In astronomy, we are generally interested in pressure applied by surrounding gas (or plasma). Ordinarily, such pressure is related to the temperature of the gas (*see* thermal pressure). In objects such as white dwarfs and neutron stars, pressure may arise from a quantum effect (*see* degeneracy pressure). Light can also exert pressure. (*See* radiation pressure.)

primary mirror The large, light-collecting mirror of a reflecting telescope.

prime focus (of a reflecting telescope) The first point at which light focuses after bouncing off the primary mirror; located in front of the primary mirror.

prime meridian The meridian of longitude that passes through Greenwich, England, defined to be longitude 0°.

primitive meteorites Meteorites that formed at the same time as the solar system itself, about 4.6 billion years ago. Primitive meteorites from the inner asteroid belt are usually stony, and those from the outer belt are usually carbon-rich.

processed meteorites Meteorites that apparently once were part of a larger object that "processed" the original material of the solar nebula into another form. Processed meteorites can be rocky if chipped from the surface or mantle, or metallic if blasted from the core.

proper motion The motion of an object in the plane of the sky, perpendicular to our line of sight.

protogalactic cloud A huge, collapsing cloud of intergalactic gas from which an individual galaxy formed.

proton–proton chain The chain of reactions by which low-mass stars (including the Sun) fuse hydrogen into helium.

protons Particles found in atomic nuclei with positive electrical charge, built from three quarks.

protoplanetary disk A disk of material surrounding a young star (or protostar) that may eventually form planets.

protostar A forming star that has not yet reached the point where sustained fusion can occur in its core.

protostellar disk A disk of material surrounding a protostar; essentially the same as a protoplanetary disk, but may not necessarily lead to planet formation.

protostellar wind The relatively strong wind from a protostar.

protosun The central object in the forming solar system that eventually became the Sun.

pseudoscience Something that purports to be science or may appear to be scientific but that does not adhere to the testing and verification requirements of the scientific method.

Ptolemaic model The geocentric model of the universe developed by Ptolemy in about 150 A.D.

pulsar A neutron star from which we see rapid pulses of radiation as it rotates.

pulsating variable stars Stars that alternately grow brighter and dimmer as their outer layers expand and contract in size.

quantum mechanics The branch of physics that deals with the very small, including molecules, atoms, and fundamental particles.

quantum state Refers to the complete description of the state of a subatomic particle, including its location, momentum, orbital angular momentum, and spin, to the extent allowed by the uncertainty principle.

quantum tunneling The process in which, thanks to the uncertainty principle, an electron or other subatomic particle appears on the other side of a barrier that it does not have the energy to overcome in a normal way.

quarks The building blocks of protons and neutrons, quarks are one of the two basic types of fermions (leptons are the other).

quasar The brightest type of active galactic nucleus.

radar mapping Imaging of a planet by bouncing radar waves off its surface, especially important for Venus and Titan where thick clouds mask the surface.

radar ranging A method of measuring distances within the solar system by bouncing radio waves off planets.

radial motion The component of an object's motion directed toward or away from us.

radial velocity The portion of any object's total velocity that is directed toward or away from us. This part of the velocity is the only part that we can measure with the Doppler effect.

radiation pressure Pressure exerted by photons of light.

radiation zone (of a star) A region of the interior in which energy is transported primarily by radiative diffusion.

radiative diffusion The process by which photons gradually migrate from a hot region (such as the solar core) to a cooler region (such as the solar surface).

radiative energy Energy carried by light; the energy of a photon is Planck's constant times its frequency, or $h \times f$.

radioactive decay The spontaneous change of an atom, in which its nucleus breaks apart or a proton turns into an electron, thereby changing into a different element. It also releases heat in a planet's interior.

radioactive element (or **radioactive isotope**) A substance whose nucleus tends to fall apart spontaneously.

radio galaxy A galaxy that emits unusually large quantities of radio waves; thought to contain an active galactic nucleus powered by a supermassive black hole.

radio lobes The huge regions of radio emission found on either side of radio galaxies. The lobes apparently contain plasma ejected by powerful jets from the galactic center.

radiometric dating The process of determining the age of a rock (i.e., the time since it solidified) by comparing the present amount of a radioactive substance to the amount of its decay product.

radio waves Light with very long wavelengths (and hence low frequencies)—longer than those of infrared light.

recession velocity (of a galaxy) The speed at which a distant galaxy is moving away from us due to the expansion of the universe.

recollapsing universe The possible fate of our universe in which the collective gravity of all its matter eventually halts and reverses the expansion. The galaxies will come crashing back together, and the universe will end in a fiery Big Crunch.

red giant A giant star that is red in color.

red-giant winds The relatively dense but slow winds from red giant stars.

redshift (Doppler) A Doppler shift in which spectral features are shifted to longer wavelengths, caused when an object is moving away from the observer.

reference frame (frame of reference) Two people (or objects) share the same reference frame if they are *not* moving relative to one another.

reflecting telescope A telescope that uses mirrors to focus light.

reflection (of light) The process by which matter changes the direction of light.

reflection nebula A nebula that we see as a result of starlight reflected from interstellar dust grains. Reflection nebulae tend to have blue and black tints.

refracting telescope A telescope that uses lenses to focus light.

resonance *See* orbital resonance

rest wavelength The wavelength of a spectral feature in the absence of any Doppler shift or gravitational redshift.

retrograde motion Motion that is backward compared to the norm; e.g., we see Mars in apparent retrograde motion during the periods of time when it moves westward, rather than the more common eastward, relative to the stars.

revolution The orbital motion of one object around another.

right ascension (RA) Analogous to longitude, but on the celestial sphere; it is the angular east-west distance between the vernal equinox and a location on the celestial sphere.

rings (planetary) Consist of numerous small particles orbiting a planet within its Roche zone.

Roche tidal zone The region within two to three planetary radii (of any planet) in which the tidal forces tugging an object apart become comparable to the gravitational forces holding it together; planetary rings are always found within the Roche tidal zone.

rocks (in solar system theory) Material common on the surface of Earth, such as silicon-based minerals, that are solid at temperatures and pressures found on Earth but typically melt or vaporize at temperatures of 500–1300 K.

rotation The spinning of an object around its axis.

rotation curve A graph that plots rotational (or orbital) velocity against distance from the center for any object or set of objects.

runaway greenhouse effect A positive feedback cycle in which heating caused by the greenhouse effect causes more greenhouse gases to enter the atmosphere, which further enhances the greenhouse effect.

saddle-shaped (or **hyperbolic**) **geometry** Refers to any case in which the rules of geometry for a saddle-shaped surface hold, such as that two lines that begin parallel eventually diverge.

Sagittarius Dwarf A small, dwarf elliptical galaxy that is currently passing through the disk of the Milky Way Galaxy.

saros cycle The period over which the basic pattern of eclipses repeats, which is about 18 years $11\frac{1}{3}$ days.

satellite Any object orbiting another object.

scattered light Light that is reflected into random directions.

Schwarzschild radius A measure of the size of the event horizon of a black hole.

science The search for knowledge that can be used to explain or predict natural phenomena in a way that can be confirmed by rigorous observations or experiments.

scientific method An organized approach to explaining observed facts through science.

scientific theory A model of some aspect of nature that has been rigorously tested and has passed all tests to date.

seafloor crust On Earth, the thin, dense crust of basalt created by seafloor spreading.

seafloor spreading On Earth, the creation of new seafloor crust at mid-ocean ridges.

search for extraterrestrial intelligence (SETI) The name given to observing projects designed to search for signs of intelligent life beyond Earth.

secondary mirror A small mirror in a reflecting telescope, used to reflect light gathered by the primary mirror toward an eyepiece or instrument.

sedimentary rock A rock that formed from sediments created and deposited by erosional processes.

seismic waves Earthquake-induced vibrations that propagate through a planet.

semimajor axis Half the distance across the long axis of an ellipse; in this text, it is usually referred to as the *average* distance of an orbiting object, abbreviated *a* in the formula for Kepler's third law.

Seyfert galaxies The name given to a class of galaxies found relatively nearby and that have nuclei much like those of quasars, except that they are less luminous.

shepherd moons Tiny moons within a planet's ring system that help force particles into a narrow ring. A variation on *gap moons*.

shield volcano A shallow-sloped volcano made from the flow of low-viscosity basaltic lava.

shock wave A wave of pressure generated by gas moving faster than the speed of sound.

sidereal day The time of 23 hours 56 minutes 4.09 seconds between successive appearances of any particular star on the meridian; essentially the true rotation period of the Earth.

sidereal month About $27\frac{1}{4}$ days, the time required for the Moon to orbit Earth once (as measured against the stars).

sidereal period (of a planet) A planet's actual orbital period around the Sun.

sidereal time Time measured according to the position of stars in the sky rather than the position of the Sun in the sky. *See also* local sidereal time

sidereal year The time required for the Earth to complete exactly one orbit as measured against the stars; about 20 minutes longer than the tropical year on which our calendar is based.

silicate rock A silicon-rich rock.

singularity The place at the center of a black hole where, in principle, gravity crushes all matter to an infinitely tiny and dense point.

Small Magellanic Cloud One of two small, irregular galaxies (the other is the Large Magellanic Cloud) located about 150,000 light-years away; it probably orbits the Milky Way Galaxy.

snowball Earth Name given to a hypothesis suggesting that, some 600–700 million years ago, the Earth experienced a period in which it became cold enough for glaciers to exist worldwide, even in equatorial regions.

solar activity Refers to short-lived phenomena on the Sun, including the emergence and disappearance of individual sunspots, prominences, and flares; sometimes called *solar weather*.

solar circle The Sun's orbital path around the galaxy, which has a radius of about 28,000 light-years.

solar day Twenty-four hours, which is the average time between appearances of the Sun on the meridian.

solar eclipse Occurs when the Moon's shadow falls on the Earth, which can occur only at new moon; may be total, partial, or annular.

solar flares Huge and sudden releases of energy on the solar surface, probably caused when energy stored in magnetic fields is suddenly released.

solar luminosity The luminosity of the Sun, which is approximately 4×10^{26} watts.

solar maximum The time during each sunspot cycle at which the number of sunspots is the greatest.

solar minimum The time during each sunspot cycle at which the number of sunspots is the smallest.

solar nebula The piece of interstellar cloud from which our own solar system formed.

solar neutrino problem Refers to the disagreement between the predicted and observed number of neutrinos coming from the Sun.

solar prominences Vaulted loops of hot gas that rise above the Sun's surface and follow magnetic-field lines.

solar sail A large, highly reflective (and thin, to minimize mass) piece of material that can "sail" through space using pressure exerted by sunlight.

solar system (or star system) Consists of a star (sometimes more than one star) and all the objects that orbit it.

solar thermostat The regulation of the Sun's core temperature that comes about because pressure and gravity must balance within the Sun.

solar wind A stream of charged particles ejected from the Sun.

solid phase The phase of matter in which atoms or molecules are held rigidly in place.

solstice *See* summer solstice *and* winter solstice

sound wave A wave of alternately rising and falling pressure.

south celestial pole (SCP) The point on the celestial sphere directly above Earth's South Pole.

spacetime The inseparable, four-dimensional combination of space and time.

spacetime diagram A graph that plots a spatial dimension on one axis and time on another axis.

special theory of relativity Einstein's theory that describes the effects of the fact that all motion is relative and that everyone always measures the same speed of light.

spectral lines Bright or dark lines that appear in an object's spectrum, which we can see when we pass the object's light through a prismlike device that spreads out the light like a rainbow.

spectral resolution Describes the degree of detail that can be seen in a spectrum; the higher the spectral resolution, the more detail we can see.

spectral type A way of classifying a star by the lines that appear in its spectrum; it is related to surface temperature. The basic spectral types are designated by a letter (OBAFGKM, with O for the hottest stars and M for the coolest) and are subdivided with numbers from 0 through 9.

spectrograph An instrument used to record spectra.

spectroscopic binary A binary star system whose binary nature is revealed because we detect the spectral lines of one or both stars alternately becoming blueshifted and redshifted as the stars orbit each other.

spectroscopy (in astronomical research) The process of obtaining spectra from astronomical objects.

spectrum (of light) *See* electromagnetic spectrum

speed The rate at which an object moves. Its units are distance divided by time, such as m/s or km/hr.

speed of light The speed at which light travels, which is about 300,000 km/s.

spherical geometry Refers to any case in which the rules of geometry for the surface of a sphere hold, such as that lines that begin parallel eventually meet.

spheroidal component (of a galaxy) The portion of any galaxy that is spherical (or football-like) in shape and contains very little cool gas; generally contains only very old stars. Elliptical galaxies have only a spheroidal component, while spiral galaxies also have a disk component.

spheroidal galaxy Another name for an elliptical galaxy.

spheroidal population Refers to stars that orbit within the spheroidal component of a galaxy. Thus, elliptical galaxies have only a spheroidal population (they lack a disk population), while spiral galaxies have spheroidal population stars in their bulges and halos. Sometimes called Population II.

spin (quantum) *See* spin angular momentum

spin angular momentum Often simply called *spin*, it refers to the inherent angular momentum of a fundamental particle.

spiral arms The bright, prominent arms, usually in a spiral pattern, found in most spiral galaxies.

spiral density waves Gravitationally driven waves of enhanced density that move through a spiral galaxy and are responsible for maintaining its spiral arms.

spiral galaxies Galaxies that look like flat, white disks with yellowish bulges at their centers. The disks are filled with cool gas and dust, interspersed with hotter ionized gas, and usually display beautiful spiral arms.

spreading centers (geological) Places where hot mantle material rises upward between plates and then spreads sideways creating new seafloor crust.

spring equinox (vernal equinox) Refers both to the point in Pisces on the celestial sphere where the ecliptic crosses the celestial equator and to the moment in time when the Sun appears at that point each year (around March 21).

spring tides The higher-than-average tides on Earth that occur at new and full moon, when the tidal forces from the Sun and Moon both act along the same line.

standard candle An object for which we have some means of knowing its true luminosity, so that we can use its apparent brightness to determine its distance with the luminosity–distance formula.

standard model (of physics) The current theoretical model that describes the fundamental particles and forces in nature.

standard time Time measured according to the internationally recognized time zones.

star A large, glowing ball of gas that generates energy through nuclear fusion in its core. The term *star* is sometimes applied to objects that are in the process of becoming true stars (e.g., protostars) and to the remains of stars that have died (e.g., neutron stars).

starburst galaxy A galaxy in which stars are forming at an unusually high rate.

star cluster *See* cluster of stars

star-gas-star cycle The process of galactic recycling in which stars expel gas into space where it mixes with the interstellar medium and eventually forms new stars.

star system *See* solar system

state (quantum) *See* quantum state

steady state theory A now-discredited theory that held that the universe had no beginning and looks about the same at all times.

Stefan–Boltzmann constant A constant that appears in the laws of thermal radiation, with value

$$\sigma = 5.7 \times 10^{-8} \frac{\text{watt}}{m^2 \times \text{Kelvin}^4}$$

stellar evolution The formation and development of stars.

stellar parallax The apparent shift in the position of a nearby star (relative to distant objects) that occurs as we view the star from different positions in the Earth's orbit of the Sun each year.

stellar wind A stream of charged particles ejected from the surface of a star.

stratosphere An intermediate-altitude layer of Earth's atmosphere that is warmed by the absorption of ultraviolet light from the Sun.

stratovolcano A steep-sided volcano made from viscous lavas that can't flow very far before solidifying.

string theory New ideas, not yet well-tested, that attempt to explain all of physics in a much simpler way than current theories.

stromatolites Large bacterial "colonies."

strong force One of the four fundamental forces; it is the force that holds atomic nuclei together.

subduction (of tectonic plates) The process in which one plate slides under another.

subduction zones Places where one plate slides under another.

subgiant A star that is between being a main-sequence star and being a giant; subgiants have inert helium cores and hydrogen-burning shells.

sublimation The process by which atoms or molecules escape into the gas phase from a solid.

summer solstice Refers both to the point on the celestial sphere where the ecliptic is farthest north of the celestial equator and to the moment in time when the Sun appears at that point each year (around June 21).

sunspot cycle The period of about 11 years over which the number of sunspots on the Sun rises and falls.

sunspots Blotches on the surface of the Sun that appear darker than surrounding regions.

superbubble Essentially a giant interstellar bubble, formed when the shock waves of many individual bubbles merge to form a single, giant shock wave.

supercluster Superclusters consist of many clusters of galaxies, groups of galaxies, and individual galaxies and are the largest known structures in the universe.

supergiants (luminosity class I) The very large and very bright stars that appear at the top of an H-R diagram.

supermassive black hole Giant black hole, with a mass millions to billions of times that of our Sun, thought to reside in the centers of many galaxies and to power active galactic nuclei.

supernova The explosion of a star.

Supernova 1987A A supernova witnessed on Earth in 1987; it was the nearest supernova seen in nearly 400 years and helped astronomers refine theories of supernovae.

supernova remnant A glowing, expanding cloud of debris from a supernova explosion.

synchronous rotation Describes the rotation of an object that always shows the same face to an object that it is orbiting because its rotation period and orbital period are equal.

synchrotron radiation A type of radio emission that occurs when electrons moving at nearly the speed of light spiral around magnetic field lines.

synodic month (or **lunar month**) The time required for a complete cycle of lunar phases, which averages about $29\frac{1}{2}$ days.

synodic period (of a planet) The time between successive alignments of a planet and the Sun in our sky; measured from opposition to opposition for a planet beyond Earth's orbit, or from superior conjunction to superior conjunction for Mercury and Venus.

tangential motion The component of an object's motion directed across our line of sight.

tangential velocity The portion of any object's total velocity that is directed across (perpendicular to) our line of sight. This part of the velocity cannot be measured with the Doppler effect. It can be measured only by observing the object's gradual motion across our sky.

tectonics The disruption of a planet's surface by internal stresses.

temperature A measure of the average kinetic energy of particles in a substance.

terrestrial planets Rocky planets similar in overall composition to Earth.

theories of relativity (*special* and *general*) Einstein's theories that describe the nature of space, time, and gravity.

theory (in science) *See* scientific theory.

theory of evolution The theory, first advanced by Charles Darwin, that explains *how* evolution occurs through the process of *natural selection*.

thermal emitter An object that produces a thermal radiation spectrum; sometimes called a blackbody.

thermal energy Represents the collective kinetic energy, as measured by temperature, of the many individual particles moving within a substance.

thermal escape The process in which atoms or molecules in a planet's exosphere move fast enough to escape into space.

thermal pressure The ordinary pressure in a gas arising from motions of particles that can be attributed to the object's temperature.

thermal pulses The predicted upward spikes in the rate of helium fusion, occurring every few thousand years, that occur near the end of a low-mass star's life.

thermal radiation The spectrum of radiation produced by an opaque object that depends only on the object's temperature; sometimes called blackbody radiation.

thermosphere A high, hot X-ray-absorbing layer of an atmosphere, just below the exosphere.

tidal force A force that is caused when the gravity pulling on one side of an object is larger than that on the other side, causing the object to stretch.

tidal friction Friction within an object that is caused by a tidal force.

tidal heating A source of internal heating created by tidal friction. It is particularly important for satellites with eccentric orbits such as Io and Europa.

time dilation Refers to the effect in which you observe time running slower in reference frames moving relative to you.

timing (in astronomical research) The process of tracking how the light intensity from an astronomical object varies with time.

torque A twisting force that can cause a change in an object's angular momentum.

total apparent brightness *See* apparent brightness. We sometimes say "total apparent brightness" to distinguish it from wavelength-specific measures such as the apparent brightness measured in visible light.

totality (eclipse) The portion of either a total lunar eclipse during which the Moon is fully within the Earth's umbral shadow or a total solar eclipse during which the Sun's disk is fully blocked by the Moon.

total luminosity *See* luminosity. We sometimes say "total luminosity" to distinguish it from wavelength-specific measures such as the luminosity emitted in visible light or the X-ray luminosity.

total lunar eclipse A lunar eclipse in which the Moon becomes fully covered by Earth's umbral shadow.

total solar eclipse A solar eclipse during which the Sun becomes fully blocked by the disk of the Moon.

transit An event in which a planet passes in front of a star (or the Sun) as seen from Earth. Only Mercury and Venus can be seen in transit of our Sun. The search for transits of extrasolar planets is an important planet detection strategy.

transmission (of light) The process in which light passes through matter without being absorbed.

transparent (material) Describes a material that transmits light.

triple-alpha reaction *See* helium fusion

Trojan asteroids Asteroids found within two stable zones that share Jupiter's orbit but lie 60° ahead of and behind Jupiter.

tropical year The time from one spring equinox to the next, on which our calendar is based.

tropic of Cancer The circle on Earth with latitude 23.5°N. It is the northernmost latitude at which the Sun ever passes directly overhead (at noon on the summer solstice).

tropic of Capricorn The circle on Earth with latitude 23.5°S. It is the southernmost latitude at which the Sun ever passes directly overhead (at noon on the winter solstice).

troposphere The lowest atmospheric layer, in which convection and weather occur.

Tully–Fisher relation A relationship among spiral galaxies showing that the faster a spiral galaxy's rotation speed, the more luminous it is; it is important because it allows us to determine the distance to a spiral galaxy once we measure its rotation rate and apply the luminosity–distance formula.

turbulence Rapid and random motion.

21-cm line A spectral line from atomic hydrogen with wavelength 21 cm (in the radio portion of the spectrum).

ultraviolet light Light with wavelengths that fall in the portion of the electromagnetic spectrum between visible light and X rays.

umbra The dark central region of a shadow.

unbound orbits Orbits on which an object comes in toward a large body only once, never to return; unbound orbits may be parabolic or hyperbolic in shape.

uncertainty principle The law of quantum mechanics that states that we can never know both a particle's position and its momentum, or both its energy and the time it has the energy, with absolute precision.

universal law of gravitation The law expressing the force of gravity (F_g) between two objects, given by the formula

$$F_g = G\frac{M_1 M_2}{d^2}$$

$$\left(\text{where } G = 6.67 \times 10^{-11} \frac{m^3}{\text{kg} \times \text{s}^2}\right)$$

universal time (UT) Standard time in Greenwich (or anywhere on the prime meridian).

universe The sum total of all matter and energy.

up quark One of the two quark types (the other is the down quark) found in ordinary protons and neutrons. Has a charge of $+\frac{2}{3}$.

velocity The combination of speed and direction of motion; it can be stated as a speed in a particular direction, such as 100 km/hr due north.

vernal equinox *See* spring equinox.

virtual particles Particles that "pop" in and out of existence so rapidly that, according to the uncertainty principle, they cannot be directly detected.

viscosity Describes the thickness of a liquid in terms of how rapidly it flows; low-viscosity liquids flow quickly (e.g., water), while high-viscosity liquids flow slowly (e.g., molasses).

visible light The light our eyes can see, ranging in wavelength from about 400 to 700 nm.

visual binary A binary star system in which we can resolve both stars through a telescope.

voids Huge volumes of space between superclusters that appear to contain very little matter.

volatiles Refers to substances, such as water, carbon dioxide, and methane, that are usually found as gases, liquids, or surface ices on the terrestrial worlds.

volcanic plains Vast, relatively smooth areas created by the eruption of very runny lava.

volcanism The eruption of molten rock, or lava, from a planet's interior onto its surface.

watt The standard unit of power in science: 1 watt = 1 joule/s.

wavelength The distance between adjacent peaks (or troughs) of a wave.

weak bosons The exchange particles for the weak force.

weak force One of the four fundamental forces; it is the force that mediates nuclear reactions; also the only force besides gravity felt by weakly interacting particles.

weakly interacting particles Particles, such as neutrinos and WIMPs, that respond only to the weak force and gravity; that is, they do not feel the strong force or the electromagnetic force.

weather Describes the ever-varying combination of winds, clouds, temperature, and pressure in a planet's troposphere.

weight The net force that an object applies to its surroundings; in the case of a stationary body on the surface of the Earth, weight = mass × acceleration of gravity.

weightless A weight of zero, as occurs during free-fall.

white dwarf limit (also called the *Chandrasekhar limit*) The maximum possible mass for a white dwarf, which is about $1.4 M_{\text{Sun}}$.

white dwarfs The hot, compact corpses of low-mass stars, typically with a mass similar to the Sun compressed to a volume the size of Earth.

white dwarf supernova A supernova that occurs when an accreting white dwarf reaches the white-dwarf limit, ignites runaway carbon fusion, and explodes like a bomb; often called a *Type Ia supernova*.

WIMPs Stands for *weakly interacting massive particles* and represents a possible form of dark matter consisting of subatomic particles that are dark because they do not respond to the electromagnetic force.

winter solstice Refers both to the point on the celestial sphere where the ecliptic is farthest south of the celestial equator and to the moment in time when the Sun appears at that point each year (around December 21).

worldline A line that represents an object on a spacetime diagram.

wormholes The name given to hypothetical tunnels through hyperspace that might connect two distant places in our universe.

X rays Light with wavelengths that fall in the portion of the electromagnetic spectrum between ultraviolet light and gamma rays.

X-ray binary A binary star system that emits substantial amounts of X rays, thought to be from an accretion disk around a neutron star or black hole.

X-ray burster An object that emits a burst of X rays every few hours to every few days; each burst lasts a few seconds and is thought to be caused by helium fusion on the surface of an accreting neutron star in a binary system.

X-ray bursts Burst of X rays coming from sudden ignition of fusion on the surface of an accreting neutron star in an X-ray binary system.

Zeeman effect The splitting of spectral lines by a magnetic field.

zenith The point directly overhead, which has an altitude of 90°.

zodiac The constellations on the celestial sphere through which the ecliptic passes.

zones (on a jovian planet) Bright bands of rising air that encircle a jovian planet at a particular set of latitudes.

credits and acknowledgments

Chapter 1

1.CO NASA/Goddard Institute for Space Studies 1.3 Jerry Lodriguss/Astropix LLC 1.5 Jeffrey Bennett 1.6 Stan Maddock 1.7 Goddard Institute for Space Studies, NASA 1.8 Akira Fujii 1.9 Jeffrey Bennett 1.10 Terraced hills: Blakeley Kim, Pearson Addison-Wesley; Pyramid: Corel; Earth: NASA; Telescope: Seth Shostak 1.13 NASA Earth Observing System 1.14 NASA Earth Observing System

Chapter 2

2.CO David Nunuk 2.1 Gordon Garradd 2.6 Richard Tauber Photography 2.14 Anthony Ayiomamitis 2.16 Husmo-foto 2.19 Akira Fujii 2.20 Photo by John Q. Waller, art by John and Judy Waller 2.22 Total: Akira Fujii; Partial: Dennis diCicco; Penumbrial: Akira Fujii 2.23 Akira Fujii 2.24 Akira Fujii 2.26 Tunç Tezel

Chapter 3

3.CO NASA/Johnson Space Center 3.2 Michael Yamashita/CORBIS 3.3 N. Pecnik/Visuals Unlimited 3.4 Kenneth Garrett Photography 3.5 William E. Woolam 3.6 Jeff Henry/Peter Arnold, Inc. 3.7 Robert Frerck/Woodfin Camp & Associates 3.8 Oliver Strewe 3.9a, b Courtesy of Carl Sagan Productions, Inc. From *Cosmos* (Random House) 3.9c Courtesy of Bibliotheca Alexandrina 3.10 Bettmann/CORBIS Page 65 (top) Giraudon/Art Resource, N.Y.; (bottom) Hulton Archive/Getty Images 3.12 The Granger Collection Page 66 Erich Lessing/Art Resource, N.Y. Page 69 Bettmann/CORBIS 3.17 Anthony Ayiomamitis

Chapter 4

4.CO NASA Earth Observing System Page 90 Bettmann/CORBIS 4.5 Spaceship: Goddard Institute for Space Studies, NASA; Baseball player: Duomo/CORBIS; Rocket: Goddard Institute for Space Studies, NASA 4.8 Bikes: Getty Images; Gas pump: Don Hammond/Design Pics/Corbis; Sunflower: Alvis Upitis/Image Bank/Getty Images 4.13 U.S. Department of Energy 4.20 Bill Bachmann/Jeff Gnass Photography

Chapter 5

5.CO Dr. N. A. Sharp, NOAO/NSO/Kitt Peak FTS/AURA/NSF 5.1 Runk/Schoenberger/Grant Heilman Photography, Inc. 5.17 Yerkes Observatory 5.18 National Optical Astronomy Observatories 5.19 Primary mirror: Russ Underwood, C.A.R.A./W. M. Keck Observatory; Overhead view: Richard J. Wainscoat 5.20 National Astronomy and Ionosphere Center's Arecibo Observatory, operated by Cornell University under contract with the National Science Foundation 5.21 NASA 5.22 NASA/JPL 5.23 Richard J. Wainscoat 5.24 Johnson Space Center, NASA 5.25 Richard J. Wainscoat 5.26 CFHT Corporation 5.27 Seth Shostak

Chapter 6

6.2 Visible-light photo: NSO Sacramento Peak, National Optical Astronomy Observatories; Ultraviolet photo: Marshall Space Flight Center, NASA 6.3 Surface: NASA/Johns Hopkins University Applied Physics Laboratory/Carnegie Institution of Washington; Planet: From the Voyage Scale Model Solar System, developed by Challenger Center for Space Science Education, the Smithsonian Institution, and NASA. Image created by ARC Science Simulations. 6.4 Surface: Marshall Space Flight Center, NASA; Planet: David P. Anderson, Southern Methodist University. From the Voyage Scale Model Solar System, developed by Challenger Center for Space Science Education, the Smithsonian Institution, and NASA. 6.5a From the Voyage Scale Model Solar System, developed by Challenger Center for Space Science Education, the Smithsonian Institution, and NASA. Image created by ARC Science Simulations. 6.5b NASA 6.6 Pano view: NASA, NSSDC & USGS; Mars full disk: NASA/JPL 6.7 From the Voyage Scale Model Solar System, developed by Challenger Center for Space Science Education, the Smithsonian Institution, and NASA. Image created by

ARC Science Simulations. 6.8 Space Science Institute, JPL, NASA 6.9 From the Voyage Scale Model Solar System, developed by Challenger Center for Space Science Education, the Smithsonian Institution, and NASA. Image created by ARC Science Simulations. 6.10 From the Voyage Scale Model Solar System, developed by Challenger Center for Space Science Education, the Smithsonian Institution, and NASA. Image created by ARC Science Simulations. 6.11 NASA, ESA, H. Weaver (JHU/APL), A. Stern (SwRI), and the HST Pluto Companion Search Team 6.12 Goddard Institute for Space Studies, NASA 6.13 Niescja Turner and Carter Emmart 6.16 M. Clampin (STScI), NASA 6.17 NASA/JPL 6.19 Robert Haag Meteorites 6.21 NASA 6.22 NASA/JPL 6.30 European Southern University 6.34 Dr. Mark Garlick 6.35 Geoffrey Bryden, JPL, NASA

Chapter 7

7.CO NASA 7.1 Mercury, Venus: NASA/JPL; Earth: NASA; Earth's moon: Akira Fujii; Mars: NASA/JPL 7.3 Richard Megna/Fundamental Photographs 7.5a Jules Bucher/Photo Researchers 7.5b Doug Duncan, University of Colorado 7.7 Don Davis 7.8 Courtesy of Brad Snowder 7.9 Paul Chesley/Getty Images 7.10 U.S. Geological Survey, Denver 7.11 NASA 7.12a Gene Ahrens/Bruce Coleman Inc. 7.12b Joachim Messerschmidt/Bruce Coleman Inc. 7.12c Craig Aurness/CORBIS 7.12d C. C. Lockwood/D. Donne Bryant Stock Photography 7.16a Akira Fujii 7.16b NASA/Johns Hopkins University Applied Physics Laboratory/Carnegie Institution of Washington 7.17 Full face: Frank Barrett, celestialwonders.com; Detail: Anthony Ayiomamitis 7.18 Lunar and Planetary Institute 7.19 NASA 7.20a NASA/Johns Hopkins University Applied Physics Laboratory/Carnegie Institution of Washington 7.20b Mark Robinson, Northwestern University 7.20c NASA/Johns Hopkins University Applied Physics Laboratory/Carnegie Institution of Washington 7.21b Mark Robinson and NASA 7.23 J. Bell (Cornell), M. Wolff (Space Science Inst.), Hubble Heritage Team (STScI/AURA), and NASA 7.24 Mars Global Surveyor, NSSDC, NASA 7.25 NASA/USGS 7.26 NASA/USGS 7.27 NASA/JPL/University of Arizona 7.28a USGS 7.28b NASA /JPL; Close up: NASA Earth Observing System 7.28c NASA Earth Observing System 7.28d R. P. Irwin III and G. A. Franz, National Air and Space Museum, Smithsonian Institution 7.29 Mars: NASA/JPL/Cornell; Earth: Marjorie A. Chan 7.30 NASA/JPL/ASI/ESA/Univ. of Rome/MOLA Science Team; Zoom out: NASA/JPL/ASI/ESA/Univ. of Rome/MOLA Science Team/USGS 7.31a The Magellan project, NASA. Additional processing by John Weiss 7.31b HiRISE, MRO, LPL (U. Arizona), NASA 7.33 NASA 7.34 Center, volcanic peaks: NASA/JPL; Round blobs: NASA; Round corona: NASA/JPL; Impact craters: NASA 7.35 ESA/VIRTIS and VMC teams 7.36 NSSDC, NASA 7.37 NASA

Chapter 8

8.CO NASA/JPL/Space Science Institute 8.1 NASA 8.8a NASA 8.8b NASA Earth Observing System 8.11 JPL/NASA 8.13 NASA/JPL; Computer-generated image of Ganymede by Björn Jónsson 8.18 Computer-generated image by Björn Jónsson 8.19 NASA/JPL; Inset: Arizona State University and NASA/JPL 8.20 Space Science Institute, JPL, NASA 8.22 NASA/JPL/USGS 8.23 Space Science Institute, JPL, NASA 8.24 NASA/JPL/Space Science Institute 8.25 NASA/JPL 8.26 Triton: NASA/USGS; Basins: NASA/JPL 8.27 Swann, Tammy Becker, and Alfred McEwen of the USGS/NASA 8.28a Lunar and Planetary Laboratory 8.28b NASA/JPL 8.28c William K. Hartmann 8.30 Jupiter: Imke de Pater and James Graham (UC Berkeley), Mike Brown (Caltech); Saturn: NASA/JPL; Uranus: Erich Karkoschka (Univ. of Arizona/LPL) and NASA; Neptune: NASA

Chapter 9

9.CO Miloslav Druckmuller, Brno University of Technology 9.1 ESO 9.2 Gaspra, Ida: NASA/JPL; Mathilde, Eros: Goddard Institute for Space Studies, NASA 9.5 Jonathan Blair/CORBIS 9.6 Robert Haag Meteorites 9.7a Peter Ceravolo 9.7b Tony & Daphne Hallas 9.10 NASA, ESA, H. Weaver (APL/JHU), M. Mutchler and Z. Levay (STScI) 9.12b Vic and Jen Winters, Icstars Astronomy 9.15 NASA/ESA and M. Brown (California Institute of Technology)

9.17a NASA, ESA, H. Weaver (JHU/APL), A. Stern (SwRI), and the HST Pluto Companion Search Team 9.17b Elliott Young, Southwest Research Institute 9.18a Hal Weaver and E. E. Smith (STScI) and NASA 9.18b MSSO, ANU/ Science Library/Photo Researchers, Inc. 9.18c Elliot Young, Southwest Research Institute 9.18d Jeffrey Bennett 9.19 Kirk Johnson, Denver Museum of Nature & Science 9.20 Virgil L. Sharpton, University of Alaska–Fairbanks 9.21 Pearson Addison-Wesley 9.22 Sovfoto/Eastfoto 9.23 La República Newspaper/ Associated Press

Chapter 10

10.CO Solar and Heliospheric Observatory (SOHO), ESA and NASA 10.1 Corel 10.9b Royal Swedish Academy of Sciences 10.10 National Optical Astronomy Observatories 10.11 Brookhaven National Laboratory 10.12 Lawrence Berkeley National Laboratory 10.13a The Royal Swedish Academy of Sciences 10.13b National Solar Observatory 10.15b NASA/JPL 10.16 Image from TRACE (Transition Region and Coronal Explorer), a mission of the Stanford-Lockheed Institute for Space Research (a joint program of the Lockheed-Martin Advanced Technology Center's Solar and Astrophysics Laboratory and Stanford's Solar Observatories Group), and part of the NASA Small Explorer Program 10.17 B. Haisch and G. Slater, Lockheed Palo Alto Research Laboratory 10.18 Goddard Space Flight Center, NASA

Chapter 11

11.CO AURA, STScI, and NASA 11.4 AURA, STScI, and NASA 11.5 Harvard College Observatory 11.14 David Malin, Anglo-Australian Observatory 11.15 NASA/JPL

Chapter 12

12.CO Michael Sherick, NASA 12.1 David Malin, Anglo-Australian Observatory 12.2 Matthew Bate, University of Exeter, UK 12.3 IPAC (Infrared Processing and Analysis Center) and JPL, NASA 12.5a JPL, NASA 12.5b NASA/JPL 12.6a Davy Kirkpatrick, Caltech IPAC/NASA 12.6b European Southern Observatory 12.11a Nordic Optical Telescope, La Palma 12.11b Andrew Fruchter and the ERO Team (Sylvia Baggett/STScI, Richard Hook/ST-ECF, Zoltan Levay/STScI), NASA 12.11c Hubble Heritage Team (STScI/AURA), NASA 12.11d R. Sahai, J. Trauger 12.20 NASA, ESA, J. Hester and A. Loll (Arizona State University) 12.21 David Malin, Anglo-Australian Observatory

Chapter 13

13.CO McGill and V. Kaspi et al., NASA 13.1 NASA/H.E. Bond and E. Helen (Space Telescope Science Institute, Baltimore Md.); M. Barstow and M. Burleigh (University of Leicester, U.K.); and J. B. Holberg (University of Arizona) 13.4b M. Shara, B. Williams, and D. Zurek (STScI), R. Gilmozzi (ESO), D. Prialnik (Tel Aviv Univ.), NASA 13.6 McGill and V. Kaspi et al., NASA 13.8 European Southern Observatory 13.17 David Malin, Anglo-Australian Observatory 13.18 S. Kulkarni, J. Bloom, P. Price, Caltech, NRAO GRB Collaboration

Chapter 14

14.CO Axel Mellinger 14.1 Pearson Benjamin Cummings 14.4 NASA/JPL 14.5 David Malin, Anglo-Australian Observatory 14.6 NASA/JPL 14.7a Tony & Daphne Hallas 14.7b NASA, ESA, the Hubble Heritage (STScI/AURA)–ESA/Hubble Collaboration, and the Digitized Sky Survey 2. Acknowledgment: J. Hester (Arizona State University) and Davide De Martin (ESA/Hubble)

14.9 John Bally, University of Colorado 14.10 Jeff Hester and Paul Scowen (Arizona State University) and NASA 14.12 NASA, ESA, M. Robberto (Space Telescope Science Institute/ESA) and the Hubble Space Telescope Orion Treasury Project Team 14.14 David Malin, Anglo-Australian Observatory 14.15 David Malin, Anglo-Australian Observatory 14.17 Hubble Heritage Team, AURA/STScI, NASA 14.20a E. Kopan, IPAC/Caltech 14.20b F. Zadeh et al., VLA/NRAO 14.20c D. A. Roberts, F. Yusef-Zadeh, and W. Goss, AUI/NRAO 14.20d European Southern Observatory 14.22 F. Bagnaoff et al., MIT and NASA

Chapter 15

15.CO H. Ford (JHU), G. Illingworth (UCO-Lick), M. Clampin and G. Hartig (STScI), the ACS Science Team, and ESA, NASA 15.1 NASA 15.2 NASA and ESA 15.3 The Hubble Heritage Team (STScI /AURA), NASA 15.4 P. Knezek (WIYN), The Hubble Heritage Team (STScI /AURA), ESA and NASA 15.5 David Malin, Anglo-Australian Observatory 15.6 Anglo-Australian Observatory and Royal Observatory, Edinburgh 15.7 Inserts: National Optical Astronomy Observatories 15.8 NASA/JPL 15.9 NASA/JPL 15.13 A. Riess (STScI), NASA 15.14 The Observatories of the Carnegie Institution of Washington 15.20 NASA/Goddard Institute for Space Studies 15.21 Springel et al. (Virgo Consortium), Max Planck Institute for Astrophysics 15.22 Brad Whitmore (STScI) and JPL, NASA 15.23 Frank Summers, Dept. of Library Services, American Museum of Natural History 15.24 Michael J. West, University of Hawaii 15.25 Subaru Telescope, National Astronomical Observatory of Japan 15.26 Hubble Heritage Team, AURA, STScI, and NASA 15.27 John Bahcall, Institute for Advanced Study and NASA 15.28 National Radio Astronomy Observatory 15.29 Ann Wehrle et al. Image by Glenn Piner 15.30 NASA/JPL 15.32 H. Ford (STScI/Johns Hopkins Univ.); L. Dressel, R. Harms, A. Kochar (Applied Research Corp.); Z. Tsvetanov, A Davidsen, G. Kriss (Johns Hopkins Univ.); R. Bohlin, G. Hartig (STScI); B. Margon (Univ. of Washington-Seattle); and NASA

Chapter 16

16.CO NASA/CXC/CfA/ M. Markevitch et al. 16.6 California Institute of Technology 16.7 NASA Earth Observing System 16.8 NASA, ESA M. J. Jee and H. Ford (Johns Hopkins University) 16.10 A. Fruchter, the ERO Team (STScI, ST-ECF), NASA 16.12 Michael Strauss, Princeton University 16.14 Andrey Kravtsov, Kavli Institute for Cosmological Physics, Dept. of Astronomy and Astrophysics, University of Chicago

Chapter 17

17.CO E. Bunn, University of Richmond 17.6 Roger Ressmeyer/CORBIS 17.9 E. Bunn, University of Richmond 17.17a John Kieffer/Peter Arnold, Inc. 17.17b Roland Gerth/zefa/Corbis

Chapter 18

18.CO Seth Shostak 18.2 Jeff Greenberg/Visuals Unlimited 18.4 Stanley M. Awramik/Biological Photo Service 18.7 Woods Hole Oceanographic Institution 18.8 Roger Ressmeyer/CORBIS 18.9 Martin Hanczyc 18.11 T. E. Adams/Visuals Unlimited 18.12 M. Di Lorenzo et al., Mars Exploration Rover Mission, Cornell U., JPL, NASA 18.13a Johnson Space Center, NASA 18.13b NASA/JPL 18.13c R. L. Folk and F. L. Lynch 18.16 Yeshe Fenner, Space Telescope Institute 18.17 Seth Shostak 18.19 Seth Shostak 18.20 National Astronomy and Ionosphere Center 18.21 NASA/JPL 18.22 NASA/JPL 18.23 NASA/JPL

index

and temperature variations in cosmic microwave background, 482–483
and understanding of gravity, 452
WIMPs as, 454–455, 482, 485
"dark side" of the Moon, 44, 103
Darwin, Charles, 505
Dawn spacecraft, 155, 259, A-25
days
length of, 35, A-2
of week, names of, t58, 59
Deep Impact mission, 266
deep-sea vents, 245, 506–507, 513
deferent, 63
degeneracy pressure
auditorium analogy for, 339
in brown dwarfs, 337–339, 385
electron, 363, 375, 385
in helium-burning stage of star life cycle, 341, 342, 343
neutron, 350, 367, 375, 384–385
supernova explosions and, 350
in white dwarfs, 343–344, 362–363, 385
Deimos (moon of Mars), 150, 169, A-16
Democritus, 114
density. *See also* critical density
of air vs. water, 96
in early universe, 427–428, 457–458
enhanced areas of in early universe, 482–483, 486
of jovian planets, 234, 235–236
of neutron stars, 368
of ordinary matter, 485, 490
of protogalactic cloud, 428–429
of seafloor vs. continental crust, 218
spiral density waves and, 400–401
of universe, 486, 488–489
of white dwarfs, 363
deoxyribonucleic acid. *See* DNA
deuterium
abundance of, 475, 490
and detection of neutrinos, 295
in early universe, 477, 484, 485
from hydrogen fusion in Sun, 291
differentiation, 191–192, 203
dinosaurs, 168, 273–275, 344, 509
Dione (moon of Saturn), 241, 248, A-16
direction, 29
disk(s)
formation of from solar nebula, 162–163, 164, 171, 180
of Milky Way, 387
spinning, 94
disk component, galactic, 413
disk population, 402, 413
disk stars, 389, 390, 401, 402
distance(s)
angular, 29–30
angular size and, 31
astronomical unit for, 5, 16, A-2
force of gravity and, 98–99
to galaxies, measurement of, 416–422
in light-years, 5, 8–10, A-2
orbital speed and, 69
in parsecs, 309, A-2
of planets from Sun, t156, 234
in stadia, 64
units of, A-2
distance chain, 416–419, 421–422
distance measurements, and age of the universe, 423–424

DNA (deoxyribonucleic acid), 505, 506, 507, 508, 524
Doppler effect, 123–125
in calculation of distances to galaxies, 420
in detection of solar vibrations, 293–294
equation for, A-3
in gas orbiting galactic centers, 436
in measuring motions of stars, 390
radial component of motion and, 125
in spectroscopic binary stars, 314–315
from stars in elliptical galaxies, 448–449
Doppler shifts, 20, 124–125, 314, 315
and interpretation of spectrum, 127
Doppler technique, for detection of extrasolar planets, 174, 175–176, 180
double-lined spectroscopic binary star systems, 315
double shell–burning stars, 342–343, 345, 385
Drake, Frank, 519
Drake equation, 519–522
dust grains, interstellar, 395
from dying low-mass star, 343
in Milky Way, 397
dust tail, of comet, 265
dwarf elliptical galaxies, 414
dwarf galaxies, 413
dwarf planets
Ceres as, 11, 155, 159, 259, 263
classification of, 5, 11, 270, 272
Eris as, 11, 155, 158, 269, 270
Pluto as, 11, 155, 158, 268–272, A-14
Dysnomia (moon of Eris), 269, 270

Eagle Nebula, 396
Earth, 149. *See also* life on Earth
age of oldest rock on, 173
angular momentum of, 93–94
atmosphere of, 131–133, 134, 196, 198–199, 216–217, 509
axis precession of, 41–42
axis tilt of, 16, 35–38, t156, 518
circumference of, 64
climate of, 218–219, 240, 301, 517–518
cosmic address of, 3
death of Sun and, 344–345
differentiation in, 191–192
erosion on, 195, 197–198
forces shaping, 189–190, 195–198
geological activity of, 190–195
greenhouse effect on, 199
habitability of, 12, 216–220, 225–226, 517–518
impact cratering on, 195–196
infrared spectrum of, 515
interior structure and temperature of, 190–193
magnetic field of, 194–195, 298
mass of, t156, A-2
Moon formation by giant impact to, 169, 170
motion of, 15–16, 20
orbit of, 16, 93–94, A-15
origin of water on, 168
plate tectonics on, 197, 217–218, 518
properties of, 10–12, t156, 158, A-2, A-14
and "rare Earth" hypothesis, 516–517, 518, 520

rotation of, 16, 31, 93–94
shape of, 62
size of, 10–12
tides on, 102–104
view of universe from, 27–30
volcanism on, 195, 196–197
earthquakes, t95, 193, 218
eccentricity
of ellipse
of ellipse, 67
orbital, A-15
eclipse
conditions for, 44–45
detection of extrasolar planets by, 176–177
lunar, 44–46, t47, 63
predicting, 47, 48, 60
solar, 44, 45, 46, 47, 48
eclipse seasons, 47
eclipsing binary star systems, 314, 315
ecliptic, 28, 34
ecliptic plane, 16, 44–45
egg balancing, and spring equinox, 76
Egypt, astronomy in ancient, 59, 60
Einstein, Albert, 96, 97
and cosmological constant, 461
equation of, 96–97, 146, 286, 291, 472
general theory of relativity of, 77, 177, 371, 372, 373–375, 379, 451, 452, 474, 488, 489, 524, 525
special theory of relativity of, 364, 371
electrical charges, in atoms, 115
electromagnet, 194
electromagnetic force, 473, 475
electromagnetic radiation, 112, 113, 114. *See also* light
electromagnetic spectrum, 112, 113–114
electromagnetic wave, 113
electromagnetism, 472
electron–antielectron pairs, 472
electron degeneracy pressure, 338–339, 363, 375, 385
electron neutrinos, 295
electrons
in atomic structure, 114–115
degeneracy pressure and, 338–339, 363, 375, 385
electrical charge of, 115
energy levels of, 119
energy level transitions of, 120–121
in era of atoms, 477
mass of, A-2
in particle era, 476
electron-volt (eV), 119, A-2
electroweak era, 474, 475–476, 485
electroweak force, 473, 478–479
elements, 114. *See also names of specific elements*
abundance patterns in cosmos of, 161, 349–350, 401–402
and Big Bang theory, 483–485
detection of by emission and absorption lines, 121
distribution of by supernova, 349
formation of, 8, 161, 349–350
galactic habitable zones and, 516–517
galactic recycling of, 161, 391–401
heavy, 391, 403–404
from high-mass stars, 347–350
periodic table of, A-13
ellipse, 66–67, 68, 99
elliptical galaxies, 412, 427
characteristics of, 412, 414

from collisions of spiral galaxies, 430
evidence of dark matter in, 448–449
formation of, 428–429
in Hubble's galaxy classes, 415
elliptical orbits, 66–68
emission, of light, 116–117
emission lines, 117, 127
emission line spectrum, 117, 118, 119–121
emission nebulae, 398
Enceladus (moon of Saturn), 241
geological activity on, 248
possibility of life on, 513
properties of, A-16
encephalization quotient (EQ), 521
end of the universe, 458–463, 492
Endurance Crater (Mars), 206
energy. *See also* dark energy
comparison of types of, t95
conservation of, 97–98, 162
gravitational potential, 95, 96, 97, 100, 162, 334
gravity of, 375–376
kinetic, 95, 96–97, 100, 162
law of conservation of, 93, 94, 97–98, 100, 162
mass-, 96–97
orbital, 100–101, 169
of photons, 120, A-3
potential, 95, 96–97
radiative, 95, 97
thermal, 95–96, 162
units of, 95, 119, A-2
energy level transitions, 119–120
epicycle, 63
EQ (encephalization quotient), 521
equations, useful, A-3
equinoxes, 38–39, 41, 76
equivalence principle, 371
eras of universe's early history, 474–477, 480
Eratosthenes, 62, 64
Eris, 270, 272
discovery of, 11, 155, 269
location of, 159
properties of, t156, 158, A-14, A-15
Eros (asteroid), 159, 260
erosion
on Earth, 195, 197–198, 219
fossils and, 502–503
lack of on Venus, 212
on Mars, 206–208
escape velocity
from black holes, 372
from Earth, 101–102, A-2
equation for, A-3
from planets, A-14
Eskimo Nebula, 343
ethane, on Titan, 152, 245–248
Europa (moon of Jupiter), 103, 104, 151, 241, 242, 244–245
orbit of, 243, 261
possibility of life on, 511, 513
properties of, A-16
European Southern Observatory, 177, 368, A-26
European Space Agency (ESA), 148, 177, 182, 209, 212, 246, 266
evaporation of black holes, 492
event horizon, 372–373, 374
evolution. *See* galaxy evolution; theory of evolution
evolutionary track (life track), 345, 347
excited states of electrons, 119